A World Compendium

The Manual of Biocontrol Agents

Fourth Edition

Editor: L. G. Copping

BCPC

Promoting the Science and Practice of Sustainable Crop Production

British Library Cataloguing in Publication Data. A catalogue record of this book is available from the British Library.

ISBN 978 1 901396 17 1

First published (as *The BioPesticide Manual*) 1998
Second edition 2001
Third edition 2004

Cover design by m360, Nottingham
Typeset by Sweetmore Publishing Projects, Witney
Printed by Latimer Trend & Co., Plymouth

Published by:
BCPC, 7 Omni Business Centre, Omega Park, Alton, Hampshire, GU34 2QD, UK
Tel +44 (0) 1420 593 200   Fax +44 (0) 1420 593 209
Email: md@bcpc.org   Web: www.bcpc.org

All BCPC publications can be purchased from:
BCPC Publications, 7 Omni Business Centre, Omega Park, Alton, Hampshire,
GU34 2QD, UK
Tel +44 (0) 1420 593 200 Fax +44 (0) 1420 593 209
Email: publications@bcpc.org
Or direct from the BCPC Online Book Shop at www.bcpc.org/bookshop

Disclaimer
Every effort has been made to ensure that all information in this edition of *The Manual of Biocontrol Agents* is correct at the time of going to press. However, the editor and the publisher do not accept liability for any error or omission in the content, or for any loss, damage or any other accident arising from the use of the products listed therein.

Before handling, storing or using any approved crop protection product, it is essential to follow the instructions on the label.

With thanks to the following for permission to reproduce cover photographs:
Top: female vetch aphid (*Megoura viciae*) releasing sex pheromone (Rothamsted Research);
Middle, left to right: the parasitic nematode *Steinernema feltiae* (NGM Hague/S Haukland, University of Reading, UK); the aphid parasitoid *Praon volucre* ovipositing (Rothamsted Research); spores of the entomopathogenic fungus *Pandora neoaphidis* (Rothamsted Research);
Bottom, left to right: seven-spot ladybird (*Cocinella septempunctata*) eating a pea aphid (*Acyrthosiphon pisum*) (Rothamsted Research); pyrethrum in flower (Roger Hiorns, courtesy John Wiley and Sons Ltd/Society of Chemical Industry: *Pest Management Science*).

Contents

The Publisher

The fourth edition of *The Manual of Biocontrol Agents* is published by BCPC, a registered charity, formed in 1967.

The principal aim of BCPC is *'to promote and encourage the science and practice of crop production for the benefit of all'*.

BCPC brings together a wide range of organisations interested in the improvement of crop protection. The members of its Board represent the interests of government departments, the agrochemical industry, farmers' and consumers' organisations, the advisory services and independent consultants, distributors, the research councils, agricultural engineers, environment interests, training and overseas development.

The corporate members of BCPC currently are:

- Agricultural Engineers Association
- Association of Applied Biologists
- Association of Independent Crop Consultants
- Biotechnology and Biological Sciences Research Council
- British Institute of Agricultural Consultants
- British Society for Plant Pathology
- Campden BRI
- Crop Protection Association
- Department for Environment, Food and Rural Affairs, represented by the Chemical Regulation Directorate (formerly Pesticides Safety Directorate)
- Department of Agriculture and Rural Development for Northern Ireland
- Environment Agency
- Imperial College, London
- Lantra
- National Association of Agricultural Contractors
- National Consumer Federation
- National Farmers' Union
- National Institute of Agricultural Botany
- Natural Environment Research Council
- Scottish Government Directorate General Environment
- Society of Chemical Industry – BioResources Group

Foreword

This fourth edition of *The Manual of Biocontrol Agents* contains many new entries compared with the third edition of 2004, showing that biological control of pests, diseases and weeds is continuing to undergo strong growth. In this new edition there are 149 entries for micro-organisms (112 in 3rd edn), 89 entries for natural products (58 in 3rd edn), 74 for semiochemicals (56 in 3rd edn) and 140 for macro-organisms (127 in 3rd edn). The strongest increase in the number of 'products' took place in the categories of micro- and macro-organisms and this, indeed, coincides with the growth of this sector of biological pest control. The increase in the other groups of 'agents' (natural products and semiochemicals) is less spectacular, understandably as many countries demand severe registration procedures for these products. Registration is often so expensive that these products do not reach the market, as the relatively small industries involved in their development are not capable of finding the necessary funding. However, strenuous efforts have been made, particularly by the registration agencies, in the interests of helping these types of crop protection to be commercially available, for example in the UK, Europe and USA we see the following arrangements. The UK Chemicals Regulation Directorate runs a biopesticide scheme for the registration of pheromone and other semiochemicals, micro-organisms, plant extracts and other novel products. An additional fee is required for European consideration. However, they still go through the same seven risk-assessment processes as pesticides. The USA is much more advanced in biopesticide registration. In 1994, the US Environmental Protection agency established the Biopesticides and Pollution Prevention Division to facilitate the registration of biocontrol agents, and they generally require fewer data to register a biopesticide than to register a conventional pesticide.

In most quarters, although sadly not in all, the realisation that current technologies will not cope with the increase in demand for agricultural and horticultural produce is driving the quest for more effective pest control systems. Although some criticisms of currently registered broad-spectrum eradicant pesticides are without scientific foundation, certainly in terms of direct impacts on the environment and human health, there are continually growing problems of pest resistance and development costs. Therefore, it is of great importance that we see an increase in the range of biocontrol agents, and this will provide evidence upon which to base expanded utilisation in the future. The main targets for these agents are horticulture and generally high-value crops, and there still need to be major new technological advances to drive the use of these agents into broad-acre cropping systems. Nonetheless, current successes, as in conservation biological control, demonstrate clear further potential, and where there is successful use even of exotic organisms these – although not sustainable in the long term – can further demonstrate the value of biological control. New developments in biotechnology will help deliver chemically based biocontrol agents, either by producing these more cheaply and sustainably from botanically resources, or by direct release from crop plants as a consequence of genetic modifications. Indeed, current negative attitudes to genetic modification are likely to relax in the socio-political perspective, although not in terms of registration, so as to use these techniques more readily in facilitating the production and use of biocontrol agents.

I cannot believe that any sector dealing with agriculture and food production, whether the hard pressed production industry, the well healed food sector, the rapidly advancing research community or the Departments of State involved in attempting to keep up with quickly changing demands and technologies, will not find this fourth edition of very great value.

John A. Pickett, CBE, DSc, FRS
Scientific Director, Rothamsted Centre for Sustainable Pest and Disease Management, and
Head, Department of Biological Chemistry, Rothamsted Research

Preface

It was a difficult decision to launch the first edition of *The BioPesticide Manual* in 1998. The use of 'biologically based products' was very limited, and the problems associated with their widespread use were seen to be restricted by a narrowness of pest/disease/weed spectrum, a lack of persistence, poor shelf life, generally poor formulations and relatively high cost. However, in the past 11 years things have changed, and the clear advantages of 'biorationals' are, at last, being realised. At a time when resistance to conventional chemical pesticides is becoming more widespread and there is a lack of new chemicals with novel modes of action, the opportunity to apply natural alternatives in sequence or in admixture with chemicals helps to reduce this risk. In addition, mixing with biorationals reduces the required rate of each chemical, thereby reducing the level of residues in the treated crop. Increasingly, the major manufacturers of crop protection chemicals based on Nature are recommending their use in mixture, and they are seeing good results from these practices. In addition, the effectiveness of biorationals in specialised situations, such as on protected crops, continues to show outstanding efficacy. In a world that is increasingly concerned about greenhouse gas emissions, chemical pollution and undesirable effects on non-target organisms, there is a clear and viable place for natural crop protection.

There are some small changes to the layout of this Fourth Edition, with the Appendix containing a piece that examines the requirements for crop protection methods in different parts of the world. This includes an examination of plant-derived products that are sold internationally, but that seldom have a crop protection use associated with them. It is hoped that *The Manual of Biocontrol Agents* will continue to be the reference text for all interested in natural pest, weed and disease control. As an aid to this, each entry is identified if it can be used somewhere in the world within organic growing systems. This aspect of the manual must be treated with care, however, because not all accepted compounds are sold in all countries, and not all organic accreditation bodies have accepted the use of all products within their area of responsibility.

The book is written in a different style from its sister publication, *The Pesticide Manual. The Manual of Biocontrol Agents* covers in detail the biological effects observed with all products listed and, wherever possible, gives information on the mode of action plus key references for readers to pursue if they wish. To ensure no information is lost, the reader is referred, where appropriate, to *The Pesticide Manual*, where comprehensive details of chemistry, synthesis and physico-chemical data of natural chemicals may be found.

A wide number of terms are used variously to describe crop protection agents that are derived from Nature. These include – Biologicals; Biorationals; Natural Products; and Biocontrol Agents.

The Manual of Biocontrol Agents is divided into four distinct sections.

- Section 1. Micro-organisms – now contains 149 different baculoviruses, bacteria, fungi, protozoa and nematodes, all of which are increasingly important in integrated crop management and organic farming, as well as conventional crop protection strategies.

- Section 2. Natural Products – the 89 compounds listed all have a claim for a crop effect and all are derived from micro-organisms, algae and marine crustaceans, as well as higher plants. Where relevant, these entries are cross-referenced to the Fifteenth Edition of *The Pesticide Manual*.

- Section 3. Macro-organisms – this section now contains 140 records of insects and mites that are predators or parasitoids of phytophagous insects and mites sold commercially for use in glasshouses, interiorscapes and in outdoor agriculture. Insects that can be used for weed control are also listed.

- Section 4. Semiochemicals – this section describes 74 different chemicals used in mating disruption, lure and kill, or insect-monitoring strategies.

This edition retains the reclassification of the *Bacillus thuringiensis* entries by biological effect rather than by the subspecies from which they derive.

Most compounds mentioned in the book are highly compatible with organic farming practices and are appropriate for use in environmentally sensitive situations. The new manual contains a brief summary of the requirements demanded by organic farming regulating authorities, with appropriate references from which the reader can gather additional information.

Each individual entry lists a number of commercial products by tradename and company. All approved names, alternative names, common names, tradenames and code numbers are listed in Index 2, and relate to the entry number. Each entry number is preceded by the number (from 1 to 4) of the section in which it is placed. The Glossary contains the scientific names of all species and genera mentioned in the main text, with authorities, classified into order and family. The English–Latin glossary is an attempt to make it easier to find a species whose scientific name is not known.

Many people have contributed significantly to the production of *The Manual of Biocontrol Agents*. Colleagues at BCPC are mentioned because of their constant help, encouragement and support. In particular, Trevor Lewis deserves special mention for his enthusiasm, encouragement and advice, while Clive Tomlin's input was essential for the successful completion of a task that would have been impossible without his help, advice and outstanding technical knowledge.

Section 1 was put together with the assistance of Helmut van Emden (University of Reading) and Roy Bateman (IPARC); Section 2 is based on information received from Clive Tomlin (BCPC) and Gary Thompson (Dow AgroSciences); Section 3 depended very heavily on guidance from Melvyn Fidgett (Novartis BCM) and Helmut van Emden (University of

Reading); Section 4 relied on help and advice from Cam Oehlschlager (ChemTica), Alan Cork (University of Greenwich) and Jack Plimmer (USA).

Others who have assisted in various ways include John Pickett (Rothamsted Research) and Bernard Blum (IOBC). Paul Lister must be acknowledged for his imaginative computer programming, and Jane Townson, David Alford, Roy Bateman and Douglas Hartley are to be congratulated for the thoroughness of their reviewing.

Finally, I must thank all those individuals from companies around the world for responding to my enquiries and providing the data that you can read in the following pages. Any errors or omissions in the transcription or interpretation of this information, gleaned from a variety of sources, are my responsibility. I urge you to bring these errors to my attention as soon as possible so that they can be corrected in time for the fifth edition.

Leonard G Copping
Editor

Guide to using *The Manual of Biocontrol Agents*

The Manual of Biocontrol Agents is divided into four sections:

1. Micro-organisms
2. Natural Products
3. Macro-organisms
4. Semiochemicals.

The general layout of each section is similar and the main headings within each section are often the same. It is intended that *The Manual of Biocontrol Agents* will allow users to find information on the nature, origin, mode of action, use, commercial availability, mammalian toxicology and environmental impact of each entry. Those products that also appear in *The Pesticide Manual* Fifteenth Edition are identified, so more detailed information on their chemistry and physicochemical properties can be found easily.

Sample Entries

The following are sample entries that illustrate how *The Manual of Biocontrol Agents* has been designed, to help the reader understand the topics that may be found under each heading. The four examples are as listed in the main volume.

Entry No. Approved name

Biological activity

The Pesticide Manual Fifteenth Edition entry number: *If the entry also appears as a major entry in* The Pesticide Manual *Fifteenth Edition.*

STRUCTURE: *If a natural product or a semiochemical, and if the entry is an identified chemical.*

TAXONOMY: *If a micro- or macro-organism, a description and classification of the organism by Order and Family.*

NOMENCLATURE: Approved name: *Approved common name of chemical plus approval authority, or scientific name plus authority if a micro- or macro-organism.* **Other names:** *Any other names under which the entry may be known.* **Common name:** *Any other common name, not necessarily approved, by which the entry is known. Trivial or English name of organism.*
CAS RN: *If a natural product or a semiochemical.* **Development code:** *If the product had a development code.*

SOURCE: *If a natural product, where is it derived from; if an organism, where is the organism found or from where was it isolated.*

TARGET PESTS: *Against which pest, disease or weed species it is recommended.*

TARGET CROPS: *In which crops its use is recommended.*

BIOLOGICAL ACTIVITY: Biology: *Details of the way the product controls its target pest; if a micro-organism, by parasitism, competition, predation or production of toxic metabolites.*
Mode of action: *Biochemical mode of action for compounds and mechanism of action if living.* **Predation, Egg laying** and **Duration of development:** *Entries for insect predators and parasitoids.* **Efficacy:** *Effectiveness of the product.* **Key reference(s):** *The most relevant reference(s) giving more details about the entry.*

COMMERCIALISATION: Formulation: *How the entry is formulated or packaged.*
Tradenames: *Names of products and manufacturers' names.* **Patent:** *Patents covering the product.*

APPLICATION: *Rate of use, timing and frequency of application.*

PRODUCT SPECIFICATIONS: Purity: *Purity of the product. Acceptable contaminants. How the effectiveness is checked.* **Storage conditions:** *How should it be stored.* **Shelf-life:** *For how long the stored product will remain effective.*

COMPATIBILITY: *Any major incompatibilities or recommendations for combination products.*

MAMMALIAN TOXICITY: *As much information on the toxicity of the active ingredient and/or the formulated product as is available, to include, where relevant or available:*

Acute oral LD$_{50}$
Acute dermal LD$_{50}$
Inhalation
Skin and eye
ADI
Toxicity class
NOEL
Other toxicological effects:
Toxicity reviews.

ENVIRONMENTAL IMPACT AND NON-TARGET TOXICITY: *As much information on the non-target toxicity and environmental impact of the active ingredient and/or the formulated product as is available to include where relevant and if known:*
Bird toxicity
Fish toxicity
Other aquatic toxicity
Effects on beneficial species
Metabolism
Behaviour in soil.

Accepted for use in organic farming: *Indicates that at least one national accreditation authority has accepted its use in organic farming. Accepted by IFOAM.*

Specific examples follow.

Section 1: Micro-organisms – sample entry

Notes

1. Number representing the section into which the entry is placed, e.g. 1 = Micro-organisms.

2. Sequential entry number.

3. Approved name.

4. Class of biological activity (and derivation).

5. Indicates an entry in *The Pesticide Manual* Fifteenth Edition.

6. Taxonomic classification by phylum, class and order.

7. The approved name (with authorities), common names, other names and development code numbers.

8. The history of the discovery of the biological activity and details of from where the organism was first isolated.

9. A summary of the method of production for commercial sale.

10. A detailed list or description of the pest species (with authorities) against which the product is recommended.

11. A list of the crops in which the product is recommended.

12. A full description of the biological activity associated with the entry, its mode of action (where known) and a summary of the efficacy of the product.

13. A list of key references that describe the discovery, effectiveness and commercialisation of the product.

❶ ❷ ❸
1:26 *Bacillus subtilis* QST 713
❹*Biological fungicide and bactericide (bacterium)*

❺ *The Pesticide Manual* **Fifteenth Edition entry number:** 56

❻ Bacterium: Schizomycetes: Eubacteriales

❼ **NOMENCLATURE: Approved name:** *Bacillus subtilis* (Ehrenberg) Cohn, strain QST 713. **Development code:** QST-713.

❽ **SOURCE:** The bacterium, *Bacillus subtilis*, is prevalent in soils and has been found in a variety of habitats world-wide. The strain selected for commercialisation was chosen because of its spectrum and level of activity against important economic fungal and bacterial pathogens.

❾ **PRODUCTION:** *Bacillus subtilis* QST 713 is produced commercially by fermentation. The fermentation medium with viable spores is used to prepare the product.

❿ **TARGET PESTS:** Used as a foliar spray for the control of a wide range of economically important fungal and bacterial diseases, including *Botrytis cinerea* Pers., *Uncinula necator* Burr., *Podosphaera leucotricha* (Ellis & Everhart) Salmon, *Erysiphe* spp., *Sphaerotheca* spp., *Leveillula taurica* Arn., *Oidium* spp., *Oidiopsis taurica* Salm., *Alternaria* spp., *Erwinia amylovora* Winsl., *Venturia* spp., *Bremia lactucae* Regel, *Peronospora* spp., *Botryosphaeria dothidea*, *Phytophthora infestans* de Bary, *Xanthomonas* spp., *Sclerotinia minor* Jagger and *Plasmopara viticola* Berl. and de T.

⓫ **TARGET CROPS:** Used on grapes, pome fruit, walnuts, stone fruit, hops, cucurbits, leafy vegetables, crucifers, peppers, tomatoes, potatoes, onions, carrots, herbs and ornamentals.

⓬ **BIOLOGICAL ACTIVITY: Mode of action:** 'Serenade Biofungicide' and 'Rhapsody Biofungicide' are effective by preventing diseases caused by a wide range of plant pathogens. They prevent plant pathogen spores from germinating, disrupt germ tube and mycelial growth and stop attachment of the plant pathogen to the leaf by producing a zone of inhibition, restricting the growth of these disease-causing pathogens. The QST 713 strain of *Bacillus subtilis* contained in 'Serenade Biofungicide' and 'Rhapsody Biofungicide' has also been shown to colonise pathogens, thus causing inhibition of the pathogen's germ tube elongation. 'Serenade Biofungicide' and 'Rhapsody Biofungicide' also activate the plant's Systemic Acquired Resistance (SAR) natural defence system. ⓭ **Key reference(s):** D R Fravel, W J Connick Jr & J A Lewis. 1998. Formulation of microorganisms to control plant diseases. In *Formulation of Microbial Biopesticides, Beneficial Microorganisms, Nematodes and Seed Treatments*, H D Burges (ed.), pp. 187–202, Kluwer Academic Publishers, Dordrecht, The Netherlands.

Notes

14   Details of the formulation types used commercially, the tradenames and the manufacturers plus selected patent numbers.

15   Details of frequency and methods of application and rate of use.

16   Information on the purity and quality of the technical material, details of the recommended storage conditions and shelf-life.

17   Information on incompatibilities and other information on the possibility of interactions between the product and other crop protection products.

18   Toxicological data to include acute oral and dermal LD_{50}, inhalation, skin and eye effects, no observed effect level, acceptable daily intake, toxicity class, toxicity reviews and any other information on adverse effects.

19   Effects on non-target organisms (birds, fish and other aquatic species, wildlife, and beneficial species), persistence, degradation rates and pathways.

20   Indicates whether at least one national accreditation authority has accepted its use in organic farming. Accepted by IFOAM.

⑭ COMMERCIALISATION: Formulation: The product is sold as a wettable powder (WP) and as an aqueous suspension (AS). **Tradenames:** 'Serenade Biofungicide' and 'Rhapsody Biofungicide' (AgraQuest). **Patents:** US 6060051 on the QST 713 strain of *Bacillus subtilis*. Worldwide patents are pending.

⑮ APPLICATION: 'Serenade Biofungicide' and 'Rhapsody Biofungicide' are applied as foliar fungicides. The wettable powder (WP) formulation is applied at a rate of 5 to 8 kg per hectare and the aqueous suspension (AS) formulation is applied at a rate of 4 to 6 litres per hectare. Both products should be applied in sufficient volume for complete foliar coverage.

⑯ PRODUCT SPECIFICATIONS: Purity: The WP formulation consists of 5×10^9 cfu/g and the AS formulation consists of 7×10^9 cfu/g. Guaranteed to be free of microbial contaminants. **Storage conditions:** Both the WP and AS formulations can be stored under dry, ambient conditions. **Shelf-life:** Stable for at least 2 years under the above storage conditions.

⑰ COMPATIBILITY: Compatible with many fungicides, bactericides, insecticides, foliar nutrients and adjuvants used in the control of crop diseases. Incompatible with strong oxidisers, acids, bases and chlorinated water.

⑱ MAMMALIAN TOXICITY: 'Serenade Biofungicide' and 'Rhapsody Biofungicide' are considered to be non-toxic products, when used according to label directions. 'Serenade Biofungicide' and 'Rhapsody Biofungicide' meet all the criteria for US-EPA Toxicity Category III with the signal word 'Caution'. **Acute oral LD$_{50}$:** rats >5000 mg/kg.
Acute dermal LD$_{50}$: rabbits >2000 mg/kg.

⑲ ENVIRONMENTAL IMPACT AND NON-TARGET TOXICITY: 'Serenade Biofungicide' and 'Rhapsody Biofungicide' have an extremely favourable environmental and ecological profile. 'Serenade Biofungicide' and 'Rhapsody Biofungicide' have low ecological risk to birds, fish and aquatic invertebrates, when used at label rates.
Effects on beneficial species: 'Serenade Biofungicide' and 'Rhapsody Biofungicide' are non-toxic to beneficials (tested on honeybees, lacewings, parasitic wasps and ladybird beetles), when used at label rates. This environmentally benign profile makes 'Serenade Biofungicide' and 'Rhapsody Biofungicide' compatible in Integrated Pest Management (IPM) programmes.

⑳ Accepted for use in organic farming: Yes.

Section 2: Natural Products – sample entry

Notes

1   Number representing the section into which the entry is placed, e.g. 2 = Natural Products.

2   Sequential entry number.

3   Approved name.

4   Class of biological activity (and derivation).

5   Indicates an entry in *The Pesticide Manual* Fifteenth Edition.

6   The approved name (with authorities), common names, other names, CAS RN and development code numbers.

7   As all compounds originate in Nature, the original source of every entry is included.

8   Some compounds are synthesised, some extracted from fermentation and some from the natural living organism. This section describes how the product is produced for commercialisation.

9   A detailed list or description of the pest species (with authorities) against which the product is recommended.

❶ ❷   ❸

2:229 spinosad

❹ *Microbial insecticide (micro-organism-derived)*

❺   *The Pesticide Manual* **Fifteenth Edition** entry number: 784

spinosyn A, R = H-

spinosyn D, R = CH$_3$-

❻   **NOMENCLATURE: Approved name:** spinosad (BSI, ANSI, ISO). **CAS RN:** *[131929-60-7]* spinosyn A; *[131929-63-0]* spinosyn D. **Development codes:** XDE-105; DE-105.

❼   **SOURCE:** The commercial product is a mixture of spinosyn A and spinosyn D. Both compounds are secondary metabolites of the soil Actinomycete, *Saccharopolyspora spinosa* Mertz & Yoa. The organism is composed of long, yellowish-pink aerial chains of spores encased in distinctive, spiny spore sheaths. The bacterium is aerobic, Gram-positive, non-acid-fast, non-motile, filamentous and differentiated into substrate and aerial hyphae. The aerial mycelium is yellowish-pink and the vegetative mycelium is yellow to yellowish-brown. The parent strain was originally isolated from an abandoned rum still in the Caribbean.

❽   **PRODUCTION:** Spinosad is obtained from a whole broth extraction, following fermentation of the organism on a feedstock of water, vegetable flours, sugar and animal fat.

❾   **TARGET PESTS:** Recommended for the control of caterpillars, leaf miners, thrips and foliage-feeding beetles. Species controlled include caterpillars (*Ostrinia nubilalis* (Hübner), *Helicoverpa zea* Boddie, *Trichoplusia ni* (Hübner), *Plutella xylostella* (L.), *Spodoptera* spp., *Heliothis* spp., *Pieris rapae* (L.), *Keiferia lycopersicella* (Walsingham), *Lobesia botrana* (Denis & Schiffermüller), *Agrotis ipsilon* (Hufnagel), *Parapediasia teterrella* (Zincken), thrips (*Frankliniella occidentalis* (Pergande), *Thrips palmi* (Karny)), flies (*Liriomyza* spp., *Ceratitis capitata* (L.)), beetles (*Leptinotarsa decemlineata* (Say)), drywood termites (*Cryptotermes brevis* (Walker), *Incisitermes snyderi* (Light)), fire ants (*Solenopsis* spp.) and grasshoppers. Under development for control of chewing and sucking lice (e.g. *Linognathus vituli*, *Bovicola ovis* (Schrank), *Solenopotes capillatus* Enderlein) and flies (e.g. *Haematobia irritans* (L.), *Lucilia cuprina* (Wiedemann)), and for control of nuisance flies (e.g. *Stomoxys calcitrans* (L.), *Musca domestica* L.).

Notes

10   A list of the crops in which the product is recommended.

11   A full description of the biological activity associated with the entry. This will include mode of action (where known) and a summary of the efficacy of the product.

12   A list of key references that describe the discovery, effectiveness and commercialisation of the product.

13   Details of the formulation type used commercially, the tradenames and the manufacturer plus selected patent numbers.

14   Details of frequency and methods of application and rate of use.

❿ TARGET CROPS: May be used on row crops (including cotton), vegetables, fruit trees, turf, vines and ornamentals. No crop phytotoxicity has been observed. Under development for use on livestock animals and in livestock premises.

⓫ BIOLOGICAL ACTIVITY: Mode of action: Spinosad effects on target insects are consistent with the activation of the nicotinic acetylcholine receptor, but at a different site than with nicotine or imidacloprid. Spinosad also affects GABA receptors, but their role in the overall activity is unclear. There is currently no known cross-resistance to other insecticide classes. **Efficacy:** The mode of action causes a rapid death of target phytophagous insects. Its moderate residual activity reduces the possibility of the onset of resistance, but it is strongly recommended that it be used within a strong, pro-active resistance management strategy. Spinosad is recommended as an ICM tool, as it shows no effects on predatory insects such as ladybirds, lacewings, big-eyed bugs or minute pirate bugs. It has reduced activity against parasitic wasps and flies. It is toxic when sprayed directly onto honeybees and other pollinators, but, once dry, residues have little effect. Spinosad is effective as a bait for fruit flies (*Ceratitis* spp., *Bactrocera* spp., etc.) and some ants (*Solenopsis* spp.).

⓬ Key references: 1) H A Kirst, K H Michel, J S Mynderse, E H Chio, R C Yao, W M Nakasukasa, L-V D Boeck, J L Occlowitz, J W Paschal, J B Deeter & G D Thompson. 1992. Discovery, isolation, and structure elucidation of a family of structurally unique, fermentation-derived tetracyclic macrolides. In *Synthesis and Chemistry of Agrochemicals III*, Chapter 20, D R Baker, J G Fenyes & J J Steffens (eds.), pp. 214–25, American Chemical Society, Washington DC, USA. 2) D J Porteus, J R Raines & R L Gantz. 1996. In *1996 Proceedings of Beltwide Cotton Conferences*, P Dugger & D Richter (eds.), pp. 875–7, National Cotton Council of America, Memphis, TN, USA. 3) T C Sparks *et al.* 1996. In *1996 Proceedings of Beltwide Cotton Conferences*. P Dugger & D Richter (eds.), pp. 692–6, National Cotton Council of America, Memphis, TN, USA. 4) V L Salgado. 1997. In *Down to Earth*, **52**:1, pp. 35–44, DowElanco, Indianapolis, IN, USA. 5) G D Thompson, R Dutton & T C Sparks. 2000. Spinosad – a case study: an example from a natural products discovery programme, *Pest Management Science*, **56**(8), 696–702.

⓭ COMMERCIALISATION: Formulation: Sold as an aqueous based suspension concentrate (SC) formulation. **Tradenames:** 'Conserve', 'Entrust', 'Success', 'SpinTor', 'Tracer', 'GF-120', 'Justice', 'Laser', 'Naturalyte' and 'Spinoace' (Dow AgroSciences), 'Racer Gold' (Agri Life), 'Extinosad' (veterinary use) (Elanco). **Patents:** US 5,202,242 (1993); EPO 375316 (1990).

⓮ APPLICATION: The compound is applied at rates of 12 to 150 grams per hectare. Apply when pest pressure demands treatment. The active ingredient does not dissolve in water and continual agitation is required to prevent the active ingredient from settling out in the spray tank. The addition of adjuvants has not been shown to improve or reduce the performance of spinosad consistently, with the exception of leaf miner control and the penetration of closed canopies, where emulsified vegetable oils have helped.

Notes

15   Information on the purity and quality of the technical material, details of the recommended storage conditions and shelf-life.

16   Information on incompatibilities and other information on the possibility of interactions between the product and other crop protection products.

17   Toxicological data to include acute oral and dermal LD_{50}, inhalation, skin and eye effects, no observed effect level, acceptable daily intake, toxicity class, toxicity reviews and any other information on adverse effects.

18   Effects on non-target organisms (birds, fish and other aquatic species, wildlife and beneficial species), persistence, degradation rates and pathways.

19   Indicates whether at least one national accreditation authority has accepted its use in organic farming. Accepted by IFOAM.

⓯ PRODUCT SPECIFICATIONS: Purity: The commercial product is composed of spinosyn A and spinosyn D. Analysis is undertaken by hplc or immunoassay (details from Dow AgroSciences). **Storage conditions:** Spinosad is stable over a wide range of temperatures. Protect from freezing. Shake well before use. **Shelf-life:** The formulated product has a shelf-life of 3 years.

⓰ COMPATIBILITY: No compatibility problems have been identified to date when tank mixing spinosad with other crop protection products, foliar fertilisers or adjuvants. A jar test for compatibility is recommended prior to use.

⓱ MAMMALIAN TOXICITY: Acute oral LD_{50}: male rats 3783, female rats >5000 mg/kg. **Acute dermal LD_{50}:** rabbits >5000 mg/kg. **Skin and eye:** Non-irritating to skin but slight irritation to eyes. **NOEL:** For dogs, mice and rats, following 13 weeks of dietary exposure to spinosad, was 5, 6 to 8 and 10 mg/kg/day, respectively.
Other toxicological effects: In acute and sub-chronic tests, spinosad did not demonstrate any neurotoxic, reproductive or mutagenic effects on dogs, mice or rats.

⓲ ENVIRONMENTAL IMPACT AND NON-TARGET TOXICITY: Bird toxicity: Spinosad is considered practically non-toxic to birds. The acute oral LD_{50} for both bobwhite quail and mallard ducks is 2000 mg/kg. **Fish toxicity:** Spinosad is considered slightly to moderately toxic to fish. The acute 96-hour LC_{50} for rainbow trout, bluegill sunfish and carp was 30, 5.9 and 5 mg/litre, respectively. **Effects on beneficial species:** Spinosad is considered highly toxic to honeybees, with less than 1 µg/bee of technical material applied topically resulting in mortality. Once residues are dry, they are non-toxic. **Metabolism:** Feeding studies produced no residues of spinosad in meat, milk or eggs. The half-life on plant surfaces ranged from 1.6 to 16 days, with photolysis as the main route of degradation. **Behaviour in soil:** Spinosad is rapidly degraded on soil surfaces by photolysis and, below the soil surface, by soil micro-organisms. **Key reference(s):** D G Saunders & B L Bret. 1997. In *Down to Earth*, **52**:1, pp. 14–21, DowElanco, Indianapolis, IN, USA. **⓳ Accepted for use in organic farming:** Yes.

Section 3: Macro-organisms – sample entry

Notes

1 Number representing the section into which the entry is placed, e.g. 3 = Macro-organisms.

2 Sequential entry number.

3 Approved name.

4 Class of biological activity.

5 Taxonomic classification by order and family.

6 The approved name (with authorities), common names and other names.

7 The history of the discovery of the insect predator and details of from where the organism was first isolated.

8 A summary of the method used to produce the predator for commercial sale.

9 A detailed list or description of the pest species (with authorities) against which the predator is recommended.

10 A list of the crop situations in which the product is recommended.

11 A full description of the biological activity of the predator to include its development, egg laying, method of predation and longevity.

3:246 *Amblyseius californicus*

❹ *Predator of mites*

❺　Predatory mite: Acari: Phytoseiidae

❻　**NOMENCLATURE: Approved name:** *Amblyseius californicus* (McGregor).
Other names: Spider mite predatory mite. **Formerly known as:** *Amblyseius chilensis* (Dosse),
Amblyseius mungeri (McGregor), *Neoseiulus californicus* (McGregor), *Typhlodromus californicus*
McGregor, *Typhlodromus mungeri* (McGregor) and *Typhlodromus marinus* (Willmann).

❼　**SOURCE:** *Amblyseius californicus* is a native of Mediterranean climates. It has also been
found in tropical and sub-tropical areas in North and South America.

❽　**PRODUCTION:** Reared on phytophagous mites, such as *Tetranychus urticae* Koch, on
beans.

❾　**TARGET PESTS:** *Amblyseius californicus* feeds on tetranychids, such as *Tetranychus urticae*,
T. cinnabarinus (Boisduval), *Panonychus ulmi* (Koch) and *Eotetranychus williamettei* Ewing, and
on tarsonemids, such as *Polyphagotarsonemus latus* (Banks). The preferred growth stages of
spider mites are the eggs and immature stages. It will survive on other small arthropods and
on pollen, but it will not reproduce in the absence of spider mites.

❿　**TARGET CROPS:** Glasshouse-grown crops, particularly those grown at relatively high
temperatures and low relative humidities, such as cucumbers, peppers and ornamentals. Also
effective in field crops grown under high temperature conditions, such as strawberries.

⓫　**BIOLOGICAL ACTIVITY: Biology:** Eggs are laid on the hairs in the axils of the midvein
and the lateral veins on the underside of leaves. The eggs are oval-shaped, white and clear
and hatch into six-legged larvae that do not feed but remain in groups near their place of
emergence. Proto- and deutonymphal stages and adults have eight legs, are mobile and feed.
Predation: At 26 °C, immature predatory mites consume, on average, 11.4 spider mite eggs
and nymphs before reaching adulthood. An adult female can consume in excess of 150 prey
over a 16-day period. *Amblyseius californicus* feeds on phytophagous mites by piercing the prey
and sucking the contents. The females eat all stages of phytophagous mites. **Egg laying:** Egg
laying varies with temperature, from less than one per day (to yield a total of 48 throughout
the female life-cycle) at 13 °C to over 3.5 per day (to yield a total of 65 throughout the
female life-cycle) at 33 °C. Duration of development, egg laying and longevity will depend
upon temperature, the type and availability of food and the ambient humidity.
Duration of development: *Amblyseius californicus* development is more rapid at high
temperatures, taking about 15 days at 15 °C, 8 days at 20 °C and 5.5 days at 25 °C. In a
study conducted at 21 °C, eggs hatched after three days, larvae matured after one day,
protonymphs and deutonymphs took 3 to 4 days to develop – giving a total development

Notes

12   A list of key references that describe the discovery, effectiveness and commercialisation of the parasite or predator.

13   Details of the development stage and carrier materials used to sell the product, together with the tradenames and the manufacturers.

14   Details of frequency and methods of application and rate of use.

15   Information on the purity and quality of the technical material, details of the recommended storage conditions and shelf-life.

16   Information on incompatibilities and other information on the possibility of interactions between the product and other crop protection products.

17   Toxicological data, as far as they are known.

18   Effects on non-target organisms.

All macro-organisms can be used in organic farming, although some have specific country restrictions applied to them.

time of 7.3 days. **Efficacy:** *Amblyseius californicus* is particularly useful where food is scarce, temperatures are high, humidity is low and when the phytophagous mites are concealed in the terminal shoots of the crop.

⑫ **Key reference(s):** G J de Moraes, J A McMurty & H A Denmark. 1986, A catalog of the mite family Phytoseiidae: references to taxonomy, synonymy, distribution and habitat, *EMBRAPA Departmento Difusao de Technologia, Brasilia DF, Brazil.*

⑬ **COMMERCIALISATION: Formulation:** Minimum of 2000 live adults and juvenile predators per unit in vermiculite or corn husks. Contains an unspecified number of predator eggs. DEFRA licence required for release in the UK. **Tradenames:** 'Amblyline cal' (Syngenta Bioline), 'Californicus-System' (Biobest), 'Spider Mite Predator' (Arbico), 'Ambsure (cal)' (Biological Crop Protection), 'Neoseiulus californicus' (Rincon-Vitova), (Biocontrol Network), (Hydro-Gardens), (Harmony) and (IPM Laboratories), 'Spical' (Koppert).

⑭ **APPLICATION:** Predatory mites are fragile and should be carefully mixed with the carrier material and sprinkled over the crop such that the carrier settles on the foliage, allowing the predators to move onto the crop. If there are hot-spots in the crop, the material can be poured into a box hanging within the crop where the phytophagous mites are most abundant. If supplied on bean leaves, lay the leaves carefully on the leaves of the crop.

⑮ **PRODUCT SPECIFICATIONS: Purity:** No phytophagous mites, no mould and no other contaminants. **Storage conditions:** Store at 8 to 10 °C, 65% to 80% r.h., out of direct sunlight. Protect from freezing and from high temperatures. **Shelf-life:** Use within 3 days of receipt, if stored under recommended conditions.

⑯ **COMPATIBILITY:** Incompatible with foliar applied insecticides and acaricides and insecticidal smokes. Can be used in conjunction with *Phytoseiulus persimilis* and with systemic soil-applied insecticides such as imidacloprid.

⑰ **MAMMALIAN TOXICITY:** Allergic reactions have been noted in production workers, but this may be due to the prey rather than the predator. There have been no reports of allergic or other adverse reactions following use in glasshouses or in the field.

⑱ **ENVIRONMENTAL IMPACT AND NON-TARGET TOXICITY:** *Amblyseius californicus* is widespread in Nature and there have been no reports of adverse environmental impact or effects on non-target organisms.

Section 4: Semiochemicals – sample entry

Notes

1   Number representing the section into which the entry is placed, e.g. 4 = Semiochemicals.

2   Sequential entry number.

3   Approved name.

4   Class of biological activity.

5   Indicates an entry in *The Pesticide Manual* Fifteenth edition.

6   The approved name (with authorities), common name, other names by which it is known and CAS RN.

7   As all compounds originate in Nature, the original source for each entry is included. The authority for the insect's scientific name is given.

8   Most pheromones are synthesised rather than being extracted from the insect.

9   A list or description of the pest species (with authorities) against which the product is recommended (usually a single species).

10   A list of the crops in which the product is recommended.

11   A full description of the biological activity associated with the entry. This will include mode of action (where known), behaviour of the insect when exposed to relatively high rates of the semiochemical and a summary of the efficacy of the product.

12   A list of key references that describe the discovery, effectiveness and commercialisation of the product.

13   Details of the formulation types used commercially, the tradenames and the manufacturers plus selected patent numbers.

4: 407 gossyplure

❹ *Pink bollworm sex pheromone*

---

❺ *The Pesticide Manual* **Fifteenth Edition entry number:** 446.

(Z,Z)-   (Z,E)-

❻ **NOMENCLATURE: Approved name:** gossyplure (name in common use); (Z,Z)- plus (Z,E)-hexadeca-7,11-dien-1-yl acetate. **Other names:** Pink bollworm sex pheromone; *Pectinophora gossypiella* sex pheromone; PBW; Z7Z11-16Ac; Z7E11-16Ac. **CAS RN:** [53042-79-8] (7Z,11E)- isomer; [52207-99-5] (Z,Z)- isomer; [122616-64-2] (7-Z,11-unspecified stereochemistry)- isomer; [50933-33-0] unspecified stereochemistry.

❼ **SOURCE:** The sex pheromone of the pink bollworm (*Pectinophora gossypiella* (Saunders)). Isolated from the abdominal tips of virgin females.

❽ **PRODUCTION:** Manufactured for use in the protection of cotton.

❾ **TARGET PESTS:** Pink bollworm (*Pectinophora gossypiella*).

❿ **TARGET CROPS:** Cotton.

⓫ **BIOLOGICAL ACTIVITY: Mode of action:** Gossyplure is the sex pheromone of the pink bollworm. Male moths locate and subsequently mate with female moths by following the trail or pheromone plume emitted by the virgin females. The indiscriminate application of high levels of gossyplure interferes with this process, as a constant exposure to high levels of pheromone makes trail following impossible (habituation/adaptation). Alternatively, the use of discrete sources of pheromone released over time presents the male with a false trail to follow (confusion). Control is subsequently achieved through the prevention of mating and laying of fertile eggs. **Efficacy:** Very low rates are required to cause mating disruption. Gossyplure is volatile and distributes throughout the crop easily.

⓬ **Key reference(s):** J R Merkl & H M Flint. 1981. Responses of male pink bollworms to various mixtures of the (Z,Z)- and (Z,E)- isomers of gossyplure, *Environ. Entomol.*, 6, 114.

⓭ **COMMERCIALISATION: Formulation:** Slow-release formulations of hollow fibres of polyacrylate resin containing gossyplure, laminated flakes covering a porous layer impregnated with gossyplure, polyamide micro-capsules containing gossyplure, twist tie dispensers and polymer bands are all commercially available. Consep has introduced a sprayable controlled-release granule formulation. **Tradenames:** 'Nomate PBW' (Scentry), 'Pectone' and 'Sirene

Notes

14   Details of frequency and methods of application, siting of the pheromone release system and rate of release.

15   Information on the purity of the technical material.

16   Information on incompatibilities and other information on the possibility of interactions between the product and other crop protection products.

17   Toxicological data to include acute oral and dermal LD_{50}, inhalation, skin and eye effects, no observed effect level, acceptable daily intake, toxicity class, toxicity reviews and any other information on adverse effects.

18   Effects on non-target organisms and the environment.

19   Indicates whether at least one national accreditation authority has accepted its use in organic farming. Accepted by IFOAM.

PBW' (plus permethrin) (Syngenta), 'PBW Rope-L' and 'PB Rope' (Shin-Etsu), 'Frustrate PBW' and 'DeCoy PBW Band' (Certis), 'Selibate PBW' (Certis) and (Monterey), 'Last Flight PBW' and 'Lost Dream' (plus methomyl plus chlorpyrifos) (Troy), 'Checkmate PBW' and 'Checkmate PBW-F' (Consep), 'Disrupt PBW' (Hercon), 'Dismate' (Russell), 'Pherocon' (Trece), 'Sirene-PBW' (plus permethrin) (IPM Technologies).

⑭ APPLICATION: Many slow-release formulations are applied by aerial spraying, with the slow-release plastic device adhering to the crop's foliage and the pheromone diffusing into the field, but the product is also suitable for Ultra-Low Volume (ULV) application or it may be applied using a back pack or trailer-mounted sprayer. Other dispensers are attached by hand to individual plants within the cotton field. Apply at the pinhead square stage of cotton crop growth, at a rate of 10 to 30 g a.i. per hectare. Do not exceed 150 g per hectare per year. Diluted mixtures should be sprayed within 12 hours of preparation.

⑮ PRODUCT SPECIFICATIONS: The product is composed of the two isomers in the ratio c. 1:1. It should be stored in a cool place until used and freezing extends the shelf-life indefinitely. **Purity:** The isomers are >95% chemically pure.

⑯ COMPATIBILITY: Gossyplure can be used alone or in combination with other chemicals, such as foliage fertilisers or with insecticides as a lure and kill strategy. When used in conjunction with a monitoring programme, the results from the monitoring traps may be used to determine the timing of applications. Typically, three spray applications will be needed between May and August. Under optimum conditions, gossyplure should suppress trap catch for two weeks. Increase in catch above tolerance level indicates the need for re-application.

⑰ MAMMALIAN TOXICITY: Gossyplure has shown no adverse toxicological effects on manufacturers, formulators, research workers or farmers. It is considered to be non-toxic. **Acute oral LD$_{50}$:** rats >5000 mg/kg. **Acute dermal LD$_{50}$:** rabbits >2000 mg/kg. **Inhalation:** LC$_{50}$ (4 h) rats >2000 mg/litre. **Toxicity class:** EPA (formulation) III.

⑱ ENVIRONMENTAL IMPACT AND NON-TARGET TOXICITY: Gossyplure is a natural insect pheromone that is specific to the pink bollworm. There is no evidence that it has caused any adverse effects on any non-target organisms nor had any adverse environmental impact. **⑲ Accepted for use in organic farming:** Yes.

Resistance

The increased use of crop protection agents with single modes of action has resulted in the development of resistance to a wide range of chemicals in a large number of insects, mites, plant pathogens and weeds. The problem is so serious that the agrochemical industry has established a number of Resistance Action Committees as Specialist Technical Groups of CropLife International (formerly the Global Crop Protection Federation). These Resistance Action Committees monitor the development of resistance and introduce industry-wide recommendations for the use of all crop protection agents in such a way as to reduce the possibility of resistance developing.

The Resistance Action Committees include the Insecticide Resistance Action Committee (IRAC), the Fungicide Resistance Action Committee (FRAC) and the Herbicide Resistance Action Committee (HRAC). Details of the recommendations and activities of these committees can be obtained from the following address:

CropLife International, Avenue Louise 143, 1050 Brussels, Belgium; www.gcpf.org

The individual Resistance Action Committees can be contacted as follows.

Fungicides (FRAC)
www.frac.info

Mr A. Leadbeater, FRAC Chairman, Syngenta Crop Protection AG, WRO-1004.4.31, Schwarzwaldallee 215, CH-4058 Basel, Switzerland
Tel: +41 61 32 34 190
Mobile: +41 79 35 84 190
Fax: +41 61 32 36 127
e-mail: Andy.Leadbeater@Syngenta.com

Members:
Dr K. Stenzel, Bayer CropScience, Germany; Vice-Chairman FRAC and Chairman – SBI Fungicides Working Group
Prof. P. E. Russell, Private Consultant; Secretary/Treasurer
Dr R. Gold, BASF, Germany; Communication and Website Officer
Dr N. Fernandes-Buzzerio, Syngenta Crop Protection, Brazil; Regional representative FRAC
Mr J.-L. Genet, DuPont, France; Chairman – Benzimidazoles Expert Forum
Dr D. Hermann, Syngenta Crop Protection, Switzerland; Chairman – Phenylamides Expert Forum
Dr G. Kemmitt, Dow Agrosciences, UK; Ordinary Member

Ms H. Lachaise, Bayer CropScience, France; Chairman – Anilinopyrimidines Working Group and Dicarboximides Expert Forum

Dr M. Merk, BASF, Germany; Chairman – CAA Fungicides Working Group

Dr G. Olaya, Syngenta Crop Protection, USA; Representative NAFRAC

Dr H. Sierotzki, Syngenta Crop Protection, Switzerland; Chairman Banana Working Group

Dr K. Tanabe Nippon Soda, Japan; Regional representative Japan

Herbicides (HRAC)
www.plantprotection.org/hrac

Harvey Glick, HRAC Chairman, Monsanto, Global Product Stewardship, St Louis, MO, USA

Tel: +1 314 694 6019

Fax: +1 314 694 1139

e-mail: harvey.l.glick@monsanto.com

Members:

David Vitolo, HRAC Treasurer, Syngenta Crop Protection,

James Bahr, FMC, Princeton, NJ, USA

Tim Obrigawitch, Dupont Crop Protection

Marvin Schultz, Dow AgroSciences

Sandra Shinn, FMC

Michelle Starke, Product Stewardship Manager, Monsanto

James Whitehead, Makhteshim Agan of North America

Les Glasgow, NA-HRAC Chairman Syngenta

Dr Kazuyuki Itoh, APAC chairman

Anne Thompson, EHRWG chairperson, Dow AgroSciences, UK

Insecticides (IRAC)
http://PlantProtection.org/IRAC

Alan Porter, IRAC Co-ordinator, Alan Porter Associates, Glentruim House, 53 Dirleton Avenue, North Berwick, EH39 4BL, Scotland

Tel: +44 1620 895 674

Fax: +44 1620 890 430

e-mail: porterapa@aol.com

IRAC is supported by 15 of the major insecticide manufacturing companies – BASF, Bayer CropScience, Belchim Crop Protection, Cheminova, Chemtura, Dow AgroSciences, DuPont, FMC, Makhteshim-Agan, Monsanto, Nihon-Nohyaku, Nufarm, Sumitomo Chemical, Syngenta and Vestergaard Frandsen; and has seven country groups – Australia, Brazil, India, South Africa, Spain, USA and South East Asia.

The use of natural products, pheromones, living systems, insect predators and parasites and the development of transgenic crops in integrated crop management systems are considered to be valuable strategies to slow the development of resistance and, thereby, improve the value and life of many crop protection agents while maintaining a level of insect, disease and weed control that the grower demands. The reason for the success of this approach is the reduced ability of target pests to develop detoxification mechanisms or to avoid the attentions of a living control agent. Today it is increasingly true that living biologicals and products based on natural agents will find increasing use in future crop protection strategies.

Organic farming

Definition

Organic farming is defined as a holistic method of farming that is designed to work with and alongside natural systems. It aims to reduce the occurrence of pest and disease damage to crops by using fundamental and sustainable farming and agronomic practices that include the following.

- Sustainable crop rotations – cropping systems that maintain fertility and control pests and diseases. Rotations remove all traces of host crops for periods of time that should be long enough to interrupt pest or disease life-cycles. In addition, rotations encourage not only more diverse and stable eco-systems than conventional, monoculture agriculture but also the establishment of populations of natural predators.

- Biodiversity – it is claimed that the exclusion of synthetic pesticides from organic farming serves to preserve and enhance biodiversity within the growing system, allowing natural enemies to thrive and, thereby, exert control on pest populations. This approach is enhanced by the conservation and improvement of natural, environmentally friendly features of the landscape such as hedgerows and ponds, the construction of beetle banks and sown wild flower strips that enable communities of beneficial organisms to flourish. (These features also are a component of Integrated Crop Management (ICM) systems.)

- Optimising crop health – the soil is the basis of environmentally sound and sustainable organic farming and the maintenance of a healthy soil is fundamental to organic farming systems. Soil micro-organisms are of primary importance because they process the organic matter, thereby providing a balance of nutrients and minerals that are used by the crop to achieve healthy and vigorous growth. Healthy plants, furthermore, are more able to withstand pest and disease attack. Good crop husbandry and hygiene and associated sound agronomic practices make a significant contribution to the health of the crop and the prevention of pest and disease problems.

- Selection of crop varieties – the selection of crop varieties that are not only appropriate for the location, but also are inherently resistant to pests and diseases, ensures that the level of inputs necessary for pest and disease control will be significantly reduced or even avoided completely.

Accreditation

The organic movement is global and is practised on every continent, with organic food being traded worldwide. In the European Union (EU) and the USA, the labelling of organic agricultural products is regulated by legislation. This legislation contains regulations applying to third countries and, therefore, opens up access to markets for organic products originating outside the EU and the USA. Proof must be provided that the production, inspection and certification processes comply with the legislation. An increasing number of countries have been accepted onto a EU provisional third-country list and, therefore, special regulations govern the export of their organic products to the EU. At present, inspections and certifications of organic products for the EU and USA markets are, to a large extent, carried out by private, international inspection and certification bodies in the EU or the USA. Local national inspection bodies in third countries have become increasingly active, and are taking over the tasks of inspection and certification in developing countries. The establishment of such local national inspection bodies is being encouraged to safeguard the independent and sustainable development of organic farming in developing countries.

Within the UK, there are several organic farming accreditation bodies, as follows.

- United Kingdom Register of Organic Food Standards (UKROFS). The standards set by UKROFS conform to the minimum standards set by International Federation of Organic Agricultural Movements (IFOAM).
- Organic Farmers and Growers (OF&G).
- Scottish Producers' Association (SOPA).
- Soil Association Certification Ltd (SA Cert). Soil Association certification has its own standards, which are more exacting than those laid down by UKROFS.
- Demeter/Bio-Dynamic Agriculture Association (BDAA).
- Irish Organic Farmers' & Growers' Association (IOFGA).
- Food Certification (Scotland) Ltd. Food Certification (Scotland) Ltd provides organic certification for farmed salmon in the UK. UK regulations permit local certification.
- Organic Trust Ltd.
- CMi Certification.

Other countries have their own organisation(s) that accredit growers within each country (in the USA, for example, one of these is the USDA and another the Organic Materials Review Institute, Box 11558, Eugene, OR 97440, USA; Tel: +1 541 343 7600; Fax: +1 541 343 8971; e-mail: info@omri.org), but each of these organisations is associated with the International Federation of Organic Agricultural Movements (IFOAM), which sets basic standards for organic agriculture and runs an international accreditation programme for organic certification bodies. IFOAM is a grassroots and democratic organisation that currently unites 750 member organisations in 108 countries. In order to achieve its mission and address the complexity of the various components of the organic agricultural movement worldwide, IFOAM has established official committees and groups with very specific purposes, from the development of standards to the facilitation of organic agriculture in developing countries. The IFOAM General Assembly serves at the foundation of IFOAM. It elects the World Board for a three-year term. The World Board appoints members to official committees, working groups and task forces based upon the recommendation of the IFOAM membership, and IFOAM member organisations also establish regional groups and sector-specific interest groups. The IFOAM World Board has established the following official structures:

- the Norms Management Committee, which includes members of the Standards Committee and the Accreditation Criteria Committee
- the FAO Liaison Office
- various working groups and temporary task forces
- IFOAM Regional Groups.

IFOAM member organisations have also established professional bodies such as the IFOAM Organic Trade Forum, the Organic Retailers Association, the IFOAM Aquaculture Group and the IFOAM Forum of Consultants and initiatives such as the Farmers' Group.

The accreditation system recognises regional differences and has provided an internationally acceptable set of standards for organic growers that are implemented by the regional groups. IFOAM standards, however, cannot be used for certification on their own – they are designed merely to provide a framework for certification systems worldwide to develop national or regional standards. These local standards take regional or local conditions into account and may be more rigorous than IFOAM basic standards.

Aims

The principal aims of organic production and processing as laid down by IFOAM (see www.ifoam.org) are based on a number of principles and ideas. All are important and the list below is not arranged in order of merit.

- To produce food of high quality and in sufficient quantity.
- To interact in a constructive and life-enhancing way with natural systems and cycles.
- To consider the wider social and ecological impact of the organic production and processing system.
- To encourage and enhance biological cycles within farming systems, involving micro-organisms, soil flora and fauna, plants and animals.
- To develop a valuable and sustainable aquatic ecosystem.
- To maintain and increase long-term fertility of soils.
- To maintain the genetic diversity in the production system and its surroundings and to protect plant and wildlife habitats.
- To promote the healthy use and proper care of water, water resources and all life therein.
- To use, as far as possible, renewable resources in locally organised production systems.
- To create a harmonious balance between crop production and animal husbandry.
- To give all livestock living conditions that have due consideration for the basic aspects of their innate behaviour.

- To minimise or eliminate all forms of pollution.
- To process organic products using renewable resources.
- To produce organic products that are fully biodegradable.
- To produce textiles that are long-lasting and of good quality.
- To allow everyone involved in organic production and processing a quality of life which meets their basic needs and allows an adequate return and satisfaction from their work, including a safe working environment.
- To progress towards an entire production, processing and distribution chain that is both socially just and ecologically responsible.

Procedures for certification

If a farmer wishes to become an organic grower, he/she is not allowed immediately to sell their produce as organic. All new organic growers are required to convert their fields to attain organic status. The rules that govern conversion are laid down by IFOAM and are administered regionally by national organic farming organisations. The establishment of an organic management system and the building of soil fertility require an interim conversion period. The conversion period may not always be of sufficient duration to improve soil fertility and re-establish the balance of the ecosystem, but it is the period in which all the actions required to reach these goals have been started.

Conversion requirements follow the principle that organic agriculture is a process which develops a viable and sustainable agro-ecosystem. For a sustainable agro-ecosystem to function optimally, diversity in crop production and animal husbandry must be arranged in such a way that all the elements of the farming management interact. Conversion may be accomplished over a period of time and a farm may be converted step by step. However, it is expected that, on a single farm, all crop production and all animal husbandry will eventually be converted to organic management, although this is not a requirement imposed by most accreditation agencies.

The national certification bodies set standards that clearly separate and document the products of different farming systems. These standards are designed to prevent the mixing of input factors and produce. These standards must all be applied to the relevant aspects from the beginning of the conversion period and progressively met during the conversion period.

Before products from a farm or project can be certified as organic, inspection must be made during the conversion period. The start of the conversion period may be calculated from the date of application to the local or regional certification bodies or standardising organisations (hereafter referred to as the certification bodies), or from the date of last application of unapproved farm inputs, but only if it can be demonstrated that standards requirements have been met from that date.

Plant products from annual crops can be certified organic when the standards requirements have been met for a minimum of 12 months before the start of the production cycle. Perennial plants (excluding pastures and meadows) can be certified organic at the first harvest after at least 18 months of management according to the standards requirements. Pastures, meadows and their products can be certified after 12 months of organic management. Where the certification body or standardising organisation requires a period of three or more years of documented non-use of prohibited materials, certification may be granted after the first of these years – that is 12 months after application. This conversion period can be extended by the certification body, depending on past use of the land and environmental conditions.

The certification body may allow plant products to be sold as 'produce of organic agriculture in process of conversion' or a similar description, when the standard requirements for certification as organic have been met for at least 12 months. A full conversion period is not required where *de facto* full standards requirements have been met for several years and where this can be verified through numerous means and sources. In such cases, inspection must be carried out with a reasonable time interval before the first harvest.

If the whole farm is not converted, the certification body must ensure that the organic and conventional parts of the farm are separate and can be inspected. Simultaneous production of conventional, in conversion and/or organic crops or animal products is allowed only where different types of production are clearly distinguished. To ensure a clear separation between organic and conventional production, the certification body must inspect the whole system from production to final market. On farms with simultaneous organic and conventional production, the use of any genetically engineered organisms is not permitted on the conventional part, even if the organic farm is not cultivating that crop.

Organic certification is based on an ongoing commitment to organic production practices and, as such, the certification body should only certify production which is likely to be maintained on a long-term basis. Converted land and animals must not be moved back and forth between organic and conventional management.

The certification body will set standards for a minimum percentage of the farm area designed to facilitate biodiversity and nature conservation to include:

- extensive grassland such as moorlands, reed land or dry land
- in general, all areas which are not under rotation and are not heavily manured: extensive pastures, meadows, extensive grassland, extensive orchards, hedges, hedgerows, edges between agriculture and forest land, groups of trees and/or bushes, and forest and woodland
- ecologically rich fallow land or arable land
- ecologically diversified (extensive) field margins
- waterways, pools, springs, ditches, floodplains, wetlands, and swamps and other water-rich areas which are not used for intensive agriculture or aquaculture production
- areas with wasteland flora.

Rules for organic crop production

- All seeds and plant material must be certified organic.

- Species and varieties cultivated must be adapted to the soil and climatic conditions and be resistant to pests and diseases.

- In the choice of varieties, genetic diversity must be taken into consideration.

- When organic seed and plant materials are available, they shall be used. The certification body shall set time limits for the requirement of certified organic seed and other plant material.

- When certified organic seed and plant materials are not available, chemically untreated conventional materials must be used as a first option.

- Where no other alternatives are available, chemically treated seed and plant material may be used. The certification body shall define conditions for exemptions and set time limits for any use of chemically treated seeds and plant materials.

- The use of genetically engineered seeds, pollen, transgenic plants or plant material is not allowed.

- Wild-harvested products shall be certified organic only if derived from a stable and sustainable growing environment. Harvesting or gathering the product shall not exceed the sustainable yield of the ecosystem, or threaten the existence of plant or animal species.

- Products can be certified organic only if derived from a clearly defined collecting area which is not exposed to prohibited substances, and which is subject to inspection.

- The collection area shall be at an appropriate distance from conventional farming, pollution and contamination.

- The operator managing the harvesting or gathering of the products shall be clearly identified and be familiar with the collecting area in question.

Where appropriate, the certification body is required to ensure that sufficient diversity is obtained, in time or place, in a manner that takes into account pressure from insects, weeds, diseases and other pests, while maintaining or increasing soil organic matter, fertility, microbial activity and general health. For non-perennial crops, this is normally, but not exclusively, achieved by means of crop rotation.

Sufficient quantities of biodegradable material of microbial, plant or animal origin should be returned to the soil to increase or at least maintain its fertility and the biological activity within it. Such material produced on organic farms should form the basis of the fertilisation programme.

All chemical inputs, including fertilisers and pest control agents, are closely controlled. Organic farming systems should be carried out in a way which ensures that losses from pests, diseases and weeds are minimised. Emphasis is placed on the use of crops and varieties well adapted to the environment, a balanced fertilisation programme, fertile soils of high biological activity, adapted rotations, companion planting and green manures. Growth and development of crops and livestock should take place in a natural manner.

Weeds, pests and diseases should be managed by a number of preventive cultural techniques which limit their development, such as suitable rotations, green manures, a balanced fertilisation programme, early and predrilling seedbed preparations, mulching, mechanical control and the disturbance of pest development cycles. The natural enemies of pests and diseases should be protected and encouraged through proper habitat management of hedges, nesting sites and other beneficial habitats. Pest management should be regulated by understanding and disrupting the ecological needs of the pests.

Products used for pest, disease and weed management, prepared at the farm from local plants, animals and micro-organisms, are allowed, but brand name products must always be confirmed as acceptable by the certification body.

Thermic weed control and physical methods for pest, disease and weed management are permitted. Thermic sterilisation of soils to combat pests and diseases is restricted to circumstances where a proper rotation or renewal of soil cannot take place. Permission may be given by the certification body only on a case by case basis.

All equipment from conventional farming systems must be properly cleaned and free from residues before being used on organically managed areas.

IFOAM establishes the standards of organic farming, but it is dependent upon national organisations within each country to monitor, regulate and accredit organic farming produce and to supervise conversion from conventional to organic systems. The vigour and enthusiasm with which these different bodies apply and regulate farmers varies, but their power for accreditation stems from membership of IFOAM and this can be withdrawn if the certification body falls below the standards expected by IFOAM. Information on national bodies is available from IFOAM.

Products that may be used for fertilisation and soil conditioning

- Farmyard manure, slurry and urine
- Guano
- Human excrement from separated sources which are monitored for contamination (not to be directly applied on edible parts)
- Vermicastings
- Blood meal, meat meal, bone, bone meal

- Hoof and horn meal, feather meal, fish and fish products, wool, fur, hair, dairy products
- Biodegradable processing by-products of microbial, plant or animal origin, e.g. by-products of food, feed, oilseed, brewery, distillery or textile processing
- Crop and vegetable residues, mulch, green manure, straw
- Wood, bark, sawdust, wood-shavings, wood-ash, wood charcoal
- Seaweed and seaweed products
- Peat (prohibited for soil conditioning), but excluding synthetic additives; permitted for seed, potting module composts
- Compost made from ingredients such as spent mushroom waste, humus from worms and insects, urban composts from separated sources which are monitored for contamination
- Plant preparations and extracts
- Worms and microbiological preparations based on naturally occurring organisms
- Biodynamic preparations
- Basic slag
- Calcareous and magnesium amendments
- Calcified seaweed
- Limestone, gypsum, marl, chalk, sugar beet lime, calcium chloride
- Magnesium rock, kieserite and Epsom salt (magnesium sulfate)
- Mineral potassium (e.g. potassium sulfate, sylvanite), if obtained by physical procedures but not enriched by chemical processes
- Natural phosphates (cadmium content should not exceed 90 mg/kg P_2O_5)
- Pulverised rock, stone meal
- Clay (e.g. bentonite, perlite, vermiculite, zeolite)
- Sodium chloride
- Trace elements
- Sulfur

Products that may be used for plant pest and disease control, weed management and growth regulation

I Plant and animal origin

- Algal preparations
- Animal preparations and oils
- Beeswax
- Chitin nematicides (natural origin)

- Coffee grounds
- Copper salts (such as sulfate, hydroxide, oxychloride, octanoate) (Copper usage has been reduced to a maximum of 8 kg/ha per year (on a rolling average basis) or less, according to national laws or private label standards)
- Corn gluten meal (weed control)
- Dairy products (such as milk, casein)
- Gelatine
- Lecithin
- Natural acids (such as vinegar)
- Neem (*Azadirachta indica* extracts where registered for use)
- Plant oils
- Plant preparations
- Plant-based repellents
- Pyrethrum [*Tanacetum* (formerly *Chrysanthemum*) *cinerariaefolium* extracts], but without the synergist piperonyl butoxide after 2006
- Quassia (*Quassia amara* extracts where registered for use)
- Rotenone (extracts of *Derris elliptica*, *Lonchocarpus* spp. and *Thephrosia* spp.)
- Ryania (*Ryania speciosa* extracts where registered for use)
- Sabadilla (where registered for use)
- Tobacco tea (pure nicotine is forbidden)

II Mineral origin
- Clay (e.g. bentonite, perlite, vermiculite, zeolite)
- Diatomaceous earth
- Lime sulfur (calcium polysulfide)
- Quicklime
- Silicates (such as sodium silicates, quartz)
- Sulfur

III Organisms used for biological pest control
- Fungal preparations
- Bacterial preparations (*Bacillus thuringiensis*)
- Release of parasites, predators and other natural enemies, and sterilised insects
- Viral preparations (polyhedro- and granuloviruses)

IV Others

- Biodynamic preparations
- Calcium hydroxide
- Carbon dioxide
- Chloride of lime
- Ethyl alcohol
- Homeopathic and ayurvedic preparations
- Light mineral oils (paraffin)
- Potassium bicarbonate
- Potassium permanganate
- Sea salt and salty water
- Soda
- Sodium bicarbonate
- Soft soap
- Sulfur dioxide

V Traps, barriers, repellents

- Physical methods (e.g. chromatic traps, mechanical traps)
- Mulches, nets
- Pheromones – in traps and dispensers only

The use of synthetic growth regulators is prohibited. Synthetic dyes may not be used for cosmetic alteration of organic products, and the use of genetically engineered organisms or products thereof is prohibited.

This summary is modified from information that is available on the IFOAM website: www.ifoam.org

1. Micro-organisms

1:01 *Acremonium diospyri*
Biological herbicide (fungus)

Fungus: anamorphic: previously classified as mitosporic Deuteromycetes: Moniliales

NOMENCLATURE: Approved name: *Acremonium diospyri* (Crandall) W. Gams.
Other names: Previously known as *Nalanthamala diospyri* (Crandall) Schroers & M. J. Wingfield.

SOURCE: Introduced into the USA from South Africa in the 1960s.

PRODUCTION: Released and self-propagating.

TARGET PESTS: Persimmon (*Diospyros virginiana* L.).

TARGET CROPS: Rangeland.

BIOLOGICAL ACTIVITY: Mode of action: Pathogenesis.

COMMERCIALISATION: Formulation: Not sold commercially.

PRODUCT SPECIFICATIONS: Specifications: Released and self-propagating.

MAMMALIAN TOXICITY: It is not expected that *Acremonium diospyri* will have any adverse effects on mammals. There is no evidence of allergenic reactions.

ENVIRONMENTAL IMPACT AND NON-TARGET TOXICITY: It is not expected that *Acremonium diospyri* will have any adverse effects on non-target organisms or the environment. **Approved for use in organic farming:** Yes.

1:02 *Adoxophyes orana* granulosis virus Swiss isolate
Insecticidal baculovirus

Virus: Baculoviridae: Granulovirus

NOMENCLATURE: Approved name: *Adoxophyes orana* granulosis virus Swiss isolate.
Other names: AoGV.

SOURCE: Originally isolated from infected summer fruit tortrix (*Adoxophyes orana* F. von R.) larvae. Occurs relatively widely in Nature.

PRODUCTION: *Adoxophyes orana* granulosis virus is produced commercially on larvae of the summer fruit tortrix moth.

TARGET PESTS: Used for control of summer fruit tortrix moths (*Adoxophyes orana*).

TARGET CROPS: Used in summer fruits.

BIOLOGICAL ACTIVITY: Mode of action: As with all insect baculoviruses, *Adoxophyes orana* GV must be ingested to exert an effect. Following ingestion, the virus enters the insect's haemolymph. It multiplies in the insect body, leading to death. **Biology:** AoGV is more active on small larvae than later larval instars. It is ingested by the feeding larva and the virus protein coat is dissolved in the insect's midgut (that is alkaline), releasing the virus particles. These pass through the peritrophic membrane and invade midgut cells by fusion with the microvilli. The virus particles invade the cell nuclei, where they are uncoated and replicated. Initial replication produces non-occluded virus particles that spread to more cells and hasten the invasion of the host insect, but later the virus particles are produced with protein coats. These remain infective when released from the dead insects. **Efficacy:** AoGV acts relatively slowly, as it has to be ingested before it exerts any effect on the insect host. It is important to ensure good cover of the foliage to effect good control. Monitoring of adult insect laying patterns and application targeted at newly hatched eggs gives better control than on a mixed population. **Key reference(s):** 1) A Schmid, O Cazellos & G Benz. 1983. *Mitteilungen der Schweizerischen Entomologischen Gesellschaft*, **56**, 225–35. 2) R R Granados & B A Federici (eds). 1986. *The Biology of Baculoviruses, Vols 1 and 2*, CRC Press, Boca Raton, Florida, USA. 3) D J Leisy & J R Fuxa. 1996. Natural and engineered viral agents for insect control. In *Crop Protection Agents from Nature: Natural Products and Analogues*, L G Copping (ed.), Royal Society of Chemistry, Cambridge, UK. 4) B A Federici. 1999. Naturally occurring baculoviruses for insect pest control. In *Biopesticides Use and Delivery*, F R Hall and J J Menn (eds), pp. 301–20, Humana Press, Totowa, NJ, USA. 5) H D Burges & K A Jones. 1998. Formulation of bacteria, viruses and Protozoa to control insects. In *Formulation of Microbial Biopesticides, Beneficial Microorganisms, Nematodes and Seed Treatments*, H D Burges (ed.), pp. 33–127, Kluwer Academic Publishers, Dordrecht, The Netherlands.

COMMERCIALISATION: Formulation: Sold as a suspension concentrate (flowable concentrate) (SC) formulation. **Tradenames:** 'Capex 2' (Andermatt).

APPLICATION: 'Capex 2' is more effective against first instar larvae and so it is recommended that adult activity is monitored and the product applied shortly after egg-laying. Ensure that the treated foliage is well covered and that the pH of the spray solution is between 6 and 8. Apply 1×10^{13} GV per hectare.

PRODUCT SPECIFICATIONS: Specifications: *Adoxophyes orana* granulosis virus is produced commercially on larvae of the summer fruit tortrix moth. **Purity:** The product is produced *in vivo* in *Adoxophyes orana* larvae. The product is tested to ensure there are no bacterial contaminants which are pathogenic to man. Product specification is checked by bioassay on *Adoxophyes orana* larvae. **Storage conditions:** Store at temperatures below 2°C, in a sealed container. **Shelf-life:** Stable for 4 weeks if stored at room temperature, but has unlimited stability if stored below 2°C.

COMPATIBILITY: Compatible with all non-copper fungicides and all pesticides which do not have a repellent effect on *Adoxophyes orana*. The spray solution should be maintained at a pH between 6 and 8. Do not use chlorinated water.

MAMMALIAN TOXICITY: There is no evidence of acute or chronic toxicity, eye or skin irritation in mammals. No allergic reactions or other health problems have been observed in research or manufacturing staff or with users of the product.

ENVIRONMENTAL IMPACT AND NON-TARGET TOXICITY: *Adoxophyes orana* granulosis virus occurs in Nature. There is no evidence that it affects any organism other than summer fruit tortrix larvae. It is unstable at extreme pH and when exposed to u.v. light. It does not persist in the environment. **Approved for use in organic farming:** Yes.

1:03 *Agrobacterium radiobacter* isolates K84, K89 and K1026 *Beneficial bacterium*

Bacterium: Eubacteriales: Rhizobiacea

NOMENCLATURE: Approved name: *Agrobacterium radiobacter* (Beijerink and van Delden) Conn. isolate K84, isolate K89 and isolate K1026. *Agrobacterium radiobacter* isolate K1026 differs from the naturally occurring bacterium *Agrobacterium radiobacter* isolate K84 in that a small portion of DNA has been removed from K84 to prevent isolate K1026 from transferring DNA to the bacterium that causes crown gall disease and reduces the likelihood of resistance.

SOURCE: Naturally occurring bacterium. Found widely in Nature. Isolate K1026 was discovered and developed in Australia by Bio-Care Technology. Isolate K84 is produced by AgBioChem.

PRODUCTION: Produced by fermentation. 'Galltrol-A' is a petri dish culture on agar.

TARGET PESTS: Used for the prevention of crown gall infections caused by *Agrobacterium tumefaciens* Conn. and *Agrobacterium rhizogenes* (Riker, Banfield, Wright, Keitt & Sagen) Conn.

TARGET CROPS: Used in a variety of crops, including fruit trees, nuts, vines, soft fruit and ornamentals under glass and outside. In the USA, 'Galltrol-A' is labelled for use in all crops that are susceptible to crown gall. It can be applied to roots, stems and cuttings of non-bearing almond, pecan, apricot, caneberries, cherry, nectarine, peach, plum, prune, walnut and ornamentals, such as euonymus and rose.

BIOLOGICAL ACTIVITY: Biology: *A. radiobacter* competes with the crown gall-causing organisms, *A. tumefaciens* and *A. rhizogenes*, for invasion sites on damaged woody stems of a wide range of crops, thereby preventing the pathogenic bacteria from becoming established. There is also evidence that the antagonist produces an antibacterial compound that is inhibitory to the growth of the plant pathogens. **Efficacy:** 'Nogall' is isolate K1026 and it is ineffective against crown gall disease in grapes, pome fruit and some ornamentals.

Key reference(s): 1) A Kerr. 1980. Biological control of crown gall through production of agrocin 84, *Plant Dis.*, **64**, 25. 2) A S Paau. 1998. Formulation of beneficial microorganisms applied to soil. In *Formulation of Microbial Biopesticides, Beneficial Microorganisms, Nematodes and Seed Treatments*, H D Burges (ed.), pp. 235–54, Kluwer Academic Publishers, Dordrecht, The Netherlands.

COMMERCIALISATION: Formulation: Sold as a water dispersible powder formulation for slurry treatment (WS) containing bacterial cells with more than 1×10^9 viable bacteria per gram. 'Galltrol-A' is a petri dish culture. 'Norbac 84C' is an aqueous suspension containing bacterial cells, methyl cellulose and phosphate buffer. **Tradenames:** 'Diegall' and 'Nogall' (Bio-Care Technology), 'Galltrol-A' (AgBioChem), 'Norbac 84C' (IPM Laboratories) and (New BioProducts).

APPLICATION: Treatment involves dipping cuttings, transplants or seeds into water-based suspensions of the bacterium and planting immediately after treatment. Control is sometimes extended by soil drench treatments. A 250 gram powder pack suspended in 12 litres of water is sufficient to treat between 200 and 5000 rooted cuttings or up to 10 000 seeds. 'Galltrol-A' is isolate K84 and the bacteria from a single petri dish per US gallon (3.8 litres) are more effective than the powder treatment.

PRODUCT SPECIFICATIONS: Specifications: Produced by fermentation. 'Galltrol-A' is a petri dish culture on agar. **Purity:** 'Galltrol-A' is 99% active pure culture with 1% water, containing only living cells of the antagonist. Purity and efficacy can be determined by plating out the formulation on agar plates and incubating in the laboratory for 24 to 48 hours and counting the colonies formed. **Storage conditions:** Store in a cool, dry situation, out of direct sunlight. Do not freeze. 'Norbac 84C' should be refrigerated. **Shelf-life:** Powders, if stored according to the manufacturers' recommendations, will remain viable for a year. 'Galltrol-A', if stored in a refrigerator, remains viable for 4 to 5 months, retaining its high vigour on agar culture.

COMPATIBILITY: It is unusual to mix *Agrobacterium radiobacter* with chemicals. Use non-chlorinated water to prepare suspensions. It should not be used with broad-spectrum fungicides such as copper-based products, or bactericides or fertilisers.

MAMMALIAN TOXICITY: There has been no record of allergic or other adverse reactions from research and manufacturing staff or from users of the product. In the USA, some or all uses of the products have been approved by the Environmental Protection Agency as

Reduced Risk applications. **Acute dermal LD$_{50}$:** No infectivity or toxicity observed in rats at 3.4 × 10^{11} spores/kg. **Inhalation:** No infectivity or toxicity at 5.4 mg/litre (2.6 × 10^7 spores/litre). **NOEL:** In a two-year study on rats, 8.4 g/kg b.w. daily; in a thirteen-week study on rats, 1.3 × 10^9 spores/kg b.w. daily.

ENVIRONMENTAL IMPACT AND NON-TARGET TOXICITY: *Agrobacterium radiobacter* occurs widely in Nature and is not expected to show any adverse effects on non-target organisms or the environment. **Bird toxicity:** In 63-day feeding trials, chickens receiving up to 5.1 × 10^7 spores/g diet showed no ill effects. **Behaviour in soil:** *Agrobacterium radiobacter* spores have a very short persistence in the environment (DT$_{50}$ 10 h), mainly due to their u.v.-light sensitivity. **Approved for use in organic farming:** Yes.

1:04 *Alternaria cassiae*

Biological herbicide (fungus)

Fungus: Ascomycota: mitosporic Pleosporaceae

NOMENCLATURE: Approved name: *Alternaria cassiae* Jurair & Khan.

SOURCE: Isolated from coffee senna (*Cassia occidentalis* L.).

PRODUCTION: Produced by fermentation.

TARGET PESTS: Sicklepod (*Senna obtusifolia* (L.) H.S. Irwin & Barneby), showy crotalaria (*Crotalaria spectabilis* Roth.) and coffee senna (*Cassia occidentalis*).

TARGET CROPS: Soybeans, peanuts and cotton.

BIOLOGICAL ACTIVITY: Mode of action: Pathogenesis. **Key reference(s):** 1) R E Hoagland. 1996. Hydroponic seedling bioassay for the bioherbicides *Colletotrichum truncatum* and *Alternaria cassiae*, *Biocontrol Science and Technology*, **5**(3), 251–60. 2) C D Boyette. 1988. Biocontrol of three leguminous weed species with *Alternaria cassiae*, *Weed Technol.*, **2**, 414–7.

COMMERCIALISATION: Formulation: No longer commercially available. **Tradenames:** 'Casst' (Mycogen).

PRODUCT SPECIFICATIONS: Specifications: Produced by fermentation.

COMPATIBILITY: It is unusual to apply *A. cassiae* with other crop protection chemicals. Do not apply fungicides within 4 weeks of use. Incompatible with strong oxidisers, acids, bases and chlorinated water.

MAMMALIAN TOXICITY: It is not expected that there will be any adverse or allergenic effects from the use of *Alternaria cassiae*.

ENVIRONMENTAL IMPACT AND NON-TARGET TOXICITY: No adverse effects are expected on any non-target organisms or on the environment.
Approved for use in organic farming: Yes.

1:05 *Alternaria cuscutacidae*
Biological herbicide (fungus)

Fungus: Ascomycota: mitosporic Pleosporaceae

NOMENCLATURE: Approved name: *Alternaria cuscutacidae* Rudakov.

SOURCE: Isolated from infected dodder (*Cuscuta campestris* Yuncker) in southern parts of the former Soviet Union.

PRODUCTION: Mass produced by fermentation.

TARGET PESTS: Dodder – especially *Cuscuta campestris*.

BIOLOGICAL ACTIVITY: Mode of action: Pathogenesis.

COMMERCIALISATION: Formulation: Distributed in Russia in the 1950s, but not sold commercially.

APPLICATION: Produced in liquid culture and applied as a suspension of mycelium and infective spores.

PRODUCT SPECIFICATIONS: Specifications: Mass produced by fermentation.

COMPATIBILITY: Incompatible with strong oxidisers, acids, bases and chlorinated water, but can be applied with conventional herbicides. Fungicides should not be used within three weeks of application.

MAMMALIAN TOXICITY: It is not expected that there will be any adverse or allergenic effects from the use of *Alternaria cuscutacidae*.

ENVIRONMENTAL IMPACT AND NON-TARGET TOXICITY: No adverse effects are expected on any non-target organisms or on the environment.
Approved for use in organic farming: Yes.

1:06 *Alternaria destruens* isolate 059
Biological herbicide (fungus)

Fungus: Ascomycota: mitosporic Pleosporaceae

NOMENCLATURE: Approved name: *Alternaria destruens* E.G. Simmons isolate 059.

SOURCE: Initially isolated from swamp dodder (*Cuscuta gronovii* Willd. ex Schult.) in 1986. It is indigenous to the USA.

PRODUCTION: Produced by fermentation.

TARGET PESTS: All species of dodder including swamp dodder (*Cuscuta gronovii*), large seed dodder (*C. indecora* Choisy), small seed dodder (*C. planiflora* Ten.) and field dodder (*C. campestris* Yuncker).

TARGET CROPS: Fruit and vegetable crops, and ornamentals.

BIOLOGICAL ACTIVITY: Mode of action: Pathogenesis. **Key reference(s):** T A Bewick, J C Porter & R C Ostrowski. 2000. Field trial results with Smolder: a bioherbicide for dodder control (abstract), *Proceedings Northeastern Weed Science Society*, **54**, 66.

COMMERCIALISATION: Formulation: *Alternaria destruens* isolate 059 is sold as a granular formulation (GR) and as a wettable powder (WP). *Alternaria destruens* isolate 059 was licensed for sale in the USA in May 2005. **Tradenames:** 'Smolder G', 'Smolder R' and 'Smolder WP' (Loveland).

APPLICATION: 'Smolder G': Granules are applied to moist soil at a rate of 50 pounds (1 bag) per acre at, or immediately prior to, dodder emergence. 'Smolder WP': The liquid sprayable product should be applied when dodder vines are beginning to reach the top of the crop canopy.

PRODUCT SPECIFICATIONS: Specifications: Produced by fermentation. **Purity:** Contains no pathogenic or allergenic micro-organisms. **Storage conditions:** Store in a cool, dry condition in a sealed container. **Shelf-life:** Use as soon after purchase as possible.

COMPATIBILITY: It is unusual to apply *A. destruens* with other crop protection chemicals. Do not apply fungicides within 4 weeks of use. Incompatible with strong oxidisers, acids, bases and chlorinated water.

MAMMALIAN TOXICITY: No harmful health effects to humans are expected from use of *Alternaria destruens* isolate 059. Appropriate tests found no evidence that the fungus is toxic to humans and other mammals. No toxicological or pathogenic effects of *A. destruens* in mammals have been reported in available public literature or in the submitted data. In addition, certain biological characteristics of *A. destruens*, which include its moisture and temperature requirements during infection, and its dependence on *Cuscuta* spp. as hosts, are further indications that this microbial pest control agent would not be pathogenic to mammals. **Acute oral LD$_{50}$:** rats >1 × 10^7 colony forming units (cfu)/animal. Toxicity Category IV for acute oral toxicity. **Acute dermal LD$_{50}$:** rats >5000 mg/kg. Toxicity Category IV. **Inhalation:** LD$_{50}$ for rats >2.03 mg/litre. Toxicity Category IV for acute inhalation toxicity. **Acute pulmonary toxicity/pathogenicity:** rats >5.0 × 10^5 cfu/animal. **Skin and eye:** rabbits: Toxicity Category III for primary eye irritation. Toxicity Category IV for primary dermal irritation.

ENVIRONMENTAL IMPACT AND NON-TARGET TOXICITY: Available studies show that no adverse environmental or non-target effects are expected when products containing *Alternaria destruens* isolate 059 are used in accordance with label instructions. *A. destruens* has not been reported to infect any organism other than *Cuscuta* spp. In addition, exposure of birds, fish, aquatic invertebrates and honeybees to 'Smolder G' and 'Smolder WP' is anticipated to be minimal because the products are applied to soil and foliage of the target weed. **Approved for use in organic farming:** Yes.

1:07 *Ampelomyces quisqualis* isolate M10
Biological fungicide (fungus)

The Pesticide Manual Fifteenth Edition entry number: 32

Fungus: anamorphic Ascomycetes: formerly Deuteromycetes: Sphaeopsidales and mitosporic Leptosphaeriaceae

NOMENCLATURE: Approved name: *Ampelomyces quisqualis* Ces. isolate M10. **Other names:** Formerly known as *Cicinnobiolum cesatii*, but renamed in 1959. Also known as A.q. and AQ10.

SOURCE: *Ampelomyces quisqualis* occurs widely in Nature. Isolate 10 was discovered in a vineyard in Israel and was selected for commercialisation following the discovery of its ability to grow and sporulate in submerged fermentation.

PRODUCTION: *Ampelomyces quisqualis* is grown commercially in semi-solid or submerged fermentation, during which process it produces spores. The spores serve as the active ingredient of the WG formulation.

TARGET PESTS: Selective fungal hyperparasite used to control powdery mildews. Although different crops are attacked by a different genus or species of the powdery mildew pathogen, *Ampelomyces quisqualis* may hyperparasitise all of them to a similar extent.

TARGET CROPS: Used in apples, cucurbits, grapes, ornamentals, strawberries and tomatoes.

BIOLOGICAL ACTIVITY: Mode of action: Germinating spores suppress the development of powdery mildew by hyperparasitism. This process requires relative humidity of at least 60% in the microenvironment of the germinating spores. Once inside hyphae of the pathogen – following a process that takes 2 to 4 hours – the hyperparasite can propagate independently of the external environment; the end result is cessation of powdery mildew development. **Biology:** *Ampelomyces quisqualis* is a well known hyperparasite of powdery mildews (Erysiphaceae). Following the discovery of isolate 10 in Israel in 1984, it was licensed to and developed by Ecogen Israel Partnership, Jerusalem (a subsidiary of Ecogen Inc.). Germinating spores of *A. quisqualis* suppress the development of powdery mildews through hyperparasitism. Once within the hyphae of the phytopathogen, *A. quisqualis* grows intracellularly in the hyphae of the powdery mildew, independently of the external environment. This leads to a cessation of the development of the powdery mildew. *A. quisqualis* propagates in the conidiophores of the host by production of pycnidia (asexual fruiting bodies). Spore germination requires a humidity of at least 60% and penetration of the powdery mildew hyphae takes 2 to 4 hours, depending upon temperature. **Efficacy:** The product works very effectively in a spray programme to control a wide range of powdery mildews. High humidity is essential for spore germination and powdery mildew hyphal invasion and this can be enhanced with the addition of mineral oil adjuvants. It is, however, recommended that the product is applied in the early morning or late afternoon when dew is present or is expected on the crop. **Key reference(s):** 1) R A Daoust & R Hofstein. 1996. *Ampelomyces quisqualis*, a new biofungicide to control powdery mildew in grapes, *Proc. Brighton Crop Protection Conference – Pests & Diseases*, **1**, 33–40. 2) R Hofstein & B Fridlender. 1994. Development of production, formulation and delivery systems, *Proc. Brighton Crop Protection Conference – Pests & Diseases*, **3**, 1273–80. 3) D R Fravel, W J Connick Jr & J A Lewis. 1998. Formulation of microorganisms to control plant diseases. In *Formulation of Microbial Biopesticides, Beneficial Microorganisms, Nematodes and Seed Treatments*, H D Burges (ed.), pp. 187–202, Kluwer Academic Publishers, Dordrecht, The Netherlands.

COMMERCIALISATION: Formulation: Sold as water dispersible granules (WG). **Tradenames:** 'AQ10' (Certis) and (Intrachem). **Patents:** US 5190754.

APPLICATION: Apply at a rate of 35 to 70 g product per hectare. Good control is achieved when disease levels are below 3%. Levels above 3% may lead to poor control. *Ampelomyces quisqualis* may give prophylactic control if applied before the powdery mildew is present in the crop, as the hyperparasite remains viable on the leaf surface for short periods. Good cover of the foliage is essential and the product should be applied when the humidity is above 60%. Addition of mineral oil at a rate of 0.3% v/v can be used to enhance the germination of the spores. Monitoring of the weather conditions that are conducive to the onset of powdery mildew infestations is recommended for the best results. The product has been employed as part of Integrated Crop Management programmes and in alternation with standard chemical mildewicides. It is commonly applied by standard spray application techniques in the presence of surfactants that are compatible with the viability of the organism. Not phytotoxic or phytopathogenic.

PRODUCT SPECIFICATIONS: Specifications: *Ampelomyces quisqualis* is grown commercially in semi-solid or submerged fermentation, during which process it produces spores. The spores serve as the active ingredient of the WG formulation. **Purity:** 'AQ10' formulation consists of 5×10^9 spores per gram. The efficacy of the formulation can be assessed using the spore germination test and/or the hyperparasitism test. The spore germination test determines the viability of the spores and involves plating out the spores on an agar plate and incubating for 48 hours, after which time all the viable spores will have germinated. The percentage germination gives an indication of spore viability of that population. The hyperparasitism test assesses performance and involves spraying cucumber plants infested with low levels of powdery mildew (*Sphaerotheca fuliginea* Poll.) and incubating the plants for ten days in the glasshouse. The hyperparasitism of the powdery mildew becomes evident and can be assessed semiquantitatively by eye. **Storage conditions:** Store in a cool, dry place, preferably under refrigeration. **Shelf-life:** 'AQ10' has a shelf-life of ≥12 months when stored at room temperature (20°C) and of ≥3 years if refrigerated.

COMPATIBILITY: Can be used concurrently with commercial biological insecticides such as *Bacillus thuringiensis*. However, it cannot be co-mixed with currently used fungicides such as systemic, sterol biosynthesis inhibitors. Incompatible with strong oxidisers, acids, bases and chlorinated water.

MAMMALIAN TOXICITY: 'AQ10' has not demonstrated evidence of toxicity, infectivity, irritation or hypersensitivity to mammals. No allergic responses or health problems have been observed by research workers, manufacturing staff or users.

ENVIRONMENTAL IMPACT AND NON-TARGET TOXICITY: *Ampelomyces quisqualis* occurs in Nature and, as such, is not expected to show any adverse effects on non-target organisms or the environment. **Bird toxicity:** No demonstrated toxicity.
Approved for use in organic farming: Yes.

1:08 *Anagrapha falcifera* nucleopolyhedrovirus
Insecticidal baculovirus

Virus: Baculoviridae: Nucleopolyhedrovirus

NOMENCLATURE: Approved name: *Anagrapha falcifera* nucleopolyhedrovirus.
Other names: Celery looper nucleopolyhedrovirus; *Anagrapha falcifera* MNPV; *Anagrapha falcifera* NPV; AfNPV; AfMNPV.

SOURCE: Naturally occurring nucleopolyhedrovirus originally isolated from the alfalfa looper (*Anagrapha falcifera* (Kirby)).

PRODUCTION: Produced for commercial sale in living caterpillars cultured under controlled conditions.

TARGET PESTS: For the control of lepidopteran larvae and, in particular, *Heliothis* and *Helicoverpa* species, including cotton bollworm (*Helicoverpa zea* (Boddie)) and tobacco budworm (*Heliothis virescens* Fabricius).

TARGET CROPS: Maize (corn), cotton, tomatoes and other vegetables, fruit crops, glasshouse crops and ornamentals.

BIOLOGICAL ACTIVITY: Biology: As with all insect baculoviruses, *Anagrapha falcifera* MNPV (AfMNPV) must be ingested to exert an effect. Following ingestion, the protective protein matrix of the polyhedral occlusion bodies (OBs) of the virus is dissolved within the midgut and the virus particles enter the insect's haemolymph. These virus particles invade nearly all cell types in the larval body, where they multiply, leading to death. Shortly after the death of the larva, the integument ruptures, releasing very large numbers of OBs.
Efficacy: AfMNPV is unusual in that it is able to infect over thirty different species of Lepidoptera, thereby overcoming, to a degree, the problem of the selectivity of baculoviruses in crop protection situations. However, AfMNPV acts relatively slowly, as it has to be ingested and then multiply before it exerts any effect on the insect host. It is important to ensure good cover of the foliage to effect good control. Monitoring of adult insect laying patterns and application targeted at newly hatched eggs gives better control than on a mixed population.
Key reference(s): 1) R R Granados & B A Federici (eds). 1986. *The Biology of Baculoviruses, Vols 1 and 2*, CRC Press, Boca Raton, Florida, USA. 2) D J Leisy & J R Fuxa. 1996. Natural and engineered viral agents for insect control. In *Crop Protection Agents from Nature: Natural Products and Analogues*, L G Copping (ed.), Royal Society of Chemistry, Cambridge, UK. 3) B A Federici. 1999. Naturally occurring baculoviruses for insect pest control. In *Biopesticides Use and Delivery*, F R Hall and J J Menn (eds), pp. 301–20, Humana Press, Totowa, NJ, USA. 4) H D Burges & K A Jones. 1998. Formulation of bacteria, viruses and Protozoa to control

insects. In *Formulation of Microbial Biopesticides, Beneficial Microorganisms, Nematodes and Seed Treatments*, H D Burges (ed.), pp. 33–127, Kluwer Academic Publishers, Dordrecht, The Netherlands.

COMMERCIALISATION: Formulation: Formulated as a liquid concentrate. **Tradenames:** 'CLV LC' (Certis).

APPLICATION: Apply in a relatively high volume to ensure good coverage of treated foliage without run-off. The virus will remain infective on the surface of leaves for 7 to 14 days.

PRODUCT SPECIFICATIONS: Specifications: Produced for commercial sale in living caterpillars cultured under controlled conditions. **Purity:** The product contains polyhedra with no bacterial contamination which is pathogenic to man. Efficacy can be checked by bioassay against a susceptible caterpillar. **Storage conditions:** Store in a tightly closed container in a cool (<21°C), dry and dark area. Shelf-life is extended if kept frozen. Unstable at temperatures above 32°C. **Shelf-life:** Stable for several weeks at 2°C and indefinitely if kept frozen.

COMPATIBILITY: Can be used with other insecticides that do not repel target species. Incompatible with strong oxidisers, acids, bases and chlorinated water.

MAMMALIAN TOXICITY: Baculoviruses are specific to invertebrates, with there being no record of any vertebrate becoming infected. The virus does not infect or replicate in the cells of mammals and is inactivated at temperatures above 32°C. There is no evidence of acute or chronic toxicity, eye or skin irritation in mammals. No allergic reactions or other health problems have been observed in research or manufacturing staff or with users of the product. Considered to be of low mammalian toxicity.

ENVIRONMENTAL IMPACT AND NON-TARGET TOXICITY: AfMNPV occurs widely in Nature and is not expected to have any adverse effects on non-target organisms or on the environment. There is no evidence of the virus infecting vertebrates or plants. It has no adverse effects on fish, birds or beneficial organisms.

Approved for use in organic farming: Yes.

1:09 *Anticarsia gemmatalis* nucleopolyhedrovirus

Insecticidal baculovirus

Virus: Baculoviridae: Nucleopolyhedrovirus

NOMENCLATURE: Approved name: *Anticarsia gemmatalis* nucleopolyhedrovirus.
Other names: *Anticarsia gemmatalis* MNPV; *Anticarsia gemmatalis* NPV; AgNPV; AgMNPV.

SOURCE: Originally isolated from the velvet bean caterpillar (*Anticarsia gemmatalis* Hübner) found in a soybean crop in the USA and introduced into Brazil, where *A. gemmatalis* is a significant problem. More recently, an isolate of *Anticarsia gemmatalis* nucleopolyhedrovirus (AgMNPV) has been identified that has good activity against the sugar cane borer (*Diatraea saccharalis* (Speyer)) and retains activity against the velvet bean caterpillar.

PRODUCTION: Produced by *in vivo* culture in *Anticarsia gemmatalis*. The newer isolate is produced in *Diatreae saccaralis*, an insect that is easier to culture than *A. gemmatalis*.

TARGET PESTS: *Anticarsia gemmatalis* (velvet bean caterpillar) and *Diatraea saccharalis* (sugar cane borer).

TARGET CROPS: Soybeans and sugar cane.

BIOLOGICAL ACTIVITY: Biology: As with all insect baculoviruses, AgMNPV must be ingested to exert an effect. Following ingestion, the protective protein matrix of the polyhedral occlusion bodies (OBs) of the virus is dissolved within the midgut and the virus particles enter the insect's haemolymph. These virus particles invade nearly all cell types in the larval body, where they multiply, leading to death. Shortly after the death of the larva, the integument ruptures, releasing very large numbers of OBs. **Key reference(s):** 1) R R Granados & B A Federici (eds). 1986. *The Biology of Baculoviruses, Vols 1 and 2*, CRC Press, Boca Raton, Florida, USA. 2) D J Leisy & J R Fuxa. 1996. Natural and engineered viral agents for insect control. In *Crop Protection Agents from Nature: Natural Products and Analogues*, L G Copping (ed.), Royal Society of Chemistry, Cambridge, UK. 3) B A Federici. 1999. Naturally occurring baculoviruses for insect pest control. In *Biopesticides Use and Delivery*, F R Hall & J J Menn (eds), pp. 301–20, Humana Press, Totowa, NJ, USA. 4) H D Burges & K A Jones. 1998. Formulation of bacteria, viruses and Protozoa to control insects. In *Formulation of Microbial Biopesticides, Beneficial Microorganisms, Nematodes and Seed Treatments*, H D Burges (ed.), pp. 33–127, Kluwer Academic Publishers, Dordrecht, The Netherlands. 5) F Moscardi. 1988. Production and use of entomopathogens in Brazil. In *Biological Pesticides and Novel Plant-Pest Resistance for Insect Pest Management*, D W Roberts & R R Granados (eds), pp. 53–60, Cornell Univ. Press, Ithaca, NY, USA.

COMMERCIALISATION: Formulation: Sold as a powder made from dried, ground, infected caterpillars. **Tradenames:** 'Polygen' (Agroggen S/A Biol Ag), 'Multigen' (EMBRAPA).

APPLICATION: A single treatment applied to soybeans at a rate of 1×10^{10} polyhedral inclusion bodies per hectare gives control of velvet bean caterpillars for the entire season.

PRODUCT SPECIFICATIONS: Specifications: Produced by *in vivo* culture in *Anticarsia gemmatalis*. The newer isolate is produced in *Diatreae saccharalis*, an insect that is easier to culture than *A. gemmatalis*. **Storage conditions:** Store in a cool, dry place in a sealed container. Do not expose to direct sunlight. **Shelf-life:** Stable for over a year, if stored under recommended conditions.

COMPATIBILITY: Compatible with many chemical insecticides, but should not be applied with strong oxidisers, acids or bases and chlorinated water.

MAMMALIAN TOXICITY: There is no evidence that AgMNPV has any allergic or other adverse effects on research workers, manufacturing or field staff. Considered to have low mammalian toxicity.

ENVIRONMENTAL IMPACT AND NON-TARGET TOXICITY: AgMNPV occurs in Nature and is not expected to have any adverse effects on non-target organisms or on the environment. **Approved for use in organic farming:** Yes.

1:10 ARF-18   *Biological nematicide (fungus)*

Fungus (yet to be classified)

NOMENCLATURE: Development code: ARF-18 (Arkansas Fungus).

SOURCE: Isolated from soybean cyst nematodes (*Heterodera glycines* Ichinohe) from fields in eastern Arkansas, but also widely scattered in surrounding states. It has also been found in the reniform nematode (*Rotylenchulus reniformis* Linford and Oliveira) from cotton fields in south-eastern Arkansas.

PRODUCTION: The fungus can be produced by fermentation, but using mycelium as the seed material. Work on large-scale production is in progress.

TARGET PESTS: Soybean cyst nematode (*Heterodera glycines*), other cyst nematodes (*Heterodera* spp.) and the reniform nematode (*Rotylenchulus reniformis*).

TARGET CROPS: Cotton and soybeans.

BIOLOGICAL ACTIVITY: Mode of action: The fungus, after penetration, fills the interior of the adult nematode life stage, third or fourth stage juveniles or egg and absorbs the internal contents, probably through the action of enzymes. **Efficacy:** The fungus activity is effective at relatively low levels on a per hectare basis. **Key reference(s):** 1) D G Kim. 1989. Biological control of soybean cyst nematode, *Heterodera glycines*, with a soil-borne fungus. Dissertation: University of Arkansas, Fayetteville, AK, USA. 2) D G Kim & R D Riggs. 1991. Characteristics and efficacy of a sterile hyphomycete (ARF18), a new biocontrol agent for *Heterodera glycines* and other nematodes, *J. Nematol.*, **23**, 275–82. 3) D G Kim & R D Riggs. 1995. Efficacy of the nematopathogenic fungus ARF18 in alginate-clay pellet formulations against *Heterodera glycines*, *J. Nematol.*, **27**, 602–8. 4) D G Kim, R D Riggs & J C Correll. 1998. Isolation, characteristics and distribution of an unnamed biocontrol fungus from cysts of *Heterodera glycines*, *Phytopathol.*, **88**, 465–71. 5) P Timper & R D Riggs. 1998. Variation in efficacy of the Fungus ARF against the soybean cyst nematode, *Heterodera glycines*, *J. Nematol.*, **30**, 461–7.

COMMERCIALISATION: Formulation: ARF-18 is under development, but alginate/clay pellet formulations have been evaluated successfully in the field.

APPLICATION: ARF-18 should be applied as a formulated product (or as chopped mycelium) as an in-furrow, in-the-row treatment. One strain is effective against the soybean cyst nematode (*Heterodera glycines*) and certain other cyst nematodes; another strain is being evaluated for reniform nematode (*Rotylenchulus reniformis*) control. The recommended application rate is 27 to 55 kg fresh weight fungus mycelium equivalent per hectare (25 to 50 lb per acre).

PRODUCT SPECIFICATIONS: Specifications: The fungus can be produced by fermentation, but using mycelium as the seed material. Work on large-scale production is in progress. **Purity:** ARF-18 can be produced as pure mycelium. **Storage conditions:** Mycelium should be stored under refrigeration, but formulations might not need refrigeration. **Shelf-life:** Very short (months) for formulations. No data are available on fresh mycelium.

COMPATIBILITY: Compatible with malathion, bentazone, sethoxydim, trifluralin and aldicarb.

MAMMALIAN TOXICITY: There are no reports of allergic or other adverse toxicological effects from research workers, manufacturers or field staff with ARF-18.

ENVIRONMENTAL IMPACT AND NON-TARGET TOXICITY: ARF-18 occurs widely in soils, particularly those infested with nematodes. There is no evidence to suggest that it will have any adverse effects on non-target organisms or on the environment.
Approved for use in organic farming: Not yet commercially available.

1:11 *Aspergillus flavus* AF36 and NRRL 21882
Beneficial fungus

Fungus: anamorphic Hypocreales: previously classified as mitosporic Deuteromycetes: Moniliales

NOMENCLATURE: Approved name: *Aspergillus flavus* Link AF36 and *Aspergillus flavus* NRRL 21882.

SOURCE: *Aspergillus flavus* AF36 was isolated from cottonseed in the Yuma Valley, Arizona, USA and *A. flavus* NRRL 21882 was isolated from a peanut in Dawson, Georgia, USA in 1991 by the United States Department of Agriculture (USDA), National Peanut Research Laboratory.

PRODUCTION: Produced by fermentation.

TARGET PESTS: Aflatoxin-producing isolates of *Aspergillus flavus*.

TARGET CROPS: Cotton (AF36) and peanuts (NRRL 21882).

BIOLOGICAL ACTIVITY: Mode of action: *Aspergillus flavus* isolates occur throughout the world in soil and can contaminate the edible portions of crops. In some cases, this contamination can lead to the production of aflatoxin, a very toxic mycotoxin associated with liver cancer. *Aspergillus flavus* AF36 and NRRL 21882 are non-aflatoxin-producing isolates and the distribution of large quantities of these isolates around growing crops builds up the population, thereby allowing them to out-compete the mycotoxin-producing isolates. **Efficacy:** *Aspergillus flavus* AF36 has been shown to be very effective in preventing the formation of aflatoxin in cotton in Texas and Arizona and NRRL 21882 has been effective in controlling aflatoxin production in peanuts in Georgia.

COMMERCIALISATION: Formulation: Sold as wheat seeds coated with the fungal mycelium. The viability of the fungus in the end-use product is 3000 cfu/g. **Tradenames:** 'Aspergillus flavus AF36' (Arizona Cotton), 'Afla-guard' (Circle One Global).

APPLICATION: The *Aspergillus flavus* AF36-coated wheat seeds are spread throughout cotton at a rate of 11 kg product/ha (10 lb product/acre) pre-blossom. It should not be incorporated and up to 5 cm (2 in) of rainfall or irrigation water should be applied within 2 days of application. 'Afla-guard' is applied at a rate of 22 kg product/ha (20 lb product/acre) pre-pegging.

PRODUCT SPECIFICATIONS: Specifications: Produced by fermentation. **Purity:** The product is checked to ensure that it contains no aflatoxin-producing isolates and that it is free from bacteria and other potentially harmful contaminants. **Storage conditions:** Store in a cool, dry, sealed container. **Shelf-life:** The product can be stored for up to one year.

COMPATIBILITY: Incompatible with broad-spectrum fungicides.

MAMMALIAN TOXICITY: There are no reports of allergic or other adverse toxicological effects from research workers, manufacturers or field staff with *Aspergillus flavus* AF36 or NRRL 21882. **Acute oral LD$_{50}$:** rats >5000 mg/kg. **Toxicity class:** EPA (formulation) IV.

ENVIRONMENTAL IMPACT AND NON-TARGET TOXICITY: *Aspergillus flavus* AF36 and NRRL 21882 occur widely in Nature and are not expected to have any adverse effects on non-target organisms or on the environment.

1:12 *Autographa californica* nucleopolyhedrovirus   *Insecticidal baculovirus*

Virus: Baculoviridae: Nucleopolyhedrovirus

NOMENCLATURE: **Approved name:** *Autographa californica* nucleopolyhedrovirus. **Other names:** *Autographa californica* MNPV; *Autographa californica* NPV; AcNPV; AcMNPV.

SOURCE: Occurs widely in Nature; originally isolated from *Autographa californica* (Speyer).

PRODUCTION: Produced commercially by *in vivo* culture in lepidopteran larvae.

TARGET PESTS: Lepidopteran larvae.

TARGET CROPS: Maize (corn), vegetables, fruit crops and ornamentals

BIOLOGICAL ACTIVITY: **Biology:** As with all insect baculoviruses, *Autographa californica* MNPV (AcMNPV) must be ingested to exert an effect. Following ingestion, the protective protein matrix of the polyhedral occlusion bodies (OBs) of the virus is dissolved within the midgut and the virus particles enter the insect's haemolymph. These virus particles invade nearly all cell types in the larval body, where they multiply, leading to death. Shortly after the death of the larva, the integument ruptures, releasing very large numbers of OBs.
Efficacy: Unlike some baculoviruses, AcMNPV will infect over thirty different species of Lepidoptera. This makes it a broader-spectrum product than most baculovirus-based products.
Key reference(s): 1) R R Granados & B A Federici (eds). 1986. *The Biology of Baculoviruses, Vols 1 and 2*, CRC Press, Boca Raton, Florida, USA. 2) D J Leisy & J R Fuxa. 1996. Natural and engineered viral agents for insect control. In *Crop Protection Agents from Nature: Natural Products and Analogues*, L G Copping (ed.), Royal Society of Chemistry, Cambridge, UK.

3) B A Federici. 1999. Naturally occurring baculoviruses for insect pest control. In *Biopesticides Use and Delivery*, F R Hall & J J Menn (eds), pp. 301–20, Humana Press, Totowa, NJ, USA.
4) H D Burges & K A Jones. 1998. Formulation of bacteria, viruses and Protozoa to control insects. In *Formulation of Microbial Biopesticides, Beneficial Microorganisms, Nematodes and Seed Treatments*, H D Burges (ed.), pp. 33–127, Kluwer Academic Publishers, Dordrecht, The Netherlands.

COMMERCIALISATION: Formulation: Formulated as a liquid concentrate.
Tradenames: 'VPN 80' (Agricola El Sol), 'Gusano' (Certis).

APPLICATION: Good coverage of the foliage is essential to ensure the targeted caterpillars consume the virus particles.

PRODUCT SPECIFICATIONS: Specifications: Produced commercially by *in vivo* culture in lepidopteran larvae. **Purity:** Formulated product contains the nucleopolyhedrovirus particles and no bacterial contaminants. **Storage conditions:** Store in a tightly closed container in a cool (<21°C), dry and dark area. Shelf-life is extended if kept frozen. Unstable at temperatures above 32°C. **Shelf-life:** Stable for several weeks if stored at 2°C. Indefinite stability if stored frozen.

COMPATIBILITY: Compatible with all insecticides, except those that repel the target caterpillars. Do not use with strong oxidising or reducing agents or with chlorinated water.

MAMMALIAN TOXICITY: There have been no reports of allergic or other adverse effects from research workers, manufacturing or field staff. Considered to be of low mammalian toxicity.

ENVIRONMENTAL IMPACT AND NON-TARGET TOXICITY: AcMNPV occurs widely in Nature and is not expected to have any adverse effects on non-target organisms or on the environment. **Approved for use in organic farming:** Yes.

1:13 *Azospirillum brasilense* isolate Cd
Biological plant growth promoter (bacterium)

Proteobacterium: 'Alphaproteobacteria': Rhodospirillales: Rhodospirillaceae. Sometimes classified as aerobic/microaerophilic, motile, helical/vibrioid Gram-negative bacterium (not in any family) within the division Gracilicutes

NOMENCLATURE: Approved name: *Azospirillum brasilense* Tarrand, Krieg & Döbereiner isolate Cd. **Other names:** Formerly known as: *Spirillum lipoferum* Beijerinck (Becking).

SOURCE: Isolated from roots of Bermuda grass (*Cynodon dactylon* Pers.).

PRODUCTION: Produced by fermentation.

TARGET CROPS: Recreational turf and lawns.

BIOLOGICAL ACTIVITY: Mode of action: *Azospirillum brasilense* colonises the roots and secretes phytohormones, mainly auxins. Indol-3-ylacetic acid (IAA) is the major hormone. **Efficacy:** Positive effects of inoculation with *Azospirillum brasilense* have been demonstrated on various root parameters, including increase in root length, particularly of the root elongation zone. **Key reference(s):** 1) S H Omay, W A Schmidt, P Martin & F Bangerth. 1993. Indoleacetic acid production by the rhizosphere bacterium (*Azospirillum brasilense*) Cd under *in vitro* conditions, *Can. J. Microbiol.*, **39**, 187–92. 2) S Mitra, M Rogers, E Pedersen, S Camyon & J Turner. 2001. Effect of using *Azospirillum brasilense* and a microbial food source on the root development of cool-season and warm-season turf grasses, *Proc. Brit. Crop Protection Conference – Weeds*, **1**, 15–22.

COMMERCIALISATION: Formulation: Liquid suspension of flocculated *Azospirillum brasilense* is applied with a liquid microbial stimulant called Lex. The product is part of the 'FreshPack' line of products. **Tradenames:** 'Recharge' (Eco Soil Systems).

APPLICATION: Bi-weekly applications to established turf.

PRODUCT SPECIFICATIONS: Specifications: Produced by fermentation.
Storage conditions: Store at room temperature. Do not expose to direct sunlight.
Shelf-life: If stored correctly, the product should encounter no loss of viability for at least 4 months.

COMPATIBILITY: Incompatible with broad-spectrum products such as copper-based fungicides.

MAMMALIAN TOXICITY: There are no reports of allergic or other adverse toxicological effects from research workers, manufacturers or field staff.

ENVIRONMENTAL IMPACT AND NON-TARGET TOXICITY: *Azospirillum brasilense* occurs widely in Nature and is not expected to have any adverse effects on non-target organisms or on the environment. **Approved for use in organic farming:** Yes.

1:14 *Bacillus cereus* isolate BP01
Biological plant growth regulator

Bacterium: Bacillales: Bacillaceae

NOMENCLATURE: Approved name: *Bacillus cereus* Frankland & Frankland isolate BP01.
OPP Chemical Code: 119802.

SOURCE: *Bacillus cereus* isolate BP01 is a widely occurring soil bacterium.

PRODUCTION: Produced by fermentation.

TARGET CROPS: Cotton.

BIOLOGICAL ACTIVITY: Mode of action: The mode of action is unknown.

COMMERCIALISATION: Formulation: 'MepPlus' is sold as a mixed formulation with mepiquat chloride. **Tradenames:** 'MepPlus' (Micro-Flo).

APPLICATION: *Bacillus cereus* isolate BP01 should be applied using conventional ground and aerial equipment as a foliar spray at rates from 0.04 to 0.4 g per hectare, depending on the cotton variety and its vigour. The maximum application level for *Bacillus cereus* isolate BP01 on cotton is 0.8 g per hectare per year, with an average of 0.25 g per hectare per year. Applications should be made from the match head square stage at 7- to 14-day intervals to within 30 days of harvest. No more than six applications should be made in a year and 'MepPlus' should not be applied to crops that are under stress.

PRODUCT SPECIFICATIONS: Specifications: Produced by fermentation. **Purity:** 'MepPlus' contains 0.05% *Bacillus cereus* isolate BP01, 4.2% mepiquat chloride plus 95.75% inert ingredients.

COMPATIBILITY: Incompatible with strong oxidisers, acids, bases and chlorinated water.

MAMMALIAN TOXICITY: *Bacillus cereus* isolate BP01 is considered to be non-toxic to mammals. Some isolates of *Bacillus cereus* have been implicated in nosocomial infections in rare instances and in food poisoning incidents. No enterotoxin was found in *Bacillus cereus* isolate BP01 using the ELISA analysis. Quality control procedures in place during manufacturing ensure that harmful levels of contaminating micro-organisms are prevented. **Acute oral LD$_{50}$:** Toxicity category IV. **Acute dermal LD$_{50}$:** Toxicity category III. **Inhalation:** Toxicity category IV. **Skin and eye:** Toxicity category IV.

ENVIRONMENTAL IMPACT AND NON-TARGET TOXICITY: Plant study data indicated that *Bacillus cereus* isolate BP01 does not induce phytotoxic responses in plants. No apparent effects were noted in germination of seeds and growth of soybean seedlings treated with 0.2 g per acre and 5.0 g per acre when compared with untreated seedlings. No observable foliar phytotoxicity was reported in any of the treatment or control groups. Therefore, risks to non-target plants are considered minimal to non-existent. **Bird toxicity:** When *Bacillus cereus* isolate BP01 was administered orally at an estimated dose of 4.44 × 10^7 colony forming units per g b.w. per day for five days, it was practically non-toxic to bobwhite quail.

Other aquatic toxicity: In a toxicity/pathogenicity study, *Bacillus cereus* isolate BP01 was toxic to *Daphnia* at concentrations ranging from 3.9 × 10^7 to 6.7 × 10^8 colony forming units per litre. **Behaviour in soil:** *Bacillus cereus* isolate BP01 is a common constituent of soil.

1:15 *Bacillus firmus* isolate N1
Biological nematicide (bacterium)

Bacterium: Bacillales: Bacillaceae

NOMENCLATURE: Approved name: *Bacillus firmus* Bredemann & Werder isolate N1.

SOURCE: The bacterium was isolated from cultivated soil in Israel.

PRODUCTION: *Bacillus firmus* is produced commercially by fermentation. The fermentation medium with viable spores is used to prepare the product

TARGET PESTS: Root-knot nematodes, particularly *Meloidogyne* species.

TARGET CROPS: Glasshouse and field-grown vegetable crops.

BIOLOGICAL ACTIVITY: Mode of action: *Bacillus firmus* colonises the egg sacs of the root-knot nematodes and subsequently destroys the eggs. Treatment of soil with the bacterium leads to a decline in the juvenile viability in soil and this is thought to be due, at least in part, to the destruction of nematode eggs. **Efficacy:** It has been shown that treatment of soil in a tomato crop gave control after 50 days that was equivalent to a chemical standard treatment and, after 85 days, nematode control was still below economic threshold levels.

Key reference(s): M Keren-Zur, J Antonov, A Bercovitz, A Husid, G Kenan, M Marcov & M Rebhun. 2000. *Bacillus firmus* formulations for the safe control of root-knot nematodes, *Proc. Brit. Crop Protection Conference – Pests & Diseases*, **1**, 47–51.

COMMERCIALISATION: Formulation: Sold as two dry powder formulations, 'BioNem' for conventional and ICM farming and 'BioSafe' for organic farming. Both formulations contain 3 × 10^9 spores per gram. **Tradenames:** 'BioNem' (Minrav), 'BioSafe' (Minrav).

APPLICATION: The products should be applied in furrow, several days prior to planting, at rates between 70 and 150 grams per metre and incorporated to a depth of 20 cm.

PRODUCT SPECIFICATIONS: Specifications: *Bacillus firmus* is produced commercially by fermentation. The fermentation medium with viable spores is used to prepare the product **Purity:** Produced by fermentation and guaranteed to be free of bacterial contaminants that are pathogenic to man. Viability of the bacterium within the formulation can be determined by plating out on agar and incubating at temperatures of about 25 °C for 48 hours, then counting the colonies formed. **Storage conditions:** Store in a sealed container under cool, dry conditions. **Shelf-life:** Stable for at least 2 years, when stored in cool, dry conditions.

COMPATIBILITY: Incompatible with soil sterilants such as methyl bromide and dazomet, but may be used after these treatments. Can be applied in combination with soil fungicides.

MAMMALIAN TOXICITY: Based on a review of the toxicology and other information submitted for soil-inhabiting *Bacillus* species, there is no evidence of risks to human health. Because the bacteria colonise plant roots, exposure to humans would be minimal once the crops are sown. There have been no reports of allergic reactions by researchers, production workers or field staff following use of 'BioNem' or 'BioSafe'.

ENVIRONMENTAL IMPACT AND NON-TARGET TOXICITY: *Bacillus firmus* isolate N1 is not harmful to non-target organisms or to the environment. Tests were conducted to help ensure that there were no risks to plants, insects or birds. Aquatic exposure is not expected. Tests have shown that this bacterium is not phytotoxic.

Approved for use in organic farming: 'BioSafe' is sold for use in organic farming.

1:16 *Bacillus licheniformis* isolate SB3086
Biological fungicide (bacterium)

Bacterium: Bacillales: Bacillaceae

NOMENCLATURE: Approved name: *Bacillus licheniformis* (Weigmann) Chester isolate SB3086.

SOURCE: *Bacillus licheniformis* isolate SB3086 is a commonly occurring soil micro-organism.

PRODUCTION: *Bacillus licheniformis* is produced commercially by fermentation. The fermentation medium with viable spores is used to prepare the product.

TARGET PESTS: The product is effective against a wide range of fungal diseases, including *Alternaria* spp., *Botrytis* spp., *Cercospora* spp., *Drechslera* spp., *Fusarium* spp., dollar spot (*Sclerotinia homeocarpa* Bennett) and other *Sclerotinia* spp., *Phytophthora* spp., powdery mildews (*Microsphaera* spp., *Erysiphe cichoracearum* DC), rusts (*Gymnosporangium* spp., *Puccinia* spp., *Uromyces* spp.), scab (*Venturia inaequalis* Wint.) and *Septoria* leafspot.

TARGET CROPS: *Bacillus licheniformis* isolate SB3086 is approved for treatment of ornamental turf, lawns, golf courses, ornamental plants, conifers and tree seedlings in outdoor, glasshouse and nursery sites. It cannot be used on crops intended for food use.

BIOLOGICAL ACTIVITY: Mode of action: *Bacillus licheniformis* isolate SB3086 is thought to produce an antifungal secondary metabolite that contributes towards the control of fungal diseases. It is also believed to produce an antifungal enzyme such as xylanase, mannase or a protease. The precise nature of these agents has not yet been identified.

COMMERCIALISATION: Formulation: *Bacillus licheniformis* is formulated as a liquid concentrate. **Tradenames:** 'Biofungicide Green-Releaf' (Novozymes).

APPLICATION: The product is diluted in water and sprayed on leaves or applied to soil. Treat lawns, golf courses or ornamental turf when disease symptoms are evident. Repeat application every 3 to 14 days, depending on level of disease. For ornamental plants, treat when conditions are favourable for disease development and repeat every 7 to 14 days until conditions are no longer favourable for disease, at a rate of 150 to 750 ml/1000 sq. m. (5 to 25 fl.oz./1000 sq. ft.).

PRODUCT SPECIFICATIONS: Specifications: *Bacillus licheniformis* is produced commercially by fermentation. The fermentation medium with viable spores is used to prepare the product. **Storage conditions:** The formulation should be stored under dry, ambient conditions. **Shelf-life:** The product is stable for at least a year under the above storage conditions.

COMPATIBILITY: Compatible with many fungicides, bactericides, insecticides, foliar nutrients and adjuvants used in the control of crop diseases. In ICM strategies, it is unusual to apply 'Biofungicide Green-Releaf' with chemical pesticides. Incompatible with strong oxidisers, acids, bases and chlorinated water.

MAMMALIAN TOXICITY: *Bacillus licheniformis* SB3086 at 1×10^8 cfu/animal was not toxic, infective, or pathogenic to rats. **Acute oral LD$_{50}$:** rats >5000 mg/kg. **Acute dermal LD$_{50}$:** rabbits >5050 mg/kg. **Inhalation:** Not toxic to rats exposed to 2.56 mg/litre. **Skin and eye:** Non-irritating to the eyes of rabbits and very slightly irritating to the skin of rats. The product is not a dermal sensitiser. **Toxicity class:** EPA (formulation) IV.

ENVIRONMENTAL IMPACT AND NON-TARGET TOXICITY: *Bacillus licheniformis* has not been known to cause any pathogenicity in plants. **Bird toxicity:** Acute oral LD$_{50}$ for mallard >4000 mg/kg. There was no pathogenicity or toxicity to young mallard ducks by oral gavage over 5 days at 9×10^9 cfu/kg b.w. per day. Neither the active ingredient nor the formulated product is harmful to mallard ducks at the proposed field use rates. **Fish toxicity:** LC$_{50}$ (30 days) for rainbow trout >1.1×10^6 cfu/ml. **Other aquatic toxicity:** LC$_{50}$ (21 days) for *Daphnia magna* 1.8×10^6 cfu/ml. **Effects on beneficial insects:** 'Biofungicide Green-Releaf' at 1.6 ×

10^6 cfu/ml had no effect on the survival of honeybee larvae. **Behaviour in soil:** The added soil density of *B. licheniformis* from the proposed use rates would be 0.42 to 1.5%. As this is a very small proportion of the naturally occurring *Bacilli* in the soil, it is not expected to add substantially to the effects of the naturally occurring *Bacillus* populations.

1:17 *Bacillus mycoides* isolate J
Biological fungicide (bacterium)

Bacterium: Bacillales: Bacillaceae

NOMENCLATURE: Approved name: *Bacillus mycoides* Flugge isolate J.
OPP Chemical Code: 006516.

SOURCE: *Bacillus mycoides* is a widespread, naturally occurring micro-organism in the soil.

PRODUCTION: By fermentation.

TARGET PESTS: Cercospora leaf spot (*Cercospora beticola* Sacc.).

TARGET CROPS: Sugar beet.

BIOLOGICAL ACTIVITY: Mode of action: Systemic acquired resistance.

COMMERCIALISATION: Formulation: *Bacillus mycoides* isolate J has been given provisional approval for experimental use as a fungicide to control Cercospora leaf spot on sugar beets in Montana, Minnesota and North Dakota. **Tradenames:** '*Bacillus mycoides* isolate J' (Montana Microbial Products).

APPLICATION: In the experimental programme, the product is mixed with water and applied to the sugar beet, using both ground and aerial spray applications beginning 4 to 6 days before the predicted disease onset and repeated at 10- to 14-day intervals thereafter for as long as the disease persists.

MAMMALIAN TOXICITY: No allergic reactions or other adverse health problems have been shown by research workers, manufacturing staff or users.

ENVIRONMENTAL IMPACT AND NON-TARGET TOXICITY: There is no evidence that the formulated product has any adverse effect on non-target organisms or on the environment. **Approved for use in organic farming:** Yes.

1:18 *Bacillus popilliae*

Biological insecticide (bacterium)

Bacterium: Bacillales: Bacillaceae

NOMENCLATURE: Approved name: *Bacillus popilliae* Newman has been reclassified. See *Paenibacillus popilliae* (entry 1:90).

1:19 *Bacillus pumilus* isolate GB34

Biological fungicide (bacterium)

Bacterium: Bacillales: Bacillaceae

NOMENCLATURE: Approved name: *Bacillus pumilus* Meyer & Gottheil isolate GB34. **OPP Chemical Code:** 006493.

SOURCE: *Bacillus pumilus* isolate GB34 is a naturally occurring bacterium that is particularly common in soil and on dead plant tissue. This isolate was selected for commercialisation because it is commonly found on the developing root system of soybean plants without damaging the crop and because of its activity against the most common soil-borne pathogens of soybeans.

PRODUCTION: *Bacillus pumilus* is produced commercially by fermentation. The fermentation medium with viable spores is used to prepare the product.

TARGET PESTS: Soil-borne pathogenic fungi and, in particular, *Pythium* spp. and *Rhizoctonia* spp.

TARGET CROPS: Soybeans.

BIOLOGICAL ACTIVITY: Mode of action: The mode of action of *Bacillus pumilus* isolate GB 34 is not fully understood. The bacterium prevents the germination of fungal spores in the root zone around developing roots. It has also been shown to grow on the surface of fungal spores.

COMMERCIALISATION: Formulation: The commercial product is sold as a slurry containing bacterial spores. **Tradenames:** 'GB34 Concentrate Biological Fungicide' (Gustafson).

APPLICATION: Soybean seeds are treated with the seed treatment at drilling.

PRODUCT SPECIFICATIONS: Specifications: *Bacillus pumilus* is produced commercially by fermentation. The fermentation medium with viable spores is used to prepare the product.
Purity: The seed treatment is guaranteed to be free of microbial contaminants.
Storage conditions: The product should be stored under dry, ambient conditions.
Shelf-life: It is recommended that the slurry is used within three months of purchase.

COMPATIBILITY: Incompatible with strong oxidisers, acids, bases and chlorinated water.

MAMMALIAN TOXICITY: 'GB34 Concentrate Biological Fungicide' is considered to be a non-toxic product when used according to label directions. It meets all the criteria for US-EPA Toxicity Category III with the signal word 'Caution'.

ENVIRONMENTAL IMPACT AND NON-TARGET TOXICITY: *Bacillus pumilus* is not harmful to non-target organisms or to the environment. Tests were conducted to help ensure that there were no risks to plants, insects or birds. Aquatic exposure is not expected. Tests have shown that this bacterium does not harm seeds or cause them to decay.
Approved for use in organic farming: Yes.

1:20 *Bacillus pumilus* isolate QST 2808
Biological fungicide (bacterium)

Bacterium: Bacillales: Bacillaceae

NOMENCLATURE: Approved name: *Bacillus pumilus* Meyer & Gottheil isolate QST 2808.
Development code: OPP Chemical Code: 006485.

SOURCE: The bacterium, *Bacillus pumilus* isolate QST 2808, is prevalent in soils and has been found in a variety of habitats worldwide. The isolate selected for commercialisation was chosen because of its spectrum and level of activity against important economic fungal and bacterial pathogens.

PRODUCTION: *Bacillus pumilus* is produced commercially by fermentation. The fermentation medium with viable spores is used to prepare the product.

TARGET PESTS: For use against a wide range of plant fungal diseases, including powdery and downy mildews and rusts.

TARGET CROPS: 'Sonata' is targeted for use in small-grain cereals, tree fruit, vegetables and vines.

BIOLOGICAL ACTIVITY: Mode of action: *Bacillus pumilus* isolate QST 2808 employs a novel mode of action that inhibits fungal pathogen development on the leaf surface and also activates the plant's immune system. Initial research has shown that 'Sonata' may have curative activity as well as preventive properties, thereby arresting the development of powdery and downy mildews, rusts and other pathogens.

COMMERCIALISATION: Formulation: 'Sonata' is based on a proprietary isolate of the bacterium *Bacillus pumilus* isolate QST 2808 and is formulated as an aqueous suspension (AS) and a wettable powder (WP). **Tradenames:** 'Sonata' (AgraQuest).

APPLICATION: 'Sonata' is applied as a foliar fungicide. The WP formulation is applied at a rate of 5 to 8 kg per hectare and the AS formulation is applied at rates of 4 to 6 litres per hectare. It should be applied in sufficient volume for complete foliar coverage. *Bacillus pumilus* isolate QST 2808 occurs naturally in the environment and will complement growers' Integrated Crop Management (ICM) programmes. With an expected zero-day post-harvest interval, 'Sonata' may be applied up to and including the day of harvest.

PRODUCT SPECIFICATIONS: Specifications: *Bacillus pumilus* is produced commercially by fermentation. The fermentation medium with viable spores is used to prepare the product. **Purity:** The WP formulation consists of 5×10^9 cfu/g and the AS formulation consists of 7×10^9 cfu/ml. Guaranteed to be free of microbial contaminants. **Storage conditions:** Both the WP and AS formulations should be stored under dry, ambient conditions. **Shelf-life:** Preliminary indications are that 'Sonata' is stable for at least 2 years under the above storage conditions.

COMPATIBILITY: Compatible with many fungicides, bactericides, insecticides, foliar nutrients and adjuvants used in the control of crop diseases. In ICM strategies, it is unusual to apply 'Sonata' with chemical pesticides. Incompatible with strong oxidisers, acids, bases and chlorinated water.

MAMMALIAN TOXICITY: 'Sonata' is considered to be a non-toxic product when used according to label directions. It meets all the criteria for US-EPA Toxicity Category III with the signal word 'Caution'.

ENVIRONMENTAL IMPACT AND NON-TARGET TOXICITY: *Bacillus pumilus* isolate QST 2808 is not harmful to non-target organisms or to the environment. Tests were conducted to help ensure that there were no risks to plants, insects or birds. Aquatic exposure is not expected. Tests have shown that this bacterium does not harm seeds or cause them to decay. **Approved for use in organic farming:** Yes.

1:21 *Bacillus sphaericus* isolate 2362

Biological insecticide (bacterium)

The Pesticide Manual **Fifteenth Edition entry number:** 55

Bacterium: Bacillales: Bacillaceae

NOMENCLATURE: Approved name: *Bacillus sphaericus* Neide isolate 2362, serotype H-5a5b. **CAS RN:** *[143447-72-7]*.

SOURCE: *Bacillus sphaericus* is found widely in Nature and this isolate (2362) was selected because of its effective control of mosquito larvae.

PRODUCTION: *Bacillus sphaericus* is produced commercially by fermentation as for *Bacillus thuringiensis* isolates for control of the *Plutella* group of Lepidoptera (entry 1:33).

TARGET PESTS: Mosquito larvae, being particularly active against *Culex* spp.

TARGET CROPS: Used as a public health insecticide. It is applied in habitats where mosquitos live and lay eggs, such as storm water and drainage systems, marine and coastal areas, freshwater bodies such as lakes and streams, and water that collects in discarded tyres.

BIOLOGICAL ACTIVITY: Mode of action: *Bacillus sphaericus* produces parasporal, proteinaceous, crystal inclusion bodies during sporulation. Upon ingestion, the crystal proteins are solubilised and the insect gut proteases convert the original pro-toxin into smaller toxins. These hydrolysed toxins bind to the insect's midgut cells at high-affinity, specific receptor binding sites where they interfere with the potassium ion-dependent, active amino acid symport mechanism. This disruption causes the formation of large cation-selective pores that increase the water permeability of the cell membrane. A large uptake of water causes cell swelling and eventual rupture, disintegrating the midgut lining. **Biology:** The crystal inclusions derived from *Bacillus sphaericus* are mosquito larva specific. Because they have to be ingested and then processed within the insect's gut, they are often slow acting (in comparison with conventional chemicals). Isolate 2362 was isolated from blackfly (*Simulium* spp.) in Africa. It is active against mosquito larvae under a wide range of conditions, including extended residual activity in highly organic, aquatic environments. *Bacillus sphaericus* has a mode of action similar to that of *Bacillus thuringiensis* subsp. *israelensis* (entry 1:31). It should be applied from first instar up to early fourth instar, with toxic symptoms often appearing within an hour of ingestion by susceptible species. The bacterium is said to recycle in the aquatic environment and this is thought to be a consequence of proliferation in susceptible insects, cannibalism and release into the water. **Efficacy:** Very effective when used against mosquito larvae in still water, even in the presence of high levels of organic matter. Light stability can cause problems if exposed to high sunlight intensities. Rapidly hydrolysed under even mild alkaline conditions.

Bacillus sphaericus is more effective than *Bacillus thuringiensis* subsp. *israelensis* for use in slow-release formulations designed to control mosquitos. **Key reference(s):** H D Burges & K A Jones. 1998. Formulation of bacteria, viruses and Protozoa to control insects. In *Formulation of Microbial Biopesticides, Beneficial Microorganisms, Nematodes and Seed Treatments*, H D Burges (ed.), pp. 33–127, Kluwer Academic Publishers, Dordrecht, The Netherlands.

COMMERCIALISATION: Formulation: Formulated as water dispersible granules (WG). **Tradenames:** 'Spherimos' (Valent BioSciences), 'VectoLex CG' (Valent BioSciences).

APPLICATION: Applied by air or by hand-held application equipment to expanses of water to be treated. Rates of application depend upon the stage of larvae to be treated and the organic content of the water. Rates between 2 and 4 kg of product per hectare are recommended, with the highest rates used against large larvae and in highly polluted water. Pesticide products containing *Bacillus sphaericus* remain active for 1 to 4 weeks after spraying, depending primarily on the species of mosquito larvae, environmental conditions, water quality and exact form of the granules.

PRODUCT SPECIFICATIONS: Specifications: *Bacillus sphaericus* is produced commercially by fermentation as for *Bacillus thuringiensis* isolates for control of the *Plutella* group of Lepidoptera (entry 1:33). **Purity:** Prepared as for *Bacillus thuringiensis* subsp. *kurstaki* (entry 1:33). The commercial product contains living spores of *Bacillus sphaericus* plus the protein endotoxin. Efficacy can be determined by bioassay on *Culex* larvae in the laboratory. **Storage conditions:** Store in dry, temperature-stable conditions. **Shelf-life:** If stored under dry, stable conditions, the formulated product will remain viable for ≥ 2 years.

COMPATIBILITY: Compatible with other insecticides. Do not use in conjunction with copper-based fungicides or algal control agents. Incompatible with strong oxidisers, acids, bases and chlorinated water.

MAMMALIAN TOXICITY: No allergic reactions or other adverse health problems have been shown by research workers, manufacturing staff or users. **Acute oral LD$_{50}$:** rats >5000 mg/kg (tech.). **Acute dermal LD$_{50}$:** rabbits >2000 mg/kg (tech.). **Inhalation: LC$_{50}$:** rats (4 h) *c.* 0.09 mg/litre (tech.). **Skin and eye:** Mild skin irritant, eye irritant (rabbits).

ENVIRONMENTAL IMPACT AND NON-TARGET TOXICITY: In approved field use, *Bacillus sphaericus* has shown no adverse effects on non-target organisms or on the environment.

1:22 *Bacillus subtilis* var. *amyloliquefaciens* isolate FZB24   *Biological fungicide (bacterium)*

The Pesticide Manual Fifteenth Edition entry number: 56

Bacterium: Bacillales: Bacillaceae

NOMENCLATURE: Approved name: *Bacillus subtilis* (Ehrenberg) Cohn var. *amyloliquefaciens* isolate FZB24.

SOURCE: *Bacillus subtilis* var. *amyloliquefaciens* is found naturally in soil and leaf litter and the commercial isolate (FZB24) was isolated from this source. FZB Biotechnik produces three different isolates of *Bacillus subtilis* var. *amyloliquefaciens*, coded FZB24, FZB13 and FZB42.

PRODUCTION: *Bacillus subtilis* var. *amyloliquefaciens* isolates are produced commercially by fermentation. The fermentation medium with viable spores is used to prepare the product.

TARGET PESTS: Specific soil fungi, such as *Rhizoctonia* and *Fusarium* species, which cause rot and wilt diseases in plants.

TARGET CROPS: *Bacillus subtilis* var. *amyloliquefaciens* isolate FZB24 is recommended for use in glasshouses and other indoor sites. It can be used at these sites on shade and forest tree seedlings, ornamentals and shrubs. 'TAE-022' and 'Taegro' are registered as a plant protection agent in the USA for use on indoor, outdoor and food crops, 'Bacillus subtilis FZB24' is registered in Switzerland by Bayer CropScience as a plant protection agent for potatoes only and in Germany by Spiess-Urania Chemicals as a plant growth strengthening agent. 'Rhizo-Plus' is recommended for home and garden use in Germany.

BIOLOGICAL ACTIVITY: Mode of action: *Bacillus subtilis* var. *amyloliquefaciens* isolate FZB24 is effective as a fungicide but is also claimed to enhance growth, strengthen plants and increase yields. The bacterium establishes itself in the rhizosphere of the treated crop and colonises the plant root system, competing with disease organisms that attack the developing root system. There is evidence that the growth of pathogenic fungi is inhibited by the production of digestive enzymes by the bacterium within the rhizosphere. It is claimed that *Bacillus subtilis* var. *amyloliquefaciens* isolate FZB24 enhances the natural resistance of plants to disease pathogens. **Biology:** The rhizosphere is an environment in which a vast number of organisms thrive. It is well aerated and contains relatively high levels of nutrients supplied by the exudate from the growing plant and from the action of bacteria and fungi that colonise the region. Many plant pathogens are attracted to this region and will invade the roots of the germinating crop. *Bacillus subtilis* var. *amyloliquefaciens* isolate FZB24, if introduced into the root zone of germinating crops as a soil treatment, persists in the rhizosphere and outcompetes fungi, allowing the crop to grow away from the pathogen.

Key reference(s): 1) D R Fravel, W J Connick Jr & J A Lewis. 1998. Formulation of microorganisms to control plant diseases. In *Formulation of Microbial Biopesticides, Beneficial Microorganisms, Nematodes and Seed Treatments*, H D Burges (ed.), pp. 187–202, Kluwer Academic Publishers, Dordrecht, The Netherlands. 2) A S Paau. 1998. Formulation of beneficial microorganisms applied to soil. In *Formulation of Microbial Biopesticides, Beneficial Microorganisms, Nematodes and Seed Treatments*, H D Burges (ed.), pp. 235–54, Kluwer Academic Publishers, Dordrecht, The Netherlands. 3) G Schmiedeknecht, H Bochow & H Jung. 1997. Biologische Kontrolle knollen- und bodenbürtiger Erkrankungen der Kartoffel, *Med. Fac. Landbouww. Gent*, **62/3b**, 1055–62. 4) M Kilian, U Steiner, B Krebs, H Junge, G Schmiedeknecht & R Hain. 2000. FZB24 *Bacillus subtilis* – mode of action of a microbial agent enhancing plant vitality, *Pflanzenschutz-Nachrichten Bayer*, **1**, 72–93.

COMMERCIALISATION: Formulation: *Bacillus subtilis* var. *amyloliquefaciens* isolate FZB24 is sold as water dispersible granules (WG). **Tradenames:** 'Bacillus subtilis FZB24' (Spiess-Urania) and (Waldstein-Wartenberg), 'Rhizo-Plus' (FZB Biotechnik), 'Rhizo-Plus Konz' (FZB Biotechnik), 'TAE-022' (isolate FZB24) (Taensa), 'Taegro' (isolate FZB24) (Taensa) and (Earth BioSciences), 'Terranal' (isolate FZB13) (Ecostyle).

APPLICATION: It is recommended that the formulated product be pre-mixed in warm water to activate the bacteria before diluting for application. The product may be applied as a drench – saturating the soil containing the plants to be treated, preferably soon after planting or transplanting, or as a dip – applying the suspension to roots or cuttings before planting or as an addition to nutrient solutions.

PRODUCT SPECIFICATIONS: Specifications: *Bacillus subtilis* var. *amyloliquefaciens* isolates are produced commercially by fermentation. The fermentation medium with viable spores is used to prepare the product. **Purity:** Produced by fermentation and guaranteed to be free of bacterial contaminants harmful to man. Viability of the bacterium within the formulation can be determined by plating out on agar and incubating at temperatures of about 25 °C for 48 hours, then counting the colonies formed. **Storage conditions:** Store in a sealed container at room temperature under dry conditions. **Shelf-life:** Stable for at least 2 years, when stored at room temperature under dry conditions.

COMPATIBILITY: Incompatible with strong oxidisers, acids, bases and chlorinated water. *Bacillus subtilis* var. *amyloliquefaciens* may be used in admixture with soluble fertilisers and many pesticides. It should not be used, however, with copper-based products or with bactericides such as streptomycin.

MAMMALIAN TOXICITY: No harmful health effects to humans are expected from use of *Bacillus subtilis* var. *amyloliquefaciens* isolate FZB24 in pesticide products. Appropriate animal tests provided no indication that this isolate of *Bacillus subtilis* is toxic or infectious to humans. Given the limited uses allowed for products containing this naturally occurring bacterium,

ENVIRONMENTAL IMPACT AND NON-TARGET TOXICITY: *Bacillus subtilis* is a naturally occurring soil bacterium and, as such, would not be expected to cause any adverse effect on non-target organisms. **Bird toxicity:** Acute oral LD_{50} for bobwhite quail >4000 mg/kg. There was no pathogenicity or toxicity to young bobwhite quail by oral gavage over 5 days at 4×10^{11} spores per kilogram. **Approved for use in organic farming:** Yes.

1:24 *Bacillus subtilis* isolate HAI-0404
Biological fungicide (bacterium)

The Pesticide Manual **Fifteenth Edition entry number:** 56

Bacterium: Bacillales: Bacillaceae

NOMENCLATURE: Approved name: *Bacillus subtilis* (Ehrenberg) Cohn isolate HAI-0404.

SOURCE: Isolated from the foliage of trees in Japan.

PRODUCTION: *Bacillus subtilis* is produced commercially by fermentation. The spores are used to prepare the product.

TARGET PESTS: Grey mould (*Botrytis cinerea* Pers.).

TARGET CROPS: Vegetables and fruit.

BIOLOGICAL ACTIVITY: Mode of action: *Bacillus subtilis* isolate HAI-0404 is effective against diseases caused by *Botrytis cinerea* by preventing its spores from germinating, disrupting germ tube and mycelial growth and stopping attachment to the leaf by producing a zone of inhibition, restricting the growth of the pathogen. It has also been shown to colonise *Botrytis*, thus causing inhibition of the pathogen's germ tube elongation.

COMMERCIALISATION: Formulation: It is being sold as a wettable powder (WP) formulation often in combination with chemical fungicides by Nippon Soda.
Tradenames: 'Agrocare' (Nippon Soda).

APPLICATION: 'Agrocare' is applied as a foliar fungicide and should be applied in sufficient volume for complete foliar coverage.

PRODUCT SPECIFICATIONS: Specifications: *Bacillus subtilis* is produced commercially by fermentation. The spores are used to prepare the product. **Purity:** Produced by fermentation and guaranteed to be free of bacterial contaminants harmful to man. Viability of the bacterium within the formulation can be determined by plating out on agar and incubating at temperatures of about 25°C for 48 hours, then counting the colonies formed.

Storage conditions: Store in a sealed container at room temperature under dry conditions.

Shelf-life: Stable for at least 2 years, when stored at room temperature under dry conditions.

COMPATIBILITY: Incompatible with strong oxidisers, acids, bases and chlorinated water. *Bacillus subtilis* isolate HAI-0404 may be used in admixture with soluble fertilisers and many pesticides. It should not be used, however, with copper-based products or with bactericides such as streptomycin.

MAMMALIAN TOXICITY: Based on a review of the toxicology and other information submitted for *Bacillus subtilis* isolate HAI-0404, there is no evidence of risks to human health.

ENVIRONMENTAL IMPACT AND NON-TARGET TOXICITY: *Bacillus subtilis* isolate HAI-0404 is a naturally occurring bacterium that was isolated from the leaves of trees and, as such, would not be expected to cause any adverse effect on non-target organisms.

Approved for use in organic farming: Yes, if used alone.

1:25 *Bacillus subtilis* isolate MBI 600
Biological fungicide (bacterium)

The Pesticide Manual **Fifteenth Edition entry number:** 56

Bacterium: Bacillales: Bacillaceae

NOMENCLATURE: Approved name: *Bacillus subtilis* (Ehrenberg) Cohn isolate MBI 600.

SOURCE: The bacterium, *Bacillus subtilis*, is prevalent in soils and has been found in a variety of habitats worldwide. The isolate selected for commercialisation was chosen because of its spectrum and level of activity against important economic fungal pathogens.

PRODUCTION: Manufactured by fermentation.

TARGET PESTS: Wilts, crown rot, root rot and other seed-borne diseases caused by the fungi *Fusarium*, *Aspergillus*, *Pythium* and *Rhizoctonia*. Also targeted in a foliar formulation for control of *Botrytis* and mildew.

TARGET CROPS: Soybeans, cotton, dry beans, grasses, barley, wheat, maize (corn), peas and peanuts.

BIOLOGICAL ACTIVITY: Mode of action: *Bacillus subtilis* MBI 600 is a spore-forming bacterium that colonises the developing root system of young plants. The bacterium thus competes with and thereby suppresses disease organisms, such as *Fusarium*, *Rhizoctonia*, *Alternaria* and *Aspergillus*, that attack root systems. As a result of this biological protection, the plant may establish a more vigorous root system. **Key reference(s):** 1) A S Paau. 1998. Formulation of beneficial microorganisms applied to soil. In *Formulation of Microbial Biopesticides, Beneficial Microorganisms, Nematodes and Seed Treatments*, H D Burges (ed.), pp. 235–54, Kluwer Academic Publishers, Dordrecht, The Netherlands. 2) M P McQuilken, P Halmer & D J Rhodes. 1998. Application of microorganisms to seeds. In *Formulation of Microbial Biopesticides, Beneficial Microorganisms, Nematodes and Seed Treatments*, H D Burges (ed.), pp. 255–85, Kluwer Academic Publishers, Dordrecht, The Netherlands. 3) D R Fravel, W J Connick Jr & J A Lewis. 1998. Formulation of microorganisms to control plant diseases. In *Formulation of Microbial Biopesticides, Beneficial Microorganisms, Nematodes and Seed Treatments*, H D Burges (ed.), pp. 187–202, Kluwer Academic Publishers, Dordrecht, The Netherlands.

COMMERCIALISATION: Formulation: Sold as a dry powder seed treatment formulation (WS). **Tradenames:** 'Epic' (Becker Underwood – discontinued), 'Histick N/T' and 'Integral' (Becker Underwood), 'Pro-mix with Biofungicide' (Premier Horticulture), 'Stimulex' (Scotts), 'Subtilex' (formerly 'Epic') (Becker Underwood).

APPLICATION: The product is used as a seed treatment in combination with chemical seed treatments as a slurry mix which is continuously agitated and applied within 72 hours. Other applications include a suspension of MBI 600 spores applied to growing media mixes where *Bacillus subtilis* colonises the mix, suppressing a range of damping-off diseases. The foliar formulation is applied in rotation with chemical fungicides for the control of *Botrytis* in tomatoes and powdery mildew in strawberries.

PRODUCT SPECIFICATIONS: Specifications: Manufactured by fermentation. **Storage conditions:** Store under dry, ambient conditions. **Shelf-life:** Stable for at least 2 years, when stored in cool, dry conditions.

COMPATIBILITY: Incompatible with broad-spectrum seed treatments such as captan and copper-based products. Recommended for use with a range of fungicides to extend the disease control achieved. May be applied with insecticidal seed treatments. Incompatible with strong oxidisers, acids, bases and chlorinated water.

MAMMALIAN TOXICITY: Based on a review of the toxicology and other information submitted for *Bacillus subtilis* MBI 600, there is no evidence of risks to human health. Because the bacterium colonises plant roots, exposure to humans would be minimal once the seeds are planted.

ENVIRONMENTAL IMPACT AND NON-TARGET TOXICITY: *Bacillus subtilis* MBI 600 is not harmful to non-target organisms or to the environment. Tests were conducted to help

ensure that there were no risks to plants, insects or birds. Aquatic exposure is not expected. Tests have shown that this bacterium does not harm seeds or cause them to decay.
Approved for use in organic farming: Yes.

1:26 *Bacillus subtilis* isolate QST 713
Biological fungicide/bactericide (bacterium)

The Pesticide Manual **Fifteenth Edition entry number:** 56

Bacterium: Bacillales: Bacillaceae

NOMENCLATURE: Approved name: *Bacillus subtilis* (Ehrenberg) Cohn isolate QST 713. **Development code:** QST-713; QRD 133 WP.

SOURCE: The bacterium, *Bacillus subtilis*, is prevalent in soils and has been found in a variety of habitats worldwide. The isolate selected for commercialisation was chosen because of its spectrum and level of activity against important economic fungal and bacterial pathogens.

PRODUCTION: *Bacillus subtilis* QST 713 is produced commercially by fermentation. The fermentation medium with viable spores is used to prepare the product.

TARGET PESTS: Used as a foliar spray for the control of a wide range of economically important fungal and bacterial diseases, including *Botrytis cinerea* Pers., *Uncinula necator* Burr., *Podosphaera leucotricha* (Ellis & Everhart) Salmon, *Erysiphe* spp., *Sphaerotheca* spp., *Leveillula taurica* Arn., *Oidium* spp., *Oidiopsis taurica* Salm., *Alternaria* spp., *Erwinia amylovora* Winsl., *Venturia* spp., *Bremia lactucae* Regel, *Peronospora* spp., *Botryosphaeria dothidea* Ces. & De Not., *Phytophthora infestans* de Bary, *Xanthomonas* spp., *Sclerotinia minor* Jagger and *Plasmopara viticola* Berl. & de T.

TARGET CROPS: Used on grapes, pome fruit, walnuts, stone fruit, hops, cucurbits, leafy vegetables, crucifers, broccoli, peppers, tomatoes, potatoes, onions, carrots, herbs and ornamentals.

BIOLOGICAL ACTIVITY: Mode of action: 'Serenade Biofungicide' and 'Rhapsody Biofungicide' are effective by preventing diseases caused by a wide range of plant pathogens. They prevent plant pathogen spores from germinating, disrupt germ tube and mycelial growth and stop attachment of the plant pathogen to the leaf by producing a zone of inhibition, restricting the growth of these disease-causing pathogens. The QST 713 isolate has been shown to produce three different groups of lipoproteins (iturins, agrastatins/plipastatins and surfactins), that have been shown to act together synergistically against several different plant pathogens. It has also been shown to colonise pathogens, thus causing inhibition of the pathogen's germ tube elongation. 'Serenade Biofungicide' and 'Rhapsody Biofungicide' also activate the plant's Systemic Acquired Resistance (SAR) natural defence system. These multiple modes of action mean that 'Serenade Biofungicide' and 'Rhapsody Biofungicide' can be used in fungicide resistance strategies. **Key reference(s):** 1) D W Edgecomb & D Manker. 2005. *Bacillus subtilis* strain QST 713, a new biological tool for integrated and organic disease control programs, *Proc. BCPC International Congress Crop Science & Technology 2005*, **1**, 69–72. 2) D R Fravel, W J Connick Jr & J A Lewis. 1998. Formulation of microorganisms to control plant diseases. In *Formulation of Microbial Biopesticides, Beneficial Microorganisms, Nematodes and Seed Treatments*, H D Burges (ed.), pp. 187–202, Kluwer Academic Publishers, Dordrecht, The Netherlands.

COMMERCIALISATION: Formulation: The product is sold as a wettable powder (WP) and as an aqueous suspension (AS). *Bacillus subtilis* QST 713 is included in EU Annex 1. **Tradenames:** 'Impression' (AgraQuest) and (SDS Biotech), 'Rhapsody Biofungicide' (AgraQuest), 'Serenade Biofungicide' (AgraQuest). **Patents:** US 6060051, on the QST 713 isolate of *Bacillus subtilis*. Worldwide patents are pending.

APPLICATION: 'Serenade Biofungicide' and 'Rhapsody Biofungicide' are applied as foliar fungicides. The wettable powder (WP) formulation is applied at a rate of 5 to 8 kg per hectare and the aqueous suspension (AS) formulation is applied at a rate of 4 to 6 litres per hectare. Both products should be applied in sufficient volume for complete foliar coverage. It has also been shown that when products containing the QST 713 isolate are used sequentially with recommended rates or in admixture with reduced rates of chemical fungicides, effective disease control is achieved with lower rates of chemical application.

PRODUCT SPECIFICATIONS: Specifications: *Bacillus subtilis* QST 713 is produced commercially by fermentation. The fermentation medium with viable spores is used to prepare the product. **Purity:** The WP formulation consists of 5×10^9 cfu/g and the AS formulation consists of 7×10^9 cfu/g. Guaranteed to be free of microbial contaminants. **Storage conditions:** Both the WP and AS formulations can be stored under dry, ambient conditions. **Shelf-life:** Stable for at least 2 years under the above storage conditions.

COMPATIBILITY: Compatible with many fungicides, bactericides, insecticides, foliar nutrients and adjuvants used in the control of crop diseases. Incompatible with strong oxidisers, acids, bases and chlorinated water.

MAMMALIAN TOXICITY: 'Serenade Biofungicide' and 'Rhapsody Biofungicide' are considered to be non-toxic products when used according to label directions. 'Serenade Biofungicide' and 'Rhapsody Biofungicide' meet all the criteria for US-EPA Toxicity Category III with the signal word 'Caution'. **Acute oral LD$_{50}$:** rats >5000 mg/kg. **Acute dermal LD$_{50}$:** rabbits >2000 mg/kg.

ENVIRONMENTAL IMPACT AND NON-TARGET TOXICITY: 'Serenade Biofungicide' and 'Rhapsody Biofungicide' have an extremely favourable environmental and ecological profile. 'Serenade Biofungicide' and 'Rhapsody Biofungicide' have low ecological risk to birds, fish and aquatic invertebrates, when used at label rates. **Bird toxicity:** Acute oral LD_{50} for bobwhite quail >5000 mg/kg. **Fish toxicity:** LC_{50} for trout 162 ppm. **Other aquatic toxicity:** EC_{50} for *Daphnia* 108 ppm. **Effects on beneficial insects:** LC_{50} for honeybee larvae >10 000 ppm; for Hymenoptera parasitic wasp >30 000 ppm; for ladybird >60 000 ppm; for lacewing larvae >60 000 ppm. 'Serenade Biofungicide' and 'Rhapsody Biofungicide' are non-toxic to beneficial insects, when used at label rates. This environmentally benign profile makes 'Serenade Biofungicide' and 'Rhapsody Biofungicide' compatible in Integrated Pest Management (IPM) programmes. **Approved for use in organic farming:** Yes.

1:27 *Bacillus thuringiensis* encapsulated delta-endotoxins for control of Coleoptera
Biological insecticide (bacterium)

The Pesticide Manual **Fifteenth Edition entry number:** 58

Bacterium: Bacillales: Bacillaceae

NOMENCLATURE: Approved name: *Bacillus thuringiensis* Berliner subsp. *morrisoni* encapsulated delta-endotoxins. **Other names:** Formerly known as *Bacillus thuringiensis* Berliner subsp. *san diego* encapsulated delta-endotoxins. **Development code:** SAN 401 I (Novartis); CGA 237218 (mixture with *Bt kurstaki*) (Novartis).

SOURCE: Produced in cells of *Pseudomonas fluorescens* Migula which have been genetically modified to produce the *Bacillus thuringiensis* endotoxin. For all 'CellCap'-based products, cells are then killed in such a way that they constitute a rigid microcapsule for the enclosed insecticidal protein.

PRODUCTION: Manufactured by fermentation, as for isolates for control of the *Plutella* group of encapsulated delta-endotoxins (entry 1:28).

TARGET PESTS: 'M-Trak' is used to control Colorado potato beetle (*Leptinotarsa decemlineata* (Say)).

TARGET CROPS: 'M-Trak' is recommended for use in potatoes.

BIOLOGICAL ACTIVITY: Mode of action: The mode of action is described in detail under the entry for *Bacillus thuringiensis* isolates for control of the *Plutella* group of Lepidoptera (entry 1:33). The products contain no live cells: the infectious role of the spores in live products is partially fulfilled by gut flora derived from bacteria that live on leaves. The absence of *Bt* spores reduces potency in some species of Coleoptera, but this is balanced by the protection given by the capsules, the net result being increased potency compared with non-encapsulated products. Encapsulation within the cell wall of *Pseudomonas fluorescens* makes formulations more robust than the standard endotoxins-plus-spores products produced by fermentation of *Bacillus thuringiensis*. **Key reference(s):** 1) W Gelernter. 1990. Targeting insecticide-resistant markets. In *Managing Resistance to Agrochemicals: From Fundamental Research to Practical Strategies*, M B Green, W K Moberg & H LeBaron (eds), American Chemical Society, New York, USA. 2) M Metz (ed.). 2004. Bacillus thuringiensis *A Cornerstone of Modern Agriculture*, 242 pp. (with index), Haworth Press Inc., Binghamton, NY, USA.

COMMERCIALISATION: Formulation: Sold as capsule suspensions (CS) (encapsulated in killed *Pseudomonas fluorescens*) and as granules (GR). **Tradenames:** 'M-Trak' (*san diego – cryIIIA*) (Ecogen).

APPLICATION: As for isolates for control of the *Plutella* group of Lepidoptera (entry 1:33).

PRODUCT SPECIFICATIONS: Specifications: Manufactured by fermentation, as for isolates for control of the *Plutella* group of encapsulated delta-endotoxins (entry 1:28).
Purity: 'M-Trak' is based on *Bt* subsp. *morrisoni* endotoxin CryIIIA. **Storage conditions:** Store under cool, dry conditions in a sealed container. Do not expose to direct sunlight.
Shelf-life: If stored under manufacturer's recommended conditions, the product will remain active for over 12 months.

COMPATIBILITY: Incompatible with strong oxidising agents, acids and bases. Can be used with most insecticides, acaricides, fungicides and spray tank additives.

MAMMALIAN TOXICITY: 'M-Trak' has shown no allergic or other toxic effects in test animals. **Other toxicological effects:** Considered to be non-toxic. Tolerance exempt in the USA on all raw agricultural commodities when applied to growing crops pre- or post-harvest.

ENVIRONMENTAL IMPACT AND NON-TARGET TOXICITY: 'M-Trak' is not expected to have any adverse effects on non-target organisms or on the environment.

1:28 *Bacillus thuringiensis* encapsulated delta-endotoxins for control of the *Plutella* group of Lepidoptera

Biological insecticide (bacterium)

1. Micro-organisms

The Pesticide Manual **Fifteenth Edition entry number:** 58

Bacterium: Bacillales: Bacillaceae

NOMENCLATURE: Approved name: *Bacillus thuringiensis* Berliner subsp. *kurstaki* encapsulated delta-endotoxins.

SOURCE: Produced in cells of *Pseudomonas fluorescens* Migula which have been genetically modified to produce the *Bacillus thuringiensis* serotype H-3a,3b endotoxin. Such products are known as 'CellCap'. For all 'CellCap'-based products, cells are then killed in such a way that they constitute a rigid microcapsule for the enclosed insecticidal protein.

PRODUCTION: A gene coding for an endotoxin protein from *Bacillus thuringiensis* subsp. *kurstaki* has been used to transform the bacterium *Pseudomonas fluorescens*. The transformed bacterium expresses the gene and produces the delta-endotoxin. The cells are produced by fermentation, killed and treated with cross-linking agents to fortify the bacterial cell wall and, thus, give an encapsulated toxin as the product. There are four practical reasons for producing the toxins commercially in *Pseudomonas fluorescens* rather than in the strains of *Bt* where they originated: *Pseudomonas fluorescens* produces higher concentrations of the toxins than *Bt*; the endotoxins remain encapsulated within the *Pseudomonas fluorescens* cells when the cells are killed, making collecting the endotoxins relatively easy; the dead *Pseudomonas fluorescens* cells protect the *Bt* endotoxins from decomposition induced by ultraviolet light, thereby lengthening the time the toxins remains active; *Pseudomonas fluorescens* does not produce the large number of additional toxins made by *Bt*, which might harm non-target species.

TARGET PESTS: Effective against Lepidoptera and some beetles. 'MVP' is recommended for diamondback moth (*Plutella xylostella* L.) and other Lepidoptera and 'M-Peril' for corn borers (*Sesamia* spp., *Ostrinia furnacalis* Guenée and *Ostrinia nubilalis* (Hübner)).

TARGET CROPS: Effective in vegetables, maize (corn), tree fruit, vines and cotton. 'MVP' is recommended for use in cruciferous crops and 'M-Peril' for maize. 'Guardjet' was developed for the Japanese crucifer market.

BIOLOGICAL ACTIVITY: Mode of action: The mode of action is described in detail under the entry for *Bacillus thuringiensis* isolates for control of the *Plutella* group of Lepidoptera

(1:33). The products contain no live cells: the infectious role of the spores in live products is partially fulfilled by gut flora derived from bacteria that live on leaves. The absence of *Bt* spores reduces potency in some species of Lepidoptera, but this is balanced by the protection given by the capsules, the net result being increased potency compared with non-encapsulated products. Encapsulation within the cell wall of *Pseudomonas fluorescens* makes formulations more robust than the standard endotoxins-plus-spores products produced by fermentation of *Bacillus thuringiensis*. **Key reference(s):** 1) W Gelernter. 1990. Targeting insecticide-resistant markets. In *Managing Resistance to Agrochemicals: From Fundamental Research to Practical Strategies*, M B Green, W K Moberg & H LeBaron (eds), American Chemical Society, New York, USA. 2) M Metz (ed.). 2004. Bacillus thuringiensis *A Cornerstone of Modern Agriculture*, 242 pp. (with index), Haworth Press Inc., Binghamton, NY, USA.

COMMERCIALISATION: Formulation: Sold as a capsule suspension (CS) (encapsulated in killed *Pseudomonas fluorescens*) and as a granule (GR). **Tradenames:** 'Guardjet' (*kurstaki – cryIA(c)*) (Kubota), 'M-Peril' (*kurstaki – cryIA(c)*) (Ecogen), 'MVP' (*kurstaki – cryIA(c)*) (Ecogen), 'MVP II' (*kurstaki – cryIA(c)*) (Ecogen).

APPLICATION: As for non-encapsulated isolates for control of the *Plutella* group of Lepidoptera (entry 1:33).

PRODUCT SPECIFICATIONS: Specifications: A gene coding for an endotoxin protein from *Bacillus thuringiensis* subsp. *kurstaki* has been used to transform the bacterium *Pseudomonas fluorescens*. The transformed bacterium expresses the gene and produces the delta-endotoxin. The cells are produced by fermentation, killed and treated with cross-linking agents to fortify the bacterial cell wall and, thus, give an encapsulated toxin as the product. There are four practical reasons for producing the toxins commercially in *Pseudomonas fluorescens* rather than in the strains of *Bt* where they originated: *Pseudomonas fluorescens* produces higher concentrations of the toxins than *Bt*; the endotoxins remain encapsulated within the *Pseudomonas fluorescens* cells when the cells are killed, making collecting the endotoxins relatively easy; the dead *Pseudomonas fluorescens* cells protect the *Bt* endotoxins from decomposition induced by ultraviolet light, thereby lengthening the time the toxins remain active; *Pseudomonas fluorescens* does not produce the large number of additional toxins made by *Bt*, which might harm non-target species. **Purity:** 'MVP' (MYX7275), 'MVP II' (MYX104) and 'M-Peril' are based on *Bacillus thuringiensis* subsp. *kurstaki* toxin CryIA(c), as microcapsule ('MVP', 'MVP II') and granular formulations. **Storage conditions:** Store under cool, dark conditions. Do not expose to direct sunlight. **Shelf-life:** If stored under manufacturer's recommendation, the products will remain active for over 12 months.

COMPATIBILITY: Incompatible with strong oxidising agents, acids and bases. Can be used with most insecticides, acaricides, fungicides and spray tank additives.

MAMMALIAN TOXICITY: Acute oral LD$_{50}$: rats >5050 mg/kg (formulation). **Acute dermal LD$_{50}$:** rabbits >2020 mg/kg (formulation). **Inhalation:** All animals survived a dose of 9.98 × 10¹⁰ cells. **Toxicity class:** EPA (formulation) III. **Other toxicological effects:** No formulation has shown any evidence of toxicity, infectivity or hypersensitivity to mammals. No allergic reactions or other health problems have been shown by research workers, manufacturing staff or users. Considered to be non-toxic. Tolerance exempt in the USA on all raw agricultural commodities when applied to growing crops pre- or post-harvest.

ENVIRONMENTAL IMPACT AND NON-TARGET TOXICITY: The formulated product, although longer lasting than conventional *Bt* products, has a short persistence due to its sensitivity to u.v. light. No adverse effects have been recorded in approved field use and none are anticipated. Formulations should not be used near water courses.

1:29 *Bacillus thuringiensis* encapsulated delta-endotoxins for control of the *Spodoptera* group of Lepidoptera
Biological insecticide (bacterium)

The Pesticide Manual **Fifteenth Edition entry number:** 58

Bacterium: Bacillales: Bacillaceae

NOMENCLATURE: Approved name: *Bacillus thuringiensis* Berliner subsp. *aizawai* encapsulated delta-endotoxins.

SOURCE: *Bacillus thuringiensis* subsp. *aizawai* occurs widely in Nature, as does the bacterium *Pseudomonas fluorescens* Migula.

PRODUCTION: Produced as for *Bacillus thuringiensis* encapsulated delta-endotoxins for the control of the *Plutella* group of Lepidoptera (entry 1:28).

TARGET PESTS: Lepidopteran larvae, such as *Spodoptera*, *Heliothis*, *Helicoverpa* and *Pieris* species, *Ostrinia nubilalis* (Hübner) and *Plutella xylostella* (L.). This product was developed because it controls *Spodoptera* spp. and some other noctuids which have only moderate susceptibility to *Btk*, because it is more active against some species recommended with *Btk* and because it controls species that have developed resistance to *Btk*. 'M/C' is recommended for control of armyworm species (*Spodoptera* spp.).

TARGET CROPS: Soft fruit, canola, maize (corn), soybeans, peanuts, cotton, tree fruits and nuts, tobacco, vegetables and vines.

BIOLOGICAL ACTIVITY: Mode of action: The mode of action is described in detail under the entry for *Bacillus thuringiensis* encapsulated delta-endotoxins for control of the *Plutella* group of Lepidoptera (1:28). **Key reference(s):** M Metz (ed.). 2004. Bacillus thuringiensis *A Cornerstone of Modern Agriculture*, 242 pp. (with index), Haworth Press Inc., Binghamton, NY, USA.

COMMERCIALISATION: Formulation: Sold as liquid concentrates. **Tradenames:** 'Mattch' (*kurstaki* + *aizawai*) (Ecogen), 'M/C' (*aizawai*, CryIC) (Ecogen).

APPLICATION: As for isolates for control of the *Plutella* group, repeating every 7 to 10 days, depending upon the intensity of insect pressure.

PRODUCT SPECIFICATIONS: Specifications: Produced as for *Bacillus thuringiensis* encapsulated delta-endotoxins for the control of the *Plutella* group of Lepidoptera (entry 1:28). **Purity:** Contains bacterial cells cross-linked for additional strength and containing *Bta* endotoxins. 'Mattch' (MYX 300) is a mixture of *kurstaki* CryIA(c) and *aizawai* CryIC toxins and 'M/C' (MYX 833) is based on *aizawai* toxin CryIC. **Storage conditions:** Store in a cool, dry place. Do not expose to direct sunlight. **Shelf-life:** Stable for over 12 months.

COMPATIBILITY: Compatible with most acaricides, insecticides and spray tank additives. Incompatible with strong oxidising agents, acids and bases.

MAMMALIAN TOXICITY: Acute oral LD$_{50}$: rats >5000 mg/kg (formulation). **Acute dermal LD$_{50}$:** rabbits >2000 mg/kg (formulation). **Inhalation: LC$_{50}$:** rats >2 × 10^{12} cells/ml. All animals exposed to 1 × 10^{11} cells by inhalation survived. **Skin and eye:** Does not penetrate the skin; mild skin irritant (rabbits). **Toxicity class:** EPA (formulation) III.
Other toxicological effects: Considered to be non-toxic. Tolerance exempt in the USA on all raw agricultural commodities when applied to growing crops pre- or post-harvest.

ENVIRONMENTAL IMPACT AND NON-TARGET TOXICITY: There is no evidence that the formulated product has any adverse effect on non-target organisms or on the environment. **Fish toxicity:** LC$_{50}$ (96 h) >250 mg/litre. **Effects on beneficial insects:** LC$_{50}$ >1000 mg/kg soil. Harmless to mite predators *Typhlodromus pyri* Scheuten and *Amblyseius californicus* (McGregor).

1:30 *Bacillus thuringiensis* isolates for control of Coleoptera

Biological insecticide (bacterium)

The Pesticide Manual **Fifteenth Edition entry number:** 57

Bacterium: Bacillales: Bacillaceae

NOMENCLATURE: Approved name: *Bacillus thuringiensis* Berliner subsp. *morrisoni* isolates Sa-10 (Certis) and NovoBtt (Valent BioSciences). **Other names:** Formerly known as *Bacillus thuringiensis* Berliner subsp. *tenebrionis*; *Bacillus thuringiensis* subsp. *san diego*; *Btm*; *Btt*. **Development code:** SAN 418 I (originally Sandoz, subsequently Novartis, then Thermo Trilogy and now Certis). **CAS RN:** *[68038-71-1]*; formerly *[12673-85-7]*, *[62628-54-0]* and *[67383-05-5]*.

SOURCE: *Bacillus thuringiensis* subsp. *morrisoni* is common in soils, particularly those rich in insects.

PRODUCTION: By fermentation, as for isolates for control of the *Plutella* group of Lepidoptera (entry 1:33).

TARGET PESTS: Used to control some Coleoptera, particularly the Colorado potato beetle (*Leptinotarsa decemlineata* (Say)).

TARGET CROPS: Solanaceous crops, mainly potatoes.

BIOLOGICAL ACTIVITY: Mode of action: The mode of action is described in detail under entry 1:33 (*Bacillus thuringiensis* isolates for control of the *Plutella* group of Lepidoptera). **Biology:** As for isolates for control of the *Plutella* group, except that *Btm* produces a single insecticidal protein of 73 kDa that does not require activation to show its full activity. The midgut environment of the target insect, *Leptinotarsa decemlineata*, is nearer to neutral pH than in Lepidoptera, being around pH 6. Ultrastructural effects on the insect midgut are similar to those of other endotoxins, with a few important exceptions. No membrane lesions or microvillar damage are observed and the first cellular response (swelling and elongation) is relatively slow. Studies on the binding affinity of a susceptible coleopteran (*Leptinotarsa decemlineata*) and a tolerant one (*Diabrotica undecimpunctata howardi* Barber – southern corn rootworm) showed that susceptibility was correlated, at least in part, with increased receptor binding affinity and pore formation. Both adults and larvae are susceptible to the product. **Key reference(s):** 1) L F Adams, C-L Liu, S C MacIntosh & R L Starnes. 1996. Diversity and biological activity of *Bacillus thuringiensis*. In *Crop Protection Agents from Nature: Natural Products and Analogues*, L G Copping (ed.), pp. 360–88, Royal Society of Chemistry, Cambridge, UK.

2) B C Carlton. 1993. Genetics of *Bt* insecticidal crystal proteins and strategies for the construction of improved strains. In *Pest Control with Enhanced Environmental Safety*, S O Duke, J J Menn & J R Plimmer (eds), pp. 326–37, ACS Symposium Series 524, American Chemical Society, Washington DC, USA. 3) M Metz (ed.). 2004. Bacillus thuringiensis *A Cornerstone of Modern Agriculture*, 242 pp. (with index), Haworth Press Inc., Binghamton, NY, USA.

COMMERCIALISATION: Formulation: Sold as a water dispersible liquid formulation (SC). **Tradenames:** 'Bt var. San Diego' (Biocontrol Network), 'Colorado Potato Beetle Beater' (Biocontrol Network), 'M-one' (Mycogen – discontinued), 'M-one Plus' (*san diego*) (Mycogen – discontinued), 'N-Cap' (Mycogen – discontinued), 'Novodor' (Valent BioSciences), 'Trident' (Thermo Trilogy – discontinued).

APPLICATION: As for isolates for control of the *Plutella* group of Lepidoptera, repeating every 7 to 10 days as necessary.

PRODUCT SPECIFICATIONS: Specifications: By fermentation, as for isolates for control of the *Plutella* group of Lepidoptera (entry 1:33). **Purity:** All formulations are produced to contain a dose of the toxin that is expressed in terms of international units (i.u.) active against a target pest per mg of product. **Storage conditions:** Store under cool, dry conditions in a sealed container. Do not expose to direct sunlight. **Shelf-life:** Stable for up to 2 years, if stored under manufacturers' recommended conditions.

COMPATIBILITY: Can be applied with conventional agrochemicals and with spray additives. Incompatible with strong oxidising agents, acids and bases.

MAMMALIAN TOXICITY: Acute oral toxicity: Oral dosing of rats showed no adverse acute effects at doses of $>2 \times 10^8$ colony forming units (cfu) per animal (technical) or >5 g/kg (WP formulation). **Toxicity class:** EPA (formulation) III. **Other toxicological effects:** Considered to be non-toxic. Tolerance exempt in the USA on all raw agricultural commodities when applied to growing crops pre- or post-harvest.

ENVIRONMENTAL IMPACT AND NON-TARGET TOXICITY: Non-toxic to the environment and neither infective nor toxic to non-target organisms.
Approved for use in organic farming: Yes.

1:31 *Bacillus thuringiensis* isolates for control of Diptera

Biological insecticide (bacterium)

The Pesticide Manual **Fifteenth Edition entry number:** 57

Bacterium: Bacillales: Bacillaceae

NOMENCLATURE: Approved name: *Bacillus thuringiensis* Berliner subsp. *israelensis* isolate EG2215. **Other names:** *Bti*. **Development code:** SAN 402 I (originally Sandoz, subsequently Novartis, then Thermo Trilogy and now Certis). **CAS RN:** *[68038-71-1]*; formerly *[12673-85-7]*, *[62628-54-0]* and *[67383-05-5]*.

SOURCE: *Bacillus thuringiensis* subsp. *israelensis* occurs naturally in soil and strain SA-3 has been selected for development as an insecticide. The original isolate for EG2215 is ONR-60A (Goldberg & Margalit) which is identical to isolate HD567 from the H. Dulmage USDA collection (Dulmage *et al.*).

PRODUCTION: Produced by fermentation, as for isolates for control of the *Plutella* group of Lepidoptera (1:33).

TARGET PESTS: Only Diptera: e.g. mosquito and blackfly larvae and sewerage flies. 'Gnatrol' can be used to control fungus gnats and mushroom fly larvae.

TARGET CROPS: Used in water bodies, sewerage filters and in glasshouses and mushroom houses.

BIOLOGICAL ACTIVITY: Mode of action: Stomach poison; the mode of action is described in detail under the entry for *Bacillus thuringiensis* isolates for control of the *Plutella* group of Lepidoptera (entry 1:33). **Biology:** As for isolates for control of the *Plutella* group of Lepidoptera, except that the crystal inclusions derived from *Bti* are the most insoluble of any *Bt* crystals, requiring a very high pH (>11) for full solubilisation. *Bti* produces five different insecticidal proteins and all have dipteran activity, although one toxin, the 27 kDa cytolytic toxin, appears to synergise the others. The effects of *Bti* on larvae are as for the *Plutella* group of Lepidoptera. Sometimes its effects are more rapid: heavily infested mosquito pools may be dramatically covered by floating, dying larvae within twenty minutes of application of *Bti* granules. Extended periods of control are achieved by virtue of the residual activity of *Bti*. The spores cause no significant increase in mortality and so spore-free products are marketed to minimise the weight carried during aerial application and to satisfy countries that do not permit addition of spores to potable water. This is in contrast to the *Bt* isolates for control of the *Plutella* group of Lepidoptera, where the action of spores is significant in some host species. **Efficacy:** *Bacillus thuringiensis* subsp. *israelensis* produces five Cry toxins with dipteran activity: Cry4A, Cry4B, Cry10A and Cry11A plus CytIA. **Key reference(s):** 1) H de Barjac & D J Sutherland (eds). 1990. *Bacterial Control of Mosquitoes and Blackflies: Biochemistry, Genetics and Applications of* Bacillus thuringiensis israelensis *and* Bacillus sphaericus, Unwin Hyman, London, UK. 2) P Fast. 1981. The crystal toxin of *Bacillus thuringiensis.* In *Microbial Control of Pests and Plant Diseases 1970–1980,* H D Burges (ed.), Academic Press, London, UK. 3) L F Adams, C-L Liu, S C MacIntosh & R L Starnes. 1996. Diversity and biological activity of *Bacillus thuringiensis.* In *Crop Protection Agents from Nature: Natural Products and Analogues,* L G Copping (ed.), pp. 360–88, Royal Society of Chemistry, Cambridge, UK. 4) P F Entwistle, J S Cory, M J Bailey & S Higgs (eds). 1993. Bacillus thuringiensis, *an Environmental Biopesticide: Theory and Practice,* 311 pp. Wiley, Chichester, UK. 5) H D Burges & K A Jones. 1989. Formulation of bacteria, viruses and Protozoa to control insects. In *Formulation of Microbial Biopesticides, Beneficial Microorganisms, Nematodes and Seed Treatments,* H D Burges (ed.), pp. 33–127, Kluwer Academic Press, Dordrecht, The Netherlands. 6) M Metz (ed.). 2004. Bacillus thuringiensis *A Cornerstone of Modern Agriculture,* 242 pp. (with index), Haworth Press Inc., Binghamton, NY, USA. 7) L J Goldberg & J Margalit. 1977. A bacterial spore demonstrating rapid larvicidal activity against *Anopheles sergentii, Uranotaenia unguiculata, Culex univitattus, Aedes aegypti* and *Culex pipiens, Mosq. News,* **37**, 355–8. 8) Dulmage, Beegle, Barjac, Reich, Donaldson & Krywienczyk. 1982. *Bacillus thuringiensis* cultures available from the US Department of Agriculture. Agricultural Research Service, New Orleans, USA.

COMMERCIALISATION: Formulation: Sold as aqueous suspensions, briquettes (BR), flowable concentrates (SC), water dispersible granules (WG), wettable powders (WP) and slow-release rings. **Tradenames:** 'Acrobe' (BASF), 'Aquabac' (Becker Microbial Products), (Biocontrol Network) and (Rincon-Vitova), 'Bactimos' (public health use) (Valent BioSciences), 'Bactis' (Caffaro), 'Gnatrol' (Valent BioSciences), 'Hil-Bti' (Hindustan), 'Liper-I' (Agri Life), 'Mosquito Dunks' (Rincon-Vitova) and (Biocontrol Network), 'Skeetal' (Valent BioSciences), 'Teknar' (Valent BioSciences), 'VectoBac' (Rincon-Vitova) and (Valent BioSciences), 'Vectocide' (Sanex).

APPLICATION: Applied by air or by hand-held application equipment to expanses of water to be treated. Rates of application increase with the age of larvae to be treated and the organic content of the water. Rates between 2 and 4 kg of product per hectare are usual. Spray into soil around germinating seedlings to control fungus gnats.

PRODUCT SPECIFICATIONS: Specifications: Produced by fermentation, as for isolates for control of the *Plutella* group of Lepidoptera (1:33). **Purity:** *Bacillus thuringiensis* subsp. *israelensis* formulations contain delta-endotoxins with or without spores.
Storage conditions: Store under cool, dry conditions. Do not expose to direct sunlight.
Shelf-life: Stable for up to 2 years, if stored under recommended conditions.

COMPATIBILITY: It is not usual to apply *Bti* with other pesticides. Incompatible with strong oxidising agents, acids and bases.

Acute oral LD$_{50}$: rats >2670 mg/kg, 1×10^{11} spores/kg; rabbits >2×10^9 spores/kg.
Acute dermal LD$_{50}$: rats >2000 mg/kg, 4.6×10^{10} spores/kg; rabbits >6.28 g/kg.
Inhalation: LC$_{50}$: 8×10^7 spores per rat. NOEL: Rats (3 months) 4 g/kg body weight daily.
Toxicity class: EPA (formulation) III ('Gnatrol' IV). Other toxicological effects: Considered to be non-toxic. Tolerance exempt in the USA on all raw agricultural commodities when applied to growing crops pre- or post-harvest.

ENVIRONMENTAL IMPACT AND NON-TARGET TOXICITY: Non-toxic to the environment and neither infective nor toxic to non-target organisms. Fish toxicity: LC$_{50}$ for water feeder guppies (*Toecilia reticulata*) >156 mg/litre (as 'Teknar').
Other aquatic toxicity: LC$_{50}$ for *Daphnia pulex* (96 h) >25 mg/litre (technical).
Approved for use in organic farming: Yes.

1:32 *Bacillus thuringiensis* isolates for control of Lepidoptera and Coleoptera
Biological insecticide (bacterium)

The Pesticide Manual Fifteenth Edition entry number: 57

Bacterium: Bacillales: Bacillaceae

NOMENCLATURE: Approved name: *Bacillus thuringiensis* Berliner subsp. *kurstaki* various isolates with some conjugates of *Bacillus thuringiensis* subsp. *kurstaki* with other subspecies, particularly *Bacillus thuringiensis* subsp. *aizawai*. This group also includes some new isolates to control caterpillars and beetles that are not very susceptible to *Btk* and related isolates. These are often effective against field populations of the *Plutella* group of Lepidoptera that have developed resistance to *Btk*. Development code: CGA 237218 (conjugate of *Btk* and *Bta* originally Ciba, subsequently Novartis, then Thermo Trilogy and now Certis) and Ecogen isolates EG7673 and EG2424 (constructed by multiple-step conjugal mating between subsp. *kurstaki* isolate EG2042 (recipient) and subspecies *morrisoni*, *kumamotoensis* and *kurstaki* isolate EG2154 (donors) (see US patent 5024837)). CAS RN: *[68038-71-1]*; formerly *[12673-85-7]*, *[62628-54-0]* and *[67383-05-5]*.

SOURCE: *Bacillus thuringiensis* isolates occur naturally in soil and other insect-rich environments.

PRODUCTION: By fermentation, as for isolates for control of the *Plutella* group of Lepidoptera (entry 1:33). The endotoxins and living spores are formulated as water dispersible liquid concentrates and as a variety of dry products.

TARGET PESTS: Ecogen isolates EG7673 and EG2424 ('Raven' and 'Jackpot', respectively) are used to control Colorado potato beetle (*Leptinotarsa decemlineata* (Say)), in addition to Lepidoptera.

TARGET CROPS: Recommended for use in various row crops, fruit and other trees, vegetables and cotton.

BIOLOGICAL ACTIVITY: Mode of action: Stomach poison; the mode of action is described in detail under the entry for *Bacillus thuringiensis* isolates for control of the *Plutella* group of Lepidoptera (entry 1:33). **Key reference(s):** 1) L F Adams, C-L Liu, S C MacIntosh & R L Starnes. 1996. Diversity and biological activity of *Bacillus thuringiensis*. In *Crop Protection Agents from Nature: Natural Products and Analogues*, L G Copping (ed.), pp. 360–88, Royal Society of Chemistry, Cambridge, UK. 2) M Metz (ed.). 2004. Bacillus thuringiensis *A Cornerstone of Modern Agriculture*, 242 pp (with index), Haworth Press Inc, Binghamton, NY, USA.

COMMERCIALISATION: Formulation: Sold as oil miscible flowable concentrates (OF), water suspension concentrates (SC) and wettable powders (WP). **Tradenames:** 'Crymax' [EG7841] (Certis), 'Lepinox' [EG2371] (Certis) and (Intrachem).

APPLICATION: As for isolates for control of the *Plutella* group of Lepidoptera and repeating every 5 to 7 days if infestations are high.

PRODUCT SPECIFICATIONS: Specifications: By fermentation, as for isolates for control of the *Plutella* group of Lepidoptera (entry 1:33). The endotoxins and living spores are formulated as water dispersible liquid concentrates and as a variety of dry products. **Purity:** The fermentation is stopped at sporulation when, in addition to spores, crystals of protein (the delta-endotoxin) are also formed. Both are included in the product. **Storage conditions:** Store under cool, dry conditions. Do not expose to direct sunlight. **Shelf-life:** If stored as recommended by the manufacturer, the formulations will have a shelf-life of 1 to 3 years or more.

COMPATIBILITY: Can be applied with a wide range of acaricides, insecticides, fungicides and spray adjuvants. Incompatible with strong oxidising agents, acids and bases.

MAMMALIAN TOXICITY: Acute oral toxicity: Rats dosed with $>1 \times 10^8$ colony forming units (cfu) per animal showed no adverse effects. **Toxicity class:** EPA (formulation) III. **Other toxicological effects:** There has been no record of allergic or other adverse effects from researchers, manufacturers or users. Tolerance exempt in the USA on all raw agricultural commodities when applied to growing crops pre- or post-harvest.

ENVIRONMENTAL IMPACT AND NON-TARGET TOXICITY: *Bacillus thuringiensis* subspp. *kurstaki* and *aizawai* are non-hazardous to birds and fish. These products are not expected to cause any adverse effects on non-target organisms. **Behaviour in soil:** *Bacillus thuringiensis* subspp. *kurstaki* and *aizawai* are common constituents of the soil microflora.

1:33 *Bacillus thuringiensis* isolates for control of the *Plutella* group of Lepidoptera
Biological insecticide (bacterium)

The Pesticide Manual Fifteenth Edition entry number: 57

Bacterium: Bacillales: Bacillaceae

NOMENCLATURE: **Approved name:** *Bacillus thuringiensis* Berliner subsp. *kurstaki* and various other isolates (*Btk* and related types). **Other names:** *Btk*. **Development code:** CGA-269941. **CAS RN:** *[68038-71-1]*; formerly *[12673-85-7]*, *[62628-54-0]* and *[67383-05-5]*.

SOURCE: *Bacillus thuringiensis* isolates are common in soil, mills, warehouses and other insect-rich environments. Isolates that are used in crop protection are selected from those isolated in Nature on the basis of their potency in test insect species, spectrum of host insects and the ease with which they can be grown in fermenters. The insecticidal activity of *Bt* was first observed in insects associated with man in 1902 in dying larvae of *Bombyx mori* Lin. (silkworm) by S. Ishiwata (cited by K. Ishikawa, *Pathology of the Silkworm*) and was characterised after isolation from larvae of *Ephestia kuehniella* Zell. by E. Berliner (1915. *Z. Angew. Entomol.*, **2**, 29). First used as the microbial insecticide 'Sporeine' against lepidopteran larvae in 1938 (S. E. Jacobs. 1950. *Proc. Soc. Appl. Bacteriol.*, **13**, 83). Developed for the control of these pests by several companies, which currently market products.

PRODUCTION: Produced by accurately controlled fermentation in deep tanks of sterilised nutrient liquid medium. The endotoxins and living spores are harvested as water dispersible liquid concentrates for subsequent formulation. *B. thuringiensis* subsp. *kurstaki* strains EG2348, EG2371, EG7841 and EG7826 were constructed at Ecogen Inc. (US patent 5080897) through conjugal mating between *B. thuringiensis* subsp. *kurstaki* strains (recipient) and *B. thuringiensis* subsp. *aizawai* strains (donor); they produce both CryI and CryII crystal proteins. Strains EG2424 and EG7673 were constructed through a multiple-step conjugal mating between subsp. *kurstaki* strain EG2042 (recipient) and subspp. *morrisoni*, *kumamotoensis* and *kurstaki* strain EG2154 (donors) (see US patent 5024837 and C. Gawron-Burke & T. Johnson in *Advances in Potato Pest Biology and Management*, G. Zehnder, (ed.)); it produces both CrylII and Cryl crystal proteins.

TARGET PESTS: Lepidopteran larvae, particularly the diamondback moth (*Plutella xylostella* (L.)) and other vegetable pests and forest insects.

TARGET CROPS: Recommended for use in vegetables, fruit, maize (corn), small-grain cereals and in forests, orchards or for general tree care.

BIOLOGICAL ACTIVITY: Mode of action: *Bacillus thuringiensis* produces parasporal, proteinaceous, crystal inclusion bodies during sporulation. Upon ingestion, these are insecticidal to larvae of the order Lepidoptera and to both larvae and adults of a few Coleoptera. Once in the insect, the crystal proteins are solubilised and the insect gut proteases convert the original pro-toxin into a combination of up to four smaller toxins. These hydrolysed toxins bind to the insect's midgut cells at high-affinity, specific receptor binding sites, where they interfere with the potassium ion-dependent, active amino acid symport mechanism. This disruption causes the formation of large cation-selective pores that increase the water permeability of the cell membrane. A large uptake of water causes cell swelling and eventual rupture, disintegrating the midgut lining. Different toxins bind to different receptors and this explains the selectivity of different *Bt* isolates in different insect species and with varying intensities; this explains species specificities. **Biology:** The crystal inclusions derived from *Btk* are generally specific to Lepidoptera. Because they have to be ingested and then processed within the insect's gut, they are often slow acting (in comparison with conventional chemicals). The toxin stops feeding and young larvae may starve to death; insects not killed by direct action of the toxin may die from bacterial infection over a period of 2 to 3 days. Different toxins have different spectra of biological activity (see other *Bacillus thuringiensis* entries). Different isolates and serotypes have been developed by different companies. For example, Novartis (*Bt*-based products now owned by Certis) developed serotypes 3a and 3b. In addition to producing the endotoxins, many isolates of *Bt* are potent insect pathogens. **Efficacy:** Very effective when used against lepidopteran species, where some damage to the crop is acceptable, such as in forestry. Stability can be a problem if exposed to strong sunlight (primarily the u.v. component). Rapidly hydrolysed under even mild alkaline conditions. **Key reference(s):** 1) P Fast. 1981. The crystal toxin of *Bacillus thuringiensis*. In *Microbial Control of Pests and Plant Diseases 1970–1980*, H D Burges (ed.), Academic Press, New York, USA. 2) L F Adams, C-L Liu, S C MacIntosh & R L Starnes. 1996. Diversity and biological activity of *Bacillus thuringiensis*. In *Crop Protection Agents from Nature: Natural Products and Analogues*, L G Copping (ed.), pp. 360–88, Royal Society of Chemistry, Cambridge, UK. 3) P F Entwistle, J S Cory, M J Bailey & S Higgs (eds). 1993. Bacillus thuringiensis, *an Environmental Biopesticide: Theory and Practice*, 311 pp. Wiley, Chichester, UK. 4) H D Burges & K A Jones. 1989. Formulation of bacteria, viruses and protozoa to control insects. In *Formulation of Microbial Biopesticides: Beneficial Microorganisms, Nematodes and Seed Treatments*, H D Burges (ed.), pp. 33–127, Kluwer Academic Press, Dordrecht, The Netherlands. 5) M Metz (ed.). 2004. Bacillus thuringiensis *A Cornerstone of Modern Agriculture*, 242 pp. (with index), Haworth Press Inc., Binghamton, NY, USA.

COMMERCIALISATION: Formulation: Sold as a combination of endotoxin crystals and living bacterial spores. Formulated as suspension concentrates (SC), granular baits (GB), ready-to-use baits (RB), suspo-emulsions (SE), granules (GR), oil miscible flowable concentrates (OF), dispersible powders (DP), water dispersible granules (WG) and wettable powders (WP). Some formulations are encapsulated *Bt* delta-endotoxins for the control of the *Plutella* group of Lepidoptera (encapsulated *Btk* endotoxins; see 1:28). **Tradenames:** 'Able' (Certis), 'Agree' (Certis), 'Agrobac' (Tecomag), 'Bacilex' (Shionogi – discontinued) and (Bayer CropScience – discontinued), 'Bactec Bt 32' (Plato), 'Bactosid K' (Sanex), 'Bactospeine' (Valent BioSciences), 'Bactospeine Koppert' (Koppert), 'Bactospeine WP' (Koppert – discontinued), 'Bactucide' (Caffaro) and (Isagro), 'Baturad' (Cequisa), 'Biobit' (Valent BioSciences), 'Biocot' (DuPont), 'Bollgard' [EG2349] (Ecogen – discontinued), 'Certan' (Steele & Brodie – discontinued), 'Collapse' (Calliope) and (NPP – discontinued), 'Cordalene' (Agrichem) and (Intrachem Bio Italia), 'CoStar' (Certis), 'Cutlass' [EG2371] (Certis), 'Delfin' (Certis) and (Kwizda), 'Delivery' [EG2371] (Certis), 'DiPel' (Biocontrol Network), (Rincon-Vitova) and (Valent BioSciences), 'Ecotech Bio' [EG2371] (Bayer CropScience), 'Ecotech Pro' [EG2348] (Bayer CropScience), 'Foil' [EG2424] (Ecogen – discontinued), 'Foray' (Valent BioSciences), 'Forwarbit' (Forward International), 'Insectobiol' (Samabiol), 'Halt' (Biostadt), 'Hil-Btk' (Hindustan), 'Insectobiol' (Samabiol), 'Jackpot' [EG2424] (Ecogen – discontinued) and (Intrachem – discontinued), 'Javelin' [EG2348] (Certis), 'Larvo-BT' (Troy Biosciences), 'Lepinox' [GC-91] (Certis), 'Lipel-K' (Agri Life), 'Novosol FC' (Ashlade – discontinued), 'Quark' (Valent BioSciences – discontinued), 'Rapax' [EG2348] (Certis) and (Intrachem), 'Scutello' (Biobest), 'Scutello 2x' (Biobest), 'Steward' ([H3a,3b]) (Novartis – discontinued), 'Thuricide' (Biocontrol Network), (Certis) and (Valent Biosciences), 'Troy-BT' (Troy Biosciences). **Patents:** Many worldwide including US 5080897; US 5024837 (both to Ecogen).

APPLICATION: May be applied using hand-held or aerial systems or through irrigation systems at rates of 100 to 300 g a.i. per hectare, ensuring that the crop is well covered with the spray suspension. Apply while insect larvae are small and repeat every 5 to 7 days if infestations are high. *Bt*-based sprays can be applied up to the day of harvest.

PRODUCT SPECIFICATIONS: Specifications: Produced by accurately controlled fermentation in deep tanks of sterilised nutrient liquid medium. The endotoxins and living spores are harvested as water dispersible liquid concentrates for subsequent formulation. *B. thuringiensis* subsp. *kurstaki* strains EG2348, EG2371, EG7841 and EG7826 were constructed at Ecogen Inc. (US patent 5080897) through conjugal mating between *B. thuringiensis* subsp. *kurstaki* strains (recipient) and *B. thuringiensis* subsp. *aizawai* strains (donor); they produce both CryI and CryII crystal proteins. Strains EG2424 and EG7673 were constructed through a multiple-step conjugal mating between subsp. *kurstaki* strain EG2042 (recipient) and subspp. *morrisoni*, *kumamotoensis* and *kurstaki* strain EG2154 (donors) (see US patent 5024837 and C. Gawron-Burke & T. Johnson in *Advances in Potato Pest Biology and Management*, G. Zehnder, (ed.)); it produces both CryIII and CryI crystal proteins. **Purity:** All formulations are standardised at a toxin content expressed in terms of international units (i.u.) active against a target pest per mg of product. Guaranteed to be free of microbial contaminants.

Storage conditions: Do not expose to direct sunlight. Keep cool, but do not freeze.

Shelf-life: If stored under cool, dark conditions, the products remain viable for 2 years or more.

COMPATIBILITY: Do not use in combination with broad-spectrum biocides such as chlorothalonil. Compatible with a wide range of acaricides, insecticides, fungicides, stickers, spreaders and wetters. Do not use water with a pH above 8.0. Incompatible with strong oxidisers, acids, bases and chlorinated water.

MAMMALIAN TOXICITY: Despite decades of the widespread use of *Bacillus thuringiensis* as a pesticide (it has been registered since 1961), there have been no confirmed reports of immediate or delayed allergic reactions to the delta-endotoxin itself despite significant oral, dermal and inhalation exposure to the microbial product. Extensive studies on *Bacillus thuringiensis*-containing pesticides demonstrate that isolates are not toxic or pathogenic. No adverse effects were observed on body weight gain, clinical effects or on necropsy. Infectivity/ pathogenicity studies show that rodents gradually eliminate *B. thuringiensis* from the body after oral, inhalation or intravenous application. Observed toxicity at high doses is attributed to the vegetative growth stage, not to the insecticidal protein or to the spores. Early formulations, produced from *B. thuringiensis* contained a non-proteinaceous, toxic β-exotoxin. (See review by US-EPA, J T McClintock, C R Schaffer & R D Sjoblad. 1995. A comparative review of mammalian toxicity of *Bacillus thuringiensis*-based pesticides, *Pestic. Sci.*, **45**, 95–105).

Acute oral LD$_{50}$: No infectivity or toxicity was observed in rats at 4.7×10^{11} spores/kg of product; rat LD$_{50}$ ('Able') >5500 mg/kg rat body weight (toxicity category IV). No adverse effects at doses of 1×10^8 up to 7×10^{12} colony forming units (cfu) per rat. Rat LD$_{50}$ (technical) >5500 mg/kg rat body weight (toxicity category IV). **Acute dermal LD$_{50}$:** rats >5000 mg/kg; rabbits >1×10^9 cfu; rabbit LD$_{50}$ ('Able' and technical) >2020 mg/kg (toxicity category III). **Inhalation:** No infectivity or toxicity at 5.4 mg/litre (2.6×10^7 spores/litre). Albino rat LC$_{50}$ (aerosolised 'Able') 5.63 mg/litre (toxicity category IV). **Skin and eye:** Some products can cause substantial but temporary eye injury (probably due to formulation ingredients). With the unformulated bacterium, there was no infectivity or toxicity observed in rats at 3.4×10^{11} spores/kg (toxicity category IV). **NOEL:** (2 years) rats 8.4 g/kg; (13 weeks) rats 1.3×10^9 spores/kg body weight daily. **Toxicity class:** EPA (formulation) III. Considered to be non-toxic. Tolerance exempt in the USA on all raw agricultural commodities when applied to growing crops pre- or post-harvest. **Other toxicological effects:** *Btk* and related isolates have not shown evidence of toxicity, infectivity or hypersensitivity to mammals.

No allergic reactions or other health problems have been reported by research workers, manufacturing staff or users.

ENVIRONMENTAL IMPACT AND NON-TARGET TOXICITY: *Btk* and related isolates have short persistence due to their sensitivity to u.v. light. No adverse effects have been recorded in approved field use and none are anticipated. *Btk* should not be used near water courses. **Bird toxicity:** In 63-day feeding trials, chickens receiving 5.1×10^7 spores/g diet showed no ill effects. In 5-day feeding trials, bobwhite quail and mallard ducklings receiving 1.6 g/kg (8 g/kg total) ('Able') by oral gavage showed no signs of toxicity. **Fish toxicity:** LC_{50} (96 h) for water gobies (*Pomatoschistus minutus*) >400 mg/litre (as 'Thuricide'). The aqueous and oral NOELs (31 days) for sheepshead minnow were $>4.9 \times 10^{10}$ colony forming units (cfu)/litre of dilution water and 3.7×10^7 cfu/g of food, respectively ('Able'). For bluegill sunfish, aqueous $LC_{50} >4.7 \times 10^{10}$ cfu/litre of dilution water and oral $LC_{50} >3.9 \times 10^7$ cfu/g of food ('Able'). **Other aquatic toxicity:** On *Daphnia*, the lowest observable effect concentration (NOEC) was 1.1×10^9 colony forming units (cfu)/litre; the reproductive NOEL was 5.6×10^8 cfu/litre and the LC_{50} was $>1.1 \times 10^9$ cfu/litre ('Able'). The oral NOEL for grass shrimp was $>3.6 \times 10^7$ cfu/g of food ('Able'). **Effects on beneficial insects:** Non-toxic to honeybees: LD_{50} >0.1 mg/bee ('Delfin'). Adult honeybees (10 days *per os*) LC_{50} 118 g/bee. The no observed effect concentration (NOEC) and lowest observed effect concentration (LOEC) were 156 and 625 ppm, respectively ('Able'). NOEL on green lacewing larvae (8 days *per os*) was 3000 ppm; the NOEL to adult *Hippodamia convergens* Guerin (21 days *per os*) was 1500 ppm; and the NOEL to adult *Nasonia vitripennis* (Walker) (15 days *per os*) was 3000 ppm ('Able'). **Approved for use in organic farming:** Yes.

1:34 *Bacillus thuringiensis* isolate for control of soil-inhabiting Coleoptera
Biological insecticide (bacterium)

The Pesticide Manual **Fifteenth Edition entry number:** 57

Bacterium: Bacillales: Bacillaceae

NOMENCLATURE: Approved name: *Bacillus thuringiensis* Berliner subsp. *japonensis* isolate *buibui*. **Other names:** *Btj*. **CAS RN:** *[68038-71-1]*; formerly *[12673-85-7]*, *[62628-54-0]* and *[67383-05-5]*.

SOURCE: *Bacillus thuringiensis* subsp. *japonensis* isolate *buibui* occurs naturally in soil and was originally isolated from soil in Japan.

PRODUCTION: As for isolates for control of the *Plutella* group of Lepidoptera (entry 1:33).

TARGET PESTS: Soil-inhabiting beetles.

TARGET CROPS: Turf, grass, landscapes and ornamentals.

BIOLOGICAL ACTIVITY: Mode of action: Stomach poison; the mode of action is described in detail under the entry for *Bacillus thuringiensis* isolates for control of the *Plutella* group of Lepidoptera (entry 1:33). **Biology:** As for isolates for control of the *Plutella* group of Lepidoptera, except that *Bacillus thuringiensis* subsp. *japonensis* isolate *buibui* is specific to beetles and is very effective at controlling phytophagous soil-inhabiting species.
Key reference(s): 1) P F Entwistle, J S Cory, M J Bailey & S Higgs (eds). 1993. Bacillus thuringiensis, *an Environmental Biopesticide: Theory and Practice*, 311 pp. Wiley, Chichester, UK. 2) M Metz (ed.). 2004. Bacillus thuringiensis *A Cornerstone of Modern Agriculture*, 242 pp. (with index), Haworth Press Inc., Binghamton, NY, USA.

COMMERCIALISATION: Formulation: Sold as powder formulations. **Tradenames:** 'KM 503' (Kubota), 'M-Press' (Mycogen – discontinued).

APPLICATION: Apply to established grass when beetle egg hatch is complete. Repeat every 7 to 10 days, depending upon the intensity of the attack.

PRODUCT SPECIFICATIONS: Specifications: As for isolates for control of the *Plutella* group of Lepidoptera (entry 1:33). **Purity:** Guaranteed to be free of bacterial contaminants. **Storage conditions:** Store in a cool, dry situation out of direct sunlight. **Shelf-life:** May be stored for 12 months.

COMPATIBILITY: Compatible with other soil-active insecticides, but do not apply with broad-spectrum fungicides such as copper-based compounds. Incompatible with strong oxidising agents, acids and bases.

MAMMALIAN TOXICITY: *Bacillus thuringiensis* subsp. *japonensis* isolate *buibui* occurs widely in Nature and there have been no reports of allergic or other adverse toxicological effects from research workers, manufacturers, formulators or field staff.

ENVIRONMENTAL IMPACT AND NON-TARGET TOXICITY: *Bacillus thuringiensis* subsp. *japonensis* isolate *buibui* is not expected to have any adverse effects on non-target organisms or on the environment.

1:35 *Bacillus thuringiensis* isolates for control of the *Spodoptera* group of Lepidoptera   *Biological insecticide (bacterium)*

The Pesticide Manual **Fifteenth Edition entry number:** 57

Bacterium: Bacillales: Bacillaceae

NOMENCLATURE: Approved name: *Bacillus thuringiensis* Berliner subsp. *aizawai* and various isolates, including some conjugates with other subspecies. This group includes strains to control insects that are less susceptible to *Btk* and related strains. These are also effective against field populations of the *Plutella* group that have developed resistance to *Btk*.
Other names: *Bta.* **Development code:** SAN 401 (originally Sandoz, subsequently Novartis, then Thermo Trilogy and now Certis); GC-91 (originally Ciba, subsequently Novartis then Thermo Trilogy and now Certis). **CAS RN:** *[68038-71-1]*; formerly *[12673-85-7]*, *[62628-54-0]* and *[67383-05-5]*.

SOURCE: *Bacillus thuringiensis* isolates occur naturally in soil and other insect-rich environments.

PRODUCTION: By fermentation, as for isolates for control of the *Plutella* group (entry 1:33). The endotoxins and living spores are formulated as water dispersible liquid concentrates and as a variety of dry products. Isolate GC-91 is a conjugant.

TARGET PESTS: Used to control Lepidoptera. Developed because it controls *Spodoptera* spp. and some other noctuids which have only moderate susceptibility to *Btk*, because it is more active against some species recommended with *Btk* and because it controls species that have developed resistance to *Btk*.

TARGET CROPS: Recommended for use in various row crops, fruit and other trees, vegetables and cotton.

BIOLOGICAL ACTIVITY: Mode of action: Stomach poison; the mode of action is described in detail under the entry for *Bacillus thuringiensis* isolates for control of the *Plutella* group of Lepidoptera (entry 1:33). **Key reference(s):** 1) L F Adams, C-L Liu, S C MacIntosh & R L Starnes. 1996. Diversity and biological activity of *Bacillus thuringiensis*. In *Crop Protection Agents from Nature: Natural Products and Analogues*, L G Copping (ed.), pp. 360–88, Royal Society of Chemistry, Cambridge, UK. 2) M Metz (ed.). 2004. Bacillus thuringiensis *A Cornerstone of Modern Agriculture*, 242 pp. (with index), Haworth Press Inc., Binghamton, NY, USA.

COMMERCIALISATION: Formulation: Sold as water dispersible suspension concentrates (SC), water dispersible granules (WG) and wettable powders (WP).

Tradenames: 'Agree' (Certis), 'Certan' (Syngenta) and (Thermo Trilogy – discontinued), 'Florbac' (Valent BioSciences), 'M-C KM' (Kubota), 'XenTari' (Valent BioSciences) and (Biobest).

APPLICATION: As for isolates for control of the *Plutella* group, repeating every 5 to 7 days if infestations are high.

PRODUCT SPECIFICATIONS: Specifications: By fermentation, as for isolates for control of the *Plutella* group (entry 1:33). The endotoxins and living spores are formulated as water dispersible liquid concentrates and as a variety of dry products. Isolate GC-91 is a conjugant. **Purity:** The fermentation is stopped at sporulation when, in addition to spores, crystals of protein (the delta-endotoxin) are also formed. Both are included in the product. **Storage conditions:** Store under cool, dry conditions. Do not expose to direct sunlight. **Shelf-life:** If stored as recommended by the manufacturer, the formulations will have a shelf-life of 1 to 3 years or more.

COMPATIBILITY: Can be applied with a wide range of acaricides, insecticides, fungicides and spray adjuvants. Incompatible with strong oxidising agents, acids and bases.

MAMMALIAN TOXICITY: Acute oral toxicity: rats dosed with $>1 \times 10^8$ colony forming units (cfu) per animal showed no adverse effects. **Toxicity class:** EPA (formulation) III. **Other toxicological effects:** There has been no record of allergic or other adverse effects from researchers, manufacturers or users. Tolerance exempt in the USA on all raw agricultural commodities when applied to growing crops pre- or post-harvest.

ENVIRONMENTAL IMPACT AND NON-TARGET TOXICITY: *Bacillus thuringiensis* subsp. *aizawai* is non-hazardous to birds and fish, but 'XenTari' is highly toxic to honeybees exposed directly to spray treatment (Editorial note – almost certainly due to formulation additives). It should not be applied when honeybees are foraging in the area to be treated. **Behaviour in soil:** In clay loam of low nutrient status (pH 7.3, pF 2, 25 °C), insecticidal activity declined rapidly in 20 days due to deterioration of the crystals; at pF 3, it declined slowly for 500 days; at higher nutrient status, there was a brief 10-fold increase of inoculum applied to soil. As a natural part of the ecosystem, it decays to complex and non-toxic organic compounds. *Bacillus thuringiensis* subspp. *kurstaki* and *aizawai* are both common constituents of the soil microflora. **Approved for use in organic farming:** Yes.

1:36 *Beauveria bassiana* isolates ATCC 74040, Bb 147, GHA and Stanes
Biological insecticide (fungus)

Fungus: anamorph of Hypocreales: Clavicipitaceae, Ascomycota form taxa: formerly named Fungi Imperfecti: mitosporic fungi or Deuteromycetes: Moniliales

NOMENCLATURE: Approved name: *Beauveria bassiana* (Balsamo) Vuillemin isolates ATCC 74040, Bb 147, GHA and Stanes. **Other names:** Formerly known as *Botrytis bassiana* Balsamo. **Common name:** White muscardine. **Development code:** ESC 170 GH (Ecoscience); F-7744 (Troy). **OPP Chemical Code:** 128818.

SOURCE: An isolate of *Beauveria bassiana* was obtained from a mycosed larva of the European corn borer (*Ostrinia nubilalis* (Hübner)) found in Beauce, France by INRA. A production process was developed by INRA and is now owned by Natural Plant Protection (NPP). The Troy isolate was obtained from a boll weevil (*Anthonomus grandis* (Boheman)) at the USDA-ARS Crop Insect Research Center, Lower Rio Grande Valley, Texas, USA. Three isolates that have been commercialised are Bb 147 – NPP; ATCC 74040 (= ARSEF 3097 = FCI 7744) – Troy; and GHA – Emerald BioAgriculture Corp. The Stanes isolate was derived from an infected tea shothole borer at the United Planters' Association of Southern India, Tamil Nadu, India.

PRODUCTION: 'Ostrinil' is produced by solid state fermentation on clay granules. Other products are produced by solid state fermentation and subsequent extraction of conidia. The Stanes isolate is produced by submerged fermentation.

TARGET PESTS: Isolate Bb 147 is recommended for use against European corn borer (*Ostrinia nubilalis*) and Asiatic corn borer (*O. furnacalis* Guenée); isolate GHA is used against whitefly, thrips, aphids and mealybugs; and isolate ATCC 74040 is effective against a range of soft-bodied coleopteran and hemipteran pests. 'Bio-Power' (Stanes isolate) is used against shothole borer, coffee berry borer, white grubs, bollworm, cutworm, brown planthopper and diamondback moth. 'Beauverin', 'Boverol' and 'Boverosil' are recommended for control of Colorado beetle. Activity is claimed for a very wide range of other insects including wireworm, mole crickets, tetranychid mites, pear psylla and Japanese beetle larvae.

TARGET CROPS: Isolate Bb 147 is recommended for use in maize (corn) in Europe; isolate GHA is used in vegetables and ornamentals; and isolate ATCC 74040 is used in turf and ornamentals as 'Naturalis-O' and 'Naturalis-T' and on all raw agricultural commodities as 'Naturalis-L'. A new formulation, 'Back-Off', is being developed for use in cotton, vegetables and ornamentals. 'Bio-Power' is recommended for use in tea, coffee, cotton, tomato, brinjal, okra and rice.

BIOLOGICAL ACTIVITY: Biology: The entomopathogen invades the insect body. Fungal conidia become attached to the insect cuticle and, after germination, the hyphae penetrate the cuticle and proliferate in the insect's body. High humidity or free water is essential for conidial germination and infection can take between 24 and 48 hours, depending on the temperature. The infected insect may live for 3 to 5 days after hyphal penetration and, after death, the conidiophores are produced on the outside of the insect's body and new conidia are released on the outside of the insect cadaver. The fungus is insect specific.

Key reference(s): 1) G Riba. 1985. Thése de Doctorat d'État, Mention Sciences, Université Pierre et Marie Curie, France. 2) J E Wright & T A Knauf. 1994. Evaluation of Naturalis-L for control of cotton insects, *Proc. Brighton Crop Protection Conference – Pests & Diseases*, **1**, 45. 3) J D Vandenberg, A M Shelton, N J Wilsey & M Ramos. 1998. Assessment of *Beauveria bassiana* sprays for control of diamondback moth on crucifers, *J. Econ. Entomol.*, **91**(3), 624–30. 4) I Mazet & D G Boucias. 1996. Effects of the fungal pathogen, *Beauveria bassiana* on protein synthesis of infected *Spodoptera exigua* larvae, *J. Insect Physiol.*, **42**, 91–9. 5) H D Burges. 1998. Formulation of mycoinsecticides. In *Formulation of Microbial Biopesticides, Beneficial Microorganisms, Nematodes and Seed Treatments*, H D Burges (ed.), pp. 131–85, Kluwer Academic Publishers, Dordrecht, The Netherlands. 6) (For classification) K H Seifert & W Gams. 2000. The taxonomy of anamorphic fungi. In *The Mycota. VII Systematics and Evolution*, *Part A*, K Esser & P A Lemke (eds), pp. 325–6, Springer, Heidelberg, Germany.

COMMERCIALISATION: Formulation: 'Ostrinil' is a clay microgranular formulation colonised by sporulating mycelia of a pyralid active isolate (MG); 'Mycotrol' and 'BotaniGard' are available as a wettable powder (WP) and an emulsifiable suspension (ES); 'Mycotrol O' is approved for use in organic crop production; 'Naturalis-L' is a suspension concentrate (SC); and 'Bio-Power' is a WP containing less than 8% moisture. **Tradenames:** 'Boverin' (ex-Soviet Union), 'Bio-Power' (Stanes), 'Boveral OF' (Intrachem), 'Boverol' (ex-Czechoslovakia), 'Boverosil' (ex-Czechoslovakia), 'Conidia' (Live Systems Technology), 'CornGuard' (Mycotech), 'Fermone Naturalis L-225' (Troy), 'Ghamycotrol' (Emerald BioAgriculture), 'Naturalis' (Intrachem), 'Naturalis-L' (Troy), 'Naturalis-O' (Troy), 'Naturalis-T' (Troy), 'Ostrinil' (NPP) and (Calliope), 'Proecol' (Probioagro), 'Racer' (SOM Phytopharma) and (Agri Life). **Patents:** EP 9040118330. The INRA patent (EP 0406103 A1) for the solid fermentation of *Beauveria bassiana* on clay granules has been granted exclusively to NPP.

APPLICATION: Used as foliar sprays through all types of applicators, with water as the carrier. Application rates depend upon the crop and the pests to be controlled. The normal application rate on commodity crops is 750 to 1000 ml of product per hectare, for ornamentals under cover or outdoors 24 to 80 ml per 10 litres and on turf and lawns 32 to 96 ml per 100 square metres. Used as a foliar spray through all types of application equipment, with water as a carrier. The normal recommendation for 'Bio-Power' is 1 kg of product in 100 litres of water. This gives a usual application rate of between 1×10^{13} and 10^{14} colony forming units (cfu) per hectare.

PRODUCT SPECIFICATIONS: Specifications: 'Ostrinil' is produced by solid state fermentation on clay granules. Other products are produced by solid state fermentation and subsequent extraction of conidia. The Stanes isolate is produced by submerged fermentation. **Purity:** 'Naturalis' contains conidia of *Beauveria bassiana* at a concentration of 2.3×10^7 spores per ml, 'Ostrinil' contains at least 5×10^8 spores per gram, 'Bio-Power' contains conidia of *Beauveria bassiana* at a concentration of 1×10^8 spores per gram, 'Mycotrol' and 'BotaniGard' contain at least 4×10^{10} spores per gram (WP) or 2×10^{10} spores per ml (ES). Viability of the spores is determined by culture on nutrient agar and counting the colonies formed, and efficacy is checked by bioassay with an appropriate insect. Detailed identification of the specific isolate in any formulation requires DNA and isoenzyme matching with the registered isolate held in a type culture collection. **Storage conditions:** Store in a cool, dry place. Do not freeze and do not allow the product to undergo thermal shock. **Shelf-life:** May be kept for up to one year, if stored below 20 °C.

COMPATIBILITY: The products may be used alone or tank mixed with other products such as sticking agents, insecticidal soaps, emulsifiable oils and insecticides or used with beneficial insects. Do not use with fungicides and wait 48 hours after application before applying fungicides. Incompatible with strong oxidisers, acids and bases and chlorinated water.

MAMMALIAN TOXICITY: No infectivity or pathogenicity was observed in rats after 21 days exposure to 1.8×10^9 colony forming units (cfu) per kg. **Acute oral LD$_{50}$:** rats $>1.8 \times 10^8$ cfu per kg. Toxicity category IV. **Acute dermal LD$_{50}$:** rats >2000 mg/kg. *Beauveria bassiana* isolate 74040 was not pathogenic, infective or toxic in rabbits dosed dermally at 2 g per animal containing 4.2×10^7 cfu/ml. Toxicity Category IV. **Inhalation: LC$_{50}$:** rats $>1.2 \times 10^8$ cfu/animal. **Skin and eye:** Possible irritant to eyes, skin and respiratory system. Rabbits displayed minimal ocular irritation when given a single 0.1 ml ocular dose of 'Naturalis-L' containing 2×10^6 cfu. Toxicity Category III. Rabbits dosed and exposed to 5 ml *Beauveria bassiana* ('Naturalis-L') containing 5.5×10^7 cfu for 4 hours showed no mortality or significant toxic effects. Toxicity Category IV. **Other toxicological effects:** Dermal, oral and inhalation studies with 'Naturalis-L' on rats indicated that the fungus is non-toxic and non-pathogenic.

ENVIRONMENTAL IMPACT AND NON-TARGET TOXICITY: Bird toxicity: Oral LD$_{50}$ (5 days) for quail >2000 mg/kg daily (by gavage). **Fish toxicity:** 'Naturalis-L' does not affect fish embryos, larvae or adults. LC$_{50}$ (31 days) for rainbow trout 7300 mg/litre. 'Bio-Power', in a 96-hour exposure study on *Tilapia* at a dose of 2×10^8 cfu per litre, had no toxic effect. There were no indications of infectivity or pathogenicity in fathead minnow (*Pimephales promelas*) dosed with 1.0×10^9 cfu/litre *Beauveria bassiana*. **Other aquatic toxicity:** EC$_{50}$ (14 days) for *Daphnia pulex* de Geer 4100 mg/litre. EC$_{50}$ for *Daphnia magna* 9.9×10^7 spores/litre. NOEC for *Daphnia magna* 7.6×10^7 spores/litre and LOEC 1.3×10^8 spores/litre. **Effects on beneficial insects:** On honeybees, 30-day dietary and contact studies indicate that 'Naturalis-L' and 'Bio-Power' have no significant effect;

LC_{50} (23 days, ingestion) 9285 µg/bee. No effects were observed on beneficial species after field application. **Behaviour in soil:** Background levels of *Beauveria bassiana* present in cotton fields were not significantly changed after repeated applications of 'Naturalis-L', indicating that the fungus does not accumulate. 'Bio-Power' is safe to the earthworm *Lampito marutii* at 10 g product per kg dry weight of soil. *Beauveria bassiana* is not pathogenic to plants. **Approved for use in organic farming:** Yes.

1:37 *Beauveria bassiana* isolate 447
Biological insecticide (fungus)

Fungus: anamorph of Hypocreales: Clavicipitaceae, Ascomycota form taxa: formerly named Fungi Imperfecti: mitosporic fungi or Deuteromycetes: Moniliales

NOMENCLATURE: **Approved name:** *Beauveria bassiana* (Balsamo) Vuillemin isolate 447. **Other names:** Formerly known as *Botrytis bassiana* Balsamo. **Common name:** White muscardine.

SOURCE: An isolate of *Beauveria bassiana* was obtained from a mycosed fire ant colony.

PRODUCTION: Produced by solid state fermentation and subsequent extraction of conidia.

TARGET PESTS: Fire ants (*Solenopsis* spp.) and other ants found indoors.

BIOLOGICAL ACTIVITY: **Biology:** The entomopathogen invades the ant's body. Fungal conidia become attached to the cuticle and, after germination, the hyphae penetrate the cuticle and proliferate in the ant's body. The infected ant may live for 3 to 5 days after hyphal penetration and, after death, the conidiophores are produced on the outside of the ant's body and new conidia are released on the outside of the cadaver. The fungus is particularly effective against ants. **Key reference(s):** 1) H D Burges. 1998. Formulation of mycoinsecticides. In *Formulation of Microbial Biopesticides, Beneficial Microorganisms, Nematodes and Seed Treatments*, H D Burges (ed.), pp. 131–85, Kluwer Academic Publishers, Dordrecht, The Netherlands. 2) (For classification) K H Seifert & W Gams. 2000. The taxonomy of anamorphic fungi. In *The Mycota. VII Systematics and Evolution*, *Part A*, K Esser & P A Lemke (eds), pp. 325–6, Springer, Heidelberg, Germany.

COMMERCIALISATION: **Formulation:** The fungus is firmly attached to the inside of a bait station. **Tradenames:** 'Baits Motel Stay Awhile – Rest Forever' (GlycoGenesys).

APPLICATION: The bait station is placed at non-food sites, such as spaces under sinks, toilets and washing machines where ants may occur. The bait stations cannot be used where they may contact food, such as in food-handling areas or on food utensils.

PRODUCT SPECIFICATIONS: **Specifications:** Produced by solid state fermentation and subsequent extraction of conidia. **Shelf-life:** May be kept for up to one year, if stored below 20°C.

MAMMALIAN TOXICITY: No adverse effects are expected on children, adults, pets or the environment when the bait stations are used as directed. No infectivity or pathogenicity was observed in rats after 21 days exposure to 1.8×10^9 colony forming units (cfu) per kg. The fungus does not grow at temperatures above 36°C. **Acute oral LD_{50}:** rats $>18 \times 10^8$ cfu per kg. **Acute dermal LD_{50}:** rats >2000 mg/kg. **Inhalation: LC_{50}:** rats $>1.2 \times 10^8$ cfu/animal. **Skin and eye:** Possible irritant to eyes, skin and respiratory system.

ENVIRONMENTAL IMPACT AND NON-TARGET TOXICITY: **Bird toxicity:** Oral LD_{50} (5 days) for quail >2000 mg/kg daily (by gavage). **Fish toxicity:** LC_{50} (31 days) for rainbow trout 7300 mg/litre. EC_{50} (14 days) for *Daphnia pulex* de Geer 4100 mg/litre.
Effects on beneficial insects: The method of use means that beneficial species will not come into contact with the product.

1:38 *Beauveria bassiana* isolate GHA
Biological insecticide (fungus)

Fungus: anamorph of Hypocreales: Clavicipitaceae, Ascomycota form taxa: formerly named Fungi Imperfecti: mitosporic fungi or Deuteromycetes: Moniliales

NOMENCLATURE: **Approved name:** *Beauveria bassiana* (Balsamo) Vuillemin isolate GHA. **Other names:** Formerly known as *Botrytis bassiana* Balsamo. **Common name:** White muscardine. **OPP Chemical Code:** 128924.

SOURCE: *Beauveria bassiana* isolate GHA was isolated from a mycosed insect.

PRODUCTION: Produced by solid state fermentation and subsequent extraction of conidia.

TARGET PESTS: *Beauveria bassiana* isolate GHA controls a wide range of adult and larval insects, including whiteflies, aphids, weevils, borers, grasshoppers and diamondback moth (*Plutella xylostella* (L.)).

TARGET CROPS: Various outdoor and indoor sites, such as rangeland, improved pastures, forests, commercial landscapes and interiorscapes, glasshouses and homes. Crops include turf, ornamentals and all food and feed uses.

BIOLOGICAL ACTIVITY: Biology: The entomopathogen invades the insect body. Fungal conidia become attached to the insect cuticle and, after germination, the hyphae penetrate the cuticle and proliferate in the insect's body. High humidity or free water is essential for conidial germination and infection can take between 24 and 48 hours, depending on the temperature. The infected insect may live for 3 to 5 days after hyphal penetration and, after death, the conidiophores are produced on the outside of the insect's body and new conidia are released on the outside of the insect cadaver. The fungus is insect specific.
Key reference(s): 1) H D Burges. 1998. Formulation of mycoinsecticides. In *Formulation of Microbial Biopesticides, Beneficial Microorganisms, Nematodes and Seed Treatments*, H D Burges (ed.), pp. 131–85, Kluwer Academic Publishers, Dordrecht, The Netherlands. 2) (For classification) K H Seifert & W Gams. 2000. The taxonomy of anamorphic fungi. In *The Mycota. VII Systematics and Evolution*, *Part A*, K Esser & P A Lemke (eds), pp. 325–6, Springer, Heidelberg, Germany.

COMMERCIALISATION: Formulation: 'BotaniGard' is a wettable powder (WP) and 'Mycotrol' and 'Organigard' are suspo-emulsions (SE). **Tradenames:** 'BotaniGard' (Emerald BioAgriculture), 'Mycotrol' (Emerald BioAgriculture) and (Mycotech), 'Organigard' (Emerald BioAgriculture).

APPLICATION: All products can be sprayed on growing plants using hand, ground or aerial equipment; they can also be applied through irrigation systems on large agricultural fields.

PRODUCT SPECIFICATIONS: Specifications: Produced by solid state fermentation and subsequent extraction of conidia. **Storage conditions:** Store in a cool, dry place. Do not freeze and do not allow the product to undergo thermal shock. **Shelf-life:** May be kept for up to one year, if stored below 20°C ('Mycotrol' and 'BotaniGard' below 27°C).

COMPATIBILITY: The products may be used alone or tank mixed with other products such as sticking agents, insecticidal soaps, emulsifiable oils and insecticides or used with beneficial insects. Do not use with fungicides and wait 48 hours after application before applying fungicides. Incompatible with strong oxidisers, acids and bases and chlorinated water.

MAMMALIAN TOXICITY: There are no expected health risks to humans who apply this fungus as a pesticide or who eat crops treated with the fungus. **Acute oral LD$_{50}$:** *Beauveria bassiana* isolate GHA was not pathogenic, infective or toxic in rats when dosed orally with 1×10^8 colony forming units (cfu)/animal. Toxicity category IV. **Acute dermal LD$_{50}$:** *Beauveria bassiana* isolate GHA was not pathogenic, infective or toxic to rabbits dosed dermally at 2 g per animal (equivalent to 1.6×10^{11} cfu/animal). There was slight to moderate dermal irritation which persisted to day 14. Toxicity Category III for dermal irritation.

Inhalation: *Beauveria bassiana* isolate GHA was not pathogenic, infective or toxic in rats when dosed intratracheally with 1×10^8 cfu/animal. Toxicity category IV.

Skin and eye: *Beauveria bassiana* isolate GHA was not pathogenic, infective or toxic in rats when dosed intraperitoneally with 1×10^7 cfu/animal. Toxicity Category IV. Toxicity Category III for primary eye irritation effects.

ENVIRONMENTAL IMPACT AND NON-TARGET TOXICITY: *Beauveria bassiana* isolate GHA occurs very widely in the environment and it demonstrates a low toxicity profile. Hence, the potential ecological risk due to exposure to this micro-organism is thought to be minimal. **Bird toxicity:** *Beauveria bassiana* isolate GHA was not toxic to American kestrels when dosed at 1 µg per g body mass. **Fish toxicity:** *Beauveria bassiana* isolate GHA showed no indications of infectivity or pathogenicity among treated fathead minnow (*Pimephales promelas*) dosed with 7.5×10^8 cfu/litre of water. **Other aquatic toxicity:** *Beauveria bassiana* isolate GHA at 9.3×10^8 cfu/litre had minor growth effects on *Daphnia*. The EC_{50} (21 days) was $>9.3 \times 10^8$ colony forming units (cfu)/litre. The NOEC was 4.7×10^8 cfu/litre and the LOEC was 9.3×10^8 cfu/litre. **Effects on beneficial insects:** When treated with the field rate of 1×10^{13} conidia/acre *Beauveria bassiana* isolate GHA, there was 37% mortality of both field-collected and laboratory-treated beneficial beetles, *Apthona flava* Guill. The LD_{50} was 2.11×10^{13} and LD_{90} was 9.4×10^{13} conidia/acre. Nymphs of *Xylocoris flavipes* treated with live conidia of *Beauveria bassiana* isolate GHA showed 16% mortality at 1×10^{14} conidia/acre and 41% at 1×10^{15} conidia/acre. The LD_{50} (10 days) was 1.55×10^{15} conidia/acre and the LD_{90} was 3.3×10^{16} conidia/acre. **Approved for use in organic farming:** Yes.

1:39 *Beauveria bassiana* isolate HF23
Biological insecticide (fungus)

Fungus: anamorph of Hypocreales: Clavicipitaceae, Ascomycota form taxa: formerly named Fungi Imperfecti: mitosporic fungi or Deuteromycetes: Moniliales

NOMENCLATURE: Approved name: *Beauveria bassiana* (Balsamo) Vuillemin isolate HF23. **Other names:** Formerly known as *Botrytis bassiana* Balsamo. **Common name:** White muscardine. **OPP Chemical Code:** 090305.

SOURCE: *Beauveria bassiana* HF23 is a naturally occurring fungus which was isolated from a domestic housefly (*Musca domestica* L.) in the USA.

PRODUCTION: Produced by solid state fermentation and subsequent extraction of conidia.

TARGET PESTS: House fly (*Musca domestica*).

TARGET CROPS: Poultry houses and chicken manure pits.

BIOLOGICAL ACTIVITY: Mode of action: Pathogenesis. **Biology:** The *Beauveria bassiana* HF23 invades the insect body. Fungal conidia become attached to the insect cuticle and, after germination, the hyphae penetrate the cuticle and proliferate in the insect's body. High humidity or free water is essential for conidial germination and infection can take between 24 and 48 hours, depending on the temperature. The infected insect may live for 3 to 5 days after hyphal penetration and, after death, the conidiophores are produced on the outside of the insect's body and new conidia are released on the outside of the insect cadaver. The fungus is specific to the housefly. **Key reference(s):** 1) A Siri, A C Scorsetti, V E Dikgolz & C C López Lastra. 2005. Natural infections caused by the fungus *Beauveria bassiana* as a pathogen of *Musca domestica*, *BioControl*, **50**(6), 937–40. 2) P E Kaufman, C Reasor, D A Rutz, J K Ketiz & J J Arends. 2005. Evaluation of *Beauveria bassiana* applications against adult housefly, *Musca domestica*, in commercial caged-layer poultry facilities in New York State, *Biological Control*, **33**, 360–7.

COMMERCIALISATION: Formulation: *Beauveria bassiana* HF23 is formulated as an emulsifiable powder (EP) containing 5.6×10^9 colony forming units per gram. **Tradenames:** 'balEnce' (JABB).

APPLICATION: The product is sprayed into chicken houses and onto chicken manure during the active housefly season. Applications should be made to walls, floors and manure piles in poultry houses. The suggested application rate of *Beauveria bassiana* HF23 for the control of housefly is 40 to 50 g (1.5 to 2 ounces) of the product per 465 m^2 (5000 square feet). The spray should be applied with conventional spray equipment to walls, floor, posts and manure in poultry houses, concentrating on areas where the greatest numbers of pests are located. Reapply at intervals of 2 to 7 days as long as pest pressure persists, or as pest eggs hatch and mature. There is no restriction on the total maximum amount which may be applied in a year.

PRODUCT SPECIFICATIONS: Specifications: Produced by solid state fermentation and subsequent extraction of conidia. **Shelf-life:** May be stored for more than 15 months at 22 to 26 °C.

COMPATIBILITY: It is unlikely that *Beauveria bassiana* HF23 will be used with other products, although use in mixture with or alternating with chemical insecticides within an Integrated Pest Management strategy is an option.

MAMMALIAN TOXICITY: Acute oral LD$_{50}$: rats $<3.20 \times 10^8$ colony forming units per animal. EPA Toxicity Category IV for acute oral toxicity/pathogenicity effects in mammals. **Acute dermal LD$_{50}$:** rabbits $<4.27 \times 10^{11}$ cfu/g . EPA Toxicity Category III. **Inhalation:** Not toxic or pathogenic via pulmonary route in rats. **Skin and eye:** Not an eye irritant to rabbits.

ENVIRONMENTAL IMPACT AND NON-TARGET TOXICITY: There is no evidence of toxicity or pathogenicity for *Beauveria bassiana* HF23 in the literature or in submitted studies on mammals, birds, insects, aquatic organisms and plants. Consequently, EPA has given it a 'NO EFFECT' finding, as it is not expected to harm endangered or threatened species. **Bird toxicity:** Oral LD_{50} (5 days) for chickens >910 mg/kg body weight and the NOEL was 910 mg/kg body weight . **Approved for use in organic farming:** Yes.

1:40 *Beauveria brongniartii* isolates Bb96, IMBST 95.031 and 95.041

Biological insecticide (fungus)

Fungus: anamorph of Hypocreales: Clavicipitaceae, Ascomycota form taxa; formerly named Fungi Imperfecti: mitosporic fungi or Deuteromycetes: Moniliales

NOMENCLATURE: Approved name: *Beauveria brongniartii* (Saccardo) Petch isolate Bb96 ('Betel', Swiss isolate), isolate IMBST 95.031 and 95.041 for 'Melocont-Pilzgerste' (both Austrian isolates). **Other names:** Formerly known as: *Beauveria tenella* (Saccardo) McLeod *sensu* McLeod.

SOURCE: *Beauveria brongniartii* for 'Betel' and 'Engerlingspilz' was isolated from a mycosed white grub (*Hoplochelus marginalis* (Fairmaine)) larva found in Madagascar, by CIRAD/ IRAT, France. The isolate was particularly virulent. A fermented-rice preparation was tested successfully by INRA and a production process was developed by them. This process is now owned by Natural Plant Protection (NPP). *Beauveria brongniartii* types for 'Melocont-Pilzgerste' were isolated from pasture soil infested with cockchafer larvae (*Melolontha melolontha* L.) by the Institute of Microbiology, Leopold-Franzens University (LFU), Innsbruck, Austria. The isolates were particularly virulent. LFU produced filamentous mycelium and conidia on barley kernels by a mass production unit using a diphasic fermentation process. This process is now owned by Agrifutur (AGF), Italy.

PRODUCTION: *Beauveria brongniartii* for 'Betel' is cultured by solid-state fermentation on clay granules. *Beauveria brongniartii* types for 'Melocont-Pilzgerste', 'Schweizer-Beauveria' and 'Engerlingspilz' are cultured by diphasic fermentation on barley kernels.

TARGET PESTS: For 'Melocont-Pilzgerste', the target pests are cockchafers (*Melolontha melolontha*, *M. hippocastani* Fabricius). For 'Betel', the targets are white grubs (*Hoplochelus marginalis* L.).

TARGET CROPS: Use in agriculture is non-restricted.

BIOLOGICAL ACTIVITY: Biology: The entomopathogen invades the insect body. Fungal conidia become attached to the insect cuticle and, after germination, the hyphae penetrate the cuticle and proliferate in the insect's body. High humidity or free water is essential for conidial germination and infection can take between 24 and 48 hours, depending on the temperature. The infected insect may live for 3 to 5 days after hyphal penetration and, after death, the conidiophores are produced on the outside of the insect's body and new conidia are released on the outside of the insect cadaver. Produces insecticidal secondary metabolites such as oosporein. The fungus is particularly effective against coleopteran pests. **Key reference(s):** 1) O Goebel. 1989. Diplôme d'Ingénieur en Agronomie Tropicale, CNEARC/ESAT, France. 2) B Vercambre. 1991. In *Rencontres Caraibes en Lutte Biologique*, Les Colloques de l'INRA, No. 58, pp. 371–8, published by INRA, France. 3) H Strasser. 1999. Beurteilung der Wirksamkeit des biologischen Pflanzenschutzpräparates Melocont-Pilzgerste zur Maikäferbekämpfung. Pflanzenschutz Beratungsservice, *Der Förderungsdienst*, **5**, 158–64. 4) H Strasser, T M Butt & A Vey. 2000. Are there any risks in using entomopathogenic fungi for pest control, with particular reference to the bioactive metabolites of *Metarhizium*, *Tolypocladium* and *Beauveria* species?, *Biocontrol Science and Technology*, **10**, 717–35. 5) H D Burges. 1998. Formulation of mycoinsecticides. In *Formulation of Microbial Biopesticides*, *Beneficial Microorganisms*, *Nematodes and Seed Treatments*, H D Burges (ed.), pp. 131–85, Kluwer Academic Publishers, Dordrecht, The Netherlands. 6) (For classification) K H Seifert & W Gams. 2000. The taxonomy of anamorphic fungi. In *The Mycota. VII Systematics and Evolution*, *Part A*, K Esser & P A Lemke (eds), pp. 325–6, Springer, Heidelberg, Germany.

COMMERCIALISATION: Formulation: *Beauveria brongniartii* is produced as a microgranule (MG) (NPP) and as inoculated barley seed (Andermatt, Agrifutur and Eric Schweizer). **Tradenames:** 'Betel' (NPP), 'Engerlingspilz' (Andermatt), 'Melocont-Pilzgerste' (Agrifutur) and (Kwizda AGRO), 'Schweizer Beauveria' (Eric Schweizer). **Patents:** The INRA patent (EP 0406103 A1) for the solid fermentation of *Beauveria* species on clay granules has been granted exclusively to NPP.

APPLICATION: In sugar cane, 'Betel' may be applied at planting on the edge of the furrow or at the foot of the ratoon canes. A dose rate of 50 kg product per hectare reduces larval populations below the damage threshold of three larvae per sugar cane. For cockchafer control, inoculated barley seed is sown in the soil at a dose of 35 to 70 kg product per hectare. This gives an application rate of 1×10^{12} colony forming units (cfu) per hectare.

PRODUCT SPECIFICATIONS: Specifications: *Beauveria brongniartii* for 'Betel' is cultured by solid-state fermentation on clay granules. *Beauveria brongniartii* types for 'Melocont-Pilzgerste', 'Schweizer-Beauveria' and 'Engerlingspilz' are cultured by diphasic fermentation on barley kernels. **Purity:** The purity of 'Betel' is measured in terms of spore count (minimum of 0.2×10^8 spores per gram) and efficacy against *Hoplochelus marginalis* larvae. The purity of 'Melocont-Pilzgerste' is measured in terms of spore count (minimum of 5×10^5 spores per kernel) and efficacy against *Melolontha melolontha* larvae. The activity of 'Engerlingspilz' is measured by bioassay with *M. melolontha*. **Storage conditions:** Store in a dry, refrigerated container. **Shelf-life:** The spores remain viable for a year if stored at 2°C.

COMPATIBILITY: Can be applied with chemical products, with the exception of fungicides.

MAMMALIAN TOXICITY: Considered to be non-toxic. **Acute oral LD$_{50}$:** rats >5000 mg/kg. In rats, there was no toxicity, infectivity or pathogenicity from a single dose of 1.1×10^9 cfu/kg. **Acute dermal LD$_{50}$:** rats >2000 mg/kg. **Skin and eye:** Mildly irritant to skin of rabbits. **Other toxicological effects:** Air sampling was performed in the incubator chambers (25°C, 80% r.h.) during 'Melocont-Pilzgerste' incubation. The number of *Beauveria* spp. (cfu/m^3) correlated with the concentration of air-borne propagules inside special care units in hospitals (one *B. brongniartii* count after 8 days and seven counts after 28 days) and, therefore, makes it without risk when exposure takes place. **Key reference:** J Rainer, U Peintner & R Pöder. 2000. Biodiversity and concentration of airborne fungi in hospital environment, *Mycopathologia*, **149**, 87–97.

ENVIRONMENTAL IMPACT AND NON-TARGET TOXICITY: The phytotoxic potential of *Beauveria brongniartii* and its main secondary metabolite, oosporein, were evaluated against seed potatoes (*Solanum tuberosum* L.), both *in vitro* and *in situ*. The results obtained suggest that *B. brongniartii* and oosporein pose no risks to potato plants or tubers, or to their consumers. Similar results were observed with cress (*Lepidium sativum* L.) and Timothy grass (*Phleum pratense* L.). **Key references:** 1) D Abendstein, B Pernfuss & H Strasser. 2000. Evaluation of *Beauveria brongniartii* and its metabolite oosporein regarding phytotoxicity on seed potatoes, *Biocontrol Science and Technology*, **10**, 789–96. 2) H Strasser, D Abendstein, H Stuppner & T Butt. 2000. Monitoring the distribution of secondary metabolites produced by the entomopathogenous fungus *Beauveria brongniartii* with particular reference to oosporein, *Mycological Research*, **10**,1227–33. **Bird toxicity:** Dietary LD$_{50}$ (5 days) for quail and mallard ducks >4000 mg/kg. **Fish toxicity:** LC$_{50}$ (30 days) for rainbow trout 7200 mg/litre, NOAEL (30 days) 3000 mg/litre. **Other aquatic toxicity:** NOAEL (21 days) for *Daphnia pulex* de Geer 500 mg/litre. **Approved for use in organic farming:** Yes.

1:41 *Brevibacillus brevis*
Biological fungicide (bacterium)

Bacterium: Bacillales: Bacillaceae

NOMENCLATURE: Approved name: *Brevibacillus brevis.* **Other names:** Formerly known as *Bacillus brevis.*

SOURCE: Isolated from soil at the University of Aberdeen, UK.

PRODUCTION: Produced by fermentation.

TARGET PESTS: A wide range of plant fungal pathogens are sensitive to *Brevibacillus brevis* but work has concentrated on *Botrytis cinerea* Pers., *Pythium* spp. and *Sphaerotheca fulginea* Poll. Under evaluation for use against other foliar diseases, stem-base and soil-borne diseases in cereals and potatoes and for use to control storage pathogens.

TARGET CROPS: Being developed for use in vegetable crops and under evaluation for use in cereals, potatoes and in post-harvest disease control.

BIOLOGICAL ACTIVITY: Mode of action: *Brevibacillus brevis* exerts its effect through two separate and distinct modes of action. It produces the antifungal metabolite, gramicidin S, that disrupts the fungal cytoplasmic membrane, particularly on germinating conidia and germ tubes. In addition, *B. brevis* produces a biosurfactant that reduces the periods of wetness on the plant surface and thereby inhibits the germination of the fungal spores and the subsequent penetration into the plant. **Efficacy:** *In vitro* biological activity is due entirely to the presence of gramicidin S and this metabolite predominates when there is an abundance of water on the plant surface. The biosurfactant is of primary importance when there is intermittent leaf wetness by reducing surface tension and allowing the leaf to dry more quickly. **Key reference(s):** B Seddon, R C McHugh & A Schmitt. 2000. *Brevibacillus brevis* – a novel candidate biocontrol agent with broad-spectrum antifungal activity, *Proc. Brit. Crop Protection Conference – Pests & Diseases*, **2**, 563–70.

COMMERCIALISATION: Formulation: Under development. Spore numbers of 1×10^8 per ml have been shown to be effective for post-harvest disease control.

APPLICATION: Applied as a suspension of spores in water. The product should be sprayed to cover the crop, but not to run-off.

PRODUCT SPECIFICATIONS: Specifications: Produced by fermentation.
Purity: Development formulations contain only *Brevibacillus brevis* spores.

COMPATIBILITY: Believed to be compatible with many fungicides, bactericides, insecticides, foliar nutrients and adjuvants used in the control of crop diseases. In ICM strategies, it is unusual to apply with chemical pesticides.

MAMMALIAN TOXICITY: *Brevibacillus brevis* is considered to be a non-toxic product when used according to label directions. There have been no reports of any allergic or other adverse effects from researchers, formulators or field workers.

ENVIRONMENTAL IMPACT AND NON-TARGET TOXICITY: *Brevibacillus brevis* occurs naturally and is not expected to be harmful to non-target organisms or to the environment.

1:42 *Burkholderia cepacia* Wisconsin isolate
Biological fungicide/nematicide (bacterium)

Bacterium: Pseudomonadales: Pseudomonadaceae

NOMENCLATURE: Approved name: *Burkholderia cepacia* (Palleroni & Holmes) Yabuuchi, Kosako, Oyaizu, Yaro, Hotta, Hashimoto, Ezaki & Arakawa Wisconsin isolate.
Other names: Formerly known as *Pseudomonas cepacia* (ex Burkholder) Palleroni and Holmes, Wisconsin isolate.

SOURCE: A common component of the rhizosphere. An aggressive coloniser of the roots of many plants. The Wisconsin isolate J82 was selected because of its ease of manufacture and effective suppression of soil-borne diseases and nematodes.

PRODUCTION: Produced by fermentation.

TARGET PESTS: Soil-colonising fungal pathogens and nematodes.

TARGET CROPS: Used as a seed treatment for many different outdoor crops, including alfalfa, barley, beans, clover, cotton, peas, grain sorghum, vegetable crops and wheat. Also used to treat transplanted crops.

BIOLOGICAL ACTIVITY: Biology: *Burkholderia cepacia* is an aggressive coloniser of the root zones of growing plants. In addition, it is antagonistic to pathogenic fungi and plant parasitic nematodes, preventing them from becoming established in the region of the crop and, thereafter, colonising the crop. **Key reference(s):** 1) A S Paau. 1998. Formulation of beneficial microorganisms applied to soil. In *Formulation of Microbial Biopesticides, Beneficial Microorganisms, Nematodes and Seed Treatments*, H D Burges (ed.), pp. 235–54, Kluwer Academic Publishers, Dordrecht, The Netherlands. 2) M P McQuilken, P Halmer & D J Rhodes. 1998. Application of microorganisms to seeds. In *Formulation of Microbial Biopesticides, Beneficial Microorganisms, Nematodes and Seed Treatments*, H D Burges (ed.), pp. 255–85, Kluwer Academic Publishers, Dordrecht, The Netherlands. 3) D R Fravel, W J Connick Jr &

J A Lewis. 1998. Formulation of microorganisms to control plant diseases. In *Formulation of Microbial Biopesticides, Beneficial Microorganisms, Nematodes and Seed Treatments*, H D Burges (ed.), pp. 187–202, Kluwer Academic Publishers, Dordrecht, The Netherlands.

COMMERCIALISATION: Formulation: Formulated as an inert powder coated with living bacteria containing 1×10^5 colony forming units per gram. It is used as a seed treatment and as a liquid suspension of live bacteria in nutrient broth. **Tradenames:** 'Blue Circle Liquid Biological Fungicide' (Stine Microbial Products), 'Deny' (CCT) and (Stine Microbial Products), 'Intercept' (Soil Technologies).

APPLICATION: Used as a seed treatment or as a powder treatment for transplants. May be applied as a drench shortly after transplanting.

PRODUCT SPECIFICATIONS: Specifications: Produced by fermentation. **Purity:** Viability of the formulated bacterium is determined by plating the product on to nutrient agar, culturing in the laboratory for 48 hours and counting the number of colonies that develop. **Storage conditions:** Store under cool, dry conditions in a sealed container. Powder formulations should be kept between −6 and 24°C and liquid formulations between 1 and 24°C. Avoid direct sunlight and do not allow liquid formulations to dry. **Shelf-life:** The powder seed treatment is stable for one year and the liquid formulation for 6 months.

COMPATIBILITY: Compatible with most chemical treatments, but do not apply with broad-spectrum fungicides such as copper-based products. Incompatible with strong oxidisers, acids, bases and chlorinated water.

MAMMALIAN TOXICITY: The product has not produced any adverse allergic reactions in research workers, manufacturing or field staff. **Skin and eye:** The product may cause eye irritation.

ENVIRONMENTAL IMPACT AND NON-TARGET TOXICITY: *Burkholderia cepacia* occurs widely in Nature and would not be expected to cause any adverse effects on non-target organisms or on the environment. **Approved for use in organic farming:** Yes.

1:43 *Candida oleophila* isolate I-82
Biological fungicide (fungus)

Fungus: anamorphic Saccharomycetales: formerly classified as mitosporic fungi and Deuteromycetes: Moniliales

NOMENCLATURE: Approved name: *Candida oleophila* Montrocher isolate I-82.
Development code: SAN 418 I (Novartis (now Syngenta)).

SOURCE: A yeast that occurs widely in Nature.

PRODUCTION: Manufactured by fermentation.

TARGET PESTS: Used to control post-harvest diseases.

TARGET CROPS: Used on citrus and pome fruit, small fruits and berries and other crops.

BIOLOGICAL ACTIVITY: Biology: Prevents the invasion of fruit in storage by storage disease organisms by competing for entry sites. **Key reference(s):** D R Fravel, W J Connick Jr & J A Lewis. 1998. Formulation of microorganisms to control plant diseases. In *Formulation of Microbial Biopesticides, Beneficial Microorganisms, Nematodes and Seed Treatments*, H D Burges (ed.), pp. 187–202, Kluwer Academic Publishers, Dordrecht, The Netherlands.

COMMERCIALISATION: Formulation: Sold as extruded granules. **Tradenames:** 'Aspire' (Ecogen) and (Elf Atochem).

APPLICATION: Applied to harvested fruit, pome fruit and citrus as a spray or dip treatment.

PRODUCT SPECIFICATIONS: Specifications: Manufactured by fermentation.
Purity: Contains only *Candida oleophila* cells. Efficacy can be determined by growing on nutrient agar in the laboratory and counting the number of colonies that develop.
Storage conditions: Store under cool (4°C), dry conditions in a sealed container. Do not allow to freeze and keep out of direct sunlight. **Shelf-life:** If stored under manufacturer's recommended conditions, it will remain effective for up to a year.

COMPATIBILITY: Compatible with several chemical post-harvest antifungal fruit treatments, such as thiabendazole and ethoxyquin. Incompatible with strong oxidisers, acids, bases and chlorinated water.

MAMMALIAN TOXICITY: There have been no records of allergic or other adverse effects from the use of *Candida oleophila* by researchers, manufacturers and field workers.
Toxicity class: EPA (formulation) III.

ENVIRONMENTAL IMPACT AND NON-TARGET TOXICITY: *Candida oleophila* is not hazardous to fish and wildlife. **Approved for use in organic farming:** Yes.

1:44 *Candida saitoana*

Biological fungicide (fungus)

Fungus: anamorphic Saccharomycetales: formerly classified as mitosporic fungi and Deuteromycetes: Moniliales

NOMENCLATURE: Approved name: *Candida saitoana* Nakase & Suzuki.

SOURCE: A yeast that occurs widely in Nature.

PRODUCTION: Produced for use in crop protection by fermentation.

TARGET PESTS: Used to control post-harvest diseases.

TARGET CROPS: Used in fruit packing houses to protect harvested fruit.

BIOLOGICAL ACTIVITY: Biology: Prevents the invasion of fruit in storage by storage disease organisms by competing for entry sites. **Key reference(s):** D R Fravel, W J Connick Jr & J A Lewis. 1998. Formulation of microorganisms to control plant diseases. In *Formulation of Microbial Biopesticides, Beneficial Microorganisms, Nematodes and Seed Treatments*, H D Burges (ed.), pp. 187–202, Kluwer Academic Publishers, Dordrecht, The Netherlands.

COMMERCIALISATION: Formulation: *Candida saitoana* is under development by Micro Flo and the USDA as a post-harvest biological fungicide. **Tradenames:** 'Biocure' (Micro Flo).

APPLICATION: Applied to harvested fruit, pome fruit and citrus as a spray or dip treatment.

PRODUCT SPECIFICATIONS: Specifications: Produced for use in crop protection by fermentation. **Storage conditions:** Store under cool (4°C), dry conditions in a sealed container. Do not allow to freeze and keep out of direct sunlight.

COMPATIBILITY: May be used with most post-harvest fungicides. Incompatible with strong oxidisers, acids, bases and chlorinated water.

MAMMALIAN TOXICITY: There have been no records of allergic or other adverse effects from the use of *Candida saitoana* by researchers, manufacturers and field workers. The fungus will not grow at temperatures ≥37°C.

ENVIRONMENTAL IMPACT AND NON-TARGET TOXICITY: *Candida saitoana* occurs widely in Nature and is not expected to have any deleterious effects on non-target organisms or on the environment. **Approved for use in organic farming:** Yes.

1:45 *Cercospora rodmanii*
Biological herbicide (fungus)

Fungus: Deuteromycetes: Hyphomycetes: Hyphales

NOMENCLATURE: Approved name: *Cercospora rodmanii* Conway. **Other names:** *Cercospora piaropi* Tharp. **Development code:** Abbott Laboratories of USA (now Valent BioSciences) developed an experimental formulation of *Cercospora rodmanii*, coded ABG-5003, against *Eichhornia crassipes*.

SOURCE: Isolated from water hyacinth (*Eichhornia crassipes* (Mart.) Solms) in Florida, USA.

PRODUCTION: Not available commercially.

TARGET PESTS: Water hyacinth (*Eichhornia crassipes*).

TARGET CROPS: Waterways, irrigation ditches and other stretches of open water.

BIOLOGICAL ACTIVITY: Mode of action: Pathogenesis. **Biology:** Symptoms caused by *Cercospora* spp. may be easily confused with those of many other foliar pathogens, including many opportunistic, weak parasites. This pathogen causes small (2–4 mm diameter) necrotic spots on laminae and petioles. The spots are characterised by pale centres surrounded by darker necrotic regions. Occasionally, the spots may appear in the shape of 'teardrops' that coalesce as the leaf matures, causing the entire leaf to turn necrotic and senescent. In fact, the senescence is accelerated by the *Cercospora* disease, and the disease can rapidly spread across water hyacinth infestations, causing large areas of the weed mat to turn brown and necrotic. Under severe infections, the plant may be physiologically stressed, lose its ability to regenerate, become waterlogged and sink or disintegrate. **Efficacy:** A *Cercospora rodmanii* formulation containing one million viable propagules per gram was applied to test tanks at rates of 0, 5, 10 and 20 g/sq m. After 6 weeks, it was shown that a treatment rate of five million viable propagules per square metre provided adequate infection of water hyacinth in a spring application. In an autumn study, treatment rates of 0, 1, 2.5 and 10 g/sq m of the *C. rodmanii* formulation containing four million viable propagules per gram were applied to replicated test tanks. After 6 weeks, all tanks that received *C. rodmanii* were infected by the fungus, but there were no significant differences in the level of infection among *C. rodmanii* treatments. Based on the results of these studies, rates of five and four million viable propagules per square metre were recommended for use.

Key reference(s): 1) R Charudattan. 1996. Pathogens for biological control of water hyacinth. In *Strategies for Water Hyacinth Control. Report of a Panel of Experts Meeting*, 11–14 September, 1995. Fort Lauderdale, Florida, USA. FAO, Rome, Italy. pp. 90–7, R Charudattan, R Labrada, T D Center & C Kelly-Begazo, (eds). 2) R Charudattan. 1986. *Cercospora rodmanii*: a biological control agent for water hyacinth, *Aquatics*, **8**(2), 21–4. 3) K E Conway. 1976. Evaluation of

Cercospora rodmanii as a biological control agent of water hyacinth, *Phytopathology*, **66**, 914–7. 4) K E Conway & T E Freeman. 1977. Host specificity of *Cercospora rodmanii*, a potential biological control agent of water hyacinth, *Pl. Dis. Reptr.*, **61**, 262–6. 5) M H Julien, M P Hill, T D Center & J Q Ding (eds). 2001. Biological and Integrated Control of Water Hyacinth, *Eichhornia crassipes, ACIAR Proceedings*, **102**.

COMMERCIALISATION: Formulation: Not sold commercially.

PRODUCT SPECIFICATIONS: Specifications: Not available commercially.

COMPATIBILITY: Incompatible with strong oxidisers, acids, bases and chlorinated water, but can be applied with conventional herbicides. Fungicides should not be used within three weeks of application.

MAMMALIAN TOXICITY: There are no reports of allergic or other adverse toxicological effects from the use of *Cercospora rodmanii*. The fungus only infects water hyacinth.

ENVIRONMENTAL IMPACT AND NON-TARGET TOXICITY: No adverse effects are expected on any non-target organisms or on the environment.

1:46 *Chondrostereum purpureum* isolates HQ1 and PFC-2139

Biological herbicide (fungus)

Fungus: Basidiomycetes: Meruliaceae

NOMENCLATURE: Approved name: *Chondrostereum purpureum* (Pers. ex. Fr.) Pouzar isolates HQ1 and PFC-2139. **Other names:** Formerly known as: *Stereum purpureum* Fr.
Common name: Silver leaf fungus.

SOURCE: The fungus is an important source of wood rot and is found widely in temperate forests. It is often the first fungus to show on the stumps of newly felled trees. *Chondrostereum purpureum* isolate PFC-2139 was isolated from a canker on a red alder (*Alnus rubra* Bong) on Vancouver Island near Duncan, British Columbia, Canada in 1994.

PRODUCTION: Produced by fermentation.

TARGET PESTS: Prevents the regrowth of undesirable forest pest trees such as the American black cherry (*Prunus serotina* Ehrh.), yellow birch (*Betula lutea* Michx.) and poplar (*Populus* spp.).

TARGET CROPS: Used in forests and rights of way as a stump rotter and as a mycoherbicide for hardwood trees.

BIOLOGICAL ACTIVITY: Biology: *Chondrostereum purpureum* invades freshly cut stumps or fresh wounds on a range of deciduous trees. The pathogen develops within the tree and spreads to the vascular system, where it blocks the vessels and leads to plant death. **Efficacy:** Trials have shown that about 95% of treated stumps are killed within 2 years of treatment. **Key reference(s):** 1) M P Greaves, P J Holloway & B A Auld. 1998. Formulation of microbial herbicides. In *Formulation of Microbial Biopesticides, Beneficial Microorganisms, Nematodes and Seed Treatments*, H D Burges (ed.), pp. 202–33, Kluwer Academic Publishers, Dordrecht, The Netherlands. 2) M D De Jong, E Sela, S F Shamoun & R E Wall. 1996. Natural occurrence of *Chondrostereum purpureum* in relation to its uses as a biological control agent in Canadian forests, *Biological Control*, **6**, 347–52. 3) E C Setliff. 2002. The wound pathogen *Chondrostereum purpureum*, its history and incidence on trees in North America, *Aust. J. Bot.*, **50**, 645–51. 4) R E Wall. 1996. Pathogenicity of the bioherbicide fungus *Chondrostereum purpureum* to some trees and shrubs of southern Vancouver Island, FDRA Report 246, 18 pp.

COMMERCIALISATION: Formulation: Sold as a suspension of fungal mycelium in water. **Tradenames:** 'Biochon' (Koppert), 'Chontrol Paste' [PFC-2139] (MycoLogic), 'ECOclear' [PFC-2139] (MycoLogic), 'Myco-Tech Paste' [HQ1] (Myco-Forestis).

APPLICATION: The product should be sprayed or spread on the fresh wound surface in late spring/early summer or in the autumn.

PRODUCT SPECIFICATIONS: Specifications: Produced by fermentation. **Storage conditions:** Store in a sealed container in a cool place out of direct sunlight. **Shelf-life:** Use as soon as possible after delivery.

COMPATIBILITY: It is not necessary to apply the product with other crop protection agents. Do not use chlorinated water.

MAMMALIAN TOXICITY: *Chondrostereum purpureum* occurs widely in Nature and has not caused allergic or other adverse effects on research workers or on production or field staff. No growth occurs at 35 °C. **Acute oral LD$_{50}$:** rats >5000 mg/kg or 1.2×10^6 cfu/kg. No clinical signs indicative of toxicity. Toxicity Category IV. **Acute dermal LD$_{50}$:** rabbits >3.4×10^4 cfu/kg. Toxicity Category III. **Skin and eye:** Toxicity Category IV. **Other toxicological effects:** There have been no confirmed reports of immediate or delayed allergic reactions to *Chondrostereum purpureum*.

ENVIRONMENTAL IMPACT AND NON-TARGET TOXICITY: *Chondrostereum purpureum* occurs very widely in Nature and it is not expected to have any adverse effects on non-target organisms or on the environment. It is applied commercially to tree stumps and this will significantly reduce the possibility of exposure to non-target organisms.

1:47 *Clonostachys rosea* f. *catenulate* isolate J1446 *Biological fungicide (fungus)*

Fungus: anamorphic Hypocreaceae: formerly mitosporic Hyphomycetales: Moniliaceae

NOMENCLATURE: Approved name: *Clonostachys rosea* f. *catenulate* (Gilman & Abbott) Schroers [see 2001. *Stud. Mycol.*, **46**, 76. for explanation of the taxonomy] isolate J1446. **Other names:** Formerly known as: *Gliocladium catenulatum* Gilman & Abbott. **OPP Chemical Code:** 021009.

SOURCE: Developed as a collaboration between the Agricultural Research Centre in Finland and Kemira Agro, the fungus was isolated from soil in Finland.

PRODUCTION: Produced by fermentation.

TARGET PESTS: Damping-off, seed rot, root and stem rot, and wilt diseases caused by species of *Rhizoctonia*, *Pythium*, *Phytophthora*, *Fusarium*, *Didymella*, *Botrytis*, *Verticillium*, *Alternaria*, *Cladosporium*, *Helminthosporium*, *Penicillium* and *Plicaria* in soil and *Botrytis* spp., *Didymella* spp. and *Helminthosporium* spp. post-harvest or as foliar pathogens.

TARGET CROPS: *Clonostachys rosea* f. *catenulate* is used in vegetables, herbs and ornamentals. In the USA, it is approved for use indoors and outdoors on a wide variety of vegetables, herbs and spices. It can also be used on turf, ornamentals and on tree and shrub seedlings.

BIOLOGICAL ACTIVITY: Mode of action: Microbial fungicide with a preventive rather than curative action. The mode of action of *Clonostachys rosea* f. *catenulate* isolate J1446 includes hyperparatisism, enzyme activity and colonisation of roots in advance of the pathogens. **Key reference(s):** 1) D R Fravel, W J Connick Jr & J A Lewis. 1998. Formulation of microorganisms to control plant diseases. In *Formulation of Microbial Biopesticides, Beneficial Microorganisms, Nematodes and Seed Treatments*, H D Burges (ed.), pp. 187–202, Kluwer Academic Publishers, Dordrecht, The Netherlands. 2) A S Paau. 1998. Formulation of beneficial microorganisms applied to soil. In *Formulation of Microbial Biopesticides, Beneficial Microorganisms, Nematodes and Seed Treatments*, H D Burges (ed.), pp. 235–54, Kluwer Academic Publishers, Dordrecht, The Netherlands.

COMMERCIALISATION: Formulation: Sold as wettable powders (WP) containing a minimum of 1×10^7 colony forming units (cfu) per gram. **Tradenames:** 'Prestop' (Verdera), 'Primastop' (Kemira).

APPLICATION: 'Prestop' can be incorporated into the growing media as an aqueous suspension at the rate of 500 g per cubic metre of soil or it can applied as a soil drench or spray at the rate of 250 grams/1000 plants. It is generally applied when plants are sown, potted or transplanted and is reapplied at intervals of a few weeks. Recommendations for 'Primastop' are 0.05–1.2 × 10^8 colony forming units (cfu)/litre of soil (0.05–1.2 g of product/litre of growth substrate), 5–25 × 10^8 cfu/m² in seedling trays or beds (5–25 g of product/m²) and 1–10 × 10^8 cfu/m² as a foliar spray or on turf (1–10 g product/m²).

PRODUCT SPECIFICATIONS: Specifications: Produced by fermentation. Purity: Contains no contaminants. Storage conditions: Store under cool (below 8°C), dry conditions, in a sealed container. Do not expose to direct sunlight. Shelf-life: Remains viable for 12 months if stored under correct conditions.

COMPATIBILITY: Intervals of 0 to 7 days are recommended between 'Prestop' application and most chemical pesticide treatments. Incompatible with strong oxidisers, acids, bases and chlorinated water.

MAMMALIAN TOXICITY: There have been no reports of allergic or other adverse toxicological effects from research and manufacturing staff, formulators or field workers. There is no evidence to suggest that the fungus is toxic or infectious to humans. Acute oral LD_{50}: rats >2000 mg/kg. Toxicity category III. Acute dermal LD_{50}: rats >2000 mg/kg. Toxicity category IV. Inhalation: LC_{50}: rats >5.57 mg/litre. Not toxic or pathogenic to rats. Toxicity category IV. Skin and eye: Mild eye irritation. May cause sensitisation by skin contact. Toxicity category IV.

ENVIRONMENTAL IMPACT AND NON-TARGET TOXICITY: *Clonostachys rosea* f. *catenulate* occurs widely in Nature and is not expected to have any adverse effects on non-target organisms or on the environment. Fish toxicity: LC_{50} (30days) for rainbow trout 3.5 × 10^8 cfu/litre (504 mg/litre). Other aquatic toxicity: EC_{50} (21 days) for *Daphnia* 5.5 × 10^6 cfu/litre (7.8 mg/litre). Approved for use in organic farming: Yes.

1:48 *Colletotrichum acutatum*
Biological herbicide (fungus)

Fungus: anamorphic Glomerellaceae: previously classified as mitosporic fungi and Deuteromycetes: Phyllachorales: Phyllachoraceae

NOMENCLATURE: Approved name: *Colletotrichum acutatum* Simmonds.
Other names: Previously known as *Colletotrichum xanthi* Halstead. Some researchers have identified the fungus in this product as an isolate of *Colletotrichum gloeosporioides* (Penz.) Sacc.

SOURCE: Isolated from infected silky wattle (*Hakea sericea* Schrad. & J.C. Wendl.) trees.

PRODUCTION: Produced by fermentation.

TARGET PESTS: Silky wattle or needlebush (*Hakea sericea*).

TARGET CROPS: Rangeland and native vegetation.

BIOLOGICAL ACTIVITY: Efficacy: An effective pathogen causing 'gummosis' symptoms in infected plants. **Key reference(s):** 1) M P Greaves. 1996. Microbial herbicides: factors in development. In *Crop Protection Agents from Nature: Natural Products and Analogues*, L G Copping (ed.), pp. 444–67, Royal Society of Chemistry, Cambridge, UK. 2) M P Greaves, P J Holloway & B A Auld. 1998. Formulation of microbial herbicides. In *Formulation of Microbial Biopesticides, Beneficial Microorganisms, Nematodes and Seed Treatments*, H D Burges (ed.), pp. 202–33, Kluwer Academic Publishers, Dordrecht, The Netherlands. 3) R E Hoagland, C D Boyette, M A Weaver & H K Abbas. 2007. Bioherbicides: Research and Risks, *Toxin Reviews*, **26**, 313–42.

COMMERCIALISATION: Formulation: The product was never registered in South Africa, but will be produced upon request. **Tradenames:** 'Hakatak'.

APPLICATION: Applied so as to give good cover of the target weed under conditions of high humidity. Do not apply fungicides for at least 3 weeks after application.

PRODUCT SPECIFICATIONS: Specifications: Produced by fermentation.
Storage conditions: Store in a sealed container under cool conditions out of direct sunlight.
Shelf-life: Use as soon as possible after receipt.

COMPATIBILITY: *Colletotrichum acutatum* is compatible with chemical herbicides, but should not be used in conjunction with a fungicide. Incompatible with strong oxidisers, acids, bases and chlorinated water.

MAMMALIAN TOXICITY: There is no report of any allergic or other adverse effects following the use of the product by researchers, formulators or users. *Colletotrichum acutatum* will not survive at temperatures $\geq 37\,^\circ$C (human body temperature).

ENVIRONMENTAL IMPACT AND NON-TARGET TOXICITY: *Colletotrichum acutatum* occurs widely in Nature and is not expected to have any adverse effects on non-target organisms or on the environment. **Approved for use in organic farming:** Yes.

1:49 *Colletotrichum coccodes*
Biological herbicide (fungus)

Fungus: anamorphic Glomerellaceae: previously classified as mitosporic fungi and Deuteromycetes: Phyllachorales: Phyllachoraceae

NOMENCLATURE: Approved name: *Colletotrichum coccodes* (Wallr.) Hughes. **Other names:** Black dot; anthracnose.

SOURCE: Isolated from velvetleaf (*Abutilon theophrasti* L.) in the USA.

PRODUCTION: Produced by fermentation.

TARGET PESTS: Velvetleaf (*Abutilon theophrasti*).

TARGET CROPS: Soybeans and peanuts.

BIOLOGICAL ACTIVITY: Mode of action: Pathogenesis. **Efficacy:** An inoculum density of 1×10^7 spores/ml applied to run-off or an application rate of 2.3×10^8 spores/m^2 was necessary to produce the most destructive levels of disease. Velvetleaf plants at all growth stages were susceptible and were reduced in vigour following inoculation. Plants inoculated at the cotyledon stage were killed. **Key reference(s):** L A Wymore, C Poirier, A K Watson & A R Gotlieb. 1988. *Colletotrichum coccodes*, a potential bioherbicide for control of velvetleaf (*Abutilon theophrasti*), *Plant Dis.* **72**, 534–8.

COMMERCIALISATION: Formulation: *Colletotrichum coccodes* is not sold commercially.

PRODUCT SPECIFICATIONS: Specifications: Produced by fermentation.

COMPATIBILITY: *Colletotrichum coccodes* is compatible with chemical herbicides, but should not be used in conjunction with a fungicide. Incompatible with strong oxidisers, acids, bases and chlorinated water.

MAMMALIAN TOXICITY: Since its discovery, there have been no reports of adverse effects, sensitivity or reactions of any type related to use or handling of *Colletotrichum coccodes*. No pathogenic or infective effects were observed in any study.

ENVIRONMENTAL IMPACT AND NON-TARGET TOXICITY: *Colletotrichum coccodes* occurs widely in Nature and is not expected to have any adverse effects on non-target organisms or on the environment. **Approved for use in organic farming:** Yes.

1:50 *Colletotrichum gloeosporioides* f. sp. *aeschynomene* *Biological herbicide (fungus)*

Fungus: anamorphic Glomerellaceae: previously classified as mitosporic fungi and Deuteromycetes: Phyllachorales: Phyllachoraceae

NOMENCLATURE: Approved name: *Colletotrichum gloeosporioides* (Penz.) Sacc. f. sp. *aeschynomene*.

SOURCE: Isolated from Northern joint vetch in the USA.

PRODUCTION: Produced by fermentation.

TARGET PESTS: Northern joint vetch (*Aeschynomene virginica* L.).

TARGET CROPS: Rice and soybeans.

BIOLOGICAL ACTIVITY: Mode of action: Pathogenesis. **Biology:** *Colletotrichum gloeosporioides* f. sp. *aeschynomene* is a pathogen of Northern joint vetch. When applied to the weed, it penetrates the plant's cuticle and invades the weed, leading to plant death. The pathogen is specific to Northern joint vetch, although there is evidence of pathogenicity to peas. **Key reference(s):** 1) M P Greaves. 1996. Microbial herbicides: factors in development. In *Crop Protection Agents from Nature: Natural Products and Analogues*, L G Copping (ed.), pp. 444–67, Royal Society of Chemistry, Cambridge, UK. 2) M P Greaves, P J Holloway & B A Auld. 1998. Formulation of microbial herbicides. In *Formulation of Microbial Biopesticides, Beneficial Microorganisms, Nematodes and Seed Treatments*, H D Burges (ed.), pp. 202–33, Kluwer Academic Publishers, Dordrecht, The Netherlands.

COMMERCIALISATION: Formulation: 'Collego' was sold as an aqueous suspension of spores, but was withdrawn in 2003. 'LockDown retro' is sold as a pressed powder or granular formulation containing 45% spore suspension with a minimum of 1×10^9 colony forming units per g of product, and 'LockDown XL' is sold as a pressed powder or granular formulation containing 30% spore suspension with a minimum of 7×10^8 cfu/g product. Both were approved for sale by the US-EPA in March 2006. **Tradenames:** 'Collego' (Encore Technologies) (withdrawn), 'LockDown retro' and 'LockDown XL' (Agricultural Research Initiatives). **Patents:** US 3849104.

APPLICATION: Applied over the top of the developing weed under conditions of high humidity. Do not apply fungicides for at least 3 weeks after application. Both 'LockDown' products are applied at a rate of 75 g of product (32 g a.i.) per acre per application (77 g a.i. per ha). Applications are made by air when *Aeschynomene virginica* is 8 to 24 inches (20 to 60 cm) high, but before rice heads appear. One or two applications are made each year.

PRODUCT SPECIFICATIONS: Specifications: Produced by fermentation.
Storage conditions: Store in a sealed container under cool conditions out of direct sunlight.
Shelf-life: Use as soon as possible after receipt.

COMPATIBILITY: *Colletotrichum gloeosporioides* f. sp. *aeschynomene* is compatible with chemical herbicides, but should not be used in conjunction with a fungicide. Incompatible with strong oxidisers, acids, bases and chlorinated water.

MAMMALIAN TOXICITY: There is no report of any allergic or other adverse effects following the use of the product. **Skin and eye:** For the end-use formulation, slight eye irritation in rabbits was observed at a dose of 0.1 ml (Toxicity Category IV) and skin irritation in rabbits was not observed at a dose of 0.5 ml (Toxicity Category IV). **Toxicity class:** Toxicity Category IV. **Other toxicological effects:** Since its discovery, no incidents of hypersensitivity have been reported by researchers, manufacturers or users.

ENVIRONMENTAL IMPACT AND NON-TARGET TOXICITY: *Colletotrichum gloeosporioides* f. sp. *aeschynomene* occurs widely in Nature and is not expected to have any adverse effects on non-target organisms or on the environment.
Approved for use in organic farming: Yes.

1:51 *Colletotrichum gloeosporioides* f. sp. *cuscutae*   *Biological herbicide (fungus)*

Fungus: anamorphic Glomerellaceae: previously classified as mitosporic fungi and Deuteromycetes: Phyllachorales: Phyllachoraceae

NOMENCLATURE: Approved name: *Colletotrichum gloeosporioides* (Penz.) Sacc. f. sp. *cuscutae*.

SOURCE: Isolated from dodder (*Cuscuta* spp.) in China.

PRODUCTION: Produced by fermentation.

TARGET PESTS: Used for the control of dodders (*Cuscuta chinensis* Semen and *C. australis* R.Br.).

TARGET CROPS: Used primarily in soybeans.

BIOLOGICAL ACTIVITY: Mode of action: Pathogenesis. **Biology:** *Colletotrichum gloeosporioides* f. sp. *cuscutae* is a pathogen of dodder (*Cuscuta* spp.). When applied to the weed, it penetrates the plant's cuticle and invades the weed, leading to plant death. The pathogen is specific to dodders. **Efficacy:** The organism will infect and kill dodder at any stage of growth, from seedling to mature plant. *Colletotrichum gloeosporioides* f. sp. *cuscutae* will invade the parasitic weed and kill it within 2 to 4 weeks. **Key reference(s):** 1) M P Greaves. 1996. Microbial herbicides: factors in development. In *Crop Protection Agents from Nature: Natural Products and Analogues*, L G Copping (ed.), pp. 444–67, Royal Society of Chemistry, Cambridge, UK. 2) M P Greaves, P J Holloway & B A Auld. 1998. Formulation of microbial herbicides. In *Formulation of Microbial Biopesticides, Beneficial Microorganisms, Nematodes and Seed Treatments*, H D Burges (ed.), pp. 202–33, Kluwer Academic Publishers, Dordrecht, The Netherlands.

COMMERCIALISATION: Formulation: Sold in China as a dry powder dispersion of spores. **Tradenames:** 'Luboa II'.

APPLICATION: Applied over the top of the developing weed under conditions of high humidity. The product is applied to crops at an early growth stage, as it is important to control dodder as early as possible.

PRODUCT SPECIFICATIONS: Specifications: Produced by fermentation. **Storage conditions:** Store in a sealed container under cool conditions out of direct sunlight.

COMPATIBILITY: *Colletotrichum gloeosporioides* f. sp. *cuscutae* is compatible with chemical herbicides, but should not be used in conjunction with a fungicide. Incompatible with strong oxidisers, acids, bases and chlorinated water.

MAMMALIAN TOXICITY: There is no report of any allergic or other adverse effects following the use of the product by researchers, formulators or users. *Colletotrichum gloeosporioides* f. sp. *cuscutae* will not survive at temperatures ≥37°C (human body temperature).

ENVIRONMENTAL IMPACT AND NON-TARGET TOXICITY: *Colletotrichum gloeosporioides* f. sp. *cuscutae* is specific to dodder and is naturally occurring. It is not expected to have any deleterious effects on non-target organisms or on the environment. **Approved for use in organic farming:** Yes.

1:52 *Colletotrichum gloeosporioides* f. sp. *malvae*

Biological herbicide (fungus)

Fungus: anamorphic Glomerellaceae: previously classified as mitosporic fungi and Deuteromycetes: Phyllachorales: Phyllachoraceae

NOMENCLATURE: Approved name: *Colletotrichum gloeosporioides* (Penz.) Sacc. f. sp. *malvae*.

SOURCE: *Colletotrichum gloeosporioides* f. sp. *malvae* is a naturally occurring fungus that is pathogenic to the weeds round-leaved mallow (*Malva rotundifolia* L.), small-flowered mallow (*Malva parviflora* L.), common mallow (*Malva neglecta* Wallr.) and velvetleaf (*Abutilon theophrasti* Medic.), all of which are members of the family Malvaceae. It was originally isolated and characterised by Dr Knud Mortensen, Agriculture Canada Research, in 1982. It had been reported as indigenous to the provinces of Saskatchewan and Manitoba, occurring as an endemic pathogen of round-leaved mallow, producing lesions on aerial parts.

PRODUCTION: Produced by fermentation.

TARGET PESTS: Malvaceous weeds, including *Malva* species and *Abutilon theophrasti*.

TARGET CROPS: *Colletotrichum gloeosporioides* f. sp. *malvae* can be used in many different crops and in grassland and pasture.

BIOLOGICAL ACTIVITY: Mode of action: Pathogenesis. **Biology:** *Colletotrichum gloeosporioides* f. sp. *malvae* is a pathogen of plants from the family Malvaceae. When applied to weeds of this family, it penetrates the plant's cuticle and invades the weed, leading to plant death. The pathogen is specific to Malvaceae. **Efficacy:** The organism will infect and kill round-leaved and small-flowered mallows at any stage of growth, from seedling to mature plant. *Colletotrichum gloeosporioides* f. sp. *malvae* causes disease lesions that will completely encircle the stems and petioles of mallow, causing the plant to collapse in 2 to 4 weeks.

Key reference(s): 1) M P Greaves. 1996. Microbial herbicides: factors in development. In *Crop Protection Agents from Nature: Natural Products and Analogues*, L G Copping (ed.), pp. 444–67, Royal Society of Chemistry, Cambridge, UK. 2) M P Greaves, P J Holloway & B A Auld. 1998. Formulation of microbial herbicides. In *Formulation of Microbial Biopesticides, Beneficial Microorganisms, Nematodes and Seed Treatments*, H D Burges (ed.), pp. 202–33, Kluwer Academic Publishers, Dordrecht, The Netherlands.

COMMERCIALISATION: Formulation: The end-use formulation is a two-component product. 'Mallet WP' Component A consists of a 16 oz (475 ml) bottle containing a water-soluble spore nutrient and rehydrating agent that activates the spores prior to application. 'Mallet WP' Component M consists of a bag containing a water-suspensible dried fungal spore formulation of *Colletotrichum gloeosporioides* f. sp. *malvae*. **Tradenames:** 'BioMal' (Philom Bios) (discontinued), 'Mallet WP' (Encore Technologies).

APPLICATION: Applied over the top of the developing weed under conditions of high humidity. The product is applied to crops at an early growth stage to control target weeds.

PRODUCT SPECIFICATIONS: Specifications: Produced by fermentation. **Storage conditions:** Store in a sealed container under cool conditions out of direct sunlight. **Shelf-life:** Use as soon as possible after receipt.

COMPATIBILITY: *Colletotrichum gloeosporioides* f. sp. *malvae* is compatible with chemical herbicides, but should not be used in conjunction with a fungicide. Incompatible with strong oxidisers, acids, bases and chlorinated water.

MAMMALIAN TOXICITY: Since its discovery in 1982, there have been no reports of adverse effects, sensitivity or reactions of any type related to use or handling of this organism. No pathogenic or infective effects were observed in any study. **Acute oral LD$_{50}$:** rats (for the active ingredient) $>6 \times 10^5$ colony forming units (cfu)/g. **Acute dermal LD$_{50}$:** rats (for the active ingredient) $>4.21 \times 10^7$ cfu/g. The acute intraperitoneal toxicity/pathogenicity in rats is $>5.7 \times 10^5$ cfu per animal. **Inhalation:** The acute pulmonary toxicity/pathogenicity in rats is $>4.55 \times 10^4$ cfu per animal. **Other toxicological effects:** *Colletotrichum gloeosporioides* f. sp. *malvae* is not pathogenic or infective to mammals. There have been no reports of toxins or secondary metabolites associated with the organism and acute toxicity studies have shown that *C. gloeosporioides* f. sp. *malvae* is non-toxic, non-pathogenic and non-irritating. Residues of *C. gloeosporioides* f. sp. *malvae* are not expected on agricultural commodities and thus exposure to the general population, from the proposed uses, is not anticipated.

ENVIRONMENTAL IMPACT AND NON-TARGET TOXICITY: *Colletotrichum gloeosporioides* f. sp. *malvae* occurs widely in Nature and is not expected to have any adverse effects on non-target organisms or on the environment. **Approved for use in organic farming:** Yes.

1:53 *Colletotrichum truncatum* isolate NRRL 18434   *Biological herbicide (fungus)*

Fungus: anamorphic Glomerellaceae: previously classified as mitosporic fungi and Deuteromycetes: Phyllachorales: Phyllachoraceae

NOMENCLATURE: Approved name: *Colletotrichum truncatum* (Schwein) isolate NRRL 18434. **Other names:** *Colletotrichum dematium* var. *truncatum* (Schwein.) v. Arx.

SOURCE: The fungal organism was isolated from hemp sesbania (*Sesbania exaltata* Cory)

seedlings in the Southern Weed Science Laboratory, Mississippi, USA. This isolate was found to be responsible for a previously unknown anthracnose disease that kills infected plants as a result of stem-girdling lesions.

PRODUCTION: Produced by fermentation.

TARGET PESTS: Hemp sesbania (*Sesbania exaltata*).

TARGET CROPS: Soybeans.

BIOLOGICAL ACTIVITY: Mode of action: Pathogenesis. **Biology:** When infecting hemp sesbania, *Colletotrichum truncatum* NRRL 18434 produces orbicular lesions with dark, concentric circles, 2–4 mm in diameter, containing pustules of truncate conidia, 18–25 by 3–5 µm, surrounded by dark setae.

COMMERCIALISATION: Formulation: Not sold commercially. **Patents:** United States Patent 5034328.

PRODUCT SPECIFICATIONS: Specifications: Produced by fermentation.

MAMMALIAN TOXICITY: Since its discovery, there have been no reports of adverse effects, sensitivity or reactions of any type related to use or handling of *Colletotricum truncatum* NRRL 18434. No pathogenic or infective effects were observed in any study.

ENVIRONMENTAL IMPACT AND NON-TARGET TOXICITY: *Colletotrichum truncatum* NRRL 18434 shows no effects on several cultivars of soybean, cotton, rice, garden bean, and tomato, as well as 19 other plant species representing a total of 7 families. Three cultivars of soybean tested sustained very slight injury, but they outgrew the injury after 3 weeks. It is not expected that *Colletotrichum truncatum* will have any adverse effects on non-target organisms or on the environment.

1:54 *Coniothyrium minitans* isolate CON/M/91-08   *Biological fungicide (fungus)*

The Pesticide Manual **Fifteenth Edition entry number:** 179

Fungus: anamorphic Leptosphaeriaceae: formerly classified as mitosporic Ascomycetes: Leptosphaeriaceae

NOMENCLATURE: Approved name: *Coniothyrium minitans* Campbell isolate CON/M/91-08. **OPP Chemical Code:** 028836.

SOURCE: Commonly occurring soil fungus and a specialised parasite of sclerotia in soil. The isolate used in the product 'Contans (WG)' was originally isolated in 1992 by the German company Prophyta. In 1995, it was formulated into the product 'Contans (WG)' which was registered by the Federal Biological Research Centre for Agriculture and Forestry on 22 December 1997 (reg. no. AP-ZA 04346-00-00) and introduced in 1998 in Germany. It is now widely registered in Europe and North America.

PRODUCTION: Manufactured by solid state fermentation. The conidia are formulated as water dispersible granules.

TARGET PESTS: Used to control the resting survival structures (sclerotia) of plant pathogens from the genus *Sclerotinia* and, in particular, *Sclerotinia sclerotiorum* de Bary and *Sclerotinia minor* Jagger.

TARGET CROPS: Oilseed rape, lettuce and other crops susceptible to plant pathogens from the genus *Sclerotinia*.

BIOLOGICAL ACTIVITY: Mode of action: *C. minitans* is a very slow-growing fungus, with its biological activity expressed predominantly by its mycoparasitic effects on sclerotia-forming fungi in the soil. **Biology:** *C. minitans* attacks the sclerotia of the target organism (*Sclerotinia* spp.) in the soil and destroys them. Sclerotial infection takes place from germinating spores of *C. minitans*. *C. minitans* penetrates through the outer pigmented rind intercellularly or via existing cracks in the surface. Growth continues inter- and intracellularly through the unpigmented internal tissues of the cortex and medulla. Penetration into the cells involves both enzymatic degradation and pressure, and can occur in all tissues of the sclerotia. The cytoplasm of invaded cells in the medulla shows plasmolysis, aggregation and vacuolation and the cell walls are gradually degraded. This may involve the production of chitinase and β-1,3-glucanase. Hyphae of *C. minitans* within rind cells frequently exhibit lysis. Subsequently, the hyphae of the mycoparasite proliferate within the sclerotia, and pycnidia form within and on the surface of the sclerotia in less than 14 days, under ideal conditions. **Key reference(s):** 1) J M Whipps & M Gerlach. 1992. Biology of *Coniothyrium minitans* and its potential for use in disease biocontrol, *Mycological Research*, **96**(11), 897–907. 2) D R Fravel, W J Connick Jr & J A Lewis. 1998. Formulation of microorganisms to control plant diseases. In *Formulation of Microbial Biopesticides, Beneficial Microorganisms, Nematodes and Seed Treatments*, H D Burges (ed.), pp. 187–202, Kluwer Academic Publishers, Dordrecht, The Netherlands. 3) A S Paau. 1998. Formulation of beneficial microorganisms applied to soil. In *Formulation of Microbial Biopesticides, Beneficial Microorganisms, Nematodes and Seed Treatments*, H D Burges (ed.), pp. 235–54, Kluwer Academic Publishers, Dordrecht, The Netherlands.

COMMERCIALISATION: Formulation: Sold as a water dispersible granule (WG) formulation containing 1×10^9 conidia (mitospores) per gram (5.3%). **Tradenames:** 'Contans (WG)' (Prophyta), (Intrachem) and (Kwizda AGRO), 'Intercept' (Encore Technologies), 'Koni'

(Biovéd). **Patents:** Patents on the active ingredient: German Patent (DE 19502065 A1, December 7 1995) and United States Patent (5,766,583, 16 June 1998) have been awarded for use of the isolate of *Coniothyrium minitans* CON/M/91-08 as a biological control agent against *Sclerotinia sclerotiorum*. PCT application (DE95/00926) was filed covering Europe and Canada. Patents on the fermentation technology needed for the product: PCT-Number DE99/01271 (International Publishing number: WO99/57239), Australian Patent No. 749402; New Zealand Patent No. 507738; European Patent No. 10093708. Other patent applications have been filed by Prophyta internationally.

APPLICATION: Soil application and incorporation 2 to 3 months before planting or 2 to 3 months before infection is likely (just before sowing or planting), in order to allow *Coniothyrium minitans* to destroy the sclerotia in the soil. Rates of application are 2 to 8 kg per hectare when applied to the soil pre-plant or 1 to 2 kg per hectare when applied post-harvest.

PRODUCT SPECIFICATIONS: Specifications: Manufactured by solid state fermentation. The conidia are formulated as water dispersible granules. **Purity:** The materials used to start the fermentation process are pure cultures of *Coniothyrium minitans*. These are obtained by plating pure lyophilised stock cultures of the isolate CON/M/91-08 on an agar medium. The formulation contains 5.3% spores of *C. minitans* and 94.7% glucose as carrier, giving 1×10^9 conidia of *C. minitans* per gram of product. **Storage conditions:** Store under cool (2 to 6°C) and dry conditions. Do not expose to extremes of temperature or direct sunlight. Do not store under wet conditions or expose to temperatures above 20°C for long periods. **Shelf-life:** Can be stored for 6 months under recommended conditions (4°C).

COMPATIBILITY: From experiences so far obtained, 'Contans (WG)' can be used in tank mix with herbicides containing trifluralin. Mixtures with fungicides, acids or bases as well as products attacking organic material should be avoided. Do not use chlorinated water.

MAMMALIAN TOXICITY: Considered to be non-toxic. There are no reports of *Coniothyrium minitans* causing allergic or other adverse toxicological effects in research and manufacturing staff, formulators or field workers. **Acute oral LD$_{50}$:** rats >2500 mg/kg. Toxicity Category III. **Acute dermal LD$_{50}$:** rats >2500 mg/kg. Toxicity Category III. **Acute intraperitoneal LD$_{50}$:** rats >2000 mg/kg. **Inhalation: LC$_{50}$:** rats >12.74 mg/litre of air for 4 hours. Toxicity Category IV. **Skin and eye:** Non-irritant to the skin and eye of the rabbit and non-sensitising to the skin of the guinea pig. Toxicity Category IV.

ENVIRONMENTAL IMPACT AND NON-TARGET TOXICITY: The naturally occurring soil fungus *Coniothyrium minitans* is a highly specific sclerotia parasite. Any adverse effect on non-target organisms or hazardous effects to the environment can be excluded. **Fish toxicity:** LC$_{50}$ (96 h) for carp >100 mg/litre. **Other aquatic toxicity:** LC$_{50}$ (48 h) for *Daphnia magna* >100 mg/litre. No toxicity to algae at 100 mg/litre. **Effects on beneficial insects:** There is no evidence of any deleterious effects on non-target

organisms living in or on the soil. **Behaviour in soil:** The spores of the naturally occurring soil fungus of *Coniothyrium minitans* can be determined in soil in the absence of sclerotia. If the host organism is present, *C. minitans* spores germinate, the fungus develops vegetatively and infects the host. The *C. minitans* population decreases when the number of sclerotia is reduced. The vegetative stage disappears and the fungus rests as spores.
Approved for use in organic farming: Yes.

1:55 *Cryphonectria parasitica*
Biological fungicide (fungus)

Fungus: Ascomycetes: Diaporthales

NOMENCLATURE: Approved name: *Cryphonectria parasitica* (Murrill) Barr.
Other names: Formerly known as: *Diaporthe parasitica* Murr., *Valsonectria parasitica* Rehm and *Endothia parasitica* Anders. and Anders.

SOURCE: A non-pathogenic isolate of the fungus that is the causal organism of chestnut blight. Isolated from a chestnut tree in France.

PRODUCTION: Produced by fermentation.

TARGET PESTS: Chestnut blight (*Cryphonectria parasitica*).

TARGET CROPS: Chestnut trees (*Castanea* spp.).

BIOLOGICAL ACTIVITY: Biology: This non-pathogenic isolate of the fungus invades possible infection sites on trees and outcompetes for these sites with the pathogen, preventing the establishment of the disease. Many avirulent isolates of this fungus carry a mycovirus (or VLP) and this virus can be transferred to virulent isolates in Nature. As a consequence, virulent isolates become avirulent. **Key reference(s):** D R Fravel, W J Connick Jr & J A Lewis. 1998. Formulation of microorganisms to control plant diseases. In *Formulation of Microbial Biopesticides, Beneficial Microorganisms, Nematodes and Seed Treatments*, H D Burges (ed.), pp. 187–202, Kluwer Academic Publishers, Dordrecht, The Netherlands.

COMMERCIALISATION: Formulation: Sold as spores suspended in a paste.
Tradenames: 'Endothia parasitica' (CNICM).

APPLICATION: Damaged or pruned areas of the tree should be treated with the product as soon as possible to allow for the establishment of the non-pathogenic isolate.

PRODUCT SPECIFICATIONS: Specifications: Produced by fermentation.
Storage conditions: Store in a cool, dry situation. Do not expose to sunlight.
Shelf-life: One year.

COMPATIBILITY: It is unusual to apply *Cryphonectria parasitica* with any other treatments. Do not use chlorinated water.

MAMMALIAN TOXICITY: There is no record of allergic or other adverse effects following the use of the product. It is considered to be of low mammalian toxicity.

ENVIRONMENTAL IMPACT AND NON-TARGET TOXICITY: *Cryphonectria parasitica* occurs in Nature and is not expected to have any adverse effects on non-target organisms or on the environment.

1:56 *Cryptococcus albidus*
Biological fungicide (fungus)

Fungus: anamorphic Tremellales: previously classified as mitosporic Basidiomycetes: Sporidiales

NOMENCLATURE: Approved name: *Cryptococcus albidus* (Saito) Skinner (CBS No. 604.94).

SOURCE: *Cryptococcus albidus* was first isolated from peach juice samples.

PRODUCTION: Produced by fermentation.

TARGET PESTS: Storage rot fungi and, in particular, *Penicillium* and *Botrytis* spp.

TARGET CROPS: Fruit such as apples and pears.

BIOLOGICAL ACTIVITY: *Cryptococcus albidus* initially outcompetes wound pathogens for nutrients and space. Thereafter, it also produces two proteins that destroy the structure of fungal cell walls and arrest any further growth. It is a contact product and can only protect where it is applied. It has no systemic action and it cannot migrate from one position to another. Efficacy will, therefore, depend on good coverage of the fruit surface.

COMMERCIALISATION: Formulation: Sold as a granular dried yeast, vacuum-packed in 5 kg blocks. **Tradenames:** 'FruitPlus' (Anchor Bio-Technologies), 'YieldPlus' (Anchor Bio-Technologies). **Patents:** RSA 96/1347.

APPLICATION: The formulation is sprayed onto the fruit immediately after harvest (or the fruit is dipped into a suspension of spores). The treated fruit is dried before entering storage.

PRODUCT SPECIFICATIONS: Specifications: Produced by fermentation. **Shelf-life:** May be kept for several months (maximum 2 years) in cool, dry conditions away from direct sunlight.

COMPATIBILITY: It is unusual to apply *Cryptococcus albidus* with any other treatments, although it has been successfully trialled in combination with iprodione and diphenylamine. Do not use chlorinated water or broad-spectrum fungicides.

MAMMALIAN TOXICITY: Some species of *Cryptococcus* have been claimed to cause superficial, non-invasive skin infections on mammals. **Acute oral LD$_{50}$:** rats >4147 mg/kg. **Acute dermal LD$_{50}$:** rats >6 750 mg/kg.

ENVIRONMENTAL IMPACT AND NON-TARGET TOXICITY: *Cryptococcus albidus* occurs widely in Nature and is not expected to cause any adverse effects on non-target organisms.

1:57 *Cydia pomonella* granulosis virus Mexican and French isolates
Insecticidal baculovirus

Virus: Baculoviridae: Granulovirus

NOMENCLATURE: Approved name: *Cydia pomonella* granulosis virus Mexican and French isolates. **Other names:** Codling moth granulosis virus; CmGV; CpGV.

SOURCE: *Cydia pomonella* granulosis virus (CpGV) has been isolated from codling moth larvae in the field.

PRODUCTION: The virus is produced by infecting codling moth larvae, harvesting the infected hosts and extracting the granular occlusion bodies by centrifugation. This technique requires very large quantities of larvae for virus production and is, thus, an expensive procedure. Trials are under way on the production of the virus in a more cost-effective way *in vivo* or through *in vitro* techniques such as insect cell culture.

TARGET PESTS: Codling moth (*Cydia pomonella* (L.)).

TARGET CROPS: Apple, pear and walnut orchards.

BIOLOGICAL ACTIVITY: Mode of action: Codling moth larvae are infected by the granulosis virus, which was first reported in 1964 following its identification in *Cydia pomonella* larvae collected in Mexico (*J. Insect Pathol.*, 1964, **6**, 373–86).

Biology: CpGV is more active on small larvae than on later larval instars that are usually inaccessible to the virus, as they burrow into the fruit. It is recommended that applications be targeted at neonate larvae by following egg-laying patterns with the use of pheromone monitoring systems. The virus is ingested by the feeding larva and the virus protein coat is dissolved in the insect's midgut (that is alkaline), releasing the virus particles. These pass through the peritrophic membrane and invade midgut cells by fusion with the microvilli. The virus particles invade the cell nuclei, where they are uncoated and replicated. Initial replication produces non-occluded virus particles that spread to new cells and hasten the invasion of the host insect, but later the virus particles are produced with protein coats and remain infective when released from the dead insects. **Efficacy:** The high level of activity shown by CpGV to first instar larvae means that, if the product is applied at the correct time, good control is achieved with very low rates of application. The virus has to be consumed to be effective and death will often take many hours, even with small larvae.

Key reference(s): 1) R R Granados & B A Federici (eds). 1986. *The Biology of Baculoviruses*, *Vols 1 and 2*, CRC Press, Boca Raton, Florida, USA. 2) D J Leisy & J R Fuxa. 1996. Natural and engineered viral agents for insect control. In *Crop Protection Agents from Nature: Natural Products and Analogues*, L G Copping (ed.), Royal Society of Chemistry, Cambridge, UK. 3) B A Federici. 1999. Naturally occurring baculoviruses for insect pest control. In *Biopesticides Use and Delivery*, F R Hall & J J Menn (eds), pp. 301–20, Humana Press, Totowa, NJ, USA. 4) H D Burges & K A Jones. 1998. Formulation of bacteria, viruses and Protozoa to control insects. In *Formulation of Microbial Biopesticides*, *Beneficial Microorganisms*, *Nematodes and Seed Treatments*, H D Burges (ed.), pp. 33–127, Kluwer Academic Publishers, Dordrecht, The Netherlands.

COMMERCIALISATION: Formulation: Sold as a liquid and a suspension concentrate (SC) formulation. **Tradenames:** 'Carposin' (Agrichem), 'Carpovirusine' (Calliope) and (Arvesta), 'CYD-X' (Certis), 'Granupom' (Bayer) and (Biobest), 'Madex' (Andermatt Biocontrol) and (Intrachem), 'Pavois' (Bayer), 'Virin-Gyap' (NPO Vector), 'Virosoft CP-4' (Biotepp).

APPLICATION: The incidence of codling moth adults should be monitored using pheromone traps, in order that the product can be applied at the optimum time for pest control. High-volume sprays of up to 1×10^{13} CpGV per hectare should be applied when adults are detected. It has been suggested that rates of one-tenth of this dose can be applied at weekly intervals with good results. It may be used in ICM programmes and used on organically grown produce. Some formulations contain the virus plus specific additives to enhance the effectiveness of the product.

PRODUCT SPECIFICATIONS: Specifications: The virus is produced by infecting codling moth larvae, harvesting the infected hosts and extracting the granular occlusion bodies by centrifugation. This technique requires very large quantities of larvae for virus production and is, thus, an expensive procedure. Trials are under way on the production of the virus in a more cost-effective way *in vivo* or through *in vitro* techniques such as insect cell culture. **Purity:** The product is monitored to ensure that it is free of bacterial contaminants pathogenic to man. It is assayed for effectiveness against codling moth larvae. **Storage conditions:** Store in sealed containers at 2°C. **Shelf-life:** Formulations stored at 2°C are stable for over 2 years. If stored at room temperature, they remain effective for one month. Infectivity is reduced by exposure to u.v. light.

COMPATIBILITY: Compatible with all crop protection agents that do not repel codling moth. Use water with a pH between 6 and 8. Incompatible with strong oxidisers, acids, bases and chlorinated water.

MAMMALIAN TOXICITY: There is no evidence of acute or chronic toxicity in mammals. No allergic reactions or other health problems have been observed in research or manufacturing staff or users of the product. **Acute oral LD$_{50}$:** rats >2000 mg/kg (>1 × 10^{10} occlusion bodies (OBs)/kg). **Acute dermal LD$_{50}$:** rabbits >1 × 10^{10} OBs/kg. **Inhalation: LC$_{50}$:** rats >2 × 10^{12} CpGV/ml. **Skin and eye:** Does not penetrate the skin, but is a mild skin irritant to rabbits. Instillation of 0.1 ml of CpGV 1 × 10^{10} OBs/ml did not produce any adverse effects in rabbits. There is no evidence of eye irritation in mammals. Toxicity category IV.

ENVIRONMENTAL IMPACT AND NON-TARGET TOXICITY: The virus occurs widely in Nature and there is no evidence that it affects any organism other than the codling moth larvae. It is unstable when exposed to u.v. light and does not persist in the environment, with no viral activity being found in soil 4 months after application. The virus particles sediment in water. **Bird toxicity:** *Cydia pomonella* granulosis virus showed no toxicity or pathogenicity to birds at 10 000 mg/kg for 5 days. **Fish toxicity:** LC$_{50}$ (96 h) for fish >250 mg/litre. **Other aquatic toxicity:** LC$_{50}$ (48 h) for *Daphnia pulex* de Geer >250 mg/litre. **Effects on beneficial insects:** LC$_{50}$ (contact) for honeybees >1 × 10^{10} CpGV per ml per bee. LC$_{50}$ for earthworms >1000 mg/kg soil. It was harmless to the mite predators *Typhlodromus pyri* Scheunten and *Amblyseius californicus* (McGregor).
Approved for use in organic farming: Yes.

1:58 *Cylindrobasidium laeve*

Biological herbicide (fungus)

Fungus: Basidiomycetes: Polyporales: Meruliaceae

NOMENCLATURE: Approved name: *Cylindrobasidium laeve* (Pers.:Fr.) Chamuris.

SOURCE: Isolated from dead wattle bushes in South Africa.

PRODUCTION: Produced by liquid fermentation.

TARGET PESTS: Black wattle (*Acacia mearnsii* de Wild.) and golden wattle (*Acacia pycnantha* Benth.).

TARGET CROPS: Rangeland.

BIOLOGICAL ACTIVITY: Mode of action: *Cylindrobasidium laeve* is a pathogen of black and golden wattle. These *Acacia* species, that were imported from Australia, invaded large areas of rangeland in South Africa and, even when cut back, regrew into large shrubs. The fungus invades the exposed vascular tissue of treated tree stumps, leading to the rapid death of the plant. **Efficacy:** Tests were conducted by the PPRI Weed Pathology Unit, Stellenbosch, on the effectiveness of *Cylindrobasidium laeve* as a biological control agent of cut wattle stumps. Results of field trials showed mortality of treated stumps of both *Acacia mearnsii* and *A. pycnantha* to be greater than 80% (reaching 90 and 100% in some cases) within 6 to 12 months of treatment. **Key reference(s):** C L Lennox, M J Morris & A R Wood. 2000. Stumpout™ – Commercial production of a fungal inoculant to prevent regrowth of cut wattle stumps in South Africa. In *Proc. X International Symposium on Biological Control of Weeds, Session 1 Abstracts*, N R Spencer (ed.), p. 140, Montana State University, Bozeman, Montana, USA.

COMMERCIALISATION: Formulation: 'Stumpout' was introduced in South Africa in 1997. It is sold as living basidiospores in an oil formulation, in small sachets. **Tradenames:** 'Stumpout' (PPRI).

APPLICATION: The product is diluted in sunflower oil and 1 to 2 ml is painted onto the fresh cut surface of the tree stump.

PRODUCT SPECIFICATIONS: Specifications: Produced by liquid fermentation. **Purity:** Contains living basidiospores in an oil formulation and no impurities. **Shelf-life:** Use as soon as possible after receipt.

COMPATIBILITY: It is unusual to use *Cylindrobasidium laeve* with any other control agent.

MAMMALIAN TOXICITY: There is no record of allergic or other adverse effects following the use of *Cylindrobasidium laeve*. It is considered to be of low mammalian toxicity.

ENVIRONMENTAL IMPACT AND NON-TARGET TOXICITY: *Cylindrobasidium laeve* occurs widely in Nature and is not expected to cause any adverse effects on non-target organisms. It appears to be specific for the black and golden wattles.
Approved for use in organic farming: Yes

1:59 *Entyloma ageratinae*
Biological herbicide (fungus)

Fungus: Basidiomycetes: Ustilaginales

NOMENCLATURE: Approved name: *Entyloma ageratinae* Barreto and Evans.
Other names: Mist flower smut. Originally misnamed as *Cercosporella* sp. and also described as *Entyloma compositarum* by Trujillo *et al.*

SOURCE: *Entyloma ageratinae* was introduced into Hawaii from Jamaica in 1975. Subsequently (in 1988), it was imported into New Zealand from Hawaii.

PRODUCTION: It can be cultured in liquid medium, although it grows very slowly. Usually grown on mist flowers (*Ageratina riparia* (Regel)).

TARGET PESTS: Mist flowers (*Ageratina riparia*).

TARGET CROPS: Forests and rangeland, as well as amenity grassland.

BIOLOGICAL ACTIVITY: Mode of action: *Entyloma ageratinae* sporulates on the underside of green, living leaves 7 to 10 days after infecting a mist flower plant. The affected leaves quickly become brown and necrotic. The fungus appears to have both biotrophic and necrotrophic stages. Necrotic leaves fall prematurely and, where climatic conditions are suitable, the fungus also causes dieback of shoots. Most plants at a site eventually become infected, leading to a decline in weed cover over wide areas. **Efficacy:** Mexican devil weed (*Ageratina adenophora* (Sprengel) King & Robinson) is the only other plant species which shows symptoms of infection by *Entyloma ageratinae*, but the smut fungus could not complete its life-cycle on this host. Within 10 years of release in Hawaii, the populations of mist flower had been substantially reduced and the desirable introduced kikuyu grass (*Pennisetum clandestinum* Hochst) and other pasture species had replaced previous weed infestations.
Key reference(s): J Frohlich, S V Fowler, A Gionotti, R L Hill, E Killgore, L Morin, L Sugiyama & C Winks. 2000. Biological control of mist flower (*Ageratina riparia*, Asteraceae): transferring a successful program from Hawai'i to New Zealand. In *Proc. X International Symposium on Biological Control of Weeds*, 4–14 July 1999. N R Spencer (ed.), pp. 51–7, Montana State University, Bozeman, Montana, USA.

COMMERCIALISATION: Formulation: The fungus is not sold commercially.

APPLICATION: *Entyloma ageratinae* is introduced either as a suspension of conidia in water (*ca.* 2×10^5 conidia per ml) painted onto several mist flower plants at the site to be treated or 3 infected mist flower plants from the glasshouse are distributed at the site.

PRODUCT SPECIFICATIONS: Specifications: It can be cultured in liquid medium, although it grows very slowly. Usually grown on mist flowers (*Ageratina riparia*).

COMPATIBILITY: It is not usual to use *Entyloma ageratinae* with other chemical agents, but its effects are enhanced by the use of a selective phytophagous insect, such as the mist flower gall fly (*Procecidochares alani* (Steyskal)).

MAMMALIAN TOXICITY: *Entyloma ageratinae* has not been reported to have any allergic or other adverse effects on researchers or growers.

ENVIRONMENTAL IMPACT AND NON-TARGET TOXICITY: *Entyloma ageratinae* has been shown to complete its life-cycle only on the mist flower. For this reason, it is not considered to offer any danger to non-target organisms or the environment.

1:60 *Entyloma compositarum*
Biological herbicide (fungus)

Fungus: Basidiomycetes: Ustilaginales

NOMENCLATURE: Approved name: *Entyloma compositarum* Trujillo *et al.*
Other names: *Entyloma compositarum* has been reclassified. See *Entyloma ageratinae* (entry 1:59).

1:61 *Erwinia carotovora*
Biological bactericide (bacterium)

Bacterium: Eubacteriales: Bacillaceae

NOMENCLATURE: Approved name: *Erwinia carotovora* Holl. **Other names:** Formerly known as *Bacillus carotovorus* Jones and *Pectobacterium carotovorum* Jones.

SOURCE: A non-pathogenic isolate of *Erwinia carotovora* isolated from Chinese cabbage.

PRODUCTION: Produced by fermentation.

TARGET PESTS: Soft rot (*Erwinia carotovora*).

TARGET CROPS: Chinese cabbage.

BIOLOGICAL ACTIVITY: Biology: This non-pathogenic isolate of the bacterium invades possible infection sites on Chinese cabbage and outcompetes for these sites with the pathogen, preventing the establishment of the disease.

COMMERCIALISATION: Formulation: Sold as a suspension of spores.
Tradenames: 'BioKeeper' (Nissan).

APPLICATION: Apply as a foliar spray, ensuring good coverage of the foliage.

PRODUCT SPECIFICATIONS: Specifications: Produced by fermentation. **Purity:** Does not contain any contaminants. **Storage conditions:** Store in a cool, dry place, out of direct sunlight. **Shelf-life:** If stored correctly, the product remains viable for 12 months.

COMPATIBILITY: It is unusual to apply the product with conventional chemicals.

MAMMALIAN TOXICITY: *Erwinia carotovora* has not been reported to cause allergic or other adverse toxicological effects on research workers, manufacturers, formulators or field staff.

ENVIRONMENTAL IMPACT AND NON-TARGET TOXICITY: *Erwinia carotovora* is not expected to have any adverse effects on non-target organisms or on the environment.

1:62 *Fusarium oxysporum* isolate Fo 47
Biological fungicide (fungus)

Fungus: anamorphic Nectriaceae: previously classified as mitosporic fungi and Deuteromycetes: Moniliales

NOMENCLATURE: Approved name: *Fusarium oxysporum* Schlechtendal isolate Fo 47.

SOURCE: *Fusarium oxysporum* isolate Fo 47 is a naturally occurring mutant isolate of the fungus that was found in suppressive soil of Chateaurenard, south-east France by INRA researchers and was selected for commercialisation, as it is not phytopathogenic and competes with other pathogenic isolates of the fungus. *Fusarium oxysporum* isolate Fo 47 is auto-incompatible and cannot hybridise with pathogenic isolates.

PRODUCTION: By solid fermentation to produce clay microgranules and by liquid fermentation to produce liquid formulation.

TARGET PESTS: For the control of vascular wilts caused by the fungal pathogens *Fusarium oxysporum* Schlechtendal and *Fusarium moniliforme* Sheldon (*Gibberella fujikuroi* Wr.).

TARGET CROPS: 'Fusaclean L' is under development for treating rock wool blocks and 'Fusaclean G' for use on field and glasshouse-grown crops.

BIOLOGICAL ACTIVITY: Mode of action: Protects crops against pathogenic *Fusarium* spp. by three distinct mechanisms – i) soil competition at root level: *Fusarium oxysporum* isolate Fo 47 is a strong root zone coloniser and is highly competitive for nutrients with other micro-organisms; ii) competition at the surface of the root system: *Fusarium oxysporum* isolate Fo 47 competes with other soil micro-organisms for access at root infection sites; iii) elicitation: *Fusarium oxysporum* isolate Fo 47 activates the plant's autoimmune system, leading to the production of phytoalexins that inhibit the production of *Fusarium* digestive enzymes and detoxify the fusaric acid produced by phytopathogenic strains. **Biology:** *Fusarium oxysporum* isolate Fo 47 is a non-pathogenic form of the fungus. It grows competitively within the root zones of crops, preventing the establishment of the pathogenic isolates of the genus.
Key reference(s): 1) D R Fravel, W J Connick Jr & J A Lewis. 1998. Formulation of microorganisms to control plant diseases. In *Formulation of Microbial Biopesticides, Beneficial Microorganisms, Nematodes and Seed Treatments*, H D Burges (ed.), pp. 187–202, Kluwer Academic Publishers, Dordrecht, The Netherlands. 2) A S Paau. 1998. Formulation of beneficial microorganisms applied to soil. In *Formulation of Microbial Biopesticides, Beneficial Microorganisms, Nematodes and Seed Treatments*, H D Burges (ed.), pp. 235–54, Kluwer Academic Publishers, Dordrecht, The Netherlands.

COMMERCIALISATION: Formulation: Sold as suspension concentrates (SC) and as microgranules (MG). **Tradenames:** 'Biofox C' (SIAPA), 'Fusaclean G' (MG, mycelium and spores, 2.5×10^8 cfu/g) (NPP), 'Fusaclean L' (SC, spores, 1×10^{11} cfu/litre) (NPP).

APPLICATION: 'Fusaclean L' is added to the rock wool blocks in which glasshouse-grown crops are propagated. 'Fusaclean G' is a microgranule formulation that is added to the soil around both field and glasshouse-grown crops.

PRODUCT SPECIFICATIONS: Specifications: By solid fermentation to produce clay microgranules and by liquid fermentation to produce liquid formulation.
Purity: The fermentation process is undertaken under carefully controlled conditions and no phytopathogenic isolates of the fungus are included in the product. Viability of the fungus can be determined by plating onto agar, incubating in the laboratory at temperatures around 25 °C and counting the colonies formed after 48 hours. **Storage conditions:** Store under cool conditions, preferably refrigerated. Do not expose to sunlight and do not freeze.
Shelf-life: Use as soon as possible after delivery. Dry formulations will last about 6 months without loss of viability.

COMPATIBILITY: Incompatible with soil-applied fungicides and with strong oxidisers, acids, bases and chlorinated water.

MAMMALIAN TOXICITY: *Fusarium oxysporum* isolate Fo 47 is not pathogenic or infective to mammals. **Acute oral LD$_{50}$:** rats >5000 mg/kg. **Acute dermal LD$_{50}$:** rats >2000 mg/kg.

Inhalation: Not toxic; not infective or pathogenic to rats after an intratracheal instillation.

ENVIRONMENTAL IMPACT AND NON-TARGET TOXICITY: *Fusarium oxysporum* isolate Fo 47 occurs in Nature and is not expected to cause any adverse reactions to any non-target organism or to have any deleterious effect on the environment. **Bird toxicity:** Oral LD_{50} (5 days) for quail >9400 mg/kg (by gavage), equivalent to 1.2×10^3 colony forming units/g. **Fish toxicity:** LC_{50} (96 h) for fish >100 mg/litre. **Approved for use in organic farming:** Yes.

1:63 *Gliocladium catenulatum*
Biological fungicide (fungus)

Fungus: anamorphic Hypocreaceae: formerly mitosporic Hyphomycetales: Moniliaceae

NOMENCLATURE: Approved name: *Gliocladium catenulatum* (Gilman & Abbott) Schroers. **Other names:** *Gliocladium catenulatum* has been reclassified. See *Clonostachys rosea* f. *catenulate* (Gilman & Abbott) (entry 1:47).

1:64 *Gliocladium virens*
Biological fungicide (fungus)

Fungus: anamorphic Hypocreaceae: formerly mitosporic Hyphomycetales: Moniliaceae

NOMENCLATURE: Approved name: *Gliocladium virens* Miller, Giddens & Foster. **Other names:** *Gliocladium virens* has been reclassified. See *Trichoderma virens* (Miller, Giddens & Foster) von Arx (entry 1:143).

1:65 *Helicoverpa armigera* nucleopolyhedrovirus
Insecticidal baculovirus

Virus: Baculoviridae: Nucleopolyhedrovirus

NOMENCLATURE: Approved name: *Helicoverpa armigera* nucleopolyhedrovirus.
Other names: *Helicoverpa armigera* MNPV; *Helicoverpa armigera* NPV; *Heliothis armigera* nucleopolyhedrovirus; *Heliothis armigera* MNPV; HaMNPV; HaNPV.

SOURCE: Originally isolated from infected *Helicoverpa armigera* (Hübner) larvae. It is found widely in Nature.

PRODUCTION: The product is produced from the culture of infected *Helicoverpa armigera* larvae under controlled conditions. Polyhedral inclusion bodies are extracted from the dead insects and formulated as a liquid concentrate.

TARGET PESTS: Particularly effective against *Helicoverpa armigera*, but also shows activity against other noctuid caterpillars.

TARGET CROPS: The product may be used in a wide range of crops, including cotton, vegetables, such as brassicae (cabbages), tomatoes and peas, and ornamentals, such as roses.

BIOLOGICAL ACTIVITY: Mode of action: As with all insect baculoviruses, *Helicoverpa armigera* MNPV must be ingested to exert an effect. Following ingestion, the protective protein matrix of the polyhedral occlusion bodies (OBs) of the virus is dissolved within the midgut and the virus particles enter the insect's haemolymph. These virus particles invade nearly all cell types in the larval body, where they multiply, leading to death. Shortly after the death of the larva, the integument ruptures, releasing very large numbers of OBs.
Efficacy: HaMNPV acts relatively slowly, as it has to be ingested before it exerts any effect on the insect host. It is important to ensure good cover of the foliage to effect good control. Monitoring of adult insect laying patterns and applications targeted at newly hatched eggs gives better control than on a mixed population. **Key reference(s):** 1) R R Granados & B A Federici (eds). 1986. *The Biology of Baculoviruses, Vols 1 and 2*, CRC Press, Boca Raton, Florida, USA. 2) D J Leisy & J R Fuxa. 1996. Natural and engineered viral agents for insect control. In *Crop Protection Agents from Nature: Natural Products and Analogues*, L G Copping (ed.), Royal Society of Chemistry, Cambridge, UK. 3) B A Federici. 1999. Naturally occurring baculoviruses for insect pest control. In *Biopesticides Use and Delivery*, F R Hall & J J Menn (eds), pp. 301–20, Humana Press, Totowa, NJ, USA. 4) H D Burges & K A Jones. 1998. Formulation of bacteria, viruses and Protozoa to control insects. In *Formulation of Microbial Biopesticides, Beneficial Microorganisms, Nematodes and Seed Treatments*, H D Burges (ed.), pp. 33–127, Kluwer Academic Publishers, Dordrecht, The Netherlands.

COMMERCIALISATION: Formulation: Sold as a liquid concentrate. **Tradenames:** 'Biovirus-H' (Biotech International), 'SamStar-Ha' (SOM Phytopharma) and (Agri Life).

APPLICATION: Apply at a rate of 1500 billion occlusion bodies per hectare (750 ml of product per hectare), ensuring good coverage of the foliage.

PRODUCT SPECIFICATIONS: Specifications: The product is produced from the culture of infected *Helicoverpa armigera* larvae under controlled conditions. Polyhedral inclusion bodies are extracted from the dead insects and formulated as a liquid concentrate. **Purity:** The liquid concentrate formulation is checked to ensure that it is free from bacteria that are harmful to man. **Storage conditions:** Store in a tightly closed container in cool (<21 °C), dry and dark conditions. Shelf-life is extended if kept frozen. Unstable at temperatures above 32 °C. **Shelf-life:** Stable for several weeks if kept cool, but has unlimited stability if kept in a freezer.

COMPATIBILITY: Can be used with other insecticides that do not repel *Helicoverpa* and *Heliothis* species. Incompatible with strong oxidisers, acids, bases and chlorinated water.

MAMMALIAN TOXICITY: Baculoviruses are specific to invertebrates, with there being no record of any vertebrate becoming infected. The virus does not infect or replicate in the cells of mammals and is inactivated at temperatures above 32 °C. There is no evidence of acute or chronic toxicity, eye or skin irritation in mammals. No allergic reactions or other health problems have been observed in research or manufacturing staff or with users of the product. Some or all of the uses in the USA have been approved by the Environmental Protection Agency as Reduced Risk application.

ENVIRONMENTAL IMPACT AND NON-TARGET TOXICITY: HaMNPV occurs widely in Nature and there is no evidence of the virus infecting vertebrates or plants. It has no adverse effects on fish, birds or beneficial organisms. **Approved for use in organic farming:** Yes.

1:66 *Helicoverpa zea* nucleopolyhedrovirus
Insecticidal baculovirus

Virus: Baculoviridae: Nucleopolyhedrovirus

NOMENCLATURE: Approved name: *Helicoverpa zea* nucleopolyhedrovirus. **Other names:** *Helicoverpa zea* MNPV; *Helicoverpa zea* NPV; HzMNPV; HzNPV.

SOURCE: The virus occurs naturally in *Heliothis* and *Helicoverpa* species.

PRODUCTION: The product is produced from the culture of infected *Helicoverpa zea* (Boddie) larvae under controlled conditions. Polyhedral inclusion bodies are extracted from the dead insects and formulated as a liquid concentrate.

TARGET PESTS: For control of *Heliothis* and *Helicoverpa* species, including cotton bollworm (*Helicoverpa zea*) and tobacco budworm (*Heliothis virescens* Fabricius).

TARGET CROPS: Vegetables, ornamentals, tomatoes and cotton.

BIOLOGICAL ACTIVITY: Mode of action: As with all insect baculoviruses, *Helicoverpa zea* MNPV must be ingested to exert an effect. Following ingestion, the protective protein matrix of the polyhedral occlusion bodies (OBs) of the virus is dissolved within the midgut and the virus particles enter the insect's haemolymph. These virus particles invade nearly all cell types in the larval body, where they multiply, leading to death. Shortly after the death of the larva, the integument ruptures, releasing very large numbers of OBs. **Efficacy:** HzMNPV acts relatively slowly, as it has to be ingested before it exerts any effect on the insect host. It is important to ensure good cover of the foliage to effect good control. Monitoring of adult insect laying patterns and applications targeted at newly hatched eggs gives better control than on a mixed population. **Key reference(s):** 1) R R Granados & B A Federici (eds). 1986. *The Biology of Baculoviruses, Vols 1 and 2*, CRC Press, Boca Raton, Florida, USA. 2) D J Leisy & J R Fuxa. 1996. Natural and engineered viral agents for insect control. In *Crop Protection Agents from Nature: Natural Products and Analogues*, L G Copping (ed.), Royal Society of Chemistry, Cambridge, UK. 3) B A Federici. 1999. Naturally occurring baculoviruses for insect pest control. In *Biopesticides Use and Delivery*, F R Hall & J J Menn (eds), pp. 301–20, Humana Press, Totowa, NJ, USA. 4) H D Burges & K A Jones. 1998. Formulation of bacteria, viruses and Protozoa to control insects. In *Formulation of Microbial Biopesticides, Beneficial Microorganisms, Nematodes and Seed Treatments*, H D Burges (ed.), pp. 33–127, Kluwer Academic Publishers, Dordrecht, The Netherlands.

COMMERCIALISATION: Formulation: Sold as a liquid concentrate. **Tradenames:** 'GemStar' (Certis) and (Rincon-Vitova), 'Elcar' (Syngenta).

APPLICATION: Apply at a rate of 1500 billion occlusion bodies per hectare (750 ml of product per hectare), ensuring good coverage of the foliage.

PRODUCT SPECIFICATIONS: Specifications: The product is produced from the culture of infected *Helicoverpa zea* (Boddie) larvae under controlled conditions. Polyhedral inclusion bodies are extracted from the dead insects and formulated as a liquid concentrate. **Purity:** The liquid concentrate formulation is checked to ensure that it is free from bacteria that are harmful to man. **Storage conditions:** Store in a tightly closed container in cool (<21 °C), dry and dark conditions. Shelf-life is extended if kept frozen. Unstable at temperatures above 32 °C. **Shelf-life:** Stable for several weeks if kept cool, but has unlimited stability if kept in a freezer.

COMPATIBILITY: Can be used with other insecticides that do not repel *Helicoverpa* and *Heliothis* species. Incompatible with strong oxidisers, acids, bases and chlorinated water.

MAMMALIAN TOXICITY: Baculoviruses are specific to invertebrates, with there being no record of any vertebrate becoming infected. The virus does not infect or replicate in the cells of mammals and is inactivated at temperatures above 32 °C. There is no evidence of acute or chronic toxicity, eye or skin irritation in mammals. No allergic reactions or other health problems have been observed in research or manufacturing staff or with users of the product.

ENVIRONMENTAL IMPACT AND NON-TARGET TOXICITY: HzMNPV occurs widely in Nature and there is no evidence of the virus infecting vertebrates or plants. It has no adverse effects on fish, birds or beneficial organisms. **Approved for use in organic farming:** Yes.

1:67 *Heterorhabditis bacteriophora*
Insect parasitic nematode

Nematode: Rhabditida: Heterorhabditidae

NOMENCLATURE: Approved name: *Heterorhabditis bacteriophora* Poinar.
Other names: Formerly known as *Heterorhabditis heliothidis*. **Development code:** SAN 402 1 (Novartis).

SOURCE: Originally found in southern and central Europe and North America, but now a commonly occurring soil-inhabiting nematode.

PRODUCTION: Produced by fermentation and sold as third stage infective juveniles.

TARGET PESTS: Effective against a wide range of insect pests, but targeted at Japanese beetles (*Popillia japonica* Newman). 'Larvanem' is recommended for control of vine weevils and grubs. 'B-Green' is recommended for control of the garden chafer (*Phyllopertha horticola* (L.)) and *Hoplia philanthus* (Foerster), as well as the vine weevil (*Otiorhynchus sulcatus* (Fabricius)).

TARGET CROPS: Used in a wide range of crops, ornamentals and turf.

BIOLOGICAL ACTIVITY: Biology: The infective juveniles are the infectious third stage larvae and only these can survive outside the host insect, as they do not require food. They enter a host through one of its natural openings or through the skin. The nematode larva then releases bacteria (*Photorhabdus luminescens* (Thomas & Poinar) Boemare, Akhurst & Mourant) into the insect's body and toxins produced by these bacteria kill the insect within

48 hours. The bacteria then digest the insect's body into material that the nematode can feed upon, and the nematode's fourth stage develops within the dead insect. These fourth stage larvae develop into hermaphrodite nematodes which lay as many as 1 500 eggs each. These eggs develop into males and females that are able to reproduce sexually. After mating, the males die and the females will lay eggs in the dead insect if there is enough food available or, if not, the first and second stage larvae develop inside her body. As soon as the larvae reach the third stage, they leave the dead insect and seek out a new host. **Efficacy:** Very effective for the control of soil-dwelling beetles and other insects. Particularly effective against phytophagous insects that are gregarious in habit. **Key reference(s):** 1) R Gaugler & H K Kaya. 1990. Entomopathogenic nematodes. In *Biological Control*, 365 pp., CRC Press, Boca Raton, Florida, USA. 2) R Georgis & H K Kaya. 1998. Formulation of entomopathogenic nematodes. In *Formulation of Microbial Biopesticides, Beneficial Microorganisms, Nematodes and Seed Treatments*, H D Burges (ed.), pp. 289–308, Kluwer Academic Publishers, Dordrecht, The Netherlands.

COMMERCIALISATION: Formulation: Sold as water dispersible clay formulations of infective third stage larvae. **Tradenames:** 'B-Green' (Biobest), 'Cruiser' (Ecogen), 'Heteromask' (BioLogic), 'Heterorhabditis bacteriophora' (Biocontrol Network) and (Rincon-Vitova), 'Larvanem' (Koppert), 'Lawn Patrol' (Hydro-Gardens), 'NemoPAK H' (Bioplanet).

APPLICATION: Apply at a rate of 5×10^5 juveniles per square metre as an overall drench when target insects are present. Ensure that soil temperatures remain above 12 °C and air temperatures should be between 12 and 30 °C for two weeks after application. Foliar applications should be at high volume, but avoiding spray run-off. Stir to prevent sedimentation.

PRODUCT SPECIFICATIONS: Specifications: Produced by fermentation and sold as third stage infective juveniles. **Purity:** Sold as infective third instar larvae with no bacterial contaminants. The efficacy of the formulation can be determined by bioassay on Japanese beetles or other susceptible insects. **Storage conditions:** Store in a refrigerator at 3 to 5 °C. Do not freeze or expose to direct sunlight. **Shelf-life:** Use as soon as possible after delivery.

COMPATIBILITY: Compatible with broad-spectrum insecticides. Incompatible with benzimidazole-based fungicides and strong oxidisers, acids and bases.

MAMMALIAN TOXICITY: There are no records of allergic or other adverse reactions in researchers, producers or users of the product. **Acute dermal LD$_{50}$:** rats >2000 mg/kg (4.6×10^{10} third stage larvae/kg), rabbits >6280 mg/kg. **Inhalation: LC$_{50}$:** 8.0×10^7 third stage larvae/ rat. **NOEL:** rats (3 months) 4 g/kg b.w. daily.

ENVIRONMENTAL IMPACT AND NON-TARGET TOXICITY: *Heterorhabditis bacteriophora* occurs in Nature and is not expected to show any adverse effects on non-target organisms or on the environment. **Fish toxicity:** LC$_{50}$ (96 h) for water feeder guppies (*Toecilia retriculata*) >156 mg/litre. **Approved for use in organic farming:** Yes.

1:68 *Heterorhabditis megidis* isolates MicroBio UK 211 and Dutch HW79
Soil insect parasitic nematode

Nematode: Rhabditida: Heterorhabditidae

NOMENCLATURE: Approved name: *Heterorhabditis megidis* Poinar, Jackson and Klein MicroBio isolate UK 211 (now Becker Underwood) and HW79 (Dutch isolate). **Other names:** Parasitic nematode.

SOURCE: Originally found in Europe, but now widespread throughout the world.

PRODUCTION: Produced by fermentation in a liquid diet or in solid-state fermentation on foam cubes. Sometimes reared on insects. Sold as third stage infective juveniles.

TARGET PESTS: Soil insects, such as the larvae and pupae of vine weevils (*Otiorhynchus sulcatus* (Fabricius)).

TARGET CROPS: Ornamentals and vegetables in glasshouses and outdoors.

BIOLOGICAL ACTIVITY: Biology: The third stage larvae are the infectious stage and only these can survive outside the host insect, as they do not require food. They enter a host through one of its natural openings or through the skin. The nematode larva then releases bacteria (*Photorhabdus temperata* Fischer-Le Saux, Viallard, Brunel, Normand & Boemare) into the insect's body and toxins produced by these bacteria kill the insect within 48 hours. The bacteria then digest the insect's body into material that the nematode can feed upon, and the nematode's fourth stage develops within the dead insect. These fourth stage larvae develop into hermaphrodite nematodes which lay as many as 1500 eggs each. These eggs develop into males and females that are able to reproduce sexually. After mating, the males die and the females lay eggs in the dead insect if there is enough food available or, if not, the first and second stage larvae develop inside her body. As soon as the larvae reach the third stage, they leave the dead insect and seek out a new host. **Efficacy:** Very effective predators that can be used to eradicate vine weevil populations. **Key reference(s):** 1) R Gaugler & H K Kaya. 1990. Entomopathogenic nematodes. In *Biological Control*, 365 pp., CRC Press, Boca Raton, Florida, USA. 2) R Georgis & H K Kaya. 1998. Formulation of entomopathogenic nematodes. In *Formulation of Microbial Biopesticides, Beneficial Microorganisms, Nematodes and Seed Treatments*, H D Burges (ed.), pp. 289–308, Kluwer Academic Publishers, Dordrecht, The Netherlands.

COMMERCIALISATION: Formulation: Sold as infective third stage larvae. **Tradenames:** 'Dickmaulrüsslernematoden' (Andermatt), 'Heterorhabditis megidis' (Neudorff) and (Rincon-Vitova), 'Heterorhabditis-System' (Biobest), 'Nemasys H' (for chafer grubs)

(Becker Underwood), 'Nemasys G' (Becker Underwood), (Westgro Sales), (Plant Products), (BCP), (Brinkman), (Bog Madsen), (Svenska Predator), (Lier Frucklager) and (Intrachem).

APPLICATION: Apply as a drench overall when target insects are present. The area to be treated must be moist at application and must not be allowed to dry out after treatment. The temperature of the soil or compost should be between 12 and 30°C at application and for 2 weeks after application. Apply at a rate of 5×10^5 juveniles per square metre, repeated every 2 weeks as necessary. Stir to prevent sedimentation.

PRODUCT SPECIFICATIONS: Specifications: Produced by fermentation in a liquid diet or in solid-state fermentation on foam cubes. Sometimes reared on insects. Sold as third stage infective juveniles. **Purity:** Product contains only third stage infective larvae. Efficacy of the product may be checked by a bioassay on vine weevils (*Otiorhyncus sulcatus*) or on larvae of *Galleria mellonella* (L.) or *Tenebrio molitor* L. Actual nematode numbers may be determined by direct counting under a microscope. **Storage conditions:** Store in a refrigerator at 3 to 5°C. Do not freeze or expose to direct sunlight. **Shelf-life:** Use as soon as possible after delivery.

COMPATIBILITY: Incompatible with soil insecticides. Soil moisture is necessary for the nematodes to survive. Incompatible with strong oxidisers, acids and bases.

MAMMALIAN TOXICITY: No allergic or other adverse reactions have been reported by researchers, production staff or users following the use of *Heterorhabditis megidis* or its associated bacterium.

ENVIRONMENTAL IMPACT AND NON-TARGET TOXICITY: It is not expected that *Heterorhabditis megidis* or its associated bacterium will have any adverse effect on non-target organisms or on the environment. **Approved for use in organic farming:** Yes.

1:69 *Hirsutella thompsonii* isolate MF(Ag)5
Biological acaricide (fungus)

Fungus: anamorphic Clavicepitaceae: previously classified as mitosporic Deuteromycetes: Hyphomycetes

NOMENCLATURE: Approved name: *Hirsutella thompsonii* Fisher isolate MF(Ag)5. **Development code:** ITCC 4962; IMI 385470.

SOURCE: Originally isolated from an eriophyid mite in Tamil Nadu, India.

PRODUCTION: Produced by fermentation.

TARGET PESTS: Eriophyid mites, particularly the coconut mite (*Aceria guerreronis* Keifer).

TARGET CROPS: Major crop use is in coconut plantations, but can be used in palmyrah palm and in arecanut.

BIOLOGICAL ACTIVITY: Mode of action: *Hirsutella thompsonii* acts through degradation of the mite's cuticle, with subsequent fungal growth in the haemolymph and tissues of the mites. Re-sporulation from dead mites leads to infection of epidemic proportions.
Biology: *Hirsutella thompsonii* is an entomopathogenic fungus specific to eriophyid mites that exerts its effect by invasion of the living mite. Spores adhere to the cuticle of the mite and, under ideal conditions, germinate, producing a germ tube that penetrates the host mite's cuticle by physical and enzymic processes and subsequently invades the haemolymph and other tissues. The fungal hyphae develop in the mite and sporulation takes place through the living and dead mite's cuticle, providing infectious spores to continue the epidemic.
Efficacy: Field investigations conducted in more than 15 locations to evaluate the performance of 'Mycohit' showed that, by the 70th day of the experiment, greater than 90% mortality of the mites was observed in coconuts sprayed twice (at 2-week intervals).

COMMERCIALISATION: Formulation: Sold as a talc-based formulation coded Formulation-T containing 2.5×10^8 colony forming units/g and with a moisture content of about 12%.
Tradenames: 'Mycohit' (Hindustan Antibiotics).

APPLICATION: 'Mycohit' is generally recommended for use as a spray when the weather is dry. It should be used at 1% concentration and about 2 litres of the spray solution is needed per tree. Up to 50 trees can be treated with 1 kg of the product. In certain situations, such as after heavy rain, dusting of the product on the bunches is enough because of the wet microclimate within the crown.

PRODUCT SPECIFICATIONS: Specifications: Produced by fermentation. **Shelf-life:** The product is stable at 4°C for up to 6 months.

COMPATIBILITY: Susceptible to some fungicides, especially dithiocarbamates. Incompatible with strong oxidisers, acids, bases and chlorinated water.

MAMMALIAN TOXICITY: No skin or eye irritation has been observed. There is no evidence of acute or chronic toxicity, infectivity or hypersensitivity to mammals. No allergic responses or health problems have been observed by research workers, manufacturing staff or users.

ENVIRONMENTAL IMPACT AND NON-TARGET TOXICITY: *Hirsutella thompsonii* is widespread in Nature and is not pathogenic to non-target species. It has not shown adverse effects on the environment.

1:70 *Lagenidium giganteum*
Biological insecticide (fungus)

Chromista: Oomycetes: Pythiales: Pythiaceae

NOMENCLATURE: Approved name: *Lagenidium giganteum* Couch.
OPP Chemical Code: 129084.

SOURCE: *Lagenidium giganteum* is a naturally occurring water mould fungus that was originally identified from mosquito larvae in North Carolina and Georgia, USA.

PRODUCTION: Produced by fermentation.

TARGET PESTS: 'Laginex' controls mosquito larvae, including *Aedes*, *Anopheles*, *Coquillettidea*, *Culex*, *Culiseta*, *Deinocerites*, *Eretmapodites*, *Haemagogus*, *Mansonia*, *Opifex*, *Orthopodomyia*, *Psorophora*, *Sabethes*, *Uranotaenia* and *Wyeomyia* species. *Culex* spp. are particularly susceptible. It does not infect any other insect species.

TARGET CROPS: Recommended for use in rice, soybeans and irrigated pasture, but can also be used on non-crop land, such as drainage ditches, flood plains, ponds, marshes, river margins and small containers such as tyres.

BIOLOGICAL ACTIVITY: Mode of action: *Lagenidium giganteum* is a parasite of mosquito larvae. It is an aquatic fungus that requires fresh water – it does not survive in marine or salty environments and is most effective in water with a temperature between 15 and 32°C (60 and 90 °F), low sodium chloride levels, a pH between 4.5 and 8.0 and with low organic content. Motile zoospores, once released into water, locate mosquito larvae, attach to and penetrate the larval cuticle and then grow within the body cavity. The mosquito larva is consumed and the fungus produces additional zoospores that are released into the water to infect other larvae. Alternatively, oospores are produced within the larvae and these can withstand adverse environmental conditions such as extended periods of drought. When favourable conditions return, oospores release infective zoospores and the cycle of infection begins again. **Biology:** 'Laginex' is effective for 3 to 4 weeks (and sometimes longer) and gives complete control of a treated population within 5 to 7 days. **Key reference(s):** H D Burges. 1998. Formulation of mycoinsecticides. In *Formulation of Microbial Biopesticides, Beneficial Microorganisms, Nematodes and Seed Treatments*, H D Burges (ed.), pp. 131–85, Kluwer Academic Publishers, Dordrecht, The Netherlands.

COMMERCIALISATION: Formulation: 'Laginex' is an aqueous suspension formulation containing 40% *Lagenidium giganteum* (California isolate) zoospores. **Tradenames:** 'Laginex' (AgraQuest).

APPLICATION: 'Laginex' is applied at rates between 0.7 and 14 litres per hectare (9 and 180 fl oz per acre), with the lower rates being used in situations where epizootics might be expected to occur (moderate populations of highly susceptible mosquito larvae in fresh, unpolluted water at a temperature between 19 and 30°C). Most treatments require 1.5 to 6 litres per hectare (20 to 80 fl oz per acre). Higher rates are recommended when the larval population is very high and the mosquito larvae are not highly susceptible or conditions are not conducive to good fungal growth.

PRODUCT SPECIFICATIONS: Specifications: Produced by fermentation. **Purity:** 'Laginex' contains 1 × 10^{10} colony forming units (cfu) per litre. **Storage conditions:** Store in a cool situation, out of direct sunlight. **Shelf-life:** Use within 2 weeks of purchase.

COMPATIBILITY: It is unusual to use 'Laginex' with other control agents. Incompatible with strong oxidisers, acids, bases and chlorinated water.

MAMMALIAN TOXICITY: 'Laginex' has been shown to be non-toxic to mammals. It has no toxic effects in acute oral, dermal, pulmonary and intravenous tests on rats.
Skin and eye: 'Laginex' is not an eye or skin irritant (rabbits).

ENVIRONMENTAL IMPACT AND NON-TARGET TOXICITY: *Lagenidium giganteum* occurs widely in Nature and has not been shown to have any adverse effects on freshwater fish, water weeds or crop plants.

1:71 *Lecanicillium lecanii* whitefly isolate or aphid isolate

Biological insecticide (fungus)

Fungus: anamorphic Hypocreales: previously classified as mitosporic Deuteromycetes: Moniliales

NOMENCLATURE: Approved name: *Lecanicillium lecanii* (Zimmerman) Gams & Zare whitefly isolate or aphid isolate. **Other names:** Formerly known as *Cephalosporium lecanii* Zimmermann and *Verticillium lecanii* (Zimmerman) Viegas whitefly isolate or aphid isolate. (See R Zare & W Gams. 2001. *Nova Hedwigia*, **71**, 329–37).

SOURCE: *Lecanicillium lecanii* occurs widely in Nature. Two commercial isolates were isolated by R A Hall, the first from the aphid, *Macrosiphoniella sanborni* (Gillette), during a natural epidemic and the second from the glasshouse whitefly, *Trialeurodes vaporariorum* (Westwood).

The two isolates had different host ranges and were initially developed by Tate and Lyle Ltd (who no longer produce or market them), although, since 1990, both isolates have been produced and marketed by Koppert. A third isolate was isolated from the bodies of cotton aphids (*Aphis gossypii* Glover) and tobacco whitefly (*Bemisia tabaci* (Gennadius)).

PRODUCTION: *Lecanicillium lecanii* isolates are cultured on a sterile undefined medium and the spores harvested by concentration and drying.

TARGET PESTS: 'Mycotal' is used for control of whitefly, with a side effect on thrips. 'Vertalec' is used to control aphids. A mutant form is under development by the USDA for control of cyst nematodes in soybeans. 'Bio-Catch' is used to control aphids, whitefly, thrips and scale insects.

TARGET CROPS: 'Mycotal' and 'Vertalec' are used on protected glasshouse crops. The USDA mutant form is under development for use in soybeans. 'Bio-Catch' is used in vegetables, ornamentals, cash crops, oil seeds and pulses.

BIOLOGICAL ACTIVITY: Mode of action: *Lecanicillium lecanii* acts through degradation of the insect's cuticle, with subsequent fungal growth in the haemolymph and tissues of insects. Re-sporulation from dead insects leads to infection of epidemic proportions.
Biology: *Lecanicillium lecanii* is an entomopathogenic fungus that exerts its effect by invasion of the living insect. Spores adhere to the cuticle of the insect and, under ideal conditions, germinate, producing a germ tube that penetrates the host insect's cuticle by physical and enzymic processes and subsequently invades the haemolymph and other tissues. The fungal hyphae develop in the insect and sporulation takes place through the living or dead insect's cuticle, providing infectious spores to continue the epidemic. **Key reference(s):** 1) R A Hall. 1976. A bioassay of the pathogenicity of *Verticillium lecanii* conidiospores on the aphid *Macrosiphoniella sanborni*, *J. Inverteb. Pathol.*, **27**, 41–8. 2) R A Hall. 1979. Effects of repeated subculturing on agar and passaging through an insect host on pathogenicity and growth rate of *Verticillium lecanii*, *J. Inverteb. Pathol.*, **36**, 216–22. 3) D L Meade. 1991. The use of *Verticillium lecanii* against submaginal instars of *Bemisia tabaci*, *J. Inverteb. Pathol.*, **57**, 296–8. 4) H D Burges. 1998. Formulation of mycoinsecticides. In *Formulation of Microbial Biopesticides, Beneficial Microorganisms, Nematodes and Seed Treatments*, H D Burges (ed.), pp. 131–85, Kluwer Academic Publishers, Dordrecht, The Netherlands.

COMMERCIALISATION: Formulation: Sold as wettable powder (WP) formulations. 'Mycotal' has 1×10^{10} and 'Vertalec' 1×10^9 colony forming units (cfu)/g. **Tradenames:** 'Bio-Catch' (Stanes), 'Mealikil' (SOM Phytopharma) and (Agri Life), 'Mycotal' (whitefly isolate) (Koppert), 'Vertalec' (aphid isolate) (Koppert).

APPLICATION: It is usual to apply the products at high volume in the presence of the target insect pests. Invasion of the pests is best under conditions of high humidity. Non-phytotoxic and non-phytopathogenic.

PRODUCT SPECIFICATIONS: Specifications: *Lecanicillium lecanii* isolates are cultured on a sterile undefined medium and the spores harvested by concentration and drying.
Purity: Products contain spores of *Lecanicillium lecanii*, the activity of which is measured in terms of spore count and efficacy against insects. Spore count is determined by plating on agar and counting the colonies formed after 48 hours' incubation in the laboratory using standard techniques. Efficacy assays are conducted against *Aphis gossypii* for 'Vertalec' and against larvae of *Trialeurodes vaporariorum* for 'Mycotal'. 'Bio-Catch' contains 2×10^8 spores of *Lecanicillium lecanii* per gram. **Storage conditions:** Refrigerate for periods of long storage. Do not allow temperature to reach 35 °C. Do not freeze. **Shelf-life:** The product is stable at 4 °C for up to 6 months.

COMPATIBILITY: Susceptible to some fungicides, especially dithiocarbamates. Incompatible with strong oxidisers, acids, bases and chlorinated water.

MAMMALIAN TOXICITY: No skin or eye irritation has been observed. There is no evidence of acute or chronic toxicity, infectivity or hypersensitivity to mammals. No allergic responses or health problems have been observed by research workers, manufacturing staff or users. *Lecanicillium lecanii* is not toxic by inhalation.

ENVIRONMENTAL IMPACT AND NON-TARGET TOXICITY: *Lecanicillium lecanii* is widespread in Nature and is not pathogenic to non-target species. It has not shown adverse effects on the environment. **Approved for use in organic farming:** Yes.

1:72 *Lymantria dispar* nucleopolyhedrovirus
Insecticidal baculovirus

Virus: Baculoviridae: Nucleopolyhedrovirus

NOMENCLATURE: Approved name: *Lymantria dispar* nucleopolyhedrovirus.
Other names: *Lymantria dispar* MNPV; *Lymantria dispar* NPV; LdMNPV; LdNPV; gypsy moth virus.

SOURCE: Isolated from infected gypsy moth (*Lymantria dispar* (L.)) larvae by the USDA. The 'expanded gypsy moth research, development and application program' was part of the USDA 'combined forest pest research and development program' co-ordinated by the Forest Insect and Disease Research Laboratory, Hampden, Connecticut, and culminated in the registration of the product in 1978 by the US-EPA.

PRODUCTION: Produced *in vivo* in gypsy moth larvae reared under controlled conditions.

TARGET PESTS: Gypsy moth (*Lymantria dispar*).

TARGET CROPS: Recommended for use in commercial forestry and on ornamental and non-commercial landscape trees. In the USA, use is limited to wide-area government-sponsored programmes.

BIOLOGICAL ACTIVITY: Mode of action: As with all insect baculoviruses, *Lymantria dispar* MNPV must be ingested to exert an effect. Following ingestion, the virus enters the insect's haemolymph and multiplies in the insect body, leading to insect death. **Biology:** Upon ingestion, the protective protein matrix of the polyhedral occlusion bodies (OBs) of the virus is dissolved within the midgut and the virus particles enter the insect's haemolymph. These virus particles invade nearly all cell types in the larval body, where they multiply. Shortly after the death of the larva, the integument ruptures, releasing very large numbers of OBs. **Efficacy:** The virus infects only gypsy moth larvae. **Key reference(s):** 1) R R Granados & B A Federici (eds). 1986. *The Biology of Baculoviruses, Vols 1 and 2*, CRC Press, Boca Raton, Florida, USA. 2) D J Leisy & J R Fuxa. 1996. Natural and engineered viral agents for insect control. In *Crop Protection Agents from Nature: Natural Products and Analogues*, L G Copping (ed.), Royal Society of Chemistry, Cambridge, UK. 3) B A Federici. 1999. Naturally occurring baculoviruses for insect pest control. In *Biopesticides Use and Delivery*, F R Hall & J J Menn (eds), pp. 301–20, Humana Press, Totowa, NJ, USA. 4) H D Burges & K A Jones. 1998. Formulation of bacteria, viruses and Protozoa to control insects. In *Formulation of Microbial Biopesticides, Beneficial Microorganisms, Nematodes and Seed Treatments*, H D Burges (ed.), pp. 33–127, Kluwer Academic Publishers, Dordrecht, The Netherlands.

COMMERCIALISATION: Formulation: Sold through the USDA Forest Service as a liquid suspension. **Tradenames:** 'Gypcheck Biological Insecticide for the Gypsy Moth' (USDA Forest Service).

APPLICATION: Two aerial applications of 2.5×10^{11} polyhedral inclusion bodies (PIBs) per hectare will typically restrict defoliation to below 55 to 60% and sometimes below 30%.

PRODUCT SPECIFICATIONS: Specifications: Produced *in vivo* in gypsy moth larvae reared under controlled conditions. **Purity:** Products contain polyhedral inclusion bodies and no contaminating bacterial pathogens of man. **Storage conditions:** Store in a refrigerator in a sealed container. Do not freeze and do not expose to direct sunlight. **Shelf-life:** Formulations can be stored for up to 6 months.

COMPATIBILITY: It is unusual to apply *Lymantria dispar* MNPV with other insecticides. Incompatible with strong oxidisers, acids, bases and chlorinated water.

MAMMALIAN TOXICITY: There are no reports of allergic or other adverse reactions to *Lymantria dispar* MNPV from research workers or manufacturing or field staff. Considered to be non-toxic to mammals.

ENVIRONMENTAL IMPACT AND NON-TARGET TOXICITY: *Lymantria dispar* MNPV occurs in Nature and is not expected to have any adverse effects on non-target organisms or on the environment.

1:73 *Mamestra brassicae* nucleopolyhedrovirus
Insecticidal baculovirus

Virus: Baculoviridae: Nucleopolyhedrovirus

NOMENCLATURE: Approved name: *Mamestra brassicae* nucleopolyhedrovirus.
Other names: *Mamestra brassicae* MNPV; *Mamestra brassicae* NPV; MbMNPV; MbNPV.

SOURCE: Occurs in Nature as a natural parasite of the cabbage moth (*Mamestra brassicae* (L.)). It was first isolated from virosed larvae of the cabbage moth, collected in France by INRA researchers. The virus isolate was developed by NPP (Natural Plant Protection), France. It is also able to infect other lepidopteran species.

PRODUCTION: MbNPV is multiplied *in vivo* on *Mamestra brassicae* infested at the larval stage. It is isolated by centrifugation techniques. The product is then formulated as a suspension containing the entomopathogenic virus particles and specific additives.

TARGET PESTS: 'Mamestrin' is registered in France for the control of *Mamestra brassicae* and may be used against *Helicoverpa armigera* (Hübner), *Phthorimaea operculella* Zeller and *Plutella xylostella* (Linnaeus).

TARGET CROPS: 'Mamestrin' can be used on a variety of crops, including vegetables, potatoes, brassicae and ornamentals.

BIOLOGICAL ACTIVITY: Mode of action: As with all insect baculoviruses, *Mamestra brassicae* NPV must be ingested to exert an effect. Following ingestion, the virus enters the insect's haemolymph and multiplies in the insect body, leading to insect death.
Biology: MbNPV is more active on small larvae than on later larval instars. The virus is ingested by the feeding larva and the virus protein matrix is dissolved in the insect's midgut (that is alkaline), releasing the virus particles. These pass through the peritrophic membrane and invade midgut cells by fusion with the microvilli. The virus particles invade the cell nuclei, where they are uncoated and replicated. Initial replication produces non-occluded virus particles which attack more cells and hasten the invasion of the host insect, but later the virus particles are produced with protein coats that remain infective when released from the

dead insects. **Efficacy:** MbNPV acts relatively slowly, as it has to be ingested before it exerts any effect on the insect host. **Key reference(s):** 1) R R Granados & B A Federici (eds). 1986. *The Biology of Baculoviruses, Vols 1 and 2*, CRC Press, Boca Raton, Florida, USA. 2) D J Leisy & J R Fuxa. 1996. Natural and engineered viral agents for insect control. In *Crop Protection Agents from Nature: Natural Products and Analogues*, L G Copping (ed.), Royal Society of Chemistry, Cambridge, UK. 3) B A Federici. 1999. Naturally occurring baculoviruses for insect pest control. In *Biopesticides Use and Delivery*, F R Hall & J J Menn (eds), pp. 301–20, Humana Press, Totowa, NJ, USA. 4) H D Burges & K A Jones. 1998. Formulation of bacteria, viruses and Protozoa to control insects. In *Formulation of Microbial Biopesticides, Beneficial Microorganisms, Nematodes and Seed Treatments*, H D Burges (ed.), pp. 33–127, Kluwer Academic Publishers, Dordrecht, The Netherlands.

COMMERCIALISATION: Formulation: Sold as liquid formulations. **Tradenames:** 'Mamestrin' (NPP) and (Calliope), 'Virin-EKS' (NPO Vector), 'Virosoft BA3' (Biotepp). **Patents:** FR 8717748; EP 90401016.

APPLICATION: 'Mamestrin' should be applied to foliage at a dose of 4 litres of product per hectare per application. It has potential for integrated crop management (ICM). Monitoring of adult insect laying patterns and application targeted at newly hatched eggs gives better control than on a mixed-age population. It is important to ensure coverage of the foliage to effect good control.

PRODUCT SPECIFICATIONS: Specifications: MbNPV is multiplied *in vivo* on *Mamestra brassicae* infested at the larval stage. It is isolated by centrifugation techniques. The product is then formulated as a suspension containing the entomopathogenic virus particles and specific additives. **Purity:** 'Mamestrin' contains 2.5×10^{12} polyhedral inclusion bodies (PIBs) per litre. The infectivity of the formulated product is assessed by bioassay against *Mamestra brassicae* larvae. **Storage conditions:** Store at 4°C. **Shelf-life:** The product has a shelf-life of 2 years, when stored under the appropriate conditions.

COMPATIBILITY: Compatible with most crop protection agents that do not repel the target insects. Do not apply with copper-based fungicides, strong oxidisers, acids, bases and chlorinated water. Use water with a neutral pH as a carrier.

MAMMALIAN TOXICITY: No allergenic symptoms or health problems have been observed with research workers, manufacturing staff or users. **Acute oral LD$_{50}$:** male rats $>10.2 \times 10^9$, female rats $>10.95 \times 10^9$ polyhedral inclusion bodies (PIB)/kg b.w. **Acute dermal LD$_{50}$:** There is no evidence of acute or chronic toxicity, eye or skin irritation to mammals, although 'Mamestrin' may induce a slight hypersensitive reaction. **Inhalation: LC$_{50}$:** rats $>7.5 \times 10^{10}$ PIB/kg b.w. No toxicity, infectivity or pathogenicity was detected after an intranasal installation in rats at 7.5×10^{10} PIB/kg b.w. **Skin and eye:** 'Mamestrin' shows no skin penetration. **NOEL:** (99 days) for mice $>1.5 \times 10^9$ PIB/kg b.w. daily.

ENVIRONMENTAL IMPACT AND NON-TARGET TOXICITY: *Mamestra brassicae* NPV occurs in Nature and is not expected to have any adverse effects on non-target organisms or on the environment. **Bird toxicity:** *Mamestra brassicae* NPV is not toxic or infective to birds. **Fish toxicity:** *Mamestra brassicae* NPV is not toxic to fish.
Effects on beneficial insects: LD_{50} (oral and contact) for honeybees >5 × 10^4 PIB/bee.
Behaviour in soil: MbNPV occurs naturally. After one year, the level of MbNPV remaining in treated soil is equivalent to natural levels.

1:74 *Mamestra configurata* nucleopolyhedrovirus
Insecticidal baculovirus

Virus: Baculoviridae: Nucleopolyhedrovirus

NOMENCLATURE: Approved name: *Mamestra configurata* nucleopolyhedrovirus.
Other names: *Mamestra configurata* MNPV; *Mamestra configurata* NPV; McMNPV; McNPV; bertha armyworm nucleopolyhedrovirus.

SOURCE: 'Virosoft' is a wild-type baculovirus isolated from *Mamestra configurata* Walker. It has been selected from the same ecozone in which it will be produced and distributed.

PRODUCTION: McNPV is multiplied *in vivo* on *Mamestra configurata* infested at the larval stage and isolated by centrifugation techniques. The product is then formulated as a suspension.

TARGET PESTS: *Mamestra configurata* (bertha armyworm).

TARGET CROPS: Canola (oilseed rape) in ecozone 3.

BIOLOGICAL ACTIVITY: Mode of action: As with all insect baculoviruses, *Mamestra configurata* NPV must be ingested to exert an effect. Following ingestion, the virus enters the insect's haemolymph, where it multiplies in the insect body, leading to death.
Biology: McNPV is more active on small larvae than on later larval instars. The virus is ingested by the feeding larva and the virus protein matrix is dissolved in the insect's midgut (that is alkaline), releasing the virus particles. These pass through the peritrophic membrane and invade midgut cells by fusion with the microvilli. The virus particles invade the cell nuclei, where they are uncoated and replicated. Initial replication produces non-occluded virus particles that invade more cells and hasten the invasion of the host insect, but later the virus particles are produced with protein coats that remain infective when released from

the dead insects. **Efficacy:** McNPV acts relatively slowly (in comparison with conventional chemical insecticides), as it has to be ingested before it exerts any effect on the insect host. **Key reference(s):** 1) R R Granados & B A Federici (eds). 1986. *The Biology of Baculoviruses, Vols 1 and 2*, CRC Press, Boca Raton, Florida, USA. 2) D J Leisy & J R Fuxa. 1996. Natural and engineered viral agents for insect control. In *Crop Protection Agents from Nature: Natural Products and Analogues*, L G Copping (ed.), Royal Society of Chemistry, Cambridge, UK. 3) B A Federici. 1999. Naturally occurring baculoviruses for insect pest control. In *Biopesticides Use and Delivery*, F R Hall & J J Menn (eds), pp. 301–20, Humana Press, Totowa, NJ, USA. 4) H D Burges & K A Jones. 1998. Formulation of bacteria, viruses and Protozoa to control insects. In *Formulation of Microbial Biopesticides, Beneficial Microorganisms, Nematodes and Seed Treatments*, H D Burges (ed.), pp. 33–127, Kluwer Academic Publishers, Dordrecht, The Netherlands.

COMMERCIALISATION: Formulation: Sold as a powder formulation. **Tradenames:** 'Virosoft' (BioTEPP).

APPLICATION: 'Virosoft' is applied either preventively directly onto the ground at seeding, or curatively on the canopy upon appearance of adult moths. This product can be used in an integrated pest management (IPM) programme because of its high specificity, which leaves the rest of the ecological system unaltered (including bees, natural predators, parasitoids and soil fauna).

PRODUCT SPECIFICATIONS: Specifications: McNPV is multiplied *in vivo* on *Mamestra configurata* infested at the larval stage and isolated by centrifugation techniques. The product is then formulated as a suspension. **Storage conditions:** Store in a sealed container under dry conditions at 4°C. **Shelf-life:** The product has a shelf-life of 2 years, when stored under the appropriate conditions.

COMPATIBILITY: Compatible with most crop protection agents that do not repel the target insects. Do not apply with copper-based fungicides, strong oxidisers, acids, bases and chlorinated water. Use water with a neutral pH.

MAMMALIAN TOXICITY: This group of insect viruses is known to be highly specific and able to infect only very closely related species. Their greatest advantage is their high host specificity. This accounts for the complete environmental safety of baculoviruses. There has been no literature which reports any adverse effects to humans. *Mamestra configurata* NPV has not demonstrated evidence of toxicity, infectivity or irritation to mammals. No allergic responses or other related adverse health problems have been observed by research workers, manufacturing staff or users.

ENVIRONMENTAL IMPACT AND NON-TARGET TOXICITY: *Mamestra configurata* NPV occurs in Nature and, as such, is not expected to show any adverse effects on non-target organisms or on the environment.

1:75 *Maravalia cryptostegiae*
Biological herbicide (fungus)

Fungus: Basidiomycetes: Uredinales

NOMENCLATURE: Approved name: *Maravalia cryptostegiae* (Cummins) Ono.

SOURCE: Introduced into Australia from Madagascar in 1995.

PRODUCTION: Reared on live host plants, the rubber-vine weed (*Cryptostegia grandiflora* (Roxb. ex R. Br.) R. Br.).

TARGET PESTS: Rubber-vine weed (*Cryptostegia grandiflora*).

BIOLOGICAL ACTIVITY: Mode of action: Pathogenesis. **Biology:** Infection results in rust-induced defoliation, producing an overall reduction in fecundity and biomass of the weed. **Efficacy:** In sites with low water tables, weed growth decreased markedly, with a reduction in plant volume reaching 90% over a 4-year period. Both rust- and drought-induced stress combined to cause up to 75% plant mortality at some sites and, at all monitored sites, seedling recruitment was virtually nil. **Key reference(s):** H C Evans. 2000. Evaluating plant pathogens for biological control of weeds: an alternative view of pest risk assessment, *Australasian Plant Pathology*, **29**(1), 1–14.

COMMERCIALISATION: Formulation: Not commercialised.

APPLICATION: The rust was applied by spraying both dry and aqueous inoculum of uredinioid teliospores from the ground using mist-blowers, as well as from the air by atomising spore suspensions.

PRODUCT SPECIFICATIONS: Specifications: Reared on live host plants, the rubber-vine weed (*Cryptostegia grandiflora*).

MAMMALIAN TOXICITY: There are no reports of allergic or other adverse toxicological effects from the use of *Maravalia cryptostegiae*. The rust is widely spread in Madagascar and there is no evidence of allergenic or other adverse reactions. It only infects the rubber-vine weed.

ENVIRONMENTAL IMPACT AND NON-TARGET TOXICITY: *Maravalia cryptostegiae* is specific to the rubber-vine weed and, as such, is not expected to have any adverse effects on non-target organisms or on the environment. **Approved for use in organic farming:** Yes.

1:76 *Metarhizium anisopliae* isolates F52 and ESF1

Biological insecticide (fungus)

The Pesticide Manual Fifteenth Edition entry number: 560

Fungus: anamorphic Nectriaceae: Hypocreales; formerly Hyphomycetes: Moniliacae, Deuteromycetes and mitosporic fungus

NOMENCLATURE: Approved name: *Metarhizium anisopliae* (Metschnikoff) Sorokin isolates F52 and ESF1. **Other names:** Formerly known as *Penicillium anisopliae* Vuill. and *Entomophthora anisopliae* Metsch. **Common name:** Green muscardine fungus. **Development code:** BIO 1020 (Bayer). **OPP Chemical Code:** 029056.

SOURCE: Commonly occurring fungus, often associated with dead insects. 'Bio-Catch M' was isolated from the dead rice brown planthopper (*Nilaparvata lugens* (Stal.)).

PRODUCTION: Produced by controlled, deep fermentation.

TARGET PESTS: Effective against a range of Coleoptera and Lepidoptera. 'BioBlast'(isolate ESF1) is used to control termites. 'Bio-Catch M' is recommended for use against the rice brown planthopper and a range of Coleoptera and Lepidoptera. 'Bio-Path' is used for cockroach control and 'Cobican' is used to control the sugarcane spittle bug.

TARGET CROPS: Used in a variety of different crops, including glasshouse-grown vegetables and ornamentals.

BIOLOGICAL ACTIVITY: Mode of action: *Metarhizium anisopliae* is a very effective entomopathogen. It attacks the target insect by penetrating its cuticle and invading the haemolymph. **Efficacy:** Applied as a foliar spray to infested crops, the entomopathogen invades and immobilises the insect within 2 days. Death occurs after 7 to 10 days. The mycosed insects remain adhered to the crop and additional spores are released to maintain a high level of infective material on the crop. Following treatment of a termite infestation, infection of as few as 5 to 10% of the population will lead to the destruction of the entire colony over a 2-week period. **Key reference(s):** 1) T M Butt. 1992. Pathogenicity of the entomogenous hyphomycete fungus, *Metarhizium anisopliae*, against the chrysomelid beetles, *Psylliodes chrysocephala* and *Phaedon cochleariae*, *Biocontrol Science and Technology*, **2**, 327–34. 2) H D Burges. 1998. Formulation of mycoinsecticides. In *Formulation of Microbial Biopesticides, Beneficial Microorganisms, Nematodes and Seed Treatments*, H D Burges (ed.), pp. 131–85, Kluwer Academic Publishers, Dordrecht, The Netherlands.

COMMERCIALISATION: Formulation: Sold as an injectable formulation for termite control and as a suspension of spores for control of foliar pests. Also used as a foliar spray for thrips control. 'Bio-Catch M' is formulated as a water dispersible powder (DP).

Tradenames: 'BioBlast Biological Termiticide' [isolate ESF1] (EcoScience) and (Terminex), 'Bio-Path' (EcoScience), 'Bio-Catch M' (Stanes), 'Cobican' (Probioagro), 'New BIO 1020', 'Taenure Granular Bioinsecticide', 'Tick-Ex G' and 'Tick-Ex EC' [isolate F52] (Earth BioSciences), 'Pacer' (SOM Phytopharma) and (Agri Life).

APPLICATION: *Metarhizium anisopliae* isolate F52 is applied as a foliar spray when insects are present on the crops or incorporated into the growth medium. Isolate ESF1 is sprayed directly on foraging termites, which take the fungus back to the nest and infect nestmates, or is injected into infested wood through drilled holes.

PRODUCT SPECIFICATIONS: Specifications: Produced by controlled, deep fermentation. **Purity:** Free from contaminants pathogenic to man. 'Bio-Catch M' contains not less than 1×10^8 colony forming units (cfu) per gram *Metarhizium anisopliae*. **Storage conditions:** Store in a cool, dry place out of direct sunlight. **Shelf-life:** If stored under recommended conditions, the formulation will remain viable for 12 months.

COMPATIBILITY: Used alone. Incompatible with fungicides. Incompatible with strong oxidisers, acids, bases and chlorinated water.

MAMMALIAN TOXICITY: There are no reports of allergic or other adverse toxicological effects from research, manufacturing, formulation or field staff. **Inhalation:** *Metarhizium anisopliae* shows no inhalation toxicity. **Skin and eye:** *Metarhizium anisopliae* shows no skin or eye irritation.

ENVIRONMENTAL IMPACT AND NON-TARGET TOXICITY: *Metarhizium anisopliae* occurs widely in Nature and is not expected to have any adverse effects on non-target organisms or on the environment. 'Bio-Catch M' has no adverse effects on fish or other aquatic organisms and is non-toxic to honeybees and silkworms. **Behaviour in soil:** Conidia of isolate F52 and ESF1 are poorly transported in a sandy loam soil; contamination of groundwater by *Metarhizium anisopliae* is, therefore, unlikely.

1:77 *Metarhizium anisopliae* var. *acridium* isolates IMI 330189 and FI-985
Biological insecticide (fungus)

The Pesticide Manual **Fifteenth Edition entry number:** 560

Fungus: anamorphic Nectriaceae: Hypocreales; formerly Hyphomycetes: Moniliacae, Deuteromycetes and mitosporic fungus

NOMENCLATURE: Approved name: *Metarhizium anisopliae* (Metschnikoff) Sorokin var. *acridium* isolates IMI 330189 and FI-985. **Other names:** Formerly known as *Metarhizium anisopliae* var. *anisopliae*; earlier names include *Entomophthora anisopliae* Metsch and *Metarhizium flavoviride* Gams & Rozsypal (see F Driver, R J Milner & J W H Trueman. 2000. A taxonomic revision of *Metarhizium* based on a phylogenetic analysis of rDNA sequence data, *Mycological Research*, **104**, 134–50). **Common name:** Green muscardine fungus.

SOURCE: Naturally occurring entomopathogen isolated only from mycosed acridoid insects. LUtte BIologique contre les LOcustes et les SAuterians (LUBILOSA) is a collaborative, multi-disciplinary research and development programme, funded by the governments of Canada, the Netherlands, Switzerland and the United Kingdom, that developed *Metarhizium anisopliae* var. *acridium* isolate IMI 330189 for control of locusts and grasshoppers in environmentally sensitive areas of Africa. A different product with a similar isolate (isolate FI-985) has been developed by CSIRO in Australia (R J Milner. 2000. Current status of *Metarhizium* for insect control in Australia, *Biocontrol News and Information*, **21**, 47N-50N).

PRODUCTION: Produced by biphasic fermentation with conidia produced by solid substrate fermentation on rice substrate.

TARGET PESTS: Grasshoppers and locusts.

BIOLOGICAL ACTIVITY: Mode of action: Both *Metarhizium anisopliae* var. *acridium* isolates IMI 330189 and FI-985 are very effective and specific entomopathogens. They attack the target insect by penetrating its cuticle and invading the haemolymph. Conditions of high humidity are essential for cuticular penetration and the oil concentrate formulation has been designed to reduce this dependence. The isolates that have been commercialised for use against grasshoppers and locusts are highly virulent under tropical conditions. These characteristics are thought to be genetically homologous, leading to their recognition as a distinct variety of *M. anisopliae*. **Efficacy:** The entomopathogen penetrates the cuticle of the target insect relatively quickly and the insect becomes lethargic and shows reduced feeding 2 to 3 days after invasion in the laboratory or 7 to 10 days in the field. Death follows after 7 to 14 days or more, depending on prevailing temperature conditions, with the sporulation of the fungus outside the cadaver potentially providing a source of spores for continued infestation. In dry conditions, conidia are produced inside the insect. The conidia can survive the dry season within the insect cadavers. Although the mycoinsecticide is slower to show any effects than conventional chemicals, over time more effective control is achieved, with a lower reinfestation rate being usual. **Key reference(s):** 1) C J Lomer, R P Bateman, D Dent, H De Groote, O-K Douro-Kpindou, J Langewald, Z Ouambama, R Peveling & M Thomas. 1999. Development of strategies for the incorporation of biological pesticides into the integrated management of locusts and grasshoppers, *Agricultural and Forest Entomology*, **1**, 71–88. 2) R P Bateman, M Carey, D Moore & C Prior. 1993. The enhanced infectivity of *Metarhizium flavoviride* in oil formulations to desert locusts at low humidities, *Annals of Applied Biology*, **122**, 145–52. 3) J Langewald, Z Ouambama, Z Mamadou, R Peveling, I Stolz, R P Bateman, S Attignon, S Blanford, S Authurs & C J Lomer. 1999. Comparisons of an organophosphate insecticide with a mycoinsecticide for the control of *Oedaleus senegalensis* (Orthoptera: Acrididae) and other Sahelian grasshoppers at an operational scale, *Biocontrol Science and Technology*, **9**, 199–214. 4) J C Scanlan, W E Grant, D M Hunter & R J Milner. 2001. Habitat and environmental factors influencing the control of migratory locusts (*Locusta migratoria*) with a biopesticide (*Metarhizium anisopliae*), *Ecological Modelling*, **136**, 223–6. 5) D M Hunter, R J Milner, J C Scanlan & P A Spurgin. 1999. Aerial treatment of the migratory locust, *Locusta migratoria* (L.) (Orthoptera: Acrididae) with *Metarhizium anisopliae* (Deuteromycotina: Hyphomycetes) in Australia, *Crop Protection*, **18**, 699–704. 6) D M Hunter, R J Milner & P A Spurgin. 2001. Aerial treatment of the Australian plague locust, *Chortoicetes terminifera* (Orthoptera: Acrididae) with *Metarhizium anisopliae* (Deuteromycotina: Hyphomycetes), *Bull. Ent. Res.*, **91**, 93–9. 7) H D Burges. 1998. Formulation of mycoinsecticides. In *Formulation of Microbial Biopesticides, Beneficial Microorganisms, Nematodes and Seed Treatments*, H D Burges (ed.), pp. 131–85, Kluwer Academic Publishers, Dordrecht, The Netherlands.

COMMERCIALISATION: Formulation: Sold as an ultra-low volume (ULV) formulation of aerial conidia (4–5 × 10^{10} per gram dry powder) of *Metarhizium anisopliae* var. *acridium* isolates IMI 330189 and FI-985 in mineral or vegetable oils. **Tradenames:** 'Green Guard' (SGB), 'Green Muscle' (NPP) and (Biological Control Products), 'Taerain' (Taensa). **Patents:** GB 22550188.

APPLICATION: Applied using controlled droplet application techniques to optimise droplet size to facilitate insect invasion, at a rate of 100 g conidia in 500 ml per hectare. For the Australian plague locust, the recommended rate of 'Green Guard' is 25 g conidia in 500 ml per hectare, but trials have shown 17 g per hectare to be effective (D Hunter. 2001. Operational use of *Metarhizium anisopliae* for locust control in Australia, *Advances in Applied Acridology*, p. 5 – electronic version). For migratory locust, the recommended rate of 'Green Guard' is 75 g per hectare in 500 ml oil.

PRODUCT SPECIFICATIONS: Specifications: Produced by biphasic fermentation with conidia produced by solid substrate fermentation on rice substrate. **Purity:** Contains no contaminants pathogenic to man. **Storage conditions:** Store in a cool, dry situation out of direct sunlight. The spores can tolerate 5 days at 50°C, 14 days at 40°C and one year at 30°C. **Shelf-life:** If stored correctly, the formulation will remain effective for 3 years (≤20°C).

COMPATIBILITY: 'Green Muscle' can be applied with low rates of conventional insecticides, as this has been shown to reduce the time taken to kill the target insects from more than a week to 3 to 4 days. However, it is usual to recommend the use of the product alone. 'Green Guard' is applied alone. Incompatible with strong oxidisers, acids, bases and chlorinated water. Do not use water with a pH above 8.0.

MAMMALIAN TOXICITY: *Metarhizium anisopliae* var. *acridium* has been recorded from many countries as a natural pathogen of acridid grasshoppers. There have been no reports of allergic or other adverse toxicological effects by research or manufacturing staff, formulators or field workers. Non-pathogenic to mammals. **Acute oral LD$_{50}$:** rats >2000 mg/kg (for 'Green Muscle'). **Acute dermal LD$_{50}$:** rats >2000 mg/kg (for 'Green Muscle'). **Inhalation:** Acute pulmonary toxicity/infectivity LC$_{50}$ for rats >4 850 mg/m^3 (for 'Green Muscle'). **Skin and eye:** Non-irritating to skin and eyes of rabbits.

ENVIRONMENTAL IMPACT AND NON-TARGET TOXICITY: *Metarhizium anisopliae* var. *acridium* is not expected to have any adverse effects on non-target organisms or the environment. There is no effect on Carabidae, Tenebrionidae, Formicidae or Epydridae. **Bird toxicity:** No adverse effects in orientation tests on quail (for 'Green Muscle'). **Other aquatic toxicity:** LC$_{50}$ (48 h) for *Daphnia* spp. >100 mg/litre (for 'Green Muscle'). **Effects on beneficial insects:** There have been no adverse effects in orientation tests on honeybees (for 'Green Muscle'). LC$_{50}$ (14 days) for earthworms >1000 mg/kg soil (for 'Green Muscle'). **Behaviour in soil:** After spraying, residual levels of *Metarhizium anisopliae* var. *acridium* isolate IMI 330189 ('Green Muscle') have been shown to decrease exponentially, with a DT$_{50}$ of 6 to 8 days under Sahelian conditions. *Metarhizium anisopliae* var. *acridium* isolate FI-985 ('Green Guard') application has not resulted in any detectable residue in the soil.

1:78 *Metarhizium anisopliae* var. *anisopliae* isolate FI-1045

Biological insecticide (fungus)

The Pesticide Manual **Fifteenth Edition entry number:** 560

Fungus: anamorphic Nectriaceae: Hypocreales; formerly Hyphomycetes: Moniliacae, Deuteromycetes and mitosporic fungus

NOMENCLATURE: Approved name: *Metarhizium anisopliae* (Metschnikoff) Sorokin var. *anisopliae* isolate FI-1045.

SOURCE: *Metarhizium anisopliae* var. *anisopliae* isolate FI-1045 was isolated from a naturally infected greyback canegrub, *Dermolepida albohirtum*, in Queensland, Australia.

PRODUCTION: Mycelium is produced in liquid fermenters and this is used to inoculate sterile broken rice for sporulation.

TARGET PESTS: The greyback canegrub (*Dermolepida albohirtum* (Waterhouse)).

TARGET CROPS: Sugar cane.

BIOLOGICAL ACTIVITY: Mode of action: The product is supplied as natural granules on which the *Metarhizium* spores have been grown. They are applied below the surface of the soil. The conidia persist and infect the target pest on contact. A systemic infection develops, killing the insect in 3 to 12 weeks and, after death, new conidia form on the outside of the cadaver, thus augmenting the original inoculum. **Efficacy:** 'BioCane' provides 50 to 60% control in the season of application and persists to give ongoing control. **Key reference(s):** 1) R J Milner. 2000. Current status of *Metarhizium* for insect control in Australia, *Biocontrol News and Information*, **21**, 47–50. 2) D P Logan, L N Robertson & R J Milner. 1999. Review of the development of *Metarhizium anisopliae* as a microbial insecticide, BioCane™, for the control of greyback canegrub *Dermolepida albohirtum* (Waterhouse) (Coleoptera: Scarabaeidae) in Queensland sugarcane, *IOBC/WPRS Bull.*, **23**, 131–7.

COMMERCIALISATION: Formulation: 'BioCane' is formulated as natural rice granules 2 to 3 mm in diameter. The product contains over 2×10^9 viable conidia/g. **Tradenames:** 'BioCane' (BioCare).

APPLICATION: 'BioCane' is applied at fill-in at a rate of 33 kg of product per hectare.

PRODUCT SPECIFICATIONS: Specifications: Mycelium is produced in liquid fermenters and this is used to inoculate sterile broken rice for sporulation. **Purity:** The formulated product contains rice and conidia and is free of other micro-organisms. **Storage conditions:** Storage should be in a dry place at 5 to 10°C. **Shelf-life:** It is recommended that the product be used immediately, but it can be stored for up to 6 months.

COMPATIBILITY: 'BioCane' is always used alone. Incompatible with strong oxidisers, acids, bases and chlorinated water.

MAMMALIAN TOXICITY: There are no reports of allergic or other adverse toxicological effects from research, manufacturing, formulation or field staff.

ENVIRONMENTAL IMPACT AND NON-TARGET TOXICITY: There are no reports of any adverse environmental or non-target effects. **Behaviour in soil:** Does not leach through soil and remains where placed in the profile.

1:79 *Metarhizium anisopliae* isolate ICIPE 30
Biological insecticide (fungus)

The Pesticide Manual **Fifteenth Edition entry number:** 560

Fungus: anamorphic Nectriaceae: Hypocreales; formerly Hyphomycetes: Moniliacae, Deuteromycetes and mitosporic fungus

NOMENCLATURE: Approved name: *Metarhizium anisopliae* (Metschnikoff) Sorokin isolate ICIPE 30. **Other names:** Formerly known as: *Penicillium anisopliae* Vuill. and *Entomophthora anisopliae* Metsch. **Common name:** Green muscardine fungus. **Development code:** ICIPE 30.

SOURCE: *Metarhizium anisopliae* isolate ICIPE 30 was isolated from the lepidopteran stemborer, *Busseola fusca* Fuller, in 1989 from Kendu Bay, on the shore of Lake Victoria, Kenya.

PRODUCTION: Produced as granules on inert/cereal substrate and as dry spore powder by accurately controlled solid substrate (maize/vermiculite) fermentation in plastic bowls.

TARGET PESTS: Macrotermitinae of the genera *Macrotermes*, *Microtermes* and *Odontotermes*.

TARGET CROPS: Recommended for use in maize (corn), cassava, passion fruit, citrus, coffee, nurseries in agroforestry and vegetables attacked by termites. Recommended for use in termite mounds, buildings and wooden structures.

BIOLOGICAL ACTIVITY: Mode of action: *Metarhizium anisopliae* produces conidia which attach to the host cuticle and germinate on its surface. The fungus then penetrates the host cuticle shortly after germination. A wide variety of enzymes is produced, which degrade the host cuticle. Following successful penetration, the fungus grows in a yeast-like phase inside the host haemocoel, forming hyphal bodies or blastospores. After host death, conidia are formed externally on the insect surface. **Biology:** Because of the different phases of infection, fungal pathogens are often slow acting (death occurring within 3 to 5 days) in comparison with conventional chemicals. The debilitating effect of infection, however, often causes a reduction in feeding and insects stop feeding and may starve to death. **Efficacy:** A single application as granules in maize cropping systems can persist for two seasons. Injection of spores into termite mounds may result in destruction of colonies within 2 to 3 months. Buildings can be protected from attack for more than 2 years. **Key reference(s):** 1) ICIPE. 1997. *ICIPE Annual Report 1996/97.* 2) ICIPE. 2000. *ICIPE Annual Report 2000.* 3) N K Maniania, S Ekesi & J Songa. 2001. Managing termites in maize cropping systems with the entomopathogenic fungus, *Metarhizium anisopliae, Insect Science and its Application*, **22**, 41–6. 4) M P McQuilken, P Halmer & D J Rhodes. 1998. Application of microorganisms to seeds. In *Formulation of Microbial Biopesticides, Beneficial Microorganisms, Nematodes and Seed Treatments*, H D Burges (ed.), pp. 255–85, Kluwer Academic Publishers, Dordrecht, The Netherlands.

COMMERCIALISATION: Formulation: Used as technical concentrate in the form of granules overgrown with conidia. **Tradenames:** 'Muchwatox' (proposed) (ICIPE).

APPLICATION: Used in development trials at the very high rate of 20 to 30 kg of granules per hectare at planting in maize, cassava, vegetables and nurseries. Used at rates of 2 to 5 g per mound (depending on the size of the mound) for termite mound destruction.

PRODUCT SPECIFICATIONS: Specifications: Produced as granules on inert/cereal substrate and as dry spore powder by accurately controlled solid substrate (maize/vermiculite) fermentation in plastic bowls. **Purity:** The formulation is standardised to have a light green colour, unclumped and free of microbial contaminants harmful to man. It has a conidial viability of >90% and a moisture content of 4 to 5% (not yet standardised). Conidia spore density is 1×10^9 conidia/g. **Storage conditions:** Do not expose to direct sunlight. Keep refrigerated (4 to 6°C). **Shelf-life:** If stored under cool, dark conditions, the product remains viable for one year.

COMPATIBILITY: Do not use in combination with broad-spectrum fungicides such as chlorothalonil, mancozeb, benomyl or dithianon. Incompatible with strong oxidisers, acids, bases and chlorinated water.

MAMMALIAN TOXICITY: Has not shown evidence of toxicity, infectivity or hypersensitivity to mammals. No allergic reactions or other related health problems have been shown by research workers, manufacturing staff or users. **Toxicity class:** EPA (formulation) III.

ENVIRONMENTAL IMPACT AND NON-TARGET TOXICITY: No adverse effects on predators and parasitoids have been recorded in approved application field trials.

1:80 *Metarhizium anisopliae* isolate ICIPE 69
Biological insecticide (fungus)

The Pesticide Manual Fifteenth Edition entry number: 560

Fungus: anamorphic Nectriaceae: Hypocreales; formerly Hyphomycetes: Moniliacae, Deuteromycetes and mitosporic fungus

NOMENCLATURE: Approved name: *Metarhizium anisopliae* (Metschnikoff) Sorokin isolate ICIPE 69. **Other names:** Formerly known as *Penicillium anisopliae* Vuill. and *Entomophthora anisopliae* Metsch. **Common name:** Green muscardine fungus. **Development code:** ICIPE 69.

SOURCE: *Metarhizium anisopliae* isolate ICIPE 69 originated from a soil sample obtained from the Democratic Republic of Congo (DRC) and isolated using the *Galleria* bait method in 1990.

PRODUCTION: Produced by accurately controlled solid substrate (long-grain rice) fermentation in plastic autoclavable containers. Conidia are harvested by sifting the substrate through a sieve.

TARGET PESTS: Thrips, particularly the legume flower thrips (*Megalurothrips sjostedti* (Trybom)), onion thrips (*Thrips tabaci* Lindeman) and the western flower thrips (*Frankliniella occidentalis* (Pergande)).

TARGET CROPS: Recommended for use in vegetables and flowers such as cowpea, French beans, snow pea, broad beans, onions, chrysanthemums, carnations and roses.

BIOLOGICAL ACTIVITY: Mode of action: *Metarhizium anisopliae* produces conidia which attach to the host cuticle and germinate on its surface. The fungus then penetrates the host cuticle shortly after germination. A wide variety of enzymes is produced, which degrade the host cuticle. Following successful penetration, the fungus grows in a yeast-like phase inside the host haemocoel, forming hyphal bodies or blastospores. After host death, conidia are formed externally on the insect surface. **Biology:** Because of the different phases of infection, fungal pathogens are often slow acting (death occurring within 3 to 5 days) in comparison with conventional chemicals. The debilitating effect of infection, however, often results in reduced feeding, causing the insects to starve to death. Adult insects, which survive infection as larvae, have reduced fecundity and longevity. **Efficacy:** Very effective when applied before thrips population build-up on the target crop. Application with reduced doses of compatible synthetic insecticides may be needed to suppress an expanding or already established thrips population. **Key reference(s):** 1) S Ekesi, N K Maniania, K Ampong-Nyarko & I Onu. 1999. Effect of intercropping cowpea with maize on the performance of *Metarhizium anisopliae* against the legume flower thrips, *Megalurothrips sjostedti* (Thysanoptera: Thripidae) and predators, *Environmental Entomology*, **28**(6), 1154–61. 2) S Ekesi, N K Maniania & I Onu. 1999. Effects of temperature and photoperiod on development and oviposition of the legume flower thrips, *Megalurothrips sjostedti*, *Entomologia Experimentalis et Applicata*, **93**(2), 149–55. 3) S Ekesi, N K Maniania & K Ampong-Nyarko. 1999. Effect of temperature on germination, radial growth and pathogenic activity of *Metarhizium anisopliae* and *Beauveria bassiana* on *Megalurothrips sjostedti*, *Biocontrol Science and Technology*, **9**(2), 177–85. 4) S Ekesi, N K Maniania, K Ampong-Nyarko & I Onu. 1998. Potential of the entomopathogenic fungus, *Metarhizium anisopliae* for the control of legume flower thrips, *Megalurothrips sjostedti* (Trybom) on cowpea in Kenya, *Crop Protection*, **17**, 661–8. 5) S Ekesi, N K Maniania, I Onu & B Löhr. 1998. Pathogenicity of entomopathogenic fungi to the legume flower thrips, *Megalurothrips sjostedti* (Trybom) (Thysanoptera: Thripidae), *Journal of Applied Entomology*, **122**, 629–34. 6) S Ekesi & N K Maniania. 2000. Susceptibility of *Megalurothrips sjostedti* developmental stages to the entomopathogenic fungus *Metarhizium anisopliae* and the effect of infection on feeding and on fecundity, egg fertility and longevity of adults surviving infection as second instar larvae, *Entomologia Experimentalis et Applicata*, **94**(3), 229–36.

COMMERCIALISATION: Formulation: Used as dry spore powder technical concentrate, tank-mixed with 0.1% nutrient agar, 0.5% molasses, 0.05% Silwet and 0.1% glycerol. **Tradenames:** 'Metathripol' (ICIPE).

APPLICATION: Used at rates of 1.0 to 1.5 kg spores per hectare against legume flower thrips and onion thrips and at rates of 2.0 to 2.5 kg spores per hectare against western flower thrips. The crop must be well covered with the spray suspension. Apply while thrips populations are between 4 to 10 thrips per 20 flowers and repeat every 5 to 7 days. The fungal-based sprays can be applied up to the day of harvest. *Metarhizium anisopliae* has a short persistence on foliage due to its sensitivity to u.v. light and is preferably administered in the evenings between 17.00 and 18.30 hours.

PRODUCT SPECIFICATIONS: Specifications: Produced by accurately controlled solid substrate (long-grain rice) fermentation in plastic autoclavable containers. Conidia are harvested by sifting the substrate through a sieve. **Purity:** The formulation is standardised to have a light green colour, unclumped and free of microbial contaminants pathogenic to man. Spores have a viability of >90% with a moisture content of 4 to 5%. Particle size (diameter) is <60 mm and spore density is 1×10^9 conidia/g. **Storage conditions:** Do not expose to direct sunlight. Keep refrigerated (4 to 6°C). **Shelf-life:** If stored under cool, dark conditions, the products remain viable for one year.

COMPATIBILITY: Do not use in combination with broad-spectrum fungicides such as mancozeb, chlorothalonil or dithianon. Compatible with a wide range of other fungicides, acaricides, insecticides, stickers, spreaders and wetters. Do not use water with a pH above 8.0. Incompatible with strong oxidisers, acids and bases and chlorinated water.

MAMMALIAN TOXICITY: 'Metathripol' has not shown evidence of toxicity, infectivity or hypersensitivity to mammals. No allergic reactions or other related health problems have been shown by research workers, manufacturing staff or users. **Toxicity class:** EPA (formulation) III.

ENVIRONMENTAL IMPACT AND NON-TARGET TOXICITY: No adverse effects on predators and parasitoids or on the environment have been recorded in approved field use.

1:81 *Metarhizium flavoviride* var. *flavoviride* isolate F001   *Biological insecticide (fungus)*

The Pesticide Manual **Fifteenth Edition entry number:** 560

Fungus: anamorphic Nectriaceae: Hypocreales; formerly Hyphomycetes: Moniliacae, Deuteromycetes and mitosporic fungus

NOMENCLATURE: Approved name: *Metarhizium flavoviride* var. *flavoviride* Gams & Rozsypal isolate F001.

SOURCE: *Metarhizium flavoviride* var. *flavoviride* isolate F001 was isolated from a naturally infected redheaded cockchafer, *Adoryphorus couloni* (Burmeister), in Victoria, Australia.

PRODUCTION: Mycelium is produced in liquid fermenters and this is used to inoculate sterile broken rice for sporulation.

TARGET PESTS: The redheaded cockchafer (*Adoryphorus couloni*).

TARGET CROPS: *Metarhizium flavoviride* var. *flavoviride* isolate F001 is used in pasture and on turf.

BIOLOGICAL ACTIVITY: Mode of action: The product is supplied as natural granules on which the *Metarhizium* spores have been grown. These granules are applied below the surface of the soil. The conidia persist and infect the target pest on contact. A systemic infection develops, killing the insect in 3 to 12 weeks and, after death, new conidia form on the outside of the cadaver, thus augmenting the original inoculum. Control has been reported for over 7 years with 'BioGreen'. **Efficacy:** 'BioGreen' may give 60 to 70% control in the first year, but builds up to give increasing levels of control in subsequent seasons. Increases of up to 25% in pasture productivity have been obtained. **Key reference(s):** 1) R J Milner. 2000. Current status of *Metarhizium* for insect control in Australia, *Biocontrol News and Information*, **21**,

47–50. 2) A C Rath, D Worledge, T B Koen & B A Rowe. 1995. Long-term field efficacy of the entomogenous fungus *Metarhizium anisopliae* against the subterranean scarab, *Adoryphorus couloni*, *Biocontrol Science and Technology*, **5**, 439–51. 3) A C Rath & G K Bullard. 1997. Persistence of *Metarhizium anisopliae* DAT F-001 in pasture soils for 7.5 years – implications for sustainable soil-pest management. In *Soil Invertebrates in 1997*, P G Allsopp, D J Rogers & L N Robertson (eds), pp. 78–80, BSES, Brisbane, Queensland, Australia.

COMMERCIALISATION: **Formulation:** 'BioGreen' is formulated as natural rice granules 2 to 3 mm in diameter. The product contains over 2×10^9 viable conidia/g. **Tradenames:** 'BioGreen' (BioCare).

APPLICATION: 'BioGreen' is applied 20 to 25 mm below the surface of the pasture or turf, at a rate of 10 kg of product per hectare.

PRODUCT SPECIFICATIONS: **Specifications:** Mycelium is produced in liquid fermenters and this is used to inoculate sterile broken rice for sporulation. **Purity:** The formulated product contains rice and conidia and is free of other micro-organisms. **Storage conditions:** Storage should be in a dry place at 5 to 10°C. **Shelf-life:** It is recommended that the product be used immediately, but it can be stored for up to 6 months.

COMPATIBILITY: 'BioGreen' is always used alone. Incompatible with strong oxidisers, acids, bases and chlorinated water.

MAMMALIAN TOXICITY: There are no reports of allergic or other adverse toxicological effects from research, manufacturing, formulation or field staff.

ENVIRONMENTAL IMPACT AND NON-TARGET TOXICITY: There are no reports of any adverse environmental or non-target effects. **Behaviour in soil:** Does not leach through soil and remains where placed in the profile.

1:82 mixed bacteria
Biological fungicide/plant growth regulator (bacterium)

NOMENCLATURE: **Approved name:** A mixture of 15 bacteria, including five Gram-negative cocci, seven Gram-positive rods and three Gram-negative rods.

SOURCE: 'Vitazyme' is a mixture of rhizosphere beneficial bacteria.

PRODUCTION: 'Vitazyme' is produced through fermentation.

TARGET CROPS: Effective in a wide range of different crops, including vegetables, maize (corn) and cotton.

BIOLOGICAL ACTIVITY: Mode of action: 'Vitazyme' has been shown to reduce the need for nitrogen inputs, to speed up seed germination and crop maturity, to improve soil structure and drainage and to increase crop yield and quality. The bacteria that are found in the commercial product 'Vitazyme' occupy the rhizosphere and improve growth by solubilisation of insoluble minerals such as phosphate; by producing antifungal antibiotics; and by fixing nitrogen. In addition, 'Vitazyme' benefits soil characteristics by increasing root growth, improving mycorrhizal activity and encouraging the activity of earthworms.
Biology: A mixture of 15 bacteria, including five Gram-negative cocci, seven Gram-positive rods and three Gram-negative rods. One of the Gram-negative rods is *Bacillus macerans*.

COMMERCIALISATION: Formulation: The product is sold as 'Vitazyme', a liquid formulation containing 2 to 7×10^6 colony forming units (cfu) per ml. **Tradenames:** 'Vitazyme' (Vital Earth Resources).

APPLICATION: 'Vitazyme' is usually applied to the soil at or shortly after planting, but can also be used as a seed treatment. The manufacturers claim that the dilution rate is not critical. It is not necessary to incorporate the product into the soil after application.

PRODUCT SPECIFICATIONS: Specifications: 'Vitazyme' is produced through fermentation.
Storage conditions: Store under dry, ambient conditions. **Shelf-life:** May be stored for up to 12 months.

COMPATIBILITY: 'Vitazyme' can be tank mixed with fertilisers and pesticides. It can be used in organic farming, where it is used alone. Incompatible with strong oxidisers, acids, bases and chlorinated water.

MAMMALIAN TOXICITY: No harmful health effects to humans are expected from use of 'Vitazyme'. There is no indication that this product is toxic or infectious to humans. Given the recommended uses for the product, additional exposure of the public is likely to be minimal or non-existent.

ENVIRONMENTAL IMPACT AND NON-TARGET TOXICITY: The bacteria found in 'Vitazyme' occur in many kinds of soil and various studies indicate that these bacteria do not pose a risk to non-target organisms or to the environment.
Approved for use in organic farming: Yes.

1:83 *Muscodor albus* isolate QST 20799
Biological fungicide (fungus)

Fungus: Ascomycetes: Xylariaceae

NOMENCLATURE: Approved name: *Muscodor albus* anam. nov. isolate QST 20799.
OPP Chemical Code: 006503.

SOURCE: *Muscodor albus* isolate QST 20799 was originally isolated from the small limbs of a cinnamon tree (*Cinnamomum zeylanicum* Nees) located in the Lancetilla Botanical Garden near La Ceiba, Honduras by Gary Strobel of Montana State University, USA.

PRODUCTION: Manufactured by fermentation.

TARGET PESTS: Effective against a range of fungal pathogens including *Pythium* spp., *Phytophthora* spp. and *Fusarium* spp. Also claimed to show activity against bacteria and nematodes.

TARGET CROPS: Recommended to control fungal root diseases in protected and field crops and for control of post-harvest decay in fresh fruits and vegetables and cut flowers. It can be used on food crops, seed and other propagules, non-food crops and cut flowers. Products containing *Muscodor albus* can be used in the field, glasshouses and warehouses.

BIOLOGICAL ACTIVITY: Mode of action: *Muscodor albus* isolate QST 20799 grows as a white sterile mycelium and does not produce asexual or sexual spores or other reproductive structures such as chlamydospores or sclerotia. When hydrated, it produces a number of volatiles, mainly alcohols, acids and esters, that inhibit and kill plant pathogenic fungi and other organisms, such as nematodes, that cause soil-borne and post-harvest diseases. **Biology:** It is proposed as a methyl bromide replacement for seed, propagule, soil and post-harvest treatments of all food or feed commodities, as well as for ornamentals and cut flowers.
Key reference(s): 1) G A Strobel, E Dirkse, J Sears & C Markworth. 2001. Volatile antimicrobials from *Muscodor albus*, a novel endophytic fungus, *Microbiology*, **147**(11), 2943–50. 2) J Worapong, G Strobel, E J Ford, J Y Li & W M Hess. 2001. *Muscodor albus* anam. gen. et sp. nov., an endophyte from *Cinnamomum zeylanicum*, *Mycotaxon.*, **79**, 67–79.

COMMERCIALISATION: Formulation: 'Andante' and 'Glissade' are sold as granules; 'Arabesque' is a liquid concentrate. **Tradenames:** 'Andante', 'Arabesque' and 'Glissade' (AgraQuest).

APPLICATION: *Muscodor albus* is formulated into pesticide products which release volatile chemicals when wet. These products are incorporated into the soil or are held in containers where they do not contact treated commodities. The products are applied to field, glasshouse and other protected crops, and stored crops, including cut flowers.

PRODUCT SPECIFICATIONS: Specifications: Manufactured by fermentation.

MAMMALIAN TOXICITY: No harmful effects are likely to occur to workers or the public from use of *Muscodor albus* on crops. Laboratory studies indicate that the active ingredient is not toxic or infective following lung, oral, eye or skin exposure in rats. In these studies, the microbe did not survive in rat tissues, indicating that it is unlikely to cause infections in mammals. During laboratory research and field trials, no workers reported adverse effects. Human exposure to *Muscodor albus* QST 20799 used as a pesticide active ingredient will be minimal. Dietary exposure to the microbe and its volatiles is not expected, because no residues remain on treated food or feed. Even if the pesticides remain on the food, the studies on rats indicate that there is a reasonable certainty that they will not harm mammals. The products' low potential for drift minimises residential and worker exposure when applied in the field. Exposure to workers is further minimised because they are required to wear appropriate personal protective equipment (PPE).

ENVIRONMENTAL IMPACT AND NON-TARGET TOXICITY: No harmful environmental effects are expected from pesticidal uses of *Muscodor albus* QST 20799. Published literature and submitted studies indicate that the active ingredient will not cause adverse effects to mammals, birds, honeybees, other non-target insects or plants. Studies show that *Muscodor albus* QST 20799 does not produce spores, and does not survive in the environment after it uses up its food sources. The volatile compounds also dissipate rapidly. Thus, run-off is not expected. Moreover, the pesticide product will not be applied to bodies of water, minimising adverse effects to aquatic organisms. There is no evidence of toxicity or pathogenicity in the literature or in submitted studies on mammals, birds, insects, aquatic organisms and plants. Consequently, *Muscodor albus* QST 20799 is not expected to harm endangered or threatened species. **Approved for use in organic farming:** Yes.

1:84 *Myrothecium verrucaria*
Biological nematicide (fungus)

The Pesticide Manual Fifteenth Edition entry number: 603

Fungus: anamorphic Hypocreales: previously classified as mitosporic fungi and Deuteromycetes: Moniliales

NOMENCLATURE: Approved name: *Myrothecium verrucaria* Ditm.

SOURCE: The fungus occurs in soil and the commercial isolate was isolated from a nematode in the USA.

PRODUCTION: Produced by fermentation.

TARGET PESTS: Plant parasitic nematodes, including root-knot (*Meloidogyne* spp.), cyst (*Heterodera* spp.), sting (*Belonolaimus longicaudatus* Rau) and burrowing (*Radopholus similis* (Cobb) Thorne) nematodes. There are also claims that *Myrothecium verrucaria* acts as a bioherbicide on kudzu (*Pueraria lobata* Ohwi. (syn. *Pueraria thunbergiana* Benth.)) and on morning glories (*Ipomoea* spp.).

TARGET CROPS: Developed for use in turf, tobacco, grapes, citrus, brassicae and bananas.

BIOLOGICAL ACTIVITY: Biology: The active ingredient is the mixture of substances that are in suspension and in solution when the fungus, *Myrothecium verrucaria*, is grown in the laboratory. To prepare the active ingredient as a dry powder, water is removed from the culture mixture and the fungus is killed by exposure to high temperatures. The pesticidal activity is apparently not due to a single identifiable component, but requires the entire mixture. Researchers do not know the mechanism of action. **Key reference(s):** 1) P Warrior, R M Beach, P A Grau, J M Conley & G W Kirfman. 1998. Commercial development and introduction of DiTera, a new nematicide. In *Abstracts of the 9th International Congress of Pesticide Chemistry*, Royal Society of Chemistry, Cambridge, UK. 2) P Warrior. 2000. Living systems as natural crop protection agents, *Pest Management Science*, **56**(8), 681–7. 3) R E Hoagland, C D Boyette & H K Abbas. 2007. *Myrothecium verrucaria* isolates and formulations as bioherbicide agents for kudzu, *Biocontrol Science and Technology*, **17**(7), 721–31.

COMMERCIALISATION: Formulation: Sold as a dry powder formulation. **Tradenames:** 'DiTera', 'ABG-9008 Biological Nematicide Wettable Powder', 'ABG-9017 Biological Nematicide Emulsifiable Suspension', 'Ditera G Biological Nematicide Granule', 'Myrothecium verrucaria Slurry' and 'Ditera DF Biological Nematicide' (Valent BioSciences).

APPLICATION: The pesticide product is incorporated into the upper 1 to 2.5 cm of soil as a dry powder or as a ground spray. It can be applied any time in the life-cycle of the plant; before planting, during planting or after planting.

PRODUCT SPECIFICATIONS: Specifications: Produced by fermentation. **Purity:** The product contains a mixture of different components produced during the fermentation of *M. verrucaria*. The active ingredient does not contain any live *M. verrucaria*, or other living organisms or spores of concern.

COMPATIBILITY: Can be applied with other crop protection agents. It is usual not to mix with chemicals, as it has a role in integrated crop management (ICM) systems.

MAMMALIAN TOXICITY: Toxicological studies indicate a very favourable acute and non-acute toxicological profile. There is no evidence of allergic reactions.

Skin and eye: Mild, reversible skin and eye irritation were seen when the active ingredient was tested on laboratory animals; based on results of standard toxicity tests, no other human health problems are expected.

ENVIRONMENTAL IMPACT AND NON-TARGET TOXICITY: *Myrothecium verrucaria* occurs widely in Nature and is not expected to have any adverse effects on non-target organisms or on the environment. The only known risk to the environment from use of products containing this active ingredient concerns possible toxicity to aquatic organisms and, consequently, it is not recommended to use the pesticide in or near bodies of water. The living fungus causes plant disease but the active ingredient does not contain living *Myrothecium verrucaria* and, therefore, cannot infect plants. **Approved for use in organic farming:** Yes.

1:85 *Neodiprion sertifer/N. lecontei* nucleopolyhedrovirus   *Insecticidal baculovirus*

Virus: Baculoviridae: Nucleopolyhedrovirus

NOMENCLATURE: Approved name: *Neodiprion sertifer* nucleopolyhedrovirus and *Neodiprion lecontei* nucleopolyhedrovirus. **Other names:** *Neodiprion sertifer* MNPV; *Neodiprion lecontei* MNPV; *Neodiprion sertifer* NPV; *Neodiprion lecontei* NPV; NsMNPV; NlMNPV; NsNPV; NlNPV; sawfly virus.

SOURCE: Naturally occurring virus isolated from sawflies in forests in the USA and Canada. Nucleopolyhedroviruses have been isolated from 25 species of sawfly.

PRODUCTION: The virus is produced from the propagation of sawflies in the field. Selected, high-density populations are treated with the virus and, several days later, the diseased insects are collected and frozen. Before application, the larvae are freeze-dried and ground into a fine powder. A liquid formulation is made by separating the polyhedral inclusion bodies (PIB) from the diseased insects, purifying the polyhedra and suspending them in an aqueous solution.

TARGET PESTS: Sawflies (*Neodiprion* spp.).

TARGET CROPS: Forests.

BIOLOGICAL ACTIVITY: Mode of action: The virus is ingested by the feeding larva and the virus protein matrix is dissolved in the insect's midgut (that is alkaline), releasing the virus particles. These pass through the peritrophic membrane and invade midgut cells by fusion with the microvilli. The virus particles invade the cell nuclei of the epithelial cells, where they are uncoated and replicated. Initial replication produces non-occluded virus particles, but later the virus particles are produced with protein coats that remain infective when released from the dead insects. **Biology:** As with all insect baculoviruses, *Neodiprion sertifer* and *Neodiprion lecontei* NPV must be ingested to exert an effect. Following ingestion, the virus acts in an unusual way, in that it infects only the epithelial cells of the larval midgut. However, the infection is sufficient to lead to larval mortality. **Efficacy:** Sawflies are gregarious forest pests and this makes them particularly vulnerable to viral epizootics. The infection of a single larva will always lead to the death of an entire colony, which gives the advantage that low rates of use are sufficient for population control. **Key reference(s):** 1) R R Granados & B A Federici (eds). 1986. *The Biology of Baculoviruses, Vols 1 and 2*, CRC Press, Boca Raton, Florida, USA. 2) D J Leisy & J R Fuxa. 1996. Natural and engineered viral agents for insect control. In *Crop Protection Agents from Nature: Natural Products and Analogues*, L G Copping (ed.), Royal Society of Chemistry, Cambridge, UK. 3) B A Federici. 1999. Naturally occurring baculoviruses for insect pest control. In *Biopesticides Use and Delivery*, F R Hall & J J Menn (eds), pp. 301–20, Humana Press, Totowa, NJ, USA. 4) H D Burges & K A Jones. 1998. Formulation of bacteria, viruses and Protozoa to control insects. In *Formulation of Microbial Biopesticides, Beneficial Microorganisms, Nematodes and Seed Treatments*, H D Burges (ed.), pp. 33–127, Kluwer Academic Publishers, Dordrecht, The Netherlands.

COMMERCIALISATION: Formulation: Sold as a powder derived from ground, freeze-dried, infected larvae. 'Neochek-S' (derived from *N. sertifer*) is sold and used only under the supervision of the US-EPA and 'Leconteivirus' (derived from *N. lecontei*) is sold and used only under the supervision of the Canadian Forestry Service. A liquid formulation ('Sertistop') is available in Finland for use in Europe against *Neodiprion sertifer*. **Tradenames:** 'Leconteivirus' (Canadian Forestry Service), 'Neochek-S' (USDA Forest Service), 'Sertistop' (Verdera), 'Virox' (Oxford Virology).

APPLICATION: The recommended rate of application is 5×10^9 polyhedral inclusion bodies (PIB) or 50 virus-killed larvae per hectare. A single application will keep a forest free of sawflies for several years.

PRODUCT SPECIFICATIONS: Specifications: The virus is produced from the propagation of sawflies in the field. Selected, high-density populations are treated with the virus and, several days later, the diseased insects are collected and frozen. Before application, the larvae are freeze-dried and ground into a fine powder. A liquid formulation is made by separating the polyhedral inclusion bodies (PIB) from the diseased insects, purifying the polyhedra and suspending them in an aqueous solution. **Purity:** The products contain ground, freeze-dried, infected larvae. **Storage conditions:** Store in a cool, dry situation. Powder formulations may be frozen. Liquid formulation must not be stored frozen. Do not expose to direct sunlight or high temperatures. **Shelf-life:** The product retains its viability for up to 2 years, if stored according to the manufacturer's recommendations.

COMPATIBILITY: It is unusual to use the product in combination with other crop protection agents. Not compatible with strong oxidisers, acids, alkalis and chlorinated water.

MAMMALIAN TOXICITY: There are no reports of allergic or other adverse effects of the product on research workers or manufacturing or field staff. Considered to be a product of low mammalian toxicity.

ENVIRONMENTAL IMPACT AND NON-TARGET TOXICITY: It is not expected that any product containing these viruses will have any adverse effects on non-target organisms or on the environment.

1:86 *Nosema locustae*
Biological insecticide (microsporidium)

Protozoa: Microsporidium

NOMENCLATURE: Approved name: *Nosema locustae* Canning. **Development code:** SHA 117001.

SOURCE: Isolated from grasshoppers in the USA.

PRODUCTION: *Nosema locustae* is produced by *in vivo* rearing in grasshoppers.

TARGET PESTS: Grasshoppers.

TARGET CROPS: Pastures, rangeland and many other crops.

BIOLOGICAL ACTIVITY: Mode of action: *Nosema locustae* has to be ingested to be effective. Second and third instar, wingless nymphs are the most susceptible stages. The microsporidial spores germinate in the grasshopper's midgut and invade the insect's body, leading to lethal infections. The response of the grasshoppers to infection varies, with some dying shortly after infection, whilst others become weakened and sluggish. In this condition, the insects do not feed, but are often cannibalised by healthy grasshoppers which subsequently become infected. Egg production is reduced by 60 to 80% in the surviving adults and, in many cases, the *Nosema locustae* parasite is passed on to new generations through the eggs. Once established in the population, the microsporidium will usually survive for at least a year. **Biology:** *Nosema locustae* is a microsporidial pathogen that infects approximately 60 different species of grasshopper, as well as Mormon crickets (*Anabrus simplex* Haldeman). Death often follows shortly after infection. **Efficacy:** *Nosema locustae* is effective against many species of grasshopper and Mormon crickets. It will not control the southern lubber grasshopper or other cricket species. **Key reference(s):** 1) W M Brooks. 1988. Entomogenous protozoa. In *Handbook of Natural Pesticides Volume V, Microbial Insecticides, Part A, Entomogenous Protozoa and Fungi*, C M Ignoffo (ed.), pp. 1–149, CRC Press, Boca Raton, Florida, USA. 2) H D Burges & K A Jones. 1998. Formulation of bacteria, viruses and Protozoa to control insects. In *Formulation of Microbial Biopesticides, Beneficial Microorganisms, Nematodes and Seed Treatments*, H D Burges (ed.), pp. 33–127, Kluwer Academic Publishers, Dordrecht, The Netherlands.

COMMERCIALISATION: Formulation: Sold as baits with *Nosema locustae* spores sprayed onto the surface of bran and thickened with 0.25% hydroxymethyl cellulose as a sticker. **Tradenames:** 'Grasshopper Control Semaspore Bait' (Beneficial Insect Company), 'Milky Spore' and 'Milky Spore Dispersal Tubes' (Biocontrol Network), 'Nolo Bait' (Biocontrol Network) and (M&R Durango), 'NOLO Bait' and 'Semaspore Bait' (Rincon-Vitova).

APPLICATION: The bait is distributed throughout the crop to be protected, either by hand or with a spreader. Best control is achieved if the bait is applied in the early morning. When grasshopper populations reach more than eight per square metre, apply 1 kg product per hectare. Repeat after 4 to 6 weeks if the population remains high.

PRODUCT SPECIFICATIONS: Specifications: *Nosema locustae* is produced by *in vivo* rearing in grasshoppers. **Purity:** Product contains *Nosema locustae* spores attached to bran bait. Efficacy can be determined by bioassay on grasshopper nymphs. **Storage conditions:** Store in a sealed container at 0 to 6 °C. Do not expose to direct sunlight. **Shelf-life:** Spores remain infective for up to 18 months, if stored according to the manufacturer's recommendations.

COMPATIBILITY: It is unusual to apply *Nosema locustae* with other pest control agents.

MAMMALIAN TOXICITY: *Nosema locustae* is considered to be of very low toxicity to mammals. It does not irritate, replicate or accumulate in rats, guinea pigs or rabbits. **Acute oral LD$_{50}$:** rats >5000 mg/kg (4.49×10^9 spores/kg). **Toxicity class:** EPA (formulation) IV.

ENVIRONMENTAL IMPACT AND NON-TARGET TOXICITY: There is no evidence of adverse effects on non-target organisms or on the environment. **Fish toxicity:** *Nosema locustae* does not replicate or accumulate in rainbow trout or bluegill sunfish. **Effects on beneficial insects:** Honeybees and other beneficial organisms are not infected by *Nosema locustae*.

1:87 *Orgyia pseudotsugata* nucleopolyhedrovirus

Insecticidal baculovirus

Virus: Baculoviridae: Nucleopolyhedrovirus

NOMENCLATURE: Approved name: *Orgyia pseudotsugata* nucleopolyhedrovirus.
Other names: *Orgyia pseudotsugata* MNPV; *Orgyia pseudotsugata* NPV; OpMNPV; OpNPV.

SOURCE: Baculovirus that occurs widely in Nature, originally isolated from the Douglas fir tussock moth (*Orgyia pseudotsugata* (McDunnough)).

PRODUCTION: Manufactured by *in vivo* methods using tussock moth larvae. The virus is separated from the insect cadavers by centrifugation.

TARGET PESTS: Douglas fir tussock moth (*Orgyia pseudotsugata*).

TARGET CROPS: Forest trees (various), ornamental and non-commercial trees. In the USA, use is limited to wide-area, government-sponsored programmes.

BIOLOGICAL ACTIVITY: Mode of action: As with all insect baculoviruses, *Orgyia pseudotsugata* NPV must be ingested to exert an effect. Following ingestion, the protective protein matrix of the polyhedral occlusion bodies (OBs) of the virus is dissolved within the midgut and the virus particles enter the insect's haemolymph. These virus particles invade nearly all cell types in the larval body, where they multiply, leading to death. Shortly after the death of the larva, the integument ruptures, releasing very large numbers of OBs.
Efficacy: OpNPV acts relatively slowly, as it has to be ingested before it exerts any effect on the insect host. It is important to ensure good cover of the foliage to effect good control. Monitoring of adult insect laying patterns and application targeted at newly hatched eggs give better control. **Key reference(s):** 1) R R Granados & B A Federici (eds). 1986. *The Biology of Baculoviruses, Vols 1 and 2*, CRC Press, Boca Raton, Florida, USA. 2) D J Leisy & J R Fuxa. 1996. Natural and engineered viral agents for insect control. In *Crop Protection Agents from Nature: Natural Products and Analogues*, L G Copping (ed.), Royal Society of Chemistry, Cambridge, UK. 3) B A Federici. 1999. Naturally occurring baculoviruses for insect pest control. In *Biopesticides Use and Delivery*, F R Hall & J J Menn (eds), pp. 301–20, Humana Press, Totowa, NJ, USA. 4) H D Burges & K A Jones. 1998. Formulation of bacteria, viruses and Protozoa to control insects. In *Formulation of Microbial Biopesticides, Beneficial Microorganisms, Nematodes and Seed Treatments*, H D Burges (ed.), pp. 33–127, Kluwer Academic Publishers, Dordrecht, The Netherlands.

COMMERCIALISATION: Formulation: First registered in the USA in 1976 and reregistered in 1998. Sold through the USDA Forest Service as a liquid suspension. **Tradenames:** 'OpNPV' (USDA Forest Service), 'TM Biocontrol-1' (USDA Forest Service).

APPLICATION: Apply as a spray, ensuring good cover of the foliage. The virus will remain infective on the surface of leaves for 7 to 14 days.

PRODUCT SPECIFICATIONS: Specifications: Manufactured by *in vivo* methods using tussock moth larvae. The virus is separated from the insect cadavers by centrifugation.
Purity: Free from bacterial contaminants pathogenic to man. **Storage conditions:** Store in sealed containers under cool, dark conditions. May be stored frozen. **Shelf-life:** Stable for several weeks if kept cool, but has unlimited stability if kept in a freezer.

COMPATIBILITY: Can be used with other insecticides that do not repel target insect species. Incompatible with strong oxidisers, acids, bases and chlorinated water.

MAMMALIAN TOXICITY: Baculoviruses are specific to invertebrates, with there being no record of any vertebrate becoming infected. The virus does not infect or replicate in the cells of mammals and is inactivated at temperatures above 32°C. There is no evidence of acute or chronic toxicity, eye or skin irritation in mammals. No allergic reactions or other health problems have been observed in research or manufacturing staff or with users of the product.

ENVIRONMENTAL IMPACT AND NON-TARGET TOXICITY: OpNPV occurs widely in Nature and there is no evidence of the virus infecting vertebrates or plants. It has no adverse effects on fish, birds or beneficial organisms.

1:88 *Paecilomyces fumosoroseus* isolate Apopka 97
Biological insecticide/acaricide (fungus)

Fungus: anamorphic Trichocomaceae: previously classified as mitosporic Deuteromycetes: Moniliales

NOMENCLATURE: Approved name: *Paecilomyces fumosoroseus* (Wize) Brown & Smith isolate Apopka 97. **Development code:** PFR 97.

SOURCE: *Paecilomyces fumosoroseus* has been isolated from a range of infected insects around the world. The Apopka 97 isolate was isolated in 1986 from the mealybug, *Phenococcus solani* Ferris, on *Gynura* (a velvet plant) growing in a conservatory in Apopka, Florida. This isolate was licensed to Thermo Trilogy Corporation (formerly W R Grace and Co and now Certis), who developed the production and formulation. European development has been undertaken by Biobest. The isolate used in 'Priority' was isolated from a red spider mite (*Tetranychus urticae* Koch).

PRODUCTION: *Paecilomyces fumosoroseus* is produced by fermentation and is sold as spores in a water dispersible granule (WG) and as a water dispersible powder (DP) formulation.

TARGET PESTS: For control of whitefly (*Trialeurodes vaporariorum* (Westwood) and *Bemisia tabaci* (Gennadius)). Also shows some activity against aphids, thrips and spider mites. 'Priority' is recommended for use against spider mites, pink mites, thrips, aphids and whitefly.

TARGET CROPS: Recommended for use on ornamentals and food crops in glasshouses and outdoors.

BIOLOGICAL ACTIVITY: Mode of action: *Paecilomyces fumosoroseus* spores germinate on the body of the target pest and it exerts its insecticidal action through penetration of the cuticle and subsequent fungal growth within the haemolymph and other tissues of the infected insects. Sporulation from dead pests leads to infections of epidemic proportions. **Biology:** *Paecilomyces fumosoroseus* cannot grow at temperatures above 32 °C. It exerts its effects on whitefly larvae by enzymic and mechanical entry through the insect cuticle and subsequently invades the haemolymph. After host death, the fungus releases spores following penetration of the exoskeleton by the conidiophores. *Paecilomyces fumosoroseus* does not produce mycotoxins. **Key reference(s):** 1) K Bolkmans & G Sterk. 1995. PreFeRal WG (*Paecilomyces fumosoroseus* strain Apopka 97), a new microbial insecticide for the biological control of whiteflies in greenhouses, *Medische Faculteit Landbouwwetenschappen Rijksuniversiteit Gent*, **60**, 707–11. 2) G Sterk, K Bolkmans & J Eyals. 1996. A new microbial insecticide, *Paecilomyces fumosoroseus* strain Apopka 97, for the control of greenhouse whitefly, *Proc. Brighton Crop Protection Conf. – Pests & Diseases*, **2**, 461–6. 3) H D Burges. 1998. Formulation of mycoinsecticides. In *Formulation of Microbial Biopesticides, Beneficial Microorganisms, Nematodes and Seed Treatments*, H D Burges (ed.), pp. 131–85, Kluwer Academic Publishers, Dordrecht, The Netherlands.

COMMERCIALISATION: Formulation: 'PreFeRal' is sold as a water dispersible granule (WG) containing 1×10^9 colony forming units per gram. 'Priority' is sold as a water dispersible powder (DP) containing no less than 2×10^8 cfu per gram. 'PreFeRal' is listed in EU Annex 1. **Tradenames:** 'Pae-Sin' (Agrobionsa), 'PFR 97' (Certis), 'PreFeRal' (Certis) and (Biobest), 'Priority' (Stanes). **Patents:** US 5968808 (to USDA).

APPLICATION: High volume sprays, ensuring good coverage of the foliage, should be applied at a dose rate of 10 g product per litre (14 oz per 100 gallons).

PRODUCT SPECIFICATIONS: Specifications: *Paecilomyces fumosoroseus* is produced by fermentation and is sold as spores in a water dispersible granule (WG) and as a water dispersible powder (DP) formulation. **Purity:** The fermentation procedure is carried out under controlled conditions and only spores of *Paecilomyces fumosoroseus* are found within the product. The viability of the spores can be determined by plating out on agar, incubating in the laboratory for 48 hours and counting the colonies formed.

Storage conditions: The formulated product should be stored in dry, refrigerated conditions at 4°C. Do not freeze. **Shelf-life:** If stored under the correct conditions, the formulated product will remain biologically active for over 6 months.

COMPATIBILITY: Cannot be tank-mixed or applied with fungicides. *Paecilomyces fumosoroseus* does not infest beneficial insects or insect parasites and predators and so it can be used in combination with other whitefly control measures, such as *Encarsia formosa* Gahan and *Macrolophus caliginosus* Wagner. Incompatible with strong oxidisers, acids, bases and chlorinated water.

MAMMALIAN TOXICITY: *Paecilomyces fumosoroseus* shows no oral toxicity, pathogenicity or infectivity to any tested animals at 1×10^6 colony forming units (cfu)/animal. No hypersensitivity has been reported in any research or manufacturing workers or users. **Acute dermal LD$_{50}$:** rats $>1 \times 10^9$ cfu/animal. **Inhalation:** No toxicity, pathogenicity or infectivity at 1×10^6 conidia spores/animal. **Skin and eye:** Slight dermal irritancy, reversible within 72 hours, at 1×10^8 cfu/animal. Practically non-irritating to the eye at $>1 \times 10^7$ cfu/animal. Not a dermal sensitiser at 1×10^9 cfu/animal.

ENVIRONMENTAL IMPACT AND NON-TARGET TOXICITY: *Paecilomyces fumosoroseus* isolate Apopka 97 is harmless to non-target organisms, including predatory and beneficial insects, mammals and birds. It is a soil fungus that is widespread in Nature and is not expected to show any adverse environmental effects. **Effects on beneficial insects:** Non-toxic to beneficial insects. **Approved for use in organic farming:** Yes.

1:89 *Paecilomyces lilacinus* isolate P 251
Biological nematicide (fungus)

Fungus: anamorphic Trichocomaceae: previously classified as mitosporic Deuteromycetes: Moniliales

NOMENCLATURE: Approved name: *Paecilomyces lilacinus* (Thom.) Samson isolate P 251.

SOURCE: The fungus occurs as a widespread saprophytic soil fungus. In 1979, *Paecilomyces lilacinus* was identified as an effective parasite of *Meloidogyne incognita* (Kofoid and White) Chitwood and *Globodera pallida* (Stone) eggs on potatoes. Further study revealed that different isolates of *P. lilacinus* differed considerably in their nematophagous potential. The commercial isolate P 251 was isolated from the egg mass of a root-knot nematode in the Philippines and has been shown to be effective at controlling nematodes by the Australian Technology Innovation Corporation, who were the first to commercialise it.

PRODUCTION: *Paecilomyces lilacinus* is manufactured by solid state fermentation. The conidia are formulated as water dispersible granules.

TARGET PESTS: Effective against plant parasitic nematodes, including root-knot (*Meloidogyne* spp.), burrowing (*Radopholus similis* (Cobb) Thorne) and cyst (*Globodera* spp. and *Heterodera* spp.) nematodes.

TARGET CROPS: Recommended for a wide range of annual crops, including vegetables, potatoes, tomatoes and cotton, in addition to perennial crops such as bananas, coffee, cacao, citrus, grapes and pineapple and non-food crops such as ornamentals and tobacco.

BIOLOGICAL ACTIVITY: Mode of action: *Paecilomyces lilacinus* isolate P 251 has a wide pH tolerance and temperature range, growth being optimal between 26 and 30°C. It directly parasitises the eggs and other stages of development of common plant-infecting nematodes. Fungal spores germinate and the mycelium invades the nematode. Repeated application of the fungus modifies the microflora of the root zone and this leads to a beneficial plant response. **Biology:** *Paecilomyces lilacinus* cannot grow at temperatures above 32°C. It exerts its effects on nematode eggs by enzymic and mechanical entry through the wall and it grows within the egg. Subsequently, the fungus releases spores and newly laid eggs are parasitised. *Paecilomyces lilacinus* does not produce mycotoxins. There are unsubstantiated reports that the fungus can enter the growing roots of nematode-infected plants. **Efficacy:** The product is particularly effective against root-knot nematodes, cyst building nematodes and the burrowing nematode. *Paecilomyces lilacinus* does not affect beneficial entomopathogenic nematodes in any way. **Key reference(s):** 1) F Cabanillas, K R Barker & M E Daykin. 1988. Histology of the interactions of *Paecilomyces lilacinus* with *Meloidogyne incognita* on tomato, *J. Nematol.*, **20**, 362–5. 2) R J Holland, K L Williams & A Khan. 1999. Infection of *Meloidogyne javanica* by *Paecilomyces lilacinus*, *Nematology*, **1**, 131–9. 3) R G Davidé & R A Zorilla. 1983. Evaluation of a fungus, *Paecilomyces lilacinus* (Thom.) Samson, for the biological control of the potato cyst nematode *Globodera rostochiensis* Woll. as compared with some nematicides, *Philippine Agriculturalist*, **66**, 397–404. 4) R A Samson. 1974. *Paecilomyces* and some allied hyphomycetes. In *Studies in Mycology No. 6*, 119 pp., Centralbureau voor Schimmelcultures, Baarn, The Netherlands.

COMMERCIALISATION: Formulation: The product is sold as water dispersible granules (WG) containing 1×10^9 colony forming units (cfu) per gram. The 'Bioact' formulation contains 6.25 g spores of *Paecilomyces lilacinus* and 93.75 g glucose as a carrier, giving 1×10^{10} conidia of *Paecilomyces lilacinus* per gram of product. **Tradenames:** 'BioAct WG' (Australian Technology Innovation Corporation) and (Prophyta), 'MeloConTM WG' (Prophyta), 'NemaChek' (Australian Technology Innovation Corporation) and (Prophyta), 'Paecil'

(Prophyta), 'Paecio' (Agri Life), 'PL Plus' (Biological Control Products). **Patents:** Patent on the product: European Patent Registration No. 0593428 B1, (PCT/AU90/00325); Spanish Patent No. ES2153817T3; German Patent No. DE69033697T2; French Patent No. 2798037 to 2798249, Italian Patent No. 90911830.9; Canadian Patent No. 2,059,642. Patent on the fermentation technology needed for the product: PCT-Number DE99/01271 (International Publishing number: WO99/57239), Australian Patent No. 749402; New Zealand Patent No. 507738; European Patent No. 10093708 (all to Prophyta).

APPLICATION: Applied as a seedling or soil drench or in drip irrigation to give between 2 and 5×10^9 viable spores to each plant, which are then distributed around the plant. Also applied as a transplanting water treatment. Overall applications at 5 kg per hectare (4 lb per acre) can be made to soil before planting, to the soil of seedlings before transplanting and as a post-plant soil drench. Repeat applications are recommended to prevent the build-up of surviving nematodes.

PRODUCT SPECIFICATIONS: Specifications: *Paecilomyces lilacinus* is manufactured by solid state fermentation. The conidia are formulated as water dispersible granules.
Purity: The fermentation procedure is carried out under controlled conditions and only spores of *Paecilomyces lilacinus* are found within the product. The viability of the spores can be determined by plating out on agar, incubating in the laboratory for 48 hours and counting the colonies formed. **Storage conditions:** 'PL Plus' should be stored between 4 and 8°C. 'BioAct WG' can be stored at −20°C. **Shelf-life:** If stored between 4 and 8°C, 'PL Plus' will remain biologically active for over 12 months. If stored at −20°C, 'BioAct WG' spores will retain their viability for over 12 months.

COMPATIBILITY: Can be tank-mixed or applied with other compounds, but not fungicides or herbicides such as linuron or metribuzin. *Paecilomyces lilacinus* does not infect beneficial insects or insect parasites and predators. There are reports of high application rates causing mortality in some species of termites. When used as part of an Integrated Crop Management (ICM) programme, it is unusual to apply *Paecilomyces lilacinus* with chemical pesticides. Incompatible with strong oxidisers, acids, bases and chlorinated water.

MAMMALIAN TOXICITY: *Paecilomyces lilacinus* isolate P 251 shows no oral toxicity, pathogenicity or infectivity to any tested animals at 1×10^6 colony forming units (cfu)/animal. The fungus does not grow well at temperatures above 35°C. Related isolates have the potential to cause respiratory sensitisation amongst asthma sufferers. **Acute oral LD$_{50}$:** rat: >2000 mg/kg (>4×10^9 colony forming units/kg). **Acute dermal LD$_{50}$:** rat: >2000 mg/kg (>4×10^9 colony forming units/kg). **Inhalation: LC$_{50}$:** rats >8×10^7 colony forming units per animal. **Skin and eye:** No dermal toxicity, pathogenicity or infectivity at 1×10^9 cfu/animal. Slight dermal irritancy, reversible within 72 hours, at 1×10^8 cfu/animal. Not a dermal sensitiser at 1×10^9 cfu/animal. Practically non-irritating to the eye at >1×10^7 cfu/animal.

ENVIRONMENTAL IMPACT AND NON-TARGET TOXICITY: *Paecilomyces lilacinus* isolate P 251 is harmless to non-target organisms, including predatory and beneficial insects, mammals, birds and beneficial soil nematodes. It is a soil fungus that is widespread in Nature and is not expected to show any adverse environmental effects. **Fish toxicity:** LC_{50} (96 h) for rainbow trout >100 mg/litre. **Other aquatic toxicity:** LC_{50} (48 h) for *Daphnia magna* >100 mg/litre. E_rC_{50} (72h) for *Scenedesmus subspicatus* ≥5.13 × 10^8 cfu/litre (nominal). **Effects on beneficial insects:** *Paecilomyces lilacinus* does not affect beneficial organisms in any way. Earthworms: 2 week NOEC 7.5 × 10^9 cfu/litre soil (no infectivity). **Behaviour in soil:** The fungus is a ubiquitous, common, saprophytic inhabitant of the soil. **Approved for use in organic farming:** Yes.

1:90 *Paenibacillus popilliae*
Biological insecticide (bacterium)

Bacterium: Bacillales: Bacillaceae

NOMENCLATURE: Approved name: *Paenibacillus popilliae* (Dutky) Pettersson. **Other names:** Formerly known as *Bacillus popilliae* Newman (*Int. J. Syst. Bacteriol.*, 1999. **49**, 531–40). **Common name:** Milky spore disease.

SOURCE: Discovered by USDA worker, Samson Robert Dutky, in the Washington DC area. Isolated from a Japanese beetle, *Popillia japonica* Newman.

PRODUCTION: *Paenibacillus popilliae* is produced commercially by *in vivo* culture in the larvae and adults of *Popillia japonica*.

TARGET PESTS: Japanese beetles (*Popillia japonica*).

TARGET CROPS: Recommended for use in a wide range of crops, including vegetable gardens, lawns and golf courses. *Paenibacillus popilliae* is not recommended for use on pastures.

BIOLOGICAL ACTIVITY: Mode of action: *Paenibacillus popilliae* is a pathogen that is specific to the Japanese beetle and a few very closely related species. It controls the insect larvae by parasitism rather than through the production of insect-specific endotoxins. Invaded larvae die some time after infection, releasing bacterial spores into the soil, where they may be eaten by newly emerged and older larvae which are subsequently invaded. **Biology:** Effects on large insect populations may be slow to develop, but, once present in the soil, *Paenibacillus* *popilliae* may remain infective for as long as twenty years, although pest recurrence may occur in a shorter time. **Efficacy:** Specific for control of Japanese beetles. Once the bacteria and spores become established in a geographic area, they greatly decrease the numbers of larvae and adult Japanese beetles, thereby reducing plant damage. **Key reference(s):** H D Burges & K A Jones. 1998. Formulation of bacteria, viruses and Protozoa to control insects. In *Formulation of Microbial Biopesticides, Beneficial Microorganisms, Nematodes and Seed Treatments*, H D Burges (ed.), pp. 33–127, Kluwer Academic Publishers, Dordrecht, The Netherlands.

COMMERCIALISATION: Formulation: Sold as a dry powder formulation containing bacterial spores. **Tradenames:** 'Milky Spore Disease' (Arbico) and (Rincon-Vitova), 'Milky Spore Powder' (Reuter).

APPLICATION: Applied to the soil as discrete spots about a metre apart to form a grid. The recommended application rate is about 10 kg product per hectare. 'Milky Spore Powder' is sold as a powder, to be spread on soil between spring and autumn (that is, when the ground is not frozen).

PRODUCT SPECIFICATIONS: Specifications: *Paenibacillus popilliae* is produced commercially by *in vivo* culture in the larvae and adults of *Popillia japonica*. **Purity:** Guaranteed to be free from other bacteria. **Storage conditions:** Store in dry conditions at ambient temperature. Do not expose to high temperatures. **Shelf-life:** If kept dry in a sealed container, the product will remain viable for over 2 years.

COMPATIBILITY: Incompatible with chlorinated water, strong acids and bases.

MAMMALIAN TOXICITY: No allergic reactions or other adverse health problems have been shown by research workers, manufacturing staff or users. A variety of tests have indicated that spores of *Paenibacillus popilliae* are not harmful to humans. The spores have been shown to develop into active bacteria only at temperatures lower than human body temperature and only in a few species of beetle.

ENVIRONMENTAL IMPACT AND NON-TARGET TOXICITY: In approved field use, *Paenibacillus popilliae* has shown no adverse effects on non-target organisms or on the environment. It will remain viable in the soil for extended periods (twenty to thirty years have been claimed), but is specific for the Japanese beetle and closely related species and, thereby, does not pose any threat to non-target organisms. The spores do not infect other insects, mammals, birds, earthworms or plants. Also, the spores have been used since the 1940s against Japanese beetle larvae, with no adverse environmental effects reported.
Approved for use in organic farming: Yes.

1:91 *Pantoea agglomerans* isolate C9-1
Biological bactericide (bacterium)

Bacterium: Enterobacteriales: Enterobacteriaceae

NOMENCLATURE: Approved name: *Pantoea agglomerans* (Ewing & Fife) Gavini *et al.* isolate C9-1. **Other names:** Formerly known as *Enterobacter agglomerans* (Beijerinck) Ewing & Fife.

SOURCE: Originally isolated in 1994 by researchers at the US Department of Agriculture, Agriculture Research Collection, from apple stem tissue. This naturally occurring bacterium has streptomycin and rifampicin resistance that is not derived through genetic engineering. *Pantoea agglomerans* is also found in soil.

PRODUCTION: Produced by fermentation.

TARGET PESTS: Fire blight (*Erwinia amylovora* Winsl.).

TARGET CROPS: Apples and pears (pome fruit).

BIOLOGICAL ACTIVITY: Mode of action: The C9-1 isolate of *Pantoea agglomerans* is antagonistic to foliar bacteria and fungi through the production of siderophores and antibiotics. **Key reference(s):** D R Fravel, W J Connick Jr & J A Lewis. 1998. Formulation of microorganisms to control plant diseases. In *Formulation of Microbial Biopesticides, Beneficial Microorganisms, Nematodes and Seed Treatments*, H D Burges (ed.), pp. 187–202, Kluwer Academic Publishers, Dordrecht, The Netherlands.

COMMERCIALISATION: Formulation: Under development for use in orchards for the control of fire blight. **Tradenames:** 'BlightBan C9-1' (Nufarm).

APPLICATION: 'BlightBan C9-1' is used to control fire blight in apples and pears through air blast spray application. It is applied at 15 to 20% bloom followed by a second application at the first petal fall or full bloom and a third application at rat-tail bloom for pear or post petal fall for apples.

PRODUCT SPECIFICATIONS: Specifications: Produced by fermentation. **Purity:** Contains no parasitic, phytotoxic or allergenic contaminants. **Storage conditions:** Store in a cool, dry situation. Do not expose to sunlight. **Shelf-life:** If stored according to the manufacturer's instructions, the product should remain viable for 12 months.

COMPATIBILITY: Incompatible with broad-spectrum compounds such as copper-based fungicides and with strong oxidisers, acids, bases and chlorinated water. Can be used in combination with streptomycin and rifampicin.

MAMMALIAN TOXICITY: *Pantoea agglomerans* has shown no allergic or other toxic effects on researchers, formulators, manufacturers or users. The bacterium is not viable at temperatures above 37°C.

ENVIRONMENTAL IMPACT AND NON-TARGET TOXICITY: *Pantoea agglomerans* occurs naturally and is not expected to have any adverse effects on non-target organisms or on the environment. **Approved for use in organic farming:** Yes.

1:92 *Pantoea agglomerans* isolate E325
Biological bactericide (bacterium)

Bacterium: Enterobacteriales: Enterobacteriaceae

NOMENCLATURE: Approved name: *Pantoea agglomerans* (Ewing & Fife) Gavini *et al.* isolate E325. **Other names:** Originally named *Erwinia herbicola* (Lohnis) Dye.

SOURCE: Isolated in 1994 by researchers at the US Department of Agriculture, Agriculture Research Collection. *Pantoea agglomerans* is also found in soil.

PRODUCTION: Produced by fermentation.

TARGET PESTS: Fire blight (*Erwinia amylovora* Winsl.).

TARGET CROPS: Apples and pears (pome fruit).

BIOLOGICAL ACTIVITY: Mode of action: The E325 isolate of *Pantoea agglomerans* is antagonistic to foliar bacteria and fungi through the production of siderophores and antibiotics. **Key reference(s):** D R Fravel, W J Connick Jr & J A Lewis. 1998. Formulation of microorganisms to control plant diseases. In *Formulation of Microbial Biopesticides, Beneficial Microorganisms, Nematodes and Seed Treatments*, H D Burges (ed.), pp. 187–202, Kluwer Academic Publishers, Dordrecht, The Netherlands.

COMMERCIALISATION: Formulation: Under development for use in orchards for the control of fire blight. **Tradenames:** 'Bloomtime Biological FD Biopesticide' (Northwest Agriculture).

APPLICATION: 'Bloomtime Biological FD Biopesticide' is used to control fire blight in apples and pears through air blast spray application. This microbial pesticide is applied at 15 to 20% bloom followed by a second application at the first petal fall or full bloom.

PRODUCT SPECIFICATIONS: Specifications: Produced by fermentation. **Purity:** Contains no parasitic, phytotoxic or allergenic contaminants. **Storage conditions:** Store in a cool, dry situation. Do not expose to sunlight. **Shelf-life:** If stored according to the manufacturer's instructions, the product should remain viable for 12 months.

COMPATIBILITY: Incompatible with broad-spectrum compounds such as copper-based fungicides and with strong oxidisers, acids, bases and chlorinated water.

MAMMALIAN TOXICITY: *Pantoea agglomerans* has shown no allergic or other toxic effects on researchers, formulators, manufacturers or users. The bacterium is not viable at temperatures above 37°C. *Pantoea agglomerans* isolate E325 has an EPA toxicity category IV.

ENVIRONMENTAL IMPACT AND NON-TARGET TOXICITY: *Pantoea agglomerans* occurs naturally and is not expected to have any adverse effects on non-target organisms or on the environment. **Approved for use in organic farming:** Yes.

1:93 *Pasteuria penetrans*
Biological nematicide (bacterium)

Bacterium: Bacillales: Alicyclobacillaceae

NOMENCLATURE: Approved name: *Pasteuria penetrans* (Thorne) Sayre and Starr.
Other names: Formerly known as *Bacillus penetrans* (Thorne) Mankau.

SOURCE: *Pasteuria penetrans* is a naturally occurring soil microbe that is a parasite of species of the nematode genus *Meloidogyne*. The isolate of interest as a biological control agent was isolated from *Meloidogyne arenaria* (Neal) Chitwood.

PRODUCTION: At present, *Pasteuria penetrans* can only be grown in the presence of plant parasitic nematodes, but trials are under way on an *in vitro* fermentation method.

TARGET PESTS: Root-knot nematodes such as *Meloidogyne* spp.

TARGET CROPS: A wide range of arable and horticultural crops, including soybeans, peanuts, tomato, cucumber, squash, melon, sweet potato and figs, in addition to protected crops.

BIOLOGICAL ACTIVITY: Mode of action: The endospores of *Pasteuria penetrans* that are formed inside the parasitised nematodes are released into soil and they attach readily to the cuticle of host nematodes on contact in soil or water. The attached endospores produce a germ tube that penetrates the nematode's cuticle. Inside the nematode pseudoceol, the germ tube develops into a vegetative, spherical colony consisting of a dichotomously branched, septate mycelium. Nematodes parasitised by *Pasteuria* spp. remain alive and reach the adult stage, but fecundity is greatly reduced or completely blocked. Two million spores have been shown to be formed within a single root-knot nematode female. **Efficacy:** *Pasteuria penetrans* and related forms are considered as potential biological control agents of phytoparasitic nematodes. Host adhesion is one of the most critical steps in the life-cycle of *Pasteuria*. The adhesion process is selective and is mediated by the external layer of parasporal fibres. These layers surround the endospore core wall and remain attached to the host cuticle until the germination process is completed. The endospore represents the parasite durable stage, as well as the infective propagule responsible for transmission. **Key reference(s):** 1) R Ahmad & S R Gowen. 1991. Studies on the infection of *Meloidogyne* spp. with isolates of *Pasteuria penetrans*, *Nematologia Mediterranea*, **19**, 233. 2) T E Hewlett, A C Schuerger & D W Dickson. 1997. Biological control of *Meloidogyne arenaria* at EPCOT, Disney World, *J. Nematol.*, **29**, 583. 3) R M Sayre & M P Starr. 1989. Genus *Pasteuria* Metchnikoff, 1888. In *Bergey's Manual of Systematic Bacteriology*, S T Williams, M E Sharpe & J G Holt (eds), pp. 2601–15, Baltimore, Maryland, USA. 4) S M Brown & D Nordmeyer. 1985. Synergistic reduction in root galling by *Meloidogyne javanica* with *Pasteuria penetrans* and nematicides, *Revue de Nematologie*, **8**, 285–6.

COMMERCIALISATION: Formulation: Sold as a wettable powder (WP) and as a granule containing 1×10^9 spores per gram. **Tradenames:** 'Pasteuria Wettable Powder' (Nematech).

APPLICATION: Spray or sprinkle a dilute suspension of spores around the crop at planting or in the spring at a rate of 1 to 3 kg product per 1 000 m^2 for annuals and 2.5 to 5 kg product per 1000 m^2 for perennial crops. The crops should be tilled into the soil after harvest to help raise the level of the bacterium in the soil for subsequent seasons.

PRODUCT SPECIFICATIONS: Specifications: At present, *Pasteuria penetrans* can only be grown in the presence of plant parasitic nematodes, but trials are under way on an *in vitro* fermentation method. **Purity:** Contains no organisms that might be harmful to the crop or to the applicator. **Shelf-life:** May be stored in a cool, dry location for several months.

COMPATIBILITY: *Pasteuria penetrans* is not compatible with chemical nematicides such as chloropicrin or methyl bromide, but is compatible with non-volatile nematicides such as carbofuran.

MAMMALIAN TOXICITY: *Pasteuria penetrans* is not known to be pathogenic, infective or toxic to humans. The specificity of the organism as a parasite only on nematodes of the genus *Meloidogyne* makes it extremely unlikely to infect any other unrelated organism. The inability to grow the bacterium in the absence of the host nematode makes standard toxicological tests impossible to perform.

ENVIRONMENTAL IMPACT AND NON-TARGET TOXICITY: *Pasteuria penetrans* is a common inhabitant of soils in areas infested with nematodes and, as such, is not expected to have any adverse effects on non-target organisms or on the environment. **Approved for use in organic farming:** Yes.

1:94 *Peniophora gigantea*
Biological fungicide (fungus)

Fungus: Basidiomycetes: Rusulales

NOMENCLATURE: Approved name: *Peniophora gigantea* (Fr.) Jul. **Other names:** *Peniophora gigantea* has been reclassified. See *Phlebiopsis gigantea* (entry 1:96).

1:95 *Phasmarhabditis hermaphrodita*
Mollusc parasitic nematode

Nematode: Rhabdita: Phasmarhabditidae

NOMENCLATURE: Approved name: *Phasmarhabditis hermaphrodita* (Schneider).
Other names: Mollusc parasitic nematode.

SOURCE: Originally found in central Europe, but now found widely in soil, although it is less common in northern countries.

PRODUCTION: Produced by fermentation and sold as third stage juveniles.

TARGET PESTS: Very effective against slugs.

TARGET CROPS: Vegetables and ornamentals in glasshouses and outdoors.

BIOLOGICAL ACTIVITY: Biology: *Phasmarhabditis hermaphrodita* third stage juveniles invade slugs through the dorsal pore (the raised area on the slug's back). Once inside the host, the nematodes release the bacterium *Moraxella osloensis* (Bovre & Henriksen) Bovre which produces toxins that prevent the slugs from feeding within 72 hours. The nematodes feed on the immobilised and dying slugs and the juvenile nematodes develop into adults that breed, producing eggs that hatch into juvenile nematodes. The mantles of the infected slugs swell during this process, with the slugs dying underground after 7 to 10 days. The nematodes may continue to reproduce in the body of the slug, or infective third stage juveniles may pass into the soil to seek more hosts, depending upon the availability of food. **Efficacy:** Very effective predator of slugs, with rates of 3×10^5 nematodes per square metre reducing slug damage to less than 1% within 15 days. **Key reference(s):** 1) R Gaugler & H K Kaya. 1990. Entomopathogenic nematodes. In *Biological Control*, 365 pp., CRC Press, Boca Raton,

Florida, USA. 2) R Georgis & H K Kaya. 1998. Formulation of entomopathogenic nematodes. In *Formulation of Microbial Biopesticides*, *Beneficial Microorganisms*, *Nematodes and Seed Treatments*, H D Burges (ed.), pp. 289–308, Kluwer Academic Publishers, Dordrecht, The Netherlands.

COMMERCIALISATION: Formulation: Sold as infective third stage juveniles in a moist, inert carrier. **Tradenames:** 'Biolug' (Andermatt), 'Nemaslug' (Becker Underwood), (Brinkman), (Bog Madsen), (BCP), (Chase Organics), (Green Gardener), (Scarletts Plant Care) and (Intrachem), 'Phasmarhabditis-System' (Biobest).

APPLICATION: For crops that are attacked during germination or immediately after emergence or planting out, apply 4 days prior to emergence or transplanting. Applications to crops that are susceptible later in the season, such as potatoes, should be made 6 to 7 weeks before harvest or when the crop is most sensitive to slug attack. Apply to moist soil or compost at a rate of 3×10^5 juveniles per square metre, as a drench, when the soil temperature is between 5 and 25 °C. A single application is effective for up to 6 weeks. For plants that require prolonged protection, a second treatment may be necessary. Apply within 4 hours of diluting the formulation and stir to prevent sedimentation. For best activity, apply in the evening. Do not use close to ponds, as this may reduce the beneficial water snail population.

PRODUCT SPECIFICATIONS: Specifications: Produced by fermentation and sold as third stage juveniles. **Purity:** Formulations contain third stage infective nematode juveniles in an inert carrier. Efficacy can be measured by bioassay on host slugs and numbers confirmed by counting under a microscope. **Storage conditions:** Store for no more than 2 days in a cool, dark place. If kept for longer than 2 days, the product must be stored in a refrigerator at temperatures around 5 °C. Do not freeze and avoid direct sunlight and temperatures above 35 °C. **Shelf-life:** The juvenile nematodes remain infective for relatively short periods after delivery.

COMPATIBILITY: Compatible with a wide range of conventional agrochemicals, but should not be used with benzimidazole-based fungicides or soil-applied insecticides. Compatible with all IPM or biological crop protection strategies. Incompatible with strong oxidisers, acids and bases.

MAMMALIAN TOXICITY: *Phasmarhabditis hermaphrodita* and its associated bacterium have not shown any allergic or other adverse reactions in researchers, production staff or users.

ENVIRONMENTAL IMPACT AND NON-TARGET TOXICITY: *Phasmarhabditis hermaphrodita* and its associated bacterium would not be expected to have any effects on non-target organisms nor any adverse environmental effects.

Approved for use in organic farming: Yes.

1:96 *Phlebiopsis gigantea*
Biological fungicide (fungus)

Fungus: Basidiomycetes: Rusulales

NOMENCLATURE: Approved name: *Phlebiopsis gigantea* (Fr.) Massee.
Other names: Formerly known as *Phlebia gigantea* (Fr.) Donk and *Peniophora gigantea* (Fr.) Jul.

SOURCE: The isolate used in the product 'Rotstop' was originally isolated in 1987 by the Finnish Forest Research Institute from a Norway spruce log left in the forest. It was used as a stump treatment agent on pine and spruce in 1988 and was re-isolated in 1989 from a Norway spruce stump. In 1991, it was formulated into the product by Kemira Oy (currently produced by Verdera Oy).

PRODUCTION: Produced by fermentation and formulated as fungal spores.

TARGET PESTS: Developed for the control of root and butt rot (*Heterobasidion annosum* (Fr.) Bref.; also known as *Fomes annosus* (Fr.) Cooke, including all three intersterility groups P, S and F).

TARGET CROPS: Stumps of pine and spruce tree species.

BIOLOGICAL ACTIVITY: Mode of action: *Peniophora gigantea* exerts its effect by competing for the entry sites of the pathogen on the cut stumps of pine and spruce trees. **Biology:** The fungus invades the stump surface of freshly cut spruce and pine trees and prevents the subsequent invasion of the stump surface by the pathogenic fungus, *Heterobasidion annosum*, the causal organism of root and butt rot in conifer tree species. **Efficacy:** Very effective if applied at tree harvest, when the temperature is above 5°C.
Key reference(s): 1) K Korhonen *et al.* 1994. Control of *Heterobasidion annosum* by stump treatment with Rotstop, a new commercial formulation of *Phlebiopsis gigantea*. In *Proc. of the 8th International Conference on Root and Butt Rots*, Wik, Sweden and Haikko, Finland, August 9th–16th, 1993, M Johanson & J Stenlid (eds), pp. 675–85, IUFRO, Uppsala, Sweden.
2) D R Fravel, W J Connick Jr & J A Lewis. 1998. Formulation of microorganisms to control plant diseases. In *Formulation of Microbial Biopesticides, Beneficial Microorganisms, Nematodes and Seed Treatments*, H D Burges (ed.), pp. 187–202, Kluwer Academic Publishers, Dordrecht, The Netherlands. 3) O Holdenrieder & B J W Greig. 1998. Biological methods of control. In: Heterobasidion annosum, *Biology, Ecology, Impact and Control*, Woodward *et al.* (eds), pp. 235–58, CAB International, Wallingford, UK. 4) J E Pratt, M Niemi & Z H Sierota. 2000. Comparison of three products based on *Phlebiopsis gigantea* for the control of *Heterobasidion annosum* in Europe. *Biocontrol Science and Technology*, **10**(4), 469–79.

COMMERCIALISATION: Formulation: Formulated as a wettable powder (WP) containing 1 × 10^6 to 1 × 10^7 colony forming units (cfu) per gram. **Tradenames:** 'Rotstop' (Verdera).

APPLICATION: Recommended application time is when *Heterobasidion annosum* is capable of spreading or growing on stumps (during the vegetative period when the temperature is between 5 and 40°C). The product is mixed with water and the suspension is sprayed on the stump surface with a spraying device on the harvester head at tree felling or manually within three hours after felling.

PRODUCT SPECIFICATIONS: Specifications: Produced by fermentation and formulated as fungal spores. **Purity:** The viability and purity of the product is checked by plate counting. An indication of the competitive efficacy of the fungus can be obtained by growing the fungus in the presence of *Heterobasidion annosum*. The efficacy of the product can also be determined by treating the upper surfaces of freshly cut pine or spruce stem pieces and exposing these to spores of *Heterobasidion annosum*. **Storage conditions:** 'Rotstop' can be stored for 1 week at room temperature. Store unopened packages at temperatures below 8°C. Opened packages and 'Rotstop' liquid suspensions must be used within one day. **Shelf-life:** One week at room temperature. Up to 12 months in unopened container at <8°C.

COMPATIBILITY: 'Rotstop' is not compatible with chemical fungicides. Incompatible with strong oxidisers, acids and bases. Certain anti-freeze agents can be used in the 'Rotstop' suspension during periods of night frost.

MAMMALIAN TOXICITY: There have been no reports of allergic or other adverse toxicological effects from research workers, manufacturing staff or users. No acute oral, acute inhalation, acute dermal or intraperitoneal pathogenicity, infectivity and toxicity was observed in animal tests. No skin irritation was observed in animal tests. The product may cause some eye irritation, and the silica carrier may irritate respiratory organs if the exposure limits are exceeded. No skin sensitisation was observed in animal tests. Inhalation of the product and skin contact should be avoided by use of standard protective equipment.

ENVIRONMENTAL IMPACT AND NON-TARGET TOXICITY: *Phlebiopsis gigantea* is a natural component of the microflora of coniferous forests and, as such, is not expected to have any effect on non-target organisms nor to show any adverse environmental properties. It is possible that the long-term use of 'Rotstop' will reduce the occurrence of the fungus in treated areas and lead to an enhanced suppression of *Heterobasidion annosum*.

1:97 *Phomopsis amaranthicola*
Biological herbicide (fungus)

Fungus: Ascomycetes: Diaporthales

NOMENCLATURE: Approved name: *Phomopsis amaranthicola*.

SOURCE: Isolated from a pigweed (*Amaranthus retroflexus* L.) plant in Florida, USA.

PRODUCTION: Produced by fermentation.

TARGET PESTS: Most species of *Amaranthus*.

TARGET CROPS: A wide range of crops including salad vegetables and many field crops.

BIOLOGICAL ACTIVITY: Mode of action: Pathogenesis. **Biology:** The highest level of control was obtained when *Phomopsis amaranthicola* was applied at 6×10^7 conidia per ml. Final mortality of all species, except *A. hybridus*, reached 100% in inoculated plots 25 days earlier than in non-inoculated control plots. Conidial suspensions were more effective in controlling the species than were mycelial suspensions. The pathogen was able to spread to non-inoculated control plots. *Phomopsis amaranthicola* causes foliar lesions and leaf abscission. **Efficacy:** The weeds are better controlled when small and it has been shown that crops such as peppers are able to outcompete treated weeds very effectively.
Key reference(s): E N Rosskopf, R Charudattan, J T Devalerio & W Stall. 2000. Field evaluation of *Phomopsis amaranthicola*, a biological control agent of *Amaranthus* spp., *Plant Disease*, **84**(11), 1225–30.

COMMERCIALISATION: Formulation: Not sold commercially.

APPLICATION: Apply at 6×10^7 conidia per ml.

PRODUCT SPECIFICATIONS: Specifications: Produced by fermentation.

MAMMALIAN TOXICITY: Since its discovery, there have been no reports of adverse effects, sensitivity or reactions of any type related to use or handling of *Phomopsis amaranthicola*. No pathogenic or infective effects were observed in any study.

ENVIRONMENTAL IMPACT AND NON-TARGET TOXICITY: *Phomopsis amaranthicola* can infect crops within the family Amaranthaceae. It is not expected that it will have any adverse effects on non-target organisms or on the environment.
Approved for use in organic farming: Yes.

1:98 *Phragmidium violaceum*
Biological herbicide (fungus)

Fungus: Basidiomycota: Urediniomycetes: Uredinales

NOMENCLATURE: Approved name: *Phragmidium violaceum* (Schultz) G. Winter. **Other names:** Violet bramble rust. Formerly known as *Puccinia violacea* Schultz.

SOURCE: Isolated from infected blackberry plants in France by the CSIRO European Laboratory in Montpellier, France.

PRODUCTION: Usually bred on blackberry plants (*Rubus fruticosus* L.).

TARGET PESTS: Specific for blackberry plants (*Rubus fruticosus*).

TARGET CROPS: Rangeland, pasture and amenity land.

BIOLOGICAL ACTIVITY: Mode of action: *Phragmidium violaceum* is an obligate pathogen of blackberry plants. **Biology:** The blackberry rust is most effective during late summer and autumn in cool, moist environments. **Efficacy:** *Phragmidium violaceum* is a very effective pathogen of blackberry, causing rapid defoliation and eventual death.
Key reference(s): 1) E B Oehrens. 1977. Biological control of blackberry through the introduction of the rust, *Phragmidium violaceum* in Chile. *FAO Plant Prot. Bull.*, **23**, 26–8. 2) L Lawrence. 2004. Better biological control of weedy blackberry, *Outlooks on Pest Management*, **15**(3), 138–9.

COMMERCIALISATION: Formulation: *Phragmidium violaceum* is not sold commercially.

APPLICATION: Mixtures of the 8 different isolates of the fungus are introduced.

PRODUCT SPECIFICATIONS: Specifications: Usually bred on blackberry plants (*Rubus fruticosus*).

COMPATIBILITY: It is unusual to use *Phragmidium violaceum* with other crop protection agents. Incompatible with strong oxidisers, acids, bases and chlorinated water.

MAMMALIAN TOXICITY: No skin or eye irritation has been observed. There is no evidence of acute or chronic toxicity, infectivity or hypersensitivity to mammals. No allergic responses or health problems have been observed by research workers, manufacturing staff or users. *Phragmidium violaceum* is not toxic by inhalation.

ENVIRONMENTAL IMPACT AND NON-TARGET TOXICITY: The new isolates of *Phragmidium violaceum* are not a threat to commercial blackberry cultivars or Australian native *Rubus* species. Thorough testing showed that they attack only weedy blackberry and, after the required consultation process, Biosecurity Australia cleared these isolates for release on blackberry. It is not expected that the fungus will have any adverse effects on non-target organisms or on the environment.

1:99 *Phytophthora palmivora*
Biological herbicide (fungus)

Chromista: Oomycetes: Pythiales

NOMENCLATURE: Approved name: *Phytophthora palmivora* (Butl.) Butl.

SOURCE: Naturally occurring soil pathogenic fungus (Oomycetes have recently been reclassified and have been placed in the Chromista Kingdom rather than fungal). Isolated from strangler vine (*Morrenia odorata* (H. & A.) Lindl.) in a Florida citrus grove. Introduced by Abbott, acquired by Valent BioSciences and licensed to Encore Technologies.

PRODUCTION: Manufactured by fermentation.

TARGET PESTS: Strangler vine or milkweed vine (*Morrenia odorata*).

TARGET CROPS: Citrus and other perennial crops.

BIOLOGICAL ACTIVITY: Biology: *Phytophthora palmivora* invades the strangler vine via the roots. It is specific to the weed and may take 6 to 10 weeks to kill it. After application, populations of *Morrenia odorata* will continue to fall, but complete control in an orchard may not be achieved for up to a year after treatment. **Key reference(s):** 1) M P Greaves. 1996. Microbial herbicides – factors in development. In *Crop Protection Agents from Nature: Natural Products and Analogues*, L G Copping (ed.), pp 444–67, Royal Society of Chemistry, Cambridge, UK. 2) W H Ridings, D J Mitchell, C L Schoulties & N E El-Gholl. 1976. Biological control of milkweed vine in Florida citrus groves with a pathotype of *Phytophthora citrophthora*. In *Proc. IV International Symposium of Biological Control of Weeds*. 3) M P Greaves, P J Holloway & B A Auld. 1998. Formulation of microbial herbicides. In *Formulation of Microbial Biopesticides, Beneficial Microorganisms, Nematodes and Seed Treatments*, H D Burges (ed.), pp. 235–54, Kluwer Academic Publishers, Dordrecht, The Netherlands.

COMMERCIALISATION: Formulation: Sold as a liquid suspension containing 6.7×10^5 living chlamydospores per ml. **Tradenames:** 'DeVine' (Encore Technologies).

APPLICATION: Apply to soil between May and September after the strangler vine has germinated. Ensure that the soil is wet at the time of application and it may be necessary to irrigate after 3 days to maintain soil moisture content. Apply at a rate of 750 ml per hectare (about 1×10^9 chlamydospores per hectare). 'DeVine' can be applied by a boom sprayer using at least 80 litres of water per hectare, by chemigation ensuring that the product is applied towards the end of the irrigation or through sprinkler irrigation. Apply only once a season.

PRODUCT SPECIFICATIONS: Specifications: Manufactured by fermentation. **Purity:** The product contains 6.7×10^5 living chlamydospores of *Phytophthora palmivora* per ml.

Storage conditions: Store under refrigeration, prevent from freezing and do not expose to direct sunlight. **Shelf-life:** Use as soon as possible after delivery.

COMPATIBILITY: Do not mix with chlorinated water or apply with any other pesticide, fertiliser or spray adjuvant. Do not use within 30 metres of susceptible plants such as cucurbit vegetables.

MAMMALIAN TOXICITY: There is no evidence that *Phytophthora palmivora* has any allergic or other adverse effects on mammals. Considered to be of low mammalian toxicity.

ENVIRONMENTAL IMPACT AND NON-TARGET TOXICITY: *Phytophthora palmivora* occurs naturally in the soil and there is no evidence that its use as a mycoherbicide has any effect on non-target organisms or on the environment. **Behaviour in soil:** There is evidence that *Phytophthora palmivora* remains biologically active in treated soil for at least a year after application.

1:100 *Plodia interpunctella* granulosis virus
Insecticidal baculovirus

Virus: Baculoviridae: Granulovirus

NOMENCLATURE: Approved name: *Plodia interpunctella* granulosis virus.
Other names: *Plodia interpunctella* GV; Indian meal moth granulosis virus; IMMGV.
OPP Chemical Code: 108896.

SOURCE: The granulosis virus was originally isolated by the USDA from infected Indian meal moth larvae.

PRODUCTION: Indian meal moth larvae are grown on a wheat bran-based diet which contains larval nutrients and vitamins. Indian meal moth eggs are added to the diet and allowed to hatch and develop into larvae. After about 10 days, these larvae are infected with IMMGV. The infection is allowed to develop for a short period, after which the entire mix of diet plus infected larvae is homogenised, dried and powdered to pass an 80-mesh sieve.

TARGET PESTS: Indian meal moth (*Plodia interpunctella* (Hübner)) larvae.

TARGET CROPS: IMMGV is used to protect stored products such as dried fruits and nuts. It can also be used to treat stored food containers and cracks and crevices where the moth breeds.

BIOLOGICAL ACTIVITY: Mode of action: As with all insect baculoviruses, *Plodia interpunctella* GV must be ingested to exert an effect. Following ingestion, the virus enters the insect's haemolymph. It multiplies in the insect body, leading to death. **Biology:** IMMGV is more active on small larvae than on later larval instars. The virus is ingested by the feeding larva and the virus protein coat is dissolved in the insect's midgut (that is alkaline), releasing the virus particles. These pass through the peritrophic membrane and invade midgut cells by fusion with the microvilli. The virus particles invade the cell nuclei, where they are uncoated and replicated. Initial replication produces non-occluded virus particles that spread to new cells and hasten the invasion of the host insect, but later the virus particles are produced with protein coats and remain infective when released from the dead insects. **Efficacy:** IMMGV has a residual effect against Indian meal moth in the treated areas, whereas the majority of other control agents do not. A single treatment with IMMGV has been shown to prevent re-infestation for up to 2 years.

COMMERCIALISATION: Formulation: The product is required to exhibit a minimum potency (LC_{50}) of 0.26 µg product per g standard bioassay diet. The product is sold as a wettable powder (WP) containing between 1.4×10^{11} and 5.7×10^{11} viral capsules per g of product (4×10^{12} and 1.6×10^{13} viral capsules per oz of product) and larval parts on milled wheat bran carrier. **Tradenames:** 'BioGuard-V' (AgriVir), 'FruitGuard-V' (AgriVir), 'NutGuard-V' (AgriVir). **Patents:** US 5,023,182, issued to the USDA.

APPLICATION: Apply as a water suspension, using low pressure spray equipment, as evenly as possible at the rate of 25 to 150 g per tonne (1 to 5 oz per ton) for dried fruit and nuts, and 60 to 300 g per 100 m^2 (2 to 10 oz per 100 square feet) for crack and crevice treatment.

PRODUCT SPECIFICATIONS: Specifications: Indian meal moth larvae are grown on a wheat bran-based diet which contains larval nutrients and vitamins. Indian meal moth eggs are added to the diet and allowed to hatch and develop into larvae. After about 10 days, these larvae are infected with IMMGV. The infection is allowed to develop for a short period, after which the entire mix of diet plus infected larvae is homogenised, dried and powdered to pass an 80-mesh sieve. **Purity:** The product contains approximately 9.35% Indian meal moth granulosis virus infected Indian meal moth larval parts, approximately 85.90% milled wheat bran carrier and approximately 4.75% other inert ingredients (largely brewer's yeast). **Shelf-life:** Formulations stored at 2°C are stable for over 2 years. If stored at room temperature, they remain effective for a few months. Infectivity is reduced by exposure to u.v. light.

COMPATIBILITY: It is unusual to use IMMGV with any other insecticide. Use water with a pH between 6 and 8. The product is incompatible with strong oxidisers, acids, bases and chlorinated water.

MAMMALIAN TOXICITY: A number of studies on the toxicity of baculoviruses, inclusive of granulosis viruses, to animals have shown that, by oral, dermal, inhalation and injection routes of exposure, the tested granulosis viruses produced no effects on overall health, gross or micro-pathology, haematology, clinical chemistry and antibody stimulation in the test animals. There is no evidence of acute or chronic toxicity, eye or skin irritation in mammals with IMMGV. No allergic reactions or other health problems have been observed in research or manufacturing staff or with users of the product. **Acute oral LD$_{50}$:** rats >1.7 × 10^8 IMMGV virion capsules/animal. Toxicity category IV. **Acute dermal LD$_{50}$:** Toxicity category III. **Inhalation:** Toxicity category IV. **Skin and eye:** Toxicity category IV.

ENVIRONMENTAL IMPACT AND NON-TARGET TOXICITY: *Plodia interpunctella* granulosis virus occurs in Nature. The commercial use pattern of the product within packing houses will not expose non-target organisms to the virus. There is no evidence that it affects any organism other than Indian meal moth larvae. It is unstable at extreme pH and when exposed to u.v. light. It does not persist in the environment.

1:101 *Pseudomonas aureofaciens* isolate Tx-1
Biological fungicide (bacterium)

Bacterium: Pseudomonadales: Pseudomonadaceae

NOMENCLATURE: Approved name: *Pseudomonas aureofaciens* (Pau) isolate Tx-1.

SOURCE: Frequently occurring soil bacterium, isolated in 1989 from turf grass crown tissue.

PRODUCTION: Produced by fermentation.

TARGET PESTS: Dollar spot (*Sclerotinia homeocarpa* Bennett), anthracnose (*Colletotrichum* spp.), *Pythium aphanidermatum* Fitzp., microdochium patch (pink snow mould) (*Microdochium nivale* (Fr.) Samuels & Hallett).

TARGET CROPS: Recreational turf including golf courses and lawns.

BIOLOGICAL ACTIVITY: Mode of action: *Pseudomonas aureofaciens* isolate Tx-1 produces antifungal metabolites, including phenazine carboxylic acid (PCA) and hydroxylated derivatives of PCA, in addition to competitive exclusion of invading pathogens. **Efficacy:** Control is achieved within 7 to 10 days after inoculation, depending upon the weather conditions and the growth of the turf grass. Best control is achieved with frequent application and optimum (high) nitrogen fertility. **Key reference(s):** D R Fravel, W J Connick Jr & J A Lewis. 1998. Formulation of microorganisms to control plant diseases. In *Formulation of Microbial Biopesticides, Beneficial Microorganisms, Nematodes and Seed Treatments*, H D Burges (ed.), pp. 187–202, Kluwer Academic Publishers, Dordrecht, The Netherlands.

COMMERCIALISATION: Formulation: Delivered through an on-site fermentation system called the BioJect. Registered by the US-EPA [EPA Registration No. 70688-1]. **Tradenames:** 'Spot-Less' (Turf Science).

APPLICATION: Apply to established turf through the irrigation system or download to spray tank.

PRODUCT SPECIFICATIONS: Specifications: Produced by fermentation. **Purity:** Contains no contaminants. **Storage conditions:** Store refrigerated. Do not expose to direct sunlight. **Shelf-life:** If stored correctly, the product should encounter no loss of viability for at least 4 months.

COMPATIBILITY: It is recommended that the product be used alone. Incompatible with broad-spectrum products such as copper-based fungicides and with strong oxidisers, acids, bases and chlorinated water.

MAMMALIAN TOXICITY: There are no reports of allergic or other adverse toxicological effects from research or manufacturing staff, formulators or field workers with *Pseudomonas aureofaciens* isolate Tx-1. No health risks to humans are expected from use of these bacterial strains in pesticide products, when label directions are followed.

ENVIRONMENTAL IMPACT AND NON-TARGET TOXICITY: *Pseudomonas aureofaciens* isolate Tx-1 occurs widely in Nature and is not expected to have any adverse effects on non-target organisms or on the environment. The bacterium is not expected to cause harm to the environment.

1:102 *Pseudomonas cepacia*
Biological fungicide (bacterium)

Bacterium: Pseudomonadales: Pseudomonadaceae

NOMENCLATURE: Approved name: *Pseudomonas cepacia* (ex Burkholder) Palleroni & Holmes. **Other names:** *Pseudomonas cepacia* has been reclassified. See *Burkholderia cepacia* (entry 1:42).

1:103 *Pseudomonas chlororaphis* isolate 63-28
Biological fungicide (bacterium)

Bacterium: Pseudomonadales: Pseudomonadaceae

NOMENCLATURE: Approved name: *Pseudomonas chlororaphis* (Guignard & Sauvageau) Bergey *et al.* isolate 63-28.

SOURCE: *Pseudomonas chlororaphis* isolate 63-28 was isolated from healthy canola plants in western Canada in 1984. It is a frequently occurring soil bacterium.

PRODUCTION: Manufactured by fermentation.

TARGET PESTS: Soil- and seed-borne fungal pathogens.

TARGET CROPS: 'Cedomon' is recommended for use in cereals. 'AtEze' is used in glasshouse ornamentals, nursery crops and vegetable transplants.

BIOLOGICAL ACTIVITY: Mode of action: *Pseudomonas chlororaphis* isolate 63-28 produces plant growth factors, such as cytokinin, which could help the plant to limit fungal damage. It also produces antibiotics, which would act directly on the fungi. In addition, the bacterium seems to contain a protein that binds iron that the fungi need for growth and reproduction, allowing it to outcompete the phytopathogenic fungi for space and other resources. **Biology:** *Pseudomonas chlororaphis* is a vigorous rhizosphere-inhabiting bacterium that rapidly colonises the root zone of treated plants. It outcompetes phytopathogenic species by depriving them of the nutrients necessary for their growth and by producing antifungal secondary metabolites. **Key reference(s):** M P McQuilken, P Halmer & D J Rhodes. 1998. Application of microorganisms to seeds. In *Formulation of Microbial Biopesticides, Beneficial Microorganisms, Nematodes and Seed Treatments*, H D Burges (ed.), pp. 255–85, Kluwer Academic Publishers, Dordrecht, The Netherlands.

COMMERCIALISATION: Formulation: 'Cedomon' is sold as a seed treatment containing 1×10^{11} colony forming units/g. 'AtEze' is a wettable powder (WP) formulation. **Tradenames:** 'AtEze' (Turf Science), 'Cedomon' (BioAgri).

APPLICATION: 'Cedomon' is used to treat seeds that are then sown. The bacterium rapidly colonises the rhizosphere and prevents the establishment of phytophagous species. 'AtEze' should be applied as a drench immediately after sowing. For longer-season crops, a second application is recommended 2 to 3 months after the first. Transplants should be treated after sowing and again just before transplanting.

PRODUCT SPECIFICATIONS: Specifications: Manufactured by fermentation. **Purity:** Contains no contaminants. **Storage conditions:** Store in a cool, dry, sealed container

out of direct sunlight. 'AtEze' should be held at refrigeration temperatures until ready to use.
Shelf-life: Remains viable for 12 months, if stored under recommended conditions.

COMPATIBILITY: 'Cedomon' should not be used with broad-spectrum fungicide seed treatments. 'AtEze' is compatible with dilute fertilisers and most fungicides.

MAMMALIAN TOXICITY: *Pseudomonas chlororaphis* has not had any reports of allergic or other adverse toxicological effects from research and manufacturing staff, formulators or field workers.

ENVIRONMENTAL IMPACT AND NON-TARGET TOXICITY: *Pseudomonas chlororaphis* is a commonly occurring soil bacterium. It is not expected that its use as a seed treatment or a soil drench will have any adverse effects on non-target organisms or on the environment.
Approved for use in organic farming: Yes.

1:104 *Pseudomonas fluorescens*
Biological fungicide/bactericide (bacterium)

Bacterium: Pseudomonadales: Pseudomonadaceae

NOMENCLATURE: Approved name: *Pseudomonas fluorescens* (Trevisan) Migula.

SOURCE: A naturally and widely occurring bacterium. An isolate that was naturally non-ice nucleating was isolated and registered in the USA for suppression of frost damage ('FrostBan'). Other isolates with antifungal or antibacterial activity have been isolated and commercialised in the USA.

PRODUCTION: Produced by fermentation.

TARGET PESTS: Fire blight (*Erwinia amylovora* Winsl.), soil-borne *Fusarium* and *Rhizoctonia* spp. and frost damage. 'Conquer' is used to control *Pseudomonas tolaasii* Paine.

TARGET CROPS: Fruit tree crops, particularly pears and apples, cotton and vegetables, as well as almond, potato and tomato crops. 'Conquer' is recommended for use in mushrooms.

BIOLOGICAL ACTIVITY: Mode of action: Bacteria on the leaves of crops often serve as nucleation sites for ice formation: ice crystals often form when they are present and the temperature falls below freezing, with resulting damage to the leaf. If these bacteria are replaced on plant leaves with competitive antagonists that lack the ice-nucleating protein, frost is prevented, even at temperatures as low as −5°C. Other isolates of *Pseudomonas fluorescens* are antagonistic to foliar or rhizosphere bacteria and fungi through the production of siderophores and antibiotics. **Key reference(s):** 1) M P McQuilken, P Halmer & D J Rhodes. 1998. Application of microorganisms to seeds. In *Formulation of Microbial Biopesticides, Beneficial Microorganisms, Nematodes and Seed Treatments*, H D Burges (ed.), pp. 255–85, Kluwer Academic Publishers, Dordrecht, The Netherlands. 2) D R Fravel, W J Connick Jr & J A Lewis. 1998. Formulation of microorganisms to control plant diseases. In *Formulation of Microbial Biopesticides, Beneficial Microorganisms, Nematodes and Seed Treatments*, H D Burges (ed.), pp. 187–202, Kluwer Academic Publishers, Dordrecht, The Netherlands. 3) A S Paau. 1998. Formulation of beneficial microorganisms applied to soil. In *Formulation of Microbial Biopesticides, Beneficial Microorganisms, Nematodes and Seed Treatments*, H D Burges (ed.), pp. 235–54, Kluwer Academic Publishers, Dordrecht, The Netherlands.

COMMERCIALISATION: Formulation: Sold as seed treatments or wettable powders (WP) of bacterial cells. 'Conquer' is a liquid suspension formulation. **Tradenames:** 'Bio AquaGuard' (plus *Trichoderma viride*) (SOM Phytopharma), 'Biocure' (Stanes), 'Biomonas' (Biotech International), 'BlightBan' (Plant Health Technologies), 'Conquer' (Mauri Foods) and (Sylvan Spawn), 'Dagger' (Ecogen − discontinued), 'Frostban A' and 'Frostban D' (both plus *Pseudomonas syringae*) (Frost Technology), 'Sheathguard' (SOM Phytopharma) and (Agri Life).

APPLICATION: Apply non-nucleating bacteria to foliage of tree crops and vegetables before the temperature falls below freezing. A single application will protect from frost damage at temperatures as low as −5°C for up to 2 months. The bacterium must be established on the foliage before freezing temperatures occur. Soil-borne pathogens are controlled with the use of seed treatments or hopper-box applications, and foliar pathogens by spray applications. Fire blight sprays should be applied before flowering. 'Conquer' is applied as a spray to mushroom beds.

PRODUCT SPECIFICATIONS: Specifications: Produced by fermentation. **Storage conditions:** Store in a cool, dry place in a sealed container. Do not expose to extremes of temperature or direct sunlight. **Shelf-life:** If stored under recommended conditions, the product remains viable for a year.

COMPATIBILITY: It is unusual to apply *P. fluorescens* with chemical treatments. Incompatible with strong oxidisers, acids, bases and chlorinated water.

MAMMALIAN TOXICITY: There are no records of allergic or other adverse effects following use of *P. fluorescens*. It is regarded as being of low mammalian toxicity.

ENVIRONMENTAL IMPACT AND NON-TARGET TOXICITY: *Pseudomonas fluorescens* occurs widely in Nature and is not expected to have any adverse effects on non-target organisms or on the environment.

1:105 *Pseudomonas gladioli* pv. *gladioli*
Bacterial herbicide (bacterium)

Bacterium: Pseudomonadales: Pseudomonadaceae

NOMENCLATURE: Approved name: *Pseudomonas gladioli* pv. *gladioli* Serenini.

SOURCE: Isolated from *Poa annua* L. in Japanese turf.

PRODUCTION: Manufactured by controlled fermentation.

TARGET PESTS: Annual meadow grass (*Poa annua*).

TARGET CROPS: Golf courses and similar areas of fine grasses.

BIOLOGICAL ACTIVITY: Mode of action: *Pseudomonas gladioli* invades the weed's xylem water-transporting system, where it multiplies, eventually blocking the tissues, leading to plant death. It does not invade the xylem of other turf grass species. **Biology:** *Pseudomonas gladioli* is a pathogen of annual meadow grass (*Poa annua*). **Key reference(s): 1)** D R Fravel, W J Connick Jr & J A Lewis. 1998. Formulation of microorganisms to control plant diseases. In *Formulation of Microbial Biopesticides, Beneficial Microorganisms, Nematodes and Seed Treatments*, H D Burges (ed.), pp. 187–202, Kluwer Academic Publishers, Dordrecht, The Netherlands. **2)** A S Paau. 1998. Formulation of beneficial microorganisms applied to soil. In *Formulation of Microbial Biopesticides, Beneficial Microorganisms, Nematodes and Seed Treatments*, H D Burges (ed.), pp. 235–54, Kluwer Academic Publishers, Dordrecht, The Netherlands.

COMMERCIALISATION: Tradenames: 'AM 301' (Japan Tobacco), 'Camperico' (Microgen).

APPLICATION: Apply to turf as a drench or through an irrigation system.

PRODUCT SPECIFICATIONS: Specifications: Manufactured by controlled fermentation. **Purity:** The product contains cells of *Pseudomonas gladioli* and no bacterial contaminants pathogenic to man. **Storage conditions:** Store in a cool, dark, dry place. Do not expose to direct sunlight or extremes of temperature. **Shelf-life:** If stored under manufacturer's recommended conditions, may be kept for up to one year.

COMPATIBILITY: Incompatible with broad-spectrum fungicides and with strong oxidisers, acids, bases and chlorinated water.

MAMMALIAN TOXICITY: There are no reports of allergic or other adverse toxicological effects from research workers, manufacturers or field staff.

ENVIRONMENTAL IMPACT AND NON-TARGET TOXICITY: *Pseudomonas gladioli* occurs widely in Nature and is not expected to have any adverse effects on non-target organisms or on the environment.

1:106 *Pseudomonas syringae* isolates ESC-10 and ESC-11
Biological fungicide (bacterium)

Bacterium: Pseudomonadales: Pseudomonadaceae

NOMENCLATURE: Approved name: *Pseudomonas syringae* Van Hall isolate ESC-10 (006441) and isolate ESC-11 (006451). **Other names:** Formerly known as: *Pseudomonas cerasi* Griff.

SOURCE: Both *Pseudomonas syringae* isolate ESC-10 and *Pseudomonas syringae* isolate ESC-11 are natural bacterial isolates that occur on many kinds of plant throughout the world. They were originally identified in and isolated from apples.

PRODUCTION: Produced by fermentation.

TARGET PESTS: Used to control fungal pathogens of stored produce.

TARGET CROPS: Apples, pears and vegetables ('BioSave 100'); and lemons, oranges and grapefruit ('BioSave 1000'); both applied after the fruit is harvested. It is intended to extend the range of crops to include bananas, cranberries and stone fruits.

BIOLOGICAL ACTIVITY: Biology: Although the exact method of disease control is unknown, these harmless bacteria probably outcompete pathogenic fungi for space and nutrients on the fruit, thereby preventing the fruit from rotting before it can be used. The bacterium competes with pathogenic fungi for occupation of the surface of the treated crop. This coverage of the treated plant surface also prevents access to entry sites.
Efficacy: *Pseudomonas syringae* covers the surface of treated harvested produce and presents a barrier to invasion by pathogenic fungi. **Key reference(s):** D R Fravel, W J Connick Jr & J A Lewis. 1998. Formulation of microorganisms to control plant diseases. In *Formulation of Microbial Biopesticides, Beneficial Microorganisms, Nematodes and Seed Treatments*, H D Burges (ed.), pp. 187–202, Kluwer Academic Publishers, Dordrecht, The Netherlands.

COMMERCIALISATION: Formulation: Sold as concentrated pellets made from frozen bacterial cells. 'BioSave 100' and 'BioSave 1000' are strain ESC-10 and 'BioSave 110' is strain ESC-11. **Tradenames:** 'BioSave 100', 'BioSave 110' and 'BioSave 1000' (EcoScience), 'Frostban A' and 'Frostban D' (both plus *Pseudomonas fluorescens*) (Frost Technology).

APPLICATION: After the fruit is harvested and cleaned, its surface is exposed to a suspension containing the ESC-10 or ESC-11 bacterium. The pesticide applicator can apply the suspension by spraying, or by dipping the fruit into the suspension. These applications are considered indoor uses, since they take place in enclosed areas.

PRODUCT SPECIFICATIONS: **Specifications:** Produced by fermentation. **Purity:** Contains no contaminants pathogenic to man. **Storage conditions:** Store under cool, dry conditions out of direct sunlight. **Shelf-life:** May be stored for 12 months.

COMPATIBILITY: It is recommended that the product be used alone. Incompatible with strong oxidisers, acids, bases and chlorinated water.

MAMMALIAN TOXICITY: Various studies show that *Pseudomonas syringae* isolate ESC-10 and *Pseudomonas syringae* isolate ESC-11 do not cause adverse effects in mammals when the bacteria are ingested, inhaled or applied to skin. Furthermore, the bacterium cannot survive at temperatures above 32°C (90°F) and, therefore, cannot grow in humans or birds, whose body temperatures are considerably higher. No health risks to humans are expected from use of these bacterial isolates in pesticide products when label directions are followed. There are no reports of allergic or other adverse toxicological effects from research or manufacturing staff, formulators or field workers.

ENVIRONMENTAL IMPACT AND NON-TARGET TOXICITY: *Pseudomonas syringae* occurs widely in Nature and is not expected to have any adverse effects on non-target organisms or on the environment. The fruits are treated in an enclosed area and thus exposures to the outside environment, including wildlife, soil and water, are not expected. **Approved for use in organic farming:** Yes.

1:107 *Pseudomonas tolassii* bacteriophage
Biological bactericide (bacteriophage)

Bacteriophage

NOMENCLATURE: **Approved name:** *Pseudomonas tolassii* bacteriophage.

SOURCE: *Pseudomonas tolassii* bacteriophage is a pathogen of *Pseudomonas tolassii* Paine and was isolated from the bacterium by NPP.

PRODUCTION: Produced by *in vivo* culture in bacterial cells.

TARGET PESTS: *Pseudomonas tolassii*.

TARGET CROPS: Edible mushrooms such as *Agaricus* and *Pleurotus* species.

BIOLOGICAL ACTIVITY: **Mode of action:** *Pseudomonas tolassii* bacteriophage is a naturally occurring pathogen of the major mushroom house bacterium, *Pseudomonas tolassii*. It controls the bacterium by pathogenesis. Following invasion and death, the infected bacterial cells rupture and new bacteriophages are released.

COMMERCIALISATION: Formulation: 'Phagus' is sold as a suspension of infected bacterial cells. **Tradenames:** 'Phagus' (NPP).

APPLICATION: 'Phagus' is applied to the developing mushrooms as a foliar spray.

PRODUCT SPECIFICATIONS: Specifications: Produced by *in vivo* culture in bacterial cells. **Purity:** Any remaining bacterial cells are checked to ensure they are not pathogenic to mushrooms and mammals. The product is monitored for the presence of other bacteria. **Storage conditions:** Use as soon after delivery as possible.

COMPATIBILITY: It is unusual to use 'Phagus' in association with any other crop protection agent.

MAMMALIAN TOXICITY: There is no evidence of any adverse or allergic effects of 'Phagus' on research workers, manufacturers or users.

ENVIRONMENTAL IMPACT AND NON-TARGET TOXICITY: *Pseudomonas tolassii* bacteriophage occurs in Nature and is a specific bacterial pathogen and, consequently, is not expected to show any adverse effects on non-target organisms or on the environment.

1:108 *Pseudozyma flocculosa* isolate PF-A22 UL
Biological fungicide (fungus)

Fungi: anamorphic Ustilaginaceae: previously Basidiomycetes: Ustilaginales

NOMENCLATURE: Approved name: *Pseudozyma flocculosa* (Traquair *et al.*) Boekhout & Traquair isolate PF-A22 UL. **Other names:** Formerly known as *Sporothrix flocculosa* Traquair, Shaw & Jarvis and *Stephanoascus flocculosus* Traquair *et al.*.

SOURCE: *Pseudozyma flocculosa* is widely distributed as a saprophytic fungal epiphyte and as a hyperparasite of powdery mildews in Canada, the USA and Europe on aerial plant surfaces in field or glasshouse agricultural ecosystems. 'Sporodex' was isolated from a leaf of red clover (*Trifolium pratense* L.) covered with *Erysiphe polygoni* DC.

PRODUCTION: *Pseudozyma flocculosa* is grown commercially in semi-solid or submerged fermentation, during which process it produces spores. The spores serve as the active ingredient of the WP formulation.

TARGET PESTS: Cucumber powdery mildew (*Sphaerotheca fuliginea* Poll.) and rose powdery mildew (*Sphaerotheca pannosa* var. *rosae*).

TARGET CROPS: Cucurbits and glasshouse grown roses.

BIOLOGICAL ACTIVITY: Mode of action: *Pseudozyma flocculosa* isolate PF-A22 UL has been shown to release fatty acid substances that destroy the membranes of the mildew. **Biology:** *Pseudozyma flocculosa* is a well known hyperparasite of powdery mildews (Erysiphaceae). **Efficacy:** The product works very effectively in a spray programme to control powdery mildews. High humidity is essential for spore germination and powdery mildew hyphal invasion and this can be enhanced with the addition of mineral oil adjuvants. It is, however, recommended that the product is applied in the early morning or late afternoon when dew is present or is expected on the crop. **Key reference(s):** D R Fravel, W J Connick Jr & J A Lewis. 1998. Formulation of microorganisms to control plant diseases. In *Formulation of Microbial Biopesticides, Beneficial Microorganisms, Nematodes and Seed Treatments*, H D Burges (ed.), pp. 187–202, Kluwer Academic Publishers, Dordrecht, The Netherlands.

COMMERCIALISATION: Formulation: Sold as a wettable powder (WP) of spores in an inert carrier. **Tradenames:** 'Sporodex' (Plant Products).

APPLICATION: Good cover of the foliage is essential and the product should be applied when the humidity is above 60%. Addition of mineral oil at a rate of 0.3% v/v can be used to enhance the germination of the spores. Monitoring of the weather conditions that are conducive to the onset of powdery mildew infestations is recommended for the best results. The product has been employed as part of Integrated Crop Management (ICM) programmes and in alternation with standard chemical mildewicides. It is commonly applied by standard spray application techniques in the presence of surfactants that are compatible with the viability of the organism. Not phytotoxic or phytopathogenic.

PRODUCT SPECIFICATIONS: Specifications: *Pseudozyma flocculosa* is grown commercially in semi-solid or submerged fermentation, during which process it produces spores. The spores serve as the active ingredient of the WP formulation. **Storage conditions:** Store in a sealed container in a cool, dry place. Do not expose to extremes of temperature or direct sunlight. **Shelf-life:** Retains viability for a year, under correct storage conditions.

COMPATIBILITY: It is recommended that the product be used alone. Incompatible with strong oxidisers, acids, bases and chlorinated water.

MAMMALIAN TOXICITY: There is no evidence of pathogenicity or infectivity of 'Sporodex' following acute oral gavage, intraperitoneal and intratracheal challenge studies in rats. No toxicity has been shown following a single oral dose in rats. No toxicity or irritation was observed following a single dermal application in rabbits. Slight toxicity was observed following a single intraperitoneal challenge in rats. However, the toxicity observed was due to normal immune response to foreign material deposited in the peritoneal cavity. Toxicity was observed in rats dosed by intratracheal challenge. Mortality was associated with the quantity of test material delivered (6×10^7 cells or 3.2×10^7 colony forming units (cfu)), which was the highest dose deliverable. In an additional study, the minimum lethal dose was shown to be higher than 6×10^7 cells, which was the highest dose deliverable. Other signs of toxicity following intratracheal challenge were associated with normal immune responses to foreign material in the lung. No reports of human toxicity have been made from those working directly with this microbe for the past 10 years. No signs of ocular irritation were observed in rabbits at or following a 48-hour scoring interval. The bioactive compounds produced by *Pseudozyma flocculosa* are not known as genotoxins.

ENVIRONMENTAL IMPACT AND NON-TARGET TOXICITY: *Pseudozyma flocculosa* occurs widely in Nature and is not thought to have any adverse effects on non-target organisms or on the environment. **Approved for use in organic farming:** Yes.

1:109 *Puccinia abrupta* var. *partheniicola*
Biological herbicide (fungus)

Fungus: Basidiomycetes: Uredinales

NOMENCLATURE: Approved name: *Puccinia abrupta* Dicks & Holw. var. *partheniicola* (Jackson) Parmelee.

SOURCE: Various isolates have been evaluated for biological efficacy. Those released in Australia were found in Mexico and those used in African control programmes were isolated in Ethiopia.

PRODUCTION: Produced on living Parthenium weed (*Parthenium hysterophorus* L.).

TARGET PESTS: Parthenium weed (*Parthenium hysterophorus*).

BIOLOGICAL ACTIVITY: Mode of action: Pathogenesis. **Biology:** Infection with the rust hastened leaf senescence of *Parthenium hysterophorus*, significantly decreased the life span and dry weight of Parthenium plants, and reduced flower production 10-fold. The individual effects of the rust on Parthenium weed morphological parameters and seed production capacity showed that the rust disease reduced mean plant height, number of leaves per plant, leaf area, number of branches, dry matter yield at maturity and number of seeds produced by 11, 22, 28, 13, 25 and 43%, respectively. **Efficacy:** May be used as a component of mixtures of biological control agents in combination with several different phytophagous insects.

Key reference(s): 1) A Parker, A N G Holden & A J Tomley. 1994. Host specificity testing and assessment of the pathogenicity of the rust, *Puccinia abrupta* var. *partheniicola*, as a biological control agent of Parthenium weed (*Parthenium hysterophorus*), *Plant Pathol.*, **43**(1), 1–16. 2) T Tessema, M Bandte, R Metz & C Büttner. 2003. Investigation of pathogens for biological control of Parthenium weed (*Parthenium hysterophorus*) in Ethiopia. In *Proc. Technological and Institutional Innovations for Sustainable Rural Development*, Göttingen, Germany.

COMMERCIALISATION: Formulation: Not sold commercially, but introduced into Australia for control of the alien invasive weed Parthenium. Also introduced into Ethiopia.

APPLICATION: Temperatures of less than 20°C and dew periods of more than 6 hours were required for abundant pustule production.

PRODUCT SPECIFICATIONS: Specifications: Produced on living Parthenium weed (*Parthenium hysterophorus* L.).

MAMMALIAN TOXICITY: There are no reports of allergic or other adverse toxicological effects from the use of *Puccinia abrupta* var. *partheniicola*.

ENVIRONMENTAL IMPACT AND NON-TARGET TOXICITY: *Puccinia abrupta* var. *partheniicola* is an obligate pathogen of Parthenium weed and, as such, is not expected to have any adverse effects on non-target organisms or on the environment. In addition, host-range studies against 120 species and varieties, with further screening against a limited number of species conducted under a range of environmental regimes, indicated the rust to be sufficiently host-specific to be considered for introduction. Additional host-range tests against sunflowers demonstrated that resistance to *P. abrupta* var. *partheniicola* was not modified by prior inoculation with *Puccinia helianthi*.

1:110 *Puccinia canaliculata*
Biological herbicide (fungus)

Fungus: Basidiomycetes: Uredinales

NOMENCLATURE: Approved name: *Puccinia canaliculata* (Schw.) Lagerh.

SOURCE: Isolated from yellow nutsedge (*Cyperus esculentus* L.).

PRODUCTION: Produced on living yellow nutsedge plants.

TARGET PESTS: Yellow nutsedge.

TARGET CROPS: Soybean, sugar cane, maize (corn), potatoes and cotton.

BIOLOGICAL ACTIVITY: Mode of action: Pathogenesis. **Biology:** When applied in the spring, *Puccinia canaliculata* inhibited nutsedge flowering and reduced the nutsedge stand and new tuber formation by 46 and 66%, respectively. It also dehydrated and killed nutsedge plants. It was less effective when applied in June. **Efficacy:** Registered, but product failed due to uneconomic production system and resistance in some weed biotypes. **Key reference(s):** S C Phatak, D R Sumner, H D Wells, D K Bell & N C Glaze. 1983. Biological control of yellow nutsedge with the indigenous rust fungus *Puccinia canaliculata*, *Science*, **219**, 1446–7.

COMMERCIALISATION: Formulation: The product consists of rust spores on finely ground leaf and stem pieces of infected yellow nutsedge plants. **Tradenames:** 'Dr BioSedge' (Tifton Innovation Corp). **Patents:** United States Patent 4731104.

APPLICATION: Apply directly to soil and young plants in early spring when new plants are beginning their growth. One application is usually sufficient, since the rust spreads by itself. In areas heavily infested with yellow nutsedge, it may be necessary to repeat the application in subsequent years.

PRODUCT SPECIFICATIONS: Specifications: Produced on living yellow nutsedge plants.

MAMMALIAN TOXICITY: There are no reports of allergic or other adverse toxicological effects from the use of *Puccinia canaliculata*. The rust only infects yellow nutsedge plants.

ENVIRONMENTAL IMPACT AND NON-TARGET TOXICITY: *Puccinia canaliculata* is specific to yellow nutsedge and, as such, is not expected to have any adverse effects on non-target organisms or on the environment. **Approved for use in organic farming:** Yes.

1:111 *Puccinia carduorum*
Biological herbicide (fungus)

Fungus: Basidiomycetes: Uredinales

NOMENCLATURE: Approved name: *Puccinia carduorum* Jacky.

SOURCE: Imported into Virginia and Maryland, USA in 1987 from Turkey.

PRODUCTION: Produced on living musk thistle (*Carduus thoermeri* Weinmann) plants.

TARGET PESTS: Musk thistle (*Carduus thoermeri*).

BIOLOGICAL ACTIVITY: Mode of action: Pathogenesis. **Biology:** *Puccinia carduorum* has been found to reduce musk thistle density by accelerating senescence of rust-infected musk thistle and reducing seed production by 20 to 57%. **Efficacy:** The rust has spread widely from its original introduction to the western states of Wyoming and California, USA.

COMMERCIALISATION: Formulation: Not sold commercially. Introduced and now widespread in the USA.

APPLICATION: Apply directly to soil and young plants in early spring when new plants are beginning their growth. One application is usually sufficient, since the rust spreads by itself. In areas heavily infested with musk thistle, it may be necessary to repeat the application in subsequent years.

PRODUCT SPECIFICATIONS: Specifications: Produced on living musk thistle (*Carduus thoermeri*) plants.

MAMMALIAN TOXICITY: There are no reports of allergic or other adverse toxicological effects from the use of *Puccinia carduorum*. The rust only infects musk thistle plants.

ENVIRONMENTAL IMPACT AND NON-TARGET TOXICITY: *Puccinia carduorum* is specific to musk thistle and, as such, is not expected to have any adverse effects on non-target organisms or on the environment. **Effects on beneficial insects:** Tests were conducted to determine the effects of *Puccinia carduorum* on feeding, oviposition, longevity, egg production, egg hatch and larval development of three established thistle beetle herbivores (*Trichosirocalus horridus* (Panzer), *Rhinocyllus conicus* Froelich and *Cassida rubiginosa* Muller). When offered both rust-infected and healthy musk thistle leaves, adults of *Cassida rubiginosa* and *Trichosirocalus horridus* consumed significantly more healthy than infected foliage. On infected leaves, feeding and oviposition were confined largely to pustule-free areas. The amount of rust-infected and healthy foliage consumed by *Cassida rubiginosa* larvae was not significantly different, but feeding on the infected leaves was again confined to rust-free leaf areas. Rust infection did not reduce oviposition by the three insects, and *Rhinocyllus conicus* oviposited only on healthy portions of the bracts. There were no significant differences in longevity, egg production, percentage of egg hatch and larval development for any of the three insects fed continuously with both healthy and rust-infected leaves. Thus, effects of *Puccinia carduorum* on the herbivores were slight, and none of the interactions found is expected to be detrimental to biological control. **Approved for use in organic farming:** Yes.

1:112 *Puccinia chondrillina*
Biological herbicide (fungus)

Fungus: Basidiomycetes: Uredinales

NOMENCLATURE: Approved name: *Puccinia chondrillina* Bubák and Sydenham.
Other names: Rush skeletonweed rust.

SOURCE: Widespread in southern Europe.

PRODUCTION: Produced on living rush skeleton weed (*Chondrilla juncea* L.) plants, but now widely introduced and endemic in many areas.

TARGET PESTS: Rush skeleton weed (*Chondrilla juncea*).

TARGET CROPS: A wide range of crops, including cereals.

BIOLOGICAL ACTIVITY: Mode of action: *Puccinia chondrillina* is tolerant of a wide range of temperature and humidity conditions, although development is minimal below 5 °C. Most severe infection results when inoculation and incubation occur at low temperatures, with 8 °C being suggested as the optimum, and during a period of at least 6 hours of darkness and when dew is present. **Biology:** *Puccinia chondrillina* is a stem and leaf rust that causes development of rust-brown pustules surrounded by yellow tissue. Infected seedlings can be killed or plants have reduced growth, vigour and reproductive potential. There is evidence of stunting, deformation, reduced branching, fewer flower buds and lower viability by seeds, where plants suffer severe rusting. **Efficacy:** There are three different biotypes of the rush skeleton weed, each with a different susceptibility to the pathogen. Initially, the susceptible type was the most common, but its population density was reduced due to biocontrol, and the two other types became more widespread. Additional rust isolates, virulent on these more resistant forms, were introduced into Australia from the Mediterranean and these isolates are exerting some degree of control of the resistant forms. Releases into the USA to control a skeleton weed biotype were only partially successful. Here, *Puccinia chondrillina* has been used along with chemical herbicides, and the insect biocontrol agents, *Cystiphora schmidti* (Rübsaamen) (chondrilla gall midge) and *Aceria chondrillae* (Canestrini) (a gall forming mite), in an integrated weed management programme to maximise its benefits. **Key reference(s):** 1) J M Cullen. 1985. Bringing the cost benefit analysis of biological control of *Chondrilla juncea* up to date. In *Proc. VI International Symposium on Biological Control of Weeds*, E S Delfosse (ed.), pp. 145–52, Agriculture Canada, Ottawa, Canada. 2) E S Delfosse, S Hasan, J M Cullen & A J Wapshere. 1985. Beneficial use of an exotic phytopathogen, *Puccinia chondrillina*, as a biological control agent for skeleton weed, *Chondrilla juncea*, in Australia. In *Pests and Parasites as Migrants*, A J Gibbs & H R C Meischke (eds), pp. 171–7, Cambridge University Press, Cambridge, UK. 3) J M Cullen & S Hasan. 1986. Pathogens for the control of weeds, *Phil. Trans. Royal Soc. London*, **318**, 213–24. 4) D M Supkoff, D B Joley & J J Marois. 1988. Effect of introduced biological control organisms on the density of *Chondrilla juncea* in California, *J. Appl. Ecol.*, **25**, 1089–95.

COMMERCIALISATION: Formulation: Introduced into regions of the world where rush skeleton weed had become established. It is now common in these regions and further introductions are not considered necessary.

APPLICATION: Infected plants are placed within the area to be treated. The spores of the fungus are distributed by wind and rain to other plants.

PRODUCT SPECIFICATIONS: Specifications: Produced on living rush skeleton weed (*Chondrilla juncea* L.) plants, but now widely introduced and endemic in many areas.

MAMMALIAN TOXICITY: There are no reports of allergic or other adverse toxicological effects from the use of *Puccinia chondrillina*.

ENVIRONMENTAL IMPACT AND NON-TARGET TOXICITY: *Puccinia chondrillina* is specific to skeleton weed and, as such, is not expected to have any adverse effects on non-target organisms or on the environment. **Approved for use in organic farming:** Yes.

1:113 *Puccinia evadens*
Biological herbicide (fungus)

Fungus: Basidiomycetes: Uredinales

NOMENCLATURE: Approved name: *Puccinia evadens* Hark.

SOURCE: Introduced into Australia from Florida, USA.

PRODUCTION: Produced on live groundsel bush (cotton seed tree) (*Baccharis halimifolia* L.), but now self-perpetuating.

TARGET PESTS: Groundsel bush (cotton seed tree) (*Baccharis halimifolia*). Groundsel bush is a native of Florida and coastal areas adjacent to the eastern side of the Gulf of Mexico. It was introduced into the Brisbane region of Australia as an ornamental plant in 1900 and has spread along the coastal areas of south-east Queensland and down the New South Wales coast. Groundsel bush is a rapid coloniser of cleared, unused land and is particularly suited to moist gullies, salt marsh areas and wetlands. It also does well on high, cleared slopes. Most germination occurs in the autumn/winter period. Plants normally do not flower in the first year of growth. Two-metre-tall plants can produce up to a million seeds a year.

BIOLOGICAL ACTIVITY: Mode of action: Pathogenesis. **Biology:** *Puccinia evadens* acts as both a leaf and a stem parasite causing development of rust-brown pustules surrounded by yellow tissue, leading to defoliation during summer and winter and stem dieback over summer. The infection process requires a moisture film on the leaf or stem surface. The dry spores are spread by wind. Infected seedlings can be killed or plants have reduced growth, vigour and reproductive potential. There is evidence of stunting, deformation, reduced branching, fewer flower buds and lower viability of seeds where plants suffer severe rusting.

COMMERCIALISATION: Formulation: Not sold commercially.

APPLICATION: Now that it is present in Australia, *Puccinia evadens* infects plants naturally.

PRODUCT SPECIFICATIONS: Specifications: Produced on live groundsel bush (cotton seed tree) (*Baccharis halimifolia*), but now self-perpetuating.

MAMMALIAN TOXICITY: There are no reports of allergic or other adverse toxicological effects from the use of *Puccinia evadens*. The rust is widely spread in Florida and there is no evidence of allergenic or other adverse reactions. It only infects the groundsel bush.

ENVIRONMENTAL IMPACT AND NON-TARGET TOXICITY: *Puccinia evadens* is specific to the groundsel bush and, as such, is not expected to have any adverse effects on non-target organisms or on the environment. **Approved for use in organic farming:** Yes.

1:114 *Puccinia jaceae* var. *solstitialis* isolate FDWSRU 84-71 *Biological herbicide (fungus)*

Fungus: Basidiomycetes: Uredinales

NOMENCLATURE: Approved name: *Puccinia jaceae* (Otth.) var. *solstitialis* isolate FDWSRU 84-71 (formerly TR 84-96).

SOURCE: The isolate was collected in 1984, east of Yarhisar and Hafik, Turkey.

PRODUCTION: Harvested from living yellow starthistle (*Centaurea solstitialis* L.) plants.

TARGET PESTS: Yellow starthistle (*Centaurea solstitialis*).

BIOLOGICAL ACTIVITY: Mode of action: Pathogenesis. **Biology:** The rust fungus *Puccinia jaceae* var. *solstitialis* isolate FDWSRU 84-71 was first released as a classical biological control for yellow starthistle in California in 2003. At that time, it was monitored to determine the best time to release the agent, to determine if the rust could complete its full life-cycle in California and to determine if it is likely to establish at two field sites in California. Subsequently, in 2005–2006, *Puccinia jaceae* was released monthly from January to June at two sites. All releases resulted in infected plants at both sites in both years. Urediniospores (infecting spores) were short-lived, and remained viable less than three weeks in the field. Teliospores (overwintering spores) emerged as the yellow starthistle senesced, and germinated as yellow starthistle seedlings emerged in the winter. Pycnia were observed shortly after teliospore germination, indicating the rust can complete its full life-cycle in California. Re-emergence, the appearance of the rust one year after inoculations, occurred only at one site. These results suggest that *Puccinia jaceae* is likely to persist in California, although establishment is expected to be limited by local environmental conditions.

Key reference(s): A J Fisher, W L Bruckart, M McMahon, D G Luster & L Smith. 2006. First report of *Puccinia jaceae* var. *solstitialis* pycnia on yellow starthistle in the United States, *Plant Disease*, **90**, 1362.

COMMERCIALISATION: Formulation: *Puccinia jaceae* var. *solstitialis* is not sold commercially.

PRODUCT SPECIFICATIONS: Specifications: Harvested from living yellow starthistle (*Centaurea solstitialis*) plants.

MAMMALIAN TOXICITY: There are no reports of allergic or other adverse toxicological effects from the use of *Puccinia jaceae* var. *solstitialis*. The rust is widely spread in Bulgaria and Turkey and there is no evidence of allergenic or other adverse reactions. It only infects the yellow starthistle.

ENVIRONMENTAL IMPACT AND NON-TARGET TOXICITY: *Puccinia jaceae* var. *solstitialis* is specific to the yellow starthistle and, as such, is not expected to have any adverse effects on non-target organisms or on the environment. **Approved for use in organic farming:** Yes.

1:115 *Puccinia myrsiphylli*
Biological herbicide (fungus)

Fungus: Basidiomycetes: Uredinales

NOMENCLATURE: Approved name: *Puccinia myrsiphylli* (Thuem) Wint. **Other names:** Bridal creeper rust fungus.

SOURCE: Isolated from bridal creeper (*Asparagus asparagoides* L. Druce) in South Africa and approved for release in Australia in 2000.

PRODUCTION: Not sold commercially.

TARGET PESTS: Bridal creeper (*Asparagus asparagoides*).

TARGET CROPS: Rangeland, woods and recreational areas.

BIOLOGICAL ACTIVITY: Mode of action: Pathogenesis. **Biology:** The rust fungus is not systemic, which means it does not spread internally throughout the plant body. Therefore, it must reinfect plants every growing season. The teliospore stage of the life-cycle can survive adverse conditions (such as summer when there is no bridal creeper) and ensures that inoculum of the rust fungus is available for a new disease cycle to be initiated the following season. **Key reference(s):** 1) C A Kleinjan, L Morin, P B Edwards & A R Wood. 2004. Distribution, host range and phenology of the rust fungus *Puccinia myrsiphylli* in South Africa, *Australasian Plant Pathology*, **33**, 263–71. 2) L Morin, A J Willis, J Armstrong & D Kriticos. 2002. Spread, epidemic development and impact of the bridal creeper rust fungus in Australia: summary of results. In *Proceedings of the 13th Australian Weeds Conference*, Perth, Western Australia, 8–13 September 2002, H Spafford-Jacob, J Dodd and J Moore (eds). 385–8.

COMMERCIALISATION: Formulation: Introduced into Australia in 2000. Subsequently, the rust has infected bridal creeper without the need for additional introductions.

PRODUCT SPECIFICATIONS: Specifications: Not sold commercially.

MAMMALIAN TOXICITY: There are no reports of allergic or other adverse toxicological effects from the use of *Puccinia myrsiphylli*. The rust only infects bridal creeper plants.

ENVIRONMENTAL IMPACT AND NON-TARGET TOXICITY: It is not expected that *Puccinia myrsiphylli* will have any adverse effects on non-target organisms or the environment. It is highly host-specific towards bridal creeper. In South Africa, it has never been reported on cultivated asparagus, a closely related species to bridal creeper, and many of the other *Asparagus* species that are found in South Africa. Experimental host-specificity testing, which involved bridal creeper and 42 plant species, including 13 cultivars of cultivated asparagus, has confirmed the extremely narrow host range of the bridal creeper rust fungus. In those tests, bridal creeper was the only species found to be highly susceptible to the rust and to develop uredinia. All other species tested were found to be immune or highly resistant to the rust. Although the rust fungus penetrated successfully some of the test plant species, no infection hyphae, haustorium mother cells or haustoria, and hence macroscopic symptoms, ever developed. These results demonstrated to the authorities that *Puccinia myrsiphylli* does not pose a threat to plant species other than bridal creeper and consequently that it was safe for release in Australia.

1:116 *Puccinia thlaspeos*
Biological herbicide (fungus)

Fungus: Basidiomycetes: Uredinales

NOMENCLATURE: Approved name: *Puccinia thlaspeos* C Schub. **Other names:** Dyer's woad rust fungus. **OPP Chemical Code:** 006489.

SOURCE: *Puccinia thlaspeos* was discovered in southern Idaho in 1979. It is now widespread in northern USA.

PRODUCTION: Produced on living dyer's woad (*Isatis tinctoria* L.) plants.

TARGET PESTS: Dyer's woad (*Isatis tinctoria*).

TARGET CROPS: Rangeland, rights of way and farmland. It is not approved for use on food crops.

BIOLOGICAL ACTIVITY: Mode of action: *Puccinia thlaspeos* was not known to be a pathogen of dyer's woad until 1978, when it was found on woad in southern Idaho. Since its discovery, woad rust has rapidly spread from Idaho into northern Utah, where it now occurs in most populations of dyer's woad. Dyer's woad rust differs from most other rust fungi used in weed biocontrol by causing systemic infections. These infections are usually asymptomatic during the first year of the woad life-cycle, with the fungus over-wintering in the tissue of infected plants. Symptoms typically appear during the second season, before the plants bolt. An additional, and very important, difference between woad rust and other rusts used in biocontrol is that a single basidiospore infection by woad rust almost entirely prevents seed production by systemically infected plants. **Efficacy:** The only host plant for *Puccinia thlaspeos* is *Isatis tinctoria* and the rust spreads very rapidly within this species following initial inoculation.

Key reference(s): B R Kropp, D R Hansen & S V Thomson. 2002. Establishment and dispersal of *Puccinia thlaspeos* in field populations of dyer's woad, *Plant Dis.*, **86**, 241–6.

COMMERCIALISATION: Formulation: The product consists of rust spores on finely ground leaf and stem pieces of infected dyer's woad (1.7×10^{10} teleospores/kg). **Tradenames:** 'Woad Warrior' (Greenville Farms).

APPLICATION: Apply as a spray or powder directly to soil and young plants in April or May when new plants are beginning their growth, at a rate of 8 kg per hectare (7 lb per acre). One application is usually sufficient, since the rust spreads by itself. In areas heavily infested with dyer's woad, it may be necessary to repeat the application in subsequent years.

PRODUCT SPECIFICATIONS: Specifications: Produced on living dyer's woad (*Isatis tinctoria* L.) plants.

MAMMALIAN TOXICITY: There are no reports of allergic or other adverse toxicological effects from the use of *Puccinia thlaspeos*. The rust only infects dyer's woad plants.

ENVIRONMENTAL IMPACT AND NON-TARGET TOXICITY: *Puccinia thlaspeos* is specific to dyer's woad and, as such, is not expected to have any adverse effects on non-target organisms or on the environment.

1:117 *Pythium oligandrum* isolate DV 74
Biological fungicide (fungus)

Chromista: Oomycetes: Pythiales

NOMENCLATURE: Approved name: *Pythium oligandrum* Drechsler isolate DV 74. **Development code:** ATCC No: 38472.

SOURCE: *Pythium oligandrum* isolate DV 74 was originally isolated in the Czech Republic, but it is widely distributed throughout the world, being common in soil and in or on plants.

PRODUCTION: Manufactured by fermentation.

TARGET PESTS: A wide range of soil-borne fungal pathogens.

TARGET CROPS: Glasshouse and outdoor vegetables, cereals and non-food tree crops.

BIOLOGICAL ACTIVITY: Mode of action: *Pythium oligandrum* outcompetes pathogenic soil fungi and also stimulates the growth of crops, rendering them less susceptible to disease attack. **Biology:** The product is applied to the soil and rapidly becomes established in the crop rhizosphere, preventing the growth and pathogenicity of soil fungi.

Key reference(s): 1) M P McQuilken, P Halmer & D J Rhodes. 1998. Application of microorganisms to seeds. In *Formulation of Microbial Biopesticides, Beneficial Microorganisms, Nematodes and Seed Treatments*, H D Burges (ed.), pp. 255–85, Kluwer Academic Publishers, Dordrecht, The Netherlands. 2) D R Fravel, W J Connick Jr & J A Lewis. 1998. Formulation of microorganisms to control plant diseases. In *Formulation of Microbial Biopesticides, Beneficial Microorganisms, Nematodes and Seed Treatments*, H D Burges (ed.), pp. 187–202, Kluwer Academic Publishers, Dordrecht, The Netherlands. 3) A S Paau. 1998. Formulation of beneficial microorganisms applied to soil. In *Formulation of Microbial Biopesticides, Beneficial Microorganisms, Nematodes and Seed Treatments*, H D Burges (ed.), pp. 235–54, Kluwer Academic Publishers, Dordrecht, The Netherlands.

COMMERCIALISATION: Formulation: Sold as wettable powders (WP).
Tradenames: 'Polygandron' and 'Polyversum' (Remeslo) and (Biopreparaty).

APPLICATION: Apply 5 g product per kilogram of seeds. For individual established plants, suspend 5 g of product in 10 litres of water and apply 5 ml per plant. A single application is usually sufficient for a season.

PRODUCT SPECIFICATIONS: Specifications: Manufactured by fermentation.
Storage conditions: Store in a sealed container in a cool, dry place. Do not expose to extremes of temperature or direct sunlight. **Shelf-life:** Retains viability for a year, under correct storage conditions.

COMPATIBILITY: Do not apply with chemical fungicides. Use non-chlorinated water to prepare suspensions. Incompatible with strong oxidisers, acids and bases.

MAMMALIAN TOXICITY: *Pythium oligandrum* has not caused allergic or other adverse effects in research workers, manufacturing or field staff. It is considered to be of low mammalian toxicity.

ENVIRONMENTAL IMPACT AND NON-TARGET TOXICITY: *Pythium oligandrum* occurs in Nature and is not thought to have any adverse effects on non-target organisms or on the environment.

1:118 *Sclerotinia minor* isolate 344141
Biological herbicide (fungus)

Fungus: Ascomycetes: Helotiales: Sclerotiniaceae

NOMENCLATURE: Approved name: *Sclerotinia minor* Jagger isolate 344141.
Development code: IMI 344141.

SOURCE: *Sclerotinia minor* isolate 344141 was obtained from a lettuce field in Sherrington, Quebec, Canada in 1983.

PRODUCTION: Produced by fermentation.

TARGET PESTS: Suppression of dandelion (*Taraxacum officinale* Weber) top growth.

TARGET CROPS: Commercial and domestic lawns, golf courses, municipal parks and turf farms.

BIOLOGICAL ACTIVITY: Mode of action: Oxalic acid, secreted by *Sclerotinia minor* isolate 344141, is thought to be the main component of its phytotoxicity to dandelion plants. The mechanisms by which oxalate secretions contribute to *Sclerotinia minor* virulence centre on three modes of action: 1) Several of the fungal enzymes secreted during invasion of plant tissues have maximal activities at low pH, thus oxalate may aid *Sclerotinia minor* virulence by lowering the apoplastic pH to a value better suited for enzymatic degradation of plant cell walls; 2) Oxalate may be directly toxic to host plants, thereby facilitating invasion; and 3) Chelation of cell wall calcium ions by the oxalate anion compromises the function of calcium dependent defence responses, thus weakening the plant cell wall. **Biology:** Disease develops quickly and complete kill of dandelion and other broadleaf weeds can be achieved within 7 days, about twice as fast as with the standard chemical herbicides. The product is compatible with normal lawn maintenance operations such as mowing, fertilisation and irrigation.
Key reference(s): A Watson. 2007. *Sclerotinia minor* – Biocontrol target or agent? In *Novel Biotechnologies for Biocontrol Agent Enhancement and Management*, M Vurro & J Gressel (eds), pp. 205–11, Springer, Dordrecht, The Netherlands.

COMMERCIALISATION: Formulation: Health Canada's Pest Management Regulatory Agency has granted conditional registration for the sale and use of Sarritor Technical Herbicide containing *Sclerotinia minor* isolate IMI 344141 and the end-use products 'Sarritor Granular Biological Herbicide (Commercial)' and 'Sarritor Domestic Granular Biological Herbicide' to suppress dandelion top growth in turf. The products are dry granules 1.4–2.0 mm in diameter containing 300 colony forming units (cfu) per g. **Tradenames:** 'Sarritor Granular Biological Herbicide (Commercial)' and 'Sarritor Domestic Granular Biological Herbicide' (Sarritor). **Patents:** Canadian Patent No. 2,292,233, Watson *et al.* (granted 24 May 2005).

APPLICATION: Applied at a rate of 40 g product/m^2 in the spring and/or the autumn when daytime high temperatures are 18 to 24°C and rainfall or irrigation will occur within 12 hours of application. The higher application rate of 60 g product/m^2 can be applied when environmental conditions are suboptimal (i.e., daily maximum temperatures are outside the optimal range of 18 to 24°C, but do not surpass 27°C) and/or when the turf is highly infested with dandelions. 'Sarritor Granular Biological Herbicide (Commercial)' and 'Sarritor Domestic Granular Biological Herbicide' should not be applied in hot and dry weather.

PRODUCT SPECIFICATIONS: Specifications: Produced by fermentation. **Purity:** The technical grade active ingredient does not contain any impurities or microcontaminants known to be Toxic Substances Management Policy (TSMP) Track-1 substances. The product meets microbiological contaminants release standards and no mammalian toxins are known to be produced by *Sclerotinia minor* isolate IMI 344141. **Shelf-life:** May be kept for 7 months if stored refrigerated and 9 months if stored frozen.

COMPATIBILITY: Compatible with normal lawn maintenance operations such as mowing, fertilisation and irrigation. Incompatible with broad-spectrum products, such as copper-based fungicides. Incompatible with strong oxidisers, acids, bases and chlorinated water.

MAMMALIAN TOXICITY: *Sclerotinia minor* is ubiquitous in nature, and globally widespread in temperate zones. Application of *Sclerotinia minor* isolate IMI 344141 is not expected to increase significantly the natural environmental background levels of this micro-organism. No adverse effects from dietary exposure have been attributed to natural populations of *Sclerotinia minor* and none were observed during acute oral toxicity testing. Furthermore, no food uses are proposed for *Sclerotinia minor* isolate IMI 344141. The establishment of a maximum residue limit (MRL) is, therefore, not required. There are no reports of any allergic or other adverse effects following the use of the product. **Skin and eye:** The products have been shown to be non-irritating to minimally irritating when applied to the skin and eyes of the rabbit. **Other toxicological effects:** Although there have been no hypersensitivity reactions reported in people during the development and testing of the products, microbial pest control agents are considered to be potential sensitisers. Exposure to allergens, including *Sclerotinia minor* isolate IMI 344141, may cause allergic reactions following repeated exposures to high concentrations. As a result, the signal words 'Potential Sensitizer' are required on the principal display panels of all product labels.

ENVIRONMENTAL IMPACT AND NON-TARGET TOXICITY: It is not expected that *Sclerotinia minor* will have any adverse effects on non-target organisms or the environment. **Approved for use in organic farming:** Yes.

1:119 *Sclerotinia sclerotiorum*
Biological herbicide (fungus)

Fungus: Ascomycetes: Helotiales: Sclerotiniaceae

NOMENCLATURE: Approved name: *Sclerotinia sclerotiorum* de Bary.

SOURCE: Isolated from a diseased thistle in Canada.

PRODUCTION: Produced by fermentation.

TARGET PESTS: *Carduus* species and *Cirsium* species.

TARGET CROPS: Pasture and rangeland.

BIOLOGICAL ACTIVITY: Mode of action: Pathogenesis. **Biology:** This isolate of *Sclerotinia sclerotiorum* is specific to thistles of the genera *Carduus* and *Cirsium*. When applied to the weed, it penetrates the plant's cuticle and invades the weed, leading to plant death.
Key reference(s): 1) M P Greaves. 1996. Microbial herbicides: factors in development. In *Crop Protection Agents from Nature: Natural Products and Analogues*, L G Copping (ed.), pp. 444–67, Royal Society of Chemistry, Cambridge, UK. 2) M P Greaves, P J Holloway & B A Auld. 1998. Formulation of microbial herbicides. In *Formulation of Microbial Biopesticides, Beneficial Microorganisms, Nematodes and Seed Treatments*, H D Burges (ed.), pp. 202–33, Kluwer Academic Publishers, Dordrecht, The Netherlands.

COMMERCIALISATION: Formulation: Sold as a wettable powder (WP) formulation of spores in an inert carrier. **Tradenames:** 'Sclerodex' (AgResearch).

APPLICATION: Applied over the top of the developing weed, under conditions of high humidity.

PRODUCT SPECIFICATIONS: Specifications: Produced by fermentation.
Storage conditions: Store in a sealed container under cool, dry conditions, out of direct sunlight. **Shelf-life:** Use as soon as possible after receipt.

COMPATIBILITY: *Sclerotinia sclerotiorum* is compatible with chemical herbicides, but should not be used in conjunction with a fungicide. Incompatible with strong oxidisers, acids, bases and chlorinated water.

MAMMALIAN TOXICITY: There are no reports of any allergic or other adverse effects following the use of the product.

ENVIRONMENTAL IMPACT AND NON-TARGET TOXICITY: *Sclerotinia sclerotiorum* occurs widely in Nature and is not expected to have any adverse effects on non-target organisms or on the environment.

1:120 *Septoria passiflorae*

Biological herbicide (fungus)

Fungus: Ascomycete: Mycosphaerellales

NOMENCLATURE: Approved name: *Septoria passiflorae* Syd.

SOURCE: *Septoria passiflorae* was discovered in 1991 at Aldana near Ipiales, Nariño, Colombia. It was transported to Hawaii in air-dried diseased banana poka (*Passiflora tarminiana* Coppens & Barney) leaves.

PRODUCTION: Produced on banana poka plants.

TARGET PESTS: Banana poka (*Passiflora tarminiana*).

TARGET CROPS: Used on the upland wet and mesic forests of the islands of Hawaii.

BIOLOGICAL ACTIVITY: Mode of action: Pathogenesis. **Biology:** Inoculum is readily airborne, with diseased vines being found more than 2.5 km from the initial inoculation sites in the opposite direction to the trade wind. **Efficacy:** Following the initial inoculations of *Septoria passiflorae* into Hawaii in 1996, nine sites had severe disease epidemics by 1998, and vine defoliation was >90%. Widespread epidemics of the disease occurred in 1999, resulting in estimated 80 to 95% biomass reductions in more than 2000 hectares of native forest infested with banana poka. **Key reference(s):** E E Trujillo, C Kadooka, V Tanimoto, S Bergfeld, G Shishido & G Kawakami. 2001. Effective biomass reduction of the invasive weed species banana poka by *Septoria* leaf spot, *Plant Disease*, **85**, 357–61.

COMMERCIALISATION: Formulation: Not sold commercially.

APPLICATION: In the initial release, banana poka vines were sprayed with 5×10^5 cfu/litre of distilled water with 0.5% gelatin and 2% sucrose. Spray was directed to the underside of the foliage and was applied to vines from 1 to 2 m from the surface of the forest floor.

PRODUCT SPECIFICATIONS: Specifications: Produced on banana poka plants.

MAMMALIAN TOXICITY: *Septoria passiflorae* occurs widely in South America and there are no reports of allergic or other adverse toxicological effects following contact with it. It only infects *Passiflora tarminiana*.

ENVIRONMENTAL IMPACT AND NON-TARGET TOXICITY: *Septoria passiflorae* is not expected to have any adverse effects on non-target organisms or on the environment. Host range studies with *S. passiflorae* conducted in 1993 and 1994 at the Hawaii Department of Agriculture Quarantine Laboratory confirmed its specificity to the banana poka vine.

1:121 *Serratia entomophila*
Biological insecticide (bacterium)

Bacterium: Eubacteriales: Enterobacteriaceae

NOMENCLATURE: Approved name: *Serratia entomophila* Grimont. **Other names:** Amber disease.

SOURCE: *Serratia entomophila* is a soil-inhabiting, non-spore-forming, aerobic bacterium that has specific activity against, and has been isolated from, the New Zealand grass grub, *Costelytra zealandica* (Coleoptera: Scarabaeidae).

PRODUCTION: *Serratia entomophila* is manufactured and formulated by Industrial Research Ltd and produced by fermentation.

TARGET PESTS: New Zealand grass grub (*Costelytra zealandica* (White)).

TARGET CROPS: Established pasture.

BIOLOGICAL ACTIVITY: Biology: The bacterium enters the insect through the mouth and adheres to the surface of the foregut, where it inhibits feeding, the biosynthesis of digestive enzymes and, finally, invades the insect haemocoel. **Efficacy:** After death, the infected larvae disintegrate, passing additional bacteria into the soil to infect additional grub larvae. The genes associated with the disease process are encoded on a single plasmid. **Key reference(s):** A S Paau. 1998. Formulation of beneficial microorganisms applied to soil. In *Formulation of Microbial Biopesticides, Beneficial Microorganisms, Nematodes and Seed Treatments*, H D Burges (ed.), pp. 235–54, Kluwer Academic Publishers, Dordrecht, The Netherlands.

COMMERCIALISATION: Formulation: Formulated as a liquid suspension containing 4×10^{13} live cells per litre. **Tradenames:** 'Invade' (Wrightson Seeds).

APPLICATION: The formulated product is applied at a rate of 1 litre per hectare.

PRODUCT SPECIFICATIONS: Specifications: *Serratia entomophila* is manufactured and formulated by Industrial Research Ltd and produced by fermentation. **Purity:** The bacterium is produced by fermentation, with checks made on the quality of the product prior to formulation. The product is evaluated for efficacy by bioassay against the New Zealand grass grub. **Storage conditions:** Store in a refrigerator. Do not freeze. **Shelf-life:** Sealed containers can be stored for up to 6 months, if kept under cool conditions.

COMPATIBILITY: The product should not be used in conjunction with broad-spectrum biocides. Incompatible with strong oxidisers, acids, bases and chlorinated water.

MAMMALIAN TOXICITY: There have been no adverse toxic reactions to the bacterium from research workers, manufacturers or users.

ENVIRONMENTAL IMPACT AND NON-TARGET TOXICITY: *Serratia entomophila* occurs in Nature and is not expected to have any adverse effects on non-target organisms nor to have any deleterious effects on the environment.

1:122 *Sphaerulina mimosae-pigrae*
Biological herbicide (fungus)

Fungus: Ascomycetes: Mycosphaerellales: Mycosphaerellaceae

NOMENCLATURE: **Approved name:** *Sphaerulina mimosae-pigrae* Evans & Carrión.
Other names: Anamorph: *Phloeospora mimosae-pigrae* Evans & Carrión.

SOURCE: The fungus was identified as occurring in Mexico, Trinidad, Venezuela, Colombia and Brazil, and samples were evaluated for control of the sensitive plant, *Mimosa pigra* L., in Australia.

TARGET PESTS: Mimosa or giant sensitive plant or giant sensitive tree (*Mimosa pigra*).

TARGET CROPS: The wetlands and floodplains of northern Australia.

BIOLOGICAL ACTIVITY: **Mode of action:** Pathogenesis. **Efficacy:** Very effective at reducing the growth and reproductive capability of *Mimosa pigra*. **Key reference(s):** H C Evans, G Carrión & G Guzman. 1993. A new species of *Sphaerulina* and its *Phloeospora* anamorph, with potential for biological control of *Mimosa pigra*, *Mycological Research*, **97**, 59–67.

COMMERCIALISATION: **Formulation:** Not sold commercially, but introduced into Australia as part of the national control strategy.

APPLICATION: Now it is present in Australia, *Sphaerulina mimosae-pigrae* infects plants naturally.

MAMMALIAN TOXICITY: There are no reports of allergic or other adverse toxicological effects from the use of *Sphaerulina mimosae-pigrae*. The pathogen is widely spread in South and Central America and there is no evidence of allergenic or other adverse reactions. It only infects *Mimosa pigra*.

ENVIRONMENTAL IMPACT AND NON-TARGET TOXICITY: *Sphaerulina mimosae-pigrae* is specific to *Mimosa pigra* and, as such, is not expected to have any adverse effects on non-target organisms or on the environment.

1:123 *Spodoptera exigua* nucleopolyhedrovirus *Insecticidal baculovirus*

The Pesticide Manual **Fifteenth Edition entry number:** 789

Virus: Baculoviridae: Nucleopolyhedrovirus

NOMENCLATURE: Approved name: *Spodoptera exigua* multicapsid nucleopolyhedrovirus. **Other names:** *Spodoptera exigua* nucleopolyhedrovirus; *Spodoptera exigua* MNPV; *Spodoptera exigua* NPV; SeMNPV; SeNPV.

SOURCE: Baculovirus that occurs widely in Nature, originally isolated from the beet armyworm, *Spodoptera exigua* (Hübner).

PRODUCTION: Produced in larvae of *Spodoptera exigua* under controlled conditions. The baculovirus is separated from the larval cadavers by centrifugation.

TARGET PESTS: Beet armyworm (*Spodoptera exigua*).

TARGET CROPS: Recommended for use in various outdoor crops, including cotton, vegetables, grapes, ornamentals and glasshouse vegetables and ornamentals.

BIOLOGICAL ACTIVITY: Mode of action: As with all insect baculoviruses, *Spodoptera exigua* MNPV must be ingested to exert an effect. Following ingestion, the virus enters the insect's haemolymph, where it multiplies in the insect body, leading to death. **Biology:** SeMNPV is more active on small larvae than on later larval instars. The virus is ingested by the feeding larva and the virus protein matrix is dissolved in the alkaline insect midgut, releasing the virus particles. These pass through the peritrophic membrane and invade midgut cells by fusion with the microvilli. The virus particles invade the cell nuclei, where they are uncoated and replicated. Initial replication produces non-occluded virus particles which invade more cells to hasten the invasion of the host insect, but later the virus particles are produced with protein coats that remain infective when released from the dead insects. **Efficacy:** SeMNPV acts relatively slowly, as it has to be ingested before it exerts any effect on the insect host. It is important to ensure good cover of the foliage to effect good control. Monitoring of adult insect laying patterns and application targeted at newly hatched eggs gives better control than on a mixed population. **Key reference(s):** 1) R R Granados & B A Federici (eds). 1986. *The Biology of Baculoviruses, Vols 1 and 2*, CRC Press, Boca Raton, Florida, USA. 2) D J Leisy & J R Fuxa. 1996. Natural and engineered viral agents for insect control. In *Crop Protection Agents from Nature: Natural Products and Analogues*, L G Copping (ed.), Royal Society of Chemistry, Cambridge, UK. 3) B A Federici. 1999. Naturally occurring baculoviruses for insect pest control. In *Biopesticides Use and Delivery*, F R Hall & J J Menn (eds), pp. 301–20, Humana Press, Totowa, NJ, USA. 4) H D Burges & K A Jones. 1998. Formulation of bacteria, viruses and Protozoa to control insects. In *Formulation of Microbial Biopesticides, Beneficial*

Microorganisms, Nematodes and Seed Treatments, H D Burges (ed.), pp. 33–127, Kluwer Academic Publishers, Dordrecht, The Netherlands.

COMMERCIALISATION: Formulation: Formulated as wettable powder formulations (WP) and liquid concentrates. 'Spod-X' is listed in EU Annex 1. **Tradenames:** 'Ness-A' (Applied Chemicals Thailand), 'Ness-E' (Applied Chemicals Thailand), 'Spod-X LC' (Certis), (Brinkman) and (Rincon-Vitova).

APPLICATION: Monitor the occurrence of adults and apply during egg laying. It is important to ensure that the foliage is well covered.

PRODUCT SPECIFICATIONS: Specifications: Produced in larvae of *Spodoptera exigua* under controlled conditions. The baculovirus is separated from the larval cadavers by centrifugation. **Purity:** Rod-shaped, elongated particles enclosed in a crystalline protein matrix (occlusion body), with no contaminating human pathogenic bacteria. Efficacy of the formulation can be determined by bioassay on *Spodoptera exigua* larvae. **Storage conditions:** Store in a cool, dry place. Keep liquid formulations refrigerated, but do not freeze. Dry formulations can be stored frozen. Do not expose to sunlight. **Shelf-life:** Wettable powder formulations are stable for up to one year. Liquid formulations should be used within 3 months.

COMPATIBILITY: Compatible with most crop protection agents that do not exert a repellent effect on *Spodoptera exigua*, but do not use with copper-based fungicides, strong oxidisers, acids, bases or chlorinated water. Should be applied in water at a neutral pH.

MAMMALIAN TOXICITY: *Spodoptera exigua* MNPV has not demonstrated evidence of toxicity, infectivity or irritation to mammals. No allergic responses or other adverse health problems have been observed by research workers, manufacturing staff or users.

ENVIRONMENTAL IMPACT AND NON-TARGET TOXICITY: *Spodoptera exigua* MNPV occurs in Nature and, as such, is not expected to show any adverse effects on non-target organisms or on the environment.

1:124 *Spodoptera litura* nucleopolyhedrovirus   *Insecticidal baculovirus*

Virus: Baculoviridae: Nucleopolyhedrovirus

NOMENCLATURE: Approved name: *Spodoptera litura* multicapsid nucleopolyhedrovirus. **Other names:** *Spodoptera litura* MNPV; *Spodoptera litura* NPV; SINPV; SIMNPV.

SOURCE: Baculovirus that occurs widely in Nature. Originally isolated from the beet armyworm (*Spodoptera litura* (Hübner)).

PRODUCTION: Produced in larvae of *Spodoptera litura* under controlled conditions. The baculovirus is separated from the larval cadavers by centrifugation.

TARGET PESTS: Beet armyworm (*Spodoptera litura*).

TARGET CROPS: Recommended for use in various outdoor crops, including cotton, vegetables, grapes, ornamentals and glasshouse vegetables and ornamentals.

BIOLOGICAL ACTIVITY: Mode of action: As with all insect baculoviruses, *Spodoptera litura* NPV must be ingested to exert an effect. Following ingestion, the virus enters the insect's haemolymph where it multiplies in the insect body, leading to death.
Biology: *Spodoptera litura* NPV is more active on small larvae than on later larval instars. The virus is ingested by the feeding larva and the protective protein matrix is dissolved in the alkaline insect midgut, releasing the virus particles. These pass through the peritrophic membrane and invade midgut cells by fusion with the microvilli. The virus particles invade the cell nuclei, where they are uncoated and replicated. Initial replication produces non-occluded virus particles which invade more cells to hasten the invasion of the host insect, but later the virus particles are produced with protein matrices and remain infective when released from the dead insects. **Efficacy:** *Spodoptera litura* NPV acts relatively slowly as it has to be ingested before it exerts any effect on the insect host. It is important to ensure good cover of the foliage to effect good control. Monitoring of adult insect laying patterns and application targeted at newly hatched eggs gives better control than on a population of mixed ages. **Key reference(s):** 1) R R Granados & B A Federici (eds). 1986. *The Biology of Baculoviruses, Vols 1 and 2*, CRC Press, Boca Raton, Florida, USA. 2) D J Leisy & J R Fuxa. 1996. Natural and engineered viral agents for insect control. In *Crop Protection Agents from Nature: Natural Products and Analogues*, L G Copping (ed.), Royal Society of Chemistry, Cambridge, UK.
3) B A Federici. 1999. Naturally occurring baculoviruses for insect pest control. In *Biopesticides Use and Delivery*, F R Hall & J J Menn (eds), pp. 301–20, Humana Press, Totowa, NJ, USA.
4) H D Burges & K A Jones. 1998. Formulation of bacteria, viruses and Protozoa to control insects. In *Formulation of Microbial Biopesticides, Beneficial Microorganisms, Nematodes and Seed Treatments*, H D Burges (ed.), pp. 33–127, Kluwer Academic Publishers, Dordrecht, The Netherlands. 5) M S Chari & N G Patel. 1983. Cotton leafworm *Spodoptera litura* (Fabr.): its biology and integrated control measures, *Cotton Development*, **13**, 7.

COMMERCIALISATION: Formulation: Formulated as wettable powder formulations (WP) and liquid concentrates. **Tradenames:** 'Spodo-lure' (Agri Life), 'Spodostar' (Agri Life).

APPLICATION: Monitor the occurrence of adults and apply during egg laying. It is important to ensure that the foliage is well covered.

PRODUCT SPECIFICATIONS: Specifications: Produced in larvae of *Spodoptera litura* under controlled conditions. The baculovirus is separated from the larval cadavers by centrifugation. Purity: Rod-shaped, elongated particles enclosed in a protein crystalline matrix (occlusion body), with no human or mammalian pathogenic bacteria. Efficacy of the formulation can be determined by bioassay on *Spodoptera litura* larvae. Storage conditions: Store in a cool, dry place. Keep liquid formulations refrigerated, but do not freeze. Dry formulations can be stored frozen. Do not expose to sunlight. Shelf-life: Wettable powder formulations are stable for up to one year. Liquid formulations should be used within 3 months.

COMPATIBILITY: Compatible with most crop protection agents that do not exert a repellent effect on *Spodoptera litura*, but do not use with copper-based fungicides or chlorinated water. Should be applied in water at a neutral pH.

MAMMALIAN TOXICITY: *Spodoptera litura* NPV has not demonstrated evidence of toxicity, infectivity, irritation or hypersensitivity to mammals. No allergic responses or other adverse health problems have been observed by research workers, manufacturing staff or users.

ENVIRONMENTAL IMPACT AND NON-TARGET TOXICITY: *Spodoptera litura* NPV occurs in Nature and, as such, is not expected to show any adverse effects on non-target organisms or on the environment.

1:125 *Steinernema carpocapsae*
Insect parasitic nematode

The Pesticide Manual Fifteenth Edition entry number: 790

Nematode: Nematoda: Steinernematidae

NOMENCLATURE: Approved name: *Steinernema carpocapsae* Weiser.
Other names: Formerly known as *Neoaplectana carpocapsae* Weiser.
Common name: Beneficial nematodes.

SOURCE: *Steinernema carpocapsae* was originally a Holarctic species found in Europe, but is now widespread throughout the world.

PRODUCTION: Produced commercially by solid-state or liquid fermentation. Sold as third instar larvae.

TARGET PESTS: Vine weevil (*Otiorhyncus sulcatus* (Fabricius)) and other soil insects such as cutworms (*Agrotis* spp.). Also used to control *Gryllotalpa gryllotalpa* (L.), *Tipula* spp., mint borer, armyworms, billbugs, root weevils, fleas, stem borers and fungus gnats. The T-14 isolate is sold for the control of termites. 'Nematac C' is recommended for the control of the cranberry girdler and other insect pests.

TARGET CROPS: Glasshouse vegetables and ornamentals and outdoor strawberries, also vegetables and blackcurrants. Recommended for use in turf. 'Nematac C' is recommended for use in cranberries.

BIOLOGICAL ACTIVITY: Biology: The third stage larvae are the infectious stage and only these can survive outside the host insect, as they do not require food. They enter a host through one of its natural openings. The nematode larva then releases bacteria (*Xenorhabdus nematophilus* (Thomas & Poinar) Akhurst & Boemare) into the insect's body and toxins produced by these bacteria kill the insect within 48 hours. The bacteria then digest the insect's body into material that the nematode can feed upon, and the nematode's fourth stage develops within the dead insect. These fourth stage larvae develop into males and females that reproduce sexually. After mating, the males die and the females lay eggs in the dead insect if there is enough food available or, if not, the first and second stage larvae develop inside her body. When larvae reach the third stage, they are able to leave the dead insect and seek out a new host. However, there may be two or more generations within the host, dependent upon the availability of food. The third stage is always the infective stage.
Efficacy: Very effective predators that can be used to eradicate vine weevil populations. Also parasitise other soil-inhabiting insects. Can be used to control pet animal fleas.
Key reference(s): R Gaugler & H K Kaya. 1990. Entomopathogenic nematodes. In *Biological Control*, 365 pp., CRC Press, Boca Raton, Florida, USA.

COMMERCIALISATION: Formulation: Supplied as third instar larvae encapsulated within granules. **Tradenames:** 'Bio Safe WG' (for turf – Japan) (Certis), 'BioVector 25' (for cranberries and mint) (Certis), 'Ecomask' (BioLogic), 'Exhibit SC-WDG' (Syngenta Bioline), 'Guardian' (Hydro-Gardens), 'Millennium' (Certis), 'Nematac C' (Becker Underwood) and (MycoTech), 'Steinernema carpocapsae' (Neudorff) and (Rincon-Vitova), 'Steinernema carpocapsae T-14 isolate' (Rincon-Vitova), 'Steinernema sp.' (species not identified) (Biocontrol Network), 'Vector TL, WG' (for turfgrass and ornamentals) (Certis).

APPLICATION: Disperse the granules in water and apply by watering evenly over soil to be treated, at a rate of about 1×10^7 infective juvenile nematodes for each 20 square metres.

PRODUCT SPECIFICATIONS: Specifications: Produced commercially by solid-state or liquid fermentation. Sold as third instar larvae. **Purity:** Products contain only third instar infectious nematode larvae. Product efficacy can be determined by bioassay against vine weevil or other coleopteran larvae. **Storage conditions:** Store under cool conditions, but do not freeze. Keep out of direct sunlight and do not allow the temperature to exceed 35°C. **Shelf-life:** Granular formulations have extended the shelf-life to 60 days at room temperature and 125 to 180 days at 5°C.

COMPATIBILITY: Will not survive in manure. High humidity and temperatures above 15°C are necessary for effective control. Incompatible with strong oxidisers, acids and bases.

MAMMALIAN TOXICITY: No allergic or other adverse reactions have been reported by researchers, production staff or users during application in glasshouse or outdoor crops from *Steinernema carpocapsae* or its associated bacterium.

ENVIRONMENTAL IMPACT AND NON-TARGET TOXICITY: It is not expected that the use of products containing *Steinernema carpocapsae* will have any effects on non-target organisms nor have any adverse effects on the environment.
Approved for use in organic farming: Yes.

1:126 *Steinernema feltiae* isolate UK 76
Sciarid fly parasitic nematode

The Pesticide Manual **Fifteenth Edition entry number:** 790

Nematode: Nematoda: Steinernematidae

NOMENCLATURE: Approved name: *Steinernema feltiae* Filipjev. MicroBio isolate UK 76.
Other names: Formerly known as *Neoaplectana bibionis* Bovien, *Neoaplectana feltiae* Filipjev, *Neoaplectana leucaniae* Hoy, *Steinernema bibionis* (Bovien). **Common name:** Beneficial nematodes.

SOURCE: *Steinernema feltiae* was originally a Holarctic species found in Europe, but is now widespread throughout the world.

PRODUCTION: Produced commercially by solid-state or liquid fermentation and sometimes bred in insects. Sold as third instar larvae.

TARGET PESTS: Sciarid flies (*Bradysia* spp., *Lycoriella* spp. and *Sciara* spp.) and other soil insects. Also sold as a preventive control for vine weevils (*Otiorhyncus sulcatus* (Fabricius)), although this market is being eroded by the use of *S. kraussei*, due to its ability to survive and infect at low temperatures. 'Exhibit sf' is recommended for garden chafer (*Phyllopertha horticola* (L.)) control.

TARGET CROPS: Glasshouse vegetables and ornamentals, mushrooms and outdoor strawberries, vegetables and turf.

BIOLOGICAL ACTIVITY: Biology: The third stage larvae are the infectious stage and only these can survive outside the host insect, as they do not require food. They enter a host through one of its natural openings. The nematode larva then releases bacteria (*Xenorhabdus bovienii* Akhurst & Boemare) into the insect's body and toxins produced by these bacteria kill the insect within 48 hours. The bacteria then digest the insect's body into material that the nematode can feed upon, and the nematode's fourth stage develops within the dead insect. These fourth stage larvae develop into males and females that reproduce sexually. After mating, the males die and the females lay eggs in the dead insect if there is enough food available or, if not, the first and second stage larvae develop inside her body. The larvae may leave the host when they reach the third stage, but two or three generations may develop within the host, depending upon the availability of food. **Efficacy:** Very effective predators that can be used to eradicate sciarid fly populations. Also parasitise other soil-inhabiting insects. May be used to control pet animal fleas. **Key reference(s):** 1) R Gaugler & H K Kaya. 1990. Entomopathogenic nematodes. In *Biological Control*, 365 pp., CRC Press, Boca Raton, Florida, USA. 2) R Georgis & H K Kaya. 1998. Formulation of entomopathogenic nematodes. In *Formulation of Microbial Biopesticides*, *Beneficial Microorganisms*, *Nematodes and Seed Treatments*, H D Burges (ed.), pp. 289–308, Kluwer Academic Publishers, Dordrecht, The Netherlands.

COMMERCIALISATION: Formulation: Supplied as third instar larvae or as granular formulations with partially dehydrated nematodes in the centre of a clay granule. **Tradenames:** 'Entonem' (Koppert), 'Exhibit sf' (Syngenta Bioline), 'HortScanmask' (for horticultural use) (BioLogic), 'Magnet' (for mushrooms) (Certis), 'NemaShield' (BioWorks), 'Nemasys' (for horticultural crops) (Becker Underwood), (Westagro), (Plant Products), (BCP), (Brinkman), (Bog Madsen), (Svenska Predator), (Lier Frucklager) and (Intrachem), 'Nemasys F' (for flower thrips and leafminers) (Becker Underwood), 'Nemasys M' (for mushrooms) (Becker Underwood), (Amycel Spawn Mate), (MIS) and (J F McKenna), 'NemoPAK S' (Bioplanet), 'Scanmask' (for 'amateur' lawn and garden use) (BioLogic) and (IPM Laboratories), 'Sciarid' (for mushrooms) (Koppert), 'Steinernema feltiae' (Neudorff) and (Rincon-Vitova), 'Steinernema sp.' (species not identified) (Biocontrol Network), 'Steinernema-System' (Biobest), 'Traunem' (Andermatt), 'X-Gnat' (for ornamentals) (Certis).

APPLICATION: Apply by watering evenly over soil to be treated, at a rate of about 1×10^5 infective juvenile nematodes per square metre. Disperse granules in water before treatment and agitate to prevent sedimentation. For best control, ensure that the soil or compost is moist and that the temperature remains between 10 and 30°C.

PRODUCT SPECIFICATIONS: Specifications: Produced commercially by solid-state or liquid fermentation and sometimes bred in insects. Sold as third instar larvae. **Purity:** Products contain only third instar infectious nematode larvae. Product efficacy can be determined by bioassay against sciarid larvae. **Storage conditions:** Store under cool conditions for a few days, but do not freeze. Keep out of direct sunlight and do not allow the temperature to exceed 35°C. **Shelf-life:** *Steinernema feltiae* can be stored for 2 weeks at 6 to 8°C. Larvae remain infectious for only a short period after delivery. Granules remain active for 60 days at room temperature and up to 6 months at 5°C.

COMPATIBILITY: Will not survive in manure. High humidity and temperatures above 15°C are necessary for effective control. Incompatible with strong oxidisers, acids and bases.

MAMMALIAN TOXICITY: No allergic or other adverse reactions have been reported by research staff, producers or users of *Steinernema feltiae* or its associated bacterium in glasshouse or outdoor crops.

ENVIRONMENTAL IMPACT AND NON-TARGET TOXICITY: It is not expected that *Steinernema feltiae* or its associated bacterium will have any adverse effects on non-target organisms nor damage the environment. **Approved for use in organic farming:** Yes.

1:127 *Steinernema glaseri* isolate B-326
White grub parasitic nematode

The Pesticide Manual **Fifteenth Edition entry number:** 790

Nematode: Nematoda: Steinernematidae

NOMENCLATURE: Approved name: *Steinernema glaseri* (Steiner) isolate B-326. **Other names:** White grub parasitic nematode. **Development code:** B-326.

SOURCE: Isolated from soil in New Jersey, USA.

PRODUCTION: Produced commercially by fermentation in 30 000- to 60 000-litre fermenters on an undefined growing medium.

TARGET PESTS: Used for the control of white grubs (Scarabaeidae).

TARGET CROPS: Turf.

BIOLOGICAL ACTIVITY: Biology: The third stage larvae are the infectious stage and only these can survive outside the host insect, as they do not require food. They measure about 1.1 mm in length and 0.42 mm in width and enter their host through one of its natural openings (the mouth, anus or spiracles). The nematode larva then releases bacteria (*Xenorhabdus poinarii* Akhurst & Boemare) into the insect's body. The bacteria are medium to long motile rods with peritrichous flagellae. They are Gram-negative facultative anaerobes that form sphaeroplasts in older cultures. They are non-spore formers, have no resistant stage and are found only in the nematodes or the insect host. The bacteria proliferate within the insect, causing septicaemia and death within 24 to 72 hours. The nematode's fourth stage develops within the dead insect. These fourth stage larvae develop into males and females that reproduce sexually. After mating, the males die and the females lay eggs in the dead insect if there is enough food available or, if not, the first and second stage larvae develop inside her body. The larvae may leave the host when they reach the third stage, but usually two or three generations develop within the host, depending upon the availability of food. **Key reference(s):** 1) R Gaugler & H K Kaya. 1990. Entomopathogenic nematodes. In *Biological Control*, 365 pp., CRC Press, Boca Raton, Florida, USA. 2) R Georgis, C T Redmond & W T Martin. 1992. *Steinernema* B-326 and B-319 (Nematoda): new biological soil insecticides, *Proc. Brighton Crop Protection Conference – Pests & Diseases*, **1**, 73–9.

COMMERCIALISATION: Formulation: Prepared as a gel polymer encasing the immobilised third stage infective juvenile nematodes. **Tradenames:** 'Steinernema glaseri' (Praxis), (Greenfire), (Intrachem) and (Integrated Pest Management).

APPLICATION: Apply at a rate of 2.5×10^9 infective juveniles per hectare to moist soil, when the temperature is between 15 and 35 °C, with the best control being achieved at temperatures of 25 to 35 °C. Applications may be made using standard irrigation systems or with common spraying systems at pressures up to 2068 kPa with nozzle sizes of 50 μm or greater. Stir to prevent sedimentation.

PRODUCT SPECIFICATIONS: Specifications: Produced commercially by fermentation in 30 000- to 60 000-litre fermenters on an undefined growing medium. **Purity:** Products contain only third instar infectious nematode larvae. Product efficacy can be determined by bioassay against white grub larvae. **Storage conditions:** Store for short periods at room temperature or for longer periods under refrigeration. Do not freeze or store at temperatures above 37 °C. The nematodes are very susceptible to desiccation. **Shelf-life:** The product may be kept for 3 months at room temperature and for 12 months if refrigerated.

COMPATIBILITY: Compatible with a wide range of biological and chemical pesticides. Incompatible with strong oxidisers, acids and bases.

MAMMALIAN TOXICITY: Tests conducted on rats, mice, rabbits and pigs showed no symptoms or mortality caused by the nematodes or their associated bacteria by oral, intradermal, subcutaneous or intraperitoneal inoculation. The nematode/bacterium complex cannot withstand body temperatures >37 °C and is eliminated by the immune system upon injection. The gut of mammals will not allow nematode penetration.

ENVIRONMENTAL IMPACT AND NON-TARGET TOXICITY: *Steinernema glaseri* isolate B-326 was isolated from soil and occurs widely in Nature. It will not survive in birds and caused no mortality or symptoms following oral, dermal, subcutaneous or intraperitoneal inoculation. It is highly sensitive to desiccation and u.v. light and shows significant mortality in exposed environments. In aquatic environments, nematode survival is poor due to low oxygen levels. **Approved for use in organic farming:** Yes.

1:128 *Steinernema kraussei* isolate L137
Insect parasitic nematode

The Pesticide Manual **Fifteenth Edition entry number:** 790

Nematode: Nematoda: Steinernematidae

NOMENCLATURE: Approved name: *Steinernema kraussei* (Steiner) Travassos isolate L137. **Other names:** Parasitic nematode.

SOURCE: *Steinernema kraussei* occurs widely in soils in temperate areas.

PRODUCTION: Produced commercially by solid state or liquid fermentation on an undefined growing medium. Can be reared on vine weevil larvae.

TARGET PESTS: Vine weevils (*Otiorhyncus sulcatus* (Fabricius)), but also effective against other soil insects.

TARGET CROPS: Glasshouse vegetables and ornamentals and outdoor strawberries, also effective in vegetables and blackcurrants.

BIOLOGICAL ACTIVITY: Biology: The third stage larvae are the infectious stage and only these can survive outside the host insect, as they do not require food. They enter a host through one of its natural openings. The nematode larva then releases bacteria (*Xenorhabdus* sp.) into the insect's body and toxins produced by these bacteria kill the insect within 48 hours. The bacteria then digest the insect's body into material that the nematode can feed upon, and the nematode's fourth stage develops within the dead insect. These fourth stage larvae develop into males and females that reproduce sexually. After mating, the males die and the females lay eggs in the dead insect if there is enough food available or, if not, the first and second stage larvae develop inside her body. When larvae reach the third stage, they are able to leave the dead insect and seek out a new host. However, there may be two or more generations within the host, depending upon the availability of food. The third stage is always the infective stage. **Efficacy:** *Steinernema kraussei* (L137) is able to survive winter field conditions including prolonged exposure to low temperatures, in contrast to *S. carpocapsae* Weiser which shows poor survival. Its ability to survive at low temperatures and the fact that it remains active at temperatures of 5°C has led to its use for vine weevil control in preference to *Steinernema carpocapsae* Weiser or *Heterorhabditis megidis* Poinar, Jackson and Klein. **Key reference(s):** 1) Z Mracek. 1994. *Steinernema kraussei* (Steiner, 1923) (Nematoda: Rhabditida: Steinernematidae): redescription of its topotype from Westphalia, *Folia Parasitological*, **41**, 59–64. 2) R Gaugler & H K Kaya. 1990. Entomopathogenic nematodes. In *Biological Control*, 365 pp., CRC Press, Boca Raton, Florida, USA. 3) R Georgis & H K Kaya. 1998. Formulation of entomopathogenic nematodes. In *Formulation of Microbial Biopesticides*,

Beneficial Microorganisms, Nematodes and Seed Treatments, H D Burges (ed.), pp. 289–308, Kluwer Academic Publishers, Dordrecht, The Netherlands.

COMMERCIALISATION: Formulation: Sold as third instar larvae in vermiculite in sealed trays. **Tradenames:** 'Exhibitline sk' (Syngenta Bioline), 'Grubsure LT' (Defenders), 'Nemasys L' (Becker Underwood).

APPLICATION: The best time to apply *Steinernema kraussei* is in the spring (March to May) and autumn (late August to November), as these are the times that most vine weevil grubs will be present in the soil. Apply by watering evenly over soil to be treated, at a rate of about 1×10^5 juvenile nematodes per square metre. The treated area should be kept moist by normal watering after application to enable the nematodes to move in the soil moisture.

PRODUCT SPECIFICATIONS: Specifications: Produced commercially by solid state or liquid fermentation on an undefined growing medium. Can be reared on vine weevil larvae. **Purity:** Products contain only third instar infectious nematode larvae. Product efficacy can be determined by bioassay against vine weevil or other coleopteran larvae. **Storage conditions:** Can be stored in a refrigerator at 5 °C. Do not freeze or expose to bright sunlight. **Shelf-life:** If stored according to the manufacturer's instructions, the product can be kept for up to 2 weeks.

COMPATIBILITY: Compatible with a wide range of biological and chemical pesticides. Incompatible with strong oxidisers, acids and bases.

MAMMALIAN TOXICITY: No allergic or other adverse reactions have been reported by researchers, production staff or users during application in glasshouse or outdoor crops from *Steinernema kraussei* or its associated bacterium.

ENVIRONMENTAL IMPACT AND NON-TARGET TOXICITY: It is not expected that the use of products containing *Steinernema kraussei* will have any effects on non-target organisms nor have any adverse effects on the environment. **Approved for use in organic farming:** Yes.

1:129 *Steinernema riobrave*
Soil insect parasitic nematode

The Pesticide Manual Fifteenth Edition entry number: 790

Nematode: Nematoda: Steinernematidae

NOMENCLATURE: Approved name: *Steinernema riobrave* Cabanillas, Poinar & Raulston. **Other names:** Soil insect parasitic nematode.

SOURCE: Originally isolated from soil.

PRODUCTION: Produced commercially by fermentation in 30 000- to 60 000-litre fermenters on an undefined medium.

TARGET PESTS: Control of large nymph and adult mole crickets (*Scapteriscus* spp.), sugar cane rootstalk borer (*Diaprepes abbreviatus* (L.)), citrus weevils (*Pachnaeus litus* Germar) and other pests of citrus.

TARGET CROPS: Recommended for use in turf, citrus and sugar cane.

BIOLOGICAL ACTIVITY: Biology: The third stage larvae are the infectious stage and only these can survive outside the host insect, as they do not require food. They enter their host through one of its natural openings (the mouth, anus or spiracles). The nematode larva then releases bacteria (*Xenorhabdus*, undescribed species) into the insect's body. The bacteria are medium to long motile rods with peritrichous flagellae. They are non-spore-forming, Gram-negative facultative anaerobes and are found only in the nematodes or the insect host. The bacteria proliferate within the insect, causing death within 24 to 72 hours. The nematode's fourth stage develops within the dead insect. These fourth stage larvae develop into males and females that reproduce sexually. After mating, the males die and the females lay eggs in the dead insect if there is enough food available or, if not, the first and second stage larvae develop inside her body. The larvae may leave the host when they reach the third stage, but usually two or three generations develop within the host, depending upon the availability of food. **Key reference(s):** 1) R Gaugler & H K Kaya. 1990. Entomopathogenic nematodes. In *Biological Control*, 365 pp., CRC Press, Boca Raton, Florida, USA. 2) R Georgis & H K Kaya. 1998. Formulation of entomopathogenic nematodes. In *Formulation of Microbial Biopesticides, Beneficial Microorganisms, Nematodes and Seed Treatments*, H D Burges (ed.), pp. 289–308, Kluwer Academic Publishers, Dordrecht, The Netherlands.

COMMERCIALISATION: Formulation: Sold as water dispersible granules (WG). **Tradenames:** 'Biovector 355 WG' (Rincon-Vitova), 'Devour WG' (Rincon-Vitova), 'Vector MC WG' (Rincon-Vitova).

APPLICATION: Apply at a rate of 2.5×10^9 infective juveniles per hectare to moist soil when the temperature is between 15 and 35 °C, with the best control being achieved at temperatures of 22 to 28 °C. Applications may be made using standard spraying systems at pressures up to 2068 kPa with nozzle sizes of 50 µm or greater. Stir to prevent sedimentation.

PRODUCT SPECIFICATIONS: Specifications: Produced commercially by fermentation in 30 000- to 60 000-litre fermenters on an undefined medium. **Purity:** Products contain third stage infective juveniles on an inert carrier. The purity of the formulation can be determined by examination under a microscope and its efficacy through bioassay on susceptible insect larvae. **Storage conditions:** Store under dry conditions at room temperature or, if stored for long periods, refrigerate. Do not freeze and do not allow the storage temperature to exceed

37°C. **Shelf-life:** If stored at room temperature, the product will remain viable for up to 2 to 3 months. If refrigerated, the product will remain active for up to a year.

COMPATIBILITY: Compatible with all biological control systems and many chemical pesticides. Do not apply in a tank-mix with copper- or benzimidazole-based fungicides, or with soil insecticides. Incompatible with strong oxidisers, acids and bases.

MAMMALIAN TOXICITY: Tests conducted on rats, mice, rabbits and pigs showed no symptoms or mortality caused by the nematodes or their associated bacteria by oral, intradermal, subcutaneous or intraperitoneal inoculation. The nematode/bacterium complex cannot withstand body temperatures >37°C and is eliminated by the immune system upon injection. The gut of mammals will not allow nematode penetration.

ENVIRONMENTAL IMPACT AND NON-TARGET TOXICITY: *Steinernema riobrave* was isolated from soil and occurs in Nature. It will not survive in birds and caused no mortality or symptoms following oral, dermal, subcutaneous or intraperitoneal inoculation. It is highly sensitive to desiccation and u.v. light and shows significant mortality in exposed environments. In aquatic environments, nematode survival is poor, due to low oxygen levels. **Approved for use in organic farming:** Yes.

1:130 *Steinernema scapterisci* isolate B-319
Mole cricket parasitic nematode

The Pesticide Manual **Fifteenth Edition entry number:** 790

Nematode: Nematoda: Steinernematidae

NOMENCLATURE: Approved name: *Steinernema scapterisci* Nguyen & Smart isolate B-319. **Other names:** Mole cricket parasitic nematode.

SOURCE: Naturally occurring nematode isolated in South America by University of Florida researchers from mole crickets during a natural epizootic. Originally reported by R Georgis *et al.* (*Proc. British Crop Protection Conference – Pests & Diseases*, 1992, **1**, 73). Introduced by Biosys (acquired by Thermo Trilogy and now Certis, who no longer manufacture it) and licensed to Ecogen (who no longer sell it). Commercialised by Becker Underwood.

PRODUCTION: Manufactured by fermentation in sterile undefined medium. Infective juveniles are harvested by concentration of the fermentation broth.

TARGET PESTS: Mole crickets (*Scapteriscus vicinus* Scudder and *Gryllotalpa* spp.).

TARGET CROPS: Turf grass, particularly golf courses.

BIOLOGICAL ACTIVITY: Biology: The third stage larvae are the infectious stage and only these can survive outside the host insect, as they do not require food. They enter a host through one of its natural openings (the mouth, anus or spiracles) and penetrate the body cavity. The nematode larva then releases its symbiotic bacteria (*Xenorhabdus*, undescribed species) into the insect's body and toxins produced by these bacteria kill the insect within 36 to 48 hours. The bacteria then digest the insect's body into material that the nematode can feed upon, and the nematode's fourth stage develops within the dead insect. These fourth stage larvae develop into males and females that reproduce sexually. After mating, the males die and the females lay eggs in the dead insect if there is enough food available or, if not, the first and second stage larvae develop inside her body. When larvae reach the third stage, they are able to leave the dead insect and seek out a new host. However, there may be two or more generations within the host, depending upon the availability of food. The third stage is always the infective stage. **Efficacy:** *Steinernema scapterisci* is a very efficient pathogen of mole crickets. **Key reference(s):** 1) R Georgis, C T Redmond & W R Martin. 1992. *Steinernema* B-326 and B-319 (Nematoda): New biological soil insecticides, *Proc. British Crop Protection Conference – Pests & Diseases*, **1**, 73–9. 2) R Gaugler & H K Kaya. 1990. Entomopathogenic nematodes. In *Biological Control*, 365 pp., CRC Press, Boca Raton, Florida, USA. 3) R Georgis & H K Kaya. 1998. Formulation of entomopathogenic nematodes. In *Formulation of Microbial Biopesticides, Beneficial Microorganisms, Nematodes and Seed Treatments*, H D Burges (ed.), pp. 289–308, Kluwer Academic Publishers, Dordrecht, The Netherlands.

COMMERCIALISATION: Formulation: Sold as a clay-based formulation containing 60 million infective third stage nematodes per pack. **Tradenames:** 'Nematac S' (Becker Underwood), 'Otinem S' (Ecogen – discontinued).

APPLICATION: Apply to moist soil at a rate of 2.2×10^9 infective juveniles per hectare. Ensure that the soil is irrigated every 3 to 4 days after treatment, unless rain occurs. Apply using standard spray application equipment at spray pressures not exceeding 2068 kPa with nozzle sizes of 50 µm or greater. Stir to prevent sedimentation.

PRODUCT SPECIFICATIONS: Specifications: Manufactured by fermentation in sterile undefined medium. Infective juveniles are harvested by concentration of the fermentation broth. **Purity:** Efficacy of the formulation can be measured by bioassay against a target pest and nematode numbers can be determined by counting juveniles under a microscope. **Storage conditions:** Store the formulation at 10°C. Keep out of direct sunlight and avoid desiccation. **Shelf-life:** Stable for 2 months under recommended storage conditions.

COMPATIBILITY: Should not be tank-mixed with benzimidazole-based fungicides or soil insecticides/nematicides. Compatible with other IPM/biological control strategies. Incompatible with strong oxidisers, acids and bases.

MAMMALIAN TOXICITY: Neither *Steinernema scapterisci* nor its bacterial symbiont have shown evidence of toxicity, infectivity, irritation or hypersensitivity to mammals. No allergic responses or related health problems have been observed by research workers, manufacturing staff or users.

ENVIRONMENTAL IMPACT AND NON-TARGET TOXICITY: *Steinernema scapterisci* and its associated bacterium have caused no adverse effects on any non-target organism or any deleterious effects on the environment.

1:131 *Streptomyces griseoviridis* isolate K 61
Biological fungicide (bacterium)

Bacterium: Actinomycetales: Streptomycetaceae

NOMENCLATURE: Approved name: *Streptomyces griseoviridis* Anderson *et al.* isolate K 61.

SOURCE: Isolated from Finnish light-coloured *Sphagnum* peat that had been reported to have disease-suppressing properties. Several isolates of *Streptomyces* shown to possess antagonistic properties against seed- and soil-borne fungal pathogens were isolated by the Department of Plant Pathology of the University of Helsinki and the isolate K 61 was selected for development.

PRODUCTION: Produced by fermentation followed by freeze-drying of the organism.

TARGET PESTS: Various seed- and soil-borne fungal pathogens and particularly *Fusarium* spp., that cause wilt, root rot and basal rots. It also shows activity against other seed- and soil-borne pathogens, such as *Alternaria* spp., *Pythium* spp., *Phytophthora* spp., *Rhizoctonia* spp. and *Botrytis cinerea* Pers.

TARGET CROPS: Glasshouse vegetables, ornamentals and herbs.

BIOLOGICAL ACTIVITY: Mode of action: *Streptomyces griseoviridis* isolate K 61 exerts its effect through a combination of factors, including root colonisation and competition for living space and essential nutrients with plant pathogenic fungi, lysis of the cell walls of pathogenic fungi by extracellular enzymes and the production of antifungal metabolites. **Efficacy:** Effective inhibition of the growth and pathogenesis of several plant pathogens is enhanced by growth stimulation of healthy crops. **Key reference(s):** 1) O Mohammadi. 1994. *Proc. 3rd International Workshop on Plant Growth-Promoting Rhizobacteria*, 282–4. 2) D R Fravel, W J Connick Jr & J A Lewis. 1998. Formulation of microorganisms to control plant diseases. In *Formulation of*

Microbial Biopesticides, Beneficial Microorganisms, Nematodes and Seed Treatments, H D Burges (ed.), pp. 187–202, Kluwer Academic Publishers, Dordrecht, The Netherlands. 3) A S Paau. 1998. Formulation of beneficial microorganisms applied to soil. In *Formulation of Microbial Biopesticides, Beneficial Microorganisms, Nematodes and Seed Treatments*, H D Burges (ed.), pp. 235–54, Kluwer Academic Publishers, Dordrecht, The Netherlands.

COMMERCIALISATION: Formulation: Sold as wettable powders (WP) containing a minimum of 1×10^8 colony forming units (cfu) per gram. **Tradenames:** 'Mycostop' (Verdera).

APPLICATION: 'Mycostop' can be applied as a dry seed treatment at 2 to 5 grams per kilogram of seed, or as an aqueous suspension for the spraying or drenching of growth substrates at the rate of 2 to 20 grams per 100 m², or via drip irrigation to give a dose of 5 grams per 100 m² or 5 to 10 grams per 1000 plants. The first treatment is recommended immediately after sowing, planting or transplanting. The application is repeated at intervals of 3 to 6 weeks. Applications should be made before the diseases become established.

PRODUCT SPECIFICATIONS: Specifications: Produced by fermentation followed by freeze-drying of the organism. **Purity:** The product contains a minimum of 1×10^8 colony forming units (cfu) per gram. Viability of the product is determined by plating out on agar and counting the number of colonies produced following incubation in the laboratory for 48 hours. Biological efficacy is tested using a bioassay with artificially infested cauliflower seeds. **Storage conditions:** Store the unopened packages below 8 °C. **Shelf-life:** The formulated material is stable for 12 months, when stored under the recommended conditions.

COMPATIBILITY: It is not recommended that 'Mycostop' be applied as a tank-mix with other pesticides or concentrated fertiliser solutions. Do not use chlorinated water.

MAMMALIAN TOXICITY: Not hazardous to animals. No allergic or health problems have been observed in research workers, manufacturing staff or users. Inhalation of the fine powder and skin contact should be avoided by using recommended protective equipment.
Acute oral LD$_{50}$: Not toxic or pathogenic to rats exposed to >15 g/kg.
Acute dermal LD$_{50}$: rabbits >2000 mg/kg. **Inhalation:** Not toxic or pathogenic to rats exposed to 1×10^8 cfu/animal. **Skin and eye:** Mild eye irritation (reversible within 2 days). No skin irritation.

ENVIRONMENTAL IMPACT AND NON-TARGET TOXICITY: As a natural component of the soil microflora, it is not expected that use of the product will lead to any adverse effects on the environment. **Bird toxicity:** NOEL for bobwhite quail and mallard ducks 2.45×10^9 colony forming units (cfu)/kg. **Fish toxicity:** Not toxic to fish, with NOEL for rainbow trout being 5000 cfu/ml. **Other aquatic toxicity:** NOEL for *Daphnia* 1×10^4 cfu/ml. **Effects on beneficial insects:** NOEL for honeybees 9.8×10^8 cfu/kg. **Behaviour in soil:** The isolate of *Streptomyces griseoviridis* used for the preparation of 'Mycostop' occurs naturally in soil. **Approved for use in organic farming:** Yes.

1:132 *Streptomyces lydicus* isolate WYEC 108
Biological fungicide (bacterium)

Bacterium: Actinomycetales: Streptomycetaceae

NOMENCLATURE: Approved name: *Streptomyces lydicus* De Boer *et al.* isolate WYEC 108.

SOURCE: 'Actinovate' is the commercial name of the beneficial soil micro-organism, *Streptomyces lydicus* isolate WYEC 108. It is a saprophytic, rhizosphere-colonising Actinomycete that was isolated from the roots of a linseed plant in the USA.

PRODUCTION: Produced by fermentation followed by freeze-drying of the organism.

TARGET PESTS: The active ingredient, *Streptomyces lydicus* WYEC 108, is intended for use as a biological fungicide for the control of soil-borne plant root-rot and damping-off fungi. Fungi controlled include *Fusarium, Rhizoctonia, Pythium, Phytophthora, Phytomatotricum, Aphanomyces, Monosprascus, Armillaria* species and other root-decaying fungi.

TARGET CROPS: Introduced for use in the professional glasshouse, nursery and landscape market for vegetable and ornamental crops and turf, but also showing promise in a wide range of agricultural crops.

BIOLOGICAL ACTIVITY: Mode of action: *Streptomyces lydicus* WYEC 108 colonises the growing root tips of plants and acts as a mycoparasite of fungal root pathogens, thereby protecting the plants. Root colonisation is a form of competitive exclusion of a pathogen from the root system. Other mechanisms of action of *S. lydicus* include the production and excretion of antifungal metabolites (antibiotics and/or low molecular weight antifungal compounds) into the rhizosphere surrounding the roots of colonised plants, and mycoparasitism of the spores and vegetative mycelium of the fungal pathogens (via colonisation of the spores and hyphae of the fungus, followed by the production of lytic enzymes such as chitinase). No deleterious effects to plants have been observed as a result of the excretion of antifungal compounds from *Streptomyces lydicus* WYEC 108. **Efficacy:** Over the past three years, trials using 'Actinovate' have been conducted on various agricultural and ornamental crops. The results of these trials have shown an increase in seedling vigour, stronger root systems and increased yield, with fewer weak and disabled plants. Growers using 'Actinovate' products have consistently reduced or completely eliminated their need for chemical fungicides. **Key reference(s):** 1) O Mohammadi. 1994. *Proc. 3rd International Workshop on Plant Growth-Promoting Rhizobacteria*, 282–4. 2) D R Fravel, W J Connick Jr & J A Lewis. 1998. Formulation of microorganisms to control plant diseases. In *Formulation of Microbial Biopesticides, Beneficial Microorganisms, Nematodes and Seed Treatments*, H D Burges (ed.), pp. 187–202, Kluwer Academic Publishers, Dordrecht, The Netherlands. 3) A S Paau. 1998. Formulation of beneficial microorganisms applied to soil. In *Formulation of Microbial*

Biopesticides, Beneficial Microorganisms, Nematodes and Seed Treatments, H D Burges (ed.), pp. 235–54, Kluwer Academic Publishers, Dordrecht, The Netherlands.

COMMERCIALISATION: Formulation: There are two main products: 'Actinovate Plus/M' – a highly concentrated, dispersible formulation of the 'Actinovate' microbe for drench, liquid feed or irrigation; and 'Actino-Iron' – a soil amendment containing a high concentration of the 'Actinovate' microbe, as well as fulvic acid and chelated iron. **Tradenames:** 'Actino-Iron', 'Actinovate' and 'Actinovate Plus/M' (Natural Industries).

APPLICATION: The products are applied as a soil mix or drench to turf grass or potted plants. They can also be applied to plant foliage in glasshouses.

PRODUCT SPECIFICATIONS: Specifications: Produced by fermentation followed by freeze-drying of the organism. **Purity:** The products contain colony forming units of *Streptomyces lydicus* WYEC 108 with no bacteria that are pathogenic to man. **Storage conditions:** Store the unopened packages below 8°C. **Shelf-life:** The formulated material is stable for 12 months, when stored under the recommended conditions.

COMPATIBILITY: 'Actinovate' can be used with crop protection chemicals. Do not use chlorinated water.

MAMMALIAN TOXICITY: The organism has never been reported as a pathogen of humans, or as causing any type of adverse effect to humans, in published literature or through commercial use. **Acute oral LD$_{50}$:** rats (for the end-use formulation) >5050 mg/kg. **Acute dermal LD$_{50}$:** A waiver is being requested for acute dermal toxicity/pathogenicity, based on the fact that there was no toxicity or pathogenicity in the pulmonary and injection studies, and no effects were observed in the skin irritation study. Dermal toxicity or pathogenicity would not be expected for this active ingredient. Since its discovery, no incidents of hypersensitivity have been reported by researchers, manufacturers or users. **Inhalation:** For the active ingredient, the acute pulmonary toxicity/pathogenicity in rats is greater than 9.1×10^8 colony forming units (cfu) per animal. **Skin and eye:** Eye irritation in rabbits was not observed at a dose of 0.1 ml. Skin irritation in rabbits was not observed at a dose of 0.5 ml. **Toxicity class:** EPA (formulation) IV. **Other toxicological effects:** The acute injection toxicity/pathogenicity in rats is greater than 9.33×10^8 cfu per animal.

ENVIRONMENTAL IMPACT AND NON-TARGET TOXICITY: As a natural component of the soil microflora, it is not expected that use of the product will lead to any adverse effects on the environment. **Approved for use in organic farming:** Yes.

1:133 *Syngrapha falcifera* nucleopolyhedrovirus *Insecticidal baculovirus*

Virus: Baculoviridae: Nucleopolyhedrovirus

NOMENCLATURE: Approved name: *Syngrapha falcifera* nucleopolyhedrovirus.
Other names: See *Anagrapha falcifera* nucleopolyhedrovirus (entry 1:08).

1:134 *Talaromyces flavus* isolate VII7b *Biological fungicide (fungus)*

Fungus: Ascomycetes: Eurotiales: Trichocomaceae

NOMENCLATURE: Approved name: *Talaromyces flavus* (Klocker) Stolk & Samson isolate V117b.

SOURCE: Isolated from soil by Prophyta.

PRODUCTION: Manufactured by solid state fermentation. The ascospores are formulated as water dispersible granules.

TARGET PESTS: Targeted for control of soil-borne diseases, such as *Verticillium dahliae* Kleb., *V. albo-atrum* Reinke and Berth. and *Rhizoctonia solani* Kuehn.

TARGET CROPS: Tomato, cucumber, strawberries and oilseed rape.

BIOLOGICAL ACTIVITY: Mode of action: *Talaromyces flavus* protects crops against pathogenic soil-borne fungi by competition in the soil and in the rhizosphere, because it is a strong root-zone coloniser, is highly competitive for nutrients with other micro-organisms and it outcompetes pathogens for access at root infection sites. There are suggestions that *Talaromyces flavus* activates the plant's auto-immune system, leading to the production of phytoalexins that inhibit the invasion of fungal pathogens.

COMMERCIALISATION: Formulation: 'Protus WG' is sold as water dispersible granules containing ascospores. **Tradenames:** 'Protus WG' (Prophyta).

APPLICATION: The product is applied as a soil or seed treatment, a soil drench or as a root dip application.

PRODUCT SPECIFICATIONS: **Specifications:** Manufactured by solid state fermentation. The ascospores are formulated as water dispersible granules. **Purity:** The fermentation procedure is carried out under controlled conditions and only spores of *Talaromyces flavus* are found within the product. The viability of the spores can be determined by plating out on agar, incubating in the laboratory for 48 hours and counting the colonies that develop. **Storage conditions:** The formulated product should be stored in dry, refrigerated conditions at 4°C. Do not freeze. **Shelf-life:** If stored under the correct conditions, the formulated product will remain biologically active for over 6 months.

COMPATIBILITY: Cannot be tank-mixed or applied with fungicides. Incompatible with strong oxidisers, acids, bases and chlorinated water.

MAMMALIAN TOXICITY: *Talaromyces flavus* shows no oral toxicity, pathogenicity or infectivity to any tested animals at 1×10^6 colony forming units (cfu)/animal. The fungus does not grow well at temperatures >35°C.

ENVIRONMENTAL IMPACT AND NON-TARGET TOXICITY: *Talaromyces flavus* isolate V117b is harmless to non-target organisms, including predatory and beneficial insects, mammals and birds. It is a soil fungus that is widespread in Nature and is not expected to show any adverse environmental effects. **Approved for use in organic farming:** Yes.

1:135 *Talaromyces flavus* isolate Y-9401
Biological fungicide (fungus)

Fungus: Ascomycetes: Eurotiales: Trichocomaceae

NOMENCLATURE: **Approved name:** *Talaromyces flavus* (Klocker) Stolk & Samson isolate Y-9401.

SOURCE: Isolated from a domestic strawberry field in Japan.

PRODUCTION: Manufactured by solid state fermentation. The ascospores are formulated as a suspension in oil.

TARGET PESTS: A range of rice diseases including rice blast (*Magnaporthe grisea* (Hebert) Barr (*Pyricularia oryzae* Cavara)), bakanae disease (*Fusarium moniliforme* Sheldon (*Gibberella fujikuroi* Wr.)) and several bacterial diseases. It is also effective against strawberry powdery mildew (*Sphaerotheca macularis* f. sp. *fragariae* (Wall. ex Fries)), grey mould (*Botrytis cinerea* Pers.) and anthracnose (*Colletotrichum* spp.).

TARGET CROPS: Rice and various vegetable crops, including strawberries.

BIOLOGICAL ACTIVITY: Mode of action: *Talaromyces flavus* isolate Y-9401 protects the leaves and fruits from the pathogen by invading its areas of infection rather than killing it directly. There are suggestions that *Talaromyces flavus* activates the plant's auto-immune system, leading to the production of phytoalexins that inhibit the invasion of fungal pathogens.

COMMERCIALISATION: Formulation: 'ToughPearl' is used for disease control in vegetable crops and 'Touch Block' is used in rice. Both are formulated as ascospores suspended in oil. **Tradenames:** 'ToughPearl' and 'Touch Block' (Idemitsu Kosan). **Patents:** EP20010980910.

APPLICATION: The products are applied as foliar sprays at the first sign of disease development. High volume applications give better results.

PRODUCT SPECIFICATIONS: Specifications: Manufactured by solid state fermentation. The ascospores are formulated as a suspension in oil. **Purity:** The fermentation procedure is carried out under controlled conditions and only spores of *Talaromyces flavus* are found within the product. The viability of the spores can be determined by plating out on agar, incubating in the laboratory for 48 hours and counting the colonies that develop.
Storage conditions: The formulated product should be stored in dry, refrigerated conditions at 4°C. Do not freeze. **Shelf-life:** If stored under the correct conditions, the formulated product will remain biologically active for over 6 months.

COMPATIBILITY: Cannot be tank-mixed or applied with fungicides. Incompatible with strong oxidisers, acids, bases and chlorinated water.

MAMMALIAN TOXICITY: *Talaromyces flavus* shows no oral toxicity, pathogenicity or infectivity to any tested animals. The fungus does not grow well at temperatures >35°C.

ENVIRONMENTAL IMPACT AND NON-TARGET TOXICITY: *Talaromyces flavus* isolate Y-9401 is harmless to non-target organisms, including predatory and beneficial insects, mammals and birds. It is a fungus that is widespread in Nature and is not expected to show any adverse environmental effects. **Approved for use in organic farming:** Yes.

1:136 Tobacco mild green mosaic virus
Biological herbicide (virus)

Tobamovirus

NOMENCLATURE: Approved name: Tobacco mild green mosaic virus.
Other names: TMGMV.

SOURCE: Isolated from tropical soda apple (*Solanum viarum* Dunal).

PRODUCTION: Tobacco (*Nicotiana tabacum* L.) cultivar Samsun, exists as two genotypes: the resistant *NN* and the susceptible *nn*. The *nn* genotype can be used to build up high titres of the virus from which it can be harvested.

TARGET PESTS: Tropical soda apple (TSA).

TARGET CROPS: May be used in a wide variety of crops, but peppers (*Capsicum* spp.) and tobacco are considered to be at risk. When only plants outside the Solanaceae were tested (a total of 226 species and subspecies), 88% were immune, 4% were resistant and 6% were asymptomatic. Thus, only 2.2% of the plants (five species) were susceptible and developed non-lethal, mild mosaic symptoms. These results support the conclusion that TMGMV can be used safely without undue risks to non-target plants.

BIOLOGICAL ACTIVITY: Mode of action: It is believed that TMGMV kills TSA by eliciting a severe hypersensitive response (HR) from the plant. The HR is mediated by a TMGMV gene that normally triggers localised necrotic leaf lesions, limiting the spread of the virus infection. In the TMGMV–TSA system, the virus provokes a weak or delayed defence response that fails to confine the virus to the local lesions. Over several days after infection, the virus elicits an intense, host-generated, autocidal HR that results in plant death. The death is probably the result of programmed cell death and massive vascular tissue dysfunction manifested by the wilt symptom. **Biology:** TMGMV has been field-tested in 20 field locations in eight Florida counties. These trials were done during different times of the year under different climatic conditions. In all of these trials, the virus has performed consistently, and there is no evidence of resistance among the TSA populations. **Efficacy:** Typically, 98–99% control is obtained with the high-pressure spray systems, followed by abrade-and-spray (73–88%) and wiper application (61–79%) methods. The survivors are presumed to be escapees that were missed during inoculation, since they were invariably killed when inoculated with TMGMV. **Key reference(s):** 1) M S Pettersen, R Charudattan, E Hiebert & F W Zettler. 2000. Tobacco mosaic virus strain U2 causes a lethal hypersensitive response in *Solanum viarum* Dunal (tropical soda apple), *WSSA Abstracts*, **40**, 84. 2) R Charudattan, M Elliott, J DeValerio, E Hiebert & M Pettersen. 2003. Biological control of the noxious weed *Solanum viarum* by tobacco mild green mosaic tobamovirus, *Poster, Proceedings of the 8th Int. Congr. Plant Pathol.*, Christchurch, New Zealand. 3) S Adkins, I Kamenova, E N Rosskopf & D J Lewandowski. 2007. Identification and characterization of a novel tobamovirus from tropical soda apple in Florida, *Plant Dis.*, **91**, 287–93. 4) R Charudattan & E Hiebert. 2007. A plant virus as a bioherbicide for tropical soda apple, *Solanum viarum*, *Outlooks on Pest Management*, **18**(4), 167–71.

COMMERCIALISATION: Formulation: The University of Florida has patented this technology and licensed a start-up biotech company, BioProdex, Inc. (Gainesville, FL, USA) to develop and register this viral bioherbicide. **Tradenames:** 'SolviNix' (BioProdex). **Patents:** US Patent No. 6,689,718 B2 (February 10, 2004).

APPLICATION: For TMGMV to act as a bioherbicide, the virus particles have to be introduced into live host cells through microscopic injuries. TMGMV particles are large macromolecules incapable of entering the target host cells by absorption and so conventional application equipment used for chemical herbicides needs to be modified for TMGMV field applications. Eight different inoculation methods using available pesticide application equipment were tested. Amongst them, the abrade-and-spray, wiper application and high-pressure spray methods were most effective and practical for use in TSA management programmes.

PRODUCT SPECIFICATIONS: Specifications: Tobacco (*Nicotiana tabacum* L.) cultivar Samsun exists as two genotypes: the resistant *NN* and the susceptible *nn*. The *nn* genotype can be used to build up high titres of the virus from which it can be harvested.
Storage conditions: Store at −20 °C. **Shelf-life:** Retains viability for more than 50 months.

COMPATIBILITY: It is unlikely that TMGMV will be used with any other crop protection agent.

MAMMALIAN TOXICITY: As a plant virus, TMGMV does not infect humans, animals or insects and does not pose any risks to these higher life forms. TMGMV does not pose any toxicological concerns nor does it elicit any toxin production in plants following infection. TMGMV is not expected to occur in beef or milk from cattle that may be exposed to the virus when used as a bioherbicide.

ENVIRONMENTAL IMPACT AND NON-TARGET TOXICITY: TMGMV is not expected to show any adverse effects on non-target organisms. However, some Solanaceous species such as peppers (*Capsicum* spp.) and tobacco are considered to be at risk.
Approved for use in organic farming: Yes.

1:137 *Trichoderma harzianum* isolate T-22
Biological fungicide (fungus)

Fungus: anamorphic Hypocreales: Hypocreaceae

NOMENCLATURE: Approved name: *Trichoderma harzianum* Tul. isolate T-22 (Rifai isolate KRL-AG2). **Other names:** Formerly known as *Trichoderma lignorum* (Tode) Harz. **Development code:** 1295–22, KRL-AG2, ATCC 20847.

SOURCE: Isolate T-22 was produced by fusing protoplasts from *Trichoderma harzianum* isolate T-95 (originally isolated from a *Rhizoctonia*-suppressive soil in Colombia) and *Trichoderma harzianum* isolate T-12 (identified as being strongly competitive in iron-limiting conditions). Isolate T-22 was selected as being the most effective and competitive new isolate.

PRODUCTION: *Trichoderma harzianum* isolate T-22 is produced commercially by fermentation.

TARGET PESTS: Soil and foliar pathogens, such as *Pythium*, *Rhizoctonia*, *Fusarium*, *Thielaviopsis*, *Cylindrocladium*, *Myrothecium*, *Botrytis* and *Sclerotinia* species.

TARGET CROPS: Used in many crops, including vegetables, ornamentals, soybeans and maize (corn).

BIOLOGICAL ACTIVITY: Mode of action: *Trichoderma harzianum* isolate T-22 exerts its effect on fungal pathogens in a variety of different ways. It is well known as a mycoparasite inhibiting the growth of pathogenic fungi by invading and parasitising hyphae. It competes in the soil for nutrients and rhizosphere dominance with phytopathogenic fungi. *T. harzianum* isolate T-22 has a dramatic effect on the growth of crop roots and in the solubilisation of various soil nutrients. It is believed that these improved root systems, together with a greater supply of soil nutrients, allow treated plants to withstand attack from phytopathogens. There are also claims that the presence of *Trichoderma harzianum* isolate T-22 on the roots of crop plants activates the plants' systemic acquired resistance response, thereby protecting them from attack by foliar pathogens. **Efficacy:** *Trichoderma harzianum* isolate T-22 has been shown to be very effective in protecting a wide variety of plants from attack by a wide range of fungal pathogens. In addition, even in the absence of disease-causing organisms, growth and yield of treated crops have been significantly enhanced. Crops sown after treated crops have also outperformed those planted into untreated areas. *T. harzianum* isolate T-22 is a preventive treatment and cannot control or eliminate existing diseases. In the presence of high disease pressure, it should be used as part of an integrated chemical–biological system. T-22 is very persistent within the rhizosphere, but will not persist in the absence of growing roots.
Key reference(s): 1) I Chet. 1987. *Trichoderma* – application, mode of action and potential as a biocontrol agent of soilborne plant pathogenic fungi. In *Innovative Approaches to Plant Disease Control*, I Chet (ed.), pp. 137–60, Wiley & Sons, New York, USA. 2) B A Bailey & R D Lumsden. 1998. Direct effects of *Trichoderma* and *Gliocladium* on plant growth and resistance to pathogens. In Trichoderma *and* Gliocladium, *Vol. 2*, G E Harman & C P Lumsden (eds), pp. 185–204, Taylor and Francis, London, UK.

COMMERCIALISATION: Formulation: The product is sold as seed treatments (ST), as granules (GR) for soil incorporation or top-dressing and as wettable powders (WP) for foliar and root applications. **Tradenames:** 'PlantShield HC' (BioWorks) and (Koppert), 'RootShield' (BioWorks) and (Koppert), 'RootShield Drench' (Koppert), 'RootShield Granules' (BioWorks) and (Koppert), 'T-22 Planter Box' (Hasel Tarim), 'T-22 PlantShield' (PHC), 'Trianum-G' (Koppert), 'Trianum-P' (Koppert).

APPLICATION: 'RootShield' is applied as a soil incorporation at 550 to 850 g per cubic metre or as a top dressing (3 g per pot). 'PlantShield', used for root protection, is applied at 0.6 to 0.75 g per litre. Both applications are sufficient for 3 months protection. A second application is recommended at half rate after 3 months. 'PlantShield', as a foliar spray, is applied at 28 g per 3.785 litres. As a seed treatment in crops such as maize (corn) or soybeans, 'T-22 Planter Box' is applied at 505 g per 100 kg of seed.

PRODUCT SPECIFICATIONS: Specifications: *Trichoderma harzianum* isolate T-22 is produced commercially by fermentation. **Purity:** Contains no pathogenic or allergenic micro-organisms. **Storage conditions:** Store in a sealed container in a cool, dry situation. **Shelf-life:** If stored between 10 and 25 °C, the product has a 6-month shelf-life. If stored between 0 °C and 10 °C, the product has a 9-month shelf-life, and if stored below 0 °C, the product has a 12-month shelf-life.

COMPATIBILITY: Compatible with broad-spectrum, soil- or seed-applied fungicides. Incompatible with strong oxidisers, acids, bases and chlorinated water.

MAMMALIAN TOXICITY: *Trichoderma harzianum* T-22 is non-infectious and non-pathogenic to mammals. **Acute oral LD$_{50}$:** rats >500 mg/kg. **Inhalation: LC$_{50}$:** rats >0.89 mg/litre. **Skin and eye:** Eye irritant, not a skin irritant. Possible skin sensitiser.

ENVIRONMENTAL IMPACT AND NON-TARGET TOXICITY: There are no known risks to non-target organisms or to the environment from the use of products containing *Trichoderma harzianum* T-22. **Bird toxicity:** Acute oral LD$_{50}$ for mallard ducks and bobwhite quail >2000 mg/kg. **Fish toxicity:** LC$_{50}$ (96 h) for zebra fish 1.23 × 10^5 colony forming units (cfu)/ml. **Other aquatic toxicity:** LC$_{50}$ (10 days) for *Daphnia pulex* de Geer 1.6 × 10^4 cfu/ml. **Effects on beneficial insects:** *Trichoderma harzianum* T-22 was not toxic to honeybees at 1000 ppm.

1:138 *Trichoderma harzianum* isolate T-39
Biological fungicide (fungus)

Fungus: anamorphic Hypocreales: Hypocreaceae

NOMENCLATURE: Approved name: *Trichoderma harzianum* Tul., variety TH11 (Harzan) isolate T-39. **Other names:** Formerly known as: *Trichoderma lignorum* (Tode) Harz. **Development code:** ABG-8007 (Makhteshim).

SOURCE: *Trichoderma harzianum* isolate T-39 occurs widely in Nature and this isolate was selected for commercialisation because of its ability to suppress plant disease through competition with phytopathogenic fungi.

PRODUCTION: *Trichoderma harzianum* isolate T-39 is produced by fermentation.

TARGET PESTS: Recommended for the control of soil-inhabiting *Botrytis* and *Sclerotinia* species. Also effective against *Botrytis cinerea* Pers. on young crop plants, as a foliar spray. Registrations for 'TRICHODEX' in various countries are for management of *Botrytis cinerea* incited diseases, but it also controls *Sclerotinia sclerotiorum* white mould and affects *Cladosporium fulvum* leaf mould and powdery mildews. 'TRICHODEX' has been found to affect other foliar and soil-borne diseases.

TARGET CROPS: Recommended for use on vines, vegetables, glasshouse crops and in the open field. In the USA, *T. harzianum* can be applied to most food crops grown in glasshouses, shade houses or in agricultural fields. No adverse effects on crops have been reported. 'TRICHODEX' must be used according to the instructions and recommendations of the registrant (Makhteshim) and according to the registration granted in each country.

BIOLOGICAL ACTIVITY: Mode of action: *Trichoderma harzianum* isolate T-39 induces resistance (local and systemic), suppresses pathogenicity enzymes of pathogens and competes with pathogens for space and nutrients. **Biology:** *Trichoderma harzianum* isolate T-39 is a widely distributed member of the microflora and it is known to antagonise phytopathogenic fungal species. **Key reference(s):** 1) G E Harman & C P Lumsden (eds). 1998. Trichoderma *and* Gliocladium, *Vol. 2*, Taylor and Francis, London, UK. 2) D R Fravel, W J Connick Jr & J A Lewis. 1998. Formulation of microorganisms to control plant diseases. In *Formulation of Microbial Biopesticides, Beneficial Microorganisms, Nematodes and Seed Treatments*, H D Burges (ed.), pp. 187–202, Kluwer Academic Publishers, Dordrecht, The Netherlands. 3) Y Elad, G Zimand, Y Zaqs, S Zuriel & I Chet. 1993. Biological and integrated control of cucumber grey mould (*Botrytis cinerea*) under commercial greenhouse conditions, *Plant Pathology*, **42**, 324–32. 4) Y Elad. 1994. Biological control of grape grey mould by *Trichoderma harzianum*, *Crop Protection*, **13**, 35–8. 5) G Zimand, L Valinsky, Y Elad, I Chet & S Manulis. 1994. The use of RPAD procedure for the identification of *Trichoderma* isolates, *Mycological Research*, **98**, 531–4. 6) G Zimand, Y Elad & I Chet. 1996. Effect of *Trichoderma harzianum* on *Botrytis cinerea* pathogenicity, *Phytopathology*, **86**, 1255–60. 7) G De Meyer, J Bigirimana, Y Elad & M Höfte. 1998. Induced systemic resistance in *Trichoderma harzianum* T39 biocontrol of *Botrytis cinerea*, *European Journal of Plant Pathology*, **104**, 279–86. 8) Y Elad, B Kirshner, Y Nitzani & D Shtienberg. 1999. Management of powdery mildew and gray mold of cucumber by *Trichoderma harzianum* T39 and *Ampelomyces quisqualis* AQ10, *BioControl*, **43**, 241–51. 9) Y Elad. 2000. *Trichoderma harzianum* T39 preparation for biocontrol of plant diseases – control of *Botrytis cinerea*, *Sclerotinia sclerotiorum* and *Cladosporium fulvum*, *Biocontrol Science and Technology*, **10**, 499–507. 10) T M O'Neill, Y Elad, D Shtienberg & A Cohen. 1996. Control of grapevine grey mould with *Trichoderma harzianum* T39, *Biocontrol Science and Technology*, **6**, 139–46.

COMMERCIALISATION: Formulation: Formulated as microgranules (MG) and as a powder. **Tradenames:** 'Bio-Fungus' (DCM), 'Harzan' (NPP), 'Promot' (Rincon-Vitova), 'Trichoderma 2000' (Mycontrol), 'TRICHODEX' (Makhteshim) and (Intrachem), 'Trichopel' (Agrimm).

APPLICATION: Apply to soil using conventional application equipment. Ensure that the soil is moist and that the temperature is at least 12°C. As a foliar spray, 'TRICHODEX' should be sprayed on crops using ground equipment only. It works best if all parts of the crop receive uniform spray coverage. Alternation with other control agents is useful.

PRODUCT SPECIFICATIONS: Specifications: *Trichoderma harzianum* isolate T-39 is produced by fermentation. **Purity:** The products contains only spores of *Trichoderma harzianum*. 'TRICHODEX' contains 22% propagules of *Trichoderma harzianum* isolate T-39. Viability of the formulation is determined by plating the product out on agar and counting the number of colonies formed after incubation in the laboratory for 48 hours. **Storage conditions:** Store under dry, stable conditions as unopened packs. Low temperatures are preferred. **Shelf-life:** Vacuum packs maintain product viability for one year if stored at 20°C and if unopened.

COMPATIBILITY: Not recommended to be used with selected fungicides. Benzimidazoles should not be mixed with *Trichoderma harzianum* isolate T-39-based products. Incompatible with strong oxidisers, acids, bases and chlorinated water.

MAMMALIAN TOXICITY: *Trichoderma harzianum* isolate T-39 is non-infectious and non-pathogenic to mammals. **Acute oral LD$_{50}$:** rats >500 mg/kg. **Acute dermal LD$_{50}$:** Not available. **Inhalation: LC$_{50}$:** rats >0.89 mg/litre. **Skin and eye:** Eye irritant, not a skin irritant. Possible skin sensitiser. **Toxicity class:** EPA (formulation) III-IV.

ENVIRONMENTAL IMPACT AND NON-TARGET TOXICITY: There are no known risks to non-target organisms or to the environment from the use of products containing *Trichoderma harzianum* isolate T-39. **Bird toxicity:** Acute oral LD$_{50}$ for mallard ducks and bobwhite quail >2000 mg/kg. **Fish toxicity:** LC$_{50}$ (96 h) for zebra fish 1.23×10^5 colony forming units (cfu)/ml. **Other aquatic toxicity:** LC$_{50}$ (10 days) for *Daphnia pulex* de Geer 1.6×10^4 cfu/ml. **Effects on beneficial insects:** *Trichoderma harzianum* isolate T-39 was not toxic to honeybees at 1000 ppm. **Approved for use in organic farming:** Yes.

1:139 *Trichoderma harzianum* Rifai isolates TH 35 and TH 315 *Biological fungicide (fungus)*

Fungi: anamorphic Hypocreales: Hypocreaceae

NOMENCLATURE: Approved name: *Trichoderma harzianum* Tul. Rifai isolates TH 35 and TH 315. **Other names:** Formerly known as *Trichoderma lignorum* (Tode) Harz.

SOURCE: *Trichoderma harzianum* isolates TH 35 and TH 315 are components of the soil microflora often occurring in the rhizosphere of germinating plants. The isolates selected for commercialisation were chosen because of their persistence in the root zone of treated plants, because of their effectiveness and the synergism between them in controlling and preventing fungal infections, and because of their effectiveness in promoting plant development due to their mycorrhitic function. Introduced as a nursery treatment in 1997 and as a soil treatment by Ofer Kleifeld, Mycontrol.

PRODUCTION: These *Trichoderma harzianum* isolates TH 35 and TH 315 ('ROOT PRO') are produced commercially by solid fermentation. Spores are developed within the final ingredients of the formulation.

TARGET PESTS: Used as a soil additive in the growth mixture in the nursery or in the field for the control of fungal pathogens, including *Pythium* spp., *Fusarium* spp., *Rhizoctonia solani* Kuehn and *Sclerotium rolfsii* Sacc.

TARGET CROPS: Effective on a wide range of crops, including vegetables, ornamentals, field crops and herbs.

BIOLOGICAL ACTIVITY: Mode of action: The fungus establishes itself in the rhizosphere of the treated crop and colonises the plant root system, antagonising (by coiling around pathogenic hyphae and digesting their inner contents) and competing with disease organisms that attack the developing root system. Fungal mycelium also helps the root system to absorb soil nutrients better (mycorrhitic effect) and promotes plant resistance against diseases.
Key reference(s): 1) G E Harman & C P Lumsden (eds). 1998. Trichoderma *and* Gliocladium, *Vol. 2*, Taylor and Francis, London, UK. 2) A S Paau. 1998. Formulation of beneficial microorganisms applied to soil. In *Formulation of Microbial Biopesticides, Beneficial Microorganisms, Nematodes and Seed Treatments*, H D Burges (ed.), pp. 235–54, Kluwer Academic Publishers, Dordrecht, The Netherlands.

COMMERCIALISATION: Formulation: Sold as a nursery soil mixture additive and as a field soil additive. **Tradenames:** 'ROOT PRO' (Mycontrol), 'ROOT PROTATO' (Mycontrol). **Patents:** US 4748021; US 4713342; US 4915944; EU 0133878; Israel 69368.

APPLICATION: Applied as a nursery soil additive at a rate of 1% (volume) and as a field soil additive at a rate of 600 litres per hectare.

PRODUCT SPECIFICATIONS: Specifications: These isolates of *Trichoderma harzianum* isolates TH 35 and TH 315 ('ROOT PRO') are produced commercially by solid fermentation. Spores are developed within the final ingredients of the formulation. **Purity:** Produced by solid fermentation and guaranteed to be free of bacterial contaminants. Quality assurance is carried out in all production stages.

COMPATIBILITY: Compatible with most fungicides and soil disinfestation treatments, except benomyl and carbendazim. It is recommended to transplant treated seedlings into pre-sanitised soil. Incompatible with strong oxidisers, acids, bases and chlorinated water.

MAMMALIAN TOXICITY: There are no known health risks from use of products containing *Trichoderma harzianum* isolates TH 35 and TH 315. The fungus is not toxic to mammals. **Acute oral LD$_{50}$:** Tests are not feasible. **Acute dermal LD$_{50}$:** rats >2000 mg/kg. **Inhalation: LC$_{50}$:** rats >4.21 mg/litre. **Skin and eye:** Non-irritant to skin of rabbits exposed to undiluted formulation. Non-irritant to eyes of rabbits exposed to undiluted formulation. Non-sensitising by contact with skin (guinea pigs).

ENVIRONMENTAL IMPACT AND NON-TARGET TOXICITY: There are no known risks to non-target organisms or to the environment from the use of products containing *Trichoderma harzianum* isolates TH 35 and TH 315. The fungus is not toxic to birds.
Approved for use in organic farming: Yes.

1:140 *Trichoderma harzianum* isolate ATCC 20475 and *T. viride* isolate ATCC 20476 *Biological fungicide (fungus)*

Fungi: anamorphic Hypocreales: Hypocreaceae

NOMENCLATURE: Approved name: *Trichoderma harzianum* Tul. isolate ATCC 20475 and *Trichoderma viride* Tul. isolate ATCC 20476. **Other names:** Formerly known as *Trichoderma harzianum/polysporum* (Link) Rifai and *Trichoderma lignorum* (Tode) Harz.

SOURCE: Both are naturally occurring soil fungi that were isolated from soil.

PRODUCTION: *Trichoderma harzianum* isolate ATCC 20475 and *Trichoderma viride* isolate ATCC 20476 are manufactured by fermentation.

TARGET PESTS: A wide range of soil and foliar pathogens, including honey fungus (*Armillaria mellea* Kumm.), *Phytophthora* spp., silver leaf (*Chondrostereum purpureum* Pouzar), *Pythium* spp., *Fusarium* spp., *Rhizoctonia* spp. and *Sclerotium rolfsii* Sacc.

TARGET CROPS: Orchards, vineyards, ornamentals, vegetables, glasshouse and horticultural use. Post-harvest fruit and vegetables.

BIOLOGICAL ACTIVITY: Mode of action: The beneficial effects of the combination of *Trichoderma harzianum* isolate ATCC 20475 and *Trichoderma viride* isolate ATCC 20476 are through competition in the soil or on newly pruned stems with pathogenic fungi. The beneficial fungi outcompete the pathogens for nutrients and rhizosphere/wound dominance, thereby preventing or significantly reducing the impact of the pathogens. *Trichoderma* species also enhance tissue development in treated crops through the enhancement of natural auxin release. **Efficacy:** It is claimed that treatment of soil with *Trichoderma* species shows greater than 90% cure of silver leaf in pome fruit and greater than 50% in stone fruit. Injections reduced the incidence of *Armillaria* in a variety of crops. Wound treatments must be allowed to become established for at least 48 hours. **Key reference(s):** 1) A Siven & I Chet. 1986. *Trichoderma harzianum:* An effective biocontrol agent of *Fusarium* spp. In *Microbial Communities in Soil*, V Jensen *et al.* (eds), Elsevier, London, UK. 2) A Tronsmo & C Dennis. 1977. The use of *Trichoderma* species to control strawberry fruit rots, *Netherlands Journal of Plant Pathol.*, **83** (Suppl. 1), 449–55. 3) G E Harman & C P Lumsden (eds). 1998. Trichoderma *and* Gliocladium, *Vol. 2*, Taylor and Francis, London, UK.

COMMERCIALISATION: Formulation: Formulated as a dry powder ('Trisan', 'Trichoseal'), in pellet form ('Trichopel'), as an injectable ('Trichoject') and as impregnated dowels ('Trichodowels'). **Tradenames:** 'Bio-Trek HB' (Wilbur-Ellis) and (Bio Works), 'Ecosom – TH' (Agri Life), 'Trichodowels' (Agrimms Biologicals), 'Trichoject' (Agrimms Biologicals), 'Trichopel' (Agrimms Biologicals), 'Trichoseal' (Agrimms Biologicals), 'Trichoseal-spray' (Agrimms Biologicals), 'Trisan' (Applied Chemicals (Thailand)).

APPLICATION: It is important to apply the products early in the growing season or immediately after pruning. *Trichoderma* species are very effective at protecting crops from the onset of disease and application before the plant pathogens are visible always gives better control. For control of soil-borne diseases, the product is applied at rates of 6 to 12 kg product per hectare or the pellets or dowels are placed in the soil around the germinating seeds or close to the established trees. The wound sealant is applied as soon as possible after pruning and, ideally, within 5 minutes. Woody crops should be injected in mid- to late spring with 20 ml of the product at ground level, with plants up to three rows away from the infection treated. Injection should be supplemented with soil treatment. Soil should be moist at time of treatment.

PRODUCT SPECIFICATIONS: Specifications: *Trichoderma harzianum* isolate ATCC 20475 and *Trichoderma viride* isolate ATCC 20476 are manufactured by fermentation.
Storage conditions: Store under cool, dry conditions in a sealed container. Do not freeze or expose to direct sunlight. **Shelf-life:** If stored under recommended conditions, the products should remain viable for up to 12 months.

COMPATIBILITY: It is unusual to apply *Trichoderma* species with other crop protection chemicals. Do not apply fungicides within 4 weeks of use.

MAMMALIAN TOXICITY: *Trichoderma harzianum* is non-infectious and non-pathogenic to mammals. **Acute oral LD$_{50}$:** rats >500 mg/kg (*Trichoderma harzianum* isolate ATCC 20475).
Inhalation: LC$_{50}$: rats >0.89 mg/litre (*Trichoderma harzianum* isolate ATCC 20475).
Skin and eye: *Trichoderma harzianum* isolate ATCC 20475 is an eye irritant, but not a skin irritant. It is a possible skin sensitiser. **Toxicity class:** EPA (formulation) III-IV (*Trichoderma harzianum* isolate ATCC 20475).

ENVIRONMENTAL IMPACT AND NON-TARGET TOXICITY: Bird toxicity: Acute oral LD$_{50}$ for mallard ducks and bobwhite quail >2000 mg/kg (*Trichoderma harzianum* isolate ATCC 20475). **Fish toxicity:** LC$_{50}$ (96 h) for zebra fish 1.23×10^5 colony forming units (cfu)/ml (*Trichoderma harzianum* isolate ATCC 20475). **Other aquatic toxicity:** LC$_{50}$ (10 days) for *Daphnia pulex* de Geer 1.6×10^4 cfu/ml (*Trichoderma harzianum* isolate ATCC 20475).
Effects on beneficial insects: *Trichoderma harzianum* isolate ATCC 20475 was non-toxic to honeybees by oral administration at 1000 ppm. **Approved for use in organic farming:** Yes.

1:141 *Trichoderma polysporum* isolate IMI 206039/ATCC 20475 and *T. harzianum* isolate IMI 206040/ATCC 20476
Biological fungicide (fungus)

Fungi: anamorphic Hypocreales: Hypocreaceae

NOMENCLATURE: Approved name: *Trichoderma polysporum* Rifai isolate IMI 206039/ATCC 20475 and *Trichoderma harzianum* Tul. isolate IMI 206040/ATCC 20476.
Other names: Formerly known as *Trichoderma harzianum/polysporum* (Link) Rifai and *Trichoderma lignorum* (Tode) Harz. The isolate ATCC 20475 was deposited by Bio-Innovation

as *Trichoderma polysporum* Bissett and ATCC 20476 was deposited as *Trichoderma harzianum* Bissett.

SOURCE: Both are naturally occurring fungi that were isolated from trees in Sweden using the method described by J L Ricard (Biological control of *Fomes annosus* Cke. in Norway spruce (*Picea abies* Karst.) with immunizing commensals, *Studia Forestalia Suecica*, 1970. **84**, 50).

PRODUCTION: *Trichoderma polysporum* isolate IMI 206039/ATCC 20475 and *Trichoderma harzianum* isolate IMI 206040/ATCC 20476 are manufactured in pure culture fermentation.

TARGET PESTS: Soil-borne fungal pathogens, especially *Botrytis cinerea* Pers. Certain Basidiomycetes, such as *Heterobasidion annosum* Bref. and *Chondrostereum purpureum* Pouzar.

TARGET CROPS: Potted plants, strawberries, pruning wounds on fruit and ornamental trees.

BIOLOGICAL ACTIVITY: Mode of action: Mycoparasitism, through enzymes and volatiles, pre-emption, plant auxin and systemic acquired resistance (SAR) induction have been reported concerning suppression of fungal pathogens by *Trichoderma* spp. **Biology:** These isolates are active over temperatures ranging from 2°C to 32°C. **Efficacy:** Formal registration tests have shown comparable efficacy of 'BINAB TF WP' and standard chemical fungicides against common *Botrytis cinerea* or greater efficacy against tolerant isolates. Commercial Swedish strawberry growers experienced similar results over the previous six open-field growing seasons. **Key reference(s):** 1) C Martinez, O Besnard & J C Baccou. 1999. Stimulation des défenses naturelles des plantes, *Phytoma*, **521**, 16–9. 2) T Ricard & H Jörgensen. 2000. BINAB's effective, economical, and environment compatible *Trichoderma* products as possible Systemic Acquired Resistance (SAR) inducers in strawberries, *DJF-Report*, **No. 12**, 67–75. 3) L Sticher, B M Mauch & J P Métraux. 1997. Systemic acquired resistance, *Annual Review Phytopathology*, **35**, 235–70. 4) B A Bailey & R D Lumsden (eds). 1998. Trichoderma *and* Gliocladium *Vol. 1, Basic Biology, Taxonomy and Genetics*, 278 pp., Taylor and Francis, Bristol, PA, USA and London, UK. 5) G E Harman & C P Kubicek, (eds). 1998. Trichoderma *and* Gliocladium. *Vol. 2. Enzymes, Biological Control and Commercial Applications*, 393 pp., Taylor and Francis, Bristol, PA, USA and London, UK. 6) D R Fravel, W J Connick Jr & J A Lewis. 1998. Formulation of microorganisms to control plant diseases. In *Formulation of Microbial Biopesticides, Beneficial Microorganisms, Nematodes and Seed Treatments*, H D Burges (ed.), pp. 187–202, Kluwer Academic Publishers, Dordrecht, The Netherlands.

COMMERCIALISATION: Formulation: Sold as wettable powders (WP) and pellets. **Tradenames:** 'BINAB TF WP' (Bio-Innovation) and (Henry Doubleday Research Association), 'BINAB T Pellets' (Bio-Innovation) and (Henry Doubleday Research Association), 'BINAB T Vector' (Bio-Innovation) and (Henry Doubleday Research Association), 'BINAB T WP' (Bio-Innovation) and (Henry Doubleday Research Association). **Patents:** US 4678669.

APPLICATION: In pasteurised potting medium, use 20 g 'BINAB TF WP' per square metre. For foliar applications to strawberries against grey mould, use 500 g 'BINAB TF WP' per hectare in 400 to 600 litres of water and, for pruning wound treatment, apply 1:5 'BINAB TF WP' with water by volume.

PRODUCT SPECIFICATIONS: Specifications: *Trichoderma polysporum* isolate IMI 206039/ ATCC 20475 and *Trichoderma harzianum* isolate IMI 206040/ATCC 20476 are manufactured in pure culture fermentation. **Purity:** A pure culture technique is used for active ingredient production. Quality assurance and control tests are made to verify the absence of bacteria pathogenic to man and unwanted fungi, including mutants and especially any that will grow at human body temperature. **Storage conditions:** Store under cool, dry conditions in the unopened original container. Do not expose to direct sunlight. **Shelf-life:** Propagule viability decreases with time, so the product should be used within one year of purchase.

COMPATIBILITY: Varies with various pesticides; generally sulfur and copper compounds are tolerated, whilst synthetic chemicals are not. After suspension in dechlorinated drinking quality water, avoid massive exposure to soil bacteria, *Bacillus subtilis* in particular. Incompatible with strong oxidisers, acids and bases.

MAMMALIAN TOXICITY: These *Trichoderma* isolates are non-infectious and non-pathogenic to mammals. No toxic effects have been observed on production workers over 25 years of continuous operations. **Acute oral LD$_{50}$:** rats >500 mg/kg. **Inhalation: LC$_{50}$:** rats >0.89 mg/litre. **NOEL:** 1×10^4 colony forming units (cfu)/g is the Swedish National Food Administration standard NOEL for dry fruit and cereals.

ENVIRONMENTAL IMPACT AND NON-TARGET TOXICITY: There are no known risks to non-target organisms or to the environment from the use of products containing these isolates of *Trichoderma polysporum* and *T. harzianum*. **Bird toxicity:** No toxic effects were observed in routine quality tests on domestic geese, wild ducks, jays and sparrows. Acute oral LD$_{50}$ for mallard ducks and bobwhite quail >2000 mg/kg. **Fish toxicity:** No toxicity was observed in routine quality tests on rainbow trout, carp, tench and minnows. LC$_{50}$ (96 h) for zebra fish 1.23×10^5 cfu/ml. **Other aquatic toxicity:** No toxicity was observed in routine quality tests on crayfish, frogs, dragonflies and plankton. LC$_{50}$ (10 days) for *Daphnia pulex* de Geer 1.6×10^4 cfu/ml. **Effects on beneficial insects:** These *Trichoderma* isolates are not toxic to bumblebees. **Behaviour in soil:** Recycled by common soil bacteria, *Bacillus subtilis* in particular. **Approved for use in organic farming:** Yes.

1:142 *Trichoderma stromaticum*
Biological fungicide (fungus)

Fungus: anamorphic Hypocreales: Hypocreaceae

NOMENCLATURE: Approved name: *Trichoderma stromaticum* Samuels & Pardo-Schultheiss.

SOURCE: Prospects of managing witches' broom disease of cacao through biological control have been envisaged and investigated for over 20 years, leading to the isolation of a new species of *Trichoderma*, *Trichoderma stromaticum*, a parasite on the mycelium and basidiocarps of *Crinipellis perniciosa* in the Bahia region of Brazil.

PRODUCTION: Manufactured by fermentation.

TARGET PESTS: Witches' broom (*Crinipellis perniciosa* (Stahel) Sing.)

TARGET CROPS: Cacao trees.

BIOLOGICAL ACTIVITY: Mode of action: Like many *Trichoderma* species, *Trichoderma stromaticum* exerts its effect through a variety of actions, including mycoparasitism, and through the production of enzymes and volatiles. *Trichoderma stromaticum* reduces new inoculum levels through the suppression of basidioma formation in *Crinipellis perniciosa*. Systemic acquired resistance (SAR) induction has been reported concerning suppression of fungal pathogens by *Trichoderma* spp. **Biology:** *Trichoderma stromaticum* is characterised by having conidia that slowly become yellow–green in agar culture and by stout, 'Pachybasium-like', phialides formed on the surface of stromatic structures. **Efficacy:** *Trichoderma stromaticum* has been shown to give approximately 99% control of *Crinipellis perniciosa* on the ground and in leaf litter and about 56% control in the plant canopy. **Key reference(s):** 1) G J Samuels, R Pardo-Schultheiss, K P Hebbar, R D Lumsden, C N Bastos & J L Bezerra. 2000. *Trichoderma stromaticum* Samuels & Pardo-Schultheiss: A parasite of the cacao witches broom pathogen, *Mycological Research*, **104**(6), 760–4. 2) J H Bowers, B A Bailey, S Sanogo & R D Lumsden. 2001. The impact of plant diseases on world chocolate production, *APSnet*, February 2001. 3) D Macagnan, R S Romeiro, A W V Pomella & M G Carvalho. 2000. Screening of bacterial residents on cocoa phylloplane for the biocontrol of witches' broom (*Crinipellis perniciosa*), www.ag.auburn.edu/argentina/pdfmanuscripts/tableofcontents.pdf

COMMERCIALISATION: Formulation: Under development in Brazil by the Brazilian Ministry of Agriculture's Cocoa Research Centre. Formulated as a powder containing 1×10^8 colony forming units (cfu)/g. **Tradenames:** 'Tricovab' (CEPLAC).

APPLICATION: Applied to foliage as a high-volume spray at 1 kg product per hectare, ensuring that the foliage is well covered. Mixed with leaf litter and soil around cacao trees to control pathogens in leaf litter.

PRODUCT SPECIFICATIONS: Specifications: Manufactured by fermentation. **Storage conditions:** Store under cool, dry conditions away from sunlight. **Shelf-life:** Use as soon after purchase as possible.

MAMMALIAN TOXICITY: There are no reports of allergic or other adverse effects by research workers, field or manufacturing staff.

ENVIRONMENTAL IMPACT AND NON-TARGET TOXICITY: *Trichoderma stromaticum* was isolated from litter around cacao trees and is relatively common in Nature. It is not expected that it will have any adverse effects on non-target organisms or on the environment.

1:143 *Trichoderma virens* isolate GL-21
Biological fungicide (fungus)

Fungus: anamorphic Hypocreales: Hypocreaceae

NOMENCLATURE: Approved name: *Trichoderma virens* (Miller, Giddens & Foster) von Arx. isolate GL-21. **Other names:** Formerly known as *Gliocladium virens* Miller, Giddens & Foster.

SOURCE: Naturally occurring soil fungus, discovered and isolated by USDA. Marketed by Certis.

PRODUCTION: Produced by fermentation.

TARGET PESTS: Soil-borne damping-off and root rot pathogens, such as *Rhizoctonia*, *Pythium*, *Fusarium*, *Thielaviopsis*, *Sclerotinia* and *Sclerotium* spp.

TARGET CROPS: Ornamentals and food crops grown in nurseries, glasshouses and interiorscapes. Outdoor use on turf and agricultural and ornamental crops.

BIOLOGICAL ACTIVITY: Mode of action: Microbial fungicide with a preventive rather than curative action. Exerts its effect in three different ways. *Trichoderma virens* produces an antibiotic, gliotoxin, that kills the plant pathogens. It also parasitises them, in addition to competing for nutrients. **Key reference(s):** 1) G E Harman & C P Lumsden (eds). 1998. Trichoderma *and* Gliocladium, *Vol. 2,* Taylor and Francis, London, UK. 2) D R Fravel, W J Connick Jr & J A Lewis. 1998. Formulation of microorganisms to control plant diseases. In *Formulation of Microbial Biopesticides, Beneficial Microorganisms, Nematodes and Seed Treatments,* H D Burges (ed.), pp. 187–202, Kluwer Academic Publishers, Dordrecht, The Netherlands. 3) A S Paau. 1998. Formulation of beneficial microorganisms applied to soil.

In *Formulation of Microbial Biopesticides, Beneficial Microorganisms, Nematodes and Seed Treatments*, H D Burges (ed.), pp. 235–54, Kluwer Academic Publishers, Dordrecht, The Netherlands.

COMMERCIALISATION: Formulation: Sold as granular formulations (GR). **Tradenames:** 'SoilGard 12G' and 'WRC-AP-1' (Certis).

APPLICATION: Applied at a rate of 0.5 to 0.75 kg per square metre or 1 to 4 kg per 10 hectolitres. The product should be incorporated into the soil. Apply before planting. Do not over-water and do not apply to plants that are already infested.

PRODUCT SPECIFICATIONS: Specifications: Produced by fermentation. **Storage conditions:** Store in a cool, dry place away from direct sunlight. **Shelf-life:** Retains biological effectiveness for a year, under correct storage conditions.

COMPATIBILITY: Fungicides should not be used at the time of incorporation of the product. Incompatible with strong oxidisers, acids, bases and chlorinated water.

MAMMALIAN TOXICITY: May cause mild, reversible irritation to the eyes and skin. No acute toxicity, infectivity or pathogenicity following inhalation.

ENVIRONMENTAL IMPACT AND NON-TARGET TOXICITY: *Trichoderma virens* occurs naturally in soil and is not expected to have any adverse effects on non-target organisms or on the environment. **Approved for use in organic farming:** Yes.

1:144 *Trichoderma viride*
Biological fungicide (fungus)

Fungus: anamorphic Hypocreales: Hypocreaceae

NOMENCLATURE: Approved name: *Trichoderma viride* Persoon (Anamorph). (Note: some isolates described as *Trichoderma viride* may be *T. asperellum* Samuels, Liechfeldt & Nirenberg; see: http://nt.ars-grin.gov/taxadescriptions/keys/frameNomenclature.cfm?gen=Trichoderma).

SOURCE: Isolated from the soil.

PRODUCTION: Produced by fermentation.

TARGET PESTS: A wide range of soil pathogens, such as *Rhizoctonia* spp., *Fusarium* spp., *Phytophthora* spp. and *Pythium* spp.

TARGET CROPS: Ornamentals, orchards, vegetables, cereals, pulses and oil seeds.

BIOLOGICAL ACTIVITY: Mode of action: Control is caused by competition in the soil. The beneficial fungi outcompete the pathogens for nutrients. **Efficacy:** *T. viride* gives between 40 and 70% control of soil- and seed-borne fungal pathogens. **Key reference(s):** 1) G Mondal, K D Srivastava & R Aggarwal. 1995. Antagonistic effect of *Trichoderma* sp. on *Ustilago segetum* var. *tritici* and their compatibility with fungicides and biocides, *Indian Phytopathology*, **48**(4). 2) G E Harman & C P Lumsden (eds). 1998. Trichoderma *and* Gliocladium, *Vol. 2*, Taylor and Francis, London, UK. 3) D R Fravel, W J Connick Jr & J A Lewis. 1998. Formulation of microorganisms to control plant diseases. In *Formulation of Microbial Biopesticides, Beneficial Microorganisms, Nematodes and Seed Treatments*, H D Burges (ed.), pp. 187–202, Kluwer Academic Publishers, Dordrecht, The Netherlands. 4) A S Paau. 1998. Formulation of beneficial microorganisms applied to soil. In *Formulation of Microbial Biopesticides, Beneficial Microorganisms, Nematodes and Seed Treatments*, H D Burges (ed.), pp. 235–54, Kluwer Academic Publishers, Dordrecht, The Netherlands.

COMMERCIALISATION: Formulation: Formulated as a powder, with less than 8% moisture. **Tradenames:** 'Bio AquaGuard' (plus *Pseudomonas fluorescens*) (SOM Phytopharma), 'Bio-cure F' (Stanes), 'EcoSOM' (SOM Phytopharma), 'Ecosom – TV' (Agri Life), 'Lycure' (SOM Phytopharma), 'Trieco' (Ecosense).

APPLICATION: *Trichoderma viride* is very effective, when applied to the soil, against a range of soil-borne diseases. Recommended rates of use are between 5 and 8 kg product per hectare.

PRODUCT SPECIFICATIONS: Specifications: Produced by fermentation. **Purity:** The formulation contains the conidia of *Trichoderma viride* at a concentration of 2×10^7 colony forming units (cfu)/g. **Storage conditions:** Store in a cool, dry condition in a sealed container. **Shelf-life:** The product may be stored for up to 8 months under ambient temperature.

COMPATIBILITY: It is unusual to apply *Trichoderma* species with other crop protection chemicals. Do not apply fungicides within 4 weeks of use. Incompatible with strong oxidisers, acids, bases and chlorinated water.

MAMMALIAN TOXICITY: In acute oral toxicity/pathogenicity tests conducted in Swiss albino mice at a dose level of 5 ml containing 1×10^9 cfu/ml, no adverse effects were observed. In acute dermal toxicity tests conducted in Himalayan albino rabbits at a dose level of 1×10^9 cfu, 'Bio-cure F' showed no adverse effects. **Inhalation:** 'Bio-cure F' is non-toxic and non-virulent to Wistar rats by inhalation. **Skin and eye:** 'Bio-cure F' is non-irritant to skin of rabbits.

ENVIRONMENTAL IMPACT AND NON-TARGET TOXICITY: *Trichoderma viride* occurs naturally in soil and is not expected to have any adverse effects on non-target organisms or on the environment. **Fish toxicity:** LC_{50} ('Bio-cure F') for *Tilapia mossambica* >200 mg/litre

(2 × 10^8 cfu/litre). **Other aquatic toxicity:** LC$_{50}$ ('Bio-cure F') for *Daphnia magna* 7.15 mg/litre. *Chlorella vulgaris* was highly tolerant of 'Bio-cure F', although some growth inhibition took place at 300 mg/litre. **Approved for use in organic farming:** Yes.

1:145 *Uromycladium tepperianum*
Biological herbicide (fungus)

Fungus: Urediniomycetes: Uredinales

NOMENCLATURE: Approved name: *Uromycladium tepperianum* (Sacc.) McAlpine.
Other names: Acacia gall rust.

SOURCE: Isolated from infected acacia trees in Australia.

PRODUCTION: Released into South Africa between 1987 and 1989 for the control of blue leaf wattle (*Acacia saligna* (Labill.) H. Wendl.). It has now spread widely throughout the Western Cape Province and is not sold commercially.

TARGET PESTS: Blue leaf wattle (*Acacia saligna*).

BIOLOGICAL ACTIVITY: Mode of action: The fungus causes extensive gall formation on branches and twigs, causing heavily infected branches to droop; the tree is eventually killed. **Efficacy:** About eight years after introduction, the rust disease had become widespread in South Africa and tree density was decreased by 90–95%. The number of seeds in the soil seed bank had also stabilised in most sites and the process of tree decline was reported to be continuing. It has been determined that the benefits of this biocontrol programme far outweigh the potential loss of social benefits, mainly as firewood, to be derived from this invasive tree species. **Key reference(s):** 1) J Gathe. 1971. Host range and symptoms in western Australia of gall rust, *Uromycladium tepperianum*, *J. Roy. Soc. W. Australia*, **54**, 114–8. 2) M J Morris. 1987. Biology of the Acacia gall rust, *Uromycladium tepperianum*, *Pl. Pathol.*, **36**, 100–6. 3) M J Morris. 1991. The use of plant pathogens for biological weed control in South Africa, *Agric. Eco-syst. Environm.*, **37**, 239–55. 4) M J Morris. 1997. Impact of the gall-forming rust fungus *Uromycladium tepperianum* on the invasive tree *Acacia saligna* in South Africa, *J. Biol. Control*, **10**, 75–82. 5) M J Morris. 1999. The contribution of the gall-forming rust fungus *Uromycladium tepperianum* (Sacc.) McAlp. to the biological control of *Acacia saligna* (Labill.) Wendl. (Fabaceae) in South Africa, *African Entomol. Mem.*, **1**, 125–8. 6) A R Wood

& M J Morris. 2007. Impact of the gall-forming rust fungus *Uromycladium tepperianum* on the invasive tree *Acacia saligna* in South Africa: 15 years of monitoring, *Biol. Control*, **41**, 68–77.

COMMERCIALISATION: Formulation: Not sold commercially.

PRODUCT SPECIFICATIONS: Specifications: Released into South Africa between 1987 and 1989 for the control of blue leaf wattle (*Acacia saligna*). It has now spread widely throughout the Western Cape Province and is not sold commercially.

MAMMALIAN TOXICITY: The organism has never been reported as a pathogen of humans, or as causing any type of adverse effect to humans, in published literature or through use. *Uromycladium tepperianum* occurs widely in Nature and there have been no reports of any adverse reaction following exposure.

ENVIRONMENTAL IMPACT AND NON-TARGET TOXICITY: It is not expected that *Uromycladium tepperianum* will cause any adverse effects on non-target organisms (although some species of *Acacia*, *Albizia* and *Racosperma* (all Fabaceae) are susceptible). It is not expected that there will be any adverse effects on the environment.

1:146 *Vairimorpha necatrix*
Biological insecticide (microsporidium)

Protozoa: Microsporidium

NOMENCLATURE: Approved name: *Vairimorpha necatrix* (Kramer) Piley.

SOURCE: *Vairimorpha necatrix*, a microsporidium with two distinctly different spore forms, was first reported from the armyworm *Pseudaletia unipuncta* (Haworth) larvae from Hawaii.

PRODUCTION: Generally, neonatal or early stage larvae are exposed to a low dose of spores placed on artificial diet or on a suitable food substrate. Several days later, depending on the rearing temperature, stage of insects and diet, mature spores are harvested from insects by various methods. *Vairimorpha necatrix* develops only in living cells. Many species of noctuid larvae, in particular *Helicoverpa zea* (Boddie) or *Trichoplusia ni* (Hübner), are suitable hosts for spore production. Spore production per infected larva is about 2×10^{10} spores per gram of host larva. Some species of microsporidia have been grown in tissue culture, but the high cost of culture media, the low yields in spores and the imperfect production techniques for mass production still limit the usefulness of this method at this present time.

TARGET PESTS: Lepidopteran insects, in particular *Helicoverpa zea* (Boddie), *Ostrinia nubilalis* (Hübner), *Spodoptera* spp. and *Trichoplusia ni* (Hübner).

TARGET CROPS: Field crops such as maize (corn), cotton and soybean, in addition to vegetable crops such as brassicae (cabbage).

BIOLOGICAL ACTIVITY: Mode of action: The insect fat bodies are the primary infection site. In advanced stages of disease, the fat body cells enlarge many times, giving the fat tissue a lobated appearance. The lobated fat tissues are filled with microsporidian spores. The abnormally large, white fat body is usually obvious through the integument. The degree of infection depends on spore dosage, temperature and larval age. Feeding by the insect is usually normal at first, decreasing to very little for the last few days before death. *Vairimorpha* is transmitted to the next generation, both on and in the eggs of infected adults. *Vairimorpha necatrix* causes two different types of disease, resulting in mortality in its hosts: death that results from gut damage followed by bacterial septicaemia; and death that results from microsporidiosis after ingestion of even a light spore dose. A low dosage of *V. necatrix* results in a chronic infection of mainly the fat bodies and some muscular tissues, whereas high dosages produce an acute infection of mainly midgut tissues. In the advanced stage of infection, the abnormally large, white fat body is usually conspicuous and a dorsal swelling may appear on the last two or three abdominal segments. **Biology:** *Vairimorpha necatrix* has two distinctly different spore forms. One form has spore sizes that are highly dependent upon temperature (for example, at 32°C, length = 2.4 to 5.2 μm and width = 2.2 μm, but at 15°C, length = 4.2 to 6.9 μm and width = 2.4 μm). The other spores are smaller and less variable (ranging from 2.82 to 3.98 μm in length by 1.74 to 2.40 μm in width). It is infective for more than 20 noctuids and many are major pests of agricultural crops. *Vairimorpha necatrix* will only infect Lepidoptera. **Efficacy:** *V. necatrix* has many features of a good potential biocontrol agent, including a high degree of virulence and a wide host range.

COMMERCIALISATION: Formulation: Under development as a biological control agent.

APPLICATION: Microsporidia must be eaten to infect an insect, although there is evidence that they can be transmitted by the action of predators and parasitic wasps. It is usual to apply the spores as a foliar spray to the leaves of the crop.

PRODUCT SPECIFICATIONS: Specifications: Generally, neonatal or early stage larvae are exposed to a low dose of spores placed on artificial diet or on a suitable food substrate. Several days later, depending on the rearing temperature, stage of insects and diet, mature spores are harvested from insects by various methods. *Vairimorpha necatrix* develops only in living cells. Many species of noctuid larvae, in particular *Helicoverpa zea* (Boddie) or *Trichoplusia ni* (Hübner), are suitable hosts for spore production. Spore production per infected larva is about 2×10^{10} spores per gram of host larva. Some species of microsporidia have been grown in tissue culture, but the high cost of culture media, the low yields in spores and the imperfect production techniques for mass production still limit the usefulness of this method at this present time. **Purity:** The spores are produced by culture in living insects. Care is taken to ensure that the formulation contains no bacteria that are pathogenic to man. Efficacy can be determined by bioassay on *Trichoplusia ni*. **Storage conditions:** Store in a sealed container at 0 to 6°C. Do not expose to direct sunlight. **Shelf-life:** Spores remain infective for up to 18 months, if stored correctly.

COMPATIBILITY: Preliminary data suggest that *Vairimorpha necatrix* is more effective when used in conjunction with other larvicides.

MAMMALIAN TOXICITY: *Vairimorpha necatrix* does not irritate, replicate or accumulate in rats, guinea pigs or rabbits. It is considered to be of very low toxicity to mammals. There are no reports of allergic responses by research workers, production staff or field workers.

ENVIRONMENTAL IMPACT AND NON-TARGET TOXICITY: It is not expected that the use of products containing *Vairimorpha necatrix* will have any effects on non-target organisms nor have any adverse effects on the environment. **Fish toxicity:** *Vairimorpha necatrix* does not replicate or accumulate in rainbow trout or bluegill sunfish.
Effects on beneficial insects: Honeybees and other beneficial organisms are not infected by *Vairimorpha necatrix*.

1:147 *Verticillium lecanii*
Biological insecticide (fungus)

Fungus: anamorphic Hypocreales: previously classified as mitosporic Deuteromycetes: Moniliales

NOMENCLATURE: Approved name: *Verticillium lecanii* (Zimm.) Viégas has been reclassified. See *Lecanicillium lecanii* (entry 1:71). (For more information on the new taxonomy see R Zare & W Gams. 2001. *Nova Hedwigia*, **71**, 329–37).

1:148 *Xanthomonas campestris* pv. *poannua*　*Biological herbicide (bacterium)*

Bacterium: Pseudomonadales: Pseudomonadaceae

NOMENCLATURE: Approved name: *Xanthomonas campestris* pv. *poannua* Dows.
Other names: Formerly known as *Bacillus campestris* Pam. and *Pseudomonas campestris* Sm.

SOURCE: Isolated from diseased annual meadow grass (*Poa annua* L.).

PRODUCTION: Produced by fermentation.

TARGET PESTS: Annual meadow grass (annual blue grass) (*Poa annua*).

TARGET CROPS: Ornamental and recreational turf, golf courses and lawns.

BIOLOGICAL ACTIVITY: Mode of action: *Xanthomonas campestris* pv. *poannua* is an aggressive pathogen of annual meadow grass and effects control by parasitism. It also produces the polysaccharide xanthan which interferes with the upward movement of water in the xylem. **Efficacy:** Control is achieved within 2 to 4 weeks, depending upon the weather conditions and the growth of the turf grass. Best control is achieved when the turf is growing well. **Key reference(s):** 1) S Mitra, J Turner, P C Bhowmik & M Elston. 2001. Biological control of *Poa annua* with *Xanthomonas campestris* pv. *Poannua*, *Proc. BCPC Conference – Weeds*, **1**, 40–54. 2) M P Greaves, P J Holloway & B A Auld. 1998. Formulation of microbial herbicides. In *Formulation of Microbial Biopesticides*, *Beneficial Microorganisms*, *Nematodes and Seed Treatments*, H D Burges (ed.), pp. 202–33, Kluwer Academic Publishers, Dordrecht, The Netherlands. 3) I Seiko, H Minoru, T Atsushi, M Kenji & F Takane. 1999. The Bioherbicide 'CAMPERICO', *J. Agric. Sci.*, **54**(5), 202–5.

COMMERCIALISATION: Formulation: Under development as a post-emergence treatment for the control of annual meadow grass in established turf. Sold as a liquid formulation of vegetative cells. **Tradenames:** 'XPo Bioherbicide' (Eco Soil Systems), 'Camperico' (Japan Tobacco).

APPLICATION: Apply to established turf post-weed emergence in a high-volume spray. Control is best when followed by conditions of high temperature and humidity.

PRODUCT SPECIFICATIONS: Specifications: Produced by fermentation.
Storage conditions: Store in a cool, dry place. Do not expose to direct sunlight. If long-term storage is required, keep at −18°C. **Shelf-life:** If stored correctly, the product should encounter no loss of viability for at least 2 months.

COMPATIBILITY: Incompatible with broad-spectrum products, such as copper-based fungicides.

MAMMALIAN TOXICITY: There are no reports of allergic or other adverse toxicological effects from research workers, manufacturers or field staff with *Xanthomonas campestris* pv. *poannua*.

ENVIRONMENTAL IMPACT AND NON-TARGET TOXICITY: *Xanthomonas campestris* pv. *poannua* occurs widely in Nature and is not expected to have any adverse effects on non-target organisms or on the environment. **Approved for use in organic farming:** Yes.

1:149 *Xanthomonas campestris* pv. *vesicatoria* and *Pseudomonas syringae* pv. *tomato* bacteriophages

Biological bactericide (bacteriophage)

Bacteriophage

NOMENCLATURE: **Approved name:** *Xanthomonas campestris* pv. *vesicatoria* and *Pseudomonas syringae* pv. *tomato* bacteriophages. **OPP Chemical Codes:** 006449 (*Xanthomonas campestris* pv. *vesicatoria* bacteriophage); 006521 (*Pseudomonas syringae* pv. *tomato* bacteriophage).

SOURCE: Both bacteriophages occur widely in Nature and were isolated from their hosts: *Xanthomonas campestris* pv. *vesicatoria* (Doidge) Dye (the causal organism of bacterial spot disease); and *Pseudomonas syringae* pv. *tomato* Van Hall (the causal organism of bacterial speck disease).

PRODUCTION: For commercial purposes, *Xanthomonas campestris* pv. *vesicatoria* and *Pseudomonas syringae* pv. *tomato* are grown in a culture of their host bacteria and then harvested when all the bacteria have been killed.

TARGET PESTS: Bacterial spot disease in peppers and tomatoes (*Xanthomonas campestris* pv. *vesicatoria*) and bacterial speck disease in tomatoes (*Pseudomonas syringae* pv. *tomato*).

TARGET CROPS: Tomato and pepper.

BIOLOGICAL ACTIVITY: **Mode of action:** Bacteriophages of *Xanthomonas campestris* pv. *vesicatoria* and *Pseudomonas syringae* pv. *tomato* attack the plant pathogenic bacteria after which they are named. They control the bacteria by pathogenesis. Following invasion and death, the infected bacterial cells rupture and new bacteriophages are released.

COMMERCIALISATION: Formulation: 'AgriPhage' is sold as a suspension of infected bacterial cells. **Tradenames:** 'AgriPhage' (OmniLytics).

APPLICATION: 'AgriPhage' can be applied to soil pre-planting and to soil and plants throughout the growing season by drench, spray or chemigation.

PRODUCT SPECIFICATIONS: Specifications: For commercial purposes, *Xanthomonas campestris* pv. *vesicatoria* and *Pseudomonas syringae* pv. *tomato* are grown in a culture of their host bacteria and then harvested when all the bacteria have been killed. **Purity:** Any remaining bacterial cells are checked to ensure they are not pathogenic to target crop plants and mammals. The product is monitored for the presence of other bacteria. **Storage conditions:** Use as soon after delivery as possible.

COMPATIBILITY: It is unusual to use 'AgriPhage' in association with any other crop protection agent.

MAMMALIAN TOXICITY: No harmful health effects to humans are expected from use of these two active ingredients because these two bacteriophages can infect only their target plant pathogenic bacteria. Bacteriophages are ubiquitous, commonly found in foods and feed, and are harmless to mammals.

ENVIRONMENTAL IMPACT AND NON-TARGET TOXICITY: No adverse environmental effects are expected from use of pesticide products containing the bacteriophages of *Xanthomonas campestris* pv. *vesicatoria* and *Pseudomonas syringae* pv. *tomato*. The bacteriophages attack only the target bacteria and biodegrade 24 to 48 hours after application. **Approved for use in organic farming:** Yes.

2. Natural Products

2:150 abamectin
Microbial insecticide and acaricide

The Pesticide Manual **Fifteenth Edition** entry number: 1

(i) R = -CH$_2$CH$_3$ (avermectin B$_{1a}$)

(ii) R = -CH$_3$ (avermectin B$_{1b}$)

NOMENCLATURE: Approved name: abamectin (BSI, draft E-ISO, ANSI); abamectine ((*f*) draft F-ISO). **Other names:** avermectin B1. **Development code:** MK-0936; C-076; L-676863. **CAS RN:** *[71751-41-2]* abamectin; *[65195-55-3]* avermectin B$_{1a}$; *[65195-56-4]* avermectin B$_{1b}$.

SOURCE: *Streptomyces avermitilis* M.S.T.D. is a naturally occurring soil Actinomycete isolated as part of a programme targeted at identifying new, biologically active secondary metabolites. Discovered in an *in vivo* screen when microbial fermentation broths were tested in mice against the nematode, *Nematospiroides dubius* Baylis, in a dual mice–nematode system.

PRODUCTION: Isolated following the fermentation of *Streptomyces avermitilis*. A mixture of two avermectins, avermectin B$_{1a}$ (i) and avermectin B$_{1b}$ (ii), was introduced as an insecticide/ acaricide by Merck Sharp and Dohme Agvet. Now owned by Syngenta.

TARGET PESTS: Recommended for the control of the motile stages of a wide range of mites, leaf miners, suckers, beetles and other insects. Also used for control of fire ants (*Solenopsis* spp.).

TARGET CROPS: Recommended for use on ornamentals, cotton, citrus fruit, pome fruit, nut crops, vegetables, potatoes and many other crops.

BIOLOGICAL ACTIVITY: Mode of action: The target for abamectin is the γ-aminobutyric acid (GABA) receptor in the peripheral nervous system. The compound stimulates the release of GABA from nerve endings and enhances the binding of GABA to receptor sites on the post-synaptic membrane of inhibitory motor neurons of nematodes and on the post-junction membrane of muscle cells of insects and other arthropods. This enhanced GABA binding results in an increased flow of chloride ions into the cell, with consequent hyperpolarisation and elimination of signal transduction resulting in an inhibition of neurotransmission. (See M J Turner & J M Schaeffer. 1989. In *Ivermectin and Abamectin*, W C Campbell (ed.), Springer-Verlag, New York, USA, p. 73). Insecticide and acaricide with contact and stomach action. It has limited plant systemic activity, but exhibits translaminar movement. **Key reference(s):** 1) W C Campbell (ed.). 1989. *Ivermectin and Abamectin*, Springer-Verlag, New York, USA. 2) M H Fisher & H Mrozik. 1984. In *Macrolide Antibiotics*, S Omura (ed.), Academic Press, New York, USA. 3) M H Fisher. 1993. Recent progress in avermectin research. In *Pest Control with Enhanced Environmental Safety*, S O Duke, J J Menn & J R Plimmer (eds), ACS Symposium Series 524, pp. 169–82, American Chemical Society, Washington DC, USA.

COMMERCIALISATION: Formulation: Formulated as an emulsifiable concentrate (EC) and a ready for use bait (RB). **Tradenames:** 'Agrimec', 'Dynamec', 'Vertimec', 'Affirm' (fire ant control), 'Agri-Mek' (citrus), 'Avid' (ornamentals), 'Clinch' and 'Zephyr' (cotton) (Syngenta), 'Abacide' (Mauget), 'Gilmectin' (Gilmore), 'Satin' (Sanonda), 'Abamex' and 'Vapcomic' (Vapco), 'Biok' (Cequisa), 'Romectin' (Rotam), 'Timectin' (Tide), 'Vibamec' (Vipesco), 'Agromec' (Chemvet), 'Apache' and 'Crater' (AFRASA), 'Belpromec' (Probelte), 'Vamectin 1.8 EC' (IQV), 'Vivid' (Florida Silvics).

APPLICATION: Rates of use are 5.6 to 28 g a.i. per hectare for mite control, 11 to 22 g a.i. per hectare for control of leaf miners. The effectiveness of the product is increased significantly by the addition of paraffinic oils to the spray tank.

PRODUCT SPECIFICATIONS: Specifications: Isolated following the fermentation of *Streptomyces avermitilis*. A mixture of two avermectins, avermectin B_{1a} (i) and avermectin B_{1b} (ii), was introduced as an insecticide/acaricide by Merck Sharp and Dohme Agvet. Now owned by Syngenta. **Purity:** A mixture containing about 80% avermectin B_{1a} (i) and 20% avermectin B_{1b} (ii).

COMPATIBILITY: Can be used with other crop protection agents, but not compatible with captan.

MAMMALIAN TOXICITY: Acutely toxic to mammals. **Acute oral LD_{50}:** rats 10 mg/kg; mice 13.6 mg/kg (in sesame oil). **Acute dermal LD_{50}:** rabbits >2000 mg/kg. **Skin and eye:** Mild eye irritant; non-irritating to skin (rabbits). **ADI:** (JMPR) 0.002 mg/kg b.w. [1997] (for sum of abamectin and 8,9-Z- isomer); 0.001 mg/kg b.w. [1995] (for residues not containing Δ-8,9-isomer). **Toxicity class:** EPA (formulation) IV. **Toxicity review:** 1) *Pesticide residues in food – 1994*, FAO Plant Production and Protection Paper, 1995 p. 127. 2) *Pesticide residues in food – 1994 evaluations. Part II – Toxicology*, World Health Organization, WHO/PCS/95.2, 1995.

3) *Pesticide residues in food – 1995*, FAO Plant Production and Protection Paper. 4) G Lankas & L R Gordon. 1989. In *Toxicology in Ivermectin and Abamectin*, W C Campbell (ed.), Springer-Verlag, New York, USA, pp. 89–112.

ENVIRONMENTAL IMPACT AND NON-TARGET TOXICITY: Bird toxicity: Acute oral LD_{50} for mallard ducks 84.6, bobwhite quail >2000 mg/kg. **Fish toxicity:** LC_{50} (96 h) for rainbow trout 3.2, bluegill sunfish 9.6 µg/litre. **Other aquatic toxicity:** EC_{50} (48 h) for *Daphnia pulex* 0.34 ppb. LC_{50} (96 h) for pink shrimp (*Panaeus duorarum*) 1.6, mysid shrimp (*Mysidopsis bahia*) 0.022; blue crab (*Callinectes sapidus*) 153 ppb. **Effects on beneficial insects:** Toxic to bees. **Metabolism:** Metabolites found in animals include 3″-demethylavermectin B_1 and 24-hydroxymethylavermectin B_1. 8,9-(*Z*)-Avermectin B_1 has been identified as a metabolite in plants. The polar degradates are the largest fraction; these are unidentified, but are non-toxic. **Behaviour in soil:** Binds tightly to soil, with rapid degradation by soil micro-organisms. No bioaccumulation.

2:151 acetic acid   *Fungal-derived herbicide*

$$CH_3-C\overset{O}{\underset{OH}{\big/\big/}}$$

NOMENCLATURE: Approved name: acetic acid. **CAS RN:** *[64-19-7]*.

SOURCE: Acetic acid is widely distributed in Nature.

PRODUCTION: Acetic acid is produced commercially by fermentation.

TARGET PESTS: Pesticide products containing acetic acid are used to control a diverse group of annual and perennial broadleaf weeds, including some grasses.

TARGET CROPS: Non-cropland areas, such as railway rights-of-way, golf courses, open space, driveways and industrial sites.

BIOLOGICAL ACTIVITY: Mode of action: To be effective, acetic acid needs to contact the plant leaves; the acidity of the spray solution damages and dries out the leaves.

COMMERCIALISATION: Formulation: Sold as an aqueous solution. **Tradenames:** 'Nature's Glory Weed and Grass Killer' and 'Nature's Glory Weed and Grass Killer Ready to Use' (Ecoval).

APPLICATION: For best control of weeds, the product must be sprayed onto weeds early in the season when there are very small. The product must contact the leaves to be effective.

PRODUCT SPECIFICATIONS: Acetic acid is produced commercially by fermentation.

COMPATIBILITY: Acetic acid is compatible with most pesticide products, although it is unusual to mix it.

MAMMALIAN TOXICITY: As a food, acetic acid is not expected to harm the public. Applicators are instructed to use protective equipment to protect against the eye and skin irritation that acids can cause.

ENVIRONMENTAL IMPACT AND NON-TARGET TOXICITY: Acetic acid is rapidly broken down to carbon dioxide and water and does not persist in the environment. It is not considered likely that its use represents any risk to non-target organisms.
Approved for use in organic farming: Yes.

2:152 4-allyl-2-methoxyphenol

Plant-derived insecticide

NOMENCLATURE: Approved name: 4-allyl-2-methoxyphenol. **Other names:** There is no ISO common name for this substance; the name eugenol has been used in the literature but has no official status. **CAS RN:** *[97-53-0]* 4-allyl-2-methoxyphenol; *[8000-34-8]* clove oil.

SOURCE: Found in a wide range of plants including laurel (*Laurus* species). Clove oil makes up 14 to 20% of the weight of cloves. It is predominantly comprised of 4-allyl-2-methoxyphenol, but also contains a small amount of acetyl 4-allyl-2-methoxyphenol.

PRODUCTION: Manufactured for use in agriculture, although use in organic farming is restricted to the naturally occurring material.

TARGET PESTS: Recommended for use against a wide range of insects, including aphids, armyworms, beetles, cutworms, grasshoppers, loopers, mites and weevils. 'Matran' is used as a total, post-emergence herbicide.

TARGET CROPS: Fruit and vegetables.

BIOLOGICAL ACTIVITY: Mode of action: 4-Allyl-2-methoxyphenol is a strong insect deterrent. At high concentrations, it is also an effective (organic farming approved) total, post-emergence herbicide.

COMMERCIALISATION: Tradenames: 'Bag-A-Bug Japanese Beetle Trap' (plus 2-phenylethyl propionate plus geraniol) and 'Japanese Beetle Combo Bait' (plus 2-phenylethyl propionate plus geraniol) (Spectrum), 'Bioganic Flying Insect Killer' (6% 4-allyl-2-methoxyphenol and 1% sesame oil), 'Bioganic Lawn and Garden Spray' (0.45% 4-allyl-2-methoxyphenol plus 0.45% thyme oil plus 0.45% sesame oil) and 'Bioganic Dust' (3.5% 4-allyl-2-methoxyphenol plus 1.5% phenethyl propionate) (Bioganic) and (Biocontrol Network), 'EcoEXEMPT HC' (21.4% 4-allyl-2-methoxyphenol plus 21.4% 2-phenethyl propionate) and 'Matran' (45.6% clove oil) (EcoSMART), 'Ecopco D Dust Insecticide', 'Ecopco AC Contact Insecticide' and 'Ecopco Jet Wasp/Hornet/Yellow Jacket Contact Insecticide' (plus 2-phenylethyl propionate) (EcoSMART), 'Ecosafe' (Arbico), 'Ecozap wasp and hornet insecticide' (plus 2-phenylethyl propionate), 'Ecozap Crawling and Flying Insecticide'(plus 2-phenylethyl propionate) and 'Bioganic Weed and Grass Killer' (2% 4-allyl-2-methoxyphenol plus 2% thyme oil plus 1% sodium lauryl sulfate plus 10% acetic acid) (Bioganic), 'Japanese Beetle Bait II' (plus 2-phenylethyl propionate plus geraniol) and 'Trece Japanese Beetle Trap' (plus 2-phenylethyl propionate plus geraniol) (Trece), 'Hercon Japanese Beetle Food Lure' (plus 2-phenethyl propionate plus geraniol) (Hercon), 'Raid Eo Ark' (plus 2-phenylethyl propionate) (Johnson), 'Ringer Japanese Beetle Bait' (Woodstream), 'Surefire Japanese Beetle Trap' (plus 2-phenylethyl propionate plus geraniol) (Suterra), 'Vizubon-D' (methyl salt plus naled) (Vipesco).

APPLICATION: Apply, in moderate volume, to foliage when insect pressure is increasing.

PRODUCT SPECIFICATIONS: Manufactured for use in agriculture, although use in organic farming is restricted to the naturally occurring material.

COMPATIBILITY: It is unusual to use 'Ecosafe' with other crop protection agents.

MAMMALIAN TOXICITY: 4-Allyl-2-methoxyphenol is an irritant and should be used with care. **Toxicity class:** EPA (formulation) III.

ENVIRONMENTAL IMPACT AND NON-TARGET TOXICITY: 4-Allyl-2-methoxyphenol is a naturally occurring plant-based phenolic and is not expected to be hazardous to non-target organisms or to the environment. **Approved for use in organic farming:** Yes.

2. Natural Products

2:153 aminoethoxyvinylglycine
Micro-organism-derived plant growth regulator

The Pesticide Manual **Fifteenth Edition entry number:** 44

NOMENCLATURE: Approved name: aminoethoxyvinylglycine.
Other names: *N*-(phenylmethyl)-1*H*-purin-6-amine; aviglycine; AVG.
Development code: ABG-3097; Ro 4468. **CAS RN:** *[49669-74-1]*, formerly *[73360-07-3]*.

SOURCE: Aminoethoxyvinylglycine is a naturally occurring amino acid.

PRODUCTION: Although originally isolated from a *Streptomyces* sp., aminoethoxyvinylglycine is now manufactured for use in crop protection.

TARGET CROPS: Apples, pears and other orchard fruit.

BIOLOGICAL ACTIVITY: Mode of action: Aminoethoxyvinylglycine is a naturally occurring amino acid which competitively inhibits the activity of 1-aminocyclopropane-1-carboxylic acid synthase (ACC synthase), thereby blocking ethylene biosynthesis. As ethylene is a gaseous plant hormone that promotes fruit ripening, the application of AVG prior to harvest has the potential to reduce preharvest fruit drop, delay harvest and improve fruit quality after storage. **Efficacy:** Aminoethoxyvinylglycine is applied 2 to 6 weeks before the optimum harvest date to reduce preharvest drop of apple cultivars and other tree fruit. Aminoethoxyvinylglycine sprays reduce the loss of fruit firmness and starch, particularly in harvests after the optimum harvest date. The development of water core is delayed and maturity cracking is delayed or prevented. Colour development is delayed for summer apple cultivars. Soluble solids concentration (SSC) is generally unaffected.
Key reference(s): 1) R E Byers. 1997. Effects of aminoethoxyvinylglycine (AVG) on preharvest fruit drop maturity of 'Delicious' apples, *J. Tree Fruit Production*, **2**(1), 53–75. 2) R E Byers. 1997. Effects of aminoethoxyvinylglycine (AVG) on preharvest fruit drop, maturity, and cracking of several apple cultivars, *J. Tree Fruit Production*, **2**(1), 77–97.

COMMERCIALISATION: Formulation: Sold as a water soluble concentrate (SL).
Tradenames: 'Retain' (hydrochloride) (Valent BioSciences).

APPLICATION: Aminoethoxyvinylglycine is applied to fruiting trees 2 to 6 weeks before the optimum harvest date, at rates between 32 and 132 mg per litre.

PRODUCT SPECIFICATIONS: Although originally isolated from a *Streptomyces* sp., aminoethoxyvinylglycine is now manufactured for use in crop protection.

MAMMALIAN TOXICITY: Not considered to be toxic. **Acute oral LD$_{50}$:** rats >5000 mg/kg. **Acute dermal LD$_{50}$:** rabbits >2000 mg/kg. **Inhalation:** LC$_{50}$ (4 h) for rats 1.13 mg/litre. **NOEL:** In a 90-day developmental toxicity study in rats by oral gavage, a NOAEL of 2.2 mg a.i. per kg b.w. per day was determined for both developmental and maternal toxicity. **ADI:** RfD 0.002 mg/kg b.w. [1997]. **Toxicity class:** EPA (formulation and technical a.i.) III, IV.

ENVIRONMENTAL IMPACT AND NON-TARGET TOXICITY: Aminoethoxyvinylglycine is not expected to have any deleterious effects on non-target organisms or on the environment. **Bird toxicity:** Aminoethoxyvinylglycine is moderately toxic to birds. Aminoethoxyvinylglycine hydrochloride acute oral LD$_{50}$ for northern bobwhite quail 121 mg/kg; dietary LC$_{50}$ (5 days) 230 ppm. **Fish toxicity:** Aminoethoxyvinylglycine hydrochloride LC$_{50}$ (96 h) for trout >139 mg/litre; NOEL (96 h) for trout 139 mg/litre; **Other aquatic toxicity:** EC$_{50}$ (48 h) for *Daphnia* >135 mg/litre; NOEL for *Daphnia* 135 mg/litre. **Effects on beneficial insects:** LD$_{50}$ (48 h, oral) for honeybees >100 μg/bee. LC$_{50}$ for earthworms >1000 ppm.

2:154 anthraquinone

Plant-derived bird repellent

The Pesticide Manual Fifteenth Edition entry number: 37

NOMENCLATURE: **Approved name:** anthraquinone. **Other names:** 9,10-anthraquinone. **CAS RN:** *[84-65-1]*.

SOURCE: Anthraquinone is found in a wide range of plants including teak (*Tectona grandis* L.), red quebracho (*Quebrachia lorentzii* (Griseb.)) and in the cuticular wax of perennial rye grass (*Lolium perenne* L.).

PRODUCTION: Manufactured for use in crop protection.

TARGET PESTS: Birds, especially geese on golf courses, any bird that is likely to endanger aircraft and seed feeders to safeguard treated crop seed. Often added to insecticide and/or fungicide treated seed to protect birds from accidental poisoning.

TARGET CROPS: Terrestrial areas at or near airports, commercial sites, industrial sites, municipal sites or in developed urban areas, golf courses, ornamental nurseries and conifer nurseries, landfills and dumpsites, building roofs, window sills and ledges and as a seed treatment on a variety of different crops.

BIOLOGICAL ACTIVITY: Mode of action: Anthraquinone has a marked repellency effect on birds, usually inducing retching. **Efficacy:** The use of the product in airports is intended to disperse birds that may pose threats to aeroplanes. Efficacy data have been submitted to the US-EPA to support claims of anthraquinone's ability to repel blackbirds, geese, cowbirds, robins, starlings, pigeons, horned lark and gulls. However, the review of the data indicated anthraquinone's repellency for geese only. Effects on the other bird species were not supported by the studies submitted. Because anthraquinone is not persistent in the environment (as demonstrated by the data), regular applications are required to keep the geese away. Anthraquinone treated seed is usually avoided by seed-feeding birds. **Key reference(s):** 1) T Robinson. 1967. *The Organic Constituents of Higher Plants*, Burgess, Minneapolis, MN, USA. 2) K K Schrader, M Q de Regt, P R Tidwell, C S Tucker and S O Duke. 1998. Selective growth inhibition of the musty odour-producing cyanobacterium *Oscillatoria* cf. *chalybea* by natural compounds. *Bull. Environ. Contam. Toxicol.*, **60**, 651–8.

COMMERCIALISATION: Formulation: Sold as dry and wettable powder, flowable and liquid concentrates, solutions and water dispersible powder seed treatments (often in mixtures – see *The Pesticide Manual* for details of typical mixed formulations) and as a liquid concentrate containing 50% anthraquinone. **Tradenames:** 'Corbit' (Bayer CropScience), 'Flight Control' (Environmental Biocontrol), 'Morkit' (Bayer CropScience).

APPLICATION: Spray applications are made when birds have been determined to be a nuisance. Repeat applications at weekly intervals. Treated seed is sown in the normal way.

PRODUCT SPECIFICATIONS: Manufactured for use in crop protection.

MAMMALIAN TOXICITY: Not considered to be toxic to mammals. **Acute oral LD_{50}:** rats >5000 mg/kg. **Acute dermal LD_{50}:** rats >5000 mg/kg. **Inhalation:** LC_{50} (4 h) for rats >1.3 mg/l air (dust). **Skin and eye:** Not irritating to eyes and skin (rabbits). **NOEL:** (90 days) for rats 15 mg/kg diet. **Toxicity class:** WHO (a.i.) U.

Bird toxicity: LD_{50} for Japanese quail >2000 mg/kg. **Fish toxicity:** LC_{50} (96 h) for rainbow trout 72, golden orfe 44 mg/l. **Other aquatic toxicity:** LC_{50} (48 h) for *Daphnia* >10 mg/l. ErC_{50} for *Scenedesmus subspicatus* >10 mg/l. **Effects on beneficial insects:** LC_{50} for worms (*Eisenia foetida*) >1000 mg/kg dry soil. **Behaviour in soil:** Anthraquinone is rapidly degraded in different soils (DT_{50} 7 to 10 days, DT_{90} 22 to 53 d (calculated from suitable laboratory trials)). Degradation is thought to be due to microbial activity. Laboratory trials have not revealed any leaching potential. In water, anthraquinone is stable to hydrolysis, but extremely sensitive to light with a DT_{50} in aqueous solution of about 9 minutes. On solid surfaces, 80% disappeared within 1 day. The low vapour pressure makes evaporation into the air unlikely.

Approved for use in organic farming: Naturally produced anthraquinone would be acceptable to IFOAM.

2:155 aureonucleomycin
Micro-organism-derived fungicide and bactericide

NOMENCLATURE: Approved name: aureonucleomycin. **Development code:** SPRI-371.

SOURCE: Derived from the Soochow isolate of *Streptomyces aureus* by the Shanghai Pesticide Research Institute.

PRODUCTION: Fermentation.

TARGET PESTS: Canker of citrus (*Xanthomonas citri* Dows.), bacterial leaf blight of rice (*Xanthomonas oryzae* Dows.) and cercospora leaf spot of rice (*Cercospora oryzae* Miyake).

TARGET CROPS: Citrus and rice.

BIOLOGICAL ACTIVITY: Mode of action: The mode of action of aureonucleomcin has not been determined. **Key reference(s):** Z M Li &Y B Zhang. 2008. The outset innovation of agrochemicals in China. *Outlooks on Pest Management*, **19**(3), 136–8.

COMMERCIALISATION: Formulation: Under development in China.

PRODUCT SPECIFICATIONS: Produced by fermentation.

MAMMALIAN TOXICITY: Studies on the mammalian toxicity of aureonucleomycin are under way.

ENVIRONMENTAL IMPACT AND NON-TARGET TOXICITY: Studies on the environmental impact and non-target effects of aureonucleomycin are under way.

2:156 azadirachtin   *Plant-derived insecticide*

The Pesticide Manual Fifteenth Edition entry number: 46

NOMENCLATURE: **Approved name:** azadirachtin. **Other names:** AZA; azad; neem. **Development code:** N-3101 (Cyclo). **CAS RN:** *[11141-17-6]*.

SOURCE: The neem tree (*Azadirachta indica* A. Juss) has been known to resist insect attack and, subsequently, it was found that extracts, particularly of the seed, were insecticidal. It is an attractive broad-leaved evergreen tree which is thought to have originated in Burma. It is now grown in the more arid subtropical and tropical zones of Southeast Asia, Africa, the Americas, Australia and the South Pacific Islands.

PRODUCTION: Azadirachtin is the principal insecticidal ingredient of neem (*Azadirachta indica*) seed extracts; these extracts also contain a variety of limonoids, such as nimbolide, nimbin and salannin. Many products also claim fungicidal activity. 'Neememulsion' is made from material containing 25% w/w azadirachtin, 30%–50% other limonoids, 25% fatty acids and 7% glycerol esters.

TARGET PESTS: A potent deterrent to many different genera of insects. Shown to be effective against whitefly, thrips, leaf miners, caterpillars, aphids, jassids, San José scale, beetles and mealybugs. 'Neemix 4.5' is recommended for use against balsam whiteflies (*Neodiprion abietis* (Harris)), yellow-headed spruce sawflies (*Pikonema alaskensis* (Rohwer)) and pine false webworms (*Acantholyda erythrocephala* (L.)). Some formulations claim effects against phytopathogenic fungi such as powdery mildews.

TARGET CROPS: Shows activity in a wide range of crops, including vegetables (such as tomatoes, cabbage, potatoes), cotton, tea, tobacco, coffee, protected crops and ornamentals. 'Neemix 4.5' is recommended for use in forestry.

BIOLOGICAL ACTIVITY: **Mode of action:** Azadirachtin has several effects on phytophagous insects. It has a dramatic antifeedant/repellent effect, with many insects avoiding treated crops, although other chemicals in the seed extract, such as salannin, have been shown to be responsible for these effects. It is thought to disrupt insect moulting by antagonising ecdysone. This leads to morphological defects in insects exposed to sprayed crops and, in some cases, the larval period is extended. This effect is independent of feeding inhibition. Azadirachtin is also believed to reduce the reproductive capabilities of phytophagous insects by disrupting normal mating behaviour and, thereby, reducing fecundity. The ecdysis inhibition also leads to effects on vitellogenesis, leading to the reabsorption of vitellarium and oviducts. Fungicidal and miticidal properties of the hydrophobic extract derive from physical smothering and desiccation. **Efficacy:** Most effective when used in spray programmes. Its mode of action means that it is slow to control insects, particularly when the populations are high.

Key reference(s): 1) H Schmutterer. 1995. *The Neem Tree; Source of Unique Natural Products for Integrated Pest Management, Medicine, Industry and Other Purposes,* VCH, Weinheim, Germany, 696 pp. 2) H Rembold. 1989. In *Focus on Phytochemical Pesticides, Vol. 1, The Neem Tree,* M Jacobsen (ed.), CRC Press, Boca Raton, Florida, USA.

COMMERCIALISATION: Formulation: Sold as an emulsifiable concentrate (EC) or as technical material. **Tradenames:** 'Align', 'Azatin', 'Bio-neem', 'Bollwhip', 'Neem', 'Neemazad', 'Neemix', 'Neemgard', 'Neem Oil 70%' (hydrophobic extract of neem oil), 'Niblecidine', 'Superneem 4.5-B', 'Triact', 'Trilogy' (hydrophobic extract of neem oil) and 'Turplex' (Certis), 'Agroneem' and 'Neem Cake' (Agro Logistics), 'Amazin 3% EC' (Amvac) and (Fortune), 'Amvac Aza 3% EC', 'Ecozin 3% EC' and 'Ornazin 3% EC' (Amvac), 'Anti-Pest-O' (Holy Terra), 'AquaNeem' (SOM Phytopharma), 'AZA-Direct' (Gowan), 'Azatrol' (PBI/Gordon), 'BioNeem' (Biocontrol Network) and (Safer), 'Blockade' (RPG), 'EI-783', 'EI-791' and 'Azatrol EC' (PBI/Gordon), 'Fortune Aza' and 'Fortune Biotech' (Fortune), 'Jawan' (J B Chemicals), 'K+ Neem' (plus insecticidal soap) (Organica), 'Kayneem' (neem oil) (Krishi Rasayan), 'Margosom' (Agri Life) and (SOM Phytopharma), 'Neemactin' (Biostadt), 'NeemAzad' (EID Parry) and (Andermatt), 'Neemazal T/S 1.2% EC' (EID Parry), 'Neememulsion' (Cyclo), 'Neemitaf' (300 mg/litre azadirachtin) and 'Neemitox' (1500 mg/litre azadirachtin) (Rallis), 'Neemolin' (seed extract) (Rallis), 'NeemPlus Liquid' (Biocontrol Network), 'Neem Suraksha', 'Proneem', 'Neem Wave' and 'AZA Technical' (Karapur Agro), 'Nimbecidine' (Stanes) and (PBT International), 'Ozoneem Aza' (25 to 35% azadirachtin), 'Ozoneem Oil' and 'Ozoneem Trishul' (Ozone Biotech), 'Trineem' (Tagros), 'Vineem' (Vipesco).

APPLICATION: Apply at rates of 100 to 500 g a.i. per hectare (0.15 to 0.65 oz a.i. per acre). Frequent applications are more effective than single sprays.

PRODUCT SPECIFICATIONS: Specifications: Azadirachtin is the principal insecticidal ingredient of neem (*Azadirachta indica*) seed extracts; these extracts also contain a variety of limonoids, such as nimbolide, nimbin and salannin. Many products also claim fungicidal activity. 'Neememulsion' is made from material containing 25% w/w azadirachtin, 30%–50% other limonoids, 25% fatty acids and 7% glycerol esters. **Purity:** Azadirachtin-based products are produced from the extraction of the seeds of the neem tree or from neem oil. In all cases, there are other components in each formulation, but the minimum claimed concentration of azadirachtin is guaranteed by the manufacturer.

COMPATIBILITY: No known incompatibilities with other crop protection agents, although, when neem was mixed with fenvalerate, its efficacy against the bollworm was reduced. Some neem formulations based on neem oil resulted in severe phytotoxicity in rice.

MAMMALIAN TOXICITY: Not considered to be acutely toxic to mammals.

Acute oral LD$_{50}$: rats >5000 mg/kg. Rats dosed once with 'Margosan-O' (since discontinued) and observed for 14 days showed no obvious effects, with the oral toxicity being above 5 ml/kg. **Acute dermal LD$_{50}$:** rabbits >2000 mg/kg. **Inhalation:** Albino rats exposed to 15.8 g of 'Margosan-O' (since discontinued) for 4 hours showed an LD$_{50}$ above 43.9 mg/litre per hour (the limit of the test). **Skin and eye:** Not a skin or eye irritant. Not a skin sensitiser.

Toxicity class: EPA (formulation) IV. **Other toxicological effects:** Neem seed oil and two components of neem oil, nimbolide and nimbic acid, showed no mutagenic effects in *Salmonella typhimurium* strains TA98 and TA100. **Toxicity review:** 1) M Jacobsen. 1986. Pharmacological and toxicological effects of neem and chinaberry on warm-blooded animals, *Neem Newsletter*, **3**(4), 39–43. 2) M Jacobsen (ed.). 1989. Pharmacology and toxicology of neem. In *Focus on Phytochemical Pesticides, Vol. 1, The Neem Tree*, pp. 133–53, CRC Press, Boca Raton, Florida, USA. 3) D Kanungo. 1993. In *Neem Research and Development*, N S Randhawa & B S Parmar (eds), pp. 250–62, Society of Pesticide Science, India.

Bird toxicity: Mallard ducks – daily oral administration of 'Margosan-O' (since discontinued) at 1–16 mg/kg induced no negative effects over a 14-day test period. Bobwhite quail fed a daily basic diet with added 'Margosan-O' at 1000 to 7000 ppm showed no negative effects over a 5-day test period and a 3-day recovery phase. Ducks fed a basic diet plus 'Margosan-O' at the same concentration for 5 days remained active and healthy throughout the test period. Dose levels of 1 to 16 ml of 'Margosan-O'/kg b.w. elicited no negative effects. **Fish toxicity:** LC$_{50}$ (96 h) for trout with 'Margosan-O' (since discontinued) 8.8 ml/litre, (24 h) for tilapia fingerlings with neem oil 1124.6 ppm, (24 h) for carp with neem oil 302.7 ppm. **Behaviour in soil:** DT$_{50}$ in soil *c*. 25 days. **Approved for use in organic farming:** Many products have US OMRI approval.

2:157 6-benzylaminopurine
Plant-derived plant growth regulator

The Pesticide Manual **Fifteenth Edition entry number:** *79*

NOMENCLATURE: Approved name: 6-benzylaminopurine. **Other names:** 6-BAP; 6-BA; BAP; 6-benzyladenine. **EC no:** 214-927-5. **OPP Chemical Code:** 116901. **CAS RN:** *[1214-39-7]*.

SOURCE: Reported as a naturally occurring plant growth regulator in higher plants by Skinner *et al.* in 1958 and introduced as a plant growth regulator in Japan by Kumiai Chemical Industry Co. Ltd in 1975.

TARGET CROPS: Fruit trees (orchards), ornamentals and cereals.

BIOLOGICAL ACTIVITY: Mode of action: 6-Benzylaminopurine stimulates RNA and protein biosynthesis, producing a number of growth-enhancing effects. These include a general increase in cell division, increased lateral bud formation in tree fruit, basal shoot formation in ornamentals, flowering in xerophytic species, fruit set in grapes, citrus and cucurbits and delayed senescence in rice. 6-Benzylaminopurine is used in combination with gibberellins A_4 and A_7 to thin apple trees in orchards and, thereby, increase the individual sizes of the fruit on each tree. **Key reference(s):** C G Skinner, F D Talbert & W Saive. 1958. Effect of 6-(substituted) purines and gibberellin on the rate of seed germination, *Plant Physiology*, **33**, 190–4.

COMMERCIALISATION: Formulation: Sold alone and as a combination product as a liquid concentrate (SL) and a paste (PA). **Tradenames:** 'Accel', 'Agtrol 6-BA', 'ABG-3164' (plus GA4 plus GA7), 'Cyclex' and 'ProShear' (Valent BioSciences), 'BA' (Kumiai), 'Beanin' (Riken Green), 'Chrysal BVB' (Pokon and Chrysal), 'Exilis Plus' (Fine Agrochemicals), 'Paturyl' (Reanal), 'Perlan' (plus GA4 with GA7) (Fine Agrochemicals) and (De Sangosse), 'Promalin' (plus GA4 plus GA7) (Valent BioSciences and Point Enterprises), 'Ritesize' and 'Typy' (plus GA4 with GA7) (Nufarm Americas).

APPLICATION: Applied as a relatively high volume foliar spray to ensure good coverage, as 6-benzylaminopurine is not well translocated in plants. Applied as a fruit thinning agent to apple trees from full blossom up to 2 weeks after petal fall.

PRODUCT SPECIFICATIONS: Purity: Contains only 6-benzylaminopurine.

COMPATIBILITY: Can be applied with other sprays.

MAMMALIAN TOXICITY: Not considered to be toxic. **Acute oral LD$_{50}$:** male rats 2125, female rats 2130, mice 1300 mg/kg. **Acute dermal LD$_{50}$:** rabbits 2900 mg/kg.
Skin and eye: Not an eye irritant. Not a skin sensitiser. **NOEL:** (2 years) for male rats 5.2, female rats 6.5, male mice 11.6 and female mice 15.1 mg/kg b.w. daily. **ADI:** 0.05 mg/kg.
Toxicity class: EPA (formulation) II; WHO (a.i.) U (company classification).
Other toxicological effects: Non-teratogenic in rats and rabbits and non-mutagenic in the Ames test.

ENVIRONMENTAL IMPACT AND NON-TARGET TOXICITY: The product does not accumulate in the environment and has no adverse effects on non-target organisms.
Fish toxicity: LC$_{50}$ (48 h) for carp >40 mg/litre. **Other aquatic toxicity:** LC$_{50}$ (24 h) for *Daphnia carinata* >40 mg/litre. EC$_{50}$ (96 h) for algae 363.1 mg/litre ('Paturyl' 10 SL formulation).
Effects on beneficial insects: LD$_{50}$ (oral) to honeybees 400 μg/bee; (contact) 57.8 litre/ha (both 'Paturyl' 10 SL formulation). **Metabolism:** In ^{14}C-metabolism studies, almost all radioactivity was excreted in the urine and faeces of test animals, with three metabolites being identified. In metabolism studies in plants, more than nine metabolites were found in studies on soybeans, grapes, maize and cocklebur (*Xanthium* sp.). **Behaviour in soil:** Sixteen days after application to soil at 22 °C, 6-benzylaminopurine had degraded to 5.3% (in a sandy loam soil) and 7.9% (in a clay loam soil) of the applied dose.
Approved for use in organic farming: Yes.

2:158 bilanafos

Micro-organism-derived herbicide

The Pesticide Manual **Fifteenth Edition entry number:** 84

NOMENCLATURE: Approved name: bilanafos (BSI, draft E-ISO, (*m*) draft F-ISO); bialaphos (JMAF). **Other names:** phosphinothricylalanyl-alanine. **Development code:** MW-801; SF-1293.
CAS RN: *[35597-43-4]* bilanafos; *[71048-99-2]* bilanafos-sodium.

SOURCE: Originally isolated from the soil-inhabiting Actinomycete, *Streptomyces hygroscopicus* (Jensen) Waksman & Henrici, and introduced by Meiji Seika. It is also produced by *Streptomyces viridochromeogenes* (Krainsky) Waksman & Henrici.

PRODUCTION: Bilanafos-sodium is produced by *Streptomyces hygroscopicus* during fermentation.

TARGET PESTS: Post-emergence control of annual weeds in crop situations and control of annual and perennial weeds in uncultivated land.

TARGET CROPS: Used post-emergence in vines, apples, brassicas, cucurbits, mulberries, azaleas, rubber and many other crops and on uncultivated land.

BIOLOGICAL ACTIVITY: Mode of action: Bilanafos is the alanylalanine amide of the biologically active acid, phosphinothricin. Phosphinothricin is a potent, irreversible inhibitor of glutamine synthetase, causing ammonia accumulation and inhibition of photophosphorylation in photosynthesis. The effects of bilanafos on plants are too rapid to be due to starvation of glutamine and other amino acids derived from glutamine and it was thought that the phytotoxic response was due to the high ammonium ion levels. However, the effects of the herbicide can be reversed by supplying the plant with glutamine and this does not reduce the levels of ammonium ions. Most of the phytotoxicity of inhibiting glutamine synthetase in C_3 plants is due to rapid cessation of photorespiration, resulting in accumulation of glyoxylate in the chloroplast and rapid inhibition of ribulose bisphosphate carboxylase. Inhibition of carbon fixation in the light leads to a series of events that ends with severe photodynamic damage. Bilanafos has no *in vitro* activity on the enzyme, but is converted to phosphinothricin within treated plants. Phosphinothricin is not metabolically degraded within higher plant tissue and is readily moved throughout treated plants in both the xylem and the phloem. The producing organism possesses an enzyme (phosphinothricin acetyl transferase (pat)) that acetylates the herbicide, rendering it non-inhibitory to glutamine synthetase and, hence, not phytotoxic. The gene that codes for this enzyme has been used to transform several crop plants to render them tolerant of over-the-top applications of bilanafos and its synthetic analogue, glufosinate.

Efficacy: Because bilanafos is converted into phosphinothricin which then interferes with glutamine synthetase, an essential enzyme of primary metabolism in higher plants, control of treated vegetation is total. However, effects often take several days to develop and death may take as long as 14 to 21 days. Regrowth of deep-rooted perennials may occur and retreatment may be necessary. Bilanafos has no effects pre-emergence.

Key reference(s): 1) S Omura, M Murata, H Hanaki, K Hinotozawa, R Oiwa & H Tanaka. 1984. Phosalacine, a new herbicidal antibiotic containing phosphinothricin. Fermentation, isolation, biological activity and mechanism of action, *J. Antibiot.*, **37**, 829. 2) E Bayer, K K Gugel, K Kaegel, H Hagenmaier, S Jessipov, W A König & H Zähner. 1972. Stoffwechselprodukte von Mikroorganismen. Phosphinothricin und Phosphinothricinyl-alanyl-alanin, *Helv. Chim. Acta*, **55**, 224. 3) Y Ogawa, H Yoshida, S Inouye & T Niida. 1973. Studies on a new antibiotic SF-1293. III. Synthesis of a new phosphorus containing amino acid, a component of antibiotic SF-1293, *Meiji Seika Kenkyu Nempo*, **13**, 49.

COMMERCIALISATION: Formulation: Sold as soluble powder (SP) and liquid formulations. **Tradenames:** 'Herbie' (Meiji Seika), (Hokko), (Sumitomo Chemical) and (Takeda).

APPLICATION: Applied post-emergence at rates of 0.5 to 1.0 kg a.i. per hectare for control of annual weeds and, at higher rates, for control of perennial weeds. Applied post-weed emergence, directed at the weed and away from the crop, in crop situations.

PRODUCT SPECIFICATIONS: Specifications: Bilanafos-sodium is produced by *Streptomyces hygroscopicus* during fermentation. **Purity:** Purity of the product is determined by nmr.

COMPATIBILITY: Compatible with most crop protection agents.

MAMMALIAN TOXICITY: Not considered to be acutely toxic to mammals. **Acute oral LD$_{50}$:** male rats 268, female rats 404 mg sodium salt/kg. **Acute dermal LD$_{50}$:** rats >5000 mg/kg. **Inhalation:** LC$_{50}$ for male rats 2.57, female rats 2.97 mg/litre. **Skin and eye:** Non-irritating to skin and eyes (rabbits). **Toxicity class:** WHO (a.i.) II. **Other toxicological effects:** In sub-acute and chronic toxicity tests, no ill-effects were observed. Non-carcinogenic, non-mutagenic and non-teratogenic. Not mutagenic in Ames and Rec assays. **Toxicity review:** A Suzuki *et al.* 1987. *J. Pestic. Sci.*, **12**, 347.

ENVIRONMENTAL IMPACT AND NON-TARGET TOXICITY: Bird toxicity: Acute oral LD$_{50}$ for chickens >5000 mg/kg. **Fish toxicity:** LC$_{50}$ (48 h) for carp 1000 mg/litre. **Other aquatic toxicity:** LC$_{50}$ (48 h) for *Daphnia pulex* 1000 mg/litre. **Effects on beneficial insects:** Not toxic to earthworms. **Metabolism:** In the mouse, the main metabolite in the faeces following oral administration was 2-amino-4-[(hydroxy)(methyl)=phosphinyl]butyric acid (A Suzuki *et al.* 1987. *J. Pestic. Sci.*, **12**, 105). Metabolised in plants to the L-isomer of glufosinate, which has a similar activity. **Behaviour in soil:** Inactivated in soil.

2:159 black pepper oil
Plant-derived animal repellent

NOMENCLATURE: Approved name: black pepper oil. **CAS RN:** *[8006-82-4]*.

SOURCE: Found in the seeds of the black pepper plant (*Piper nigrum* L.).

PRODUCTION: Oil of black pepper, a pale yellow irritating liquid with a sharp peppery odour, is obtained by steam distillation of the unripe dried fruit (peppercorns) of the black pepper plant.

TARGET PESTS: It is used as a repellent for a wide variety of small mammals, including dogs, cats, groundhogs, squirrels, skunks and raccoons.

TARGET CROPS: Oil of black pepper is used indoors, but only in non-living areas (attics, basements, cellars, storage areas, garages, sheds and barns) and outdoors on lawns, garden paths, flower beds, ornamental plants, trees, shrubs and garbage bags.

BIOLOGICAL ACTIVITY: Mode of action: The sharp and distinctive odour of the pepper oil is irritating to most target animal species.

COMMERCIALISATION: Formulation: Black pepper oil is sold as a granular formulation. **Tradenames:** 'Animal Repellent Granular' (Woodstream).

APPLICATION: The granules are distributed by hand or with a spreader.

PRODUCT SPECIFICATIONS: Specifications: Oil of black pepper, a pale yellow irritating liquid with a sharp peppery odour, is obtained by steam distillation of the unripe dried fruit (peppercorns) of the black pepper plant. **Purity:** The black pepper oil contains many different components, including chemicals commonly found in other plant oils, such as piperine and capsaicin. Oil of black pepper has many non-food uses, including aromatherapy and as a component of therapeutic skin products.

MAMMALIAN TOXICITY: No adverse effects to humans are expected from use of oil of black pepper in small mammal repellent products, because oil of black pepper is considered Generally Recognised As Safe for use in food (GRAS) by the Food and Drug Administration (FDA), and is widely used as a flavouring agent in foods; there is widespread exposure to oil of black pepper without any reported adverse effects to human health; therapeutically, the oil is mixed with other ingredients and applied to human skin with no apparent adverse effects; and very low levels of oil of black pepper are present in the registered end product, so exposure is expected to be minimal.

ENVIRONMENTAL IMPACT AND NON-TARGET TOXICITY: No toxic effects have been identified in mammals, birds, or fish. No adverse effects are expected based on the widespread use of oil of black pepper and the lack of reported adverse effects. Oil of black pepper is intended to repel small mammals, and, therefore, is not expected to harm target or non-target organisms. **Approved for use in organic farming:** Yes: it has US OMRI approval.

2:160 blasticidin-S

Micro-organism-derived fungicide

The Pesticide Manual **Fifteenth Edition entry number:** Superseded entry 957

blasticidin S

benzylaminobenzenesulfonic acid

NOMENCLATURE: Approved name: blasticidin-S (JMAF). **Development code:** BcS-3; -BAB; -BABS. **CAS RN:** *[2079-00-7]*; formerly *[11002-92-9]* and *[12767-55-4]*.

SOURCE: Isolated from the soil Actinomycete, *Streptomyces griseochromogenes* Fukunaga, in 1955 by K Fukunaga *et al.* Its fungicidal properties were first described by T Misato *et al.* in 1959.

PRODUCTION: By the fermentation of *Streptomyces griseochromogenes.* It is sold as the benzylaminobenzenesulfonate salt.

TARGET PESTS: Control of rice blast (*Pyricularia oryzae* Cavara; perfect stage *Magnaporthe grisea* (Hebert) Barr) by foliar application.

TARGET CROPS: Rice.

BIOLOGICAL ACTIVITY: Mode of action: Blasticidin-S inhibits protein biosynthesis by binding to the 50S ribosome in prokaryotes (at the same site as gougerotin), leading to the inhibition of peptidyl transfer and protein chain elongation. It is a contact fungicide with protective and curative action. **Efficacy:** Blasticidin-S exhibits a wide range of inhibitory activity on the growth of bacterial and fungal cells. In addition, it has been shown to have antiviral and anti-tumour activity. It inhibits spore germination and mycelial growth of *Pyricularia oryzae* in the laboratory at rates below 1 µg per ml. **Key reference(s):** 1) S Takeuchi, K Hirayama, K Ueda, H Sasaki & H Yonehara. 1958. Blasticidin S, a new antibiotic, *J. Antibiot. Ser. A*, **11**, 1. 2) T Misato, I Ishii, M Asakawa, Y Okimoto & K Fukunaga. 1959. Antibiotics as protectant fungicides against rice blast. II. The therapeutic action of blasticidin S, *Ann. Phytopathol. Soc. Jpn*, **24**, 302. 3) K T Huang, T Misato & H Suyama. 1964. Effect of blasticidin S on protein biosynthesis of *Pyricularia oryzae*, *J. Antibiot. Ser. A*, **17**, 65. 4) K Fukunaga *et al*. 1955. Blasticidin, a new anti-phytopathogenic fungal substance. Part I. *Bull. Agric. Chem. Soc. Jpn*, **19**, 181–8.

COMMERCIALISATION: Formulation: Sold as dispersible powder (DP), emulsifiable concentrate (EC) and wettable powder (WP) formulations. **Tradenames:** 'Bla-S' (Kaken), (Kumiai) and (Nihon Nohyaku).

APPLICATION: Applied at a rate of between 100 and 300 g a.i. per hectare for the control of rice blast (*Pyricularia oryzae*) by foliar application. Damage can be caused to alfalfa (lucerne), aubergines, clover, potatoes, soybeans, tobacco and tomatoes. Excessive application produces yellow spots on rice leaves.

PRODUCT SPECIFICATIONS: Specifications: By the fermentation of *Streptomyces griseochromogenes*. It is sold as the benzylaminobenzenesulfonate salt. **Purity:** Sold as the benzylaminobenzenesulfonate salt to reduce the possibility of crop damage. Sometimes sold in admixture with calcium acetate to reduce the incidence of eye irritation.

COMPATIBILITY: Incompatible with alkaline materials.

MAMMALIAN TOXICITY: Blasticidin-S is relatively toxic to mammals. **Acute oral LD$_{50}$:** male rats 56.8, female rats 55.9, male mice 51.9, female mice 60.1, rats 53.3 mg/kg (benzylaminobenzenesulfonate (BABS) salt). **Acute dermal LD$_{50}$:** rats >500 mg/kg. **Skin and eye:** Severe eye irritant. **NOEL:** (2 years) for rats 1 mg/kg diet. **Toxicity class:** WHO (a.i.) 1b; EPA (formulation) II. **Other toxicological effects:** Non-mutagenic in bacterial reversion tests.

ENVIRONMENTAL IMPACT AND NON-TARGET TOXICITY: Fish toxicity: LC$_{50}$ (48 h) for carp >40 mg/litre. **Other aquatic toxicity:** LC$_{50}$ (3 h) for *Daphnia pulex* >40 mg/litre. **Metabolism:** Almost all of ^{3}H-blasticidin-S administered to mice was excreted in the urine and faeces within 24 hours. In rice plants, cytomycin and deaminohydroxy blasticidin-S were identified as the main metabolites. **Behaviour in soil:** In soil, DT$_{50}$ <2 days (two soil types, o.c. 2.53%, 9.6%; moisture 42.6%, 87%, respectively; pH 6.0, 25 °C).

2:161 3-[N-butyl-N-acetyl]-aminopropionic acid, ethyl ester *Plant-derived insect repellent*

$$CH_3(CH_2)_3-N\begin{matrix} O=C-CH_3 \\ | \\ \end{matrix}-CH_2-CH_2-C(=O)-O-CH_2CH_3$$

NOMENCLATURE: Approved name: ethyl 3-[N-butyl-N-acetyl]-aminopropionate.
Other names: 3-[N-butyl-N-acetyl]-aminopropionic acid, ethyl ester.
OPP Chemical Code: 113509. **CAS RN:** *[52304-36-6]*.

SOURCE: 3-[N-butyl-N-acetyl]-aminopropionic acid, ethyl ester is structurally related to β-alanine, which occurs naturally.

PRODUCTION: Manufactured for commercial use.

TARGET PESTS: Mosquitoes, deer ticks, body lice and biting flies.

TARGET CROPS: Human skin.

BIOLOGICAL ACTIVITY: Mode of action: 3-[N-butyl-N-acetyl]-aminopropionic acid, ethyl ester has been known as a repellent of biting insects for many years. Its exact mode of action, however, is unknown.

COMMERCIALISATION: Tradenames: 'IR3535' (EM Industries).

APPLICATION: Apply to skin according to the labelled instructions.

PRODUCT SPECIFICATIONS: Manufactured for commercial use.

MAMMALIAN TOXICITY: 3-[N-butyl-N-acetyl]-aminopropionic acid, ethyl ester has been used as an insect repellent in Europe for 20 years, with no substantial adverse effects. Toxicity tests show that 'IR3535' is not harmful when ingested, inhaled, or used on skin. Eye irritation could occur if the chemical enters a person's eyes.

ENVIRONMENTAL IMPACT AND NON-TARGET TOXICITY: Because the active ingredient is used only in products applied to human skin, no risks to the environment are expected.

2:162 canola oil — *Plant-derived insecticide*

NOMENCLATURE: Approved name: canola oil.

SOURCE: Canola oil is an edible refined vegetable oil obtained from the seeds of two species of rape plants, *Brassica napus* L. and *B. campestris* L., of the family Cruciferae (mustard family).

PRODUCTION: Extracted from the seeds of oilseed rape plants.

TARGET PESTS: Canola oil is active against a wide range of insect pests.

TARGET CROPS: Canola oil can be used on a wide range of plants, including citrus, maize (corn), fruit and nut trees, sugar beet, soybeans, tomatoes, vegetables, figs, melons, olives, soft fruit, alfalfa (lucerne), bedding plants, ornamentals and houseplants.

BIOLOGICAL ACTIVITY: Mode of action: It is believed that canola oil repels insects by altering the outer layer of the leaf surface or by acting as an insect irritant. **Efficacy:** Canola oil is primarily an insect repellent and, as such, can be sprayed onto the foliage of crops or applied in irrigation water around the growing crop.

COMMERCIALISATION: Formulation: Sold as an oil formulation. **Tradenames:** 'Canola oil' (Neudorff).

APPLICATION: The products are applied either with spray or irrigation systems.

PRODUCT SPECIFICATIONS: Extracted from the seeds of oilseed rape plants.

MAMMALIAN TOXICITY: Canola oil is considered safe for human consumption. No harmful health effects to humans are expected from the use of canola oil to repel insects. Information available from published studies indicates that canola oil's nutritional and toxicological profiles are similar to those of other vegetable oils that are used as food.

ENVIRONMENTAL IMPACT AND NON-TARGET TOXICITY: Adverse effects to the environment or to organisms other than insects are not anticipated because of the low toxicity of canola oil and its rapid decomposition in the environment. In addition, canola pesticide products must not be applied directly to bodies of water; therefore, exposure of aquatic organisms should be extremely limited. **Approved for use in organic farming:** Yes: it has US OMRI approval.

2:163 capsaicin
Plant-derived insect and small mammal repellent

NOMENCLATURE: Approved name: capsaicin. **Other names:** Hot pepper extract; oleoresin. **CAS RN:** *[404-86-4]*.

SOURCE: Capsaicin is a common ingredient of solanaceous plants of the genus *Capsicum*. It is the compound responsible for the hotness of these chilli peppers. 'Armorex' is a mixture of extract of chilli (capsaicin) and essential oil of mustard (allyl isothiocyanate).

PRODUCTION: Capsaicin is obtained by grinding dried, ripe chilli peppers (*Capsicum frutescens* L.) into a fine powder. The oleoresin is derived by distilling the powder in a solvent and evaporating the solvent. The resulting highly concentrated liquid has little odour, but has an extremely pungent taste. Often sold in combination with extracts of garlic, onion, mustard and various herbs.

TARGET PESTS: A general insect and mite repellent. Also claimed to reduce transpiration in treated crops and to repel larger animals. 'Armorex' is used as a soil drench before planting and will control a wide range of fungi (including *Pythium*, *Rhizoctonia*, *Phytophthora*, *Pyrenochaeta*, *Sclerotium*, *Armillaria* and the clubroot organism, *Plasmodiophora*), soil insects (such as wireworms, cutworms, June beetle, June beetle larvae and white grubs), molluscs, nematodes (including root-knot species *Tylenchus*, *Pratylenchus*, *Xiphinema*, *Criconemoides* and *Pratylenchus*) and some weeds (seeds, roots, stolons and bulbs of broad-leaved weeds and grasses, including couch grass (*Elytrigia repens* (L.) Desv.), annual meadow grass (*Poa annua* L.), broomrape (*Orobanche* spp.), fat hen (*Chenopodium album* L.), torpedo grass (*Panicum repens* L.) and Bermuda grass (*Cynodon dactylon* Pers.)). It is not effective against mallow (*Malva* spp.), dodder (*Cuscuta* spp.) and some species of clover (*Trifolium* spp.). 'Nemastroy' is used to control soil-borne nematodes and to repel phytophagous insects. 'Valoram' is recommended for the control of many phytophagous insects and mites. 'Hot Pepper Wax Animal Repellent' is recommended for use against rabbits and squirrels.

TARGET CROPS: May be used on a wide range of different crops. 'Nemastroy' is recommended for use in landscapes, on golf courses and on ornamentals. 'Valoram' is recommended for use in a very wide range of vegetables and fruit and in turf. Capsaicin repellents can be used indoors to protect carpets and upholstered furniture.

BIOLOGICAL ACTIVITY: Mode of action: Capsaicin is claimed to disrupt insect metabolism and to affect the insect central nervous system. The pepper extract tends to damage the cell membranes of the insect, causing punctures or holes to form. The mustard extract is a neurotoxin, which will penetrate the damaged membrane and exoskeleton of the insect and kill the insect through its neurotoxic effects. Animal repellency effects are due to the pungent odour of the extract and an associated irritancy. **Efficacy:** Unlike other insect repellents, capsaicin also has some insecticidal activity and can be applied to crops with low levels of insect infestation.

COMMERCIALISATION: Formulation: Sold as a paraffinic concentrate. 'Armorex' is a liquid emulsifiable concentrate (EC). **Tradenames:** 'Armorex', 'Nemastroy' and 'Valoram' (Soil Technologies), 'Bonide Hot Pepper Wax' (Bonide), 'Dazitol' (Champon), 'Hot Pepper Wax' (Hot Pepper Wax), 'Hot Pepper Wax Animal Repellent' (Biocontrol Network). **Patents:** 'Armorex' is covered by US patent 6051233.

APPLICATION: Apply the product to the foliage of the crops, following the manufacturer's recommendations for dilution. May be applied as a curative spray or as a preventive treatment. Three treatments per season are usually sufficient for long-term control. 'Armorex' is a pre-plant or re-plant treatment. A waiting period of at least 3 days should be observed between application and planting and the soil should be covered with a non-porous cover to obtain the best control of soil-borne organisms.

PRODUCT SPECIFICATIONS: Capsaicin is obtained by grinding dried, ripe chilli peppers (*Capsicum frutescens* L.) into a fine powder. The oleoresin is derived by distilling the powder in a solvent and evaporating the solvent. The resulting highly concentrated liquid has little odour, but has an extremely pungent taste. Often sold in combination with extracts of garlic, onion, mustard and various herbs.

COMPATIBILITY: All products are compatible with most surfactants and conventional pesticides. The most effective ambient temperature range for application is 16 to 30 °C (60 to 85 °F). Capsaicin is toxic to beneficial insects and must not be used in combination with them.

MAMMALIAN TOXICITY: 'Hot Pepper Wax' is composed of compounds extracted from plants used for culinary purposes and is not considered to be toxic to mammals. The active ingredients of 'Armorex', 'Nemastroy' and 'Valoram' (allyl isothiocyanate (from essential oil of mustard) and capsaicin and related capsaicinoids (oleoresin of capsicum)) have been granted exemptions from the requirement for a tolerance for residues in or on all raw agricultural commodities, when applied according to approved EPA labelling and good agricultural practice. The active components are considered to be 'generally recognised as safe'. **Skin and eye:** Capsaicin is a severe eye irritant.

ENVIRONMENTAL IMPACT AND NON-TARGET TOXICITY: It is rapidly degraded in the environment. **Effects on beneficial insects:** The products are toxic to beneficial insects including honeybees. **Approved for use in organic farming:** Yes: it has US OMRI approval.

2:164 changchuanmycin
Micro-organism-derived fungicide

NOMENCLATURE: **Approved name:** changchuanmycin. **Development code:** Antibiotic SPRI-2098.

SOURCE: Discovered in a new isolate of *Streptomyces hygroscopicus* (Jensen) Waksman & Henrici in China by the Shanghai Pesticide Research Institute.

PRODUCTION: Fermentation.

TARGET PESTS: Powdery mildew (*Blumeria* spp., *Erysiphe* spp. and *Leveillula* spp.), early blight (*Alternaria solani* Jones & Grout) and black spot (*Cercospora* spp.).

TARGET CROPS: A wide variety of fruit and vegetables.

BIOLOGICAL ACTIVITY: **Mode of action:** The mode of action of changchuanmycin has not been determined. **Key reference(s):** Z M Li & Y B Zhang. 2008. The outset innovation of agrochemicals in China. *Outlooks on Pest Management*, **19**(3), 136–8.

COMMERCIALISATION: **Formulation:** Under development in China.

PRODUCT SPECIFICATIONS: Produced by fermentation.

MAMMALIAN TOXICITY: Studies on the mammalian toxicity of changchuanmycin are under way.

ENVIRONMENTAL IMPACT AND NON-TARGET TOXICITY: Studies on the environmental impact and non-target effects of changchuanmycin are under way.

2:165 cinnamaldehyde
Plant-derived fungicide, insect attractant and animal repellent

NOMENCLATURE: Approved name: cinnamaldehyde. **Other names:** Cinnamic aldehyde; cinnamylaldehyde; 3-phenyl-2-propenal. **CAS RN:** *[104-55-2]*.

SOURCE: Cinnamaldehyde occurs as the major component of oil found in cassia plants (*Cassia tora* L. (synonym *Cassia obtusifolia* L.)).

PRODUCTION: The oil is extracted from fresh seeds. In some situations, it is manufactured rather than extracted.

TARGET PESTS: Dry bubble (*Verticillium fungicola* (Preuss) Hassebrauk), other *Verticillium*, *Rhizoctonia* and *Pythium* species; dollar spot (*Sclerotinia homeocarpa* Bennett); and pitch canker disease (*Fusarium moniliforme* var. *subglutinans* Wr. & Reinking). It is used as an attractant for corn rootworm (*Diabrotica* spp.) and as a repellent for cats and dogs.

TARGET CROPS: Mushrooms, row crops, horticultural crops, turf and pine forests.

BIOLOGICAL ACTIVITY: Mode of action: The mode of action of cinnamaldehyde is not understood. Its specificity to particular genera of plant pathogens suggests that it is more than a simple disruption of the fungal membranes. Its repellent and attractant properties are based on its strong odour.

COMMERCIALISATION: Formulation: Cinnamaldehyde is sold as a 30% wettable powder (WP) formulation. **Tradenames:** 'Vertigo Wettable Powder Fungicide' (Monterey), 'Cinnacure A3005' (Proguard).

APPLICATION: Apply as a preventive treatment, at the rate of 0.2% to 0.5% cinnamaldehyde. Spray around areas to be protected from cat and dog invasions.

PRODUCT SPECIFICATIONS: Specifications: The oil is extracted from fresh seeds. In some situations, it is manufactured rather than extracted. **Purity:** As an extract of *Cassia tora*, the product will contain other, inert plant components.

COMPATIBILITY: It is not recommended that cinnamaldehyde be used with other crop protection agents.

MAMMALIAN TOXICITY: Not considered to be acutely toxic to mammals.
Acute oral LD$_{50}$: rats 2.25 or 3.35 g/kg, guinea pigs 1.15 g/kg ('Vertigo').
Acute dermal LD$_{50}$: rats >1.2 g/kg ('Vertigo'). **Inhalation:** LC$_{50}$ for rats >2.03 mg/litre
('Cinnacure A3005'). EPA (inhalation) IV ('Cinnacure A3005'). **Skin and eye:** Moderate eye
and skin irritant ('Cinnacure A3005'). **Toxicity class:** EPA III ('Vertigo' and 'Cinnacure A3005').

ENVIRONMENTAL IMPACT AND NON-TARGET TOXICITY: Cinnamaldehyde is not
soluble in water and is degraded rapidly in the soil and is not expected to pose any hazard to
non-target organisms or to the environment. **Other aquatic toxicity:** The US-EPA has ruled
that the following statement be carried on the labels of cinnamaldehyde-based products
– 'Environmental Hazards: Do not apply directly to water, or to areas where surface water is
present or to intertidal areas below the mean high water mark. Do not contaminate water by
cleaning of equipment or disposal of equipment wash waters'.
Approved for use in organic farming: Yes.

2:166 citric acid   *Plant-derived insecticide*

$$HO-\overset{\displaystyle CH_2CO_2H}{\underset{\displaystyle CH_2CO_2H}{C}}-CO_2H$$

NOMENCLATURE: Approved name: citric acid.

SOURCE: Component of all citrus fruit.

PRODUCTION: Extracted from citrus fruit for use in organic farming.

TARGET PESTS: A wide range of insects, including ants, aphids, beetles, caterpillars,
leafhoppers, mealybugs, mites and whitefly.

TARGET CROPS: Ornamentals, vegetables, shrubs and fruit trees.

BIOLOGICAL ACTIVITY: Mode of action: The mode of action of citric acid is not identified
with certainty. Relatively high rates have a rapid effect on a wide range of phytophagous
insects.

COMMERCIALISATION: Formulation: Sold as an aqueous concentrate.
Tradenames: 'SharpShooter' (St. Gabriel).

APPLICATION: Spray when insects first appear, covering as many insects as possible to
ensure control. If foliage is thick, ensure good coverage by spraying upper and lower foliage

to run-off. Repeat weekly or as needed to control infestation. Avoid spraying plants in hot midday sun. Spray can be applied up to the day of harvest.

PRODUCT SPECIFICATIONS: Extracted from citrus fruit for use in organic farming.

COMPATIBILITY: Can be used with other insecticides.

MAMMALIAN TOXICITY: Citric acid is not considered hazardous to mammals.

ENVIRONMENTAL IMPACT AND NON-TARGET TOXICITY: Citric acid occurs widely and is not expected to have any adverse effects on non-target organisms.
Approved for use in organic farming: Yes: it has US OMRI approval.

2:167 citric acid and mint oil
Plant-derived fungicide and bactericide

NOMENCLATURE: Approved name: citric acid and mint oil. **Other names:** Plant essential oils.

SOURCE: The product is a mixture of mint oil and citric acid. It is prepared by extraction from harvested plant material.

PRODUCTION: Plant essential oils are extracted from the appropriate crop plant (these plants must be fully described for all products in the USA).

TARGET PESTS: Effective against plant pathogenic fungi including *Phytophthora* spp. and other downy mildews, *Sclerotinia* spp. and *Botrytis* spp. and many bacteria such as *Pseudomonas*.

TARGET CROPS: 'Fungastop' can be used on a wide range of crops.

BIOLOGICAL ACTIVITY: Mode of action: It is believed to act on fungi and bacteria by alteration of their cell membrane and by the inhibition of cellular respiration.

COMMERCIALISATION: Formulation: Sold as an aqueous concentrate (AC). Additional components in the formulation include citrus pulp, fish oil, glycerol and vitamin C.
Tradenames: 'Fungastop' (Soil Technologies) and (Agriculture Solutions).

APPLICATION: 'Fungastop' is diluted 50 to 1 with water and applied in high volume to target crops when conditions are favourable for outbreaks of disease and repeated every 7–15 days depending upon disease pressure.

PRODUCT SPECIFICATIONS: Specifications: Plant essential oils are extracted from the

appropriate crop plant (these plants must be fully described for all products in the USA).
Storage conditions: If stored under cool, dry conditions 'Fungastop' has a shelf-life of 2 years.

MAMMALIAN TOXICITY: The product consists of plant essential oils that are used in foodstuffs. As a consequence, it is considered to be 'Generally Regarded As Safe' (GRAS) and is not considered to be toxic to mammals.

ENVIRONMENTAL IMPACT AND NON-TARGET TOXICITY: Specific data regarding toxicity to fish or other aquatic organisms are not available for this product, but it is believed that it should have no adverse effects on non-target organisms or on the environment.
Approved for use in organic farming: Yes: it has US OMRI approval.

2:168 citronella *Plant-derived insect repellent*

$$\begin{array}{cc} CH_3 & CH_2-CH_2 \\ \quad\ \ C{=}CH & \quad\ CH{-}CH_2 \\ CH_3 & CH_3\quad CHO \end{array}$$

NOMENCLATURE: Approved name: citronella.

SOURCE: The insecticidal properties of this oil were discovered in 1901.

TARGET PESTS: Used widely as a repellent for mosquitoes, gnats and other public hygiene insects.

BIOLOGICAL ACTIVITY: Mode of action: In the USA, citronella is a popular botanical ingredient in insect repellent formulations. Candles and incense containing oil of citronella are sold as insect repellents and it was used for a time as a hair dressing for the control of fleas and lice. The major component of citronella is citronellal.

COMMERCIALISATION: Formulation: Sold as candles containing citronella oil and as lotions for application to skin or clothing. **Tradenames:** 'Bio Bug Insect Repellent' (Nutribiotic), 'Bug Block Sunscreen and Insect Repellent' (Young), 'Buzz Away' (containing citronella, cedarwood, eucalyptus and lemongrass oils) (Quantum), 'Cardinal Insecticide Repellent Spray for Horses' (Cardinal), 'Cutter Insect Repellent RDCO31RN', 'Cutter Insect Repellent Icarus' and 'Cutter Insect Repellent Promethius' (Spectrum), 'Green Ban' (containing citronella, cajuput, lavender, safrole-free sassafras, peppermint and bergaptene-free bergamot oils) (Golden Harvest Organics), (Natural Living Store) and (Gaines), 'OFF! Citronella Candle' (Johnson), 'Scent-Off Twist-ons' and 'Scent-Off Pellets' (Scent-Off), 'Skin-So-Soft' (containing various 'oils and stearates') (Avon).

APPLICATION: Apply before entering an insect-favourable environment.

PRODUCT SPECIFICATIONS: Purity: The plant extract contains several minor components including alpha-citronellal, citronellol and alpha-citronellol.

COMPATIBILITY: It is unusual to use citronella oil with other chemical insecticides.

MAMMALIAN TOXICITY: Citronella occurs in Nature and has not shown any allergic or other adverse effects on mammals.

ENVIRONMENTAL IMPACT AND NON-TARGET TOXICITY: Citronella oil is a naturally occurring compound and it is not expected that it will have any adverse effects on non-target organisms or on the environment. It has a non-cidal effect on target insects, repelling rather than killing them.

2:169 cottonseed oil plus maize (corn) oil plus garlic oil *Plant-derived fungicide*

NOMENCLATURE: Approved name: cottonseed oil plus maize (corn) oil plus garlic oil. **Other names:** Plant essential oils. **CAS RN:** *[8001-29-4]* cottonseed oil; *[55512-33-9]* maize (corn) oil; *[8000-78-0]* garlic oil.

SOURCE: The vegetable oils are derived from the seeds of cotton and maize (corn) and the bulbs of garlic.

PRODUCTION: The vegetable oils are obtained from crushing the seeds of cotton and maize (corn) and the bulbs of garlic.

TARGET PESTS: Powdery mildews.

TARGET CROPS: 'SaferGro Mildew Cure' can be used on a very wide range of crops.

BIOLOGICAL ACTIVITY: Mode of action: The mode of action is not completely understood, but it is likely that the combination of oils prevents the powdery mildew pathogens from penetrating the cuticles of the treated plants. Membrane disruption of the germinating mildew spore is also thought to be involved.

COMMERCIALISATION: Formulation: 'SaferGro Mildew Cure' contains 30% cottonseed oil, 30% maize (corn) oil and 23% garlic oil and is formulated as an aqueous concentrate (AC). **Tradenames:** 'SaferGro Mildew Cure' (J H Biotech).

APPLICATION: The product should be mixed at a rate of 10 g per litre (1.5 oz. per gallon) and sprayed, as needed, onto the infected plants to cover the entire leaf surface as soon as disease is noticed.

PRODUCT SPECIFICATIONS: The vegetable oils are obtained from crushing the seeds of cotton and maize (corn) and the bulbs of garlic.

MAMMALIAN TOXICITY: The product consists of plant essential oils that are used in foodstuffs. As a consequence, it is considered to be 'Generally Regarded As Safe' (GRAS) and is not considered to be toxic to mammals.

ENVIRONMENTAL IMPACT AND NON-TARGET TOXICITY: Specific data regarding toxicity to fish or other aquatic organisms are not available for this product, but it is believed that it should have no adverse effects on non-target organisms or on the environment.
Approved for use in organic farming: Yes: it has US OMRI approval.

2:170 cytokinins (mixed)
Plant-derived plant growth regulator and nematicide

The Pesticide Manual **Fifteenth Edition entry number:** 225

NOMENCLATURE: Approved name: mixed cytokinins. **CAS RN:** *[308064-23-5]* cytokinins; *[2365-40-4]*, formerly *[13255-47-5]* and *[5122-37-2]* 6-isopentenylaminopurine; *[525-79-1]*, formerly *[525-80-4]* and *[33446-70-7]* kinetin; *[1637-39-4]*, formerly *[10052-59-2]* and *[129900-07-8]* zeatin.

SOURCE: Cytokinins occur naturally in plants and are involved in the initiation of plant cell division.

PRODUCTION: These products are extracted from seaweed.

TARGET PESTS: 'Suppress' is used for suppression of nematodes.

TARGET CROPS: Avocados, berries, citrus, coffee, cucurbits, herbs, hops, kiwi fruit, mangoes, pineapples, pome fruit, stone fruit, tree nuts, tropical fruit and vegetables.

BIOLOGICAL ACTIVITY: Mode of action: Cytokinins are naturally occurring plant growth regulators associated with cell division, and subsequently increased plant growth and the delay of senescence in treated organs. **Efficacy:** 'Soil Triggrr' is claimed to increase early seedling vigour and improve resistance to environmental stress. 'Foliar Triggrr' is claimed to increase flowering, fruit and seed set and to improve crop quality, uniformity, size and yield.

COMMERCIALISATION: Formulation: Sold as aqueous concentrates and granules. **Tradenames:** 'Agriblend' (Aqua-10), 'Culbac Foli-Veg' and 'PGR Plus' (Transagra), 'Cytex' (Atlantic and Pacific Research), 'Cytokin Bioregulator Concentrate', 'Cytogro Hormone Biostimulant', 'Xtra Set Blossom Spray', 'Burst Yield Booster/Plant Growth Regulator', 'Cytoplex HMS' and 'Cyzer' (PBT), 'Cytoplex HMS' (kinetin plus 4-indol-3-ylbutyric acid plus gibberellic acid) (Plant BioTech), 'Early Harvest' (cytokinins plus gibberellic acid plus 4-indol-3-ylbutyric acid) (Griffin) and (Nutrachem), 'Foli-Zyme GA' (cytokinins plus gibberellic acid plus 4-indol-3-ylbutyric acid), 'Stimulate Yield Enhancer', 'Stimulate Plus Yield Enhancer', 'Stoller X-Cyte' and 'Vigor S' (Stoller), 'Green Sol 48' and 'Green Sol 70' (kinetin plus gibberellic acid) (Frit), 'Happygro', 'Goldengro TM R', 'MegaGro L' (LT Biosyn), 'Nitrozyme' (cytokinins) (Atlantic Labs) and (Agri-Growth), 'Stimplex' (cytokinins) (Acadian), 'Super Lagniappe' (cytokinins plus gibberellic acid plus 4-indol-3-ylbutyric acid) and 'Maxon' (Agriliance), 'Suppress', 'Foliar Triggrr', 'Foliar Triggrr MFG', 'Organic Triggrr' (liquid seaweed), 'Soil Triggrr' and 'Soil Triggrr MFG' (cytokinins) (Westbridge), 'X-Cyte' (Stoller).

APPLICATION: Most products are applied as sprays to crop plants from an early growth stage. 'Soil Triggrr' and 'Suppress' are applied to the soil as granules or through chemigation techniques.

PRODUCT SPECIFICATIONS: These products are extracted from seaweed.

COMPATIBILITY: Fully compatible with other products.

MAMMALIAN TOXICITY: Not considered to be acutely toxic to mammals. **Acute oral LD$_{50}$:** rats >5000 mg/kg. **Acute dermal LD$_{50}$:** rabbits >2000 mg/kg. **Skin and eye:** Slight skin and eye irritant (rabbits).

ENVIRONMENTAL IMPACT AND NON-TARGET TOXICITY: Cytokinins occur widely in Nature and are not expected to have any adverse effects on non-target organisms or on the environment. **Behaviour in soil:** Rapidly degraded in soil. **Approved for use in organic farming:** Yes.

2:171 dihydroazadirachtin
Plant-derived insecticide

NOMENCLATURE: Approved name: dihydroazadirachtin. **Other names:** DAZA.

SOURCE: The active ingredient, dihydroazadirachtin (DAZA), is a reduced (hydrogenated) form of the naturally occurring azadirachtin (AZA) obtained from the seed kernels of the neem tree, *Azadirachta indica* A. Juss.

TARGET PESTS: A wide range of insect pests.

TARGET CROPS: Recommended for use on horticultural and ornamental plants, trees, shrubs and on turfgrass, fibre crops, forage and fodder crops and other agricultural crops.

BIOLOGICAL ACTIVITY: Mode of action: DAZA is structurally similar to azadirachtin (see entry 2:156) and the two compounds are functionally identical in their anti-pupation properties. DAZA has both antifeedant and insect growth regulator properties.

COMMERCIALISATION: Formulation: 'DAZA' is registered as a technical powder and an end-use product for indoor and outdoor use. **Tradenames:** 'DAZA' (SOM Phytopharma).

APPLICATION: The recommendation for use is foliar application by ground or aerial equipment, at a rate of 50 g per hectare, no more than seven times a season.

MAMMALIAN TOXICITY: Not considered to be acutely toxic to mammals.
Acute oral LD_{50}: rats ('DAZA') >5000 mg/kg. Toxicology Category IV.
Acute dermal LD_{50}: rabbits ('DAZA') >2000 mg/kg. Toxicology Category III.
Inhalation: LC_{50} for rats ('DAZA') >2.9 mg/litre. Toxicology Category IV. **Skin and eye:** The primary eye irritation of 'DAZA' in rabbits shows slight conjunctival irritation which clears in 24 hours (non-irritating). Toxicology Category IV. The acute dermal irritation of 'DAZA' in rabbits shows barely perceptible erythema in all test sites at 24 hours and no irritation at 72 hours (non-irritating). Toxicology Category IV. Not a dermal sensitiser.
Other toxicological effects: No detectable mutations were observed in the Salmonella/

Mammalian Microsomal Reverse Mutation Assay with five tester strains (TA98, TA100, TA1535, TA1537 and TA1538 with additional mutations) at levels of 6.6 to 5000 μg/plate of DAZA with ethanol vehicle controls, positive mammalian controls or no microsome controls.

ENVIRONMENTAL IMPACT AND NON-TARGET TOXICITY: Risk to the environment is not expected because, under current use conditions, 'DAZA' is not persistent, is relatively short lived in the environment (in the order of days) and is metabolised by ubiquitous micro-organisms in the soil and aquatic environments. **Bird toxicity:** The avian acute oral LD_{50} for 'DAZA' administered by way of gelatin capsule to bobwhite quail is >816 mg/kg (slightly toxic). Avian dietary LD_{50} for bobwhite quail is >1875 mg/kg, making it slightly toxic. **Fish toxicity:** 'DAZA' is slightly toxic to freshwater fish, with an LC_{50} for rainbow trout of 17.65 mg/litre and a 96-hour LC_{50} for bluegill sunfish of 17.65 mg/litre.
Other aquatic toxicity: 'DAZA' is slightly toxic to freshwater invertebrates, with an LC_{50} for *Daphnia magna* of 11.625 mg/litre. **Effects on beneficial insects:** 'DAZA' has been classified as relatively non-toxic to honeybees, as demonstrated by a 96-hour acute dust exposure to 'DAZA'. Being a broad spectrum insecticide, 'DAZA' is slightly toxic to beneficial insects (green lacewing, immature and mature mites, whitefly parasitoids and ladybirds).

2:172 (*E*)-3,7-dimethyl-2,6-octadien-1-ol
Plant-derived insecticide

NOMENCLATURE: Approved name: (*E*)-3,7-dimethyl-2,6-octadien-1-ol.
Other names: Geraniol; guaniol; lemonol; geranyl alcohol. **CAS RN:** *[106-24-1].*

SOURCE: Geraniol is found widely as a chief constituent in essential oils including ilang-ilang oil, palmarosa oil, geranium oil, orange flower oil, lemongrass oil, hops oil and lavender oil.

PRODUCTION: Extracted from plants for use as an insect repellent.

TARGET PESTS: Geraniol is effective in repelling mosquitoes, house flies, stable flies, horn flies, head lice, cockroaches, fire ants, gnats, dog ticks, lone star ticks and 'no-see-ums'.

TARGET CROPS: Used in public and animal health.

BIOLOGICAL ACTIVITY: Mode of action: (*E*)-3,7-dimethyl-2,6-octadien-1-ol is a strong insect deterrent.

COMMERCIALISATION: Formulation: Sold as ready-to-use formulations. Tradenames: 'Bag-a-Bug Japanese Beetle Trap' (2-phenylethyl propionate plus eugenol) and 'Japanese Beetle Combo Bait' (plus 2-phenylethyl propionate plus eugenol) (Spectrum), 'Biomite' (NPP), 'Hercon Japanese Beetle Food Lure' (plus eugenol plus 2-phenethyl propionate) (Aberdeen Road), 'Japanese Beetle Bait II' (plus 2-phenylethyl propionate plus eugenol) and 'Trece Japanese Beetle Trap' (plus 2-phenylethyl propionate plus eugenol) (Trece), 'Surefire Japanese Beetle Trap' (plus 2-phenylethyl propionate plus eugenol) (Suterra). Patents: US Patent No. 5,753,686; US Patent No. 5,633,236; US Patent 5,635,173 and US Patent 5,665,781.

APPLICATION: Apply liberally to areas to be protected from flying and biting insects.

COMPATIBILITY: Incompatible with strong oxidising agents.

MAMMALIAN TOXICITY: (*E*)-3,7-dimethyl-2,6-octadien-1-ol is not considered to be hazardous to mammals.

ENVIRONMENTAL IMPACT AND NON-TARGET TOXICITY: (*E*)-3,7-dimethyl-2,6-octadien-1-ol occurs widely in Nature and is not expected to show any adverse effects on non-target organisms or the environment. Approved for use in organic farming: Yes: it has US OMRI approval.

2:173 DMDP   *Plant-derived nematicide*

NOMENCLATURE: Approved name: (2*R*,5*R*)-dihydroxymethyl-(3*R*,4*R*)-dihydroxypyrrolidine. Other names: DMDP.

SOURCE: The naturally occurring sugar analogue, DMDP, has been isolated from tropical legumes in the genera *Lonchocarpus* and *Derris*.

PRODUCTION: The production process for DMDP is under development.

TARGET PESTS: Many species of nematode.

TARGET CROPS: DMDP has shown activity in a wide range of crops, including vegetables, cotton, tobacco, coffee and bananas.

BIOLOGICAL ACTIVITY: Mode of action: The mode of action of DMDP is under investigation. **Efficacy:** DMDP is unusual in that it can be applied as a foliar spray to crops to protect them from attack by soil-inhabiting nematodes. It has been shown that the sugar analogue is phloem-mobile, which means that it can be applied to the leaves and will be translocated to all parts of the plant including the roots. DMDP has shown good activity against nematodes when applied as a foliar spray. Excellent control has been observed, although field trials have yet to be completed. **Key reference(s):** N Birch *et al.* 1993. DMDP – A plant-derived sugar analogue with systemic activity against plant parasitic nematodes, *Nematologica*, **39**, 521–35.

COMMERCIALISATION: Formulation: Trials are in progress globally to evaluate the potential opportunities for DMDP as a foliar-applied nematicide. **Patents:** GB 2250439 A

APPLICATION: Applied as a spray to the foliage of crops. Rates and timings of application are under investigation.

PRODUCT SPECIFICATIONS: The production process for DMDP is under development.

MAMMALIAN TOXICITY: Preliminary toxicity tests have indicated that DMDP is of low toxicity to mammals. Although further tests will be needed to meet registration requirements around the world, there is a precedent that data requirements will be substantially less than those required for synthetic chemicals.

ENVIRONMENTAL IMPACT AND NON-TARGET TOXICITY: The environmental impact of DMDP is expected to be low. Data are awaited.

2. Natural Products

2:174 emamectin benzoate

Micro-organism-derived insecticide and acaricide

B_{1a} R = CH_3CH_2-

B_{1b} R = CH_3-

NOMENCLATURE: Approved name: emamectin benzoate. **Development code:** MK 244.
CAS RN: [155569-91-8], formerly [137512-74-4] and [179607-18-2].

SOURCE: *Streptomyces avermitilis* M.S.T.D. is a naturally occurring soil Actinomycete isolated as part of a programme targeted at identifying new, biologically active secondary metabolites. Abamectin was discovered when microbial fermentation broths were tested in an *in vivo* mouse/nematode screen. Emamectin benzoate is synthesised from a naturally occurring insecticide/acaricide. It consists of two homologues, emamectin B_{1a} and emamectin B_{1b}.

TARGET PESTS: Particularly effective against caterpillar pests (Lepidoptera), with suppressive activity against mites, leaf miners and thrips.

TARGET CROPS: Recommended for use in vegetables, maize (corn), tea, cotton, peanuts and soybeans. Recommended for injection into pine trees.

BIOLOGICAL ACTIVITY: Mode of action: The target for emamectin is the γ-aminobutyric acid (GABA) receptor in the peripheral nervous system. The compound stimulates the release of GABA from nerve endings and enhances the binding of GABA to receptor sites on the post-synaptic membrane of inhibitory motor neurons of nematodes and on the post-junction membrane of muscle cells of insects and other arthropods. This enhanced GABA binding results in an increased flow of chloride ions into the cell, with consequent hyperpolarisation and elimination of signal transduction resulting in an inhibition of neurotransmission. (See M H Fisher. 1993. Recent progress in avermectin research. In *Pest*

Control with Enhanced Environmental Safety, S O Duke, J J Menn & J R Plimmer (eds), ACS Symposium Series 524, pp. 169–82. American Chemical Society, Washington DC, USA). Insecticide with contact and stomach action. It has limited plant systemic activity, but exhibits translaminar movement. Emamectin benzoate irreversibly paralyses treated Lepidoptera, preventing subsequent crop damage. The lepidopteran stops feeding within hours of ingestion and dies 2 to 4 days after treatment. **Efficacy:** Emamectin benzoate gives excellent control of a wide range of lepidopteran insect pests, with additional good effects against thrips. It causes a rapid cessation of feeding, shows translaminar activity and is taken up through the roots. **Key reference(s):** 1) H Mrozik, P Eskola, B O Linn, A Lusi, T L Shih, M Tishler, F S Wakmunski, M J Wyvratt, N J Hilton, T E Anderson, J R Babu, R A Dybas, F A Preiser & M H Fisher. 1989. Discovery of novel avermectins with unprecedented insecticidal activity, *Experientia*, **45**, 315–6. 2) R A Dybas, N J Hilton, J R Babu, F A Preiser & G J Dolce. 1998. Novel second-generation avermectin insecticides and miticides for crop protection. In *Novel Microbial Products for Medicine and Agriculture*, A L Demain, G A Somkuti, J C Hunter-Cevera & H W Rossmoore (eds), Elsevier, Amsterdam, The Netherlands. 3) M H Fisher. 1993. Recent progress in avermectin research. In *Pest Control with Enhanced Environmental Safety*, S O Duke, J J Menn & J R Plimmer (eds), ACS Symposium Series 524, pp. 169–82, American Chemical Society, Washington DC, USA. 4) D M Dunbar, D S Lawson, S M White, N Ngo, P Dugger & D Richter. 1998. Emamectin benzoate: control of the heliothine complex and impact on beneficial arthropods, *Proc. Beltwide Cotton Conf.*, San Diego, California, USA, **2**, 1116–8. 5) R K Jansson & R A Dybas. 1996. Avermectins: biochemical mode of action, biological activity and agricultural importance. In *Insecticides with Novel Modes of Action: Mechanisms and Application*, I Ishaaya (ed.), Springer-Verlag, New York, USA.

COMMERCIALISATION: Formulation: Emamectin benzoate is sold as emulsifiable concentrate (EC) and soluble granule (SG) formulations. **Tradenames:** 'Proclaim', 'Affirm', 'Shot-One', 'Arise' and 'Denim' (Syngenta).

APPLICATION: It is used at rates between 5 and 25 g a.i. per hectare. The effectiveness of the product is increased significantly by the addition of paraffinic oils to the spray tank. The products 'Shot-One' and 'Arise' are used in Japan as pine tree injection treatments for control of pine wood nematodes.

COMPATIBILITY: May be used with other crop protection products.

MAMMALIAN TOXICITY: The formulated product is classed as slightly hazardous. **Acute oral LD$_{50}$:** rats (technical) 70 mg/kg; ('Proclaim 019EC') 2646 mg/kg; ('Proclaim 05G') 1516 mg/kg. **Acute dermal LD$_{50}$:** rats (technical) >2000 mg/kg; rabbits ('Proclaim 019EC') >2000 mg/kg; ('Proclaim 05G') >2000 mg/kg. **Inhalation:** LC$_{50}$ (4 h) for rats (technical) 2.12 to 4.4 mg/litre; ('Proclaim 019EC') 9.6 mg/litre. **Skin and eye:** For rabbits (technical) severe eye and slight skin irritation; ('Proclaim 019EC') severe eye and mild to moderate skin irritation; ('Proclaim 05G') mild eye (reversible within 96 hours) and slight skin irritation.

Neither technical material nor 'Proclaim 019EC' is a skin sensitiser. **NOEL:** (1 year) for dogs 0.25 mg/kg b.w. **ADI:** 0.0025 mg/kg. **Toxicity class:** WHO (technical material) II; ('Proclaim 019EC' and 'Proclaim 05G') III. **Other toxicological effects:** Not tumorigenic.

ENVIRONMENTAL IMPACT AND NON-TARGET TOXICITY: Bird toxicity: Acute oral LD_{50} for mallard ducks 46; bobwhite quail 264 mg/kg. Dietary LC_{50} (8 days) for mallard ducks 570; bobwhite quail 1318 ppm. **Fish toxicity:** LC_{50} (96 h) for rainbow trout 174 µg/litre; sheepshead minnow 1430 µg/litre. 'Proclaim' is toxic to fish and must not be sprayed near water courses. **Other aquatic toxicity:** LC_{50} for *Daphnia* 0.99 µg/litre.

Effects on beneficial insects: Will harm most beneficial arthropods; toxic to honeybees. Avoid spraying crops during flowering or when bees are active. LC_{50} for worms >1000 mg/kg dry soil. **Metabolism:** Emamectin benzoate is partially metabolised in animals, but rapidly cleared (DT_{50} following oral dosing 34 to 51 hours), indicating that it has no potential for bioaccumulation. Metabolism in plants has been investigated in lettuce, cabbage and sweetcorn. It is non-systemic, and rapidly degrades in sunlight to various complex residues in which undegraded parent is the only significant residue. The residues are very low.

Behaviour in soil: Binds tightly to soil, with rapid degradation by soil micro-organisms. No bioaccumulation.

2:175 extract of *Chenopodium ambrosioides*
Plant-derived insecticide

NOMENCLATURE: Approved name: extract of *Chenopodium ambrosioides* near *ambrosioides*. **Other names:** Extract of American wormseed; extract of Mexican tea; American wormseed oil; *Chenopodium* oil. **Development code:** QRD 400. **OPP Chemical Code:** 599995. **CAS RN:** *[89997-47-7]*.

SOURCE: The product is a blended extract of *Chenopodium ambrosioides* L. near *ambrosioides*.

TARGET PESTS: Aphids, mites, whitefly, thrips and leaf miners.

TARGET CROPS: 'Requiem' may be used on a wide range of crops including brassicas, bulb vegetables, leafy vegetables , fruiting vegetables , legumes, cucurbits, potatoes, citrus, grapes, pome fruit and tree nuts.

BIOLOGICAL ACTIVITY: Mode of action: Being lipophilic, *Chenopodium ambrosioides* near *ambrosioides* is concentrated in the outer surface of the target pest. 'Requiem' breaks down the insect's exoskeleton. This degradation of the body and joints causes a loss of fluid and inhibits the pest's ability to move. It clogs the trachea, interrupting the insect's respiratory system, preventing respiration and causing suffocation and also disrupts the insect's ability to navigate, blinding it from finding sources of food. Without the ability to locate food, the pest stops its destruction of crops and starves. These three unique modes of action means that 'Requiem' will find a place in insect resistance management when used in combination with other products.

COMMERCIALISATION: Formulation: Sold as a 25% emulsifiable concentrate (EC). **Tradenames:** 'Requiem' (AgraQuest).

APPLICATION: Apply as a 0.5 to 1% emulsion as soon as thrips are seen or when other insect pests reach damaging levels. Use a high volume spray and repeat every 7 to 14 days.

PRODUCT SPECIFICATIONS: Storage conditions: Store in a cool, dry place.

COMPATIBILITY: 'Requiem' is compatible with most biological control agents, but new introductions should not be made within 2 hours of application.

MAMMALIAN TOXICITY: 'Requiem' is not considered to be toxic to mammals. **Acute oral LD$_{50}$:** rats >5000 mg/kg. US-EPA Toxicity Category III. **Acute dermal LD$_{50}$:** rabbits >5000 mg/kg. US-EPA Toxicity Category IV. **Inhalation:** LC$_{50}$ for rats >2.02 mg/litre. (This was the maximum attainable aerosol concentration for this test due to the physical nature of the product. No deaths or toxic effects were observed in the test animals during the study period). US-EPA Toxicity Category IV. **Skin and eye:** US-EPA Toxicity Category III.

ENVIRONMENTAL IMPACT AND NON-TARGET TOXICITY: Data indicate that the pesticide should pose no significant risk to the environment if used in accordance with label directions. **Bird toxicity:** Practically non-toxic to birds. The acute oral LD$_{50}$ for bobwhite quail is >2250 mg/kg. **Fish toxicity:** 'Requiem' is toxic to aquatic vertebrates. LC$_{50}$ (96 h) for fathead minnows 20 mg/litre. **Other aquatic toxicity:** LC$_{50}$ (24 h) for *Daphnia* 3.3 mg/litre. **Approved for use in organic farming:** Yes: it has US OMRI approval.

2:176 fatty acids (oleic acid)
Plant-derived herbicide, fungicide and insecticide

The Pesticide Manual Fifteenth Edition entry number: 629

$$CH_3(CH_2)_7CH=CH(CH_2)_7CO_2M$$

$$M = H, Na \text{ or } K$$

NOMENCLATURE: Approved name: fatty acids, often oleic acid. Other names: (Z)-9-octadecenoic acid. Development code: MYX-6121 (Mycogen); JT-201 (Japan Tobacco) for potassium salt; OK-8905 (Otsuka) for sodium salt. CAS RN: *[112-80-1]* (Z)- isomer; *[112-79-8]* (E)- isomer; *[2027-47-6]* unspecified stereochemistry; *[143-18-0]* (Z)- isomer, potassium salt; *[84776-33-0]* ammonium salt.

SOURCE: Naturally occurring fatty acids extracted from plant and animal sources. Oleic acid is also a major constituent of neem oil (*q.v.*, under azadirachtin).

PRODUCTION: Extracted from plant and animal sources. Often manufactured for use in crop protection.

TARGET PESTS: 'Naturell' and other insecticide formulations are effective against aphids, thrips and scale insects.

TARGET CROPS: Insecticide uses include vegetables, fruit and ornamentals. Fungicide uses include grapes, roses and other crops. Herbicide uses include total weed control and moss control in lawns.

BIOLOGICAL ACTIVITY: Mode of action: Fatty acid extracts interfere with the cell membrane constituents of the target organism, leading to a breakdown of the integrity of the membrane and subsequent death. Different fatty acids are effective as insecticides, fungicides, total herbicides or as moss killers. Efficacy: 'M-Pede' and 'Naturell' are effective at controlling soft-bodied insects such as aphids and may also be applied to the soil to control soil-inhabiting insects. They also give curative control of powdery mildew pathogens. 'Scythe' and 'Naturell WK Herbicide' are effective total, non-residual herbicides and 'DeMoss' and 'Naturell WK Mosskiller' control moss in lawns and moss and liverworts on fences, roofs and glasshouses.

COMMERCIALISATION: Formulation: Sold as a liquid concentrate (SL).
Tradenames: 'M-Pede' (a mixture of oleic and linoleic acids, as their potassium salts) and 'Scythe' (Greenfire), 'Hinder' (ammonium salts) (Amvac), 'Neo-Fat' (Nufarm), 'Naturell', 'Naturell WK Mosskiller' and 'Naturell WK Herbicide' (Russell), 'Oleate' (sodium salt) (Otsuka), 'Savona' (potassium salts) (Koppert), 'Neu 1128' (potassium salts) (Neudorff).

APPLICATION: Apply as a foliar spray to crops, ensuring good coverage of the pest. The products may also be used to control soil-inhabiting insects when applied as a soil drench. They are suitable for use prior to the release of natural predators, such as *Encarsia formosa* Gahan.

PRODUCT SPECIFICATIONS: Extracted from plant and animal sources. Often manufactured for use in crop protection.

MAMMALIAN TOXICITY: Generally considered to be non-toxic. **Acute oral LD$_{50}$:** rats and mice >5000 mg/kg. **Acute dermal LD$_{50}$:** rats >2 000 mg/kg. **Inhalation:** LC$_{50}$ for rats >2000 mg/kg. **Toxicity class:** EPA (formulation) II ('Neo-Fat' III).

ENVIRONMENTAL IMPACT AND NON-TARGET TOXICITY: Non-toxic to non-target organisms or to the environment. Derived from edible-grade natural oils. Fully biodegradable with no harm to the environment. Sprayable up to the day of harvest. **Bird toxicity:** Dietary LC$_{50}$ for mallard ducks >5620 ppm. **Fish toxicity:** LC$_{50}$ (48 h) for carp 59.2 ppm. **Other aquatic toxicity:** LC$_{50}$ (3 h) for *Daphnia similis* >100 ppm. **Effects on beneficial insects:** LD$_{50}$ (contact) for honeybees >25 µg/bee. **Behaviour in soil:** Rapidly degraded in soil. **Approved for use in organic farming:** Yes: it has US OMRI approval.

2:177 formic acid   *Miticide*

NOMENCLATURE: Approved name: formic acid. **OPP Chemical Code:** 214900. **CAS RN:** *[64-18-6]*.

SOURCE: Formic acid is an irritating pungent liquid at room temperature. Ants are known to produce formic acid, which helps protect them from predators. Some birds take advantage of the ant's chemical defence by placing live ants in their feathers to rid themselves of mites.

PRODUCTION: Formic acid is manufactured for use in crop protection.

TARGET PESTS: Varroa (*Varroa destructor* Anderson & Trueman) and tracheal mites.

TARGET CROPS: Bee hives.

BIOLOGICAL ACTIVITY: Mode of action: Formic acid is a severe irritant and acts by directly killing the mites, while not disrupting bee behaviour or life span substantially.

COMMERCIALISATION: Formulation: The formic acid is formulated as a gel contained in a pre-packaged perforated pouch. The pouch is vented by removing all or part of an adhesive strip, allowing slow release of formic acid within the sealed beehive over a period 30 days. **Tradenames:** 'Formic Acid Gel' (Apicure), 'For-Mite' (Mann Lake), 'Mite Away II' (Nod Apiary).

APPLICATION: 'Mite-Away II' is used as part of an Integrated Pest Management (IPM) programme when mite thresholds are exceeded. Outside temperature should be between 10 and 25 °C (50 and 79 °F) at the time of application. Pads must be removed from the hives in the event of a heat wave within the first 7 days of treatment. Treatment can be resumed at the end of the heat wave. (Failure to remove the pads during a heat wave can cause excessive brood mortality and absconding). Treat all bee colonies in the apiary simultaneously, allowing a one month minimum period between applications.

PRODUCT SPECIFICATIONS: Formic acid is manufactured for use in crop protection.

MAMMALIAN TOXICITY: Formic acid is mildly acutely toxic via the oral and inhalation routes (EPA Toxicity Category III), a severe eye irritant (EPA Toxicity Category I), corrosive to the skin (EPA Toxicity Category I) and highly irritating to the respiratory tract. Ames tests for mutagenic potential are negative.

ENVIRONMENTAL IMPACT AND NON-TARGET TOXICITY: Formic acid is only registered for use in bee hives and, as a consequence, is not expected to accumulate in the environment or to be hazardous to non-target organisms.

2:178 garlic extract

Plant-derived insect repellent

NOMENCLATURE: Approved name: garlic oil extract. **CAS RN:** *[8000-78-0]*.

SOURCE: It has been known for centuries that garlic is an effective deterrent for insects, but its odour was always a problem for its widespread use in covered horticultural crops.

PRODUCTION: Extracted from garlic (*Allium sativum* L.).

TARGET PESTS: A general deterrent to all phytophagous insect species. It is also recommended as a bird repellent.

TARGET CROPS: May be used in vegetables, fruits, nuts, grains, all horticultural crops and applied to the structure of glasshouses.

BIOLOGICAL ACTIVITY: Mode of action: The extract of the garlic plant has been shown to repel insect pests (and, to a lesser extent, small animals). It is thought that it is the sulfur-containing secondary metabolites that contain the repellent properties. **Efficacy:** Garlic extract must be used before the pests arrive, as, once the pests become established, the repellent is ineffective in displacing them. The rationale is that the insect pests are directed to a site other than the one that has been treated.

COMMERCIALISATION: Formulation: Sold as an aqueous concentrate or as a dust, sometimes in combination with red pepper extracts. **Tradenames:** 'Garlic Barrier' (Green Spot) and (Garlic Research), 'Mosquito Barrier' and 'Garlic Grow' (Garlic Research), 'Allityn Insect Repellent' (Helena), 'CropGuard' and 'RepellEX AG' (American Biodynamics), 'Envirepel' (Cal Crop), 'Guardian Spray' (Guardian), 'Repeller' (Natural Resources), 'Biorepel' (JH Biotech), 'Brálic' (Ingeniería Industrial).

APPLICATION: The garlic extract should be mixed with horticultural oil (or fish oil for organic growers), diluted according to the manufacturer's instructions and then applied in high volume over the plants to be protected. Best results are achieved if applications are made late in the day. Repeat applications every 10 days, beginning early in the growing season. Do not apply when crops are in flower, as pollinating insects may also be repelled.

PRODUCT SPECIFICATIONS: Extracted from garlic (*Allium sativum* L.).

COMPATIBILITY: It is unusual to use garlic extract insect repellent with other crop protection products.

MAMMALIAN TOXICITY: Garlic extract is used as a food supplement and in cooking and is not considered to be hazardous. Garlic is 'Generally Recognised As Safe' (GRAS) by the US Environment Protection Agency.

ENVIRONMENTAL IMPACT AND NON-TARGET TOXICITY: Garlic extract is repellent to many animals including bees and wasps, but is not toxic to them. It is not persistent in the environment. **Approved for use in organic farming:** Yes: it has US OMRI approval.

2:179 gibberellic acid
Fungal-derived plant growth regulator

The Pesticide Manual **Fifteenth Edition entry number:** 441

NOMENCLATURE: Approved name: gibberellic acid (BSI, draft E-ISO, accepted in lieu of a common name); acide gibbérellique (draft F-ISO). **Other names:** Gibberellin A_3; GA_3 (ambiguous). **CAS RN:** *[77-06-5]*.

SOURCE: Originally isolated from rice infected with *Gibberella fujikuroi* Wr. (*Fusarium moniliforme* Sheldon). Infected rice in Japan was a common sight and the seedlings grew well above the height of uninfected plants, leading to the name *bakanae* disease (foolish seedling). Discovered by E Kurosawa in 1926, who called it gibberellin A. Later, ICI Plant Protection Ltd (now Syngenta) isolated a compound with similar biological properties and chemical structure, which was called gibberellic acid. It and over 70 other known members of the gibberellin group of plant growth regulators have been shown to occur naturally in a wide variety of plant species. It was introduced by ICI Plant Protection, but is no longer sold by Syngenta.

PRODUCTION: Produced from the fermentation of *Gibberella fujikuroi*.

TARGET CROPS: Plant growth regulator, used in a variety of applications, e.g. to improve fruit setting of clementines and pears (especially William pears); to loosen and elongate clusters and increase berry size in grapes; to control fruit maturity by delaying development of the yellow colour in lemons; to reduce rind stain and retard rind ageing in navel oranges; to counteract the effects of cherry yellows virus diseases in sour cherries; to produce uniform seedling growth in rice; to promote elongation of winter celery crop; to induce uniform bolting and increase seed production in lettuce for seed; to break dormancy and stimulate sprouting in seed potatoes; to extend the picking season by hastening maturity in artichokes; to increase the yield in forced rhubarb; to improve the malting quality of barley; to produce brighter-coloured, firmer fruit, and to increase the size of sweet cherries; to increase yields and aid harvesting of hops; to reduce internal browning and increase yields of Italian prunes; to increase fruit set and yields of tangelos and tangerines; to improve fruit setting in blueberries; to advance flowering and increase the yield of strawberries; and also a variety of applications on ornamentals.

BIOLOGICAL ACTIVITY: Mode of action: A naturally occurring plant growth regulator, which is part of the system that regulates plant growth and development. It acts as a plant growth regulator because of its physiological and morphological effects in extremely low concentrations. Translocated. Generally affects only the plant parts above the soil surface. **Efficacy:** Gibberellic acid is very effective at low rates of use. It exerts a wide range of effects on many different plant processes. **Key reference(s):** 1) E Kurosawa. 1926. Experimental studies on the secretion of *Fusarium heterosporum* on rice plants, *Trans. Nat. Hist. Soc. (Formosa)*, **16**, 213. 2) J F Grove. 1961. The gibberellins, *Q. Rev. Chem. Soc.*, **15**, 56–70.

COMMERCIALISATION: Formulation: Sold as soluble powder (SP), crystal, water soluble granule (SG), emulsifiable concentrate (EC) and tablet (TB) formulations.
Tradenames: 'Ceku-Gib' (Cequisa), 'GIB' (Burlington), 'Gibbex' (Griffin), 'Kri-Gibb' and 'Rasayanic Acid' (Krishi Rasayan), 'Gibrel', 'ProGibb', 'Release', 'RyzUp' and 'Ralex' (Valent BioSciences), 'Strong' (Sanonda), 'Uvex' (Productos OSA), 'GibGro' (Agtrol Chemical Products), (Nufarm GmbH) and (Nufarm Americas), 'Pol-Gibrescol' (Ciech SA), 'Forgibbs' (Forward International), 'Point Acigib' (Point Enterprises), 'Ro-Gibb' (Rotam Group), 'Tigibb' (Tide International) , 'Falgro' (Fine Agrochemicals), 'Activol', 'Berelex', 'Grocel' and 'Regulex' (Abbott), 'Vimogreen' (Vipesco), 'Agro-Gibb' (AgroSan), 'N-Large' (Stoller), 'Novagib' (Fine Agrochemicals) and (De Sangosse), 'Fengib' (plus MCPA-thioethyl) (Hokko), 'Boll-Set' (plus 4-indol-3-ylbutyric acid) (Micro Flo), 'PGR-IV' (plus 4-indol-3-ylbutyric acid) (Marman) and (Micro Flo), 'Cytoplex HMS' (plus kinetin plus 4-indol-3-ylbutyric acid) (Plant BioTech), 'Early Harvest' (plus 4-indol-3-ylbutyric acid plus cytokinins) (Griffin), 'Foli-Zyme GA' (plus 4-indol-3-ylbutyric acid plus cytokinins) (Stoller), 'Green Sol 48' (plus kinetin) (Frit), 'Maxon II' (plus 4-indol-3-ylbutyric acid) and 'Super Lagniappe' (plus 4-indol-3-ylbutyric acid plus cytokinins) (Agriliance).

PRODUCT SPECIFICATIONS: Specifications: Produced from the fermentation of *Gibberella fujikuroi*. **Purity:** The product quality is confirmed by hplc analysis.

COMPATIBILITY: Incompatible with alkaline materials and solutions containing chlorine.

MAMMALIAN TOXICITY: Not considered to be toxic. **Acute oral LD_{50}:** rats and mice >15000 mg/kg. **Acute dermal LD_{50}:** rats >2000 mg/kg. **Inhalation:** No ill effect on rats subjected to 400 mg/litre for 2 hours per day for 21 days. **Skin and eye:** Non-irritating to skin and eyes. **NOEL:** (90 days) for rats and dogs >1000 mg/kg diet (6 days per week). **Toxicity class:** WHO (a.i.) U; EPA (formulation) III.

ENVIRONMENTAL IMPACT AND NON-TARGET TOXICITY: Gibberellic acid has no adverse effects on non-target organisms or on the environment. **Bird toxicity:** Acute oral LD_{50} for bobwhite quail >2250 mg/kg. Acute oral LC_{50} >4640 mg/kg. **Fish toxicity:** LC_{50} (96 h) for rainbow trout >150 ppm. **Effects on beneficial insects:** Not toxic to honeybees. **Behaviour in soil:** Rapidly degraded in soil.

2:180 gibberellin A₄ with gibberellin A₇
Fungal-derived plant growth regulator

The Pesticide Manual **Fifteenth Edition entry number:** 442

NOMENCLATURE: **Approved name:** gibberellin A_4 plus gibberellin A_7. **CAS RN:** *[468-44-0]* gibberellin A_4 (i); *[510-75-8]* gibberellin A_7 (ii); *[8030-53-3]*, formerly *[8071-03-2]* gibberellin A_4 plus A_7.

SOURCE: Produced from the fermentation of *Gibberella fujikuroi* Wr.

TARGET CROPS: Plant growth regulator used to reduce russetting in apples, increase fruit set in pears and seed germination and yield in celery.

BIOLOGICAL ACTIVITY: **Mode of action:** Naturally occurring plant growth regulators, which are part of the system that regulates plant growth and development. The mixture acts as a plant growth regulator because of its physiological and morphological effects in extremely low concentrations. Translocated. Generally affects only the plant parts above the soil surface.

COMMERCIALISATION: **Formulation:** Sold as a liquid concentrate (SL).
Tradenames: 'ProVide' and 'Procone' (Valent BioSciences), 'Typrus' (Nufarm Americas) and (Nufarm SA), 'Novagib' (De Sangosse) and (Fine Agrochemicals), 'Regulex' (Syngenta), 'Everest' (Rocca), 'Promalin', 'ABG-3164' and 'Accel' (all plus 6-benzylaminopurine) (Valent BioSciences), 'Perlan' (plus 6-benzylaminopurine) (Fine Agrochemicals) and (De Sangosse), 'Ritesize' and 'Typy' (plus 6-benzylaminopurine) (Nufarm Americas).

PRODUCT SPECIFICATIONS: **Purity:** The product quality is confirmed by hplc analysis.

COMPATIBILITY: Incompatible with alkaline materials and solutions containing chlorine.

MAMMALIAN TOXICITY: The combination of gibberellin A_4 and gibberellin A_7 is non-toxic to mammals.

ENVIRONMENTAL IMPACT AND NON-TARGET TOXICITY: Gibberellin A_4 and gibberellin A_7 occur naturally in plants and are not expected to cause any adverse effects on non-target organisms or the environment. **Effects on beneficial insects:** Not toxic to honeybees. **Behaviour in soil:** Rapidly degraded in soil.

2:181 L-glutamic acid plus γ-aminobutyric acid

Plant growth regulator, fungicide and plant metabolic primer

H_2N-CH_2-CH_2-CH_2-CO_2H HO_2C-CH_2-CH_2-$CH(NH_2)$-CO_2H

gamma-aminobutyric acid   **L-glutamic acid**

NOMENCLATURE: Approved name: L-glutamic acid plus γ-aminobutyric acid (GABA).
CAS RN: [56-12-2] γ-aminobutyric acid; [56-86-0] L-glutamic acid.

SOURCE: L-Glutamic acid and γ-aminobutyric acid (GABA) are found in virtually all living organisms. In their pure form, they are powders.

TARGET PESTS: Growth enhancer that both increases yield and improves quality. Prevents powdery mildew on grapes and suppresses certain other crop diseases.

TARGET CROPS: Used in certain fruits and vegetables, tree nuts, peanuts, grains, animal feed crops, lawn and turfgrasses and ornamentals. 'AuxiGro WP' has received US-EPA registration for use on snap beans, cucumbers, navy beans, pinto beans, grapes, bulb onions, peppers, strawberries, watermelons, celery, lettuce, peanuts, potatoes and tomatoes. It has an 'Exemption From Tolerance' for all crop applications.

BIOLOGICAL ACTIVITY: Mode of action: L-Glutamic acid is one of the major amino acids in plant and animal proteins and is also involved in many physiological functions. Both active ingredients act as neurotransmitters in the brain. **Biology:** The active ingredients in 'AuxiGro WP' are naturally occurring, are present in all life forms and are environmentally safe. A proprietary component of 'AuxiGro WP' is γ-aminobutyric acid (GABA), a natural product found in bacteria, plants and all animals, including humans. The other active ingredient in 'AuxiGro WP' is the immediate precursor to GABA, L-glutamic acid, a naturally occurring amino acid found in all living organisms. These naturally occurring products are neither fertilisers, pesticides nor conventional plant growth regulators, but form a new category for which the term Plant Metabolic Primer has been introduced to describe their biological function.

COMMERCIALISATION: Formulation: Sold as a wettable powder (WP) formulation containing 29.2% GABA and 36.5% L-glutamic acid. **Tradenames:** 'AuxiGro WP' (Auxein Corporation).

APPLICATION: Pesticide products containing these active ingredients can be applied by spraying from the ground or the air, by drenching the soil or through certain irrigation systems. To prevent powdery mildew, the pesticide product must be applied before the disease develops.

MAMMALIAN TOXICITY: No risks to human health are expected from the use of L-glutamic acid and GABA as pesticide-active ingredients. Both substances occur naturally in plants and animals. Humans normally ingest large and variable amounts of L-glutamic acid, but rapidly metabolise it so that plasma levels remain constant. The US Food and Drug Administration (FDA) classifies L-glutamic acid as 'GRAS' (generally recognised as safe for human consumption). Toxicity tests in animals and humans showed no adverse effects from GABA or L-glutamic acid. **Acute oral LD$_{50}$:** rats >5000 mg/kg ('Auxigrow WP').
Acute dermal LD$_{50}$: rabbits >5000 mg/kg ('Auxigrow WP'). **Toxicity class:** Toxicity category IV.

ENVIRONMENTAL IMPACT AND NON-TARGET TOXICITY: No risks to the environment are expected from use of these active ingredients, because they occur naturally and do not persist in the environment. They are not toxic to mammals or other organisms tested and they are not likely to be toxic to plants, given that they enhance growth of many kinds of plants. Products with these active ingredients are not approved for application directly to water or to areas where surface water is present. **Fish toxicity:** 'AuxiGro WP' is practically non-toxic to freshwater fish (>100 mg/litre). **Other aquatic toxicity:** 'AuxiGro WP' is practically non-toxic to freshwater invertebrates (EC$_{50}$ >100 mg/litre).
Approved for use in organic farming: Yes.

2:182 harpin aβ protein
Micro-organism-derived elicitor of systemic acquired resistance

NOMENCLATURE: Approved name: harpin aβ protein. **Development code:** EBC-351.

SOURCE: Harpin aβ is a protein that consists of four fragments from other harpin proteins found in certain bacteria that cause diseases in plants.

PRODUCTION: For commercial production of the proteins, the DNA sequences coding for harpin aβ were put into a weakened strain of *Escherichia coli* Escherich (*E. coli* K-12). This genetically modified *E. coli* K-12 produces large amounts of harpin aβ, which is then isolated and purified from the bacterial growth medium.

TARGET PESTS: Harpin aβ is effective in controlling a wide range of fungal, bacterial and viral plant pathogens in the growing plant, with these effects continuing post-harvest. It also reduces infestation by selected insect and nematode pests. It provides significant plant growth enhancement and, thereby, suppression of some insects. Growth enhancements may include improved germination, increased overall plant vigour, accelerated flowering and fruit set, advanced maturity and increased yield and quality of the final harvest.

TARGET CROPS: Harpin aβ proteins can be used on a broad range of crops both outdoors and protected, including traditional field crops, minor use crops, turf and ornamentals.

BIOLOGICAL ACTIVITY: Mode of action: Harpin aβ does not act directly on disease organisms, nor does it permanently alter the DNA of treated plants. It activates the natural defence mechanism in plants, a response known as systemic acquired resistance (SAR). Harpin aβ protects against certain bacterial, viral, and fungal diseases; soil-borne pathogens; and harmful nematodes and insects. Harpin aβ protein also enhances plant growth and vigour, and increases the yield for a variety of crops, including vegetables, trees and ornamentals.
Efficacy: Harpin aβ does not directly act on plant pests and consequently it is argued that the pests are unlikely to become resistant to this active ingredient. By decreasing the use of conventional pesticide products, harpin aβ proteins can play an important role in reducing risks to workers, the public and the environment.

COMMERCIALISATION: Formulation: 'ProAct' is an end-use product formulation containing 1% Harpin aβ protein, formulated as a wettable granule. **Tradenames:** 'ProAct' (Plant Health Care).

APPLICATION: 'ProAct' is applied using conventional ground or aerial foliar equipment or as a pre-plant dip; as a seed treatment; by application via conventional sprinkler, drip or chemigation systems; or as a glasshouse drench or transplant application. Use rates are generally 0.35 to 10 g a.i. per ha (0.14 to 3.69 g a.i. per acre) at approximately 14-day intervals. Glasshouse applications are made after seedling emergence and repeated every 14 to 21 days. It can be applied 5 to 7 days before transplant and/or as a drench or pre-plant dip at transplant. Foliar applications in the field are recommended at planting and at approximately 14-day intervals up to harvest. Sprays begin at the appearance of the first true leaf in newly seeded crops.

PRODUCT SPECIFICATIONS: Specifications: For commercial production of the proteins, the DNA sequences coding for harpin aβ were put into a weakened strain of *Escherichia coli* (*E. coli* K-12). This genetically modified *E. coli* K-12 produces large amounts of harpin aβ, which is then isolated and purified from the bacterial growth medium. **Purity:** Harpin aβ is an acidic protein that is glycine rich, and contains few or no cysteine residues.

COMPATIBILITY: Harpin aβ protein is an elicitor of the plant's natural defence mechanism that renders it resistant to a wide range of insects, mites and nematodes, as well as fungal, bacterial and viral pathogens. Consequently, it is unlikely that harpin will be applied with other compounds. Chlorinated water should not be used to prepare the spray mixture.

MAMMALIAN TOXICITY: Harpin aβ protein is classified as a Toxicity Category IV product via the oral and dermal route, and a Toxicity Category IV eye and skin irritant. It is classified as Toxicity Category III via the inhalation route because the limit dose could not be reached. Researchers and workers have worked with harpin aβ in its production and application for several years, and there has been no indication of any toxicity or hypersensitivity associated with its use. Because of the lack of demonstrated adverse health effects, low rates of application and rapid degradation in the field, no residues are expected on treated crops with no attendant dietary risks expected. Because of the lack of demonstrable toxicity, no adverse effects are expected to applicators, handlers and other workers. There is reasonable certainty that no harm to adults, infants or children will result from aggregate exposure to harpin aβ residues. **Acute oral LD$_{50}$:** rats ('ProAct') 5050 mg/kg body weight. Toxicity Category IV. **Acute dermal LD$_{50}$:** rabbits ('ProAct') >5050 mg/kg. Toxicity Category IV. **Inhalation:** LC$_{50}$ for rats >0.832 mg/litre ('ProAct') (the maximum concentration attainable). Toxicity Category III. **Skin and eye:** Rabbits no positive effects in any eyes, 24 h after treatment. 'ProAct' is rated minimally irritating. Toxicity Category IV. 'ProAct' is rated slightly irritating to the skin of rabbits. Toxicity Category IV. **Other toxicological effects:** Harpin aβ is commercially produced in *Escherichia coli* by transfer of DNA fragments encoding harpin aβ protein derived from multiple bacterial sources to the cell production strain, *E. coli* K-12. The harpin aβ producing strain is considered a debilitated strain of *E. coli*, which cannot grow in the human digestive tract, or survive in the environment.

ENVIRONMENTAL IMPACT AND NON-TARGET TOXICITY: Evidence from environmental impact studies suggests that the amounts of harpin aβ required to elicit acute toxicities in non-target organism populations are unlikely to be achieved by intended exposures to recommended applications. 14-d seedling emergence (ten agronomically important plants) NOEC 1430 litre/ha. **Bird toxicity:** Avian acute oral toxicity LD$_{50}$ >4000 mg/kg body weight; NOEC 4000 mg/kg body weight; avian dietary toxicity LC$_{50}$ >5620 mg/kg body weight. **Fish toxicity:** Rainbow trout acute toxicity LC$_{50}$ >3270 mg/litre; NOEC 378 mg/litre. **Other aquatic toxicity:** Acute toxicity *Daphnia* EC$_{50}$ 1473 mg 'ProAct'/litre; NOEC 626 mg 'ProAct'/litre. **Effects on beneficial insects:** Acute contact toxicity (48 h) honeybee LD$_{50}$ >13 μg harpin aβ/bee. **Metabolism:** It is likely that the natural proteolysis taking place in soil and on plant surfaces will ensure that harpin-derived proteins do not persist in the environment. **Behaviour in soil:** Non-persistent.

2:183 harpin protein
Micro-organism-derived elicitor of systemic acquired resistance

Deduced amino acid sequence of harpin (from Z-M Wei *et al.* 1992.)

MSLNTSLGASTMQISTGGAGGNNGLLGTSRQNAGLGGNSALGLGGGNQNDTVNQLAGLLTGMM
MMMSMMGGGGLMGGGLGGGLGNGLGGSGGLGEGLSNALNDMLGGSLNTLGSKGGNNTTSTTN
SPLDQALGINSTSQNDDSTSGTDSTSDSSDPMQQLLKMFSEIMQSLFGDGGDTQGSSSGGKQPTEGE
QNAYKKGVTDALSGLMGNGLSQLLGNGGLGGGQGGNAGTGLDGSSLGGKGLRGLSGPVDYQQL
GNAVGTGIGMKAGIQALNDIGTHRHSSTRSFVNKGDRAMAKEIGQFMDQYPEVFGKPQYQKGPGQ
EVKTDDKSWAKALSKPDDDGMTPASMEQFNKAKGMIKRPMAGDTGNGNLHDAVPVVLRWVLMP

NOMENCLATURE: Approved name: harpin protein.

SOURCE: In Nature, harpin is produced by *Erwinia amylovora* Winsl., a bacterium that causes the disease fire blight in apples and pears. Harpin is an acidic, heat-stable, cell envelope associated protein with a molecular weight of about 40 kilodaltons. The protein consists of 403 amino acid residues with no cysteine.

PRODUCTION: Harpin is produced commercially in a weakened strain of *Escherichia coli* Escherich by transfer of the DNA fragment encoding harpin protein from *Erwinia amylovora* to the cell production strain, *E. coli* K-12. This harpin producing strain is considered to be a debilitated, non-pathogenic, nutritionally deficient bacterium strain of *E. coli*, which cannot grow in the human digestive tract or survive in the environment. *E. coli* K-12 cells are killed and lysed at the end of the fermentation process. Harpin protein and other cell constituents are then extracted from the growth medium and concentrated for formulation into the end-use product.

TARGET PESTS: A broad range of fungal, bacterial and viral disease organisms are controlled, including bacterial leaf spot (*Xanthomonas campestris* Dows.), bacterial speck (*Pseudomonas syringae* Van Hall), bacterial wilt (*Pseudomonas solanacearum* (Smith) Smith), Fusarium wilt, Phytophthora root rot, rice stem rot (*Magnaporthe salvinii* (Cattaneo) Krause & Webster), sheath blight (*Rhizoctonia solani* Kuehn), apple scab (*Venturia inaequalis* Wint.), fire blight (*Erwinia amylovora*), Botrytis bunch rot, black rot (*Guignardia bidwellii* Viala & Rivas), black leaf spot (*Diplocarpon rosae* Wolf), cucumber mosaic virus, root-knot nematodes (*Meloidogyne* spp.), tobacco cyst nematode (*Globodera solanacearum* Miller & Gray), and tobacco mosaic virus (TMV). It provides significant plant growth enhancement and, thereby, suppression of some insects. Growth enhancements may include improved germination, increased overall plant vigour, accelerated flowering and fruit set, advanced maturity, and increased yield and quality of the final harvest.

TARGET CROPS: Harpin protein can be used on a broad range of crops both outdoors and protected, including traditional field crops, minor use crops, turf and ornamentals. Harpin protein has been approved for use on tomatoes to enhance uniformity, size, and yield; and on citrus to increase fruit set and yield.

BIOLOGICAL ACTIVITY: Mode of action: Harpin does not act directly on the disease organism, but instead activates a natural defence mechanism in the host plant, referred to as systemic acquired resistance (SAR). Harpin is effective against certain viral diseases for which there are no other controls or resistant plant varieties. It also protects against soil-borne pathogens and pests, such as certain nematodes and fungal diseases, which have few effective controls except for methyl bromide, which has adverse human health and environmental impacts. In addition to its ability to protect plants against diseases, harpin protein also reduces infestations of selected insects and enhances plant growth, general vigour and yield of many crops, including vegetables, traditional agronomic crops and ornamentals.

Efficacy: Harpin exhibits no direct inhibitory or toxic effect on plant pathogens and thus, it is claimed, cannot exert the selection pressure that would promote the development of resistance in pest populations. By decreasing the use of conventional pesticides, harpin is expected to be an important tool in resistance management programmes.

Key reference(s): Z-M Wei, R J Laby, C H Zumoff, D W Bauer, S Y He, A Collmer & S V Beer. 1992. Harpin, elicitor of the hypersensitive response produced by the plant pathogen *Erwinia amylovora*, *Science*, **257**, 85–8.

COMMERCIALISATION: Formulation: 'Messenger' is an end-use product formulation containing 3% harpin protein, formulated as a wettable granule. **Tradenames:** 'Messenger' (Plant Health Care).

APPLICATION: The end use product may be applied pre-planting or as a foliar spray with conventional ground or aerial spray equipment or by conventional irrigation/chemigation systems. In addition, it may be used as a seed treatment or in glasshouses as a soil drench. Use rates are very low, generally 5 to 25 g a.i. per hectare (2 to 11.5 g a.i. per acre), applied at 14-day intervals. Glasshouse drench applications should be made 3 weeks after seeding with a second application 5 to 7 days before transplanting. Field foliar applications are recommended at planting and at 14-day intervals until harvest. For newly seeded crops, sprays should begin at the appearance of the first true leaf.

PRODUCT SPECIFICATIONS: Specifications: Harpin is produced commercially in a weakened strain of *Escherichia coli* Escherich by transfer of the DNA fragment encoding harpin protein from *Erwinia amylovora* to the cell production strain, *E. coli* K-12. This harpin producing strain is considered to be a debilitated, non-pathogenic, nutritionally deficient bacterium strain of *E. coli*, which cannot grow in the human digestive tract or survive in the environment. *E. coli* K-12 cells are killed and lysed at the end of the fermentation process. Harpin protein and other cell constituents are then extracted from the growth medium and concentrated for formulation into the end-use product. **Purity:** 'Messenger' is a fine granule of pH 7.86 (1% solution) at 22 °C; no foreign matter has been noted.

COMPATIBILITY: Harpin protein is an elicitor of the plant's natural defence mechanism that renders it resistant to a wide range of insects, mites and nematodes, as well as fungal, bacterial and viral pathogens. Consequently, it is unlikely that harpin will be applied with other compounds. Chlorinated water should not be used to prepare the spray mixture.

MAMMALIAN TOXICITY: Harpin is not considered to be a toxic pesticide. Harpin protein is classified as a Toxicity Category IV product via the oral, dermal and inhalation routes, and a Toxicity Category IV eye and skin irritant. Researchers and workers have worked with harpin in its production and application for over six years and there has been no indication of any toxicity or hypersensitivity associated with this protein. Because of the lack of demonstrated adverse health effects, low rates of application and rapid degradation in the field, no residues are expected on treated crops and with no attendant dietary risks. Because of the lack of demonstrable toxicity, no adverse effects are expected to applicators, handlers or other workers. There is reasonable certainty that no harm to adults, infants or children will result from aggregate exposure to harpin residues. The strain of *E. coli* used to produce harpin is a nutritionally deficient, environmentally debilitated laboratory strain that does not survive for extended periods, nor reproduce in the open environment. Even though some bacteria may survive the manufacturing process, results from controlled studies to detect viable cells on treated plants suggest that the cell production strain would not survive the common mixing and field application process for this biopesticide. **Acute oral LD$_{50}$:** rats (0.3 % harpin protein technical) LD$_{50}$ >2000 mg/kg body weight. Toxicity Category III. Rats ('Messenger') >>5000 mg/kg. Toxicity category IV. **Acute dermal LD$_{50}$:** rats >6000 mg/kg ('Messenger'). Toxicity Category IV. **Inhalation:** LC$_{50}$ for rats >2 mg/litre ('Messenger'). Toxicity Category IV. **Skin and eye:** 'Messenger' is not irritating to skin or eyes. No corneal opacity, iridal lesions, conjunctival chemosis or erythema were noted. 'Messenger' is not considered an ocular irritant of rabbits, although conjunctival redness was noted in five out of six rabbits, which cleared within 72 h. Toxicity Category IV. **Toxicity class:** EPA IV.

Other toxicological effects: Harpin has been exempted from the requirement of a tolerance on all food crops by the US-EPA. **Toxicity review:** United States Environmental Protection Agency (US-EPA). 2000. Messenger: a promising reduced risk biopesticide, *Pesticide Environmental Stewardship Program (PESP)*, **3**, Number 1.

ENVIRONMENTAL IMPACT AND NON-TARGET TOXICITY: Harpin is not expected to cause any harm to the environment because it is applied at low rates and degrades rapidly after application; it poses little or no concern as a ground or surface water contaminant. In addition, it has no demonstrable adverse effects on birds, fish, aquatic invertebrates, honeybees, non-target plants or algae. Therefore, risks to wildlife and beneficial insects are expected to be minimal. Classified as practically non-toxic by the US-EPA. 14-day seedling emergence (ten agronomically important plants) NOEC 1430 litre/ha (143 g harpin/ha).

Bird toxicity: LC$_{50}$ (dietary) for bobwhite quail ('Messenger') 100 000 ppm; acute oral LD$_{50}$ for bobwhite quail >4000 mg/kg. Fish toxicity: LC$_{50}$ for rainbow trout ('Messenger') >3720 mg/litre; NOEC 378 mg ('Messenger')/litre. Other aquatic toxicity: EC$_{50}$ for *Daphnia magna* ('Messenger') 1173 mg/litre. EC$_{50}$ for algae ('Messenger') >182 mg/litre; NOEC 120 mg ('Messenger')/litre. Effects on beneficial insects: LD$_{50}$ for honeybees ('Messenger') >1258 µg/bee; NOEC 39 µg harpin/bee. Metabolism: Harpin protein is non-persistent and does not accumulate in the environment. It is readily degraded by natural sunlight and micro-organisms. Harpin protein degraded rapidly on the surface of leaves within 3 to 4 days after application and degraded quickly when reconstituted with pond water. Behaviour in soil: Harpin is rapidly broken down in the soil.

2:184 indol-3-ylacetic acid
Plant-derived plant growth regulator

The Pesticide Manual Fifteenth Edition entry number: 490

NOMENCLATURE: Approved name: indol-3-ylacetic acid. Other names: IAA; AIA (France). CAS RN: *[87-51-4]*.

SOURCE: Naturally occurring plant growth regulator.

PRODUCTION: Indol-3-ylacetic acid is manufactured for use in agriculture, rather than being extracted from higher plants.

TARGET CROPS: Used on herbaceous and woody ornamentals to stimulate rooting of cuttings.

BIOLOGICAL ACTIVITY: Mode of action: Plant growth regulator, which affects cell division and cell elongation. Indol-3-ylacetic acid is very effective at initiating the formation of roots in cuttings through causing cell division in the cambial tissue.

COMMERCIALISATION: Formulation: Sold as tablet (TB) and dispersible powder (DP) formulations. Tradenames: 'Rhizopon A' (Fargro) and (Rhizopon).

APPLICATION: Cuttings are dipped in the product prior to planting.

PRODUCT SPECIFICATIONS: Specifications: Indol-3-ylacetic acid is manufactured for use in agriculture, rather than being extracted from higher plants. **Purity:** Product quality is checked by u.v. spectrophotometry.

COMPATIBILITY: It is unlikely that indol-3-ylacetic acid would be used with any other crop protection chemical, with the exception of broad-spectrum fungicides with which it is compatible.

MAMMALIAN TOXICITY: Indol-3-ylacetic acid occurs in all higher plants and has not been shown to cause any adverse effects on mammals. **Acute oral LD$_{50}$:** mice 1000 mg/kg.

ENVIRONMENTAL IMPACT AND NON-TARGET TOXICITY: Indol-3-ylacetic acid occurs naturally in plants and is not expected to show any adverse effects on non-target organisms or on the environment. **Effects on beneficial insects:** Not toxic to honeybees.
Behaviour in soil: Rapidly degraded in soil. **Approved for use in organic farming:** Yes.

2:185 trans-α-ionone

Plant-derived insecticide

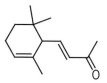

NOMENCLATURE: Approved name: trans-α-ionone. **Other names:** 4-(2,6,6-trimethyl-1-cyclohex-2-enyl)but-3-en-2-one (IUPAC); alpha-ionone; beta-ionone. **CAS RN:** *[127-41-3]*.

SOURCE: Found in many tropical and subtropical plants including copaiba (*Copaifera officinalis* (Jacq.) L.).

PRODUCTION: Extracted from plants for use as an insect repellent.

TARGET PESTS: Flying and biting insects such as mosquitoes.

TARGET CROPS: Used in public and animal health.

BIOLOGICAL ACTIVITY: Mode of action: Trans-α-ionone is a strong insect repellent.

COMMERCIALISATION: Formulation: Sold as ready-to-use formulations.
Tradenames: 'Scent-Off Twist-ons' (plus geraniol) and 'Scent-Off Pellets' (plus geraniol) (Scent-Off).

APPLICATION: Apply when mosquitoes and other flying insects are active.

PRODUCT SPECIFICATIONS: Extracted from plants for use as an insect repellent.

MAMMALIAN TOXICITY: The chemical, physical and toxicological properties of trans-α-ionone have not been thoroughly investigated, but it is naturally occurring and not considered likely to have significant effects on mammals. **Skin and eye:** Irritant, skin and respiratory sensitiser.

ENVIRONMENTAL IMPACT AND NON-TARGET TOXICITY: The non-target toxicological properties of trans-α-ionone have not been thoroughly investigated, but it is naturally occurring and not considered likely to have significant effects on non-target organisms or the environment. **Approved for use in organic farming:** Yes: it has US OMRI approval.

2:186 jojoba oil
Plant-derived fungicide and insecticide

NOMENCLATURE: Approved name: jojoba oil.

SOURCE: Derived from jojoba seeds (*Simmondsia californica* Nutt.).

PRODUCTION: Extracted from jojoba seeds and formulated for use in crop protection. Jojoba oil has been recognised as effective against whitefly for some time and it was first registered in the USA in 1994. In 1998, it was found that jojoba wax controls powdery mildew on a number of plant species.

TARGET PESTS: 'Detur' is used for whitefly (*Bemisia* spp. and *Trialeurodes vaporariorum* (Westwood)) control. 'E-RASE' and 'ECO E-RASE' are recommended for powdery mildew control.

TARGET CROPS: 'Detur' is used in melons, vegetables, cotton, ornamentals and other crops. 'ECO E-RASE' is recommended for use in ornamentals, grapes, vegetables and many other crops. 'PERMATROL' is recommended for use in grapes and in ornamentals grown in the field, in glasshouses, in shadehouses and in nurseries. 'ECO E-RASE' and 'PERMATROL' can be used in organic farming.

BIOLOGICAL ACTIVITY: Mode of action: The mode of action and the reason for jojoba oil specificity are not totally understood. Activity against whitefly eggs and immatures is through suffocation. Activity against adults is by repellency. Against powdery mildew, fungal spores and mycelium are killed by blocking access to oxygen.

COMMERCIALISATION: Formulation: Sold as a 97.5% jojoba oil formulation.
Tradenames: 'Detur', 'E-RASE' and 'ECO E-RASE' (IJO Products), 'PERMATROL' (Soil Technologies).

APPLICATION: Apply as a foliar spray, ensuring good coverage, but avoiding run-off.

PRODUCT SPECIFICATIONS: Extracted from jojoba seeds and formulated for use in crop protection. Jojoba oil has been recognised as effective against whitefly for some time and it was first registered in the USA in 1994. In 1998, it was found that jojoba wax controls powdery mildew on a number of plant species.

COMPATIBILITY: 'Detur', 'ECO E-RASE' and 'PERMATROL' can be used with other crop protection agents; however, the most restrictive label requirements of the products should apply.

MAMMALIAN TOXICITY: Jojoba oil is considered safe for human consumption. No harmful health effects to humans are expected from the use of jojoba oil in crop protection. Information available from published studies indicates that jojoba oil's nutritional and toxicological profiles are similar to those of other vegetable oils that are used as food. It is widely used in the cosmetics industry.

ENVIRONMENTAL IMPACT AND NON-TARGET TOXICITY: Adverse effects to the environment or to organisms other than insects are not anticipated because of the low toxicity of jojoba oil and its rapid decomposition in the environment. In addition, jojoba pesticide products are not allowed to be applied directly to bodies of water; therefore, exposure of aquatic organisms should be extremely limited.
Approved for use in organic farming: Yes, it has US OMRI approval.

2:187 karanjin

Plant-derived insecticide and acaricide

The Pesticide Manual Fifteenth Edition entry number: 513

NOMENCLATURE: **Approved name:** karanjin. **Other names:** 3-methoxy-2-phenyl-4*H*-furo[2,3-*H*]-1-benzopyran-4-one. **CAS RN:** *[521-88-0]*.

SOURCE: Extract of pongam or Indian beech (*Derris indica* (Lam.) Bennet; synonym *Pongamia pinnata* (L.) Pierre).

PRODUCTION: Extracted from the seeds of pongam or a wettable powder is produced by grinding the seeds.

TARGET PESTS: Mites, scales, chewing and sucking insect pests and some fungal diseases. A potent deterrent to many different genera of insects and mites. Shown to be effective against whitefly, thrips, leaf miners, caterpillars, aphids, jassids, San José scale, beetles and mealybugs.

TARGET CROPS: Shows activity in a wide range of crops, including vegetables (such as tomatoes, cabbage, potatoes), cotton, tea, tobacco, coffee, protected crops and ornamentals.

BIOLOGICAL ACTIVITY: **Mode of action:** Karanjin has a dramatic antifeedant/repellent effect, with many insects avoiding treated crops. It suppresses ecdysone hormone and, thereby, it acts as an insect growth regulator (IGR) and antifeedant. There are claims that it inhibits cytochrome P_{450} in susceptible insects and mites. There are reports that it is a nitrification inhibitor.

COMMERCIALISATION: **Formulation:** Sold as a 20 g per litre emulsifiable concentrate (EC). **Tradenames:** 'Derisom' (Agri Life) and (SOM Phytopharma).

APPLICATION: Apply at rates of 100 to 500 g a.i. per hectare. Frequent applications are more effective than single sprays. At high volume, apply 1 to 2 ml of product per litre of water.

PRODUCT SPECIFICATIONS: **Specifications:** Extracted from the seeds of pongam or a wettable powder is produced by grinding the seeds. **Purity:** Karanjin-based products are produced from the extraction of the seeds. In all cases, there are other, inert, components in each formulation.

COMPATIBILITY: It is not usual to apply karanjin with other crop protection compounds. It is incompatible with sulfur.

MAMMALIAN TOXICITY: There is no evidence of allergic or other adverse effects on producers, formulators or users.

ENVIRONMENTAL IMPACT AND NON-TARGET TOXICITY: It is not expected that karanjin-based products will have any adverse effects on non-target organisms or on the environment.

2:188 kasugamycin
Micro-organism-derived fungicide and bactericide

The Pesticide Manual **Fifteenth Edition entry number:** 515

NOMENCLATURE: Approved name: kasugamycin (JMAF). **CAS RN:** *[6980-18-3]*.

SOURCE: Isolated from the soil Actinomycete, *Streptomyces kasugaensis* Hamad *et al.* First described by H Umezawa *et al.* in 1965 and introduced by Hokko Chemical Industry Co. Ltd.

PRODUCTION: Produced by fermentation of *Streptomyces kasugaensis* and isolation of the secondary metabolite from the fermentation broth. Sold as the hydrochloride.

TARGET PESTS: Rice blast (*Pyricularia oryzae* Cavara; perfect stage *Magnaporthe grisea* (Herbert) Barr); leaf spot (*Cercospora* spp.) in sugar beet and celery; bacterial disease in rice and vegetables; and scab (*Venturia* spp.) in apples and pears.

TARGET CROPS: Rice, top fruit and vegetables.

BIOLOGICAL ACTIVITY: Mode of action: Inhibition of protein biosynthesis by interfering with the binding of aminoacyl-tRNA to both the mRNA-30S and the mRNA-70S ribosomal subunit complexes, thereby preventing the incorporation of amino acids into proteins. Kasugamycin is a systemic fungicide and bactericide with both protectant and curative properties. **Efficacy:** Kasugamycin controls rice blast at concentrations as low as 20 mg per litre. Resistance to kasugamycin was detected within three years of its introduction in 1965 and, by 1972, it had become a serious problem in Japanese rice fields. Today, mixtures of kasugamycin with other fungicides with different modes of action are used. Resistant strains of *Pyricularia oryzae* are less fit than susceptible strains and, once applications of kasugamycin are discontinued in the field, the level of resistance declines very rapidly. Kasugamycin inhibits hyphal growth of *Pyricularia oryzae* on rice, preventing lesion development; it is a comparatively weak inhibitor of spore germination, appressorium formation and penetration into the epidermal cells. In contrast, against *Cladosporium fulvum* on tomatoes, inhibition of sporulation is high, but its effects on hyphal growth are poor. Kasugamycin is taken up by plant tissue and is translocated. **Key reference(s):** 1) H Umezawa, Y Okami, T Hashimoto, Y Suhara & N Otake. 1965. A new antibiotic, kasugamycin, *J. Antibiot. Ser. A*, **18**, 101. 2) M Hamada, T Hashimoto, S Takahashi, M Yoneyama, T Miyake, Y Takeuchi, Y Okami & H Umezawa. 1965. Antimicrobial activity of kasugamycin, *J. Antibiot. Ser. A*, **18**, 104. 3) N Tanaka, H Yamaguchi & H Umezawa. 1966. Mechanism of kasugamycin action on polypeptide synthesis, *J. Biochem.*, **60**, 429.

COMMERCIALISATION: Formulation: Sold as the hydrochloride as wettable powder (WP), dispersible powder (DP), ultra-low volume (UL), soluble concentrate (SL) and granule (GR) formulations. **Tradenames:** 'Kasugamin' and 'Kasumin' (Hokko), 'Kasu-rab-valida-sumi' (plus fthalide plus validamycin plus fenitrothion) (Hokko). **Patents:** JP 42006818; BE 657659; GB 1094566.

APPLICATION: Applied as a foliar spray, a dust or a seed treatment, at rates from 20 g a.i. per litre.

PRODUCT SPECIFICATIONS: Specifications: Produced by fermentation of *Streptomyces kasugaensis* and isolation of the secondary metabolite from the fermentation broth. Sold as the hydrochloride. **Purity:** The product is analysed by cup assay with *Pseudomonas fluorescens* (NIHJ B-254).

COMPATIBILITY: Often sold or recommended for use in combination with other rice blast fungicides to counteract the onset of resistance. Incompatible with compounds that are strongly alkaline. There has been evidence of slight phytotoxicity on crops such as peas, beans, soybeans, grapes, citrus and apples. No injury has been found on rice, tomatoes, sugar beet, potatoes and many other vegetables.

MAMMALIAN TOXICITY: As with other aminoglycoside antibiotics, kasugamycin is not considered to be toxic to mammals.

ENVIRONMENTAL IMPACT AND NON-TARGET TOXICITY: Kasugamycin is not expected to have any adverse effects on non-target organisms or on the environment. **Bird toxicity:** Acute oral LD_{50} for male Japanese quail >4000 mg/kg. **Fish toxicity:** LC_{50} (48 h) for carp and goldfish >40 mg/litre. **Other aquatic toxicity:** LC_{50} (6 h) for *Daphnia pulex* >40 mg/litre. **Metabolism:** Kasugamycin administered orally to rabbits was almost completely excreted in the urine within 24 hours. When injected intravenously to dogs, it was almost completely excreted within 8 hours. After oral administration to rats at 200 mg/kg, no residues were found in eleven organs or in the blood and 96% of the administered dose remained in the digestive tract one hour after administration. Plants metabolise kasugamycin to kasugamycinic acid and kasuganobiosamine and eventually to ammonia, oxalic acid, carbon dioxide and water. **Behaviour in soil:** Metabolism in soil is identical to that in plants, with the final products being ammonia, water and carbon dioxide.

2:189 kasugamycin hydrochloride hydrate
Micro-organism-derived fungicide and bactericide

The Pesticide Manual **Fifteenth Edition entry number:** 515

NOMENCLATURE: Approved name: kasugamycin hydrochloride hydrate. **CAS RN:** *[19408-46-9]*.

SOURCE: Isolated from the soil Actinomycete, *Streptomyces kasugaensis* Hamad *et al.*

PRODUCTION: By fermentation of *S. kasugaensis*.

TARGET PESTS: Control of rice blast (*Pyricularia oryzae* Cavara; perfect stage *Magnaporthe grisea* (Herbert) Barr) and some other diseases (particularly bacterial grain rot, bacterial seedling blight and bacterial brown stripe caused by *Pseudomonas* spp.). Also used to control other plant diseases, e.g. leaf mould and *Venturia* sp. in apples; *Cercospora* leaf spot in peanuts and other legumes; *Colletotrichum lagenarium* Ell. in melons and watermelons; *Cladosporium fulvum* Cke. in tomatoes (at 20 to 40 g/ha); bacterial soft rot; bacterial canker; bean halo blight; bacterial spot; and other bacterial diseases, such as *Pseudomonas syringae* Van Hall pv. *lachrymans* in cucumbers (at 30 to 60 g/ha); *Corynebacterium michiganense* Jens. in tomatoes (at 20 to 40 g/ha); *Mycovellosiella* spp. in aubergines; *Erwinia carotovora* Holl. subsp. *carotovora* in potatoes, carrots and onions; *Pseudomonas syringae* Van Hall pv. *coronafaciens* (halo blight) in kidney beans (at 40 to 60 g/ha); *Xanthomonas* spp. in paprika; and *Pseudomonas marginalis* Stevens pv. *marginalis* in kiwifruit.

TARGET CROPS: Recommended for use in rice, tomatoes, aubergines, paprika, beans and other legumes, pome fruit (apples and pears), sugar beet, celery, potatoes, carrots, onions, cucumbers and other cucurbits, citrus fruit, kiwifruit and ornamentals.

BIOLOGICAL ACTIVITY: Mode of action: Protein biosynthesis inhibitor. Inhibits binding of met-RNA to the mRNA-30S complex, thereby preventing amino acid incorporation. Systemic fungicide and bactericide with protective and curative action. Inhibits hyphal growth of *Pyricularia oryzae* on rice, preventing lesion development; comparatively weak inhibitory action to spore germination, appressoria formation on the plant surface or penetration into the epidermal cell. Rapidly taken up into plant tissue and translocated. In contrast, against *Cladosporium fulvum* on tomatoes, inhibition of sporulation is strong, but inhibition of hyphal growth is weak. Non-phytotoxic to rice, tomatoes, sugar beet, potatoes and other vegetables, but slight injury has been noted to peas, beans, soybeans, grapes, citrus and apples. **Efficacy:** Controls rice blast at concentrations as low as 20 mg per litre. Resistance to kasugamycin hydrochloride hydrate was detected within three years of its introduction in 1965 and, by 1972, it had become a serious problem in Japanese rice fields. Today, mixtures of kasugamycin hydrochloride hydrate with other fungicides with different modes of action are used. Resistant strains of *Pyricularia oryzae* are less fit than susceptible strains and, once applications of kasugamycin hydrochloride hydrate are discontinued in the field, the level of resistance declines very rapidly. Kasugamycin hydrochloride hydrate inhibits hyphal growth of *Pyricularia oryzae* on rice, preventing lesion development; it is a comparatively weak inhibitor of spore germination, appressorium formation and penetration into the epidermal cells. In contrast, against *Cladosporium fulvum* on tomatoes, inhibition of sporulation is high but its effects on hyphal growth are poor. Kasugamycin is taken up by plant tissue and is translocated. **Key reference(s):** 1) H Umezawa, Y Okami, T Hashimoto, Y Suhara & N Otake. 1965. A new antibiotic, kasugamycin, *J. Antibiot. Ser. A*, **18**, 101. 2) M Hamada, T Hashimoto, S Takahashi, M Yoneyama, T Miyake, Y Takeuchi, Y Okami & H Umezawa. 1965. Antimicrobial activity of kasugamycin, *J. Antibiot. Ser. A*, **18**, 104. 3) N Tanaka, H Yamaguchi & H Umezawa. 1966. Mechanism of kasugamycin action on polypeptide synthesis, *J. Biochem.*, **60**, 429.

COMMERCIALISATION: Formulation: Sold as wettable powder (WP), dispersible powder (DP), granules (GR), ultra-low volume (UL) and soluble concentrate (SL) formulations. **Tradenames:** 'Kasugamin' (Dong Bang), 'Kasumin' (Hokko), 'Kasai' (plus fthalide), 'Kasumin Bordeaux' (plus copper oxychloride), 'Kasu-rabcide' (plus fthalide), 'Kasu-rab-validatrebon' (plus fthalide plus validamycin plus etofenprox) and 'Kasu-ran' (plus copper oxychloride) (Hokko), 'Kasu-rab-sumibassa' (plus fthalide plus fenitrothion plus fenobucarb) (Hokko), (Mitsui) and (Sumitomo Chemical Takeda). **Patents:** JP 42006818; BE 657659; GB 1094566.

APPLICATION: Applied as a foliar spray, a dust or a seed treatment, at rates from 20 g a.i. per litre.

PRODUCT SPECIFICATIONS: Specifications: By fermentation of *S. kasugaensis*. Purity: The product is analysed by cup assay with *Pseudomonas fluorescens* (NIHJ B-254).

COMPATIBILITY: Incompatible with pesticides that are strongly alkaline.

MAMMALIAN TOXICITY: Not considered to be acutely toxic to mammals. Acute oral LD_{50}: male rats >5000 mg/kg. Acute dermal LD_{50}: rabbits >2000 mg/kg. Inhalation: LC_{50} (4 h) for rats >2.4 mg/litre. Skin and eye: Non-irritating to eyes and skin of rabbits. NOEL: (2 years) for rats 300, dogs 800 mg/kg diet. Non-mutagenic and non-teratogenic in rats and without effect on reproduction. Toxicity class: WHO (a.i.) U; EPA (formulation) IV.

ENVIRONMENTAL IMPACT AND NON-TARGET TOXICITY: Bird toxicity: Acute oral LD_{50} for male Japanese quail >4000 mg/kg. Fish toxicity: LC_{50} (48 h) for carp and goldfish >40 mg/litre. Other aquatic toxicity: LC_{50} (6 h) for *Daphnia pulex* >40 mg/litre. Effects on beneficial insects: LD_{50} to honeybees (contact) >40 μg per bee. Metabolism: Kasugamycin hydrochloride hydrate, orally administered to rabbits, was mostly excreted in the urine within 24 hours. When injected intravenously to dogs, it was mostly excreted within 8 hours. After oral administration to rats at 200 mg/kg, no residues were detected in 11 organs or blood; 96% of the administered dose remained in the digestive tract one hour after administration. In plants, it is degraded to kasugamycinic acid and kasuganobiosamine and finally to ammonia, oxalic acid, carbon dioxide and water. Behaviour in soil: Degradation proceeds as in plants.

2:190 laminarine   *Algal-derived fungicide*

The Pesticide Manual Fifteenth Edition entry number: 519

M- series   n = 20–30

NOMENCLATURE: Approved name: laminarine.

SOURCE: Laminarine is a storage polysaccharide of the marine brown algae, *Laminaria digitata* (Hudson). It is a hydrophilic β-1,3-glucan with occasional β-1,6-linked branches.

PRODUCTION: Laminarine is extracted from kelps.

TARGET PESTS: Fungal pathogens of cereals, particularly septoria and powdery mildews.

TARGET CROPS: Cereals, particularly wheat.

BIOLOGICAL ACTIVITY: Mode of action: Laminarine has no anti-fungal activity in its own right, but it stimulates the plant's natural defence mechanism, rendering it much less susceptible to attack. It is claimed that it acts as a systemic acquired resistance (SAR) inducer.

COMMERCIALISATION: Formulation: Laminarine is sold as a 40% wettable powder (WP) formulation. **Tradenames:** 'Iodus 40' (Goemar).

APPLICATION: The end use product may be applied as a foliar spray with conventional ground or aerial spray equipment. It is not essential to give complete crop cover, but applications should be made before foliar diseases become established.

PRODUCT SPECIFICATIONS: Laminarine is extracted from kelps.

COMPATIBILITY: Laminarine is compatible with most crop protection agents, although it is not recommended to apply in admixture with other products.

MAMMALIAN TOXICITY: Laminarine is considered to be non-toxic to mammals.

ENVIRONMENTAL IMPACT AND NON-TARGET TOXICITY: Laminarine is derived from naturally occurring seaweeds and is considered unlikely to have any adverse effects on non-target organisms or on the environment. **Approved for use in organic farming:** Yes.

2:191 linalool   *Plant-derived insecticide*

NOMENCLATURE: Approved name: linalool. **Other names:** 3,7-Dimethyl-1,6-octadien-3-ol; linalyl alcohol; allo-ocimenol; 2,6-dimethyl-2,7-octadien-6-ol; licareol (L-linalool); coriandrol (D-linalool). **CAS RN:** *[78-70-6]* DL-linalool; *[126-90-9]* D-linalool; *[126-91-0]* L-linalool.

SOURCE: Linalool is a naturally occurring terpene alcohol chemical with many commercial applications, the majority of which are based on its pleasant scent (floral, with a touch of spiciness). It is found in many flowers and spice plants, such as coriander seeds.

PRODUCTION: Extracted from plants for use as an insect repellent.

TARGET PESTS: Mosquitoes.

TARGET CROPS: Used in public and animal health.

BIOLOGICAL ACTIVITY: Mode of action: Linalool is a strong mosquito deterrent.

COMMERCIALISATION: Formulation: Sold as ready-to-use formulations and as components of insect repellent candles. **Tradenames:** 'Conceal Candle' and 'Mosquito Cognito' (Biosensory), 'Mosquitol-L' (Bedoukian).

APPLICATION: Apply generously when mosquitoes are active.

PRODUCT SPECIFICATIONS: Extracted from plants for use as an insect repellent.

MAMMALIAN TOXICITY: Not considered to be acutely toxic to mammals.
Acute oral LD$_{50}$: rats 2790 mg/kg; mice 3120 mg/kg. **Acute dermal LD$_{50}$:** rats 5610 mg/kg; rabbits >5000 mg/kg. **Inhalation:** LC$_{50}$ for mice 2.95 mg/litre. **Skin and eye:** From animal studies, linalool is regarded as a skin irritant and should be seen as mildly irritant for man. Linalool is, at most, a moderate eye irritant; moreover, in about a third of human subjects it did not cause any eye irritation at 320 ppm. Linalool is considered not to be a sensitiser. **NOEL:** Rats NOAEL 160 mg/kg per day; LOAEL 400 mg/kg per day.

ENVIRONMENTAL IMPACT AND NON-TARGET TOXICITY: Bird toxicity: Bobwhite quail (*Colinus virginianus*) LC$_{50}$ >5620 ppm. **Fish toxicity:** Rainbow trout (*Oncorhynchus mykiss*) NOEC <3.5 mg/litre; LC$_{50}$ 27.8 mg/litre; bluegill (*Lepomis macrochirus*) LC$_{50}$ 36.8 mg/litre; golden orfe (*Leuciscus idus*) NOEC = 22 mg/litre; LC$_{50}$ <46 mg/litre.
Other aquatic toxicity: *Daphnia magna* NOEC 25 mg/litre; EC$_{50}$ 60 mg/litre; *Scenedesmus subspicatus* EC$_{50}$ 88.3 mg/litre.

2:192 *Macleaya* extract *Plant-derived fungicide*

NOMENCLATURE: Approved name: *Macleaya* extract. **CAS RN:** *[112025-60-2]*.

SOURCE: The pink plume poppy (*Macleaya cordata* R. Br.) is widely distributed.

PRODUCTION: The product is extracted from *Macleaya cordata*.

TARGET PESTS: Foliar fungal diseases, such as powdery mildew, alternaria leaf spot and septoria leaf spot.

TARGET CROPS: Ornamental protected crops.

BIOLOGICAL ACTIVITY: Mode of action: The mode of action has not been confirmed, but it is thought likely that, when applied to crops, 'Qwel' often induces increased amounts of naturally occurring phenolic substances in the treated plants. These phenolics act as phytoalexins and have been shown to prevent the attack of several commercially important plant diseases, including powdery mildew and grey mould in ornamental plants such as roses. In this way, 'Qwel' induces systemic acquired resistance (SAR).

COMMERCIALISATION: Formulation: Sold as a 1.5% aqueous extract. **Tradenames:** 'Qwel' (Camas).

APPLICATION: Applied to the foliage of the crops as soon as there are signs of disease.

PRODUCT SPECIFICATIONS: The product is extracted from *Macleaya cordata*.

COMPATIBILITY: It is unusual to apply 'Qwel' with any other crop protection chemical.

MAMMALIAN TOXICITY: There have been no adverse or allergic effects seen in researchers, formulators or users from *Macleaya* extract. It is considered to be non-toxic to mammals.

ENVIRONMENTAL IMPACT AND NON-TARGET TOXICITY: *Macleaya* extract is a naturally occurring substance and is considered not to be toxic to non-target organisms and to be without adverse effects on the environment. **Approved for use in organic farming:** Yes.

2:193 maple lactone

Plant-derived insect attractant

NOMENCLATURE: Approved name: maple lactone. **Other names:** 2-hydroxy-3-methyl-2-cyclopenten-1-one.

SOURCE: Maple lactone is a naturally occurring chemical found in several species of the genus *Acer*.

PRODUCTION: Extracted from trees for use.

TARGET PESTS: Cockroaches (*Periplaneta americana* L. and *Blatella germanica* (L.)).

TARGET CROPS: Indoors in dark or humid areas where cockroaches are usually found, such as near plumbing, under and behind sinks, in corners and cracks, behind kitchen appliances, in bathrooms, around laundry machines and in similar situations.

BIOLOGICAL ACTIVITY: Mode of action: Maple lactone has an odour typical of stale beer, which attracts cockroaches.

COMMERCIALISATION: Formulation: Sold in tablet form or as a part of cockroach traps. **Tradenames:** 'Babolna Insect Attractant Trap' (Babolna Bioenvironmental), 'MGK Roach Trap' (MGK).

APPLICATION: Maple lactone is prepared in tablet form for use inside a trap to attract cockroaches. The trap can be used for monitoring the numbers of cockroaches in an area, or for controlling them. The tablet is placed inside the trap, which is constructed from a piece of cardboard folded into a 'house-like' shape. The tablet is placed in the centre of the adhesive-coated inside surface. When the attractant lures cockroaches into the trap, they are caught by the adhesive surface. For monitoring, the traps are placed at a density of 1 to 2 traps per 10 square metres. For cockroach control, the traps are placed at a density of 3 to 5 traps per 10 square metres.

PRODUCT SPECIFICATIONS: Extracted from trees for use.

MAMMALIAN TOXICITY: Maple lactone is not likely to produce adverse health effects in humans. It is essentially non-toxic to humans.

ENVIRONMENTAL IMPACT AND NON-TARGET TOXICITY: No harmful environmental effects are expected because there is virtually no exposure to any species except the cockroach; the traps are approved only for use indoors; and maple lactone has been shown not to be toxic to any organisms against which it has been tested.

2:194 *p*-menthane-3,8-diol
Plant-derived insect repellent

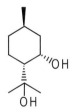

NOMENCLATURE: Approved name: *p*-menthane-3,8-diol. **CAS RN:** *[42822-86-6]*.

SOURCE: *p*-Menthane-3,8-diol occurs naturally in the lemon eucalyptus plant (*Eucalyptus citriodora* Hook).

PRODUCTION: For commercial use, the active ingredient is chemically synthesised. *p*-Menthane-3,8-diol is structurally similar to menthol.

TARGET PESTS: Repels specific insects, including mosquitoes, biting flies and gnats.

TARGET CROPS: *p*-Menthane-3,8-diol is used on humans and their clothing to repel insects.

BIOLOGICAL ACTIVITY: Efficacy: 'Quwenling' is a popular eucalyptus-based repellent product containing a mixture of *p*-menthane-3,8-diol, isopulegone and citronellol. 'Quwenling' has largely replaced dimethyl phthalate as the insect repellent of choice in China. Lemon eucalyptus has also been used for some years in China as an insect repellent. A particularly effective extract of lemon eucalyptus has been marketed as 'Mosiguard' over the past few years. Published laboratory data indicate equivalence to 20% DEET, but field tests are fewer in number and scale than those that have been carried out with DEET (*N,N*-dimethyl-*m*-toluamide). There are limited data or direct studies comparing the product with higher concentrations of DEET or with long-acting preparations. However, of all the natural-based products, lemon eucalyptus is probably among the most effective. **Key reference(s):** 1) J K Trigg & N Hill. 1996. Laboratory evaluation of a eucalyptus-based repellent against four biting arthropods, *Phytother. Res.*, **10**, 313–6. 2) J K Trigg. 1996. Evaluation of a eucalyptus-based repellent against *Anopheles* spp. in Tanzania, *J. Am. Mosquito Control Assoc.*, **12**, 243–6.

COMMERCIALISATION: Formulation: Products containing *p*-menthane-3,8-diol are sold as liquids. **Tradenames:** 'Mosiguard Natural' (Johnson), 'Quwenling' (produced in China).

APPLICATION: Products containing *p*-menthane-3,8-diol are sprayed on skin or clothing. Lotions are rubbed onto skin.

PRODUCT SPECIFICATIONS: Specifications: For commercial use, the active ingredient is chemically synthesised. *p*-Menthane-3,8-diol is structurally similar to menthol. **Purity:** Lemon eucalyptus oil is manufactured and is guaranteed to be pure.

COMPATIBILITY: It is not recommended that the product be used with any other agent.

MAMMALIAN TOXICITY: In studies using laboratory animals, *p*-menthane-3,8-diol shows no adverse effects except for eye irritation. Therefore, special precautions are put on the label to prevent the product from contacting users' eyes. If used according to label instructions, *p*-menthane-3,8-diol is not expected to pose health risks to humans, including children and other sensitive populations.

ENVIRONMENTAL IMPACT AND NON-TARGET TOXICITY: Based on laboratory animal studies, *p*-menthane-3,8-diol poses minimal or no risks to wildlife. Because of the low toxicity and limited uses of *p*-menthane-3,8-diol, it is not harmful to the environment.

2:195 MI8

Micro-organism-derived fungicide

2. Natural Products

COOH

NOMENCLATURE: Approved name: MI8.

SOURCE: It was discovered in a fluorescent pseudomonad isolate originating from the soil of a Shanghai suburb.

PRODUCTION: Fermentation.

TARGET PESTS: Fusarium wilt (*Fusarium oxysporum* Schlect.), late blight (*Phytophthora infestans* De Bary) and gummy stem blight (*Didymella bryoniae* (Fuckel) Rehm).

TARGET CROPS: Fruit and vegetables such as water melons, desert melons and peppers.

BIOLOGICAL ACTIVITY: Mode of action: The mode of action of MI8 has not been determined. **Key reference(s):** Z M Li & Y B Zhang. 2008. The outset innovation of agrochemicals in China. *Outlooks on Pest Management*, **19**(3), 136–8.

COMMERCIALISATION: Formulation: Under development in China.

MAMMALIAN TOXICITY: Studies on the mammalian toxicity of MI8 are under way.

ENVIRONMENTAL IMPACT AND NON-TARGET TOXICITY: Studies on the environmental impact and non-target effects of MI8 are under way.

2:196 milbemectin
Micro-organism-derived acaricide and insecticide

The Pesticide Manual **Fifteenth Edition entry number:** 596

milbemycin A₃: R = –CH₃
milbemycin A₄: R = –CH₂CH₃

NOMENCLATURE: Approved name: milbemectin (BSI, pa E-ISO).
Development code: B-41; E-187; SI-8601 (Sankyo). **CAS RN:** *[51596-10-2]* A₃; *[51596-11-3]* A₄.

SOURCE: Isolated from the soil Actinomycete, *Streptomyces hygroscopicus* (Jensen) Waks. & Henrici subsp. *aureolacrimosus*.

PRODUCTION: Produced by the fermentation of *S. hygroscopicus* subsp. *aureolacrimosus* and isolation from the fermentation broth.

TARGET PESTS: Control of citrus red mites and pink citrus rust mites, Kanzawa spider mites and other spider mites. Also recommended for control of leaf miners.

TARGET CROPS: Recommended for use on citrus fruit, tea and aubergines. 'Ultiflora' is recommended for use on protected ornamentals.

BIOLOGICAL ACTIVITY: Mode of action: The target for milbemectin is the γ-aminobutyric acid (GABA) receptor in the peripheral nervous system. The compound stimulates the release of GABA from nerve endings and enhances the binding of GABA to receptor sites on the post-synaptic membrane of inhibitory motor neurons of mites and other arthropods. This enhanced GABA binding results in an increased flow of chloride ions into the cell, with consequent hyperpolarisation and elimination of signal transduction resulting in an inhibition of neurotransmission. (See M J Turner & J M Schaeffer. 1989. In *Ivermectin and Abamectin*, W C Campbell (ed.), Springer-Verlag, New York, USA, p. 73). Acaricide with contact and stomach action. It has limited plant systemic activity, but exhibits translaminar movement.

Efficacy: Milbemectin has been shown to have very high insecticidal and acaricidal activity, but shows no antimicrobial effects. **Key reference(s):** 1) Y Takiguchi, H Mishima, M Okuda, M Terao, A Aoki & R Fukuda. 1980. Milbemycins, a new family of macrolide antibiotics: Fermentation, isolation and physicochemical properties, *J. Antibiot.*, **33**, 1120. 2) M Mishima. 1983. Milbemycin: a family of macrolide antibiotics with insecticidal activity. In *IUPAC Pesticide Chemistry*, J Miyamoto & P C Kearney (eds), Pergamon Press, Oxford, UK, Vol. 2, p. 129.

COMMERCIALISATION: Formulation: Sold as emulsifiable concentrate (EC) and wettable powder (WP) formulations. **Tradenames:** 'Milbeknock' and 'Koromite' (Sankyo Agro), 'Matsuguard' (for pine wood nematode control) (Sankyo Agro), 'Ultiflora' and 'Mesa' (Gowan) and (Sankyo Agro).

APPLICATION: Rates of use are 5.6 to 28 g a.i. per hectare for mite control. The effectiveness of the product is increased significantly by the addition of paraffinic oils to the spray tank.

PRODUCT SPECIFICATIONS: Specifications: Produced by the fermentation of *S. hygroscopicus* subsp. *aureolacrimosus* and isolation from the fermentation broth. **Purity:** A mixture of the homologues milbemycin A_3 (methyl) and milbemycin A_4 (ethyl) in the ratio 3 to 7.

COMPATIBILITY: Compatible with most crop protection chemicals.

MAMMALIAN TOXICITY: Considered to be relatively toxic to mammals. **Acute oral LD_{50}:** male mice 324, female mice 313, male rats 762, female rats 456 mg/kg. **Acute dermal LD_{50}:** male and female rats >5000 mg/kg. **Inhalation:** LC_{50} (4 h) for male rats 1.9, female rats 2.8 mg/kg. **NOEL:** for male rats 6.81, female rats 8.77, male mice 18.9, female mice 19.6 mg/kg. **ADI:** 0.03 mg/kg (Japan). **Other toxicological effects:** Not carcinogenic, not teratogenic, not mutagenic.

ENVIRONMENTAL IMPACT AND NON-TARGET TOXICITY: Milbemectin is not persistent in the environment and is not thought to pose any threat to non-target organisms. **Bird toxicity:** LD_{50} for male chickens 660, female chickens 650, male Japanese quail 1005, female Japanese quail 968 mg/kg. **Fish toxicity:** LC_{50} (96 h) for rainbow trout 4.5, carp 17 μg/litre. **Other aquatic toxicity:** LC_{50} (3 h) for *Daphnia pulex* >100 ppm. NOEL (72 h) for algae ≥7.3 mg/litre. **Effects on beneficial insects:** LD_{50} (oral) for honeybees 0.46 μg/bee; (contact) 0.025 μg/bee. LC_{50} (14 days) for earthworms 61 ppm. **Behaviour in soil:** Soil DT_{50} 16 to 33 days.

2:197 mildiomycin
Micro-organism-derived fungicide

The Pesticide Manual Fifteenth Edition entry number: 597

NOMENCLATURE: Approved name: mildiomycin (JMAF). **Development code:** antibiotic B-98891;TF-138 (Takeda). **CAS RN:** *[67527-71-3]*; formerly *[57497-78-6]* and *[67983-11-3]*.

SOURCE: Produced by the soil Actinomycete *Streptoverticillium rimofaciens* strain B-98891.

PRODUCTION: Manufactured by fermentation of *Streptoverticillium rimofaciens* B-98891 and extraction from the culture medium.

TARGET PESTS: Powdery mildews (*Erysiphe* spp., *Uncinula necator* Burr., *Podosphaera* spp., *Sphaerotheca* spp.).

TARGET CROPS: Ornamentals.

BIOLOGICAL ACTIVITY: Mode of action: Mildiomycin is believed to inhibit protein biosynthesis in fungi by blocking peptidyl-transferase. It is effective as an eradicant, with some systemic activity. **Efficacy:** Mildiomycin is specifically active against the pathogens that cause powdery mildew and is much less effective against bacteria. **Key reference(s):** 1) S Harada & T Kishi. 1978. Isolation and characterisation of mildiomycin, a new nucleoside antibiotic, *J. Antibiot.*, **31**, 519. 2) Y Om, I Yamaguchi & T Misato. 1984. Inhibition of protein biosynthesis by mildiomycin, an anti-mildew substance, *J. Pestic. Sci.*, **9**, 317. 3) T Kusaka, K Suetomi, T Iwasa & S Harada. 1979. TF-138: a new fungicide, *Proc. 1979 Brit. Crop Protection Conference – Pests & Diseases*, 2, 589–95.

COMMERCIALISATION: Formulation: Sold as a wettable powder (WP) formulation. **Tradenames:** 'Mildiomycin' (Sumitomo Chemical Takeda).

APPLICATION: Used as a foliar spray to eradicate the diseases and, subsequently, protect crops from attack by powdery mildews, at rates of 5 to 10 g per hectolitre.

PRODUCT SPECIFICATIONS: Manufactured by fermentation of *Streptoverticillium rimofaciens* B-98891 and extraction from the culture medium.

MAMMALIAN TOXICITY: Not considered to be acutely toxic to mammals.
Acute oral LD$_{50}$: male rats 4300, female rats 4120, male mice 5060, female mice 5250 mg/kg.
Acute dermal LD$_{50}$: male and female rats and mice >5000 mg/kg. **Skin and eye:** No irritation has been observed to the cornea and the skin of rabbits at 1000 μg per ml over 10 days.
NOEL: In 30-day feeding studies, there were no significant adverse effects to mice or rats fed 200 mg per kg daily. In a 3-month sub-acute feeding study in rats, the maximum no-effect level was 50 mg per kg daily. **Other toxicological effects:** Negative results were obtained from the Ames test for mutagenicity conducted with or without rat liver homogenate.

ENVIRONMENTAL IMPACT AND NON-TARGET TOXICITY: Mildiomycin is not expected to have any adverse effects on non-target organisms or on the environment.
Fish toxicity: LC$_{50}$ (72 h) for carp >40 mg/litre and (168 h) for killifish >40 mg/litre.
Other aquatic toxicity: LC$_{50}$ (6 h) for *Daphnia pulex* >20 mg/litre; EC$_{50}$ (72 h) for algae (*Selenastrum* spp.) 8.05 mg/litre. **Effects on beneficial insects:** LD$_{50}$ (48 h, oral and contact) to honeybees >100 μg/bee.

2:198 milsana   *Plant-derived fungicide*

The Pesticide Manual **Fifteenth Edition entry number:** 764

NOMENCLATURE: Approved name: milsana. **Other names:** REYSA; *Reynoutria sachalinensis* extract. **Development code:** BAS 114 UBF; KHHUBF; KHHCN; KHHCN-01-01.

SOURCE: Milsana is the partially refined extract of the giant knotweed plant (*Reynoutria sachalinensis* (Fr. Schm.) Nakai). Originally developed and sold by BASF AG in Europe; now under development by KHH BioSci Inc., for use in the Americas, Asia and Oceania.

PRODUCTION: Extracted from the giant knotweed plant (*Reynoutria sachalinensis*), which is grown commercially for extraction in south-eastern USA. The technical grade a.i. consists entirely of the dried and ground plant material. The product is a 5% ethanolic extract suspended in calcium nitrate solution.

TARGET PESTS: The extract is recommended for use against a wide range of fungal pathogens, including *Botrytis* spp. and powdery mildews. Activity against some bacterial diseases, such as *Xanthomonas* spp., is also claimed. It is suggested that the high level of phenolics within treated plants also gives some protection against attack by phytophagous insects.

TARGET CROPS: Milsana is being developed for use in ornamental and glasshouse crops, such as roses, and in vegetable and fruit crops, such as cucurbits, peppers and grapes.

BIOLOGICAL ACTIVITY: Mode of action: The extract, applied to growing plants at onset of disease, prevents disease development by raising the plant's natural defence system. When applied to ornamental, vegetable and fruit crops, milsana will induce increased amounts of naturally occurring phenolic substances in the treated plants. These phenolics act as phytoalexins and have been shown to prevent the attack of several commercially important plant diseases, including powdery mildew and grey mould in ornamental plants, such as roses, and in vegetable and fruit crops, such as cucumbers, peppers and grapes.

Key reference(s): 1) F Daayf, A Schmitt & R R Bélanger. 1997. Evidence of phytoalexins in cucumber leaves infected with powdery mildew following treatment with leaf extracts of *Reynoutria sachalinensis*, *Plant Physiol.*, **113**, 719–27. 2) New company developing products for growers as alternative to fungicides, *Phytopathol.*, November 1998. 3) B Fofana, D J McNally, C Labbé, R Boulanger, N Benhamou, A Séguin & R R Bélanger. 2002. Milsana-induced resistance in powdery mildew-infected cucumber plants correlates with the induction of chalcone synthase and chalcone isomerase, *Physiological and Molecular Plant Pathol.*, **61**, 121–32.

COMMERCIALISATION: Formulation: The end product, 'Milsana Bioprotectant Concentrate', contains 5% of the ethanolic extract of *Reynoutria sachalinensis* suspended in a calcium nitrate solution. **Tradenames:** 'Milsana Bioprotectant Concentrate' and 'Milsana' (KHH BioScience).

APPLICATION: 'Milsana Bioprotectant Concentrate' should be applied at 0.5% v/v in 500 to 1000 litres of water per hectare, sprayed at 7 to 10 day intervals, and with the addition of 0.02% v/v of an anionic wetting agent, to the foliage at the 4- to 6-leaf stage of crops to be protected. Treated plants, such as roses and cucurbits, will also be darker green and senescence will be delayed. A second treatment may be applied after 14 to 21 days.

PRODUCT SPECIFICATIONS: Extracted from the giant knotweed plant (*Reynoutria sachalinensis*), which is grown commercially for extraction in south-eastern USA. The technical grade a.i. consists entirely of the dried and ground plant material. The product is a 5% ethanolic extract suspended in calcium nitrate solution.

COMPATIBILITY: It is not expected that milsana will be applied with other fungicidal products.

MAMMALIAN TOXICITY: There is no evidence of allergic or other negative responses from researchers, manufacturers, formulators or users. Milsana has been commercialised in Germany for over twelve years, with no reports of any adverse effects.

Acute oral LD$_{50}$: rats >5000 mg/kg. **Acute dermal LD$_{50}$:** rabbits >2000 mg/kg. **Inhalation:** LC$_{50}$

for rats (of concentrate) >2.6 mg/litre. **Skin and eye:** Moderately irritating to the eyes, but not irritating to the skin of rabbits. Not a skin sensitiser (guinea pigs). **Toxicity class:** EPA IV.

ENVIRONMENTAL IMPACT AND NON-TARGET TOXICITY: Milsana is a naturally occurring plant extract and is not expected to have any adverse effects on non-target organisms or on the environment. In the USA, the US-EPA waived data requirements because it is a natural plant extract and already exists in the environment. **Approved for use in organic farming:** Yes: it has US OMRI approval.

2:199 natamycin

Micro-organism-derived fungicide

The Pesticide Manual **Fifteenth Edition entry number:** Superseded entry 1282

NOMENCLATURE: Approved name: natamycin (BAN). **Other names:** Myprozine; pimaricin; tennectin (traditional names). **CAS RN:** *[7681-93-8]*.

SOURCE: Secondary metabolite of the Actinomycetes, *Streptomyces natalensis* Struyk, Hoette, Drost, Waisvisz, Van Eek & Hoogerheide and *S. chattanoogensis* Burns & Holtman.

PRODUCTION: Manufactured by fermentation.

TARGET PESTS: Fungal diseases, especially basal rots caused by *Fusarium oxysporum* Schlecht.

TARGET CROPS: Ornamental bulbs such as daffodils.

BIOLOGICAL ACTIVITY: Efficacy: Used as a dip, usually in combination with hot water treatment. Gives effective control of various fungal diseases of ornamental bulbs prior to planting. **Key reference(s): 1)** B T Golding, R W Rickards, W E Meyer, J B Patrick & M Barber. 1966. The structure of the macrolide antibiotic pimaricin, *Tetr. Letters*, 3551–7.

2) G J Levinskas, W E Ribelin & C B Shaffer. 1966. Acute and chronic toxicity of pimaricin. *Toxicol. Appl. Pharmacol.*, **8**, 97–109. 3) A P Struyk, I Hoette, G Drost, J M Waisvisz, J van Eek & J C Hoogerheide. 1958. In *Antibiotics Annual, 1957–1958*, p. 878, H Welch & F Marti-Ibanez (eds), Medical Encyclopaedia Inc., New York, USA.

COMMERCIALISATION: Formulation: Sold as a wettable powder (WP), now withdrawn. **Tradenames:** 'Delvolan' (withdrawn) (Gist-Brocades). **Patents:** GB 712547; GB 844289; US 3892850.

APPLICATION: Bulbs are dipped in a solution of the product. It is usual to accompany this treatment with a hot water treatment.

PRODUCT SPECIFICATIONS: Specifications: Manufactured by fermentation. **Purity:** Product purity is checked by bioassay with a suitable fungal pathogen such as *Fusarium oxysporum*.

COMPATIBILITY: It is unusual to apply natamycin in combination with any other crop protection agent.

MAMMALIAN TOXICITY: Not considered to be acutely toxic to mammals. **Acute oral LD$_{50}$:** rats 2730–4670 mg/kg. **Skin and eye:** No acute toxicity, even at high doses. Not a skin sensitiser. Not a skin or eye irritant (rabbits). **Toxicity class:** WHO (a.i.) III.

ENVIRONMENTAL IMPACT AND NON-TARGET TOXICITY: Natamycin is not toxic to fish and is readily biodegradable. No adverse effects have been observed on non-target organisms or on the environment.

2:200 nicotine   *Plant-derived insecticide*

The Pesticide Manual **Fifteenth Edition entry number:** 614

NOMENCLATURE: Approved name: nicotine (BSI, E-ISO, F-ISO, ESA, in lieu of a common name); nicotine sulfate (E-ISO (from 1984), JMAF, for sulfate salt); sulfate de nicotine (F-ISO). **EC no:** 200-193-3. **CAS RN:** *[54-11-5]* (S)- isomer; *[22083-74-5]* (RS)- isomers; *[75202-10-7]* unstated stereochemistry; *[65-30-5]* nicotine sulfate, (S)- isomer.

SOURCE: Nicotine is the main bioactive component of the tobacco plants *Nicotiana tabacum* L., *N. glauca* Graham and particularly the species *N. rustica* L., and is also present in a number of other plants belonging to the families of Lycopodiaceae, Crassulaceae, Leguminosae, Chenopodiaceae and Compositae. The average nicotine content of the leaves of *N. tabacum* and *N. rustica* is 2% to 6% dry weight.

PRODUCTION: Once prepared from extracts of the tobacco plant, but now manufactured and sold as either technical nicotine or nicotine sulfate. Nicotine is often obtained from waste of the tobacco industry and its use is mainly confined to small-scale or glasshouse application.

TARGET PESTS: Used for the control of a wide range of insects, including aphids, thrips and whitefly.

TARGET CROPS: Recommended for use on protected ornamentals and field-grown crops, including orchard fruit, vines, vegetables and ornamentals.

BIOLOGICAL ACTIVITY: Mode of action: Non-systemic insecticide that binds to the cholinergic acetylcholine nicotinic receptor in the nerve cells of insects, leading to a continuous firing of this neuroreceptor. Active predominantly through the vapour phase, but also slight contact and stomach action. **Efficacy:** Nicotine has been used for many years as a fumigant for the control of many sucking insects. It can be used to give partial control of OP (organophosphate) and pyrethroid resistant whitefly. Nicotine has a rapid vapour action and is highly toxic to mammals (see below). The laevorotatory isomer, (S)-nicotine, is two to three times more active than the other enantiomer. Illustrative topical LD_{50} values are 4 and 140 µg/g for *Bombyx mori* L. and male *Periplaneta americana* L., respectively; in a fumigant assay, it has an LC_{50} value of 0.0032 mg/litre for *Aphis gossypii* Glover. **Key reference(s): 1)** I Schmeltz. 1971. *Naturally Occurring Insecticides*, M Jacobson & D G Crosby (eds), Marcel Dekker, New York, USA. **2)** I Ujváry. 1999. Nicotine and other insecticidal alkaloids. In *Nicotinoid Insecticides and the Nicotinic Acetylcholine Receptor*, I Yamamoto & J E Casida (eds), pp. 29–69, Springer-Verlag, Tokyo, Japan. **3)** K Chamberlain, A A Evans & R H Bromilow. 1996. 1-Octanol/water partition coefficient (K_{ow}) and pKa for ionisable pesticides measured by a pH-metric method, *Pestic. Sci.*, **47**, 265–71. **4)** W O Negherbon. 1959. Nicotine. In *Handbook of Toxicology. Vol. III: Insecticides. A Compendium*, W B Saunders (ed.), pp. 508–19, Philadelphia, USA. **5)** A Brossi & X-F Pei. 1998. Biological activity of unnatural alkaloid enantiomers. In *The Alkaloids: Chemistry and Pharmacology*, **50**, G A Cordell (ed.), pp. 109–39, Academic Press, San Diego, California, USA.

COMMERCIALISATION: Formulation: Sold as dispersible powder (DP), soluble concentrate (SL) or as fumigant formulations. **Tradenames:** 'Stalwart' (United Phosphorus Ltd), 'No-Fid' (Hortichem), 'XL-All Nicotine' (Vitax), 'Nicotine 40% Shreds' (Dow AgroSciences), 'Tobacco Dust' (Bonide).

APPLICATION: Applied as a foliar spray to cover the undersides of leaves and repeated as necessary. Best results are achieved when the temperature is above 16 °C. If used as a fumigant, the temperature must be above 16 °C. The maximum number of treatments in protected crops is three. In the UK, nicotine is subject to regulation under the Poisons Act.

PRODUCT SPECIFICATIONS: Specifications: Once prepared from extracts of the tobacco plant, but now manufactured and sold as either technical nicotine or nicotine sulfate. Nicotine is often obtained from waste of the tobacco industry and its use is mainly confined to small-scale or glasshouse application. **Purity:** The predominant component of the crude alkaloid extract is (S)-(−)-nicotine; small amounts of related alkaloids may be present. Manufactured nicotine may be the racemic mixture.

COMPATIBILITY: May be used with most crop protection agents. Test for crop phytotoxicity before treating large areas.

MAMMALIAN TOXICITY: Intravenous LD_{50} value of the (S)- and (R)- stereoisomer in mice is 0.38 and 2.75 mg/kg, respectively. **Acute oral LD_{50}:** rats 50–60; mice 24 mg/kg. **Acute dermal LD_{50}:** rabbits 50 mg/kg. **Inhalation:** Toxic to man by inhalation. **Skin and eye:** Readily absorbed through the skin. Toxic to man by skin contact. **Toxicity class:** WHO (a.i.) Ib; EPA (formulation) I. **Other toxicological effects:** Lethal oral dose for man is stated to be 40 to 60 mg.

ENVIRONMENTAL IMPACT AND NON-TARGET TOXICITY: A compilation of acute toxicological data for insects and other animals has been published (Negherbon, 1959), while the pharmacology of nicotine isomers has been summarised by Brossi and Pei (1998). **Bird toxicity:** Toxic to birds. **Fish toxicity:** LC_{50} for larval rainbow trout 4 mg/litre. **Other aquatic toxicity:** LC_{50} for *Daphnia pulex* 0.24 mg/litre. **Effects on beneficial insects:** Toxic to bees, but has a repellent effect. **Metabolism:** Nicotine decomposes relatively quickly under the influence of light and air.

2:201 oct-1-en-3-ol

Plant-derived insect attractant

$$H_2C \underset{CH}{\overset{}{=}} \overset{\overset{OH}{|}}{CH} (CH_2)_4 CH_3$$

NOMENCLATURE: Approved name: oct-1-en-3-ol. **Other names:** amylvinylcarbinol; octenol; 1-octen-3-ol. **OPP Chemical Code:** 069037. **CAS RN:** [3391-86-4].

SOURCE: Oct-1-en-3-ol is a component of a large number of plant species and is a major insect attractant. It is found in many edible fruits and vegetables.

PRODUCTION: Manufactured for use as an insect attractant. Found naturally in clover, lucerne (alfalfa) and other plants.

TARGET PESTS: Mosquitoes, dragonflies and biting flies.

TARGET CROPS: Oct-1-en-3-ol is intended for outdoor non-food use and is used in dispensers placed on electric bug killer stations. Registered for use on non-agricultural sites only.

BIOLOGICAL ACTIVITY: Efficacy: Tests on the effectiveness of oct-1-en-3-ol in attracting mosquito and biting fly species gave variable results in that some species were repelled, some species were attracted and some species did not appear affected. The label claims state that the products make electric insect killers more effective in killing certain mosquitoes and biting flies. The statements do not claim that they control mosquitoes, only that they make the electric insect killers more effective.

COMMERCIALISATION: Tradenames: 'Hercon Mosquito Attractant' (Hercon), 'Flowtron Octenol Mosquito Attractant' (Armitron), 'Bedoukian Octenol Technical' (Bedoukian), 'The Dragonfly Octenol Lure' (Biosensory), 'Mosquito Magnet Octenol Biting Insect Attractant' (American Biophysics).

APPLICATION: Oct-1-en-3-ol is used in electric insect traps to increase the attractiveness of the traps to insects.

PRODUCT SPECIFICATIONS: Manufactured for use as an insect attractant. Found naturally in clover, lucerne (alafalfa) and other plants.

MAMMALIAN TOXICITY: Not considered to be toxic to mammals. **Acute oral LD$_{50}$:** rats 340 mg/kg (Toxicity Category II). **Acute dermal LD$_{50}$:** rats 3300 mg/kg (Toxicity Category III). **Inhalation:** LC$_{50}$ for rats 3.72 mg/litre (Toxicity Category IV). **Skin and eye:** Dermal and eye contact may cause irritation, but oct-1-en-3-ol was not irritating when applied full strength to abraded and unabraded rabbit skin (using an occlusive material) for 24 hours and no irritation occurred in humans after a 48-hour closed-patch test of oct-1-en-3-ol at a concentration of 10%. A maximisation test on 22 human volunteers produced very slight sensitisation in 2 individuals when tested at a concentration of 10%. Oct-1-en-3-ol is a component of some perfumes up to levels of 1% and there is no evidence of eye or respiratory effects caused by the evaporation of the perfume. Use of oct-1-en-3-ol as an insect attractant is expected to result in less exposure to oct-1-en-3-ol than if wearing a perfume containing oct-1-en-3-ol.

ENVIRONMENTAL IMPACT AND NON-TARGET TOXICITY: It is unclear what non-target insects would be attracted by the oct-1-en-3-ol that were not already attracted to the electronic bug killer. Non-target insect exposure to the electric shock of the bug killer would likely be limited to small flying insects due to the trap design. It is believed that oct-1-en-3-ol will have no effect on listed endangered species. **Bird toxicity:** There is minimal risk to non-target wildlife from the uses of oct-1-en-3-ol due to the low exposure from the products. The exposure to birds, fish, aquatic invertebrates and plants is expected to be minimal to non-existent for oct-1-en-3-ol, because it is attached to electric insect traps.

2:202 oxytetracycline
Micro-organism-derived bactericide

The Pesticide Manual **Fifteenth Edition entry number:** 649

NOMENCLATURE: Approved name: oxytetracycline (ISO, BSI, JMAF, BAN).
Other names: Terramitsin (Russia). **CAS RN:** *[79-57-2]* oxytetracycline; *[2058-46-0]* oxytetracycline hydrochloride.

SOURCE: Oxytetracycline is produced by the fermentation of *Streptomyces rimosus* Sobin *et al.*

PRODUCTION: Produced by fermentation and usually sold as the hydrochloride.

TARGET PESTS: Bacterial diseases such as fire blight (*Erwinia amylovora* Winsl.) and diseases caused by *Pseudomonas* and *Xanthomonas* species. Also effective against diseases caused by mycoplasma-like organisms.

TARGET CROPS: Orchard stone and pome fruit and turf grass.

BIOLOGICAL ACTIVITY: Mode of action: Oxytetracycline is a potent inhibitor of protein biosynthesis in bacteria. It binds to the 30S and 50S bacterial ribosomal subunits and inhibits the binding of aminoacyl-tRNA and the termination factors RF1 and RF2 to the A site of the bacterial ribosome. It is much less active in mammalian systems. **Efficacy:** Oxytetracycline is rapidly taken up by plant leaves, particularly through stomata, and is readily translocated to other plant tissues. It is an effective antibacterial and is often mixed with streptomycin to prevent the development of streptomycin resistance. **Key reference(s):** 1) A C Finlay, G L Hobby, S Y Pan, P P Regna, J B Routier, D B Seeley, G M Shull, B A Sobin, I A Solomons, J W Vinson & J H Kane. 1950. Terramycin, a new antibiotic, *Science*, **111**, 519. 2) T Ishii, Y Doi, K Yora & H Asuyama. 1967. Suppressive effects of antibiotics of the tetracycline group on symptom development of mulberry dwarf disease, *Ann. Phytopathol. Soc. Jpn*, **33**, 267. 3) C T Caskey. 1973. Inhibitors of protein biosynthesis. In *Metabolic Inhibitors*, R M Hochster, M Kates & J H Quastel (eds), Vol. IV, p. 131, Academic Press, New York, USA.

COMMERCIALISATION: Formulation: Sold as a water soluble powder (SP). **Tradenames:** 'Mycoshield' (calcium salt) (Nufarm Americas), 'Cuprimicina Agrícola', 'Cuprimicín 100' (plus streptomycin) and 'Cuprimicín 500' (plus streptomycin plus copper sulfate (tribasic)) (Ingeniería Industrial), 'Mycoject' (calcium salt) (Mauget), 'Phytomycin' (plus streptomycin sulfate) (Ladda).

APPLICATION: Applied as a foliar spray to infected plants.

PRODUCT SPECIFICATIONS: Specifications: Produced by fermentation and usually sold as the hydrochloride. **Purity:** Efficacy of the formulation can be checked by bioassay against a suitable bacterium.

COMPATIBILITY: Compatible with most crop protection agents. It is often applied in combination with streptomycin sulfate.

MAMMALIAN TOXICITY: Considered to be non-toxic to mammalian systems.

ENVIRONMENTAL IMPACT AND NON-TARGET TOXICITY: Oxytetracycline is not expected to show any adverse effects on non-target organisms or on the environment.

2. Natural Products

2:203 pelargonic acid
Plant-derived herbicide and plant growth regulator

The Pesticide Manual **Fifteenth Edition entry number:** 619

$$CH_3(CH_2)_7CO_2H$$

NOMENCLATURE: Approved name: pelargonic acid. **Other names:** Nonoic acid; nonanoic acid; fatty acids. **Development code:** MYX 6121 (Mycogen); JT-101 (Japan Tobacco). **CAS RN:** *[112-05-0]*.

SOURCE: Pelargonic acid occurs naturally in members of the family Geraniaceae.

PRODUCTION: May be extracted from plants, but usually manufactured for use in crop protection.

TARGET PESTS: Contact, broad-spectrum herbicide.

TARGET CROPS: Apples and other orchard tree fruit.

BIOLOGICAL ACTIVITY: Mode of action: Pelargonic acid is a mild phytotoxicant and, when applied to trees shortly after petal fall, causes thinning of the fruit, thereby allowing larger, more regular-shaped fruit to develop.

COMMERCIALISATION: Formulation: Sold as a liquid concentrate (SL).
Tradenames: 'Scythe' (Dow AgroSciences) and (Extremely Green), 'Grantico' (Japan Tobacco), 'Weedol Max' (Scotts), 'Advanced 3 Hour Weedkiller' and 'Advanced Moss Killer' (3% ammonium octanoate and ammonium decanoate) (Bayer Garden).

APPLICATION: Apply to orchard trees at blossom or within 2 weeks of petal fall.

PRODUCT SPECIFICATIONS: May be extracted from plants, but usually manufactured for use in crop protection.

MAMMALIAN TOXICITY: Not considered to be acutely toxic to mammals.
Acute oral LD$_{50}$: rats and mice >5000 mg/kg. **Acute dermal LD$_{50}$:** rats >2000 mg/kg.
Inhalation: LC$_{50}$ (4 h) for rats >5.3 mg/litre. **Toxicity class:** EPA (formulation) III.

ENVIRONMENTAL IMPACT AND NON-TARGET TOXICITY: Pelargonic acid occurs in Nature and is not expected to show any adverse effects on non-target organisms or on the environment. **Bird toxicity:** Dietary LC$_{50}$ for mallard ducks >5620 ppm. **Fish toxicity:** LC$_{50}$ (48 h) for carp 59.2 ppm. **Other aquatic toxicity:** LC$_{50}$ (3 h) for *Daphnia similis* >100 ppm. **Effects on beneficial insects:** LC$_{50}$ (contact) to honeybees >25 µg/bee.
Behaviour in soil: Rapidly degraded in soil. **Approved for use in organic farming:** Yes: it has US OMRI approval.

2:204 2-phenylethyl propionate
Plant-derived insecticide/attractant

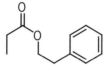

NOMENCLATURE: Approved name: 2-phenylethyl propionate. **CAS RN:** *[122-70-3]*.

SOURCE: Found in a wide variety of plant species, particularly peppermint and peppermint oil.

PRODUCTION: Synthesised for use in crop protection. Some products are derived from plants.

TARGET PESTS: Mosquitoes, dragonflies, biting flies, wasps, ants, earwigs, millipedes and yellow jackets, with some applications for control of Japanese beetles (*Popillia japonica* Newman).

TARGET CROPS: Used in public and animal health and some uses in ornamental crops. Landscaped areas and perimeters around residential structures.

BIOLOGICAL ACTIVITY: Mode of action: 2-Phenylethyl propionate is a strong deterrent of many insects, but may also be used in combinations as an insect attractant.

COMMERCIALISATION: Formulation: Sold as lures, baits and ready-to-use formulations. **Tradenames:** 'Hercon Japanese Beetle Food Lure' (plus eugenol plus geraniol) (Aberdeen Road), 'Ecozap Wasp & Hornet Insecticide' (plus eugenol) and 'Ecozap Crawling and Flying Insecticide' (plus eugenol) (Biorganic), 'Ecopco Jet Wasp/Hornet/Yellow Jacket Contact Insecticide' (plus eugenol), 'Ecopco Acu' (plus eugenol), 'Bioganic Flying Insect Killer' (plus eugenol) and 'Biorganic Granular Insecticide' (plus thymol and eugenol) (Ecosmart), 'Raid Eo Ark' (plus eugenol) (Johnson), 'Bag-a-Bug Japanese Beetle Trap' (plus eugenol plus geraniol) and 'Japanese Beetle Combo Bait (plus eugenol plus geraniol)' (Spectrum), 'Surefire Japanese Beetle Trap' (plus eugenol plus geraniol) (Suterra), 'Japanese Beetle Bait II' (plus eugenol plus geraniol) and 'Trece Japanese Beetle Trap (plus eugenol plus geraniol)' (Trece).

APPLICATION: Apply liberally to areas to be protected from flying and biting insects. Place lures and traps amongst the crops to be protected as soon as the target insects are sighted. Treat paths and rights of way liberally once insect activity is noted. Repeat as necessary.

PRODUCT SPECIFICATIONS: Synthesised for use in crop protection. Some products are derived from plants.

MAMMALIAN TOXICITY: 2-Phenylethyl propionate occurs naturally in a wide range of crops, many of which are used in the culinary and perfume industries. It is not expected that it will have any adverse effects on mammals.

ENVIRONMENTAL IMPACT AND NON-TARGET TOXICITY: 2-Phenylethyl propionate occurs naturally in a wide range of crops, many of which are used in the culinary and perfume industries. It is not expected that it will have any adverse effects on non-target organisms or the environment. **Approved for use in organic farming:** The naturally occurring compound can be used in organic farming.

2:205 plant-derived porphyrin-derivatives
Plant-derived plant growth regulator and nematicide

NOMENCLATURE: Approved name: porphyrin-derivatives. **Other names:** Plant extract 620.

SOURCE: Extracts of plant tissues and, in particular, of *Quercus falcata* Michaux (red oak), *Opuntia lindheimerii* Engelmann (prickly pear cactus), *Rhus aromatica* Aiton (sumac) and *Rhizophora mangle* L. (mangrove).

PRODUCTION: Prepared as crude aqueous extracts of the harvested plants. 'Sincocin' is plant extract 620 with a fatty acid solution. It is blended with balancers and stabilisers for use.

TARGET PESTS: 'Sincocin' and 'Nemastop' are used to control plant parasitic nematodes.

TARGET CROPS: 'Agrispon' is recommended for use as a plant growth regulator in a wide range of crops. 'Sincocin' and 'Nemastop' are used to control plant parasitic nematodes in many food crops and in orchards.

BIOLOGICAL ACTIVITY: Mode of action: It is claimed that the porphyrin-based plant extracts possess cytokinin-like activity and that their application to soil or growing plants promotes plant growth. This is claimed to happen either directly on the crop or indirectly by increasing the population of beneficial micro-organisms that raise the levels of soil nutrients. Nematode effects are thought to be caused by the plant extracts masking the exudates of susceptible plant roots, confusing parasitic nematodes so that they cannot locate the roots. The fatty acids in the formulated product have been shown to suppress female feeding. **Efficacy:** 'Agrispon' and 'Sincocin' are water-based products containing trace minerals and plant extract 620. 'Agrispon' is a plant growth regulator that may be applied to turf and agricultural crops to stimulate root growth and increase a plant's ability to withstand pests and environmental stresses. 'Sincocin' is used to control plant parasitic nematodes by reducing the feeding vigour of nematodes. 'Sincocin' is effective as a preventive nematicide and, when used in rotation with non-biological or biochemical treatments, has given good population control and reduced species that are resistant to knock-down products.

COMMERCIALISATION: Formulation: Sold as liquid concentrate (SL).
Tradenames: 'Agrispon', 'Sincocin' and 'Plant Extract 620' (Agriculture Sciences), 'Nemastop' (Soil Technologies).

APPLICATION: The products are mixed with enough water to cover evenly the desired area at the recommended rate of application and to allow penetration to the growing roots. The maximum recommended application rate for any use pattern should not exceed 150 g of plant extract per hectare (2 oz per acre) per application; the maximum application rate for food crops should not exceed 40 g of plant extract per hectare (0.6 oz per acre) per application. 'Agrispon' is diluted with water to cover evenly the desired area at an application rate of 925 ml per hectare (13 fl oz per acre) for annuals and glasshouses. The recommended timing and frequency of application depend on the plant growth cycle length. A single application is recommended for plants with a growth cycle of 60 days or less. A second application, 45 to 60 days after the first, is recommended for plants with a growing cycle of 60 to 120 days. For long-season plants, or those having a growth cycle longer than 120 days, 'Agrispon' may be applied every 45 to 60 days during the period when the plant is growing vigorously. 'Agrispon' is applied to the soil surface under trees at a rate of 925 ml per hectare (13 fl oz per acre), with an additional 450 ml per hectare (6 fl oz per acre) applied to the tree canopy. For evergreens, applications are recommended every 60 days. Deciduous trees should first be treated at bud break or leaf flush in the spring. Subsequent applications are recommended every 60 days until dormancy occurs. 'Sincocin' is applied to food crops and orchards at a rate of 1.85 litres per hectare (26 fl oz per acre). For both food crops and orchards, the first application should be made during initial root flush, with subsequent applications every 60 days during active growth. The application rate for turf and ornamentals is 25 litres per hectare (2.75 US gallons (87 fl oz) per acre). Golf greens and tee boxes should be treated every day for root pathogens or every 30 days for nematode control. Golf fairways should be treated every 30 days. Ornamentals should first be treated at root flush, with subsequent applications every 30 to 60 days during active growth.

PRODUCT SPECIFICATIONS: Specifications: Prepared as crude aqueous extracts of the harvested plants. 'Sincocin' is plant extract 620 with a fatty acid solution. It is blended with balancers and stabilisers for use. **Purity:** Plant extract from *Quercus falcata*, *Rhus aromatica*, *Rhizophora mangle* and *Opuntia lindheimerii* is the sole active ingredient in the end-use products 'Agrispon' and 'Sincocin'.

COMPATIBILITY: Both 'Agrispon' and 'Sincocin' are compatible with liquid plant protection products and fertilisers.

MAMMALIAN TOXICITY: Plant-derived porphyrins are not considered to be toxic to mammals. **Acute oral LD$_{50}$:** rats >5000 mg/kg. **Acute dermal LD$_{50}$:** rats >5000 mg/kg. **Inhalation:** LC$_{50}$ for rats >2.04 ml/litre. **Skin and eye:** In rabbits, there was minimal primary eye irritation at 0.1 ml. This was reversible within 2 days of treatment. **Toxicity class:** EPA (formulation) IV.

ENVIRONMENTAL IMPACT AND NON-TARGET TOXICITY: Not considered to be hazardous to non-target organisms or to the environment. **Bird toxicity:** LD$_{50}$ for bobwhite quail >2500 mg/kg. **Fish toxicity:** LC$_{50}$ (96 h) for fathead minnow 879.2 ppm ('Agrispon') and 249 ppm ('Sincocin'). **Other aquatic toxicity:** LC$_{50}$ (48 h) for *Daphnia pulex* 53.76 ppm ('Agrispon') and 53 ppm ('Sincocin'). **Effects on beneficial insects:** Not toxic to honeybees. **Behaviour in soil:** Rapidly degraded in soil. **Approved for use in organic farming:** Yes.

2:206 plant essential oils 1

Plant-derived insecticide

NOMENCLATURE: **Approved name:** plant essential oils. **Other names:** Hexa-hydroxyl. **CAS RN:** *[97-53-0]* eugenol; *[8007-46-3]* thyme oil.

SOURCE: Products are a mixture of plant essential oils and are prepared by extraction from harvested plant material.

TARGET PESTS: A wide range of insects, including ants, beetles, caterpillars and bugs.

TARGET CROPS: Recommended for use in landscaped areas and in areas around residential homes and business premises.

BIOLOGICAL ACTIVITY: Mode of action: The mode of action is not completely understood, but it is thought to act as an irritant to the insects and may prevent gas exchange (suffocation) and water loss by covering the insects' bodies.

COMMERCIALISATION: Formulation: Sold as a granular formulation (GR) containing 2.90% eugenol and 0.60% thyme oil as the active ingredients. **Tradenames:** 'Eco Exempt G Granular Insecticide' (EcoSMART). **Patents:** US 6004569. Other US and international patents are pending.

APPLICATION: Apply 1 to 2.5 kg product per 100 square metres (2 to 5 lb per 1000 square feet) for general insect control and apply directly into ants' nests.

MAMMALIAN TOXICITY: The product is considered to be 'Generally Regarded As Safe' (GRAS) and is not considered to be acutely toxic to mammals. **Acute oral LD$_{50}$:** rats >5000 mg/kg. **Acute dermal LD$_{50}$:** Low dermal toxicity. **Inhalation:** Inhalation of the dust may cause irritation of the nasal passages and lungs. **Skin and eye:** Prolonged exposure to this product may cause dry skin, eye irritation and drying of the nasal passages.

ENVIRONMENTAL IMPACT AND NON-TARGET TOXICITY: Specific data regarding toxicity to fish or other aquatic organisms are not available for this product, but it is believed that it should have no adverse effects on non-target organisms or on the environment. **Approved for use in organic farming:** Yes.

2:207 plant essential oils 2
Plant-derived insecticide, fungicide and animal repellent

NOMENCLATURE: Approved name: plant essential oils. **CAS RN:** *[57-06-7]* oil of mustard; *[404-86-4] Capsicum* oleoresin; *[8007-70-3]* oil of anise; *[8001-22-7]* soybean oil; *[8000-48-4]* eucalyptus oil.

SOURCE: Plant essential oils are defined as any volatile oil that gives distinctive odour or flavour to a plant, flower or fruit. They were first registered as pesticide active ingredients in the USA in 1947.

PRODUCTION: Plant essential oils are extracted from the appropriate crop plant (these plants must be fully described for all products in the USA).

TARGET PESTS: Oil of lemongrass is used to repel cats and dogs. Oil of eucalyptus is an effective insecticide and miticide, but is now only used to control fleas. Oil of mustard is used on lawns, flowers, bushes, shade trees and refuse containers. Soybean oil is used as an insecticide and miticide. Oil of anise is used to repel cats and dogs.

TARGET CROPS: Oil of lemongrass was first registered in the USA in 1962 for use on ornamentals, shade trees, patio furniture and garbage cans. Oil of eucalyptus was first registered in the USA in 1948 and is now used as a herbal flea collar for pets. Oil of mustard (allyl isothiocyanate) was first registered in the USA in 1962 as a dog repellent and is used outdoors either to repel cats and dogs or to kill insects. Soybean oil was first registered in the USA in 1959 for use on citrus fruits and a variety of ornamentals. Oil of anise was first registered in the USA in 1952 for use as a liquid spray for use on soil near lawns, gardens and flower beds.

BIOLOGICAL ACTIVITY: Mode of action: The mode of action of plant essential oils is not completely understood. It is believed that the animal repellent properties are associated with the strong odours associated with these oils and the subsequent irritant effect. Insect repellency is similarly poorly understood, but they are thought to act as irritants to the insects. Insect death may be a result of the oils covering the insects' bodies and thereby preventing gas exchange (leading to suffocation) and water loss.

COMMERCIALISATION: Formulation: Plant essential oils are formulated as liquid sprays, crystals and pellets. **Tradenames:** 'Bugitol', 'Nemitol' and 'Knokitol' (mixed plant oils) (Champon), 'Golden Pest Spray Oil' (soybean oil) (Stoller), 'Poison-Free Ant Killer' (mint oil), 'Outdoor animal repellent' (mustard oil) and 'Pesticidal Spray Oil' (canola oil) (Woodstream).

APPLICATION: To repel animals, apply freely around the area to be protected. Repeat after heavy rain. As an insecticide and miticide, apply diluted product to the target pests ensuring that the whole plant is covered.

PRODUCT SPECIFICATIONS: Plant essential oils are extracted from the appropriate crop plant (these plants must be fully described for all products in the USA).

MAMMALIAN TOXICITY: All products consist of plant essential oils that are used in foodstuffs. As a consequence, all products are considered to be 'Generally Regarded As Safe' (GRAS) and are not considered to be toxic to mammals.

ENVIRONMENTAL IMPACT AND NON-TARGET TOXICITY: All products are derived from the essential oils of a variety of different plants and are not considered to be a danger to non-target organisms or to the environment. **Approved for use in organic farming:** Yes.

2:208 poly-D-glucosamine
Crustacean-derived plant defence booster

NOMENCLATURE: Approved name: poly-D-glucosamine. **Other names:** Chitosan.
CAS RN: *[9012-76-4]*.

SOURCE: Derived from crustacean exoskeletons.

PRODUCTION: Produced from dried, crushed crustacean exoskeletons.

TARGET PESTS: Powdery mildews (various species) and *Botrytis* species.

TARGET CROPS: Fruit, such as grapes, strawberries, cherries and apples. Vegetables, such as cucumbers, squash, pumpkins, peas, peppers and tomatoes. Glasshouse and nursery plants, such as flowers (roses, etc.), potted plants and protected vegetables

BIOLOGICAL ACTIVITY: Mode of action: 'Elexa 4' is an elicitor that stimulates the natural defence response system in treated plants. Poly-D-glucosamine, the active ingredient in 'Elexa 4', binds to fungal receptor sites, mimicking an attack by fungal spores. This, in turn, results in signals being sent to the nuclei of the plant cells. These signals elicit multiple genetic and biological responses, including the production of phytoalexins (anti-microbial compounds produced in plants), aimed at inhibiting fungal infections.

COMMERCIALISATION: Formulation: 'Elexa 4' is a 4% aqueous suspension.
Tradenames: 'Elexa 4' (SafeScience). **Patents:** US 6060429 and US 5965545.

APPLICATION: Depending on disease pressure and canopy size, 'Elexa 4' can be applied at varying rates. For lower spray volumes, use 1 part 'Elexa 4' to 19 parts water (a final tank-mix ratio of 1:20). For very high spray volumes up to 900 litres per hectare (100 gallons per acre), use 1 part 'Elexa 4' to 39 parts water (a final tank-mix ratio of 1:40). 'Elexa 4' is recommended for use as an early- to mid-season treatment when disease pressure is light to moderate. The use of a non-ionic commercial surfactant (one that is cleared for food use) is highly recommended. Apply 'Elexa 4' on a 7- to 14-day schedule, depending on disease pressure.

PRODUCT SPECIFICATIONS: Produced from dried, crushed crustacean exoskeletons.

COMPATIBILITY: 'Elexa 4' is very effective when used as part of an IPM programme in rotation with conventional fungicides such as sulfur, sterol biosynthesis inhibitors, strobilurin chemistry-based products and biorational products.

MAMMALIAN TOXICITY: 'Elexa 4' has an excellent toxicity profile. It is an 'Exempt from Tolerance' product. 'Elexa 4' is registered by the US-EPA as a 'reduced risk' pesticide and carries a 'Caution' signal word. Its Pre-Harvest Interval (PHI) is 0 days, meaning that it may be used up to and including the day of harvest. 'Elexa 4' has a Restricted Entry Interval (REI) of 4 hours.

ENVIRONMENTAL IMPACT AND NON-TARGET TOXICITY: 'Elexa 4' has not shown any adverse effects on non-target organisms, nor does it accumulate in the environment. **Approved for use in organic farming:** Yes.

2:209 polynactins

Micro-organism-derived acaricide

The Pesticide Manual **Fifteenth Edition entry number:** 693

dinactin: $R_1, R_3 = CH_3-; R_2, R_4 = CH_3CH_2-$
trinactin: $R_1 = CH_3-; R_2, R_3 R_4 = CH_3CH_2-$
tetranactin: $R_1, R_2, R_3, R_4 = CH_3CH_2-$

NOMENCLATURE: Approved name: polynactins. **Other names:** Tetranactin; trinactin; dinactin. **CAS RN:** *[35396-61-5]* tetranactin; *[7561-71-9]* trinactin; *[20261-85-2]* dinactin; *[39285-04-6]* polynactin (formerly *[56485-45-1]* and *[56573-63-8]*).

SOURCE: Secondary metabolites from the Actinomycete, *Streptomyces aureus* Waksman & Henrici isolate S-3466. A mixture of tetranactin, trinactin and dinactin.

PRODUCTION: Manufactured by fermentation.

TARGET PESTS: Spider mites, such as carmine spider mite (*Tetranychus cinnabarinus* (Boisduval)), two-spotted mite (*T. urticae* (Koch)) and European red mite (*Panonychus ulmi* Koch).

TARGET CROPS: Orchard fruit trees.

BIOLOGICAL ACTIVITY: Mode of action: The polynactins are very effective at controlling spider mites under wet conditions. It is thought that the mode of action is through a leakage of basic cations (such as potassium ions) through the lipid layer of the membrane in the mitochondrion. Water is considered to be an essential component of this toxic effect by either assisting penetration or accelerating ion leakage. **Efficacy:** Following application of polynactins to apple trees, the proliferation of mites is stopped for a period of 32 days or more. The polynactins are usually sold as combination products to avoid the possibility of the development of resistance. **Key reference(s):** 1) K Ando, H Oishi, S Hirano, T Okutomi, K Suzuki, H Okazaki, M Sawanda & T Sagawa. 1971. Tetranactin, a new miticidal antibiotic. I. Isolation, characterisation and properties of tetranactin, *J. Antibiot.*, **24**, 347. 2) K Ando, T Sagawa, H Oishi, K Suzuki & T Nawata. 1974. Tetranactin, a pesticidal antibiotic, *Proc. 1st Intersect. Congr. IAMS (Sci. Counc. Jpn)*, **3**, 630.

COMMERCIALISATION: Formulation: Sold as emulsifiable concentrates (EC) mixed with other acaricides. **Tradenames:** 'Mitecidin' (plus fenobucarb) and 'Mitedown' (plus fenbutatin oxide) (Eikou Kasei).

APPLICATION: Applied as a foliar spray when conditions are wet. Polynactins should not be used under very dry conditions.

PRODUCT SPECIFICATIONS: Manufactured by fermentation.

COMPATIBILITY: Compatible with most other chemical acaricides. Ineffective under dry conditions.

MAMMALIAN TOXICITY: The polynactins are not considered to be toxic to mammals. **Acute oral LD$_{50}$:** mice >15 000 mg/kg. **Acute dermal LD$_{50}$:** mice >10 000 mg/kg. **Skin and eye:** Mild skin and eye irritant. **Toxicity class:** EPA (formulation) IV.

ENVIRONMENTAL IMPACT AND NON-TARGET TOXICITY: Relatively non-toxic to beneficial insects. **Fish toxicity:** High toxicity to fish, with a median tolerance limit to carp of 0.003 ppm.

2:210 poly-N-acetyl-D-glucosamine
Crustacean-derived nematicide

NOMENCLATURE: **Approved name:** poly-N-acetyl-D-glucosamine. **Other names:** Chitin. **CAS RN:** *[1398-61-4]*.

SOURCE: Poly-N-acetyl-D-glucosamine is present in the shells of all crustaceans and insects and in certain other organisms, including many fungi, algae and yeasts.

PRODUCTION: Commercially, poly-N-acetyl-D-glucosamine is isolated from the shells of crustaceans, after the edible parts have been removed.

TARGET PESTS: Plant pathogenic nematodes.

TARGET CROPS: Most field crops, ornamentals and turf grown in fields, home gardens and nurseries.

BIOLOGICAL ACTIVITY: **Mode of action:** Poly-N-acetyl-D-glucosamine appears to control plant pathogenic nematodes by stimulating the growth of certain naturally occurring micro-organisms in soil, which, in turn, release chemicals that kill the pathogenic nematodes and their eggs.

COMMERCIALISATION: **Formulation:** 'Clandosan' is formulated as pellets and granules. **Tradenames:** 'Clandosan' (Igene).

APPLICATION: 'Clandosan' is applied in the field 2 to 4 weeks before planting, so that it is concentrated 15 to 25 cm (6 to 10 inches) below the soil surface. For use in glasshouses and nurseries, the product is mixed with soil. After planting, 'Clandosan' can be used on short grasses, such as on lawns, turf and golf courses.

PRODUCT SPECIFICATIONS: Commercially, poly-N-acetyl-D-glucosamine is isolated from the shells of crustaceans, after the edible parts have been removed.

MAMMALIAN TOXICITY: No risks to humans are expected when products containing chitin are used according to label directions. Chitin is closely related structurally to the active ingredient chitosan (poly-D-glucosamine), which shows no toxicity to mammals and is approved by the Food and Drug Administration (FDA) as a food additive.

ENVIRONMENTAL IMPACT AND NON-TARGET TOXICITY: Risks to the environment are not expected, because a structurally similar active ingredient has not shown any toxicity and because chitin is abundant in Nature. **Approved for use in organic farming:** Yes.

2:211 polyoxin B

Micro-organism-derived fungicide

The Pesticide Manual **Fifteenth Edition entry number:** 694

NOMENCLATURE: Approved name: polyoxin B. **CAS RN:** *[19396-06-6]* polyoxin B; *[11113-80-7]* polyoxins.

SOURCE: Isolated from the soil Actinomycete, *Streptomyces cacaoi* var. *asoensis* Isono *et al.* Polyoxin B was first isolated by K Isono *et al.* in 1965 and was introduced by Hokko Chemical Industry Company Ltd, Kaken Pharmaceutical Company Ltd, Kumiai Chemical Industry Company Ltd and Nihon Nohyaku Company Ltd.

PRODUCTION: Polyoxin B is produced by fermentation of *S. cacaoi* var. *asoensis.*

TARGET PESTS: Various plant pathogenic fungi, such as *Sphaerotheca* spp. and other powdery mildews, *Botrytis cinerea* Pers., *Sclerotinia sclerotiorum* De Bary, *Corynespora melonis* Lindau, *Cochliobolus miyabeanus* Drechs. and *Alternaria alternata* (Fr.) Keissl. and other *Alternaria* species.

TARGET CROPS: Vines, apples, pears, vegetables and ornamentals.

BIOLOGICAL ACTIVITY: Mode of action: Polyoxins cause a marked abnormal swelling on the germ tubes of spores and hyphal tips of the pathogen, rendering the treated fungus non-pathogenic. In addition, the incorporation of $[^{14}C]$-glucosamine into cell-wall chitin of *Cochliobolus miyabeanus* was inhibited. It is suggested that polyoxins exert their effects through an inhibition of cell wall biosynthesis. They are systemic fungicides with protective action. **Efficacy:** Polyoxin B is effective at controlling a variety of fungal pathogens, but is ineffective against bacteria and yeasts. It is particularly effective against pear black spot and apple cork spot (*Alternaria* spp.), grey moulds (*Botrytis cinerea*) and other sclerotia-forming plant pathogens. Resistance to polyoxin B has been found in *Alternaria alternata* in some orchards in Japan, following intensive treatment. This resistance is due to lowered penetration of polyoxin B through the fungal cell membrane and, therefore, to the site of action. **Key reference(s):** 1) S Suzuki, K Isono, J Nagatsu, T Mizutani, Y Kawashima & T Mizuno. 1965. A new antibiotic, polyoxin A, *J. Antibiot. Ser. A*, **20**, 109. 2) K Isono, J Nagatsu, Y Kawashima & S Suzuki. 1965. Studies on polyoxins, antifungal antibiotics. Part 1. Isolation and characterisation of polyoxins A and B, *J. Antibiot. Ser. A*, **18**, 115. 3) J Eguchi, S Sasaki, N Ohta, T Akashiba, T Tsuchiyama & S Suzuki. 1968. Studies on polyoxins, antifungal antibiotics. Mechanism of action on the diseases caused by *Alternaria* spp., *Ann. Phytopathol. Soc. Jpn*, **34**, 280.

COMMERCIALISATION: Formulation: Sold as wettable powder (WP), emulsifiable concentrate (EC) and soluble granule (SG) formulations. **Tradenames:** 'Polyoxin AL' (Kaken), (Kumiai), (Nihon Nohyaku) and (Hokko), 'Polybelin' (plus iminoctadine tris(albesilate)) (Kumiai). **Patents:** JP 577960.

APPLICATION: Applied as foliar sprays when disease becomes evident. Rates up to 200 g per hectolitre are recommended.

PRODUCT SPECIFICATIONS: Polyoxin B is produced by fermentation of *S. cacaoi* var. *asoensis*. **Purity:** Polyoxin complex consists of polyoxin B and several other polyoxins of lower potency.

COMPATIBILITY: Incompatible with alkaline products. May be mixed with various other fungicides with different modes of action in order to reduce or delay the onset of resistance.

MAMMALIAN TOXICITY: Polyoxin B is not considered to be toxic to mammals. **Acute oral LD$_{50}$:** male rats 21 000, female rats 21 200, male mice 27 300, female mice 22 500 mg/kg. **Acute dermal LD$_{50}$:** rats >2000 mg/kg. **Inhalation:** LC$_{50}$ (6 h) for rats 10 mg/litre. **Skin and eye:** Non-irritant to mucous membranes and skin (rats). **NOEL:** (2 years) for rats and mice >48 000 mg/kg diet. **Toxicity class:** EPA (WP formulation) IV.

ENVIRONMENTAL IMPACT AND NON-TARGET TOXICITY: Polyoxin B is not expected to have any adverse effects on non-target organisms or on the environment, because of the

specific mode of action against chitin biosynthesis and inactivity against crustaceans.
Fish toxicity: LC_{50} (48 h) for carp >40 mg/litre. Japanese killifish were unaffected by 100 mg/litre for 72 hours. **Other aquatic toxicity:** LC_{50} (3 h) for *Daphnia pulex* >40 mg/litre. LC_{50} (3 h) for water flea (*Moina macrocopa*) >40 mg/litre. EC_{50} (72 h) for algae (*Selenastrum capricornutum*) >100 mg/litre. **Behaviour in soil:** In upland conditions at 25 °C, DT_{50} <2 days (two soils, o.c. 6.2%, pH 6.3, moisture 23.3% and o.c. 1.1%, pH 6.8, moisture 63.6%). In water, DT_{50} 15 days (pH 7.0, 20 °C) and 4.2 days (pH 9.0, 35 °C).

2:212 polyoxorim
Micro-organism-derived fungicide

The Pesticide Manual Fifteenth Edition entry number: 694

NOMENCLATURE: Approved name: polyoxorim (BSI, pa ISO); polyoxin D (JMAF).
CAS RN: *[22976-86-9]* polyoxorim; *[146659-78-1]* zinc salt of polyoxorim; *[11113-80-7]* polyoxins.

SOURCE: Polyoxorim (polyoxin D) was isolated by S Suzuki *et al.* in 1965. It is one of the secondary metabolites produced by the fermentation of the soil Actinomycete, *Streptomyces cacaoi* var. *asoensis*. The zinc salt was introduced as a fungicide by Kaken Pharmaceutical Company Ltd, Kumiai Chemical Industry Company Ltd and Nihon Nohyaku Company Ltd.

PRODUCTION: Polyoxorim is produced from the fermentation of *S. cacaoi* var. *asoensis*. It is isolated as the zinc salt.

TARGET PESTS: Used for the control of rice sheath blight (*Rhizoctonia solani* Kuehn (*Pellicularia sasakii* Ito)). Also effective against apple and pear canker (*Nectria galligena* Bres. (*Diplodia pseudodiplodia* Fckl.)) and *Drechslera* spp., *Bipolaris* spp., *Curvularia* spp. and *Helminthosporium* spp.

TARGET CROPS: The major use is in rice, but it also has applications in pome fruit and for disease control in turf.

BIOLOGICAL ACTIVITY: Mode of action: Polyoxins cause a marked abnormal swelling on the germ tubes of spores and hyphal tips of the pathogen, rendering the treated fungus non-pathogenic. In addition, the incorporation of [^{14}C]-glucosamine into cell-wall chitin of *Cochliobolus miyabeanus* Drechs. was inhibited. This suggested that polyoxorim exerts its effect by disruption of cell wall biosynthesis by mimicking UDP-*N*-acetylglucosamine, the natural substrate for the enzyme, chitin synthase. The polyoxins are systemic fungicides with protective action. **Efficacy:** Polyoxorim has good activity against rice sheath blight when applied as a foliar spray, at rates of about 200 g per hectolitre. High volume sprays on turf are effective at controlling many fungal pathogens, with the exception of members of the Phycomycetes (no chitin in the cell wall). Polyoxorim can be used to control apple and pear canker when applied as a paste. It is ineffective against bacteria and yeasts.

Key reference(s): 1) K Isono & S Suzuki. 1979. The polyoxins: pyrimidine nucleoside peptide antibiotics inhibiting cell wall biosynthesis, *Heterocycles*, **13**, 333. 2) K Isono, J Nagatsu, K Kobinata, K Sasaki & S Suzuki. 1967. Studies on polyoxins, antifungal antibiotics. Part V. Isolation and characterisation of polyoxins, C, D, E, F, G, H and I, *Agric. Biol. Chem.*, **31**, 190.

COMMERCIALISATION: Formulation: Sold as wettable powder (WP) and paste (PA) formulations. **Tradenames:** 'Endorse', 'Polyoxin Z' (zinc salt) and 'Stopit' (Kaken). **Patents:** JP 577960.

APPLICATION: Applied as the zinc salt at rates of 200 g a.i. per hectare for the control of sheath blight. Applications should be made when disease first appears or when conditions favour the onset of disease.

PRODUCT SPECIFICATIONS: Specifications: Polyoxorim is produced from the fermentation of *S. cacaoi* var. *asoensis*. It is isolated as the zinc salt. **Purity:** The product contains a mixture of polyoxins, of which polyoxorim is the main fraction.

COMPATIBILITY: Polyoxorim should not be used with alkaline materials. It is sold in admixture with other fungicides with different modes of action, in order to delay the possibility of resistance developing. It has not shown any phytotoxicity to crops at rates above those recommended for use.

MAMMALIAN TOXICITY: The mode of action of polyoxorim means that it is unlikely to have any effects on mammals. **Acute oral LD$_{50}$:** male and female rats >9600 mg/kg. **Acute dermal LD$_{50}$:** rats >750 mg/kg. **Inhalation:** LC$_{50}$ (4 h) for male rats 2.44, female rats

2.17 mg/litre. **Skin and eye:** Non-irritant to mucous membranes and skin (rats).
NOEL: (2 years) for rats >50 000, mice >40 000 mg/kg diet. **Toxicity class:** EPA
(WP formulation) III.

ENVIRONMENTAL IMPACT AND NON-TARGET TOXICITY: Polyoxorim is unlikely
to produce any adverse effects on non-target organisms or on the environment because
of its mode of action and inactivity against crustacea. **Bird toxicity:** LD_{50} for mallard ducks
>2150 mg/kg. **Fish toxicity:** LC_{50} (96 h) for carp >100 mg/litre. **Other aquatic toxicity:** LC_{50}
(3 h) for *Daphnia pulex* >40 mg/litre. EC_{50} (72 h) for algae (*Selenastrum capricornutum*)
>100 mg/litre. LC_{50} (3 h) for water flea (*Moina macrocopa*) >40 mg/litre.
Effects on beneficial insects: LD_{50} (96 h) for honeybees >28.774 µg/bee. Tests indicate it is
unlikely to present an increased risk to terrestrial insects. **Behaviour in soil:** In flooded soils
at 25 °C, the DT_{50} of polyoxorim is <10 days. In upland soils at 25 °C, the DT_{50} is <7 days. In
water at pH 5.5 at 24 °C, the DT_{50} is 4 hours and, at pH 5.8 at 26.5 °C, it is 8 hours.

2:213 pyrethrins (pyrethrum)
Plant-derived insecticide and acaricide

The Pesticide Manual **Fifteenth Edition entry number:** 737

R = -CH$_3$ (chrysanthemates) or -CO$_2$CH$_3$ (pyrethrates)
R$_1$ = -CH=CH$_2$ (pyrethrin) or -CH$_3$ (cinerin) or -CH$_2$CH$_3$ (jasmolin)

NOMENCLATURE: Approved name: pyrethrins (BSI, E-ISO, ESA, JMAF); pyrèthres (F-ISO).
CAS RN: *[8003-34-7]*.

SOURCE: The dried, powdered flower of *Chrysanthemum cinerariaefolium* Vis. has been used
as an insecticide from ancient times. Recent taxonomic revisions have transferred this species
from the genus *Chrysanthemum* to the genus *Tanacetum*. In some circles, it is still recorded
as *Pyrethrum*. The species was identified in antiquity in China. It spread west via Iran (Persia),

probably via the Silk Routes in the Middle Ages. The dried, powdered flower heads were known as 'Persian Insect Powder'. Records of use date from the early 19th century when it was introduced to the Adriatic coastal regions of Croatia (Dalmatia) and some parts of the Caucasus. Subsequently, it was grown in France, the USA and Japan. It is now widely grown in East African countries, especially Kenya (1930), in Ecuador and Papua New Guinea (1950) and in Australia (1980).

PRODUCTION: Pyrethrum is extracted from the flower of *Tanacetum cinerariaefolium* (Trev.) Schultz-Bip. (syns. *Chrysanthemum cinerariaefolium* Vis., *Pyrethrum cinerariaefolium* Trev.). The extract is refined using methanol (Pyrethrum Board of Kenya and MGK) or supercritical carbon dioxide (Botanical Resources and Agropharm).

TARGET PESTS: Control of a wide range of insects and mites.

TARGET CROPS: Recommended for use on fruit, vegetables, field crops, ornamentals, glasshouse crops and house plants, as well as in public health, stored products, animal houses and on domestic and farm animals.

BIOLOGICAL ACTIVITY: Mode of action: Pyrethrins have been shown to bind to the sodium channels in insects, prolonging their opening and thereby causing knockdown and death. They are non-systemic insecticides with contact action. Initial effects include paralysis, with death occurring later. They have some acaricidal activity. **Key reference(s):** 1) J E Casida & G B Quistad (eds). 1994. *Pyrethrum Flowers; Production, Chemistry, Toxicology and Uses*, Oxford University Press, Oxford, UK. 2) J E Casida (ed.). 1973. *Pyrethrum, the Natural Insecticide*, Academic Press, New York, USA. 3) C B Gnadinger. 1936. In *Pyrethrum Flowers*, 2nd edn, McLaughlin Gormley King Co., Minneapolis, Minnesota, USA.

COMMERCIALISATION: Formulation: Formulated as aerosol dispensers (AE), dispersible powders (DP), emulsifiable concentrates (EC), fogging concentrates, pressurised liquid CO_2, wettable powders (WP) and ultra-low volume liquids (UL). **Tradenames:** 'Alfadex' (Syngenta), 'Evergreen', 'Pyrocide', 'Premium Pyganic 175', 'Pyganic Crop Protection EC 1.4' and 'Pyganic Crop Protection EC 5.0' (MGK), 'Pyronyl' (mixture with piperonyl butoxide), 'Prentox Pyrethrum Extract' and 'ExciteR' (Prentiss), 'Milon' (Frunol Delicia), 'Pycon' (concentrated mixture with piperonyl butoxide) (Agropharm), 'Hash' (plus piperonyl butoxide) (Kemio), 'CheckOut' (Consep), 'Py-rin Growers Spray' (Wilbur-Ellis), 'Diatect II', 'Diatect III' and 'Diatect V' (all plus piperonyl butoxide and silica) (Diatect), 'Bonide Liquid Rotenone/Pyrethrin Spray' (rotenone 1.1% plus pyrethrum 0.8%), 'Garden Dust' (pyrethrins 0.03%, rotenone 0.5%, cube resins 1%, copper 5% and sulfur 25%) and 'Earth Friendly Fruit Tree Spray/Dust' (rotenone plus pyrethrum plus sulfur plus copper) (Bonide), 'Safer Yard and Garden Insect Killer' (pyrethrum plus insecticidal soap) (Woodstream), 'Ecozone Pyrethrum Insect Powder' (Natural Animal Health Products).

APPLICATION: Normally applied in combination with synergists, e.g. piperonyl butoxide, which inhibit detoxification. Good cover of crop foliage is essential for effective control. Many combination products with other insecticides are available.

PRODUCT SPECIFICATIONS: Specifications: Pyrethrum is extracted from the flower of *Tanacetum cinerariaefolium* (Trev.) Schultz-Bip. (syns. *Chrysanthemum cinerariaefolium* Vis., *Pyrethrum cinerariaefolium* Trev.). The extract is refined using methanol (Pyrethrum Board of Kenya and MGK) or supercritical carbon dioxide (Botanical Resources and Agropharm). **Purity:** Pyrethrum extract is defined as a mixture of three naturally occurring, closely related insecticidal esters of chrysanthemic acid, pyrethrins I, and the three corresponding esters of pyrethrin acid, pyrethrins II. In the USA, it is standardised as 45% to 55% w/w total pyrethrins, but samples may be 20%; the ratio of pyrethrins I to II is typically 0.8–2.8:1; the ratio of individual esters (pyrethrins: cinerins: jasmolins) is 71:21:7. In Europe, pyrethrum extract is 25±0.5% pyrethrins. The three components of pyrethrins I are pyrethrin I ($R = CH_3$, $R_1 = CH=CH_2$); jasmolin I ($R = CH_3$, $R_1 = CH_2CH_3$); cinerin I ($R = CH_3$, $R_1 = CH_3$); the components of pyrethrins II correspond ($R = CO_2CH_3$).

COMPATIBILITY: Incompatible with alkaline substances.

MAMMALIAN TOXICITY: Not considered to be toxic to mammals. **Acute oral LD_{50}:** male rats 3920, female rats 1280, mice 273–796 mg/kg. **Acute dermal LD_{50}:** rats >1500, rabbits >2000 mg/kg. **Inhalation:** LC_{50} (4 h) for rats 3.4 mg/litre. **Skin and eye:** Slightly irritating to skin and eyes. Constituents of the flowers may cause dermatitis to sensitised individuals, but are removed during the preparation of refined extracts. **NOEL:** for rats 100 ppm. **ADI:** 0.04 mg/kg b.w. [1972]. **Toxicity class:** WHO (a.i.) II; EPA (formulation) III.

Other toxicological effects: There is no evidence that synergists increase toxicity of pyrethrins to mammals. **Toxicity review:** 1) *Pesticide Residues in Food.* FAO Agricultural Studies, No. 90. 2) WHO Technical Report Series, No. 525. 1973. *1972 Evaluations of Some Pesticide Residues in Food.* AGP:1972/M/9/1. 3) WHO Pesticide Residue Series No. 2. 1973.

ENVIRONMENTAL IMPACT AND NON-TARGET TOXICITY: Bird toxicity: Acute oral LD_{50} for mallard ducks >10 000, bobwhite quail >2000 mg a.i./kg. **Fish toxicity:** Highly toxic to fish. LC_{50} (96 h) (static tests) for coho salmon 39, channel catfish 114 mg/litre. LC_{50} for bluegill sunfish 10, rainbow trout 5.2 mg/litre. **Other aquatic toxicity:** LC_{50} for *Daphnia pulex* 12 µg/litre. **Effects on beneficial insects:** Highly toxic to honeybees, but exhibits a repellent effect. LD_{50} (oral) 22 ng/bee, (contact) 130 to 290 ng/bee. **Metabolism:** In mammals, rapidly degraded in the stomach by hydrolysis of the ester bond to harmless metabolites.

Behaviour in soil: In the environment, degradation, promoted by sunlight and u.v. light, begins at the alcohol group and involves the formation of numerous unknown cleavage products. **Approved for use in organic farming:** Yes.

2:214 pyrethrins (chrysanthemates)

Plant-derived insecticide and acaricide

The Pesticide Manual Fifteenth Edition entry number: 737

R = -CH$_3$
R$_1$ = -CH=CH$_2$ or -CH$_3$ or -CH$_2$CH$_3$

NOMENCLATURE: Approved name: pyrethrins (chrysanthemates). CAS RN: *[121-21-1]* pyrethrin I; *[25402-06-6]* cinerin I; *[4466-14-2]* jasmolin I.

BIOLOGICAL ACTIVITY: Mode of action: See entry 2:213 – pyrethrins (pyrethrum).

PRODUCT SPECIFICATIONS: Purity: The three components of pyrethrins I are pyrethrin I (R = CH$_3$, R$_1$ = CH=CH$_2$); jasmolin I (R = CH$_3$, R$_1$ = CH$_2$CH$_3$); cinerin I (R = CH$_3$, R$_1$ = CH$_3$).

Approved for use in organic farming: Yes.

2:215 pyrethrins (pyrethrates)

Plant-derived insecticide and acaricide

The Pesticide Manual **Fifteenth Edition entry number:** 737

R = -CO₂CH₃

R = -CO$_2$CH$_3$
R$_1$ = -CH=CH$_2$ or -CH$_3$ or -CH$_2$CH$_3$

NOMENCLATURE: Approved name: pyrethrins (pyrethrates). **CAS RN:** *[121-29-9]* pyrethrin II; *[121-20-0]* cinerin II; *[1172-63-0]* jasmolin II.

BIOLOGICAL ACTIVITY: Mode of action: See entry 2:213 – pyrethrins (pyrethrum).

PRODUCT SPECIFICATIONS: Purity: The three components of pyrethrins II are pyrethrin II (R = CO$_2$CH$_3$, R$_1$ = CH=CH$_2$); jasmolin II (R = CO$_2$CH$_3$, R$_1$ = CH$_2$CH$_3$); cinerin II (R = CO$_2$CH$_3$, R$_1$ = CH$_3$).

Approved for use in organic farming: Yes.

2:216 rosemary oil

Plant-derived natural insecticide

NOMENCLATURE: Approved name: rosemary oil. **CAS RN:** *[8000-25-7]*.

SOURCE: A component of the herb rosemary (*Rosmarinus officinalis* L.).

PRODUCTION: Extracted from rosemary plants.

TARGET PESTS: 'Sporan' is recommended for control of a wide range of plant pathogens, 'EcoEXEMPT' is effective against many pests both indoors and outside, including aphids, ants, bees, centipedes, cockroaches, crickets, beetles, fleas, flies, mites, millipedes, spiders, ticks and wasps. 'EcoTrol' is used against a wide range of agricultural pests.

TARGET CROPS: 'Sporan' is recommended for use in a wide range of crops including vegetables, soft fruit and nut and fruit trees. 'EcoEXEMPT' is used in a wide range of horticultural and garden crops and in and around domestic property. 'EcoTrol' is used in fruit, nuts and vegetables.

BIOLOGICAL ACTIVITY: Mode of action: Rosemary oil has been shown to kill growing fungal hyphae and to prevent spores from germinating. The insecticidal mode of action of rosemary oil is not completely understood, but it is thought to act as an irritant to the insects and may prevent gas exchange (suffocation) and water loss by covering the insects' bodies.

COMMERCIALISATION: Formulation: 'Sporan' is an oil concentrate (17.6% rosemary oil and 82.4% wintergreen oil), 'EcoEXEMPT' is a liquid emulsifiable concentrate (EC) formulation (10% rosemary oil plus 10% to 70% oil of wintergreen and 10% to 70% mineral oil) and 'EcoTrol' is an EC formulation containing 10% rosemary oil. 'HEXACIDE' contains 5.0% rosemary oil plus the inert ingredients 5% to 60% oil of wintergreen, 5% to 50% mineral oil, 25% to 80% water and 0.1% to 5% lecithin. **Tradenames:** 'Sporan', 'EcoEXEMPT', 'EcoTrol' and 'HEXACIDE' (EcoSMART), 'Bioganic Crawling Insect Killer' (5% rosemary oil) (Bioganic).

APPLICATION: 'Sporan' should be applied at high volume (with added surfactant), at a rate of 2.5 to 10 litres of product per hectare (0.3 to 1 gallon per acre). For indoor use, apply as a space spray through conventional fogging equipment by mixing 100 to 300 g (4 to 10 oz) of 'EcoEXEMPT' with sufficient oil to equal 4.5 litres (1 gallon) of diluted spray. Apply at a rate of 30 to 90 ml (1 to 3 fl oz) per 28 000 litres (1000 cubic feet), filling the room with mist (or fog, if thermal equipment is used). Keep area closed for at least 15 minutes. Vacate treated area. Ventilate before reoccupying. Retreat if re-infestation occurs. For outdoor use, using compressed air sprayers, dilute 100 to 250 ml (4 to 8 fl oz) of 'EcoEXEMPT' per 4.5 litres (gallon) of water and apply at the rate of 9 litres (2 gallons) per 100 square metres (1000 square feet) or until area is thoroughly wet. Also apply a band of soil and vegetation 2 to 3 metres (6 to 10 feet) wide around and adjacent to the building. Treat the building foundation to a height of about 1 metre (3 feet) with sufficient water for coverage. For power sprayers, mix approximately 50 to 200 ml (2 to 6 fl oz) per 4.5 litres (gallon) of water and apply until area is sufficiently covered. 'EcoTrol' should be applied in relatively high volume sprays at a rate of 0.75 to 6 litres per hectare (8 to 64 fl oz per acre).

PRODUCT SPECIFICATIONS: Specifications: Extracted from rosemary plants.
Purity: 'EcoEXEMPT' contains 10% rosemary oil plus 10% to 70% oil of wintergreen and 10% to 70% mineral oil.

COMPATIBILITY: May be applied with other insecticides to increase the speed of insect knockdown.

MAMMALIAN TOXICITY: Not considered to be toxic to mammals. **Acute oral LD$_{50}$:** rats 5000 mg/kg. **Acute dermal LD$_{50}$:** rabbits 10 000 mg/kg. **Inhalation:** May be irritating to some individuals. **Skin and eye:** Moderate skin and eye irritation to rabbits at 500 mg/kg for 24 hours. **Other toxicological effects:** Prolonged exposure to this product may cause skin irritation, eye/nasal irritation, dizziness, headache or nausea. The State of California requires that any exempt product containing 8.5% or more of rosemary oil must, at a minimum, bear the signal word 'Caution', the phrase 'Keep Out of Reach of Children', appropriate precautionary language and a requirement for protective eyewear and gloves.

ENVIRONMENTAL IMPACT AND NON-TARGET TOXICITY: Data on effects on wildlife and aquatic organisms have not been determined, but it is not thought that rosemary oil will have any effect on non-target organisms. **Approved for use in organic farming:** Yes: it has US OMRI approval.

2:217 rosemary and clove oils
Plant-derived fungicide

NOMENCLATURE: Approved name: rosemary and clove oils. **Other names:** Plant essential oils. **CAS RN:** *[8000-25-7]* rosemary oil; *[8000-34-8]* clove oil.

SOURCE: The product is a mixture of plant essential oils and is prepared by extraction from harvested plant material.

PRODUCTION: Plant essential oils are extracted from the appropriate crop plant (these plants must be fully described for all products in the USA).

TARGET PESTS: Powdery mildews and bacterial spot.

TARGET CROPS: 'Phyta-Guard EC' can be used on a very wide range of crops.

BIOLOGICAL ACTIVITY: Mode of action: The mode of action is not completely understood, but it is likely that the combination of oils prevents the powdery mildew pathogens from penetrating the cuticles of the treated plants. Membrane disruption of the germinating mildew spore is also thought to be involved.

COMMERCIALISATION: Formulation: Sold as an emulsifiable concentrate (EC). **Tradenames:** 'Phyta-Guard EC' (Biocontrol Network).

APPLICATION: The product should be sprayed, as needed, onto the infected plants to cover the entire leaf surface as soon as disease is noticed.

PRODUCT SPECIFICATIONS: Plant essential oils are extracted from the appropriate crop plant (these plants must be fully described for all products in the USA).

MAMMALIAN TOXICITY: The product consists of plant essential oils that are used in foodstuffs. As a consequence, it is considered to be 'Generally Regarded As Safe' (GRAS) and is not considered to be toxic to mammals.

ENVIRONMENTAL IMPACT AND NON-TARGET TOXICITY: Specific data regarding toxicity to fish or other aquatic organisms are not available for this product, but it is believed that it should have no adverse effects on non-target organisms or on the environment. **Approved for use in organic farming:** Yes: it has US OMRI approval.

2:218 rosemary, thyme and clove oil
Plant-derived fungicide and bactericide

NOMENCLATURE: Approved name: rosemary, thyme and clove oil. **Other names:** Plant essential oils. **CAS RN:** *[8000-25-7]* rosemary oil; *[8007-46-3]* thyme oil; *[8000-34-8]* clove oil.

SOURCE: The vegetable oils used in 'Sporatec' are extracted from the foliage or flowers of the plants.

PRODUCTION: Plant essential oils are extracted from the appropriate crop plant (these plants must be fully described for all products in the USA).

TARGET PESTS: Anthracnose (*Colletotrichum* spp.), *Botrytis* spp., downy mildews, powdery mildews, leaf spots, rusts and bacterial spot.

TARGET CROPS: 'Sporatec' can be used on a wide range of agricultural, horticultural and ornamental crops and turf.

BIOLOGICAL ACTIVITY: Mode of action: 'Sporatec' disrupts cell membrane integrity at different stages of disease development, whilst also acting as a fungistat and preventing fungal attachment.

COMMERCIALISATION: Formulation: Sold as an emulsifiable concentrate (EC) formulation. **Tradenames:** 'Sporatec' (Brandt Consolidated).

APPLICATION: The product should be sprayed, as needed, onto the infected plants to cover the entire leaf surface as soon as disease is noticed.

PRODUCT SPECIFICATIONS: **Specifications:** Plant essential oils are extracted from the appropriate crop plant (these plants must be fully described for all products in the USA). **Storage conditions:** Store in original container only in a cool, well-ventilated, dry place at temperatures between 4 and 34 °C (40 and 95 °F).

MAMMALIAN TOXICITY: The product consists of plant essential oils that are used in foodstuffs. As a consequence, it is considered to be 'Generally Regarded As Safe' (GRAS) and is not considered to be toxic to mammals. **Inhalation:** No adverse effects are expected, but inhalation may cause irritation of nasal passages and/or dizziness. **Skin and eye:** May cause eye irritation.

ENVIRONMENTAL IMPACT AND NON-TARGET TOXICITY: Specific data regarding toxicity to fish or other aquatic organisms are not available for this product, but it is believed that it should have no adverse effects on non-target organisms or on the environment. **Approved for use in organic farming:** Yes: it has US OMRI approval.

2:219 rotenone

Plant-derived insecticide and acaricide

The *Pesticide Manual* **Fifteenth Edition entry number:** 766

NOMENCLATURE: **Approved name:** rotenone (BSI, E-ISO, F-ISO, ESA, accepted in lieu of a common name); derris (JMAF). **Other names:** Derris root; tuba-root; aker-tuba (for the plant extract). Barbasco; cubé; haiari; nekoe; timbo (for the plants). **Development code:** ENT 133. **EC no:** 201-501-9. **CAS RN:** *[83-79-4]*.

SOURCE: Rotenone and related rotenoids were obtained from *Derris*, *Lonchocarpus* and *Tephrosia* species and were used originally in Asia and South America as fish poisons.

PRODUCTION: Produced by extraction of *Derris* roots and stabilised by phosphoric acid.

TARGET PESTS: Control of a wide range of arthropod pests, including aphids, thrips, suckers, moths, beetles and spider mites. Also used for the control of fire ants and of mosquito larvae when applied to pond water. Recommended for the control of lice, ticks and warble flies on animals and for insect control in premises. Also used to control fish populations.

TARGET CROPS: Recommended for use in fruit and vegetable cultivation, in premises and for use on animals. Also used in fish management.

BIOLOGICAL ACTIVITY: Mode of action: Inhibitor of Site I respiration within the electron-transport chain. Selective non-systemic insecticide with contact and stomach action. Secondary acaricidal activity. **Key reference(s):** 1) J Fukami. 1976. Insecticides as inhibitors of respiration. In *Insecticide Biochemistry and Physiology*, C F Wilkinson (ed.), pp. 353–96, Plenum Press, New York, USA. 2) H Fukami & M Nakajima. 1971. In *Naturally Occurring Insecticides*, M Jacobson & D G Crosby (eds), Marcel Dekker, New York, USA.

COMMERCIALISATION: Formulation: Sold as dispersible powder (DP), emulsifiable concentrate (EC) and wettable powder (WP) formulations. **Tradenames:** 'Chem Sect', 'Cube Root', 'Chem-Fish' and 'Rotenone Extract' (Tifa), 'Vironone' (Vipesco), 'PB-Nox' (plus piperonyl butoxide) (Penick), 'Pyrellin' (plus pyrethrins (pyrethrum)) (Wright Webb), 'Prenfish' (mixture), 'Noxfish', 'Nusyn-Noxfish' (plus piperonyl butoxide) and 'Synpren Fish' (plus piperonyl butoxide) (Prentiss), 'PB-Nox' (plus piperonyl butoxide) (Penick), 'Chem Fish Synergized' (plus piperonyl butoxide) (Tifa), 'Rotenone 5% Organic Insecticide', 'Bonide Liquid Rotenone/Pyrethrin Spray' (rotenone 1.1% plus pyrethrum 0.8%), 'Garden Dust' (rotenone 0.5%, pyrethrins 0.03%, cube resins 1%, copper 5% and sulfur 25%), 'Rotenone-Copper Dust' and 'Earth Friendly Fruit Tree Spray/Dust' (rotenone plus pyrethrum plus sulfur plus copper) (Bonide).

APPLICATION: Applied as an overall spray to give good cover of the foliage. Often used as a component of mixtures. Can be used in organic systems in extreme conditions of insect attack.

PRODUCT SPECIFICATIONS: Specifications: Produced by extraction of *Derris* roots and stabilised by phosphoric acid. **Purity:** Product purity is checked by i.r. spectrometry, by rplc or by solvent extraction and crystallisation.

COMPATIBILITY: Not compatible with alkaline substances.

MAMMALIAN TOXICITY: Considered to be toxic to mammals. **Acute oral LD$_{50}$:** white rats 132–1500, white mice 350 mg/kg. **Acute dermal LD$_{50}$:** rabbits >5000 mg/kg.

Inhalation: LC_{50} for male rats 0.023, female rats 0.0194 mg/litre. **Toxicity class:** WHO (a.i.) II; EPA (formulation) III, (EC formulation) I). **Other toxicological effects:** Estimated lethal dose for humans 300 to 500 mg/kg; more toxic when inhaled than when ingested. Very toxic to pigs.

ENVIRONMENTAL IMPACT AND NON-TARGET TOXICITY: Fish toxicity: (for formulated product) LC_{50} (96 h) for rainbow trout 31, bluegill sunfish 23 µg/litre; (for 98% pure extract) LC_{50} (96 h) for rainbow trout 1.9, bluegill sunfish 4.9 µg/litre.

Effects on beneficial insects: Rotenone is not toxic to bees, but combinations with pyrethrum are very toxic. **Metabolism:** In rat liver and in insects, the furan ring is enzymically opened and cleaved, leaving behind a methoxy group. The principal metabolite is rotenonone. An alcohol has been found as a further metabolite, this being formed via oxidation of a methyl group of the isopropenyl residue (I Yamamoto. 1969. *Residue Rev.*, **25**, 161).

Approved for use in organic farming: Yes, in cases of extreme insect attack.

2:220 ryania extracts   *Plant-derived insecticide*

The Pesticide Manual **Fifteenth Edition entry number:** 1360 (superseded)

NOMENCLATURE: Approved name: ryania extract; ryanodine. **Development code:** SHA 071502 (ryanodine). **CAS RN:** *[8047-13-0]* ryania; *[15662-33-6]* ryanodine; *[94513-55-0]* 9,21-didehydroryanodine.

SOURCE: Alkaloids from the stem of *Ryania* species, particularly *R. speciosa* Vahl, represent the first successful discovery of a natural insecticide. The collaboration between Rutgers University and Merck in the early 1940s followed the lead from the use of *Ryania* species in South America for euthanasia and as rat poisons. This collaborative work revealed that *Ryania* alkaloid extracts were insecticidal.

PRODUCTION: Ground stem wood of *R. speciosa*.

TARGET PESTS: Codling moth (*Cydia pomonella* L.), European corn borer (*Ostrinia nubilalis* (Hübner)) and citrus thrips.

TARGET CROPS: Maize (corn), apples, pears and citrus.

BIOLOGICAL ACTIVITY: Mode of action: Ryanodine and related alkaloids affect muscles by binding to the calcium channels in the sarcoplasmic reticulum. This causes calcium ion flow into the cells and death follows very rapidly. **Efficacy:** Ryania extracts have limited use as insecticides, but do give effective control of selected species. The size and complexity of the natural product means that it can be used only to treat infested crops and it has no systemic activity. The rapidity of its effect is an advantage in the control of boring insects. **Key reference(s):** 1) J E Casida, I N Pessah, J Seifert & A L Waterhouse. 1987. In *Naturally Occurring Pesticides*, R Greenlaugh & T R Roberts (eds), Blackwell Scientific Publishers, Oxford, UK, p. 177. 2) I Ujváry. 1999. Nicotine and other insecticidal alkaloids. In *Nicotinoid Insecticides and the Nicotinic Acetylcholine Receptor*, I Yamamoto & J E Casida (eds), pp. 29–69, Springer-Verlag, Tokyo, Japan.

COMMERCIALISATION: Formulation: Sold as a water dispersible powder (DP). All US uses were cancelled voluntarily in 1997. **Tradenames:** 'Natur-Gro R-50' and 'Natur-Gro Triple Plus' (AgriSystems International), 'Ryan 50' (Dunhill Chemical). **Patents:** US 2400295.

APPLICATION: Applied when insects are attacking the crop, at rates of 10 to 72 kg ryania per hectare (20 to 145 g alkaloid per hectare). Good coverage is essential. Avoid spraying near water courses.

PRODUCT SPECIFICATIONS: Ground stem wood of *R. speciosa*.

COMPATIBILITY: Compatible with most crop protection chemicals.

MAMMALIAN TOXICITY: Not considered to be toxic to mammals. **Acute oral LD$_{50}$:** rats 1200 mg/kg. **Toxicity class:** EPA (formulation) III.

ENVIRONMENTAL IMPACT AND NON-TARGET TOXICITY: Fish toxicity: Ryania extracts are toxic to fish. **Approved for use in organic farming:** Yes.

2:221 sabadilla   *Plant-derived insecticide*

The *Pesticide Manual* **Fifteenth Edition** entry number: 769

cevadine (i) R =

veratridine (ii) R =

NOMENCLATURE: Approved name: sabadilla powder. **Other names:** Veratrine; cevadine (for major component); veratridine (for second component). **CAS RN:** *[8051-02-3]* sabadilla or veratrine mixture; *[5876-23-3]* veracevine; *[28111-33-3]* cevacine; *[62-59-9]* cevadine; *[124-80-1]* sabadine; *[187237-90-7]* 3-O-vanilloylveracevine; *[71-62-5]* veratridine.

SOURCE: An insecticidal preparation from the crushed seeds of the liliaceous plant, *Schoenocaulon officinale* Gray (formerly *Veratrum sabadilla* Retr.) was used by local Indian tribes of South and Central America as an insecticide for many years. Preparations from various hellebore plants (*Veratrum* spp.) were once used commercially. Sabadilla has been used since the 1970s. Its practical use in modern agriculture was revived by the work of Allen *et al.* in the 1940s (see below).

PRODUCTION: The seeds of *S. officinale* contain a mixture of alkaloids – veratrine, consisting of an approximately 2:1 mixture of cevadine and veratridine, in combination with many minor components, all of which are esters of the alkamine, veracine. The product is produced by grinding the seeds of the plant and subsequent concentration. The seeds contain between 2% and 4% alkaloids.

TARGET PESTS: Thrips (*Frankliniella* spp. and *Thrips* spp.).

TARGET CROPS: Citrus and avocado.

BIOLOGICAL ACTIVITY: Mode of action: Cevadine, veratridine and related ceveratrum alkaloids have a mode of action that is similar to the pyrethrins in that they activate the voltage-sensitive sodium channels of nerve, heart and skeletal muscle cell membranes, although the binding site appears to be different from that of the pyrethroids. They are non-systemic insecticides with contact action. Initial effects include paralysis, with death occurring later. **Efficacy:** Sabadilla powder is a contact insecticide that is often mixed with other botanical insecticides for improved biological efficacy. Some formulations contain sugar as a feeding stimulant. **Key reference(s):** 1) T C Allen, K P Link, M Ikawa & L K Brunn. 1945. The relative effectiveness of the principal alkaloids of sabadilla seed, *J. Econ. Entomol.*, **38**, 293–6. 2) I Ujváry. 1999. Nicotine and other insecticidal alkaloids. In *Nicotinoid Insecticides and the Nicotinic Acetylcholine Receptor*, I Yamamoto & J E Casida (eds), pp. 29–69, Springer-Verlag, Tokyo, Japan. 3) D G Crosby. 1971. Minor insecticides of plant origin. In *Naturally Occurring Insecticides*, M Jacobsen and D G Crosby (eds), pp. 177–239, Marcel Dekker, New York, USA. 4) D H R Barton, O Jeger, V Prelog & R B Woodward. 1954. The constitutions of cevine and some related alkaloids, *Experientia*, **10**, 81–90. 5) W A Catterall. 1980. Neurotoxins that act on voltage-sensitive sodium channels in excitable membranes, *Ann. Rev. Pharmacol. Toxicol.*, **125**, 987–94.

COMMERCIALISATION: Formulation: Sabadilla powder is sold as wettable powder (WP) and water soluble concentrate (SL) formulations. Some formulations contain sugar as an insect feeding stimulant. The alkaloid content in these formulations varies between 0.2% and 25%. **Tradenames:** 'Veratran D' and 'Veratran' (Dunhill Chemical).

APPLICATION: Recommended use rates are between 20 and 100 g per hectare of total alkaloid. Veratridine persists longer than cevadine, but both are degraded in air and sunlight. Consequently, frequent applications and good foliage cover enhance the biological effects. No residual activity against thrips was observed 7 days after treatment.

PRODUCT SPECIFICATIONS: Specifications: The seeds of *S. officinale* contain a mixture of alkaloids – veratrine, consisting of an approximately 2:1 mixture of cevadine and veratridine, in combination with many minor components, all of which are esters of the alkamine, veracine. The product is produced by grinding the seeds of the plant and subsequent concentration. The seeds contain between 2% and 4% alkaloids. **Purity:** The alkaloidal components of sabadilla powder are not standardised.

COMPATIBILITY: Compatible with most compounds used in citrus and avocados. Often applied in combination with other botanical insecticides and/or sugar. Can be used in pest management programmes that include the release of beneficial insects.

MAMMALIAN TOXICITY: Not considered to be toxic to mammals. **Acute oral LD$_{50}$:** rats 4000 mg/kg. **Inhalation:** Sabadilla is an irritant to mucous membranes, causing sternutatory reactions. **Skin and eye:** Irritant to eyes, causing lachrymation and inflammation. **Toxicity class:** WHO (a.i.) O. **Other toxicological effects:** LD$_{50}$ (i.p.) for mice >100 mg/kg (veracevine); 7.5 mg/kg (veratrum derivatives); 3.5 to 5.8 mg/kg (cevadine); 1.35 to 9 mg/kg (veratridine). **Toxicity review:** 1) E D Swiss & R O Bauer. 1951. Acute toxicity of veratrum derivatives, *Proc. Soc. Exp. Biol. Med.*, **76**, 847–9. 2) I Ujváry, B K Eya, R L Grendell, R F Toia & J E Casida. 1991. Insecticidal activity of various 3-acyl and other derivatives of veracevine relative to the *Veratrum* alkaloids veratridine and cevadine, *J. Agric. Food Chem.*, **39**, 1875–81.

ENVIRONMENTAL IMPACT AND NON-TARGET TOXICITY: Readily degraded in air and sunlight and not considered to be a hazard to non-target organisms.

Effects on beneficial insects: Sabadilla powder is not toxic to beneficial insects (T S Bellows Jr & J G Morse. 1993. Toxicity of insecticides in citrus to *Aphytis melinus* DeBach (Hymenoptera: Aphelinidae) and *Rhizobius laphanthae* (Blasid.) (Coleoptera: Coccinellidae), *Can. J. Entomol.*, **125**, 987–94). **Metabolism:** Sabadilla powder breaks down rapidly in air and sunlight, with little residual activity 7 days after treatment. It is readily absorbed through animal skin.

Approved for use in organic farming: Yes.

2:222 saponin
Plant-derived fungicide and nematicide

NOMENCLATURE: Approved name: saponin **OPP Chemical Code:** 097094 (saponins of *Chenopodium quinoa*). **CAS RN:** *[404589-23-7]* (saponins of *Chenopodium quinoa*); *[68990-67-0]* (saponins from *Quillaja saponaria*).

SOURCE: Saponins of *Chenopodium quinoa* Willd. are derived by an extraction from seeds of *C. quinoa*. The major saponin constituents in the extract of *C. quimoa* seeds contain 49.65% saponins of approximately equimolar amounts of the triterpene bidesmosidic glycosides of oleanolic acid, hederagenin and phytolaccagenic acid and are sold as 'Heads Up Plant Protectant'. 'Nema-Q' saponins are the aqueous extract of *Quillaja saponaria* (Quillay).

PRODUCTION: 'Heads Up Plant Protectant' is extracted from the seeds of *Chenopodium quinoa*. 'Nema-Q' is the aqueous extract of *Quillaja saponaria*.

TARGET PESTS: 'Heads Up Plant Protectant' is used against fungi, bacteria and viral plant diseases. 'Nema-Q' is used to control nematodes.

TARGET CROPS: 'Heads Up Plant Protectant' is used on tubers (potato seed pieces), and on legume and cereal seeds and for root dip or foliar application to tomato seedlings prior to transplanting. 'Nema-Q' is recommended for a wide range of fruit and vegetable crops such as strawberries, brassicas, cucurbits, peppers, tomato and leafy vegetables as well as orchard, citrus and nut crops, vineyards and golf courses.

BIOLOGICAL ACTIVITY: Mode of action: Saponins are found in many plants where they act as insect antifeedants and protect against attack by fungal, bacterial and viral pathogens. Their activity is thought to be due to the surfactant properties of the compounds which disrupt the membrane structure of the target pest.

COMMERCIALISATION: Formulation: 'Heads Up Plant Protectant' is sold as a soluble powder (SP) formulation. 'Nema-Q' is sold as an aqueous extract. **Tradenames:** 'Heads Up Plant Protectant' (Heads Up Plant Protectants), 'Nema-Q' (Monterey).

APPLICATION: 'Heads Up Plant Protectant' is prepared as a dip or spray at a rate of 1 g per litre (0.035 ounces per 34 ounces of water) for application to 170 kg (375 pounds) of legume or cereal seed or 50–100 kg (110–220 pounds) of tuber. Seeds or roots are dipped (submerged for 60 seconds) or sprayed with solution until thoroughly covered. For foliar application, spay until the leaves are completely wet, once to immature tomato seedlings, up to 15 cm tall, pre-flowering, up to 4 days before transplanting. A 1% solution of 'Nema-Q' may be applied using ground equipment with a band sprayer, soil fertiliser shanks, injected through drip irrigation or through above ground sprinkler systems between 1 and 6 times per season. Application should be made to soil that has been dampened just prior to or during root flushes. For row crops, the irrigation zone is 30–50% of the row area. For orchards or around trees, apply from tree trunk to drip-line to spray at least 50% of soil area or the area under the canopy of the tree, whichever is greater. 'Nema-Q' can be applied using chemigation techniques.

PRODUCT SPECIFICATIONS: Specifications: 'Heads Up Plant Protectant' is extracted from the seeds of *Chenopodium quinoa*. 'Nema-Q' is the aqueous extract of *Quillaja saponaria*. **Storage conditions:** Saponins are stable under normal storage conditions. Avoid freezing.

COMPATIBILITY: Saponins are compatible with most common pesticides and can be tank mixed with fungicides, insecticides, herbicides, fertilisers or nematicides. Do not mix with products that will give a final pH above 8 or below 3.

MAMMALIAN TOXICITY: Saponins can cause irritation, but they do occur widely in plants and are often components of food. **Acute oral LD$_{50}$:** rats: >3000 mg/kg **Acute dermal LD$_{50}$:** rats: >4000 mg/kg. **Inhalation:** LC$_{50}$ (4 h) for rats >12.14 mg/litre. **Skin and eye:** Saponins can be damaging to the eye. Moderate skin irritants.

ENVIRONMENTAL IMPACT AND NON-TARGET TOXICITY: No long-term, adverse effects have been identified in mammals, birds, or fish exposed to saponins. They degrade within 3 to 5 days in the environment. **Approved for use in organic farming:** Yes: it has US OMRI approval.

2:223 sesame oil
Plant-derived insecticide and fungicide

NOMENCLATURE: Approved name: sesame oil. **Other names:** Plant essential oils.

SOURCE: 'Organocide' is a mixture of sesame oil and fish oil. It is prepared by extraction from harvested plant material. 'Dragonfire-CPP' is extracted from wild sesame (*Sesamum indicum* L.) CPP seeds by cold pressing.

PRODUCTION: Plant essential oils are extracted from the appropriate crop plant (these plants must be fully described for all products in the USA).

TARGET PESTS: 'Organocide' is recommended for use against a wide range of insect and fungal pests including mites (Acari), leaf miners (*Agromyza* spp.), leaf rollers (*Homona* spp.), scale (Margarodidae), aphids (Aphididae), whitefly (Aleyrodidae), powdery mildew (Erysiphaceae) and rose black spot (*Diplocarpon rosae* Wolf). 'Dragonfire-CPP' is recommended for the control of nematodes (Nematoda) including sting, lance, stubby-root, root-knot and spiral.

TARGET CROPS: 'Organocide' is approved for use in a wide range of crops including vegetables, fruit and nut orchards, ornamentals and turf. 'Dragonfire-CPP' is approved for use on all agricultural crops, fruit and nut trees, bananas and forage, as well as on vegetable gardens, golf courses, sports fields and lawns.

BIOLOGICAL ACTIVITY: Mode of action: The mode of action is not completely understood, but it is thought to act as an irritant to the insects and may prevent gas exchange (suffocation) and water loss by covering the insects' bodies. Disease control is thought to be due to disruption of the cellular membranes. Plant parasitic nematodes are deterred from feeding by 'Dragonfire-CPP'; if the nematodes do feed, the acidic content dries them out, killing them. Survivors become lethargic and will not reproduce. The root nematode population decreases, and the plants have a chance to recover and grow.

COMMERCIALISATION: Formulation: 'Organocide' is sold as an aqueous concentrate (AC). 'Dragonfire-CPP' is sold as a concentrate. **Tradenames:** 'Organocide' (Agriculture Solutions), 'Dragonfire-CPP' (Poulenger) and (Arbico).

APPLICATION: Apply 15 g of 'Organocide' per litre of water (2 ounces per gallon) in high volume to the eggs, larvae and nymphs of insects, as well as to the adults of soft-bodied insects and to powdery mildews when symptoms first appear. 'Dragonfire-CPP' is applied at 4 to 8 litres per hectare (2.5 to 5 gallons per acre) depending upon the level of nematode infestation. It should be thoroughly mixed with water at a minimum ratio of 1 part 'Dragonfire-CPP' to 10 parts water. It can be applied by spraying, hose end applicator, hand drenching or by sub-surface injection. Irrigate immediately after application if applied to flowering plants.

2. Natural Products

PRODUCT SPECIFICATIONS: Plant essential oils are extracted from the appropriate crop plant (these plants must be fully described for all products in the USA).

MAMMALIAN TOXICITY: The product consists of plant essential oils that are used in foodstuffs. As a consequence, it is considered to be 'Generally Regarded As Safe' (GRAS) and is not considered to be toxic to mammals.

ENVIRONMENTAL IMPACT AND NON-TARGET TOXICITY: Specific data regarding toxicity to fish or other aquatic organisms are not available for this product, but it is believed that it should have no adverse effects on non-target organisms or on the environment.
Approved for use in organic farming: Yes: it has US OMRI approval.

2:224 sesame meal

Plant-derived nematicide

NOMENCLATURE: Approved name: sesame meal

SOURCE: Derived from the air-dried, wild sesame (*Sesamum indicum* L.) CPP seeds.

TARGET PESTS: All nematodes.

TARGET CROPS: Recommended for use on all forms of turf, field crops, vegetable crops and fruit and nut trees.

BIOLOGICAL ACTIVITY: Mode of action: 'Ontrol' deters turf and plant nematodes from feeding on plants once it is absorbed by their roots.

COMMERCIALISATION: Formulation: 'Ontrol' is sold as a granule (GR).
Tradenames: 'Ontrol' (Poulenger). **Patents:** US Patent: 6,599,539.

APPLICATION: 'Ontrol' can be applied by tilling into the soil, as a top dressing or by spreading, drilling, spraying, hand drenching or pressure sub-surface injection.

MAMMALIAN TOXICITY: The product consists of plant essential oils that are used in foodstuffs. As a consequence, it is considered to be 'Generally Regarded As Safe' (GRAS) and is not considered to be toxic to mammals.

ENVIRONMENTAL IMPACT AND NON-TARGET TOXICITY: Specific data regarding toxicity to fish or other aquatic organisms are not available for this product, but it is believed that it should have no adverse effects on non-target organisms or on the environment.
Approved for use in organic farming: Yes: it has US OMRI approval.

2:225 sesame stalks   *Plant-derived nematicide*

NOMENCLATURE: Approved name: sesame stalks.

SOURCE: Sesame (*Sesamum indicum* L.) is a crop that has been grown for many years for its seeds that are widely used in cooking.

PRODUCTION: Once the sesame seeds have been harvested, the remaining stalks, that consist primarily of long chain polymers such as cellulose and fibre, are ground into small pieces suitable for adding to soil.

TARGET PESTS: Plant parasitic nematodes (Nematoda).

TARGET CROPS: A very wide range of both indoor and outdoor, food and non-food crops, including ornamentals and turf.

BIOLOGICAL ACTIVITY: Mode of action: The mode of action of sesame stalks is not understood. **Efficacy:** Under ideal conditions, excellent control of a wide range of plant parasitic nematodes has been reported.

COMMERCIALISATION: Formulation: The commercial product consists of finely ground fragments that can be added to the soil. **Tradenames:** 'Nematrol' (Sesaco).

APPLICATION: The ground stalks are mixed into soil before planting, or applied as a mulch around growing plants.

PRODUCT SPECIFICATIONS: Once the sesame seeds have been harvested, the remaining stalks, that consist primarily of long chain polymers such as cellulose and fibre, are ground into small pieces suitable for adding to soil.

COMPATIBILITY: It is unusual to apply sesame stalks at the same time as any other nematicidal treatment. Compatible with most other crop protection agents.

MAMMALIAN TOXICITY: In view of sesame's long history as a crop, no adverse effects are expected from the use of sesame stalks. No harmful effects have been reported in humans, in livestock that feed on leftover stalks or as a result of mixing the stalks into soil after harvesting the seeds. The US Food and Drug Administration categorises sesame as GRAS (Generally Recognised As Safe for food use).

ENVIRONMENTAL IMPACT AND NON-TARGET TOXICITY: It is not expected that sesame stalks will pose any risk to wildlife or to the environment if label directions are followed. **Approved for use in organic farming:** Yes: it has US OMRI approval.

2:226 soft soap

Plant-derived insecticide

NOMENCLATURE: Approved name: soft soap. **Other names:** Potassium salts of plant oils.

SOURCE: Certain plants contain fatty acids that are effective insecticides. These extracted acids are sold as the potassium salts.

TARGET PESTS: Insecticidal soaps are used against aphids, caterpillars, earwigs, flea beetles, lace bugs, thrips, spider mites, mealybugs, leafhoppers, psyllids, sawfly larvae, scale crawlers, squash bugs and whitefly.

TARGET CROPS: Insecticidal soaps are recommended for use on a wide range of different crops, including bedding plants, flowers, foliage plants, fruits and nuts, herbs and spices, ornamentals, pot plants, trees and shrubs, vegetables and many others.

BIOLOGICAL ACTIVITY: Mode of action: Insecticidal soap is a contact toxin, causing a breakdown of the target pest's cuticle, leading to dehydration and, ultimately, death. **Efficacy:** Insecticidal soap causes the rapid knockdown of phytophagous insects, but, because it is broken down rapidly once sprayed, it will not prevent subsequent re-invasion. It is often used in conjunction with insect predators, being used to bring the populations down to manageable levels prior to release. It should not be applied over the top of crops that have been treated with natural predators, as these are equally susceptible. Insecticidal soaps may be phytotoxic to some plants, so it is important to test this before overall application.

COMMERCIALISATION: Formulation: Sold as an aqueous concentrate. **Tradenames:** 'Safer Insecticidal Soap' (Safer).

APPLICATION: Use 20 g of product per litre of water to give good coverage of the foliage. Applications should be repeated 2 to 3 times at weekly intervals or as needed.

COMPATIBILITY: Insecticidal soap is recommended for organic agriculture and, as such, should not be used with chemical pesticides.

MAMMALIAN TOXICITY: Insecticidal soap is composed of compounds extracted from plants and is not considered to be toxic to mammals. The active components are considered to be 'generally recognised as safe' (GRAS).

ENVIRONMENTAL IMPACT AND NON-TARGET TOXICITY: It is rapidly degraded in the environment and is not expected to cause any long-term adverse effects.
Approved for use in organic farming: Yes: it has US OMRI approval.

2:227 soybean oil   *Plant-derived insecticide*

NOMENCLATURE: Approved name: soybean oil. **Other names:** Plant essential oil. **CAS RN:** *[8001-22-7].*

SOURCE: Extracted from soybean (*Glycine max* Merr.) seeds.

TARGET PESTS: Kills soft-bodied insects and egg masses on contact. May be used as a barrier treatment in silking corn (0.5 ml per ear) to prevent entry of corn earworm (*Helicoverpa zea* Boddie).

TARGET CROPS: Recommended for use on a wide range of crops.

BIOLOGICAL ACTIVITY: Mode of action: The mode of action is not completely understood, but it is thought to act as an irritant to the insects and may prevent gas exchange (suffocation) and water loss by covering the insect's bodies. Insecticidal activity was noticed when it was being used as a spray adjuvant.

COMMERCIALISATION: Formulation: Sold as 93% soybean oil formulation. **Tradenames:** 'Natur'l Oil' (Stoller).

APPLICATION: Dilute with water and spray at high volume to contact the insect pests.

MAMMALIAN TOXICITY: The product is 93% soybean oil and, as such, is not expected to show mammalian toxicity.

ENVIRONMENTAL IMPACT AND NON-TARGET TOXICITY: It is not expected that 'Natur'l Oil' will have any adverse effects on the environment or on non-target organisms. **Approved for use in organic farming:** Yes, unless formulated with an emulsifier.

2:228 spinetoram

Micro-organism-derived insecticide

The Pesticide Manual Fifteenth Edition entry number: 783

(i)

(ii)

NOMENCLATURE: Approved name: spinetoram (BSI, ANSI, ISO).
Development code: XDE-175; DE-175; XR-175; X574175. CAS RN: [187166-40-1] spinosyn J; [187166-15-0] spinosyn L.

SOURCE: The commercial product is a mixture of chemically modified spinosyn J ((i) – major component) and spinosyn L (ii). Both compounds are derived from the soil Actinomycete, *Saccharopolyspora spinosa* Mertz & Yoa. The organism is composed of long, yellowish-pink aerial chains of spores encased in distinctive, spiny spore sheaths. The bacterium is aerobic, Gram-positive, non-acid-fast, non-motile, filamentous and differentiated into substrate and aerial hyphae. The aerial mycelium is yellowish-pink and the vegetative mycelium is yellow to yellowish-brown. The parent strain was originally isolated from an abandoned rum still in the Caribbean.

PRODUCTION: Spinetoram is obtained from a whole broth extraction, following fermentation of the organism on a feedstock of water, vegetable flours, sugar and animal fat, followed by synthetic modification.

TARGET PESTS: Applications for spinetoram products include control of crop-damaging pests such as codling moth (*Cydia pomonella* L.), leafminers (*Liriomyza* spp.), apple maggot (*Rhagoletis pomonella* (Walsh)), pear psylla (*Cacopsylla pyri* (L.) and *Cacopsylla pyricola* (Förster)), oriental fruit moth (*Carposina niponensis* Walsingham), navel orangeworm (*Amylois transitella* (Walker)), diamond-back moth (*Plutella xylostella* (L.)), armyworm (*Spodoptera* spp.), thrips (*Thrips* spp.), tomato fruit worm (*Heliothis zea* (Boddie)), loopers (*Anagrapha* spp., *Anticarsia* spp., *Autographa* spp., *Chrysodeixis* spp., *Syngrapha* spp. and *Trichoplusia ni* (Hübner)), bollworm (*Heliothis* spp and *Helicoverpa* spp.), cutworm (*Agrotis* spp.), grape berry moth (*Eupoecilia ambiguella* (Hübner)) and others.

TARGET CROPS: Crops that can be treated with spinetoram include apples and pears, stone fruit, tree nuts, crucifers, leafy vegetables, fruiting vegetables, citrus, grapes, honeydew and cantaloupe melons, maize (corn), cotton, soybeans, caneberries and bushberries.

BIOLOGICAL ACTIVITY: Mode of action: Spinetoram's effects on target insects are consistent with the activation of the nicotinic acetylcholine receptor, but at a different site than nicotine or the neonicotinoids. Spinetoram also affects GABA (γ-aminobutyric acid) receptors, but their role in the overall activity is unclear. There is currently no known cross-resistance to other insecticide classes. O-ethylation of the rhamnose sugar, as in spinetoram, has been shown to increase insecticidal activity. Reduction of the 5,6-double bond is associated with improved residuality in the field. **Efficacy:** The mode of action causes a rapid death of target phytophagous insects. Its moderate residual activity reduces the possibility of the onset of resistance, but it is strongly recommended that it be used within a strong, pro-active resistance management strategy. Spinetoram is recommended as an ICM tool, as it shows no effects on predatory insects such as ladybirds, lacewings, big-eyed bugs or minute pirate bugs. It has reduced activity against parasitic wasps and flies. It is toxic when sprayed directly onto honeybees and other pollinators, but, once dry, residues have little effect. Due to its low effective use rate of 10–100 g/ha (0.5–2.0 oz/acre), safety toward the environment and safety toward mammals and beneficial insects, spinetoram was accepted for review and registration under the Reduced Risk Pesticide Program of the US Environmental Protection Agency (EPA). **Key reference(s):** 1) A Chloridis, P Downard, J E Dripps, K Kaneshi, L C Lee, Y K Min & L A Pavan. 2007. Spinetoram (XDE-175): a new spinosyn, *Proc. XVI Int. Plant Protection Congress*, Glasgow, **1**, 68–73. 2) Spinetoram Technical Bulletin, Dow AgroSciences LLC, Form No. Y47-343-001 (11/06) BOD, November 2006, pages 2–4. 3) Spinetoram Pesticide Tolerance, US Environmental Protection Agency, Federal Register Docket ID No. DOCID:fr21mr07-4, Federal Register: March 21, 2007 (Volume 72, Number 54), Rules and Regulations pages 13170–2.

COMMERCIALISATION: Formulation: Spinetoram is formulated as a water dispersible (WG) granule or as a suspension concentrate (SC). **Tradenames:** 'Delegate WG', 'Exalt SC' and 'Radiant SC' (Dow AgroSciences).

2. Natural Products

APPLICATION: The compound is applied at rates of 5 to 100 g per hectare. It should be applied when pest pressure demands treatment. The active ingredient does not dissolve in water and continual agitation is required to prevent the active ingredient from settling out in the spray tank. The addition of adjuvants has not been shown to improve or reduce the performance of spinetoram consistently, with the exception of leaf miner control and the penetration of closed canopies, where emulsified vegetable oils have helped.

PRODUCT SPECIFICATIONS: Specifications: Spinetoram is obtained from a whole broth extraction, following fermentation of the organism on a feedstock of water, vegetable flours, sugar and animal fat, followed by synthetic modification. **Purity:** The commercial product is composed of spinosyns J and L. Analysis is undertaken by hplc or immunoassay (details from Dow AgroSciences). **Storage conditions:** Spinetoram is stable over a wide range of temperatures. Protect from freezing. Shake well before use. **Shelf-life:** The formulated product has a shelf-life of 3 years.

COMPATIBILITY: No compatibility problems have been identified to date when tank-mixing spinetoram with other crop protection products, foliar fertilisers or adjuvants. A jar test for compatibility is recommended prior to use.

MAMMALIAN TOXICITY: Not considered to be toxic to mammals. **Acute oral LD_{50}:** rats >5000 mg/kg. **Inhalation:** LC_{50} for rats >5.5 mg/litre. **Skin and eye:** Acute percutaneous LD_{50} >5000 mg/kg. Eye contact with spinetoram formulations may cause slight irritation. **Other toxicological effects:** No indications of mutagenicity, teratogenicity, or oncogenicity (Ames and chromosomal aberration tests, mutation and mouse bone marrow micronucleus assays).

ENVIRONMENTAL IMPACT AND NON-TARGET TOXICITY: The acute and chronic toxicity of spinetoram to mammals, birds, fish, earthworms and aquatic plants is low. Spinetoram is toxic to aquatic invertebrates from chronic exposure. However, because of its rapid degradation in aquatic systems, spinetoram is not expected to impact these species negatively. **Bird toxicity:** Acute oral LD_{50} for mallard ducks and bobwhite quail >2250 mg/kg. Acute dietary LC_{50} for mallard ducks and bobwhite quail >5620 ppm. **Fish toxicity:** LC_{50} (96 h) for rainbow trout >3.46, bluegill sunfish 2.69 mg/litre. **Other aquatic toxicity:** EC_{50} (48 h) for *Daphnia* >3.17 mg/litre. **Effects on beneficial insects:** Toxic to honeybees, but residues aged 3 hours or longer are practically non-toxic. LC_{50} (96 h) for worms (*Eisenia foetida*) >1000 mg/kg soil. Toxic to predatory mites and insect parasitoids in laboratory tests, but effects under field conditions are much less significant. Not toxic to predatory insects, such as coccinellids and lacewings. **Behaviour in soil:** Rapidly degraded in soil, field dissipation DT_{50} 3–5 days; aquatic field dissipation DT_{50} <1 day.

2:229 spinosad *Micro-organism-derived insecticide*

The Pesticide Manual Fifteenth Edition entry number: 784

spinosyn A, R = H-

spinosyn D, R = CH₃-

2. Natural Products

NOMENCLATURE: Approved name: spinosad (BSI, ANSI, ISO). **Development code:** XDE-105; DE-105. **CAS RN:** *[168316-95-8]* spinosad; *[131929-60-7]* spinosyn A; *[131929-63-0]* spinosyn D.

SOURCE: The commercial product is a mixture of spinosyn A and spinosyn D. Both compounds are secondary metabolites of the soil Actinomycete, *Saccharopolyspora spinosa* Mertz & Yoa. The organism is composed of long, yellowish-pink aerial chains of spores encased in distinctive, spiny spore sheaths. The bacterium is aerobic, Gram-positive, non-acid-fast, non-motile, filamentous and differentiated into substrate and aerial hyphae. The aerial mycelium is yellowish-pink and the vegetative mycelium is yellow to yellowish-brown. The parent strain was originally isolated from an abandoned rum still in the Caribbean.

PRODUCTION: Spinosad is obtained from a whole broth extraction, following fermentation of the organism on a feedstock of water, vegetable flours, sugar and animal fat.

TARGET PESTS: Recommended for the control of caterpillars, leaf miners, thrips and foliage-feeding beetles. Species controlled include caterpillars (*Ostrinia nubilalis* (Hübner), *Helicoverpa zea* Boddie, *Trichoplusia ni* (Hübner), *Plutella xylostella* (L.), *Spodoptera* spp., *Heliothis* spp., *Pieris rapae* (L.), *Keiferia lycopersicella* (Walsingham), *Lobesia botrana* (Denis & Schiffermüller), *Agrotis ipsilon* (Hufnagel), *Parapediasia teterrella* (Zincken), thrips (*Frankliniella occidentalis* (Pergande), *Thrips palmi* (Karny)), flies (*Liriomyza* spp., *Ceratitis capitata* (L.)), beetles (*Leptinotarsa decemlineata* (Say)), drywood termites (*Cryptotermes brevis* (Walker), *Incisitermes snyderi* (Light)), fire ants (*Solenopsis* spp.) and grasshoppers. Under development for control of chewing and sucking lice (e.g. *Linognathus vituli*, *Bovicola ovis* (Schrank), *Solenopotes capillatus* Enderlein) and flies (e.g. *Haematobia irritans* (L.),

Lucilia cuprina (Wiedemann)), and for control of nuisance flies (e.g. *Stomoxys calcitrans* (L.), *Musca domestica* L.).

TARGET CROPS: May be used on row crops (including cotton), vegetables, fruit trees, turf, vines and ornamentals. No crop phytotoxicity has been observed. Under development for use on livestock animals and in livestock premises.

BIOLOGICAL ACTIVITY: Mode of action: Spinosad's effects on target insects are consistent with the activation of the nicotinic acetylcholine receptor, but at a different site than nicotine or the neonicotinoids. Spinosad also affects GABA (γ-aminobutyric acid) receptors, but their role in the overall activity is unclear. There is currently no known cross-resistance to other insecticide classes. **Efficacy:** The mode of action causes a rapid death of target phytophagous insects. Its moderate residual activity reduces the possibility of the onset of resistance, but it is strongly recommended that it be used within a strong, pro-active resistance management strategy. Spinosad is recommended as an ICM tool, as it shows no effects on predatory insects such as ladybirds, lacewings, big-eyed bugs or minute pirate bugs. It has reduced activity against parasitic wasps and flies. It is toxic when sprayed directly onto honeybees and other pollinators, but, once dry, residues have little effect. Spinosad is effective as a bait for fruit flies (*Ceratitis* spp., *Bactrocera* spp., etc.) and some ants (*Solenopsis* spp.). **Key reference(s):** 1) H A Kirst, K H Michel, J S Mynderse, E H Chio, R C Yao, W M Nakasukasa, L-V D Boeck, J L Occlowitz, J W Paschal, J B Deeter & G D Thompson. 1992. Discovery, isolation, and structure elucidation of a family of structurally unique, fermentation-derived tetracyclic macrolides. In *Synthesis and Chemistry of Agrochemicals III*, Chapter 20, D R Baker, J G Fenyes & J J Steffens (eds), pp. 214–25, American Chemical Society, Washington DC, USA. 2) D J Porteus, J R Raines & R L Gantz. 1996. In *1996 Proceedings of Beltwide Cotton Conferences*, P Dugger & D Richter (eds), pp. 875–7, National Cotton Council of America, Memphis, TN, USA. 3) T C Sparks *et al.* 1996. In *1996 Proceedings of Beltwide Cotton Conferences*, P Dugger & D Richter (eds), pp. 692–6, National Cotton Council of America, Memphis, TN, USA. 4) V L Salgado. 1997. In *Down to Earth*, **52**:1, pp. 35–44, DowElanco, Indianapolis, IN, USA. 5) G D Thompson, R Dutton & T C Sparks. 2000. Spinosad – a case study: an example from a natural products discovery programme, *Pest Management Science*, **56**(8), 696–702.

COMMERCIALISATION: Formulation: Sold as an aqueous based suspension concentrate (SC) formulation. **Tradenames:** 'Conserve', 'Entrust', 'Success', 'SpinTor', 'Tracer', 'GF-120', 'Justice', 'Laser', 'Naturalyte' and 'Spinoace' (Dow AgroSciences), 'Racer Gold' (Agri Life), 'Extinosad' (veterinary use) (Elanco). **Patents:** US 5,202,242 (1993); EPO 375316 (1990).

APPLICATION: The compound is applied at rates of 5 to 150 g per hectare. Apply when pest pressure demands treatment. The active ingredient does not dissolve in water and continual agitation is required to prevent the active ingredient from settling out in the spray tank. The addition of adjuvants has not been shown to improve or reduce the performance of spinosad consistently, with the exception of leaf miner control and the penetration of closed canopies, where emulsified vegetable oils have helped.

PRODUCT SPECIFICATIONS: Specifications: Spinosad is obtained from a whole broth extraction, following fermentation of the organism on a feedstock of water, vegetable flours, sugar and animal fat. **Purity:** The commercial product is composed of spinosyn A and spinosyn D. Analysis is undertaken by hplc or immunoassay (details from Dow AgroSciences).
Storage conditions: Spinosad is stable over a wide range of temperatures. Protect from freezing. Shake well before use. **Shelf-life:** The formulated product has a shelf-life of 3 years.

COMPATIBILITY: No compatibility problems have been identified to date when tank-mixing spinosad with other crop protection products, foliar fertilisers or adjuvants. A jar test for compatibility is recommended prior to use.

MAMMALIAN TOXICITY: Not considered to be toxic to mammals. **Acute oral LD_{50}:** male rats 3783, female rats >5000 mg/kg. **Acute dermal LD_{50}:** rabbits >5000 mg/kg. **Inhalation:** LC_{50} (4 h) for rats >5.18 mg/litre. **Skin and eye:** Non-irritating to skin, but slight irritation to eyes (rabbits). Not a skin sensitiser (guinea pigs). **NOEL:** for dogs, mice and rats, following 13 weeks of dietary exposure to spinosad, was 5, 6 to 8 and 10 mg/kg/day, respectively. **ADI:** 0.02 mg/kg b.w. [2001]; (US) 0.027 mg/kg b.w.; (Japan, Australia) 0.024 mg/kg b.w. **Toxicity class:** WHO (a.i.) U; EPA (formulation) IV (tech.), IV ('Tracer').
Other toxicological effects: In acute and sub-chronic tests, spinosad did not demonstrate any neurotoxic, reproductive or mutagenic effects on dogs, mice or rats.

ENVIRONMENTAL IMPACT AND NON-TARGET TOXICITY: The acute and chronic toxicity of spinosad to mammals, birds, fish, earthworms and aquatic plants is low.
Bird toxicity: Spinosad is considered practically non-toxic to birds. The acute oral LD_{50} for both bobwhite quail and mallard ducks is >2000 mg/kg. Acute dietary LC_{50} for bobwhite quail and mallard ducks >5156 ppm. **Fish toxicity:** Spinosad is considered slightly to moderately toxic to fish. The acute 96-hour LC_{50} for rainbow trout, bluegill sunfish, common carp, Japanese carp and sheepshead minnow were 30, 5.9, 5, 3.5 and 7.9 mg/litre, respectively.
Other aquatic toxicity: EC_{50} (48 h) for *Daphnia* 14 ppm. EC_{50} (96 h) for Eastern oyster 0.3, grass shrimp >9.76 ppm. EC_{50} for *Lemna gibba* 10.6 ppm. EC_{50} for algae – *Selenastrum capricornutum* >105.5, *Skeletonema costatum* 0.2, *Navicula pelliculosa* 0.09, *Anabaena flos-aquae* 8.9 ppm. **Effects on beneficial insects:** Spinosad is considered highly toxic to honeybees, with less than 1 µg/bee of technical material applied topically resulting in mortality. Once residues are dry, they are non-toxic. **Metabolism:** Feeding studies produced no residues of spinosad in meat, milk or eggs. The half-life on plant surfaces ranged from 1.6 to 16 days, with photolysis as the main route of degradation. **Behaviour in soil:** Spinosad is rapidly degraded on soil surfaces by photolysis and, below the soil surface, by soil micro-organisms.
Key reference: D G Saunders & B L Bret. 1997. In *Down to Earth*, **52**:1, pp. 14–21, DowElanco, Indianapolis, IN, USA. **Approved for use in organic farming:** Yes.

2:230 streptomycin
Micro-organism-derived bactericide

The Pesticide Manual Fifteenth Edition entry number: 792

NOMENCLATURE: Approved name: streptomycin (BSI, E-ISO, BAN, JMAF); streptomycine (F-ISO); no name (Denmark). **CAS RN:** *[57-92-1]* streptomycin; *[3810-74-0]* streptomycin sesquisulfate.

SOURCE: Isolated from the soil Actinomycete, *Streptomyces griseus* (Krainsky) Waksman & Henrici. First reported by A Schatz *et al.* (*Proc. Soc. Exp. Biol. Med.*, **55**, 66) in 1944, its structure was elucidated in 1947 (F A Kuehl Jr, R L Peck, C E Hoffnine Jr, E W Peel & K Folkers, *J. Am. Chem. Soc.*, **69**, 1234). Discovered and first commercialised by Meiji Seika Kaisha Ltd.

PRODUCTION: Streptomycin is obtained by fermentation of *S. griseus*, and is isolated and sold as the sesquisulfate.

TARGET PESTS: Control of bacterial shot-hole, bacterial rots, bacterial canker, bacterial wilts, fire blight and other bacterial diseases (especially those caused by Gram-positive species of bacteria). Streptomycin is particularly effective against *Xanthomonas oryzae* Dows., *X. citri* Dows., *Pseudomonas tabaci* Stevens and *P. lachrymans* Carsner.

TARGET CROPS: Recommended for use in pome fruit, stone fruit, citrus fruit, olives, vegetables, potatoes, tobacco, cotton and ornamentals.

BIOLOGICAL ACTIVITY: Mode of action: Streptomycin inhibits protein biosynthesis by binding to the 30S ribosomal sub-unit and causing a misreading of the genetic code in protein biosynthesis. It is a bactericide with systemic action. **Efficacy:** Resistance to streptomycin is widespread, reducing the value of the compound in crop protection.
Key reference(s): 1) A Schatz, E Bugie & S A Waksman. 1944. Streptomycin, a substance exhibiting antibiotic activity against Gram-positive and Gram-negative bacteria, *Proc. Soc. Exp. Biol. Med.*, **55**, 66. 2) T E Likover & C G Kurland. 1967. The contribution of DNA to

translation errors induced by streptomycin *in vitro*, *Proc. Natl Acad. Sci. USA*, **58**, 2385.
3) F A Kuehl Jr, R L Peck, C E Hoffhine Jr, E W Peel & K Folkers. 1947. Streptomyces antibiotics. XIV. The position of the linkage of streptobiosamine to streptidine in streptomycine. *J. Am. Chem. Soc.*, **69**(5), 1234. 4) M L Wolfrom, M J Cron, C W DeWalt & R M Husband. 1954. Configuration of the glycosidic unions in streptomycin, *J. Am. Chem. Soc.*, **76**(14), 3675–7.

COMMERCIALISATION: Formulation: Sold as a wettable powder (WP) and liquid concentrate (SL). **Tradenames:** 'Streptrol' (streptomycin) (Agtrol) and (Nufarm Americas), 'AAstrepto' (streptomycin) (Bayer CropScience), 'Paushamycin' (streptomycin) (Paushak), 'Plantomycin' (streptomycin) (Aries), 'Agrept' (streptomycin sesquisulfate) (Meiji Seika), 'Agrimycin 17' (streptomycin sesquisulfate) and 'AS-50' (streptomycin sesquisulfate) (Syngenta), 'Bac-Master' (streptomycin sesquisulfate) (Amvac), 'Cuprimicín 17' (streptomycin), 'Cuprimicín 100' (streptomycin plus oxytetracycline) and 'Cuprimicín 500' (streptomycin plus copper sulfate (tribasic) plus oxytetracycline) (Ingeniería Industrial), 'Dustret' (streptomycin sesquisulfate plus maneb) (Agsco), 'Seed Treatment For Potatoes' (streptomycin sesquisulfate plus maneb) (Helena).

APPLICATION: Applied as a foliar spray, at rates of 200 g a.i. per hectolitre.

PRODUCT SPECIFICATIONS: Specifications: Streptomycin is obtained by fermentation of *S. griseus*, and is isolated and sold as the sesquisulfate. **Purity:** Isolated and sold as the sesquisulfate. The product is analysed by bioassay against a suitable bacterium.

COMPATIBILITY: Can cause chlorosis to rice, grapes, pears, peaches and some ornamentals and these symptoms can be relieved by the addition of iron chloride or iron citrate to the spray tank. Incompatible with pyrethrins and other alkaline products. Often used in collaboration with a bactericide with a different mode of action (such as oxytetracycline), to reduce the likelihood of resistance.

MAMMALIAN TOXICITY: As with all aminoglycoside antibiotics, streptomycin has very low mammalian toxicity. **Acute oral LD_{50}:** mice >10 000 mg/kg. Streptomycin sesquisulfate – rats 9000, mice 9000, hamsters 400 mg/kg. **Acute dermal LD_{50}:** male mice 400, female mice 325 mg/kg. **Skin and eye:** May cause allergic skin reaction. **NOEL:** In chronic toxicity studies on rats, NOEL was 125 mg/kg. **Toxicity class:** EPA (formulation) IV.
Other toxicological effects: Acute i.p. LD_{50} for male mice 340, female mice 305 mg/kg.

ENVIRONMENTAL IMPACT AND NON-TARGET TOXICITY: Streptomycin is not considered to be hazardous to non-target organisms or to the environment.
Bird toxicity: Practically non-toxic to birds. **Fish toxicity:** Slightly toxic to fish.
Effects on beneficial insects: Practically non-toxic to honeybees.

2:231 sugar octanoate esters
Plant-derived insecticide/nematicide

NOMENCLATURE: Approved name: sugar octanoate esters. **Other names:** Sucrose octanoate esters and sorbitol octanoate. **OPP Chemical Codes:** 035300 and 035400. **CAS RN:** *[42922-74-7]* and *[58064-47-4]* sucrose octanoate esters; *[108175-15-1]* sorbitol octanoate.

SOURCE: Sucrose octanoate esters are made from a caprylic fatty acid ester derived from an edible oil or fat and sucrose. Sorbitol octanoate is derived as a sugar ester synthesised via the condensation of sorbitol with octanoic acid.

TARGET PESTS: Sucrose octanoate targets soft-bodied insects and mites, sciarid flies (*Bradysia* spp.) and varroa mites (*Varroa destructor* Anderson & Trueman). Sorbitol octanoate is used against soft-bodied insects and mites.

TARGET CROPS: Sucrose octanoate can be used on a wide range of crops, on mushrooms and adult honey bees. Sorbitol octanoate can be used on a range of glasshouse, nursery and field crops.

BIOLOGICAL ACTIVITY: Mode of action: The mode of action of sucrose octanoate is physical and non-toxic; the surfactant effect of the esters de-waxes the cuticle of the target pest, causing it to desiccate. The mode of action of sorbitol octanoate is a physical, surfactant effect that results in rapid suffocation and/or de-waxing of the cuticle of the target pests, subsequently causing desiccation via loss of body fluids. There are no neurological and/or physiological interactions with the target pest.

COMMERCIALISATION: Formulation: 'Avachem Sucrose Octanoate' is a 85.43% aqueous concentrate (AC). 'Avachem Sucrose Octanoate' is a 40% aqueous concentrate. **Tradenames:** 'Avachem Sucrose Octanoate' (withdrawn) and 'Avachem Sucrose Octanoate' (AVA Chemical).

APPLICATION: Apply 'Avachem Sucrose Octanoate' by ground spray equipment. Shake or stir before use, adding the appropriate quantity to water, with agitation. Maintain gentle agitation during application. Application rates of 0.5% volume/volume (v/v) are recommended for all use sites.

PRODUCT SPECIFICATIONS: Storage conditions: Stable for at least a year if stored as recommended.

MAMMALIAN TOXICITY: Sugar octanoate esters are considered to have low mammalian toxicities. **Acute oral LD_{50}:** rats and mice – EPA Toxicity Category IV. **Acute dermal LD_{50}:** rats

and mice – EPA Toxicity Category IV. **Inhalation:** rats – EPA Toxicity Category IV.
Skin and eye: Primary eye irritation in rabbits – EPA Toxicity Category I.

ENVIRONMENTAL IMPACT AND NON-TARGET TOXICITY: It is not expected that
sugar octanoate esters will have any adverse effects on the environment or on non-target
organisms. **Effects on beneficial insects:** LC_{50} for honeybees >80 µg/bee.

2:232 synergol   *Plant-derived insecticide/acaricide*

NOMENCLATURE: Approved name: synergol

SOURCE: Synergol is made from the essential oils of three plant species – thyme (*Thymus
vulgaris* L.), wintergreen (*Gaultheria procumbens* L.) and African marigold (*Tagetes erecta* L.)) in
a vegetable oil formulation. The technology was developed by Plant Impact plc and is patent
protected.

PRODUCTION: The product is derived from steam-distilled essential oils of *Thymus vulgaris*,
Gaultheria procumbens and *Tagetes erecta* encapsulated in an emulsifiable vegetable oil base.

TARGET PESTS: Synergol is effective against a wide range of insect and mite species including
whitefly (*Bemisia* spp. and *Trialeurodes vaporarorium* (Westwood)), aphids (*Aphis gossypii*
Glover), thrips (*Frankliniella* spp.), red spider mite (*Panonychus ulmi* Koch), two-spotted
spider mite (*Tetranychus urticae* Koch), Pacific mite (*T. pacificus* McGregor), Kanzawa mite (*T.
kanzawi* Kishida), citrus rust mite (*Phyllocoptruta oleivora* (Ashm.)) and citrus red mite (*P. citri*
(McGregor)).

TARGET CROPS: Vegetables (including tomato, pepper, cucumber, courgette (zucchini),
eggplant and melon); top fruit (apples and pears); citrus; and tea.

BIOLOGICAL ACTIVITY: Mode of action: Synergol has been shown to affect target species
in three different ways. Its insecticidal mechanism is unknown, but bioassays have shown that
it is unaffected by known resistance mechanisms (Q-type esterase and non-specific esterase
E4) in *Myzus persicae* and *Bemisia tabaci*. Synergol has repellent effects by acting as an insect
irritant, and the volatile oils repel insects and mites. It has also been shown to cover the target
insects and mites, thereby suffocating them. **Efficacy:** When used as a foliar spray on fruit and
vegetable crops it gives rapid knockdown, and good repellency (up to 42 days in some crops).
It gives broad spectrum control of the adult, immature forms and eggs of mites. It controls
whitefly adults and scales and is effective against adult aphids and thrips.

COMMERCIALISATION: Formulation: Synergol is sold as a vegetable oil formulation. **Tradenames:** 'BugOil' (Plant Impact). **Patents:** Pending W2005/053395, GB 0512866.5.

APPLICATION: Can be applied throughout the season whenever target pests are present. Rates between 5 and 20 ml/l product applied to runoff as a foliar spray.

PRODUCT SPECIFICATIONS: The product is derived from steam-distilled essential oils of *Thymus vulgaris*, *Gaultheria procumbens* and *Tagetes erecta* encapsulated in an emulsifiable vegetable oil base.

COMPATIBILITY: It is unusual to apply 'BugOil' with any other insect control chemicals.

MAMMALIAN TOXICITY: All materials used in products based on synergol chemistry are either used as food or food additives at levels higher than those found in the product. There are no reports of any allergic or negative responses to the chemicals contained (at the concentrations present) in scientific literature. Products based on synergol chemistry are EPA exempt in the USA. **Toxicity class:** Non-hazardous.

ENVIRONMENTAL IMPACT AND NON-TARGET TOXICITY: The essential oils used in synergol are all present at lower levels than in the plants from which they are derived. As such they are naturally present in the environment and are not expected to cause any adverse environmental effects. **Effects on beneficial insects:** GLP studies show the technology to be safe on many beneficial organisms including bees, ladybirds, earthworms and lacewings. **Approved for use in organic farming:** Yes: it has US OMRI approval.

2:233 thyme oil
Plant-derived fungicide, insecticide and nematicide

NOMENCLATURE: Approved name: thyme oil. **Other names:** Plant essential oil. **CAS RN:** *[8007-46-3]*.

SOURCE: Derived from the foliage of thyme plants (*Thymus vulgaris* L.).

TARGET PESTS: 'Promax' is an effective soil applied nematicide and fungicide. 'Proud 3' is a foliar-applied, contact insecticide, miticide and fungicide. As an insecticide, it works best on immature, soft bodied insects.

TARGET CROPS: 'Promax' and 'Proud 3' are registered as biopesticides for a wide variety of crops, ornamental plants and turf.

BIOLOGICAL ACTIVITY: Mode of action: The mode of action is not completely understood, but it is thought to act as an irritant to the insects and may prevent gas exchange (suffocation) and water loss by covering the insects' bodies. Disease control is thought to be due to disruption of the cellular membranes. Plant parasitic nematodes are deterred from feeding.

COMMERCIALISATION: Formulation: Both products are sold as 3.5% solutions. **Tradenames:** 'Promax' and 'Proud 3' (Bio Huma Netics).

APPLICATION: Apply 'Proud 3' as a foliar spray, ensuring good coverage, but avoiding run-off. 'Promax' should be applied to the soil.

MAMMALIAN TOXICITY: The product consists of a plant essential oil that is used in foodstuffs. As a consequence, it is considered to be 'Generally Regarded As Safe' (GRAS) and is not considered to be toxic to mammals. Prolonged exposure to these products may cause skin irritation, eye/nasal irritation, dizziness, headache or nausea.

ENVIRONMENTAL IMPACT AND NON-TARGET TOXICITY: Specific data regarding toxicity to fish or other aquatic organisms are not available for this product, but it is believed that it should have no adverse effects on non-target organisms or on the environment. **Approved for use in organic farming:** Yes: it has US OMRI approval.

2:234 thymol   *Miticide*

NOMENCLATURE: Approved name: thymol. **Other names:** 5-Methyl-2-isopropyl-1-phenol. **OPP Chemical Code:** 080402. **CAS RN:** *[89 83 8]*.

SOURCE: Thymol is a constituent of oil of thyme, a naturally occurring mixture of compounds in the thyme plant (*Thymus vulgaris* L.).

PRODUCTION: Thymol is manufactured for use in agriculture.

TARGET PESTS: Varroa mite (*Varroa destructor* Anderson & Trueman).

TARGET CROPS: Beehives.

BIOLOGICAL ACTIVITY: Mode of action: Thymol has been known to possess acaricidal activity for many years. It is very volatile and its application in a gel above the brood frames in beehives has been shown to be an effective method of control of varroa mites, with mortalities of over 90% common. The exact mechanism of action is not known. The movement of bees through the dispenser causes the bees to pick up the thymol impregnated Entostat powder and this powder is further distributed throughout the hive as a consequence of normal bee activity. The powder increases the grooming behaviour of the bees and this, in addition to the presence of thymol, causes the varroa mites to fall from the bees' bodies. The presence of a varroa screen prevents the mites from re-entering the hive. **Key reference(s):** A Imdorf, S Bogdanov, R Ibanez Ochoa & N Calderone. 1999. Use of essential oils for the control of *Varrroa destructor* (Oud.) in honey bee colonies. *Apidologie* **30**, 209–28.

COMMERCIALISATION: Formulation: 'Apiguard' is sold as a gel formulation. 'Exomite Apis' consists of Entostat powder impregnated with thymol and placed inside a dispenser which is positioned at the entrance to the hive. **Tradenames:** 'Apiguard' (Vita), 'Exomite Apis' (Exosect).

APPLICATION: 'Apiguard' is placed in the hive on a piece of wax sheet, cardboard or plastic sheet centrally on top of the brood frames as a treatment tray. Fifty grams of gel are removed from the tub using the dosing tools (scoop and spatula) and placed on the tray. After two weeks, a second dose is applied following the same procedure. The product is left in the colony until it disappears from the tray. Small and wintering bee colonies may require a single dose of 25 g gel only. Treatments must not be made during honey flow; when the temperature is below 25 °C; or when the colony activity is very low. The efficacy of the end-use product is maximised if the product is used in late summer after the honey harvest (when the number of brood present is diminishing). However, in the case of severe infestations, the end-use product can also be used during the Spring, when temperatures are above 25 °C. For 'Exomite Apis', two applications are required at two-week intervals when the drone brood is at its smallest and during periods before and after honey flow (Spring and Autumn). Each application lasts 12 days and they are applied immediately after each other in order to ensure the mites, which would emerge over a prolonged period, are exposed to the powder. 'Exomite Apis' should be used as part of an Integrated Pest Management approach and can form part of a resistance management strategy.

PRODUCT SPECIFICATIONS: Thymol is manufactured for use in agriculture.

COMPATIBILITY: It is unusual to use varroa mite products containing thymol with any other treatment.

MAMMALIAN TOXICITY: Not considered to be toxic to mammals. **Acute oral LD$_{50}$:** mice 640 mg/kg. US-EPA Toxicity Category III. **Acute dermal LD$_{50}$:** rats 1049 mg/kg. US-EPA Toxicity Category II. **Inhalation:** LC$_{50}$ for rats >5000 mg/kg. US-EPA Toxicity Category IV.

Skin and eye: At high concentrations, thymol is corrosive to skin and eye and it is also a skin sensitiser. However, its use in beehives means that it is unlikely that the public would be exposed to it. Workers are required to wear appropriate personal protective equipment to minimise their exposure. US-EPA Toxicity Category I.

ENVIRONMENTAL IMPACT AND NON-TARGET TOXICITY: Because of the use pattern of thymol, it is unlikely that any non-target organisms will be exposed. In addition, it has low mammalian toxicity and is a common component of a widely used foodstuff.

Fish toxicity: Thymol is hazardous to aquatic organisms and must not be allowed to come into contact with waterways. **Approved for use in organic farming:** Yes.

2:235 trypsin modulating oostatic factor
Mosquito larvicide

$$H_2N–Tyr–Asp–Pro–Ala–(Pro)_6–COOH$$

NOMENCLATURE: Approved name: Trypsin Modulation Oostatic Factor.
Other names: TMOF. **OPP Chemical Code:** 105403.

SOURCE: Trypsin modulating oostatic factor occurs naturally in mosquitoes.

PRODUCTION: The gene coding for this peptide was isolated and engineered into *Pichia pastoris* Guillierm, Phaff (as *pastori*) Antonie van Leeuwenhoek yeast. TMOF is produced by the fermentation of this engineered yeast. The yeast cells are then killed by exposure to extremely high temperatures.

TARGET PESTS: Mosquito larvae. TMOF has been shown to be effective against several mosquito species (*Aedes aegypti* (L.), *A. albopticus*, *A. taneorhynchus*, *Culex fatigans* Wiedemann, *C. nigripalpus* Theobald, *Anopheles quadrimaculatus* (Say.), *A. albimanus* Wiedemann) and the sand fly (*Lutzomyia anthophora* (Addis)).

BIOLOGICAL ACTIVITY: Mode of action: TMOF is used to down-regulate digestive trypsin-like protease translation and transcription after passing into the haemolymph from the midgut, causing larvae to succumb to starvation. After application to water, TMOF is consumed by mosquito larvae which are aquatic particulate filter feeders. As the yeast is broken down by digestive processes, TMOF is released into the insect gut. TMOF then passes through the peritropic membrane and midgut wall into the haemolymph. TMOF then binds to specific receptors which, once activated, trigger a halt in trypsin-like enzyme biosynthesis and ultimately mRNA synthesis. The biochemical process of receptor activation is not well understood, although affected insects are unable to produce digestive proteases needed to provide essential amino acids. **Biology:** TMOF interferes with the production of trypsin in mosquito larvae, a critical enzyme needed by the larval digestive system. The mode of action of TMOF is hormonal disruption of transcription and translation of trypsin, resulting in reduced digestion of mosquito diet, ultimately leading to starvation of mosquito larvae.

COMMERCIALISATION: Formulation: TMOF is under evaluation by the US-EPA. The product is a solid that can be dispersed onto bodies of water. **Tradenames:** 'Skeetercide' (Insect Biotechnology).

APPLICATION: TMOF is applied directly to water or other places mosquito larvae may be found.

PRODUCT SPECIFICATIONS: The gene coding for this peptide was isolated and engineered into *Pichia pastoris* Guillierm, Phaff (as *pastori*) Antonie van Leeuwenhoek yeast. TMOF is produced by the fermentation of this engineered yeast. The yeast cells are then killed by exposure to extremely high temperatures.

COMPATIBILITY: It is not expected that TMOF will be used with other compounds.

MAMMALIAN TOXICITY: TMOF protein is broken down rapidly in the human gut and does not have the opportunity to inhibit trypsin synthesis. **Acute oral LD$_{50}$:** mice >2000 mg/kg. US-EPA Toxicity Category III. **Acute dermal LD$_{50}$:** rabbits >5000 mg/kg. US-EPA Toxicity Category IV **Inhalation:** LC$_{50}$ for rats >2.4 mg/litre. US-EPA Toxicity Category IV. **Skin and eye:** A mild irritant with a US-EPA Toxicity Category III. Classified as US-EPA Toxicity Category III for eye irritation.

ENVIRONMENTAL IMPACT AND NON-TARGET TOXICITY: Bird toxicity: NOEC 14-day old mallard ducks (*Anas platyrhynchos*) was >125 mg TMOF yeast/kg body weight. **Fish toxicity:** The NOEC of TMOF to sheepshead minnow (*Cyprinodon variegatus*) was >1.038 × 10^6 cells/ml. **Other aquatic toxicity:** NOEC of TMOF against *Daphnia magna* (flow-through) was 1.026 × 10^6 cells/ml. The NOEC of TMOF yeast to mysid shrimp (*Americamysis bahia*) was <1.037 × 10^6 cells/ml and the LOEC was 1.037 × 10^6 cells/ml.

2:236 validamycin
Micro-organism-derived fungicide

The Pesticide Manual Fifteenth Edition entry number: 897

NOMENCLATURE: Approved name: validamycin (JMAF); validamycin A (Japanese Antibiotics Research Association). **CAS RN:** *[37248-47-8]* validamycin; *[38665-10-0]* validoxylamine A (metabolite).

SOURCE: Originally isolated from the soil Actinomycete, *Streptomyces hygroscopicus* (Jensen) Waks. & Henrici var. *limoneus.*

PRODUCTION: By the fermentation of *S. hygroscopicus* var. *limoneus.* Validamycin A is the most active component of a mixture of seven closely related compounds known as validamycins A to G.

TARGET PESTS: Control of *Rhizoctonia solani* Kuehn and other *Rhizoctonia* species.

TARGET CROPS: Recommended for use in rice, potatoes, vegetables, strawberries, tobacco, ginger, cotton, sugar beet and other crops.

BIOLOGICAL ACTIVITY: Mode of action: Non-systemic with fungistatic action. Validamycin shows no fungicidal action against *Rhizoctonia solani*, but causes abnormal branching of the tips of the pathogen followed by a cessation of further development. It has been shown that validamycin has a potent inhibitory activity against trehalase in *Rhizoctonia solani* AG-1, without any significant effects on other glycohydrolytic enzymes tested. Trehalose is well known as a storage carbohydrate in the pathogen and trehalase is believed to play an essential role in the digestion of trehalose and transport of glucose to the hyphal tips.
Efficacy: Low rates of use give excellent control of *Rhizoctonia solani* in various crops. Rates of 30 g per hectolitre gave effective control of rice sheath blight. **Key reference(s): 1)** S Horii, Y Kameda & K Kawahara. 1972. Studies on validamycins, new antibiotics. VIII. Validamycins C, D, E and F, *J. Antibiot.*, **25**, 48. 2) K Matsuura. 1983. Characteristics of validamycin A in controlling *Rhizoctonia* diseases. In *IUPAC Pesticide Chemistry*, J Miyamoto & P C Kearney (eds), Vol. 2, p. 301, Pergamon Press, Oxford, UK. 3) R Shigemoto, T Okuno & K Matsuura. 1989. Effect of

validamycin A on the activity of trehalase of *Rhizoctonia solani* and several sclerotial fungi, *Ann. Phytopathol. Soc. Jpn*, **55**, 238.

COMMERCIALISATION: Formulation: Sold as dispersible powder (DP), soluble concentrate (SL), powder seed treatment (DS) and liquid formulations.
Tradenames: 'Validacin' and 'Valimun', 'Dantotsupadanvalida' (plus cartap hydrochloride plus clothianidin) and 'Hustler' (plus cartap hydrochloride plus clothianidin plus ferimzone plus phthalide) (Sumitomo Chemical Takeda), 'Sheathmar' (Sumitomo Chemical Takeda) and (Dhanuka), 'Mycin' (Sanonda), 'Rhizocin' (Nagarjuna Agrichem), 'Solacol' (Bayer CropScience), 'Valida' (Nichimen), 'Kasu-rab-valida-sumi' (plus fenitrothion plus kasugamycin plus phthalide) and 'Kasu-rab-validatrebon' (plus etofenprox plus kasugamycin hydrochloride hydrate plus phthalide) (Hokko), 'Vivadamy' and 'Vimix' (plus 1-naphthylacetic acid plus 2-naphthyloxyacetic acid) (Vipesco).

APPLICATION: Applied as a foliar spray, a soil drench, a seed treatment or by soil incorporation. Rates from 30 g per hectolitre give good control.

PRODUCT SPECIFICATIONS: By the fermentation of *S. hygroscopicus* var. *limoneus*. Validamycin A is the most active component of a mixture of seven closely related compounds known as validamycins A to G.

COMPATIBILITY: Validamycin can be used in conjunction with many other agrochemicals. Concentrations as high as 1000 mg per litre showed no phytotoxicity to over 150 different target crops.

MAMMALIAN TOXICITY: Validamycin is not considered to be toxic to mammals.
Acute oral LD$_{50}$: rats and mice >20000 mg/kg. **Acute dermal LD$_{50}$:** rats >5000 mg/kg.
Inhalation: LC$_{50}$ (4 h) for rats >5 mg/litre air. **Skin and eye:** Non-irritating to skin (rabbits). Not a skin sensitiser (guinea pigs). **NOEL:** In 90-day feeding trials, rats receiving 1000 mg/kg of diet and mice receiving 2000 mg/kg of diet showed no ill-effects. In 2-year feeding trials, NOEL for rats was 40.4 mg/kg b.w. daily. **Toxicity class:** WHO (a.i.) U; EPA (formulation) IV.
Other toxicological effects: Non-mutagenic in bacterial reversion assay systems. Non-teratogenic.

ENVIRONMENTAL IMPACT AND NON-TARGET TOXICITY: Validamycin has no adverse effects on non-target organisms or on the environment. **Bird toxicity:** No effect on chickens or quail at 12.5 g/kg administered orally. **Fish toxicity:** LC$_{50}$ (72 h) for carp >40 mg/litre.
Other aquatic toxicity: LC$_{50}$ (24 h) for *Daphnia pulex* >40 mg/litre. **Metabolism:** In animals, orally administered validamycin is readily decomposed to carbon dioxide and amine residues that are excreted. **Behaviour in soil:** Rapid microbial degradation in soil; DT$_{50}$ ≤5 hours.

2:237 yeast extract hydrolysate

Fungal-derived fungicide

NOMENCLATURE: Approved name: yeast extract hydrolysate.

SOURCE: Yeast extract hydrolysate is a by-product of the fermentation of brewers yeast (*Saccharomyces cerevisiae* Meyen).

PRODUCTION: Yeast extract hydrolysate is prepared from hydrolysed brewer's yeast protein, with subsequent oxidation of its constituent amino acids.

TARGET PESTS: Yeast extract hydrolysate is used to control many plant diseases, such as post-bloom fruit drop and greasy spot diseases of citrus, and bacterial leaf spot disease of tomatoes.

TARGET CROPS: All food crops, ornamentals and turf.

BIOLOGICAL ACTIVITY: Mode of action: The mode of action is thought to be due to a stimulation of the treated plant's natural defence mechanisms. It is also said to increase the yield of treated crops and lengthen the shelf-life of the harvested produce.

COMMERCIALISATION: Formulation: 'KeyPlex 350' is an end-use product formulation containing 0.063% yeast hydrolysate, with chelated micronutrients, in a liquid formulation. **Tradenames:** 'KeyPlex 350' (Morse Enterprises).

APPLICATION: 'KeyPlex 350' may be applied by conventional ground or aerial foliar application, or by 'fertigation' (low volume irrigation). Application rates for most crops and turf range from 2 to 6 litres per hectare (1 to 3 quarts per acre), with applications generally repeated at 14- to 21-day intervals. For citrus, foliar spray applications are made pre-bloom, at petal fall plus 1 to 2 summer sprays, tank mixed with summer oil. For vegetables, including tomato and pepper, foliar spray applications should begin at the 4- to 6-leaf stage, with two applications pre-bloom and four applications post-bloom. For turf and most other crops, spray applications are made pre-bloom plus two applications post-bloom; for annuals, the first spray is applied at the 4- to 6-leaf stage.

PRODUCT SPECIFICATIONS: Specifications: Yeast extract hydrolysate is prepared from hydrolysed brewer's yeast protein, with subsequent oxidation of its constituent amino acids. **Purity:** 'KeyPlex 350' contains no impurities of toxicological significance.

COMPATIBILITY: 'KeyPlex 350' is compatible with most crop protection agents. It is recommended that 400 to 600 g of urea or potassium nitrate be added per 100 litres (3 to 5 lb per 100 gallons) of water to aid leaf penetration.

MAMMALIAN TOXICITY: All products based on yeast extract hydrolysate have been deemed by the US Food and Drug Administration (FDA) to be Generally Regarded As Safe

(GRAS). **Acute oral LD$_{50}$:** rats >5000 mg/kg. **Acute dermal LD$_{50}$:** rats >5000 mg/kg.

Skin and eye: A primary dermal irritation study has shown 'KeyPlex 350' to be very slightly to non-irritating, placing the product in Toxicity Category IV. An eye irritation study indicated that the product is slightly irritating to rabbit eyes, with no corneal or iridial damage and only slight conjunctival irritation, which cleared within 7 days, placing the product in Toxicity Category III for primary eye irritation.

ENVIRONMENTAL IMPACT AND NON-TARGET TOXICITY: Because of the lack of toxicity of 'KeyPlex 350', the common occurrence of yeast extract hydrolysate in a large number of food products and its extremely low use rates, no environmental risk is anticipated by its use. **Approved for use in organic farming:** Yes: it has US OMRI approval.

2:238 zeatin
Plant-derived plant growth regulator

The Pesticide Manual **Fifteenth Edition entry number:** 224

NOMENCLATURE: Approved name: zeatin. **CAS RN:** *[1637-39-4]*, formerly *[10052-59-2]* and *[129900-07-8]*.

SOURCE: Plant-derived growth regulator, often extracted from seaweeds. Involved in the initiation of plant cell division.

PRODUCTION: Extracted from plant tissue or seaweed.

TARGET CROPS: Recommended for use on a wide range of crops, including citrus, cucumber, stone fruit crops, peppers, pine, potatoes and tomatoes.

BIOLOGICAL ACTIVITY: Mode of action: Zeatin is a naturally occurring plant growth regulator from the cytokinin group. It is associated with cell division, and subsequently increased plant growth, and also delays senescence in treated organs.

COMMERCIALISATION: Formulation: Sold as a water soluble concentrate (SL). **Tradenames:** 'Cytex' (Atlantic and Pacific Research).

APPLICATION: Apply as a foliar spray to growing crops. It can also be applied as a fluid-drilling gel at the time of planting.

PRODUCT SPECIFICATIONS: Extracted from plant tissue or seaweed.

MAMMALIAN TOXICITY: Naturally occurring plant growth regulator that has not shown any allergic or other adverse effects on producers or users. **Acute oral LD$_{50}$:** rats >5000 mg/kg. **Acute dermal LD$_{50}$:** rabbits >2000 mg/kg. **Skin and eye:** Slight skin and eye irritant (rabbits).

ENVIRONMENTAL IMPACT AND NON-TARGET TOXICITY: Cytokinins occur widely in Nature and are not expected to have any adverse effects on non-target organisms or on the environment. **Effects on beneficial insects:** Not toxic to honeybees. **Behaviour in soil:** Rapidly degraded in soil. **Approved for use in organic farming:** Yes: it has US OMRI approval.

3. Macro-organisms

3:239 *Aceria malherbae*

Phytophagous mite

Phytophagous mite: Acari: Eriophyidae

NOMENCLATURE: Approved name: *Aceria malherbae* Nuzzaci. **Other names:** Bindweed gall mite. Formerly known as: *Aceria convolvuli* (Nalepa).

SOURCE: The mite originated in Greece, France and Italy, but its native distributions now include central and southern Europe and northern Africa.

PRODUCTION: Reared on field bindweed (*Convolvulus arvensis* L.) plants.

TARGET PESTS: The plant species attacked by *Aceria malherbae* are bindweeds (*Convolvulus arvensis* and *Calystegia* spp.).

BIOLOGICAL ACTIVITY: Biology: Eggs are spherical and transparent and are deposited in galls, swellings of the stems of the field bindweed produced when attacked by *Aceria malherbae*. There are two nymph stages, which are similar in appearance to the adults, but are smaller and lacking in external genitalia. The adults are yellow-white in colour and their bodies are soft, with no exoskeleton. The adults are active from May to November. They are small and worm-like, 0.2 mm in length. The adult has a circular body with two pairs of legs found on the joined head and thorax. Adult and nymph stages overwinter, living on root buds under the soil. **Predation:** The mite causes galls in the plant. In a heavily damaged plant, the shoots are stunted in growth and they are misshapen. They attack along the midvein of the plant and produce a 'fuzzy' texture and galls also occur on the leaves, petioles and stems. The adult and nymph stages are the most damaging. **Efficacy:** The mite causes stunted growth and the plant does not produce flowers. Mites produce galls on actively growing leaves and stem buds. When the stem buds are attacked by the mites, they fail to grow and, therefore, grow in a close bunch of distorted and shortened leaves. **Key reference(s)**: A McClay. 1999. *Aceria malherbae* Nuzzaci (Acari: Eriophyidae), www.arc.ab.ca/crop/weed/BindweedAgent.html.

COMMERCIALISATION: Formulation: Galls containing adults and nymphs. **Tradenames:** 'Aceria malherbae' (Praxis).

APPLICATION: Both adult and nymph stages can be transferred to growing field bindweed plants and both stages are found in the galls. Mites in galls will live as long as the galls are kept damp and cool. Galls containing *Aceria malherbae* are distributed throughout the weed, ensuring that the growing plants are in contact with the galls. The mite is slow-moving and its efficacy is improved if the galls are moved into the weedy area as they are produced.

PRODUCT SPECIFICATIONS: Purity: Commercial products consist of field bindweed galls containing adults and nymphs of *Aceria malherbae* and no contaminants harmful to man.

3. Macro-organisms

Storage conditions: Keep galls moist and cool. Shelf-life: Adults and nymphs will remain alive with galls that are stored as recommended.

COMPATIBILITY: Incompatible with acaricides.

MAMMALIAN TOXICITY: There have been no reports of allergic or other adverse reactions from research workers, manufacturing staff or from the field release of *Aceria malherbae*.

ENVIRONMENTAL IMPACT AND NON-TARGET TOXICITY: *Aceria malherbae* occurs widely in Nature and is not expected to have any adverse effects on non-target organisms or on the environment. It has not been known to feed on any plants other than bindweeds. **Approved for use in organic farming:** Yes.

3:240 *Adalia bipunctata*   *Predator of aphids*

Predatory ladybird: Coleoptera: Coccinellidae

NOMENCLATURE: **Approved name:** *Adalia bipunctata* (L.). **Other names:** Two-spot ladybird; ladybug beetle.

SOURCE: Originally found in the Palaearctic, but now occurs throughout the Holarctic in a variety of habitats and is very common throughout central and northern Europe.

PRODUCTION: Usually collected in the wild.

TARGET PESTS: Preys on a wide variety of aphids.

TARGET CROPS: Can be used in a variety of protected and open field crops, including vegetables, ornamentals, shrubs, trees, roses, vineyards and orchards.

BIOLOGICAL ACTIVITY: **Biology:** The adult beetle is very variable in colour, but the typical form is red, with one round black mark on each wing cover. Another common form is mainly black, with two red marks on each wing cover. Adults are 3.5 to 5.5 mm long. There are four larval instars before pupation. **Predation:** Both adults and larvae are predacious on aphids, actively searching plants for aphid colonies. Newly hatched larvae search for prey immediately and their survival depends upon the presence of prey close to the hatching site. When prey is scarce, the larvae become cannibalistic. **Egg laying:** Adults lay elongated, yellowish-orange eggs, which are usually laid in batches of 20 to 50 on the surface of the plant close to an aphid colony. Hatching occurs in 4 to 8 days, depending upon temperature. The presence of abundant conspecific larvae can inhibit egg laying. **Duration of development:** Larval development, as well as adult longevity and fecundity, is influenced by diet as well as by temperature. The ladybirds perform much better when fed on some aphid species than when fed on others. Under laboratory conditions, larvae completed development in about 9 days when fed on the damson-hop aphid, *Phorodon humuli* (Schrank). By contrast, development took 14 to 15 days when fed on the black bean aphid, *Aphis fabae* Scopoli, and mortality was also very high. The pupal stage lasts about 8 days at 20°C. Egg to adult development takes around 3 weeks at 20 to 25°C. **Efficacy:** Very active and voracious predators in both the larval and adult stages. The adult is an active flier and so can disperse throughout a crop very quickly, but can also rapidly leave open field crops if prey is scarce.

COMMERCIALISATION: Formulation: Sold as adults, often in popcorn. 'Adalia250L' contains L2 and L3 larvae, in inert materials with some food for transport.
Tradenames: 'Adalia-System' (Biobest), 'Adalline b' (Syngenta Bioline), 'Adalia250L' (Bioplanet).

APPLICATION: Release into crops when aphids are present. It is very important to guarantee a careful distribution around and within infested sites as the beetle needs large quantities of food.

PRODUCT SPECIFICATIONS: Shelf-life: Release as soon after receipt as possible.

COMPATIBILITY: It is not usual to use *Adalia bipunctata* with chemical agents.

MAMMALIAN TOXICITY: There have been no reports of allergic or other adverse toxicological effects arising from contact with *Adalia bipunctata* from research staff, producers or users.

ENVIRONMENTAL IMPACT AND NON-TARGET TOXICITY: *Adalia bipunctata* is widespread in Nature and is not thought to be damaging to non-target species nor to the environment. **Approved for use in organic farming:** Yes.

3:241 *Agapeta zoegana*
Phytophagous caterpillar

Phytophagous caterpillar: Lepidoptera: Cochylidae

NOMENCLATURE: Approved name: *Agapeta zoegana* (L.). **Other names:** Sulfur knapweed root moth.

SOURCE: *Agapeta zoegana* is a root-boring moth from Europe that was first released into the United States in 1984.

PRODUCTION: The adult moths can be collected from knapweed sites in the USA. They have a requirement for knapweed plants and can be reared successfully on both spotted and diffuse knapweed.

TARGET PESTS: Spotted knapweed (*Centaurea maculosa* Monnet De La Marck) and diffuse knapweed (*C. diffusa* Monnet De La Marck).

TARGET CROPS: Used in rangeland.

BIOLOGICAL ACTIVITY: Biology: *Agapeta zoegana* is a small, bright yellow moth, 10 mm in length, with brown wing bands. The adults may be found resting vertically on the knapweed stems or under the leaves. They have the appearance of dead or dying knapweed leaves. **Predation:** *A. zoegana* is considered host-specific to spotted and diffuse knapweed. It has been tested on 51 plant species, including closely related native species and species of economic importance, such as safflower. **Egg laying:** Adult moths emerge from knapweed roots from early July until early September. Mating takes place within 24 hours of emergence and the mated female begins laying eggs after 24 hours. The eggs are laid in the stem crevices and on the leaves of the spotted and diffuse knapweed plants. The larvae hatch from the eggs in 7 to 10 days and move immediately to the root crown and mine into the root. Larvae are capable of killing small rosettes and then moving a few centimetres to another knapweed plant to feed. The larvae pupate in the root in midsummer. Adults live from 11 to 14 days, with each female laying between 21 and 28 eggs in her lifetime. The moths are strong fliers and will invade new knapweed patches. **Efficacy:** It is expected that this moth will be an effective control agent in conjunction with other root-boring and seed-feeding insects. Spotted and diffuse knapweed patches are hard to find in Europe, because they are kept under control with the native insects and diseases. **Key reference(s):** 1) H Muller. 1989. Growth pattern of diploid and tetraploid spotted knapweed, *Centaurea maculosa* Lam. (Compositae), and the effects of the root-mining moth *Agapeta zoegana* (L.) (Lep.: Cochylidae), *Weed Research*, **29**, 103–11. 2) H Muller, D Schroeder & A Gassmann. 1988. *Agapeta zoegana* (L.) (Lepidoptera: Cochylidae), a suitable prospect for biological control of spotted and diffuse knapweed, *Centaurea maculosa* Monnet De La Marck (Compositae) and *Centaurea diffusa* Monnet De La Marck (Compositae) in North America, *Can. Entomol.*, **120**, 109–24. 3) J M Story, K W Boggs & W R Good. 1990. First report of the establishment of *Agapeta zoegana* L. (Lepidoptera: Cochylidae) on spotted knapweed, *Centaurea maculosa* La Marck, in the United States. Montana Agricultural Experiment Station, Western Agricultural Research Center, Corvallis, MT 59828, USA.

COMMERCIALISATION: Formulation: *Agapeta zoegana* has been released in Arizona, California, Colorado, Idaho, Minnesota, Montana, Nebraska, Nevada, North Dakota, Oregon, South Dakota, Utah, Washington, Wisconsin and Wyoming. The moth has established in these states. In some states, *A. zoegana* adults may be obtained at no cost from state weed management agencies. Several commercial suppliers can also provide *A. zoegana* adults. **Tradenames:** 'Agapeta zoegana' (Praxis).

APPLICATION: For best establishment results, *A. zoegana* should be released in areas where there are green knapweed rosettes in August and in the area of the knapweed patch where the knapweed is abundant, but not a monoculture.

PRODUCT SPECIFICATIONS: Shelf-life: Adults should be released as soon as possible after receipt.

COMPATIBILITY: It is not usual to use *Agapeta zoegana* with chemical agents.

MAMMALIAN TOXICITY: There have been no reports of allergic or other adverse toxicological effects arising from contact with *Agapeta zoegana* from research staff, producers or users.

ENVIRONMENTAL IMPACT AND NON-TARGET TOXICITY: *Agapeta zoegana* occurs widely in Europe, where it invades only knapweed plants. In the USA, it has been tested on 51 plant species, including closely related native species and species of economic importance, such as safflower. It is not expected that its release will have any adverse effects on non-target species or on the environment. **Approved for use in organic farming:** Yes.

3:242 *Agasicles hygrophila*

Phytophagous beetle

Phytophagous beetle: Coleoptera: Chrysomelidae

NOMENCLATURE: Approved name: *Agasicles hygrophila* Selman & Vogt.
Other names: Alligatorweed flea beetle.

SOURCE: Alligatorweed was introduced into the USA in the 1890s from South America. The alligatorweed flea beetle, *Agasicles hygrophila*, is a native of South America, specifically southern Brazil and northern Argentina, and was introduced into the USA in 1964 to combat alligatorweed which had invaded United States waterways from Virginia to southern Florida and along coast waterways to Texas and in California.

PRODUCTION: Once introduced, the beetles survive on alligatorweed and further introductions are unnecessary. Initial introductions are through collection and redistribution of adults.

TARGET PESTS: Alligatorweed (*Alternanthera philoxeroides* (Mart.) Griseb.).

TARGET CROPS: Water courses.

BIOLOGICAL ACTIVITY: Biology: Adults are black, about 4 to 6mm long, with two longitudinal yellow stripes. The larvae are black, up to 6mm long when mature third instars; eggs are yellowish, 1.25mm long and are laid in clusters on the undersides of leaves. Adults have greatly enlarged hind femora for jumping, hence the name flea beetle. **Predation:** Adults overwinter among the roots and stems of alligatorweed along the margins of waterways. Eggs are laid on the undersides of leaves. Larvae eat the leaves during their development, often leaving transparent windows of uneaten epidermis. Pupation takes place inside the hollow plant stems. The life-cycle takes less than one month, and adults live about 48 days. **Efficacy:** The alligatorweed flea beetle is so successful that it is often used as the symbol of biological weed control. Within 4 years, alligatorweed was practically eliminated at the two northern Florida sites where the flea beetle was first introduced. The beetle is less effective in southern Florida and in northern areas of Alabama, Louisiana, Mississippi and in North and South Carolina. Additional beetles from a colder region in Argentina were released in the Carolinas in 1979, apparently without success. **Key reference(s):** 1) G R Buckingham. 1994. Biological control of aquatic weeds. In *Pest Management in the Subtropics: Biological Control – a Florida Perspective.* D Rosen, F D Bennett & J L Capinera (eds), Intercept Ltd, Andover, UK, 737 pp. 2) D M Maddox. 1968. Bionomics of an alligatorweed flea beetle, *Agasicles* sp., in Argentina, *Ann. Entomol. Soc. America,* **61**, 1299–2305. 3) G B Vogt, P C Quimby Jr & S H Kay. 1992. Effects of weather on the biological control of alligatorweed in the lower Mississippi Valley region, 1973–83. *USDA Tech. Bull. 1766,* 143 pp.

COMMERCIALISATION: Formulation: *Agasicles hygrophila* is not available commercially, but was introduced into the USA for control of alligatorweed.

APPLICATION: Adult flea beetles are introduced to infestations of alligatorweed, ideally in early summer.

PRODUCT SPECIFICATIONS: Shelf-life: Adults should be released as soon as possible after receipt.

COMPATIBILITY: Susceptible to topically applied parathion-methyl.

MAMMALIAN TOXICITY: There have been no reports of allergic or other adverse toxicological effects arising from contact with *Agasicles hygrophila* from research staff, producers or users.

ENVIRONMENTAL IMPACT AND NON-TARGET TOXICITY: Host-range studies in Argentina and Uruguay limited to 14 plant species in 8 families determined that only alligatorweed was suitable as a host plant for *A. hygrophila.* No additional testing has been done in North America. It is not expected that its release will have any adverse effects on non-target species or on the environment.

3:243 *Agrilus hyperici* *Phytophagous beetle*

Phytophagous beetle: Coleoptera: Buprestidae

NOMENCLATURE: Approved name: *Agrilus hyperici* (Creutzer). **Other names:** St John's wort beetle.

SOURCE: *Agrilus hyperici* is native to southern, central and eastern Europe. Commercial samples were first collected in France. In North America, the beetle is found, for the most part, in the mountain areas. In Europe, it tends to favour drier regions as, in damp sites, larvae often succumb to fungal attack. This particular beetle has also been established in Australia where it disperses widely and, in most cases, is thought lost for years after the release, but then the population grows extremely rapidly and damage to the weed is obvious. *Agrilus hyperici* will also attack plants in the shade that are not attacked by *Chrysolina hyperici*.

PRODUCTION: Produced on living St John's wort (*Hypericum perforatum* L.).

TARGET PESTS: *Agrilus hyperici* is used to control St John's wort (*Hypericum perforatum*) and *Hypericum montanum*. The beetle has also been found on *Hypericum concinnum* in California.

TARGET CROPS: The beetle is used in pasture and grassland.

BIOLOGICAL ACTIVITY: Biology: Eggs are laid on stems from just above soil level to a height of approximately 20 cm (8 inches) in July and August. The larvae are long and white and have flattened, dark brownish heads. They feed within the roots from July until May or June of the following year. The larvae pupate inside the damaged root during early May to June. Initially, pupae are a creamy white, but they darken with age. This stage lasts 9 to 15 days under laboratory conditions. Reddish-bronze adults appear in July or early August. They are usually active during the heat of the day. They disperse readily by flight and feed on the St John's wort foliage, causing only minor damage to the plant. They are about 5 mm in length and somewhat tapered and flat towards the rear. The beetle overwinters as larvae within the host root. **Efficacy:** When larvae feed within the root of a plant, the tissue is entirely consumed, leading to severe damage to the plant. Flower production is reduced, because stems produced from an infested root crown are stunted. Most infested plants are killed.
Key reference(s): 1) D T Briese. 1991. Current status of *Agrilus hyperici* (Coleoptera: Buprestidae) release in Australia in 1940 for the control of St John's wort: Lessons for insect introductions, *Biocontrol Science and Technology*, **1**, 207–15. 2) D T Briese. 1991. Demographic processes in the root-borer *Agrilus hyperici* (Coleoptera: Buprestidae): a Biological Control Agent of St John's wort, *Biocontrol Science and Technology*, **1**, 195–206. 3) G L Piper. 1986. The St John's wort root borer, *Agrilus hyperici*, in Washington, *Pacific Northwest Weed Topics*, **86-3**, 3–4.

COMMERCIALISATION: Formulation: First introduced into California in 1950, but now established in Washington, Oregon, Montana, Idaho and California. Only adults can be

3. Macro-organisms

introduced into the weed areas. **Tradenames:** 'Agrilus hyperici' (Praxis), (BioCollect) and (CalTec).

APPLICATION: The beetles should be released in July and August as soon as possible after delivery.

PRODUCT SPECIFICATIONS: Purity: Fecund adults, with no contaminants.
Storage conditions: *Agrilus hyperici* can be stored for 1 to 3 days in a refrigerator with some fresh foliage in the container. They can be transported in a cooler with an ice pack to the release site. **Shelf-life:** Adults should be released as soon as possible after receipt.

COMPATIBILITY: It is not usual to use *Agrilus hyperici* with chemical agents.

MAMMALIAN TOXICITY: There have been no reports of allergic or other adverse reactions from research workers, manufacturing staff or from the field release of *Agrilus hyperici*.

ENVIRONMENTAL IMPACT AND NON-TARGET TOXICITY: *Agrilus hyperici* occurs widely in Nature and is not expected to have any adverse effects on non-target organisms or on the environment. It has not been shown to feed on plants other than St John's Wort.

3:244 *Amblyseius andersoni*
Predator of mites

Predatory mite: Acari: Phytoseiidae

NOMENCLATURE: Approved name: *Amblyseius andersoni* (Chant). **Other names:** Spider mite predator.

SOURCE: Predatory mite that occurs widely in Europe, particularly in orchards and boxwood plantations.

PRODUCTION: Reared under controlled conditions on a diet mainly of spider mites and thrips, but also rust mites, gall mites and boxwood bud mites (*Eurytetranychus buxi* Garman).

TARGET PESTS: Spider mites, thrips, gall mites, fruit tree red spider mite (*Panonychus ulmi* (Koch)), bud mites and eriophiid (rust) mites.

TARGET CROPS: Tree nurseries, such as apple and pear orchards and boxwood plantations, glasshouse ornamentals, such as roses, and open-field roses.

BIOLOGICAL ACTIVITY: Biology: Eggs are laid on the hairs in the axils of the midvein and the lateral veins on the underside of leaves. The eggs are oval-shaped, white and clear, and hatch into six-legged larvae that do not feed, but remain in groups near their place of emergence. Proto- and deutonymphal stages and adults have eight legs, are mobile and feed. **Predation:** Immature predatory mites consume both spider mite eggs and nymphs. An adult female can consume in excess of 150 prey over a 16-day period. *Amblyseius andersoni* feeds on phytophagous mites and thrips by piercing the prey and sucking the contents and will also consume rust mites, gall mites and boxwood bud mites (*Eurytetranychus buxi*). **Egg laying:** Egg laying varies with temperature, from less than one per day (to yield a total of 48 throughout the female life-cycle) at 13 °C, to over 3.5 per day (to yield a total of 65 throughout the female life-cycle) at 33 °C. **Duration of development:** Development, egg laying and longevity are dependent upon temperature, the type and availability of food and the ambient humidity. **Efficacy:** A very effective predator, as it will consume thrips in addition to a wide variety of mites. **Key reference(s):** 1) C Duso. 1989. Role of the predatory mites *Amblyseius aberrans* (Oud.), *Typhlodromus pyri* Scheuten and *Amblyseius andersoni* (Chant) (Acari, Phytoseiidae) in vineyards. 1. The effect of single or mixed phytoseiid population releases on spider mite densities (Acari, Tetranychidae). *J. Appl. Ent.*, **107**, 474–92. 2) C Duso. 1992. Role of the predatory mites *Amblyseius aberrans* (Oud.), *Typhlodromus pyri* Scheuten and *Amblyseius andersoni* (Chant) (Acari, Phytoseiidae) in vineyards. III. Influence of variety characteristics on the success of *A. aberans* and *T. pyri* releases. *J. Appl. Ent.*, **114**, 455–62.

COMMERCIALISATION: Formulation: 1 000 live adults and juvenile predators per unit in vermiculite or corn husks. Contains an unspecified number of predator eggs.
Tradenames: 'Amblyseius andersoni' (Rincon-Vitova), 'Amblyseius andersoni' (under development) (Biobest), 'Anderline aa' (Syngenta Bioline).

APPLICATION: The introduction of 100 *Amblyseius andersoni* per m² gives excellent results.

PRODUCT SPECIFICATIONS: Purity: Commercial product contains only *Amblyseius andersoni*, with no mould or other contaminants. **Storage conditions:** Store at 8 to 10 °C, 65 to 80% r.h. and out of direct sunlight. Protect from freezing and high temperatures.
Shelf-life: Use as soon as possible after delivery.

COMPATIBILITY: Very sensitive to insecticides and benzimidazole-based fungicides. Do not release if methomyl or a synthetic pyrethroid has been used.

MAMMALIAN TOXICITY: There have been no reports of allergic or other adverse reactions following laboratory development or field use.

ENVIRONMENTAL IMPACT AND NON-TARGET TOXICITY: *Amblyseius andersoni* is widespread in Europe and there have been no reports of adverse environmental impact or effects on non-target organisms. It will feed on other mites and thrips, but will not reproduce in the absence of spider mites and this limits its longevity when the spider mite population is well controlled. **Approved for use in organic farming:** Yes.

3:245 *Amblyseius barkeri*   *Predator of thrips*

Predatory mite: Acari: Phytoseiidae

NOMENCLATURE: Approved name: *Amblyseius barkeri* (Hughes). **Other names:** Thrips predator; thrips predatory mite. Formerly known as: *Amblyseius mackenziei* Schuster & Pritchard and *Neoseiulus barkeri* Hughes.

SOURCE: *Amblyseius barkeri* is cosmopolitan and occurs widely in Nature, being found in Europe, northern and equatorial Africa, California and Israel. It thrives under conditions of high humidity, but enters diapause when days shorten and temperatures fall below 15 °C.

PRODUCTION: Reared on living, storage prey mites, such as *Tyrophagus* spp. or *Acarus siro* L., in a ratio of 1:1 to 1:5 predator:prey, excluding eggs. Small quantities of bran are provided as food for the prey mites.

TARGET PESTS: Main prey are thrips (Thysanoptera), including *Frankliniella occidentalis* (Pergande), *F. intosa* (Trybom), *Thrips tabaci* Lindeman, *T. palmi* Karny and *Parthenothrips dracaenae* (Hegeer), although the predator can survive on pollen.

TARGET CROPS: Glasshouse-grown crops, such as tomatoes, cucumbers, peppers and ornamentals. Also used for thrips control in interiorscapes.

BIOLOGICAL ACTIVITY: Biology: Eggs (oval and about 0.14 mm in diameter) are laid in the axils of the midvein and the lateral veins on the underside of leaves. The first active stage (larva) has six legs and does not feed, but remains near the place of emergence. Proto- and deutonymphs and adults have eight legs, are very mobile and consume food actively. **Predation:** Predatory mites pierce their prey and suck the contents. They consume between 1 and 5 thrips per day, depending on prey-instar, temperature and humidity, giving a total of about 85 in their lifetime. They do not feed on adults. **Egg laying:** Female predatory mites lay between 22 (at 15 to 16 °C) and 47 (at 25 to 26 °C) eggs throughout their life. **Duration of development:** Eggs hatch after 2 or 3 days, followed by 4 days for immature development at 25 °C. Adults live up to 30 days, depending on the temperature. **Efficacy:** Because *Amblyseius barkeri* can survive on pollen, it can be introduced before the thrips populations build up. However, in crops that produce large quantities of pollen, its effectiveness is reduced because of this alternative food source, so higher populations must be introduced. *Amblyseius barkeri* consumes first instar thrips nymphs more readily than later instars. **Key reference(s):** 1) R J Jacobson. 1997. Integrated pest management (IPM) in greenhouses. In *Thrips as Crop Pests*, T Lewis (ed.), pp. 639–66, CABI, Wallingford, UK. 2) J Riudavets. 1995. Predators of *Frankliniella occidentalis* (Perg.) and *Thrips tabaci* Lind.: A review. In *Biological Control of Thrips Pests*, A J M Loomans *et al.* (eds), **95-1**, 44–87, Wageningen Agricultural University Press, The Netherlands.

COMMERCIALISATION: Formulation: Sold as a vermiculite formulation with 10 000 to 50 000 predatory mites per litre and as a specially formulated slow-release breeding colony of the predatory mite. Sometimes sold in bran in combination with *Amblyseius cucumeris* (Oudemans). **Tradenames:** 'Thripex' (Koppert), 'Broad Mite Biocontrol' (IPM Laboratories), 'Thrips Predator' (Arbico), 'Amblyseius barkeri' (plus *Amblyseius cucumeris*) (Neudorff).

APPLICATION: In glasshouses, sprinkle culture/carrier carefully on the leaves of the crop in early spring at a rate of 50 to 120 predatory mites per plant and repeat every 2 to 3 weeks. Release on peppers immediately after the appearance of the first flowers. Often a high-level early release gives the best results. In interiorscapes, a release rate of 1000 to 10 000 predators per 50 square metres every 2 weeks is recommended.

PRODUCT SPECIFICATIONS: Purity: No phytophagous mites, no mould or other contaminants. **Storage conditions:** Store at 15 to 20 °C. Do not refrigerate. Do not place containers in direct sunlight. **Shelf-life:** Will remain viable for 5 days, if received within one day of shipping and if stored under recommended conditions.

COMPATIBILITY: Very sensitive to insecticides and benzimidazole-based fungicides. Do not release if methomyl or a synthetic pyrethroid has been used. Most effective on smooth-leaved plants.

MAMMALIAN TOXICITY: Allergic reactions have been noted in production workers, but this may be due to the prey rather than the predator. There have been no reports of allergic or other adverse reactions following field use.

ENVIRONMENTAL IMPACT AND NON-TARGET TOXICITY: *Amblyseius barkeri* is now widespread in Nature and there have been no reports of any adverse impact on the environment or on non-target organisms. It does not feed on beneficial species.
Approved for use in organic farming: Yes.

3:246 *Amblyseius californicus*   *Predator of mites*

Predatory mite: Acari: Phytoseiidae

NOMENCLATURE: Approved name: *Amblyseius californicus* (McGregor).
Other names: Spider mite predatory mite. Formerly known as: *Amblyseius chilensis* (Dosse), *Amblyseius mungeri* (McGregor), *Neoseiulus californicus* (McGregor), *Typhlodromus californicus* McGregor, *Typhlodromus mungeri* (McGregor) and *Typhlodromus marinus* (Willmann).

SOURCE: *Amblyseius californicus* is a native of Mediterranean climates. It has also been found in tropical and subtropical areas in North and South America.

PRODUCTION: Reared on phytophagous mites, such as *Tetranychus urticae* Koch., on beans.

TARGET PESTS: *Amblyseius californicus* feeds on tetranychid mites, such as *Tetranychus urticae*, *T. cinnabarinus* (Boisduval), *Panonychus ulmi* (Koch) and *Eutetranychus* spp., and on tarsonemid mites, such as *Polyphagotarsonemus latus* (Banks). The preferred growth stages of spider mites are the eggs, larvae and nymphs. It will survive on other small arthropods and on pollen, but it will not reproduce in the absence of spider mites.

TARGET CROPS: Glasshouse-grown crops, particularly those grown at relatively high temperatures and low relative humidities, such as cucumbers, peppers and ornamentals. Also effective in field crops grown under high temperature conditions, such as strawberries.

BIOLOGICAL ACTIVITY: Biology: Eggs are laid on the hairs in the axils of the midvein and the lateral veins on the underside of leaves. The eggs are oval-shaped, white and clear, and hatch into six-legged larvae that do not feed, but remain in groups near their place of emergence. Proto- and deutonymphal stages and adults have eight legs, are mobile and feed. **Predation:** At 26 °C, immature predatory mites consume, on average, 11.4 spider mite eggs and nymphs before reaching adulthood. An adult female can consume in excess of 150 prey over a 16-day period. *Amblyseius californicus* feeds on phytophagous mites by piercing the prey and sucking the contents. The females eat all stages of phytophagous mites. **Egg laying:** Egg laying varies with temperature, from less than one per day (to yield a total of 48 throughout the female life-cycle) at 13 °C, to over 3.5 per day (to yield a total of 65 throughout the female life-cycle) at 33 °C. Development, egg laying and longevity are dependent upon temperature, the type and availability of food and the ambient humidity. **Duration of development:** *Amblyseius californicus* development is more rapid at high temperatures, taking about 15 days at 15 °C, 8 days at 20 °C and 5.5 days at 25 °C. In a study conducted at 21 °C, eggs hatched after 3 days, larvae matured after one day, protonymphs and deutonymphs took 3 to 4 days to develop – giving a total development time of 7.3 days. **Efficacy:** *Amblyseius californicus* is particularly useful where food is scarce, temperatures are high, humidity is low and when the phytophagous mites are concealed in the terminal shoots of the crop.

Key reference(s): G J de Moraes, J A McMurty & H A Denmark. 1986. A catalog of the mite family Phytoseiidae: references to taxonomy, synonymy, distribution and habitat, EMBRAPA Departmento Difusao de Technologia, Brasilia DF.

COMMERCIALISATION: Formulation: Minimum of 2000 live adults and juvenile predators per unit in vermiculite, corn husks, corn grit or on bean leaves. Contains an unspecified number of predator eggs. DEFRA licence required for release in the UK.

Tradenames: 'Amblyline cal' (Syngenta Bioline), 'Californicus-System' (Biobest), 'Spider Mite Predator' (Arbico), 'Ambsure (cal)' (Biological Crop Protection), 'Neoseiulus californicus' (Rincon-Vitova), (Biocontrol Network), (Hydro-Gardens), (Harmony) and (IPM Laboratories), 'Spical' (Koppert).

APPLICATION: Predatory mites are fragile and should be carefully mixed with the carrier material and sprinkled over the crop such that the carrier settles on the foliage, allowing the predators to move onto the crop. If there are hot spots in the crop, the material can be poured into a box hanging within the crop where the phytophagous mites are most abundant. If supplied on bean leaves, lay the leaves carefully on the leaves of the crop.

PRODUCT SPECIFICATIONS: Purity: No phytophagous mites, no mould and no other contaminants. **Storage conditions:** Store at 8 to 10 °C, 65 to 80% r.h. and out of direct sunlight. Protect from freezing and high temperatures. **Shelf-life:** Use within 3 days of receipt, if stored under recommended conditions.

COMPATIBILITY: Incompatible with foliar-applied insecticides and acaricides and insecticidal smokes. Can be used in conjunction with *Phytoseiulus persimilis* and with systemic soil-applied insecticides, such as imidacloprid.

MAMMALIAN TOXICITY: Allergic reactions have been noted in production workers, but this may be due to the prey rather than the predator. There have been no reports of allergic or other adverse reactions following use in glasshouses or in the field.

ENVIRONMENTAL IMPACT AND NON-TARGET TOXICITY: *Amblyseius californicus* is widespread in Nature and there have been no reports of adverse environmental impact or effects on non-target organisms. It will feed on other mites and insect species, but will not reproduce in the absence of spider mites and this limits its longevity when the spider mite population is well controlled. **Approved for use in organic farming:** Yes.

3:247 *Amblyseius cucumeris*   *Predator of thrips*

Predatory mite: Acari: Phytoseiidae

NOMENCLATURE: Approved name: *Amblyseius cucumeris* (Oudemans).
Other names: Thrips predator; thrips predatory mite. Formerly known as: *Neoseiulus cucumeris* (Oudemans).

SOURCE: *Amblyseius cucumeris* is cosmopolitan and occurs widely in Nature and is found in Europe, North Africa, California and Australia. It thrives under conditions of high humidity, but some strains enter diapause when days shorten to 12.5 hours, day temperatures fall below 22 °C and night temperatures drop to 17 °C. A strain that does not enter diapause has been selected for commercial application.

PRODUCTION: Reared on living, storage prey mites, such as *Tyrophagus* spp. or *Acarus farris* (Oudemans) in a ratio of 1:1 to 1:5 predator:prey, excluding eggs. Small quantities of bran are provided as food for the prey mites. Up to 100000 phytoseiids per litre can be produced in this way.

TARGET PESTS: Main prey are thrips, including *Frankliniella occidentalis* (Pergande), *F. tritici* (Fitch), *Thrips tabaci* Lindeman and *T. obscuratus* Crawford, but the predator will eat spider mites, *Tetranychus urticae* Koch, and can survive on pollen.

TARGET CROPS: Glasshouse crops, such as tomatoes, cucumbers, peppers and ornamentals. Also used for thrips control in interiorscapes and can be used outdoors in hot climates.

BIOLOGICAL ACTIVITY: Biology: Adult mites are a pale colour with a pear shaped body. The eggs hatch into six-legged larvae that do not feed, but remain aggregated near their place of emergence. Proto- and deutonymphs and adults have eight legs, are very mobile and consume food actively. **Predation:** Predatory mites pierce their prey and suck the contents. This species feeds only on first instar thrips, consuming between 1 and 5 thrips per day, depending on temperature and humidity. **Egg laying:** Eggs are laid in the axils of the midvein and the lateral veins on the underside of leaves. Female predatory mites lay between 22 (at 15 to 16 °C) and 47 (at 25 to 26 °C) eggs throughout their life. Eggs are laid close to recently hatched thrips nymphs. **Duration of development:** Eggs hatch after 2 or 3 days, larvae mature after 0.5 to 1.5 days, protonymphs and deutonymphs take 2 to 3.5 days to develop and adults live for 6 to 11 days, depending on the temperature. The lower threshold temperature for larval development is 7.7 °C. Development can be very quick (1 to 2 weeks) when climatic parameters are suitable, e.g. temperature between 18 and 20 °C and r.h. above 60%. **Efficacy:** Because *Amblyseius cucumeris* can survive on pollen, it can be introduced before the thrips appear, allowing the predator population to build up. However, in crops that produce large quantities of pollen, its effectiveness is reduced because of this alternative food source and, consequently, higher populations must be introduced. *Amblyseius cucumeris* feeds only on first instar thrips and, as a consequence, it may take a few months to control *Thrips tabaci*. **Key reference(s):** 1) R J Jacobson. 1997. Integrated pest management (IPM) in greenhouses. In *Thrips as Crop Pests*, T Lewis (ed.), pp. 639–66, CABI, Wallingford, UK. 2) M W Sabelis & P J C van Rijn. 1997. Predation by insects and mites. In *Thrips as Crop Pests*, T Lewis (ed.), pp. 259–354, CABI, Wallingford, UK.

COMMERCIALISATION: Formulation: Sold as a vermiculite or bran formulation with 10000 to 50000 predators per litre and as a specially formulated slow-release breeding colony of the predatory mite. The controlled-release sachets are commercially more important for crops such as cucumbers and peppers. The loose product is more important in ornamentals. Sometimes sold in bran in combination with *Amblyseius barkeri*. 'Amblyseius-Vermiculite-System' is sold in tubes of 10000 and 25000 predators in a mixture of vermiculite (minimum 80%) and bran for use with an air-sprayer. **Tradenames:** 'AmbliPAK50000' (Bioplanet), 'Ambsure (abs)' and 'Ambsure (cv)' (Biological Crop Protection), 'Amblyline flo', 'Amblyline crs-wp' and 'Amblyline cu' (Syngenta Bioline), 'Amblyseius barkeri' (plus *Amblyseius cucumeris*) (Neudorff), 'Amblyseius-C' (Applied Bio-Nomics), 'Amblyseius cucumeris' (non-diapause strain) (Biocontrol Network) and (Sautter & Stepper), 'Amblyseius System', 'Amblyseius-Vermiculite-System' and 'Amblyseius Breeding System' (Biobest), 'Amblyseius Thrips Predators' (Biowise), 'Neoseiulus cucumeris' (thrips predator) (M&R Durango), 'Neoseiulus cucumeris', 'Amblyseius cucumeris mite' (in bags), 'Amblyseius cucumeris' (in tubes) and 'Amblyseius cucumeris' (in controlled-release sachets) (Rincon-Vitova), 'Thrips Biocontrol' (*Neoseiulus cucumeris*) (IPM Laboratories), 'Thrips Destroyer' (Nature's Alternative Insectary), 'Thripex' and 'Thripex Plus' (Koppert), 'Thrips Predator' (Arbico).

APPLICATION: In glasshouses, place bran carefully on the leaves of the crop in early spring at a rate of 500 mites per plant and repeat every 2 to 3 weeks. Release on peppers immediately after the appearance of the first flowers. Often a high-level early release gives the best results. In glasshouses and interiorscapes, a release rate of 1000 predators per 50 square metres weekly is recommended. For cucumbers, an initial application of 250 predators per plant is recommended, followed by subsequent applications of 50 per plant every 15 days. Field applications of 150 million predators per hectare per season are recommended. The use of controlled-release sachets gives longer periods of control in crops such as peppers and cucumbers.

PRODUCT SPECIFICATIONS: Purity: No phytophagous mites, no mould or other contaminants. **Storage conditions:** Store at 10 to 15 °C; do not refrigerate. Do not place containers in direct sunlight. **Shelf-life:** Use within 5 days, if received within one day of shipping and if stored under recommended conditions. *Amblyseius cucumeris* cannot survive temperatures below 0 °C, but can be stored at 9 °C with only low mortality.

COMPATIBILITY: Very sensitive to insecticides and benzimidazole-based fungicides. Do not release if methomyl or a synthetic pyrethroid has been used. Most effective on smooth-leaved plants.

MAMMALIAN TOXICITY: Allergic reactions have been noted in production workers, but this may be due to the prey rather than the predator. There have been no reports of allergic or other adverse reactions following field use.

ENVIRONMENTAL IMPACT AND NON-TARGET TOXICITY: *Amblyseius cucumeris* is now widespread in Nature and there is no evidence of adverse environmental effects or adverse effects on non-target organisms. It can survive on pollen and this allows it to survive periods when the thrips and spider mites have been well controlled.

Approved for use in organic farming: Yes.

3:248 *Amblyseius degenerans*
Predator of thrips

Predatory mite: Acari: Phytoseiidae

NOMENCLATURE: Approved name: *Amblyseius degenerans* (Berlese). **Other names:** Thrips predator; thrips predatory mite. Formerly known as: *Iphiseius degenerans* Berlese.

SOURCE: *Amblyseius degenerans* occurs naturally in Africa and the Mediterranean. It thrives under conditions of high humidity, but tolerates lower humidities better than *Amblyseius cucumeris* (Oudemans). It will not enter diapause under glasshouse vegetable growing conditions (16 to 25 °C).

PRODUCTION: Usually reared on pollen.

TARGET PESTS: Main prey are thrips *Frankliniella occidentalis* (Pergande). *Thrips tabaci* Lindeman are less favoured as prey. The predator will eat spider mites, *Tetranychus urticae* Koch, and can survive on pollen.

TARGET CROPS: Glasshouse crops, such as peppers and ornamentals. Also used for thrips control in interiorscapes.

BIOLOGICAL ACTIVITY: Biology: Eggs (oval and about 0.14 mm in diameter) are laid in the axils of the midvein and the lateral veins on the underside of leaves. The eggs hatch into six-legged larvae that do not feed, but remain aggregated near their place of emergence. Proto- and deutonymphs and adults have eight legs. These stages are very mobile and consume food actively. **Predation:** Predatory mites pierce their prey and suck the contents. They consume between 1 and 5 thrips per day, depending on temperature and humidity. Larger and more aggressive than *Amblyseius cucumeris*. **Egg laying:** A female predatory mite lays between 22 (at 15 to 16 °C) and 47 (at 25 to 26 °C) eggs throughout her life. Females require multiple matings for egg laying. **Duration of development:** Eggs hatch after 2 or 3 days, larvae mature after 0.5 to 1.5 days, protonymphs and deutonymphs take 2 to 3.5 days to develop and adults live for over 20 days, depending on the temperature. **Efficacy:** Because *Amblyseius degenerans* can survive on pollen, it can be introduced before the thrips appear, allowing the predator population to build up. *Amblyseius degenerans* consumes first stage thrips nymphs preferentially over later stages. **Key reference(s):** 1) R J Jacobson. 1997. Integrated pest management (IPM) in greenhouses. In *Thrips as Crop Pests*, T Lewis (ed.), pp. 639–66, CABI, Wallingford, UK. 2) M W Sabelis & P J C van Rijn. 1997. Predation by insects and mites. In *Thrips as Crop Pests*, T Lewis (ed.), pp. 259–354, CABI, Wallingford, UK.

COMMERCIALISATION: Formulation: Sold as a vermiculite formulation with 10 000 to 50 000 predators per litre. Also sold in small 30 to 60 ml vials. **Tradenames:** 'Amblyline d' (Syngenta Bioline), 'Degenerans L' (Applied Bio-Nomics), 'Iphiseius degenerans' (Rincon-

Vitova), 'Degenerans-System' (Biobest), 'Thripans' (Koppert), 'Amblyseius degenerans' (Biotactics), (International Technology Services) and (IPM Laboratories).

APPLICATION: Place predators carefully on the leaves of the crop in early spring, ensuring that they are well distributed, at a rate of one mite per plant and repeat every 2 to 3 weeks. Release on peppers immediately after the appearance of the first flowers. Often a high-level early release gives the best results. As the females require multiple matings, it is recommended that introductions are concentrated on a few plants per row rather than evenly distributed through the crop. In glasshouses and interiorscapes, a release rate of 1000 predators per 100 square metres weekly is recommended. Normally only one or two introductions are made on peppers.

PRODUCT SPECIFICATIONS: Purity: No phytophagous mites, no mould or other contaminants. **Storage conditions:** Store at 15 to 20 °C. Do not refrigerate. Do not place containers in direct sunlight. **Shelf-life:** Viable for 5 days, if received within one day of shipping and if stored under recommended conditions.

COMPATIBILITY: Very sensitive to insecticides and benzimidazole-based fungicides. Do not release if methomyl or a synthetic pyrethroid has been used. Most effective on smooth-leaved plants.

MAMMALIAN TOXICITY: There have been no reports of allergic or other adverse reactions following field use.

ENVIRONMENTAL IMPACT AND NON-TARGET TOXICITY: *Amblyseius degenerans* is widespread in Nature and has not been shown to cause any adverse effects on non-target organisms or on the environment. It can survive on pollen and this allows it to survive periods when the thrips and spider mites have been well controlled.
Approved for use in organic farming: Yes.

3:249 *Amblyseius fallacis*   *Predator of mites*

Predatory mite: Acari: Phytoseiidae

NOMENCLATURE: Approved name: *Amblyseius fallacis* (Garman). **Other names:** Spider mite predator. Formerly known as: *Neoseiulus fallacis* Garman.

SOURCE: Predatory mite that occurs widely in Nature, particularly in orchards.

PRODUCTION: Bred on spider mites (*Tetranychus urticae* Koch) under controlled conditions.

TARGET PESTS: Spider mites, in particular *Tetranychus urticae* and *Panonychus ulmi* (Koch).

TARGET CROPS: Orchard crops, especially apples and pears, strawberries, ornamentals and protected vegetable crops.

BIOLOGICAL ACTIVITY: Biology: Eggs are laid on the hairs in the axils of the midvein and the lateral veins on the underside of leaves. The eggs hatch into six-legged larvae that do not feed, but remain in groups near their place of emergence. Proto- and deutonymphal stages and adults have eight legs, are mobile and feed. **Predation:** At 26 °C, immature predatory mites consume an average of 11.4 spider mite eggs and nymphs before reaching adulthood. An adult female can consume in excess of 150 prey over a 16-day period. *Amblyseius fallacis* feeds on phytophagous mites by piercing the prey and sucking the contents.
Duration of development: *Amblyseius fallacis* development is more rapid under high temperature conditions, taking about 15 days at 15 °C, 8 days at 20 °C and 5.5 days at 25 °C. It prefers conditions of high humidity. **Efficacy:** *Amblyseius fallacis* is particularly useful where food is scarce, temperature and humidity are high and when the phytophagous mites are concealed in the terminal shoots of the crop.

COMMERCIALISATION: Formulation: Minimum of 2000 live adults and juvenile predators per unit in vermiculite or corn husks or on bean leaves. Contains an unspecified number of predator eggs. **Tradenames:** 'Amblyseius fallacis' (Applied Bio-Nomics), (IPM Laboratories) and (Praxis), 'Neoseiulus fallacis' (Rincon-Vitova), 'TriPak' (*Phytoseiulus persimilis* plus *Amblyseius fallacis* plus *Stethorus punctillum*) (Rincon-Vitova).

APPLICATION: Distribute leaves on which *A. fallacis* has been bred carefully within the crop.

PRODUCT SPECIFICATIONS: Purity: Commercial product contains only *Amblyseius fallacis*, with no mould or other contaminants. **Storage conditions:** Keep between 8 and 10 °C and out of direct sunlight. Do not freeze. **Shelf-life:** Use within 2 days of delivery.

COMPATIBILITY: Compatible with most agrochemicals. Do not apply with acaricides or persistent insecticides.

MAMMALIAN TOXICITY: *Amblyseius fallacis* has caused no allergic or other adverse effects on research workers, manufacturing or field staff. It is considered to be of low mammalian toxicity.

ENVIRONMENTAL IMPACT AND NON-TARGET TOXICITY: *Amblyseius fallacis* occurs in Nature and is not expected to have any adverse effects on non-target organisms or on the environment. It is specific to pest mites and no other insect or mite species.
Approved for use in organic farming: Yes.

3:250 *Amblyseius montdorensis*
Predator of thrips

Predatory mite: Acari: Phytoseiidae

NOMENCLATURE: Approved name: *Amblyseius montdorensis* Schicha. **Other names:** Thrips predator. Formerly known as: *Typhlodromus montdorensis* Schicha.

SOURCE: Reported by Schicha from New Caledonia, Fiji, New Hebrides, Tahiti and Queensland, Australia. Since found to be widely distributed in subtropical areas in Queensland with an anomalous recording in South Australia.

PRODUCTION: Can be reared on pollen.

TARGET PESTS: *Amblyseius montdorensis* will feed on thrips, spider mites, Eriophyidae, Tenuipalpidae and Tarsonemidae.

TARGET CROPS: Protected strawberries and ornamentals.

BIOLOGICAL ACTIVITY: Biology: Eggs are laid in the axils of the midvein and the lateral veins on the underside of leaves. The eggs hatch into six-legged larvae that do not feed, but remain aggregated near their place of emergence. Proto- and deutonymphs and adults have eight legs, are very mobile and consume food actively. *Amblyseius montdorensis* does not establish or breed well in crops with a minimum temperature below 15 °C, even when the daytime temperature is high. Failure of *A. montdorensis* to establish on chrysanthemums in Spain with a minimum temperature of 17 °C is thought to be humidity-related. At 25 °C, a humidity of around 70% is required for 50% egg hatch. **Predation:** Predatory mites pierce their prey and suck the contents. **Egg laying:** When fed on pollen, *A. montdorensis* females lay 3 eggs per day at 25 °C, falling to 1.8 eggs per day at 20 °C. When fed on thrips at 25 °C, females will lay over 3.5 eggs per day. **Duration of development:** The life-cycle of *A. montdorensis* from egg to egg is 7.8 days at 25 °C. **Efficacy:** At 25 °C, *Amblyseius montdorensis* will consume 14.8 first instar thrips plus 1.6 second instar thrips, on average, per day. Dispersal both within and between crop plants is very good.

COMMERCIALISATION: Formulation: Under development by Syngenta Bioline. The mites are supplied as units of 2000 adults in vermiculite in 250 cm^2 cardboard tubes. Each unit contains an unspecified number of nymphal stages and viable eggs. **Tradenames:** 'Amblyline m' (Syngenta Bioline).

APPLICATION: Each unit is sufficient to treat 200 square metres to give between 5 and 10 mites per square metre. Release should be repeated at least twice.

3. Macro-organisms

PRODUCT SPECIFICATIONS: Purity: Product contains nymphal stages and eggs plus small quantities of *Tyrophagus putrescentiae* added as a source of food. **Storage conditions:** Do not store at temperatures below 4°C. Keep out of direct sunlight. **Shelf-life:** Use as soon as possible after delivery.

COMPATIBILITY: Very sensitive to insecticides and benzimidazole-based fungicides. Do not release if methomyl or a synthetic pyrethroid has been used. Most effective on smooth-leaved plants.

MAMMALIAN TOXICITY: There have been no reports of allergic or other adverse reactions following field use.

ENVIRONMENTAL IMPACT AND NON-TARGET TOXICITY: *Amblyseius montdorensis* is widespread in Nature and has not been shown to cause any adverse effects on non-target organisms or on the environment. It can survive on pollen and this allows it to survive periods when the thrips and spider mites have been well controlled.
Approved for use in organic farming: Yes.

3:251 *Amblyseius swirskii*   *Predator of whitefly*

Predatory mite: Acari: Phytoseiidae

NOMENCLATURE: Approved name: *Amblyseius swirskii* Athias-Henriot.
Other names: *Typhlodromips swirskii* Athias-Henriot

SOURCE: *Amblyseius swirskii* originates from the Nile Delta of Egypt and adjacent areas of Israel and the Middle East.

PRODUCTION: Can be reared on pollen.

TARGET PESTS: *Amblyseius swirskii* will reduce, but not eradicate, whitefly populations on protected crops to which pollen has been added. It has also been studied for control of other pests, such as thrips. *A. swirskii* needs a satisfactory food source and prefers whitefly and thrips., but also feeds on pollen and to a certain degree on spider mites and tarsonemid mites.

TARGET CROPS: Glasshouse crops, such as pepper, cucumber, eggplant, strawberry and some ornamental crops.

BIOLOGICAL ACTIVITY: Biology: Adult female mites lay single eggs onto leaf hairs. The mites go through three immature stages, larva, protonymph and deutonymph, before becoming adults. At 25°C, the entire cycle from egg to adult can take less than 7 days. **Predation:** All mobile stages are predatory. *Amblyseius swirskii* feeds by piercing small arthropod prey with its mouthparts, and draining the contents. **Duration of development:** In optimal climate and feeding conditions, its total life-cycle takes 5 to 6 days. *Amblyseius swirskii* does not have a diapause period and so can also be applied during short days. **Efficacy:** *Amblyseius swirskii* mainly predates young stages and so it is recommended that ichneumon wasps are introduced for older larval stages of whitefly and *Orius* spp. for older thrips larvae and adults. *Phytoseiulus persimilis* is a recommended addition when there are spider mite infestations.

COMMERCIALISATION: Formulation: 'Swirskiline' is supplied in cardboard shaker tubes containing 25 000 mites or bulk bags containing 125 000 mites. **Tradenames:** 'Swirskiline as' (Syngenta Bioline), 'Swirskii-System' (Biobest).

APPLICATION: Apply *Amblyseius swirskii* to crops as soon as night-time temperatures reach 15 °C, and daytime temperatures are higher than 20 °C. It is well suited to use in crops grown in warm conditions. Earlier application, before temperatures are sufficiently high, can reduce or prevent establishment. It can be released onto the crop over several weeks, and can be introduced before the prey is present. *Amblyseius swirskii* should not be used as the sole means of control when whitefly populations are already very high, or when whitefly adults are invading the crop in large numbers. It is not suitable for use in tomatoes.

PRODUCT SPECIFICATIONS: Storage conditions: Keep in the shade until use at 15 °C to 20 °C; use as soon as possible after receipt. Do not refrigerate.

COMPATIBILITY: As a general rule, neonicotinoid insecticides applied for control of whitefly have no direct effect upon Phytoseiid mites.

MAMMALIAN TOXICITY: There have been no reports of allergic or other adverse reactions following use.

ENVIRONMENTAL IMPACT AND NON-TARGET TOXICITY: *Amblyseius swirskii* cannot be released without appropriate permits in some countries. It is widespread in Nature and has not been shown to cause any adverse effects on non-target organisms or on the environment. It can survive on pollen and this allows it to survive periods in the absence of prey insects. **Approved for use in organic farming:** Yes.

3:252 Amitus spiniferus

Parasitoid of woolly aphids

Parasitoid wasp: Hymenoptera: Platygasteridae

NOMENCLATURE: Approved name: *Amitus spiniferus* (Brethès).

SOURCE: Originally found in Central America, but now introduced widely.

PRODUCTION: Reared under controlled conditions on citrus woolly aphid (*Aleurothrixus floccosus* (Maskell)).

TARGET PESTS: Citrus woolly aphid.

TARGET CROPS: Citrus, especially mandarins.

BIOLOGICAL ACTIVITY: Biology: The female is very small (0.75 mm long) and is shiny black in colour. The female antenna is 0.65 mm long, with ten segments (the last three closely united and forming a club); its wings are shiny and its hind tarsus is five-segmented. The male is similar to the female, but the antenna is ten-segmented and filiform, with all flagellar segments longer than they are wide. They are covered with short erect hairs.
Predation: *Amitus spiniferus* is non-polyembryonic and bi-parental in reproduction. It lays eggs in all three nymphal stages of the host, with a preference for the first stage. Females have a life span of 4 to 5 days, and males live 3 to 4 days. Life-cycle from egg to adult varies from 45 to 60 days under laboratory conditions (at 27 °C) but, in the field, reproduction is well synchronised with the host. A female can produce more than 60 offspring, but has a poor searching capability and survives only 4 to 5 days under field conditions. **Efficacy:** *Amitus spiniferus* is a very effective parasitoid of citrus woolly aphid.

COMMERCIALISATION: Formulation: Sold as pupae. **Tradenames:** 'Amitus spiniferus'.

APPLICATION: Place pupae within trees close to woolly aphid colonies.

PRODUCT SPECIFICATIONS: Shelf-life: Release as soon as possible after receipt.

COMPATIBILITY: *Amitus spiniferus* can be used in conjunction with other beneficial insects. It is not compatible with broad-spectrum, residual insecticides.

MAMMALIAN TOXICITY: No allergic or other adverse reactions have been reported from the use of *Amitus spiniferus* under glasshouse or field conditions.

ENVIRONMENTAL IMPACT AND NON-TARGET TOXICITY: *Amitus spiniferus* is not expected to have any adverse effects on non-target organisms or on the environment. It is very specific to citrus woolly aphid. **Approved for use in organic farming:** Yes.

3:253 *Anagrus atomus* *Parasitoid of leafhoppers*

Parasitoid wasp: Hymenoptera: Mymaridae

NOMENCLATURE: Approved name: *Anagrus atomus* (L.). **Other names:** Leafhopper egg parasitoid.

SOURCE: *Anagrus atomus* was originally Palaearctic, but now occurs widely in Nature. It is a very effective parasitoid of leafhoppers.

PRODUCTION: *Anagrus atomus* is reared under controlled conditions on leafhopper eggs.

TARGET PESTS: For control of leafhoppers (*Hauptidia maraccana* Melichar and *Empoasca decipiens* Paoli).

TARGET CROPS: Recommended for use on glasshouse-grown tomatoes, peppers and ornamentals.

BIOLOGICAL ACTIVITY: Biology: The adults are small (2 mm long). The females search for leafhopper eggs, laying their eggs inside the leafhopper eggs. Parasitised eggs are bright red in colour. The wasp develops in the leafhopper egg. Following pupation, the adults emerge, mate and the females seek out new host eggs. **Duration of development:** The adult females are short-lived (2 to 3 days). **Efficacy:** The females are mobile and are very effective at seeking out host eggs.

COMMERCIALISATION: Formulation: Sold as pupae close to emergence on leaf pieces, in units of 50 to 100. **Tradenames:** 'Anagrus atomus native Species' (Biowise), 'Anagsure (a)' (Biological Crop Protection), 'Anagline a' (Syngenta Bioline).

APPLICATION: The leaf pieces bearing the *Anagrus atomus* pupae are laid out within the crop and the adults are allowed to emerge and seek host eggs. The parasitoid may be applied to protected edible and ornamental crops at any time of the year.

PRODUCT SPECIFICATIONS: Purity: Products contain only parasitised leafhopper eggs with *Anagrus atomus* pupae close to emergence. **Storage conditions:** Keep cool, out of direct sunlight. Do not allow to freeze. **Shelf-life:** Use immediately upon receipt.

COMPATIBILITY: Incompatible with residual insecticides.

MAMMALIAN TOXICITY: *Anagrus atomus* has not demonstrated evidence of toxicity, infectivity, irritation or hypersensitivity in mammals. No allergic responses or other adverse health problems have been observed by research workers, manufacturing staff or users.

ENVIRONMENTAL IMPACT AND NON-TARGET TOXICITY: *Anagrus atomus* occurs in Nature and, as such, is not expected to cause any adverse effects on non-target organisms. It is specific to leafhoppers. **Approved for use in organic farming:** Yes.

3. Macro-organisms

3:254 Anagrus epos   *Parasitoid of leafhoppers*

Parasitoid wasp: Hymenoptera: Mymaridae

NOMENCLATURE: Approved name: *Anagrus epos* Girault. **Other names:** Grape leafhopper parasitoid.

SOURCE: *Anagrus epos* occurs widely in Nature and is a very effective parasitoid of grape leafhoppers.

PRODUCTION: Raised on the eggs of the grape leafhopper (*Erythroneura elegantula* (Osborn)).

TARGET PESTS: Grape leafhopper (*Erythroneura elegantula*).

TARGET CROPS: Grape vines.

BIOLOGICAL ACTIVITY: Biology: The adults are small (2 mm long) and the females search for leafhopper eggs, laying their eggs inside the leafhopper eggs. Parasitised eggs are bright red in colour. The wasp develops in the leafhopper egg. Following pupation, the adults emerge from the parasitised eggs, mate and the females seek out new host eggs. The wasp overwinters in the eggs of the blackberry leafhopper, *Dikrella californica* (Lawson), a non-economic species whose eggs are present throughout the year on wild blackberries (*Rubus ursinus* Bailey and *R. procerus* Muell.). Overwintering wasp populations tend to be largest in blackberries along rivers that have an overstory of sheltering trees. When the blackberries produce new foliage in February, heavy oviposition by *Dikrella* leafhoppers is stimulated. The parasitoid population increases enormously on the eggs, so that, by late March and early April, there is widespread dispersal of newly produced *Anagrus* wasps.
Duration of development: The adult females are short-lived. **Efficacy:** The females are mobile and are very effective at seeking out host eggs.

COMMERCIALISATION: Formulation: Sold as pupae close to emergence on leaf pieces, in units of 50 to 100. **Tradenames:** 'Anagrus epos' (Bio Agronomics), (CalTec), (Integrated Pest Management) and (Praxis).

APPLICATION: The leaf pieces bearing the *Anagrus epos* pupae are laid out within the crop and the adults are allowed to emerge and seek host eggs. The parasitoid may be applied to protected edible and ornamental crops at any time of the year. Planting of blackberries in the location of the vineyards helps to establish high wasp populations. Vineyards located within 5.6 km of an established blackberry refuge will benefit immediately from the immigrant parasitoids. Vineyards beyond this distance are rarely colonised by wasps until mid-season.

PRODUCT SPECIFICATIONS: Purity: Products contain only parasitised leafhopper eggs with *Anagrus epos* pupae close to emergence. **Storage conditions:** Keep cool, out of direct sunlight. Do not allow to freeze. **Shelf-life:** Use immediately upon receipt.

COMPATIBILITY: Incompatible with residual, broad-spectrum insecticides. Can withstand normal applications of sulfur.

MAMMALIAN TOXICITY: *Anagrus epos* has not demonstrated evidence of toxicity, infectivity, irritation or hypersensitivity in mammals. No allergic responses or other adverse health problems have been observed by research workers, manufacturing staff or users.

ENVIRONMENTAL IMPACT AND NON-TARGET TOXICITY: *Anagrus epos* occurs in Nature and, as such, is not expected to cause any adverse effects on non-target organisms. It is specific to leafhoppers. **Approved for use in organic farming:** Yes.

3:255 *Anagyrus fusciventris*
Parasitoid of mealybugs

Parasitoid wasp: Hymenoptera: Encyrtidae

NOMENCLATURE: Approved name: *Anagyrus fusciventris* (Girault). **Other names:** Long-tailed mealybug parasitoid.

SOURCE: Originated in Australia, but now widely distributed.

PRODUCTION: Bred under controlled conditions on long-tailed mealybugs (*Pseudococcus longispinus* (Targioni)).

TARGET PESTS: Mealybugs (Pseudococcidae), but particularly *Pseudococcus longispinus*, the long-tailed mealybug.

TARGET CROPS: The long-tailed mealybug has a wide host range, including apples, asparagus, avocado, begonia, coconut and other palms, coffee, citrus, cycads, dracaena, gardenia, guava, heliconia, hibiscus, lilies, macadamia, mango, orchids, pears, philodendron, pineapple and other bromeliads, potato, soybean and sugar cane.

BIOLOGICAL ACTIVITY: Biology: *Anagyrus fusciventris* is approximately 3 mm in length. Males are black and females are yellow-grey, with vivid blue eyes. The proximal parts of the antennae of females are widened. **Predation:** *Anagyrus fusciventris* parasitises mostly larger stages of the prey, which turn into mummies, being a little darker in colour than unparasitised ones. A minimum temperature of 18 °C is needed for effective control, but parasitism occurs at both high and low relative humidity. **Efficacy:** *Anagyrus fusciventris* is a very effective long-tailed mealybug parasitoid.

COMMERCIALISATION: Formulation: Sold as mummies, with 25 per capsule.
Tradenames: 'Anagyrus fusciventris' (Entocare).

APPLICATION: *Anagyrus fusciventris* mummies are distributed within the crop to be treated at the time of mealybug invasion.

PRODUCT SPECIFICATIONS: Purity: Contains only parasitised mummies.
Storage conditions: Store under cool, dry conditions. **Shelf-life:** Use as soon as possible after receipt.

COMPATIBILITY: *Anagyrus fusciventris* is not compatible with broad-spectrum insecticides, although it may be used with other beneficial insects.

MAMMALIAN TOXICITY: *Anagyrus fusciventris* has not demonstrated evidence of toxicity, infectivity, irritation or hypersensitivity to mammals. No allergic responses or other adverse health problems have been observed by research workers, manufacturing staff or users.

ENVIRONMENTAL IMPACT AND NON-TARGET TOXICITY: *Anagyrus fusciventris* occurs in Nature and, as such, is not expected to cause any adverse effects on non-target organisms. It is a specific mealybug parasitoid. **Approved for use in organic farming:** Yes.

3:256 *Anagyrus pseudococci*
Parasitoid of mealybugs

Parasitoid wasp: Hymenoptera: Encyrtidae

NOMENCLATURE: Approved name: *Anagyrus pseudococci* Girault (Noyes & Hayat).

SOURCE: Originated in Mediterranean regions, where it was found in citrus orchards as a parasitoid of the citrus mealybug.

PRODUCTION: Produced commercially on citrus mealybug.

TARGET PESTS: Citrus mealybug (*Planococcus citri* (Risso) and *Pseudococcus affinis* (Maskell)).

TARGET CROPS: Citrus orchards.

BIOLOGICAL ACTIVITY: Predation: *Anagyrus pseudococci* will parasitise all growth stages of the mealybug, but parasitised younger mealybugs are more likely to die than be successfully parasitised. The time required for mortality other than mummification is 1 to 2, 2 to 3 and just over 3 days for first, second and third instar *Planococcus citri*, respectively. Hosts parasitised in the first instar do not produce any mummies, but the second and third instar

nymphs and adult stages of the mealybugs mummify, on average, 8 to 11 days after attack. Parasitism causes cessation of normal fecundity, as well as induction of early egg maturation of the mealybugs. Mealybugs parasitised in the adult stage produce a few eggs, but the fecundity of unparasitised mealybugs is much higher, being about 40 times greater than that of parasitised mature adults. Parasitised pre-ovipositing and mature adults lay eggs within 24 hours of attack. **Efficacy:** The wasps are very motile and seek prey very effectively.
Key reference(s): K S Islam, H A S Perera & M J W Copland. 1997. The effects of parasitism by an encyrtid parasitoid, *Anagyrus pseudococci* on the survival, reproduction and physiological changes of the mealybug, *Planococcus citri* (Risso), *Entomologia Experimentalis et Applicata*, **84**(1), 77–83.

COMMERCIALISATION: Formulation: Sold as parasitised mummies. **Tradenames:** 'Anagyrus pseudococci' (Rincon-Vitova) and (Praxis), 'Anagyrline' (Syngenta Bioline).

APPLICATION: The mummies are distributed within the citrus orchard at the time of mealybug invasion.

PRODUCT SPECIFICATIONS: Purity: Contains only parasitised mummies.
Storage conditions: Store under cool, dry conditions. **Shelf-life:** Use as soon as possible after receipt.

COMPATIBILITY: Do not use with broad-spectrum insecticides.

MAMMALIAN TOXICITY: *Anagyrus pseudococci* has not demonstrated evidence of toxicity, infectivity, irritation or hypersensitivity to mammals. No allergic responses or other adverse health problems have been observed by research workers, manufacturing staff or users.

ENVIRONMENTAL IMPACT AND NON-TARGET TOXICITY: *Anagyrus pseudococci* occurs in Nature and, as such, is not expected to cause any adverse effects on non-target organisms. It is a specific mealybug parasitoid. **Approved for use in organic farming:** Yes.

3:257 *Anaphes flavipes*
Parasitoid of cereal leaf beetles

Parasitoid wasp: Hymenoptera: Mymaridae

NOMENCLATURE: Approved name: *Anaphes flavipes* (Foerster). **Other names:** Cereal leaf beetle parasitoid.

SOURCE: When the cereal leaf beetle, *Oulema melanopus* (L.), became established in the USA, local natural enemies did not exist to control this invasive pest. USDA collected parasitoids from the beetle's native areas in Europe and Asia, which were then studied at the ARS laboratory in Sevres, France to determine their control potential and host specificity. Four hymenopterous parasitoids were selected and now represent the established complex that is successfully controlling this beetle in most of the United States.

PRODUCTION: Produced on eggs of the cereal leaf beetle (*Oulema melanopus*). Techniques have been developed using *Lema trilineata* Olivier eggs for laboratory rearing of *Anaphes flavipes*.

TARGET PESTS: Cereal leaf beetle (*Oulema melanopus*) is the main host; however, *Anaphes flavipes* is also known to parasitise *O. gallaeciana* (Heyden), *Lema collaris* Say and *L. trilineata* Olivier (three-lined potato beetle) if the primary host is not available. *Anaphes flavipes* will oviposit in eggs of other Chrysomelidae, but development does not proceed much beyond embryogenesis.

TARGET CROPS: Cereals, such as wheat and barley.

BIOLOGICAL ACTIVITY: Biology: *Anaphes flavipes* is a minute (approximately 0.75 mm long) mymarid wasp, with greatly reduced venation in the forewings. The antennae are long and slender and are used to distinguish the sex of the wasp. The males are characterised by filiform, twelve-segmented antennae, whereas the females have clavate, nine-segmented antennae. The tarsi are four- or five-segmented. Females are generally more active than males, walking briskly with their antennae in alternating, but constant, tapping motions.
Egg laying: *Oulema* eggs (at 21 °C) are susceptible to parasitisation up to 115 hours after deposition. Between 118 to 120 hours, *Anaphes flavipes* will lay eggs, but neither parasitoid nor host develops. The parasitoid will develop successfully in host eggs that are 115 to 118 hours old, but the time for development and emergence is retarded. If the development in the host egg has progressed to where the mandibular structures of the beetle embryo have sclerotised, the parasitoid will retract its ovipositor and not parasitise the egg.
Duration of development: *Anaphes flavipes* can have, on average, two generations per year, although as many as eight have been observed on *Oulema*. Its life-cycle consists of the egg, larvae, pre-pupa, pupa and adult stages. The egg stage consists of the internal cleavage of the parasitoid and its growth into a larva. The larval stage consists of the consumption of the host's egg yolk. As a pre-pupa, the parasitoid is completely motionless; the excretion of the faecal matter signifies the end to this stage. In the early pupal stage, the red compound eyes are the first feature visible. The body begins to darken and faint adult characteristics become visible. In the late pupal stage, the body is dark and the ovipositor is visible on the female. Just before emergence, the head and legs show movement as the parasitoid chews its way through the egg chorion. Under ideal conditions (21 to 25 °C), the adults emerge from the host egg in 10 to 11 days. A minimum of 174 hours (at 32.2 °C) to a maximum of 1089 hours (at 2.7 °C) was determined for the development of *Anaphes* from egg to

adult. Within one hour after emergence, adult female *Anaphes* will attack suitable hosts and commence egg deposition. On average, females deposit 20 eggs during an average 2 to 3 day post-emergence period. Females will deposit fertilised eggs (resulting in female offspring) and non-fertilised eggs (which result in male offspring). **Efficacy:** An active searcher for host eggs. *Anaphes flavipes* is a multivoltine egg parasitoid of the cereal leaf beetle. This is the preferred beneficial wasp for controlling the beetle, because it eliminates a large proportion of the beetle population and has a high dispersal rate. **Key reference(s)**: 1) R C Anderson. 1968. The biology and ecology of *Anaphes flavipes* (Foerster) (Hymenoptera: Mymaridae), an exotic egg parasite of the Cereal Leaf Beetle (Coleoptera: Chrysomelidae), *Ph.D. Thesis, Purdue University*, 148 pp. 2) R C Anderson & J D Paschke. 1968. The biology and ecology of *Anaphes flavipes*, an exotic egg parasite of the cereal leaf beetle, *Ann. Entomol. Soc. Amer.*, **61**, 1–5. 3) C L Staines. 1984. Cereal leaf beetle, *Oulema melanopus* (L.) (Coleoptera: Chrysomelidae): Density and parasitoid synchronization study in Washington County, Maryland 1977–1979, *Proc. Entomol. Soc. Wash.*, **86**, 435–8.

COMMERCIALISATION: Formulation: Sold as adults. **Tradenames:** 'Anaphes flavipes' (Bio Agronomics), (CalTec), (Integrated Pest Management Services) and (Praxis).

APPLICATION: Apply adults when cereal leaf beetles are first seen.

PRODUCT SPECIFICATIONS: Purity: Products contain only adult *Anaphes flavipes*. **Storage conditions:** It is recommended that *Anaphes flavipes* is used immediately upon receipt. If this is not possible, application should be made within 2 to 3 days. **Shelf-life:** If *Anaphes flavipes* must be stored, keep at a temperature between 6 and 10°C.

COMPATIBILITY: Pesticide applications are detrimental to successful parasitoid establishment.

MAMMALIAN TOXICITY: *Anaphes flavipes* has shown no evidence of toxicity, infectivity, irritation or hypersensitivity to mammals. No allergic responses or other adverse health problems have been observed by research workers, manufacturing staff or users.

ENVIRONMENTAL IMPACT AND NON-TARGET TOXICITY: *Anaphes flavipes* occurs in Nature and, as such, is not expected to cause any adverse effects on non-target organisms, although eggs of some beneficial insects may be parasitised in the absence of sufficient numbers of prey eggs.

3:258 Anaphes iole

Parasitoid of plant bugs

Parasitoid wasp: Hymenoptera: Mymaridae

NOMENCLATURE: Approved name: *Anaphes iole* Girault. **Other names:** Lygus bug parasitoid. Formerly known as: *Anaphes ovijentatus* (Crosby and Leonard).

SOURCE: Relatively common in North America.

PRODUCTION: Produced by parasitising lygus and/or tarnished plant bugs. Not easy to produce in large quantities.

TARGET PESTS: Tarnished, lygus and four-lined plant bugs, including *Lygus hesperus* Knight (Western tarnished plant bug), *Lygocoris pabulinus* L. (common green capsid), *Lygus lineolaris* (Palisot) (Southern tarnished plant bug) and, probably, *Lygus desertinus* Knight. Also parasitises mymarids.

TARGET CROPS: Recommended for use in high-value crops, such as strawberries and on glasshouse crops.

BIOLOGICAL ACTIVITY: Biology: The wasps are very active and seek eggs of prey actively. Best control is achieved when the plant bug population is high and this is always at the time of crop flowering. **Predation:** A wide range of *Lygus* (and related) species is parasitised and it is reported that eggs of some beneficial insects can be parasitised in the absence of sufficient numbers of prey eggs. **Egg laying:** The adult female wasps lay their eggs in the eggs of the plant bugs. **Duration of development:** *Anaphes iole* develops well in warm, slightly humid conditions, both in glasshouses and in outdoor crops. At a temperature of 30 °C, it will complete its life-cycle from egg to egg in about 15 days. **Efficacy:** An active searcher for host eggs.

COMMERCIALISATION: Formulation: Sold as adults. **Tradenames:** 'Anaphes iole' (Rincon-Vitova), (Integrated Pest Management) and (The Green Spot).

APPLICATION: *Anaphes iole* should be released at the commencement of crop flowering, as this is the time that the bugs are attracted to the crops. Relatively large quantities are required, with the recommendation for outdoor crops, such as strawberries, for 35 000 adults to be released per hectare (15 000 per acre) every week for 4 weeks from start of flowering, with another release encouraged some 6 to 7 weeks later. These releases should only be made when *Lygus* populations are high, with bug egg laying noticed in the flowers and flower buds. Similar application strategies are recommended for other crops, including protected crops.

PRODUCT SPECIFICATIONS: Purity: Products contain only adult *Anaphes iole*.

Storage conditions: It is recommended that *Anaphes iole* is used immediately upon receipt. If this is not possible, application should be made within 2 to 3 days. **Shelf-life:** If *Anaphes iole* must be stored, keep at a temperature between 6 and 10°C.

COMPATIBILITY: *Anaphes iole* is compatible with most fungicide sprays, although it is recommended that release is delayed for 2 to 3 days after fungicide application. Incompatible with most chemical insecticides.

MAMMALIAN TOXICITY: *Anaphes iole* has not demonstrated evidence of toxicity, infectivity, irritation or hypersensitivity to mammals. No allergic responses or other adverse health problems have been observed by research workers, manufacturing staff or users.

ENVIRONMENTAL IMPACT AND NON-TARGET TOXICITY: *Anaphes iole* occurs in Nature and, as such, is not expected to cause any adverse effects on non-target organisms, although eggs of some beneficial insects may be parasitised in the absence of sufficient numbers of prey eggs. **Approved for use in organic farming:** Yes.

3:259 *Anaphes nitens*
Parasitoid of eucalyptus snout beetle

Parasitoid wasp: Hymenoptera: Mymaridae

NOMENCLATURE: Approved name: *Anaphes nitens* (Girault). **Other names:** Eucalyptus snout beetle parasitoid. Formerly known as: *Anaphoidea nitens* (Girault); *Anaphes gonipteri* Debauche; and *Patasson gonipteri* Heqvist.

SOURCE: *Anaphes nitens* is a native of Australia, but is now widely distributed.

PRODUCTION: *Anaphes nitens* is reared under controlled conditions on the eucalyptus snout beetle (*Gonipterus scutellatus* Gyllenhal).

TARGET PESTS: Eucalyptus snout beetle (*Gonipterus scutellatus*) and *Gonipterus gibberus* Boisduval.

TARGET CROPS: Eucalyptus plantations.

BIOLOGICAL ACTIVITY: Biology: *Anaphes nitens* is a solitary parasitoid of the egg capsules of the eucalyptus snout beetle. The eggs of both *Gonipterus scutellatus* and *G. gibberus* are laid within an egg capsule, forming a rigid structure when dry and thereby protecting the eggs. Beetle egg numbers vary from 3 to 16 per egg capsule. A female *Anaphes nitens* finds these egg capsules and locates the position of the eggs within the capsule by using her antennae.

She then lays an egg adjacent to each beetle egg. When the wasp egg hatches, the larva enters the egg capsule and feeds on the developing *Gonipterus* embryo, undergoing three instars within the egg capsule before pupation. The life-cycle of the wasp is between 17 and 33 days, depending upon the temperature and food source. Parasitised eggs become dark and cloudy, compared with the clear yellow colour of healthy eggs. **Efficacy:** *Anaphes nitens* is a very effective parasitoid of *Gonipterus scutellatus*, with females being very efficient in locating egg capsules.

COMMERCIALISATION: Formulation: Sold as adults. **Tradenames:** 'Anaphes nitens'.

APPLICATION: Introduce adults throughout the plantation as soon as snout beetles are first seen.

PRODUCT SPECIFICATIONS: Purity: Products contain only adult *Anaphes nitens*. **Storage conditions:** It is recommended that *Anaphes nitens* is used immediately upon receipt. If this is not possible, application should be made within 2 to 3 days.

COMPATIBILITY: Insecticide applications are detrimental to successful parasitoid development.

MAMMALIAN TOXICITY: *Anaphes nitens* has not demonstrated evidence of toxicity, infectivity, irritation or hypersensitivity to mammals. No allergic responses or other adverse health problems have been observed by research workers, manufacturing staff or users.

ENVIRONMENTAL IMPACT AND NON-TARGET TOXICITY: *Anaphes nitens* occurs in Nature and, as such, is not expected to cause any adverse effects on non-target organisms. It is specific to snout beetles.

3:260 *Anastatus tenuipes*
Parasitoid of cockroaches

Parasitoid wasp: Hymenoptera: Eupelmidae

NOMENCLATURE: Approved name: *Anastatus tenuipes* Bolivar & Pieltain.

SOURCE: Common, where cockroaches abound.

PRODUCTION: Reared on the egg cases of the brown-banded cockroach (*Supella longipalpa* (Fabricius)).

TARGET PESTS: The brown-banded cockroach and the German cockroach (*Blattella germanica* (L.))

TARGET CROPS: Industrial and domestic premises.

BIOLOGICAL ACTIVITY: Biology: *Anastatus tenuipes* attacks the egg capsules (oothecae) of only two cockroach species, the brown-banded and the German cockroach. Females of *Anastatus tenuipes* lay their eggs inside cockroach oothecae. The eggs hatch and the developing parasitoid larvae consume the developing cockroach nymphs. The larvae continue to mature and eventually pupate inside the cockroach oothecae. *Anastatus tenuipes* reaches the adult stage inside the cockroach ootheca. Adults emerge by chewing a hole in the oothecal shell. The wasps mate immediately upon emergence. After mating, the females search for more cockroach egg capsules. Fertilised eggs develop into females and unfertilised eggs develop into males. **Predation:** A female adult will attack 1 to 2 hosts in her lifetime. **Duration of development:** The developmental time from egg to adult is 42 to 48 days, dependent on temperature and number of individuals developing in the host. On average, there are 7 female progeny per host. The female lives for between 4 and 6 days. **Key reference(s):** 1) L M Lebeck. 1991. A review of the hymenopterous natural enemies of cockroaches with emphasis on biological control, *Entomophaga*, **36**(3), 335–52. 2) F Agudelo-Silva. 1995. Biological control of roaches, *IPM Practitioner*, **17**(5/6), 12–13.

COMMERCIALISATION: Formulation: Sold as adults as part of a cockroach control programme. **Tradenames:** 'Anastatus tenuipes' (Rincon-Vitova).

APPLICATION: The wasps should be released where there is evidence of cockroach activity.

PRODUCT SPECIFICATIONS: Storage conditions: Release immediately.

COMPATIBILITY: Incompatible with residual insecticides.

MAMMALIAN TOXICITY: There have been no reports of allergic or other adverse toxicological effects arising from contact with *Anastatus tenuipes* from research staff, producers or users.

ENVIRONMENTAL IMPACT AND NON-TARGET TOXICITY: *Anastatus tenuipes* has a wide distribution in Nature and there is no evidence that it has any adverse effects on non-target organisms or on the environment. It is a specific cockroach parasitoid.

3. Macro-organisms

3:261 *Anisopteromalus calandrae*

Parasite of weevils

Parasitic wasp: Hymenoptera: Pteromalidae

NOMENCLATURE: Approved name: *Anisopteromalus calandrae* Howard.

SOURCE: A widely distributed species.

PRODUCTION: Reared under controlled conditions on host weevils.

TARGET PESTS: Many grain feeding Coleoptera and Lepidoptera, including snout beetles (*Sitophilus* spp.), drugstore beetle (*Stegobium paniceum* L.), cigarette beetle (*Lasioderma serricorne* (F.)), lesser grain borer (*Rhyzopertha dominica* (F.)), spider beetle (*Gibbium psylloides* (Czempinski)), Australian spider beetle (*Ptinus tectus* Boieldieu), white-marked spider beetle (*Ptinus fur* L.), grain moth (*Sitotroga cerealella* (Oliver)) and bean weevil (*Callosobruchus* spp.).

TARGET CROPS: Stored grain.

BIOLOGICAL ACTIVITY: Biology: The females of the parasitic wasp *Anisopteromalus calandrae* actively search for and parasitise hidden larvae of a variety of beetles and caterpillars infesting stored products. She lays her eggs inside them. The wasp larvae consume the parasitised host, pupate within the dead insect's body and emerge as adult wasps, mate and seek new larvae to parasitise. **Efficacy:** Female wasps are very effective searchers of larvae.

COMMERCIALISATION: Formulation: Supplied as parasitised snout beetles.
Tradenames: 'Anisopteromalus calandrae' (Andermatt), (Biofac) and (Caltec).

APPLICATION: Apply to grain stores when filled.

PRODUCT SPECIFICATIONS: Purity: Products contain no contaminants.
Storage conditions: May be stored for short periods, under cool, dry conditions.

MAMMALIAN TOXICITY: No allergic or other adverse reactions have been reported with *Anisopteromalus calandrae* under use conditions.

ENVIRONMENTAL IMPACT AND NON-TARGET TOXICITY: Under its conditions of use it is not expected that *Anisopteromalus calandrae* would come into contact with non-target species. **Approved for use in organic farming:** Yes.

3:262 *Anthocoris nemoralis*
Predator of psyllids and aphids

Predatory bug: Hemiptera: Anthocoridae

NOMENCLATURE: Approved name: *Anthocoris nemoralis* (Fabricius). **Other names:** Flower bug; pirate bug.

SOURCE: Palaearctic species, common in many parts of Europe from Finland to the Mediterranean. Breeds chiefly on trees infested with psyllids. Has successfully colonised parts of North America, following accidental or intentional introductions. Released in British Columbia in 1963 to control pear psylla.

PRODUCTION: Reared in insectaries. Grows and reproduces best on a diet of psyllids, but can be reared on some aphids and on *Ephestia kuehniella* L. eggs.

TARGET PESTS: Psyllids such as *Cacopsylla pyri* (L.) and *C. pyricola* (Förster) and aphids, mainly on trees. Will also prey on thrips.

TARGET CROPS: Orchard fruit crops, especially apples and pears.

BIOLOGICAL ACTIVITY: Biology: A small, brownish-black bug, 3.3 to 4 mm long, occurring predominantly in deciduous trees. *Anthocoris nemoralis* has five nymphal instars and it is usual to have three generations per year. Immediately after emergence, the larvae begin to feed close to their place of emergence on psyllid eggs and larvae. They become very mobile and search actively for prey within the plant canopy. It is a polyphagous predator that can be found in many crops or wild plants other than apple and pear orchards. **Predation:** The bugs search trees actively for prey, which they pierce with their needle-like mouthparts and suck dry. Each adult can prey on more than 200 psylla larvae in its lifetime. **Egg laying:** Eggs are laid after a three-day pre-oviposition period and are deposited in plant tissue, particularly in leaves, but sometimes in stems. The eggs are elongated and barely visible. Females lay, on average, 100 to 200 eggs at an average rate of 5 to 6 per day. **Duration of development:** Nymphs take 15 to 21 days to develop from hatching to adult at temperatures of 20 to 23 °C. *Anthocoris nemoralis* can overwinter as adult females in sheltered locations. There are between 2 and 3 generations a year, depending on climatic conditions. **Efficacy:** *Anthocoris nemoralis* must come into direct physical contact with prey in order to perceive its presence. Most time is spent searching leaf midribs and major veins, where immature psyllids are often found. On hops, nymphs killed an average of 174 aphids during development, whilst adults killed an average of 33 aphids per day. **Key reference(s):** J Péricart. 1972. *Hémipteres, Anthocoridae, Cimicidae, Microphysidae de L'Ouest Paléartique*, Masson et Cie, Paris, France.

COMMERCIALISATION: Formulation: Sold as adults in buckwheat and vermiculite.

Tradenames: 'Anthocoris System' (Biobest), 'Antholine n' (Syngenta Bioline), 'AntoPAK 200' (Bioplanet).

APPLICATION: Released into trees or vines as psyllids and aphids become active (in the spring). The bugs mixed with the carrier material are placed in cardboard containers in 5 to 10 release points per hectare. The adults can mate and/or fly into the crop within hours. A preferred release quantity is between 1000 and 1500 adults/hectare.

PRODUCT SPECIFICATIONS: Purity: Residue of food source, but no mites or other contaminants. **Storage conditions:** Store at 8 to 10 °C, 65% to 80% r.h., out of direct sunlight. Protect from freezing and from high temperatures. **Shelf-life:** Use within 2 days of receipt, if stored under recommended conditions.

COMPATIBILITY: Incompatible with most foliar-applied insecticides and acaricides and insecticidal smokes.

MAMMALIAN TOXICITY: There have been no reports of adverse allergic or other reactions from research workers, manufacturing staff or from the field release of *Anthocoris nemoralis*.

ENVIRONMENTAL IMPACT AND NON-TARGET TOXICITY: *Anthocoris nemoralis* occurs widely in Nature and, although it will consume insects from various different families, is not expected to have any long-term adverse effects on non-target organisms or on the environment. **Approved for use in organic farming:** Yes.

3:263 *Aphelinus abdominalis*

Parasitoid of aphids

Parasitoid wasp: Hymenoptera: Aphelinidae

NOMENCLATURE: Approved name: *Aphelinus abdominalis* (Dalman). **Other names:** Aphid parasitic wasp.

SOURCE: Its original distribution was Europe, where it is now widespread.

PRODUCTION: Reared in insectaries on aphids such as *Macrosiphum euphorbiae* (Thomas).

TARGET PESTS: Aphids, with particular activity against 'large' aphids, such as *Macrosiphum euphorbiae* and *Aulacorthum solani* (Kaltenbach).

TARGET CROPS: Protected vegetable and ornamental crops.

BIOLOGICAL ACTIVITY: Biology: Eggs are laid within the aphid and the wasp passes through four larval stages after egg hatch, during which time the aphid continues to feed. The adult emerges from its pupal stage within the mummified aphid, which is always black. **Predation:** Female wasps are most effective against larger aphids. Eggs are laid within aphid colonies and the female can parasitise between 10 and 15 aphids a day. The adult wasps also predate aphids. **Egg laying:** Adults mate within a day of emergence. Unfertilised females lay only male eggs, whilst fertilised females lay eggs that can develop into either males or females, dependent upon the size of the aphid to be parasitised. As a rule, eggs laid in larger aphids are fertilised and develop into females, whilst those laid in smaller aphids are not fertilised and develop into males. **Duration of development:** An adult will emerge between 13 and 15 days from parasitism, depending upon the temperature. The pupal stage takes about 5 days to hatch. **Efficacy:** A very efficient flyer and seeker of aphids. The relatively large ovipositor restricts the size of aphid that can be parasitised. **Key reference(s):** 1) E B Hågvar & T Hofsvang. 1991. Aphid parasitoids (Hymenoptera, Aphidiidae): Biology, host selection and use in biological control, *Biocontrol News and Information*, **12**(1), 13–41. 2) P Stary. 1988. Aphelinidae. In *Aphids: their Biology, Natural Enemies and Control*, A K Minks & P Harrewijn (eds), Vol. B, pp. 185–8, Elsevier, Amsterdam, The Netherlands.

COMMERCIALISATION: Formulation: Supplied as mummies that hatch during and after shipment and sometimes as adults. **Tradenames:** 'Aphelline ab' (Syngenta Bioline), 'Aphiline ace Mix' (50% *Aphidius colemani*, 25% *Aphidius ervi* and 25% *Aphelinus abdominalis*) (Syngenta Bioline) and (Rincon-Vitova), 'Aphelinus-System' (Biobest), 'Aphelinus' (Sautter & Stepper), (IPM Laboratories) and (Praxis), 'Aphelsure (a)' (Biological Crop Protection), 'Aphilin' (Koppert), 'Aphelinus abdominalis' (Rincon-Vitova).

APPLICATION: Allow adult wasps to fly from the container close to the aphid colonies. Rest bottles or vials in the crop so that the remaining adults can enter the crop as they emerge from the mummies.

PRODUCT SPECIFICATIONS: Purity: Adult wasps with no contaminants or pure mummies with no carrier. **Storage conditions:** Store in the dark at 8 to 10 °C. **Shelf-life:** Use as soon as possible. *Aphelinus abdominalis* will remain active for 2 to 3 days, if stored as recommended.

COMPATIBILITY: Incompatible with residual insecticides. Ants will protect aphids from parasitism.

MAMMALIAN TOXICITY: No allergic or other adverse reactions have been noted with research and production staff or users.

ENVIRONMENTAL IMPACT AND NON-TARGET TOXICITY: *Aphelinus abdominalis* is not expected to have any adverse effects on non-target organisms or on the environment. It is specific to aphids. **Approved for use in organic farming:** Yes,

3:264 *Aphelinus mali* *Parasitoid of woolly aphids*

Parasitoid wasp: Hymenoptera: Aphelinidae

NOMENCLATURE: Approved name: *Aphelinus mali* (Haldeman). **Other names:** Woolly aphid parasitoid wasp.

SOURCE: Widespread in Europe.

PRODUCTION: Reared in insectaries on woolly aphids.

TARGET PESTS: The woolly apple aphid (*Eriosoma lanigerum* (Hausmann)).

TARGET CROPS: Orchards, particularly apples.

BIOLOGICAL ACTIVITY: Biology: Eggs are laid within the aphid and the wasp passes through four larval stages after egg hatch, during which time the aphid continues to feed. The adult emerges from its pupal stage within the mummified aphid, which is always black. **Predation:** Eggs are laid within aphid colonies and the female can parasitise between 10 and 15 aphids a day. The adult wasps also predate aphids. **Egg laying:** Adults mate within a day of emergence. Unfertilised females lay only male eggs, whilst fertilised females lay eggs that can develop into either males or females, dependent upon the size of the aphid to be parasitised. As a rule, eggs laid in larger aphids are fertilised and develop into females, whilst those laid in smaller aphids are not fertilised and develop into males. **Efficacy:** A very efficient flyer and seeker of aphids. **Key reference(s):** E B Hågvar & T Hofsvang. 1991. Aphid parasitoids (Hymenoptera, Aphidiidae): Biology, host selection and use in biological control, *Biocontrol News and Information*, **12**(1), 13–41.

COMMERCIALISATION: Formulation: Supplied as mummies that hatch during and after shipment and sometimes as adults. **Tradenames:** 'Aphelinus mali' (Praxis).

APPLICATION: Allow adult wasps to fly from the container close to the aphid colonies. Rest bottles or vials in the crop so that the remaining adults can enter the crop as they emerge from the mummies.

PRODUCT SPECIFICATIONS: Purity: Adult wasps with no contaminants or pure mummies with no carrier. **Storage conditions:** Store in the dark at 8 to 10 °C. **Shelf-life:** Use as soon as possible. *Aphelinus mali* will remain active for 2 to 3 days, if stored as recommended.

COMPATIBILITY: Incompatible with residual insecticides. Ants will protect aphids from parasitism.

MAMMALIAN TOXICITY: No allergic or other adverse reactions have been noted with research and production staff or users.

ENVIRONMENTAL IMPACT AND NON-TARGET TOXICITY: *Aphelinus mali* is not expected to have any adverse effects on non-target organisms or on the environment. It is specific to aphids. **Approved for use in organic farming:** Yes.

3:265 *Aphidius colemani*   *Parasitoid of aphids*

Parasitoid wasp: Hymenoptera: Aphidiidae

NOMENCLATURE: Approved name: *Aphidius colemani* Viereck. **Other names:** Aphid parasitoid. Formerly known as: *Aphidius platensis* (Bréthes); *Aphidius transcaspicus* (Telenga).

SOURCE: This parasitic wasp was thought to have originated in India, but the species *Aphidius platensis* and *A. transcaspicus* were reclassified as junior synonyms of *A. colemani* and this extended its geographical range from Central Asia to the Mediterranean. The wasp has been introduced into Australia, Africa, Central America, California, England, Norway and the Netherlands. Traditionally, *Aphidius matricariae* Haliday was the principal commercial agent, but recently, strains selected from commercial colonies for their superior performance against *Aphis gossypii* Glover were shown to be *Aphidius colemani* contaminants and this species has generally superseded *A. matricariae*.

PRODUCTION: *Aphidius colemani* is reared on *Aphis gossypii* Glover or *Myzus persicae* (Sulzer) under controlled glasshouse conditions. Mummies of known age are collected, packaged in bottles or vials and despatched to the customer.

TARGET PESTS: Aphids. The primary hosts are *Myzus persicae*, *Myzus nicotianae* Blackman and *Aphis gossypii*. *Aulacorthum solani* (Kaltenbach) and *Rhopalosiphum padi* (L.) are also parasitised.

TARGET CROPS: Glasshouse-grown vegetable crops, such as tomatoes, cucumbers, peppers, aubergines and ornamentals. Also effective in interiorscapes.

BIOLOGICAL ACTIVITY: Biology: After egg hatch, the wasp larva progresses through four stages inside the aphid before pupating within the aphid's body, the mummy stage. On pupating, the parasitoid larva spins a silk cocoon within the aphid cuticle. The cuticle of the aphid host then hardens and the body swells, providing a protective case from which the adult wasp emerges. The adult leaves the mummy through a small, round hole cut in the dorsum of the mummified host. Adults are 2 to 3 mm long with dark bodies and long antennae. **Predation:** Female wasps can parasitise 100 to 200 aphids within 7 days, although this will vary with host. Parasitised aphids continue to feed and can transmit viruses. The presence of *Aphidius colemani* will often disturb the aphid colony, leading to the production of an aphid alarm pheromone that causes aphids to migrate from or fall off the leaf.
Egg laying: Females mate within one day of emergence and begin to lay eggs within a few hours of mating. Female adults lay single eggs within an aphid. Fertilised females can lay both unfertilised eggs, that develop into males, and fertilised eggs, that develop into females. Females can lay viable eggs without mating, but these always develop into males.
Duration of development: It takes between 13 and 15 days from parasitism or egg laying to adult emergence under suitable climatic condition (between 16 and 22 °C), but when the

temperature rises above 28 to 32 °C, activity sharply decreases. It is usual for the mummy to exist for 5 to 6 days at 20 °C. **Efficacy:** Particularly effective against *Aphis gossypii*, *Myzus persicae* and *Myzus nicotianae*, although it will parasitise over 40 different aphid species. Very effective searchers that work well against small, well-dispersed populations as adult females frequently explore the sites preferred by the pest. Can be affected by hyperparasitic wasps.

Key reference(s): 1) E B Hågvar & T Hofsvang. 1991. Aphid parasitoids (Hymenoptera, Aphidiidae): Biology, host selection and use in biological control, *Biocontrol News and Information*, **12**(1), 13–41. 2) P Stary. 1988. Aphidiidae. In *Aphids: their Biology*, *Natural Enemies and Control*, A K Minks & P Harrewijn (eds), Vol. B, pp. 171–84, Elsevier, Amsterdam, The Netherlands. 3) P M J Ramakers. 1989. Biological control in greenhouses. In *Aphids: their Biology*, *Natural Enemies and Control*, A K Minks & P Harrewijn (eds), Vol. C, pp. 199–208, Elsevier, Amsterdam, The Netherlands.

COMMERCIALISATION: Formulation: Most frequently sold as freshly collected aphid mummies of known age with no carrier or with wood chips in bottles with a feeder ring in the cap. Sometimes supplied as newly emerged adult wasps. Some manufacturers produce rearing systems which contain aphids that act as a food source for *Aphidius colemani*, but do not prey on glasshouse crops. Such a system is 'Cereal Aphids' from Biobest, which contains barley infested with cereal aphids, *Rhopalosiphum padi*. **Tradenames:** 'Aphiline c' (Syngenta Bioline), 'Aphiline CE Mix' (50% *Aphidius colemani* and 50% *Aphidius ervi*) (Syngenta Bioline), 'Aphiline Ace Mix' (50% *Aphidius colemani*, 25% *Aphidius ervi* and 25% *Aphelinus abdominalis*) (Syngenta Bioline) and (Rincon-Vitova), 'Aphipar' (Koppert), 'Aphisure (c)' (Biological Crop Protection), 'Aphid Destroyer' (Nature's Alternative Insectary), 'Aphidius System' and 'Aphidius Mix System' (Biobest), 'Aphidius aphid parasite' (M&R Durango), 'Aphidius colemani' (IPM Laboratories), (Harmony), (Praxis), (Neudorff) and (Rincon-Vitova), 'Aphidius' (Sautter & Stepper), 'AAP 2539' (BioSafer), 'AphidiPAK' (Bioplanet).

APPLICATION: Release near infested areas early in the season. Less effective at high aphid population and should be used in combination with other beneficials when aphid infestations are high. Release throughout the infested area by allowing parasitoids to escape from the shipping vial. Treat every other week at a rate of between 2 and 3 wasps per square metre. *Aphidius colemani* can be used at lower rates as a preventive treatment.

PRODUCT SPECIFICATIONS: Purity: Aphid mummies with no non-parasitised aphids. Absence of hyperparasitoids is essential. Percentage hatch should be greater than 70%. Adults of known age with no hyperparasitoids. **Storage conditions:** Maintain at 5 to 10 °C and do not expose to direct sunlight. **Shelf-life:** Use as soon as possible and within 5 days, if received within one day of despatch and stored under recommended conditions. Adults must be released immediately. Release in the early morning or in the late evening when glasshouse vents are closed.

COMPATIBILITY: Do not use yellow sticky traps, as these attract *Aphidius colemani*. Blue sticky traps can be used. Ants will reduce the effectiveness of *A. colemani*.

MAMMALIAN TOXICITY: There have been no reports of acute or chronic toxicity, eye or skin irritation or allergic or other adverse reactions to *Aphidius colemani* in research, production or horticultural staff.

ENVIRONMENTAL IMPACT AND NON-TARGET TOXICITY: *Aphidius colemani* has no effects on non-target organisms and no adverse environmental effects are expected from its use. It is specific to aphids. **Approved for use in organic farming:** Yes.

3:266 *Aphidius ervi*   *Parasitoid of aphids*

Parasitoid wasp: Hymenoptera: Aphidiidae

NOMENCLATURE: Approved name: *Aphidius ervi* Haliday. **Other names:** Aphid parasitoid.

SOURCE: Palaearctic. This parasitoid wasp is native to Europe and Asia. Introduced and established in Chile, Canada, USA and Australia.

PRODUCTION: Reared on aphids under controlled conditions.

TARGET PESTS: Aphids, including *Acyrthosiphon pisum* (Harris), *A. kondoi* Shinji, *Aulacorthum solani* (Kaltenbach), *Sitobion avenae* (Fabricius) and *Macrosiphum euphorbiae* (Thomas).

TARGET CROPS: Glasshouse-grown vegetable crops, such as peppers. May be used in ornamentals under glass in the spring. *Aphidius ervi* was widely released in the past in classical biological control programmes and is used in cereal and legume crops in South America.

BIOLOGICAL ACTIVITY: Biology: Female adults lay single eggs within an aphid. Parasitised aphids continue to feed and can transmit viruses. After egg hatch, the wasp larva progresses through four stages inside the aphid before pupating within the aphid's body, to form the mummy stage. The adult leaves the mummy through a small, round hole. **Predation:** Female wasps can parasitise 100 to 200 aphids within 7 days, although this varies with the host aphid. The presence of *Aphidius ervi* will often disturb the aphid colony, leading to the production of an aphid alarm pheromone that causes aphids to migrate from or fall off the leaf.

Egg laying: Females mate within one day of emergence and begin to lay eggs within a few hours of mating. Fertilised females can lay both unfertilised eggs, that develop into males, and fertilised eggs, that develop into females. Females can lay viable eggs without mating, but these all develop into males. **Duration of development:** It takes between 13 and 15 days from

egg hatch to adult emergence. It is usual for the mummy to exist for 5 to 6 days at 20 °C.
Efficacy: Particularly effective against *Acyrthosiphon pisum*, *Aulacorthum solani* and *Macrosiphum euphorbiae*, although it will parasitise many different aphid species. Very effective searchers that work well against small, well-dispersed populations. Can be attacked by hyperparasitic wasps. **Key reference(s):** 1) E B Hågvar & T Hofsvang. 1991. Aphid parasitoids (Hymenoptera, Aphidiidae): Biology, host selection and use in biological control, *Biocontrol News and Information*, **12**(1), 13–41. 2) P Stary. 1988. Aphidiidae. In *Aphids: their Biology, Natural Enemies and Control*, A K Minks and P Harrewijn (eds), Vol. B, pp. 171–84, Elsevier, Amsterdam, The Netherlands. 3) P M J Ramakers. 1989. Biological control in greenhouses. In *Aphids: their Biology, Natural Enemies and Control*, A K Minks & P Harrewijn (eds), Vol. C, pp. 199–208, Elsevier, Amsterdam, The Netherlands.

COMMERCIALISATION: Formulation: Supplied as freshly emerged adults or as mummies of known age or mixed mummies and adults. Some manufacturers produce rearing systems which contain aphids that act as a food source for *Aphidius ervi*, but do not attack glasshouse crops. Such a system is 'Cereal Aphids-E' from Biobest, which contains barley infested with cereal aphids, *Sitobion avenae* (Fabricius). **Tradenames:** 'Aphiline e' (Syngenta Bioline), 'Aphiline CE Mix' (50% *Aphidius colemani* and 50% *Aphidius ervi*) (Syngenta Bioline), 'Aphiline Ace Mix' (50% *Aphidius colemani*, 25% *Aphidius ervi* and 25% *Aphelinus abdominalis*) (Syngenta Bioline) and (Rincon-Vitova), 'Ervi-M-System' (mummies) and 'Banker-E-System' (Biobest), 'Aphidius ervi' (Rincon-Vitova), (IPM Laboratories), (Praxis) and (Neudorff), 'Ervipar' (Koppert).

APPLICATION: Release near infested areas early in the season. Less effective at high aphid population and should be used in combination with other beneficials when aphid infestations are high. Release throughout the infested area by allowing parasitoids to escape from the shipping vial. Treat every other week at a rate of between 2 and 3 wasps per square metre.

PRODUCT SPECIFICATIONS: Purity: Adult wasps with no non-parasitised aphids or mummies with no carrier material. Absence of hyperparasitoids is essential.
Storage conditions: Maintain at 5 to 10 °C and do not expose to direct sunlight.
Shelf-life: Five days, if received within one day of despatch and stored under recommended conditions.

COMPATIBILITY: Do not use yellow sticky traps, as these attract *Aphidius ervi*. Blue sticky traps can be used.

MAMMALIAN TOXICITY: There have been no reports of allergic or other adverse reactions following use.

ENVIRONMENTAL IMPACT AND NON-TARGET TOXICITY: *Aphidius ervi* occurs in Nature and is not expected to parasitise non-target species or cause any adverse environmental effects. It is specific to aphids. **Approved for use in organic farming:** Yes.

3:267 *Aphidius matricariae*

Parasitoid of aphids

Parasitoid wasp: Hymenoptera: Aphidiidae

NOMENCLATURE: Approved name: *Aphidius matricariae* Haliday. **Other names:** Aphid parasitoid.

SOURCE: Holarctic and now widespread in temperate regions. Generally replaced as an aphid parasitoid by *Aphidius colemani* Viereck which was isolated from a commercial strain of *Aphidius matricariae* that had been selected for its more effective control of aphids.

PRODUCTION: Reared under controlled conditions on aphids such as *Myzus persicae* (Sulzer).

TARGET PESTS: Aphids, particularly *Myzus persicae* and *Myzus nicotianae* Blackman.

TARGET CROPS: Glasshouse-grown vegetable crops, such as tomatoes, aubergines, cucumbers, peppers and ornamentals. Also effective in interiorscapes.

BIOLOGICAL ACTIVITY: Biology: Female adults lay single eggs within an aphid. Parasitised aphids continue to feed and can transmit viruses. After egg hatch, the wasp larva progresses through four stages inside the aphid before pupating within the aphid's body, forming the mummy stage. The adult leaves the mummy through a small, round hole. **Predation:** Female wasps can parasitise 100 to 200 aphids within 7 days. The presence of *Aphidius matricariae* will often disturb the aphid colony, leading to the production of an aphid alarm pheromone that causes aphids to migrate from or fall off the leaf. **Egg laying:** Females mate within one day of emergence and begin to lay eggs within a few hours of mating. Fertilised females can lay both unfertilised eggs, that develop into males, and fertilised eggs, that develop into females. Females can lay viable eggs without mating, but these develop only into males. It is usual for females to lay twice as many 'female' eggs as 'male'. **Duration of development:** It takes between 13 and 15 days from egg hatch to adult emergence, depending upon the host species and the temperature. It is usual for the mummy to exist for 5 to 6 days at 20 °C. **Efficacy:** Particularly effective against *Myzus persicae*, although it will parasitise several different species of aphid. Less effective against *Macrosiphum euphorbiae* (Thomas) and ineffective against *Aphis gossypii* Glover. Very effective searcher that works well against low-density, well-dispersed populations. Can be affected by hyperparasitoid wasps. Less effective and narrower host range than *Aphidius colemani*. **Key reference(s):** 1) E B Hågvar & T Hofsvang. 1991. Aphid parasitoids (Hymenoptera, Aphidiidae): Biology, host selection and use in biological control, *Biocontrol News and Information*, **12**(1), 13–41. 2) P Stary. 1988. Aphidiidae. In *Aphids: their Biology, Natural Enemies and Control*, A K Minks & P Harrewijn (eds), Vol. B, pp. 171–84, Elsevier, Amsterdam, The Netherlands. 3) P M J Ramakers. 1989. Biological control

3. Macro-organisms

in greenhouses. In *Aphids: their Biology*, *Natural Enemies and Control*, A K Minks & P Harrewijn (eds), Vol. C, pp. 199–208, Elsevier, Amsterdam, The Netherlands.

COMMERCIALISATION: Formulation: Sold as adults. **Tradenames:** 'Aphidius matricariae' (Hydro-Gardens), (Applied Bio-Nomics) and (Rincon-Vitova), 'Aphid Parasite' (Biocontrol Network).

APPLICATION: Release near infested areas early in the season. Less effective at high aphid populations and should be used in combination with other beneficials when aphid infestations are high. Release throughout the infested area by allowing parasites to escape from the shipping vial. Treat every other week at a rate of between 2 and 3 wasps per square metre. Can be used at lower rates as a preventive treatment.

PRODUCT SPECIFICATIONS: Purity: Adults only, with no hyperparasitoids.
Storage conditions: Maintain at 5 to 10 °C and do not expose to direct sunlight.
Shelf-life: Can be stored for up to 5 days under recommended conditions, although the parasitoid is more effective if used immediately.

COMPATIBILITY: Do not use yellow sticky traps, as these attract *Aphidius matricariae*. Blue sticky traps can be used. Control ants, as these will protect aphids from parasitism.

MAMMALIAN TOXICITY: No allergic or other adverse reactions have been reported following release of *Aphidius matricariae* by research, manufacturing or field staff.

ENVIRONMENTAL IMPACT AND NON-TARGET TOXICITY: *Aphidius matricariae* occurs widely in Nature and is not expected to have any adverse effects on non-target organisms or on the environment. It is specific to aphids. **Approved for use in organic farming:** Yes.

3:268 *Aphidoletes aphidimyza*
Predator of aphids

Predatory midge: Diptera: Cecidomyiidae

NOMENCLATURE: Approved name: *Aphidoletes aphidimyza* Rondani. **Other names:** Gall midge; aphid midge; aphid gall-midge.

SOURCE: Native to northern parts of North America and Europe. *Aphidoletes aphidimyza* was originally developed as an aphid control measure in Finland and Russia in the 1970s, culminating in its commercial sale in Finland in 1978. Subsequently, commercial trials were conducted in northern Europe and Canada alongside new breeding strategies. This led to the release of commercial products from the mid-1980s.

PRODUCTION: Raised on *Aphis gossypii* Glover or other suitable aphid and sold as pupae within a vermiculite carrier or moist cotton. Often raised on *Myzus persicae* (Sulzer) or *Aphis fabae* Scopoli on sweet peppers, aubergines or beans.

TARGET PESTS: Aphids, all genera and species.

TARGET CROPS: Glasshouse vegetables, such as tomatoes, peppers and cucumbers. Recommended for use on ornamentals and within interiorscapes. In Canada, it has been used successfully in parks on flowers, trees and shrubs. It is also used in home gardens and apple orchards.

BIOLOGICAL ACTIVITY: Biology: *Aphidoletes aphidimyza* adults are active at night, and mating and egg laying occur at night and at dusk. In the day, the adults seek refuge within the crop, usually hanging under leaves or in spider webs low over the ground. Oval eggs (0.3 × 0.1 mm) are laid near or under the aphids. Newly emerged larvae are 0.3 mm long and reach 2.5 mm in length when fully grown. They attack aphids by injecting them with a paralysing toxin through the knee joints. Within 10 minutes of the injection, the body contents of the aphid are dissolved and the midge sucks them out. *Aphidoletes aphidimyza* will consume more aphids as it grows and, if prey are plentiful, will inject and kill more aphids than it can eat. Larvae can consume aphids much larger than themselves and may kill 4–65 aphids per day, many more aphids than they eat when aphid populations are high. *Aphidoletes aphidimyza* enters diapause in the autumn with falling temperatures and shortening days. In the autumn, fully grown larvae tunnel into the soil, where they hibernate about 2 cm underground in a cocoon. In glasshouses or interiorscapes, winter diapause can be avoided with low-level supplemental lighting (60 watts for 20 square metres). **Predation:** *Aphidoletes aphidimyza* larvae will eat between 10 and 100 aphids from egg hatch to pupation. The preferred temperature for predation is between 19 and 28 °C at high relative humidity. The adults feed on honeydew. **Egg laying:** The number of eggs laid by a female depends on the weather, the amount of food consumed as a larva and the amount of honeydew absorbed as an adult. Most eggs are laid in the first 2 to 4 days of adulthood. Females live for about 10 days and lay up to 250 eggs. **Duration of development:** Development time is very dependent on temperature. At 21 °C, the egg stage takes 2 to 3 days, the larval stage takes 7 to 14 days and the pupal stage takes 14 days. Female adults live for about 10 days and males for about 7 days. **Efficacy:** The searching behaviour of the adult female midge is very effective, with infested plants being found easily within uninfested plants. Large aphid populations are preferred for egg laying. This allows for the rapid spread of the midges throughout the glasshouse. Larvae can move about 6 cm without feeding and can detect aphids from a distance of 2.5 cm. **Key reference(s)**: P M J Ramakers. 1989. Biological control in greenhouses. In *Aphids: their*

3. Macro-organisms

Biology, Natural Enemies and Control, A K Minks & P Harrewijn (eds), Vol. C, pp. 199–208, Elsevier, Amsterdam, The Netherlands.

COMMERCIALISATION: Formulation: Pupae stored in vermiculite in bottles, in damp cotton or in moist peat. Sometimes sold as adults. **Tradenames:** 'Aphidoline a' (Syngenta Bioline), 'Aphidoletes-V' (in vermiculite in bottles) (Applied Bio-Nomics), 'Aphidoletes Aphid parasites' (Biowise), 'Aphidoletes Gallmücken' (Neudorff) and (Sautter & Stepper), 'Aphidoletes aphidimyza midge AA250' and 'Aphidoletes aphidimyza midge AA1' (Rincon-Vitova), 'Aphidoletes aphidimyza' (Harmony), (Hydro-Gardens), (IPM Laboratories) and (Praxis), 'Aphidoletes aphidimyza predatory gall midge' (M&R Durango), 'Aphidoletes-System' (Biobest), 'Aphid Predator' (Arbico) and (Biocontrol Network), 'Aphidoletes aphidimyza' (adults) (Applied Bio-Nomics) and (Nature's Control), 'Aphidosure (a)' (Biological Crop Protection), 'Aphidend' (Koppert).

APPLICATION: Apply 1 to 6 midges per plant weekly for 2 to 4 weeks. A guide for release is about one midge for 10 aphids. Keep packing material moist for 2 weeks following opening. For overall application, release 1 to 4 midges per square metre, but use higher rates to control established colonies. Distribute the carrier onto the growing medium or into release boxes. The bottles can also be opened and left in a shady situation where the midges will escape. Early application is recommended in aphid hot spots. Pupae survive best when organic matter is present for a pupation site. Keep conditions moist but not wet. For best results, use in conjunction with *Hippodamia* and *Chrysoperla* spp. Low release rates may be used for preventative control. Two to four releases are usually necessary to achieve control in heavy infestations. Additional releases are not necessary as *Aphidoletes* reproduces well.

PRODUCT SPECIFICATIONS: Purity: No live aphids, only viable pupae.
Storage conditions: Can be stored for up to 5 days in a refrigerator. Will emerge from pupae within a week at room temperature. **Shelf-life:** *Aphidoletes aphidimyza* will survive for up to 5 days in a refrigerator (10 °C). Cold storage reduces female egg laying and delays emergence. They should be used as soon after delivery as possible.

COMPATIBILITY: Avoid residual pesticides. Do not hose down the crop, as this will dislodge the larvae. Use in conjunction with other aphid parasitoids.

MAMMALIAN TOXICITY: No allergic or other adverse effects have been recorded with research or manufacturing staff or through its use in the field. There is no evidence of acute or chronic toxicity, eye or skin irritation or hypersensitivity to mammals.

ENVIRONMENTAL IMPACT AND NON-TARGET TOXICITY: No adverse environmental or non-target effects have been recorded from the use of *Aphidoletes aphidimyza*. It is specific to aphids. **Approved for use in organic farming:** Yes.

3:269 *Aphthona flava*   *Phytophagous beetle*

Phytophagous beetle: Coleoptera: Chrysomelidae

NOMENCLATURE: Approved name: *Aphthona flava* Guill. **Other names:** Copper leafy spurge flea beetle; amber spurge flea beetle.

SOURCE: *Aphthona flava* is native to Europe and was introduced into the USA from Italy and Hungary in 1985. It is now widely established in North America.

PRODUCTION: Collected from leafy spurge plants in the field, but may also be reared on living plants.

TARGET PESTS: The host range of *Aphthona flava* appears restricted to spurge plants in the subgenus *Esula* of the genus *Euphorbia*. In Europe, this beetle feeds on leafy spurge (*Euphorbia esula* L.) and several other closely related spurge species.

TARGET CROPS: Used in pasture and grassland.

BIOLOGICAL ACTIVITY: Biology: *Aphthona flava* is the largest of the spurge-feeding flea beetle species and is copper in colour. Depending on location, adults emerge from the soil beginning in June to August and are present for several weeks to several months. *Aphthona flava* adults feed on leafy spurge foliage and flowers and high populations may defoliate spurge plants. Newly hatched larvae burrow into the soil and begin feeding on very small leafy spurge roots. Larvae feed on progressively larger roots and root buds as they develop. *A. flava* larvae are up to 6 mm long, with short legs, yellow heads and creamy-white bodies. The larvae overwinter, resume feeding in the spring and then pupate in a soil cell in late spring to early summer. There is one generation per year. **Egg laying:** Females lay groups of 20 to 30 eggs at, or just below, the soil surface, near the base of a leafy spurge stem. **Duration of development:** The eggs are deposited in June until early autumn, generally on the plant stem at, or below, the soil surface and sometimes on or in the soil, but near the plant stem. The larvae are active from July until early spring of the following year. The young larvae begin feeding in/on the root hairs; as they become older and larger, they migrate to the larger roots. They are difficult to observe, except under a microscope. The more mature larvae are whitish and worm-like and can be observed with the naked eye in freshly extracted roots. Pupation occurs in a soil cell from late spring to early summer. Adults will emerge in June until early autumn, depending on degree-days. Adults have the characteristic flea beetle appearance and jump when disturbed. Adult males are about 3.4 mm long; females are about 3.6 mm long. **Efficacy:** Feeding on the foliage reduces photosynthesis and flower consumption slightly reduces flowering ability. Feeding within the roots reduces the plant's ability to absorb moisture and nutrients. Light populations reduce plant height and retard flowering, while high populations reduce plant density and cause what is often referred to as 'a hole in the spurge'. At one research site in the USA, this species reduced the aerial portion of leafy spurge in a

212 by 167 m area in six years from 57% to less than 2%.

Key reference(s): 1) R W Pemberton & N E Rees. 1990. Host specificity and establishment of *Aphthona flava* Guill. (Chrysomelidae), a biological control agent for leafy spurge (*Euphorbia esula*) in the United States, *Proc. Entomol. Soc. Wash.*, **92**, 351–7. 2) R W Pemberton. 1995. Leafy spurge (*Euphorbia esula* L. complex). In *Biological Control in the U.S. Western Region: Accomplishments and Benefits of Regional Research Project W-84 (1964–1989)*, J R Nechols, L A Andres, J W Beardsley, R D Goeden & C G Jackson (eds), Univ. California, Berkeley, USA, pp. 149–55.

COMMERCIALISATION: Formulation: Supplied as fecund adults. **Tradenames:** 'Aphthona flava' (Praxis).

APPLICATION: Release adults into leafy spurge infested pasture in June/July.

PRODUCT SPECIFICATIONS: Purity: Only fecund adults, with no contaminants.

COMPATIBILITY: It is not usual for *Aphthona flava* to be used with other pest control measures.

MAMMALIAN TOXICITY: No allergic or other adverse reactions have been noted with workers in the field. There is no evidence of acute or chronic toxicity, eye or skin irritation or hypersensitivity to mammals.

ENVIRONMENTAL IMPACT AND NON-TARGET TOXICITY: *Aphthona flava* occurs widely in Nature and its host range is restricted to plants in the subgenus *Esula* of the genus *Euphorbia*. As such, it is not expected to have any effects on non-target organisms or to have any adverse effects on the environment.

3:270 *Aphytis holoxanthus*
Parasitoid of red scale

Parasitoid wasp: Hymenoptera: Aphelinidae

NOMENCLATURE: Approved name: *Aphytis holoxanthus* (DeBach) **Other names:** Red scale parasitoid.

SOURCE: *Aphytis holoxanthus* was originally collected from *Chrysomphalus aonidum* (L.) in Hong Kong and introduced into the USA in 1956. Following outstanding control of *C. aonidum* there, the parasitoid was introduced into Israel, where its effect was equally spectacular. In Australia, it was released in New South Wales and south-eastern Queensland in 1974. It spread throughout the citrus orchards and, since 1977, *C. aonidum* has become extremely uncommon and difficult to find in the field. Chemical control of red scale is no longer required.

PRODUCTION: Reared on Florida red scale insects and sold as adult wasps fed on honey for 24 hours prior to shipment.

TARGET PESTS: Citrus red scale (*Chrysomphalus aonidum*).

TARGET CROPS: Citrus.

BIOLOGICAL ACTIVITY: Biology: *Aphytis holoxanthus* is a bi-parental, external parasitoid of the adult female *Chrysomphalus aonidum*. **Predation:** After boring through the hard scale covering with its ovipositor, the parasitoid deposits an egg on the soft tissues of the host scale. After hatching, parasitoid larvae become attached to the scale, where they feed on the body fluids of the host. Pupation by the parasitoid occurs on the host's remains beneath the scale cover. Adult parasitoids cut an irregular-shaped hole in the scale cover to emerge. **Efficacy:** *Aphytis holoxanthus* can move as far as six trees from the point of introduction, attracted by the scale sex pheromone. It is a very effective parasitoid, particularly under conditions of high temperature. Adults also feed on nectar and the presence of nectar-producing plants aids parasitism.

COMMERCIALISATION: Formulation: Introduced by the USDA in Florida to control the introduced red scale in citrus.

APPLICATION: Apply by hanging container on a twig of the infested plant on the shaded side of the tree, one capsule to every 9 to 12 trees. Begin release before flowering and continue until the autumn. Most effective if released before male scale flight. Recommended release rate 12 000 to 25 000 wasps per hectare. Young trees represent poor candidates for biological control, because they offer little natural shelter for the beneficial wasp.

PRODUCT SPECIFICATIONS: Purity: Adult wasps, with no contaminants.
Storage conditions: Can be stored in a cooled, insulated container at 15 to 19 °C, away from direct sunlight. **Shelf-life:** Adults will survive in storage for about 26 days after emergence.

COMPATIBILITY: Extreme cold or low humidity will limit activity. Ants interfere with the parasitoids and reduce their performance. Can be released following application of copper or foliar fertilisers.

MAMMALIAN TOXICITY: No allergic or other adverse reactions have been noted with workers in the field. There is no evidence of acute or chronic toxicity, eye or skin irritation or hypersensitivity to mammals.

ENVIRONMENTAL IMPACT AND NON-TARGET TOXICITY: *Aphytis holoxanthus* occurs widely in Nature and is not expected to parasitise non-target organisms or have any adverse effects on the environment. It is specific to scale insects.
Approved for use in organic farming: Yes.

3:271 *Aphytis lepidosaphes*
Parasitoid of citrus mussel scale

Parasitoid wasp: Hymenoptera: Aphelinidae

NOMENCLATURE: Approved name: *Aphytis lepidosaphes* Compère. **Other names:** Citrus mussel scale parasitoid; purple scale parasitoid.

SOURCE: *Aphytis lepidosaphes* was introduced from China into California in 1948 and was first recorded in Australia as a parasitoid of *Lepidosaphes beckii* (Newman) in 1967, although not intentionally introduced. It is possible that *A. lepidosaphes* became established accidentally in Australia in the 1960s, at a time when the parasitoid was being considered for introduction from the USA.

PRODUCTION: Established in Australia and the USA and no longer produced commercially.

TARGET PESTS: Citrus mussel scale (citrus purple scale) (*Lepidosaphes* (*Cornuaspis*) *beckii*).

TARGET CROPS: Citrus.

BIOLOGICAL ACTIVITY: Biology: *Aphytis lepidosaphes* is a bi-parental, gregarious ectoparasitoid, specific to *Lepidosaphes* species. **Predation:** Adult females are preferred as hosts, but nymphs and male pre-pupae may also be attacked. From one to eight individual parasitoids may develop in one parasitised scale host. Adult females of *Aphytis lepidosaphes* outnumber males and feed from wounds caused by oviposition and probing, leading to considerable mortality of scales in addition to those killed by parasitisation. **Efficacy:** *Aphytis lepidosaphes* is very mobile and moves well between citrus trees. It has been shown to give 100% control of citrus mussel scale.

COMMERCIALISATION: Formulation: Not sold commercially.

APPLICATION: Release around scale-attacked mature trees where there is plenty of shade. Adults do feed on nectar and so the presence of nectar-producing plants enhances parasitism.

COMPATIBILITY: Extreme cold or low humidity will limit activity. Ants interfere with the parasitoids and reduce their performance. Can be released following application of copper or foliar fertilisers.

MAMMALIAN TOXICITY: No allergic or other adverse reactions have been noted with workers in the field. There is no evidence of acute or chronic toxicity, eye or skin irritation or hypersensitivity to mammals.

ENVIRONMENTAL IMPACT AND NON-TARGET TOXICITY: *Aphytis lepidosaphes* occurs widely in Nature and does not parasitise non-target organisms or have any adverse effects on the environment. It is specific to scale insects.

3:272 *Aphytis lignanensis* *Parasitoid of scale*

Parasitoid wasp: Hymenoptera: Aphelinidae

NOMENCLATURE: Approved name: *Aphytis lignanensis* Compère. **Other names:** Armoured scale parasitoid; red scale parasitoid; scale parasitoid; golden chalcid.

SOURCE: Native of India and Pakistan.

PRODUCTION: Reared on scale insects and sold as adult wasps fed on honey for 24 hours prior to shipment.

TARGET PESTS: Scale insects, particularly of citrus, olives, nut crops and passion fruit, including Florida red scale (*Aonidiella aurantii* (Maskell)), oriental yellow scale (*A. orientalis* (Newstead)) and oleander scale (*Aspidiotus nerii* (Bouché)).

TARGET CROPS: Citrus, ornamentals, orchard crops and passion fruit.

BIOLOGICAL ACTIVITY: Biology: The female wasp lays eggs under the scale cover and onto the body of second instar and unmated female scales. The unmated female scale releases a pheromone to attract the males and this also attracts the parasitoid female wasp. The eggs develop into legless larvae that feed on the scale by sucking its body fluids. All stages of scale, except the crawlers, are parasitised by the larvae. The larvae pupate within the scale and later emerge to mate and continue the cycle. Adults also feed on scales. **Predation:** One parasitoid will destroy an average of 30 scales in its lifetime. **Egg laying:** The female wasp lays her eggs under the female scale when it loosens to allow the male to mate. Between 1 and 5 eggs are laid, dependent upon the size of the scale. Approximately 6 scales will be attacked per day. **Duration of development:** Egg to adult takes about 12 to 13 days at 27 °C and the adult has a life span of 26 days in the presence of an adequate food source. **Efficacy:** *Aphytis lignanensis* can move as far as six trees from the point of introduction, attracted by the scale sex pheromone. Less effective under conditions of high or low temperature or low humidity. Adults also feed on nectar and the presence of nectar-producing plants aids parasitism. More effective parasitoid in hot, humid conditions than *Aphytis melinus* DeBach.

COMMERCIALISATION: Formulation: Supplied in boxes lined with wet newspaper, packed in ice or in sealed plastic capsules. **Tradenames:** 'Aphytis lignanensis' (Bugs for Bugs).

APPLICATION: Apply by hanging container on a twig of the infested plant on the shaded side of the tree, one capsule to every 9 to 12 trees. Begin release before flowering and continue until the autumn. Most effective if released before male scale flight. Recommended release rate 12 000 to 25 000 wasps per hectare. Young trees represent poor candidates for biological control, because they offer little natural shelter for the beneficial wasp.

PRODUCT SPECIFICATIONS: Purity: Adult wasps, with no contaminants.
Storage conditions: Can be stored in a cooled, insulated container at 15 to 19 °C, away from

direct sunlight. **Shelf-life:** Adults will survive in storage for about 26 days after emergence.

COMPATIBILITY: Extreme cold or low humidity will limit activity. Ants interfere with the parasitoids and reduce their performance. Generally, copper and foliar feeds will not harm *Aphytis lignanensis*, but organophosphates are toxic. If an OP has been applied, a minimum of 4 weeks should elapse before release of the parasitoids.

MAMMALIAN TOXICITY: No allergic or other adverse reactions have been noted with workers in the field. There is no evidence of acute or chronic toxicity, eye or skin irritation or hypersensitivity to mammals.

ENVIRONMENTAL IMPACT AND NON-TARGET TOXICITY: *Aphytis lignanensis* occurs widely in Nature and is not expected to parasitise non-target organisms or have any adverse effects on the environment. It is specific to scale insects.
Approved for use in organic farming: Yes.

3:273 *Aphytis melinus*   *Parasitoid of scale*

Parasitoid wasp: Hymenoptera: Aphelinidae

NOMENCLATURE: Approved name: *Aphytis melinus* DeBach. **Other names:** Armoured scale parasitoid; red scale parasitoid; scale parasitoid; golden chalcid.

SOURCE: Native to India and Pakistan.

PRODUCTION: Reared on scale insects and sold as adult wasps fed on honey for 24 hours prior to shipment.

TARGET PESTS: Red scale and oriental scale insects, particularly of citrus, olives, nut crops and passion fruit, including *Aonidiella aurantii* (Maskell), *A. orientalis* (Newstead), *Chrysiomphalus dictyospermi* Morgan and *Aspidiotus nerii* (Bouché).

TARGET CROPS: Citrus, ornamentals, orchard and nut crops and passion fruit.

BIOLOGICAL ACTIVITY: Biology: Adult females are yellow, very mobile and 1 mm. long. The female wasp lays eggs under the scale cover and onto the body of second instar and unmated female scales. The unmated female scale releases a pheromone to attract the males and this also attracts the parasitoid female wasp. The eggs develop into legless larvae that feed on the scale by sucking its body fluids. All stages of scale, except the crawlers, are parasitised by the larvae. The larvae pupate within the scale and later emerge to mate and continue the cycle. Adults also feed on scales. **Predation:** One parasite will destroy an average of 30 scales in its lifetime. **Egg laying:** The female wasp lays her eggs under the female scale when it loosens to allow the male to mate. Between 1 and 5 eggs are laid, dependent upon the size of the scale. Approximately 6 scales will be attacked per day. **Duration of development:** Egg to adult takes about 12 to 13 days at 27 °C and the adult has a life span of about 26 days in the presence of an adequate food source. **Efficacy:** *Aphytis melinus* can move as far as six trees from the point of introduction, attracted by the scale sex pheromone. Less effective under conditions of high or low temperature or low humidity. Adults also feed on nectar and the presence of nectar-producing plants aids parasitism. *Aphytis melinus* is more suited to cooler, less humid regions than *Aphytis lignanensis* Compère.

COMMERCIALISATION: Formulation: Supplied as adults in boxes lined with wet newspaper, packed in ice or in sealed plastic capsules. **Tradenames:** 'Aphytis melinus' (Biological Services), (Biocontrol Network), (Arbico), (IPM Laboratories), (Praxis), (Hydro-Gardens), (Sespe Creek) and (Rincon-Vitova), 'MeliPAK5000' (Bioplanet).

APPLICATION: Apply by hanging container on a twig of the treated plant on the shaded side of the tree, one capsule to every 9 to 12 trees. Begin release before flowering and continue until the autumn. Most effective if released before male scale flight. Recommended release rate 12 500 to 25 000 wasps per hectare. Young trees represent poor candidates for biological control, because they offer little natural shelter for the beneficial wasp.

PRODUCT SPECIFICATIONS: Purity: Adult wasps, with no contaminants.
Storage conditions: Can be stored in a cooled, insulated container at 15 to 19 °C, away from direct sunlight. **Shelf-life:** Adults will survive in storage for about 26 days after emergence.

COMPATIBILITY: Extreme cold or low humidity will limit activity. Ants interfere with the parasitoids and reduce their performance. Very effective in dry conditions. Do not release within 4 weeks of an OP spray. Can be released following application of copper or foliar fertilisers.

MAMMALIAN TOXICITY: No allergic or other adverse reactions have been noted with workers in the field. There is no evidence of acute or chronic toxicity, eye or skin irritation or hypersensitivity to mammals.

ENVIRONMENTAL IMPACT AND NON-TARGET TOXICITY: *Aphytis melinus* occurs widely in Nature and is not expected to parasitise non-target organisms or have any adverse effects on the environment. It is specific to scale insects.
Approved for use in organic farming: Yes.

3. Macro-organisms

3:274 Apion fuscirostre   *Phytophagous weevil*

Phytophagous weevil: Coleoptera: Apionidae

NOMENCLATURE: Approved name: *Apion fuscirostre* Fabricius. **Other names:** Scotchbroom seed weevil. Formerly known as: *Exapion fuscirostre* (Fabricius).

SOURCE: Native of western Europe, but introduced in Australasia and the USA.

PRODUCTION: Reared on living scotchbroom (*Sarothamnus scoparius* Wimmer).

TARGET PESTS: Scotchbroom (*Sarothamnus scoparius*).

TARGET CROPS: Rangeland and pasture.

BIOLOGICAL ACTIVITY: Biology: The adult weevils are small. The females lay eggs on the seed pods and the emerging larvae bore into the pods and feed on the seeds.
Predation: Larvae are white, squat, legless grubs, 2.0 to 2.5 mm long, with a very small, brown head. Pupae are found inside the seed pod. **Efficacy:** The adults disperse readily on warm winter days and throughout summer by flying from bush to bush.

COMMERCIALISATION: Formulation: Sold as fecund adults. **Tradenames:** 'Apion fuscirostre' (Praxis).

APPLICATION: Adults should be released in the scotchbroom bushes as the pods are forming.

PRODUCT SPECIFICATIONS: Purity: Only fecund adults, with no contaminants.
Storage conditions: Can be stored for short periods at low temperatures (in a refrigerator).
Shelf-life: Use as soon as possible, but, if stored following the producer's instructions, the weevils can be kept for several days.

COMPATIBILITY: It is unusual to use any other products in combination with *Apion fuscirostre*.

MAMMALIAN TOXICITY: No allergic or other adverse reactions have been noted with workers in the field. There is no evidence of acute or chronic toxicity, eye or skin irritation or hypersensitivity to mammals.

ENVIRONMENTAL IMPACT AND NON-TARGET TOXICITY: *Apion fuscirostre* occurs widely in Nature and has never been shown to parasitise any plants other than *Sarothamnus* species. For this reason, it is not expected to have any effect on non-target organisms or to have any adverse effects on the environment.

3:275 *Apion ulicis*　　　　　*Phytophagous weevil*

Phytophagous weevil: Coleoptera: Apionidae

NOMENCLATURE: Approved name: *Apion ulicis* (Forster). **Other names:** Gorse seed weevil. Formerly known as: *Exapion ulicis* (Forster).

SOURCE: *Apion ulicis* is a native of western areas of Europe and Great Britain. Between 1926 and 1932, 321 000 weevils were imported from England to New Zealand. From 1931, approximately 240 000 weevils were released throughout New Zealand at 240 sites. The species established readily in all areas and is now extremely common. The weevil was introduced into the USA in 1956 and is now common there.

PRODUCTION: *Apion ulicis* is reared on gorse (*Ulex europaeus* L.).

TARGET PESTS: Gorse (*Ulex europaeus*).

TARGET CROPS: Rangeland and pasture.

BIOLOGICAL ACTIVITY: Biology: The adult, a grey, hard-bodied, pear-shaped weevil with a long, curved snout, is only 1.8 to 2.5 mm long. A covering of minute scales gives the weevil its grey colour, but, if these scales are rubbed off, the underlying colour is much darker. The gender of adults can be determined by the relative length of the rostrum (snout). In males, the rostrum length is approximately the same as the dorsal length of the thorax, whereas, in females, it is 1.5 times that length. Generally females are larger than males. **Predation:** Larvae are white, squat, legless grubs, 2.0 to 2.5 mm long, with a very small, brown head. Pupae are found inside the seed pod, and, when newly formed, are white. As the pupae develop, their colour changes to grey. **Egg laying:** The glistening yellow eggs are laid through a hole chewed by the female, singly or in clusters of up to 20 inside young, green gorse pods.

Duration of development: Development from egg to adult takes approximately 16 weeks. The eggs take 4 weeks to hatch, the larvae take 6 to 8 weeks to pupate and the pupae take 4 weeks to complete development within the pods. There is only one generation each year. In warmer climates, weevils complete development between spring and early summer, but, in cooler areas (for example at higher altitudes), development time is extended into midsummer. **Efficacy:** The adults disperse readily on warm winter days and throughout summer by flying from bush to bush. Populations are commonly found on isolated gorse plants.

COMMERCIALISATION: Formulation: Available as fecund adults. **Tradenames:** 'Apion ulicis' (Praxis).

APPLICATION: Release adults in the gorse bush.

PRODUCT SPECIFICATIONS: Storage conditions: Can be stored for short periods at low temperatures (in a refrigerator). **Shelf-life:** Use as soon as possible, but, if stored following the producer's instructions, the weevils can be kept for several days.

3. Macro-organisms

COMPATIBILITY: It is unusual to use any other products in combination with *Apion ulicis*.

MAMMALIAN TOXICITY: No allergic or other adverse reactions have been noted with workers in the field. There is no evidence of acute or chronic toxicity, eye or skin irritation or hypersensitivity to mammals.

ENVIRONMENTAL IMPACT AND NON-TARGET TOXICITY: *Apion ulicis* occurs widely in Nature and has never been shown to parasitise any plants other than *Ulex* species. For this reason, it is not expected to have any effect on non-target organisms or to have any adverse effects on the environment.

3:276 *Aprostocetus hagenowii*
Hyperparasite of cockroaches

Parasitic wasp: Hymenoptera: Euliophidae

NOMENCLATURE: Approved name: *Aprostocetus hagenowii* Ratzeburg.

SOURCE: A commonly occurring wasp; present where cockroaches are found.

PRODUCTION: *Aprostocetus hagenowii* is bred on the egg cases of the American cockroach (*Periplaneta americana* (L.)) under controlled conditions.

TARGET PESTS: Cockroaches, but especially the American cockroach (*Periplaneta americana*), the smokybrown cockroach (*P. fuliginosa* (Serville)), the Australian cockroach (*P. australasiae* (Fabricius)), the brown cockroach (*P. brunnea* Burmeister), the oriental cockroach (*Blatta orientalis* L.), the harlequin cockroach (*Neostylopyga rhombifolia* (Stoll)) and the Florida cockroach (*Eurycotis floridana* (Walker)).

TARGET CROPS: Used in warehouses, industrial properties and domestic houses.

BIOLOGICAL ACTIVITY: Biology: *Aprostocetus hagenowii* attacks the oothecae (egg capsules) of eight species of cockroaches. Females lay their eggs inside cockroach oothecae. The cockroach eggs hatch and the developing larvae of the parasite consume the developing cockroach nymphs. The larvae continue to mature and eventually pupate inside the cockroach oothecae. *Aprostocetus hagenowii* reaches the adult stage inside the cockroach egg cases. Adults emerge from the egg case by chewing a hole in the shell. Male and female adults mate immediately upon emerging from the cockroach ootheca. After mating, *Aprostocetus hagenowii* females search for more cockroach oothecae. Fertilised eggs produce female progeny and unfertilised eggs produce male progeny. Female wasps live 7 to 10 days. **Egg laying:** Females deposit eggs in host's oothecae early in life (3 to 5 days of adult life). Females attack 1 to 2 (occasionally 3) hosts in their lifetime. **Duration of development:** It takes between 32 and 40 days for *Aprostocetus hagenowii* to develop from egg to adult, depending upon the temperature and the number of individuals developing within the host. The average number of females that develop from each host is dependent upon the host and varies between 45 and 70. **Efficacy:** *Aprostocetus hagenowii* will not attack cockroach nymphs or adults. These stages need to be controlled by other methods compatible with natural enemies, such as trapping or baiting. *Aprostocetus hagenowii* only attacks cockroach oothecae.

Key reference(s): 1) L M Lebeck. 1991. A review of the hymenopterous natural enemies of cockroaches with emphasis on biological control, *Entomophaga*, **36**(3), 335–52. 2) F Agudelo-Silva. 1995. Biological control of roaches, *IPM Practitioner*, **17**(5/6), 12–13.

COMMERCIALISATION: Formulation: Sold as adults as part of a cockroach control programme. **Tradenames:** 'Aprostocetus hagenowii' (Rincon-Vitova), 'Periplan' (Entocare).

APPLICATION: The wasps should be released where there is evidence of cockroach activity.

PRODUCT SPECIFICATIONS: Storage conditions: Release immediately.

COMPATIBILITY: Incompatible with residual insecticides.

MAMMALIAN TOXICITY: There have been no reports of allergic or other adverse toxicological effects arising from contact with *Aprostocetus hagenowii* from research staff, producers or users.

ENVIRONMENTAL IMPACT AND NON-TARGET TOXICITY: *Aprostocetus hagenowii* has a wide distribution in Nature and there is no evidence that it has any adverse effects on non-target organisms or on the environment. It is a specific cockroach hyperparasite.

3:277 *Atheta coriaria*   *Predatory beetle*

Predatory beetle: Coleoptera: Staphylinidae

NOMENCLATURE: Approved name: *Atheta coriaria* Kraatz. **Other names:** Rove beetle.

SOURCE: Widespread soil-inhabiting beetle.

PRODUCTION: Reared under controlled conditions on sciarid flies (*Bradysia* spp.).

TARGET PESTS: Soil-inhabiting insects such as sciarid flies (fungus gnats), shoreflies (*Scatella* spp.), springtails, mites, thrips pupae and nematodes..

TARGET CROPS: Glasshouse-grown ornamentals and vegetables.

BIOLOGICAL ACTIVITY: Biology: *Atheta coriaria* adults are dark brown to shiny black, 3 to 4mm long and are completely covered with hair. They spend most of their time in the soil, but they can fly and spread quickly throughout the glasshouse. They have 3 larval stages that vary in colour from creamy white to brown as they age. The adult and the larvae all move rapidly in the soil and are highly predaceous on almost anything that moves, including sciarid flies and shoreflies (eggs and larvae) and thrips (pupae). **Predation:** *Atheta coriaria* adult beetles and larvae mainly search for eggs, young larvae and pupae of the sciarid fly and the shorefly. The adults also predate the second larval stage of the Western flower thrips (*Frankliniella occidentalis*) when in the soil prior to pupation. **Egg laying:** An adult female will lay 8 eggs per day for the first 2 weeks after mating. **Duration of development:** Total development time from egg to adult takes about 3 weeks at 25 °C. After that period, the adult predatory beetle will be active for approximately 3 weeks. **Efficacy:** *Atheta coriaria* adults and larvae are voracious predators.

COMMERCIALISATION: Formulation: Sold as adult beetles in peat, vermiculite or similar carrier. **Tradenames:** 'Staphyline ac' (Syngenta Bioline), 'Atheta-System' (Biobest), 'Atheta coriaria' (Rincon-Vitova).

APPLICATION: *Atheta coriaria* can be used in all different growing media. Introduce at least 2 predatory beetles per square metre, evenly throughout the crop.

PRODUCT SPECIFICATIONS: Purity: Contains adult beetles, with no phytophagous mites, mould or other contaminants. **Storage conditions:** *Atheta coriaria* can be stored for short periods at 10 to 15 °C and r.h. >85%. **Shelf-life:** Use as soon as possible after receipt.

COMPATIBILITY: *Atheta coriaria* may consume other soil-inhabiting beneficials in the absence of prey.

MAMMALIAN TOXICITY: No allergic or other adverse reactions have been reported following the use of *Atheta coriaria* in glasshouse crops by producers, research workers or growers.

ENVIRONMENTAL IMPACT AND NON-TARGET TOXICITY: *Atheta coriaria* occurs widely in Nature and is not expected to have any adverse effects on non-target organisms or on the environment. It is a polyphagous soil-inhabiting beetle and is not expected to move long distances from its point of introduction. **Approved for use in organic farming:** Yes.

3:278 *Bangasternus orientalis*

Phytophagous weevil

Phytophagous weevil: Coleoptera: Curculionidae

NOMENCLATURE: Approved name: *Bangasternus orientalis* Capiomont.
Other names: Yellow star-thistle bud weevil.

SOURCE: *Bangasternus orientalis* is native to southern Eurasia and the Mediterranean basin.

PRODUCTION: The weevil is raised on yellow star-thistle plants.

TARGET PESTS: Yellow star-thistle (*Centaurea solstitialis* L.) and purple star-thistle (*Centaurea calcitrapa* L.).

TARGET CROPS: Grassland and pasture.

BIOLOGICAL ACTIVITY: Biology: There is one generation of *Bangasternus orientalis* per year. The weevil overwinters as an adult, outside the host plant. A female may produce up to 470 eggs, laid in late spring to early summer. The single eggs are covered with a dark mucilage which is laid on or near scale leaves beneath the immature head buds at the tips of flowering shoots. Hatched larvae tunnel through the scale leaf, the flowering stalk or peduncle and into the flower head, where they feed on receptacle tissue and developing seeds. Larvae pupate into adults within a chamber constructed of chewed seeds within the seed heads. The adults are brown, with yellow to whitish hairs that give a somewhat mottled appearance. They are 4 to 6 mm long, not including the snout. Adults exit from the pupal chambers in the heads in late summer to overwinter outside the host plant. **Predation:** The larval is the destructive stage. They tunnel through the scale leaf, the flowering stalk or peduncle and into the flower head. **Efficacy:** Larval feeding reduces seed production. Preliminary data indicate that a single larva destroys 50% to 60% of the seeds in a seed head. **Key reference(s):** N Rees *et al.* (eds). 1996. *Biological Control of Weeds in the West*, Western Society of Weed Science, in cooperation with USDA-ARS, Montana Department of Agriculture and Montana State University, Color World Printers, Bozeman, Montana, USA.

COMMERCIALISATION: Formulation: The original source for release in the USA was northern Greece. The first releases in the USA were made in 1985. *Bangasternus orientalis* is now established in California, Idaho, Oregon and Washington. **Tradenames:** 'Bangasternus orientalis' (Praxis).

APPLICATION: Release mating adults amongst the star-thistles in late spring to early summer.

PRODUCT SPECIFICATIONS: Storage conditions: Can be stored for short periods at low temperatures (in a refrigerator). **Shelf-life:** Use as soon as possible, but, if stored following the producer's instructions, the weevils can be kept for several days.

COMPATIBILITY: It is unusual to use any other products in combination with *Bangasternus orientalis.*

MAMMALIAN TOXICITY: No allergic or other adverse reactions have been noted with workers in the field. There is no evidence of acute or chronic toxicity, eye or skin irritation or hypersensitivity to mammals.

ENVIRONMENTAL IMPACT AND NON-TARGET TOXICITY: *Bangasternus orientalis* occurs widely in Nature and has never been shown to parasitise any plants other than *Centaurea* species. For this reason, it is not expected to have any effect on non-target organisms or to have any adverse effects on the environment.

3:279 *Bathyplectes curculionis*
Parasitoid of weevils

Parasitoid wasp: Hymenoptera: Ichneumonidae

NOMENCLATURE: Approved name: *Bathyplectes curculionis* (Thompson).
Other names: Alfalfa weevil parasitoid.

SOURCE: *Bathyplectes curculionis* occurs throughout Europe.

PRODUCTION: *Bathyplectes curculionis* is raised on alfalfa weevil.

TARGET PESTS: Alfalfa weevil (*Hypera postica* (Gyllenhal)).

TARGET CROPS: Alfalfa (lucerne).

BIOLOGICAL ACTIVITY: Biology: *Bathyplectes* are small, non-stinging wasps that are parasitoids of the alfalfa weevil. Unlike generalist predators that are found in many agriculture habitats and feed on numerous prey species, *Bathyplectes* are very specific natural enemies that occur only in and around alfalfa fields and attack only the alfalfa weevil, *Hypera postica*. There are two common species of *Bathyplectes*, *B. curculionis* and *B. anurus* (Thompson), often referred to simply as *Bathyplectes* spp. The two species are similar in appearance and habits. Adults are about 3 mm long, with black, robust bodies. Both lay their eggs in alfalfa weevil larvae, preferring to oviposit in the early instars. The wasp larva that hatches from the egg feeds internally and slowly devours the weevil larva, ultimately killing its host after the weevil has finished spinning its cocoon. The parasitoid larva then emerges from the weevil and spins a cocoon of its own. Only one parasitoid can successfully develop in a host weevil.

Bathyplectes cocoons are about 3.5 mm long, brown and oval in shape, with a white band around the middle. It is actually the cocoons of the two species that are the most visible sign of the parasitoid and also are the easiest method of separating the species – in *B. anurus*, the white band is raised and the cocoon has the unusual habit of 'jumping' when disturbed, whereas, in *B. curculionis*, the white band is not raised and the cocoons do not jump.

Predation: The alfalfa weevil typically has just one generation per year, with larvae present during the spring. The alfalfa weevil adults emerge from pupae during late spring to early summer, feed for several weeks and then spend the remainder of the summer in diapause. Diapause is completed by late summer or autumn and the adults become active whilst the weather remains favourable, then hibernate during the winter and resume feeding and laying eggs the following spring. The adult flight activity of both *Bathyplectes* species is synchronised with the spring activity period of the alfalfa weevil larvae. The flight lasts up to several weeks and peak parasitism levels occur 1 to 2 weeks prior to the peak in numbers of weevil larvae. *B. anurus* has just one generation a year, with all parasitoid pupae produced by spring parasitism undergoing diapause and not emerging as adults until the following spring, when weevil larvae are again abundant. *B. curculionis*, however, has a partial second generation – many of the parasitoid pupae from spring parasitism are in diapause, but some develop and emerge as adults which then must find and parasitise weevil larvae during the summer.

Egg laying: *Bathyplectes curculionis* females lay up to 200 eggs. **Efficacy:** In laboratory studies, adult wasps caged with alfalfa infested with pea aphids fed on the aphid honeydew and lived about 50% longer than wasps caged on alfalfa alone. In another experiment, newly emerged unmated *B. curculionis* wasps were fed for 2 days on aphid honeydew, sucrose dissolved in water or just water. Egg production was significantly greater (11 to 15% increase) when wasps fed on aphid honeydew or sucrose dissolved in water, than when wasps only had access to water. These results suggest that the availability of pea aphid honeydew in alfalfa fields may increase the longevity and fecundity of *B. curculionis*. Thus, the presence of moderate levels of pea aphids may be beneficial for biological control of alfalfa weevils. Other studies have shown that aphid honeydew may be most important to wasp parasitoids when flower nectar is not available, early spring or late autumn, depending on the flowering period of local plants.

Key reference(s): 1) S England & E W Evans. 1997. Effects of pea aphid honeydew on longevity and fecundity of the alfalfa weevil parasitoid, *Bathyplectes curculionis*, *Environ. Entomol.*, **26**, 1437- 41. 2) R I Sailer. 1979. Progress report on importation of natural enemies of insect pests in the U.S.A. In *Proceedings of the Joint American-Soviet Conference on Use of Beneficial Organisms in the Control of Crop Pests*, published by Entomological Society of America, p. 23.

COMMERCIALISATION: Formulation: Available as pupae. **Tradenames:** 'Bathyplectes curculionis' (Praxis).

APPLICATION: Distribute pupae throughout the crop in late spring, as weevils begin to oviposit.

PRODUCT SPECIFICATIONS: Storage conditions: Can be stored for short periods at low temperatures (in a refrigerator).

COMPATIBILITY: Populations of *Bathyplectes curculionis* are reduced by applications of broad-spectrum insecticides.

MAMMALIAN TOXICITY: No allergic or other adverse reactions have been noted with workers in the field. There is no evidence of acute or chronic toxicity, eye or skin irritation or hypersensitivity to mammals.

ENVIRONMENTAL IMPACT AND NON-TARGET TOXICITY: *Bathyplectes curculionis* occurs widely in Nature and has never been shown to parasitise any insects other than *Hypera postica*. For this reason, it is not expected to have any effect on non-target organisms or to have any adverse effects on the environment. **Approved for use in organic farming:** Yes.

3:280 *Bracon hebetor*   *Parasitoid of grain moths*

Parasitoid wasp: Hymenoptera: Braconidae

NOMENCLATURE: Approved name: *Bracon hebetor* (Say).

SOURCE: Cosmopolitan, associated with stored product moths. Not injurious to stored grain.

PRODUCTION: Raised on eggs of grain moths, such as *Ephestia kuehniella* L.

TARGET PESTS: *Bracon hebetor* parasitises several of the common grain moths, such as the Indian meal moth (*Plodia interpunctella* (Hübner)) in the late larval stage.

TARGET CROPS: Used in grain stores.

BIOLOGICAL ACTIVITY: Biology: Females paralyse and lay eggs in late instar moth larvae. On average, 8 larvae develop in one host. **Egg laying:** Each fertilised female lays approximately 100 eggs. **Duration of development:** The time from egg lay to adult is between 9 and 10 days at 30 °C. Adult female longevity is about 23 days.

COMMERCIALISATION: Tradenames: 'Bracon hebetor' (Rincon-Vitova), (Biofac), (BugLogical Control Systems), (CalTec), (Panhandle), (Arbico) and (Praxis).

APPLICATION: Release adults into grain store.

PRODUCT SPECIFICATIONS: Purity: No contaminating insects. **Storage conditions:** Can be stored in a cooled, insulated container at 15 to 19 °C, away from direct sunlight.

COMPATIBILITY: It is unusual to use *Bracon hebetor* with chemical treatments.

MAMMALIAN TOXICITY: No allergic or other adverse reactions have been noted with workers in the field. There is no evidence of acute or chronic toxicity, eye or skin irritation or hypersensitivity to mammals.

ENVIRONMENTAL IMPACT AND NON-TARGET TOXICITY: *Bracon hebetor* occurs widely in Nature and is not expected to parasitise non-target organisms or have any adverse effects on the environment. *Bracon hebetor* is used in grain stores rather than released into growing crops and this restricts its contact with the environment.

3:281 *Cactoblastis cactorum*
Phytophagous caterpillar

Phytophagous lepidopteran: Lepidoptera: Pyralidae

NOMENCLATURE: Approved name: *Cactoblastis cactorum* (Berg). **Other names:** Prickly pear moth; previously known as *Zophodia cactorum* Berg.

SOURCE: *Cactoblastis cactorum* was originally found in Argentina, but was introduced into Australia in 1926 by the release of 2 000 000 eggs near the town of Chinchilla in order to control the prickly pear which was invading much of the state of Queensland. Subsequently, it was introduced into other countries including the Caribbean Islands, Hawaii and South Africa.

PRODUCTION: Collected in the wild and introduced as eggs. Once established, the moth population stabilises in relation to the population of the host plant.

TARGET PESTS: Prickly pear cactus (*Opuntia stricta* (Haw.) Haw.).

TARGET CROPS: Rangeland.

BIOLOGICAL ACTIVITY: Biology: The caterpillars live and feed communally inside the tough, leathery (and prickly) skin of the host plant. They are initially pinkish-cream coloured, with dark red dots on the back of each segment. Later instars become orange, and the dots expand and then fuse to become a dark band across each segment. The caterpillars grow to length of about 1.5 cm. They normally pupate in white cocoons amongst the ground debris. The adult is fawn with faint dark dots and lines on the wings, and has a long nose. It normally rests with its wings wrapped around its body. The moth has a wingspan of about 2 cm.
Egg laying: The eggs are laid on top of each other, the first one being anchored to a cactus spine. The stack can contain over 100 eggs with a length of up to 2 cm. When the eggs hatch,

the young caterpillars walk down the stick and start burrowing into the cactus.

Efficacy: *Cactoblastis cactorum* is very effective at controlling the prickly pear cactus and populations in Australia have been reduced to a few isolated areas. In Queensland, the prickly pear population has been reduced by 99%.

COMMERCIALISATION: Formulation: *Cactoblastis cactorum* is not sold commercially.

MAMMALIAN TOXICITY: There is no record of allergic or other adverse toxicological effects in research workers or production or field staff. Considered to be non-toxic to mammals.

ENVIRONMENTAL IMPACT AND NON-TARGET TOXICITY: The primary host of *Cactoblastis cactorum* is the prckly pear cactus, but some adults have spread to the mainland of the USA where they have become a pest on native cacti.

Approved for use in organic farming: Yes.

3:282 *Cales noacki*   *Parasitoid of whiteflies*

Parasitoid wasp: Hymenoptera: Aphelinidae

NOMENCLATURE: Approved name: *Cales noacki* Howard. **Other names:** Citrus woolly whitefly parasitoid.

SOURCE: Originated in South America, but now widespread.

PRODUCTION: Reared under controlled conditions on the citrus woolly whitefly, *Aleurothrixus floccosus* Maskell.

TARGET PESTS: Citrus woolly whitefly (*Aleurothrixus floccosus*).

TARGET CROPS: Citrus plantations (mandarins are particularly susceptible to the citrus woolly whitefly).

BIOLOGICAL ACTIVITY: Biology: *Cales noacki* is a parasitoid wasp, 0.6 mm in length and yellow/orange in colour. It has long legs. It parasitises the second, third and fourth instars of whitefly nymphs, developing inside the nymphs, where it feeds, thereby destroying them. Its only disadvantage is its great sensitivity to high temperatures, which can lead to a seasonal balance in favour of control in the autumn and winter. Very good results can be achieved if it is used in conjunction with another parasitoid such as *Amitus spiniferus* (Brèthes), which is better adapted to summer heat.

COMMERCIALISATION: Formulation: Sold as adults in Peru by the National Program of Biological Control (NPBC). **Tradenames:** 'Cales noacki' (NPBC).

APPLICATION: Release between 100 and 500 adults per hectare, depending upon the level of infestation.

PRODUCT SPECIFICATIONS: Purity: Contains only *Cales noacki* adults.

COMPATIBILITY: *Cales noacki* is intolerant of high temperatures. It should not be used in conjunction with residual insecticides.

MAMMALIAN TOXICITY: *Cales noacki* has not demonstrated evidence of toxicity, infectivity, irritation or hypersensitivity to mammals. No allergic responses or other adverse health problems have been observed by research workers, manufacturing staff or users.

ENVIRONMENTAL IMPACT AND NON-TARGET TOXICITY: *Cales noacki* occurs in Nature and is now widely spread. As such, is not expected to show any adverse effects on non-target organisms or on the environment. It is specific to whiteflies.
Approved for use in organic farming: Yes.

3:283 *Calosoma sycophanta*   *Predatory beetle*

Predatory ground beetle: Coleoptera: Carabidae

NOMENCLATURE: Approved name: *Calosoma sycophanta* (L.) **Other names:** Gypsy moth predator.

SOURCE: A native of northern and central Europe, introduced into the USA between 1905 and 1910 and now widely distributed.

PRODUCTION: Reared on lepidopteran larvae.

TARGET PESTS: Gypsy moth (*Lymantria dispar* L.).

TARGET CROPS: Forests, woodlands and amenity trees.

BIOLOGICAL ACTIVITY: Biology: Mature adults reach a length of 3 cm and may live for up to 4 years. Eggs are laid in the soil at the foot of trees and the larvae emerge and feed on lepidopteran larvae and, in particular, larvae of the pine beauty moth (*Panolis flammea* (Denis & Schiffermüller)), the pale tussock moth (*Dasychira pudibunda* (L.)), the black arches nun moth (*Lymantria monacha* (L.)) and the gypsy moth (*Lymantria dispar*). Both adults and larvae are excellent predators, seeking out suitable food in bushes and trees. The larvae live for about 3 weeks, consuming approximately 40 caterpillars in this time. Adults will consume between 300 and 400 caterpillars a year. **Efficacy:** *Calosoma sycophanta* is an excellent predator of caterpillars, in particular gypsy moth larvae. Both adults and larvae seek out food and are very mobile. **Key reference(s)**: A T Drooz (ed.). 1989. *Insects of Eastern Forests*, Misc. Publ. 1426, Washington, DC, USDA Forest Service, USA, 608 pp.

COMMERCIALISATION: Formulation: Sold as fecund adults. **Tradenames:** 'Calosoma sycophanta' (Praxis).

APPLICATION: Release adults amongst trees when gypsy moth adults are flying.

PRODUCT SPECIFICATIONS: Purity: Supplied as fecund adults, with no contaminants. **Storage conditions:** *Calosoma sycophanta* adults can be kept for 2 weeks in a cool place such as a refrigerator. They can be transported over several days (in a cooler with an ice pack) without any adverse effects. **Shelf-life:** Adults remain alive for several weeks, if stored as directed.

COMPATIBILITY: *Calosoma sycophanta* should not be released into trees that have been treated with broad-spectrum insecticides.

MAMMALIAN TOXICITY: There have been no reports of allergic or other adverse reactions from research workers, manufacturing staff or users from the field release of *Calosoma sycophanta*.

ENVIRONMENTAL IMPACT AND NON-TARGET TOXICITY: *Calosoma sycophanta* occurs in Nature and is now widely spread. As such, it is not expected to show any adverse effects on non-target organisms or on the environment. Both adults and larvae will eat a variety of caterpillars that feed upon the trees it is released to protect.
Approved for use in organic farming: Yes.

3:284 *Cheyletus eruditus*   *Predatory mite*

Predatory mite: Acari: Cheyletidae

NOMENCLATURE: Approved name: *Cheyletus eruditus* (Schrank). **Other names:** Storage mite predator.

SOURCE: Abundant in grain stores, particularly when infested with storage mites.

PRODUCTION: Reared on storage mites, particularly *Acarus siro* (L.).

TARGET PESTS: Storage mites, such as *Acarus siro* and *Lepidoglyphus* (*Glycyphagus*) *destructor* (Schrank), and spider mites, such as *Tetranychus urticae* Koch and *T. cinnabarinus* (Boisduval).

TARGET CROPS: Stored grain and also effective on glasshouse-grown crops and in interiorscapes.

BIOLOGICAL ACTIVITY: Biology: *Cheyletus eruditus* develops at humidities as low as 55% r.h. Its development increases one- to fourfold each week between 10 °C and 30 °C, but it can survive for 6 months at 0 °C. Higher temperature and lower moisture requirements mean it usually peaks in summer during prolonged storage. Its natural presence indicates pest problems in the store. *Cheyletus eruditus* is naturally very tolerant to the OP insecticides used to control grain pests. **Key reference(s):** 1) V I Volgin. 1989. Acarina of the family Cheyletidae of the world, *Keys to the Fauna of the USSR*, **101**, 532 pp. 2) M Emekci & S Toros. 1994. Studies on the biology of *Cheyletus eruditus* (Schrank) (Acari:Cheyletidae) under laboratory conditions, *Turkiye III. Biyolojik Mucadele Kongresi Bildirileeri*.

COMMERCIALISATION: Tradenames: 'Cheyletin' (Biopreventa), 'Cheyletus eruditus' (Precision Herbs).

APPLICATION: Introduce before pest problems are noticed.

PRODUCT SPECIFICATIONS: Purity: Contains no contaminants. **Storage conditions:** *Cheyletus eruditus* must be stored at 0 to 4 °C.

COMPATIBILITY: Very tolerant of OP insecticides.

MAMMALIAN TOXICITY: Cases of allergic reactions to farmers have been reported.

ENVIRONMENTAL IMPACT AND NON-TARGET TOXICITY: *Cheyletus eruditus* occurs widely in Nature and is not expected to have any adverse effects on non-target organisms or on the environment.

3:285 *Chilocorus* species
Predator of scale insects

Predatory beetle: Coleoptera: Coccinellidae

NOMENCLATURE: Approved name: *Chilocorus* species. **Other names:** There are several species of *Chilocorus* that are effective predators of scale insects, but the most important species are *Chilocorus stigma* (Say.) (a predator of white louse scale, oleander scale, oriental scale and red scale), *Chilocorus nigritus* (Fabricius) (a predator of armoured scales), *Chilocorus bipustulatus* (L.) (a predator of the olive scale or black scale (*Saissetia oleae* (Olivier))) and *Chilocorus kuwanae* Silvestri (a predator of euonymus scale).

SOURCE: *Chilocorus kuwanae* is common throughout China, Korea and Japan, but is now widely dispersed throughout the USA following its introduction. *Chilocorus stigma* is common in North America. *Chilocorus bipustulatus* originated in the South Palaearctic, but is now widespread throughout south and central Europe.

PRODUCTION: Raised on scale insects under controlled conditions.

TARGET PESTS: In North America, *Chilocorus kuwanae* is an effective predator of euonymus scale (*Unaspis euonymi* (Comstock)) and other scales, such as San José scale (*Quadraspidiotus perniciosus* (Comstock)). In China, Korea and Japan, where it is common, *C. kuwanae* helps to keep several species of armoured scale under control in citrus groves and on landscape shrubs. *Chilocorus stigma* is used against white louse scale, oleander scale, oriental scale and red scale. *Chilocorus nigritus* is effective against armoured scales.

TARGET CROPS: *Chilocorus stigma* and *Chilocorus nigritus* are used in protected and orchard crops. *Chilocorus kuwanae* is effective against euonymus scale, but will feed on other scale insects in crops such as citrus.

BIOLOGICAL ACTIVITY: Biology: Beetles spend the winter as adults in leaf litter at the base of scale-infested plants. They become active and feed on scale when the temperature exceeds 10 °C in the spring. Adults lay bright orange eggs, singly or in small groups under the scale covers. Eggs hatch into brown larvae covered with black spines. Larvae will turn over or chew through scale covers to feed on the fleshy scale body. After three larval stages, larvae move to the underside of a leaf or to a crack or crevice on a twig or branch. There they become immobile and the larval skins split to form pupae. In the laboratory, it takes about a month to mature from egg to adult. There are usually three generations a year. *Chilocorus kuwanae* is a black ladybird, about 3 mm long, with red spots. It is quite similar to *Chilocorus stigma*, but it can be distinguished by the colour, shape and location of the spots on its wings. Spots of *C. kuwanae* tend to be deep red and rectangular and located near the centre of the wing. In contrast, spots of *C. stigma* tend to be more orange-yellow, round and oriented more towards the head of the beetle. **Predation:** All *Chilocorus* species are very effective predators. *C. kuwanae* is noted for its voracious appetite. Each larval *C. kuwanae* must feed on over 100 scale to become an adult. In the USA, this appetite has caused the beetle to deplete rapidly the euonymus scale population at landscape release sites. Like many other predatory beetles, *Chilocorus* species fly away when food gets scarce. However, other scale, such as San José scale and oystershell scale, can be used by adults as a food source.

COMMERCIALISATION: Formulation: Several species of *Chilocorus* are commercially available, being sold as fecund adults. **Tradenames:** 'Chilocorus kuwanae' (Praxis), 'Chilocorus stigma' (Bugs for Bugs), 'Chilocorus nigritus' (Entocare).

APPLICATION: *Chilocorus stigma* should be used in orchard crops at a rate of 600 to 1500 adults per hectare and in protected crops at a rate of 30 adults per 50 square metres. Similar rates are recommended for other *Chilocorus* species.

PRODUCT SPECIFICATIONS: Purity: Adults consist of fecund females. All scale supplied are parasitised. **Storage conditions:** Adults may be kept under normal conditions in the presence of a suitable food source. Very active in direct sunlight. **Shelf-life:** Adults may be kept for a few days, if fed on honey solution.

COMPATIBILITY: Incompatible with foliar insecticides.

MAMMALIAN TOXICITY: *Chilocorus* species have shown no allergic or other adverse effects on research workers, manufacturing or field staff. They are considered to be safe to mammals.

ENVIRONMENTAL IMPACT AND NON-TARGET TOXICITY: *Chilocorus* species occur widely in Nature and are not expected to have any adverse effects on non-target organisms or on the environment. They are specific to scale insects.
Approved for use in organic farming: Yes.

3:286 *Chrysolina hyperici*   *Phytophagous beetle*

Phytophagous beetle: Coleoptera: Chrysomelidae

NOMENCLATURE: Approved name: *Chrysolina hyperici* (Forster). **Other names:** St John's wort beetle; klamath weed beetle.

SOURCE: *Chrysolina hyperici* originated in northern and central Europe and western Asia. The beetle was introduced into Canada from stock from Britain via Australia and California. It was released in British Columbia in 1952.

PRODUCTION: Bred on living St John's wort (*Hypericum perforatum* L.).

TARGET PESTS: *Chrysolina hyperici* is used to control St John's wort. In no-choice tests, the beetle developed on all the *Hypericum* spp. tested, but on no other plants.

TARGET CROPS: Pastures and grassland.

BIOLOGICAL ACTIVITY: Biology: Adult *Chrysolina hyperici* are oval, metallic green or bronze beetles, 5.3 to 6.1 mm long. They emerge in late autumn, overwinter well, tend to lay eggs in the spring, and their dormancy is not disrupted by summer rain. The beetles require summer moist habitats and are well adapted to maritime habitats. In the autumn, eggs are laid on the leaves of St John's wort. This only happens in their native countries in Europe and where they were released in California and Oregon. The eggs are laid in the spring in other parts of the north-western United States. The females deposit hundreds of eggs, either one at a time or in a cluster. The eggs, which are long and orange in colour, do not hatch for 6 or 7 days.

Reproductive diapause is determined by day length. In a 16-hour day, feeding is minimal, there is no egg development and testes are small. In an 8-hour day, reproduction starts at day 60 and all adults are breeding by day 100. The overwintering stage is primarily eggs on winter foliage, but adults and larvae can also survive mild winters or under snow. There is one generation per year. **Predation:** After hatching, the larvae are humped, initially orange and later greyish. The larvae start feeding 45 minutes before sunrise and disappear about three hours later, the small ones hiding in leaf buds and larger ones under the plant or in the soil. Feeding resumes around sunset, the larvae eat the leaf buds, consuming all the leaves. They then move to another plant before they mature. Finally, larvae move into the soil, make cells and then pupate. The pupae are fat and C-shaped, becoming greyish-pink in colour. *Chrysolina hyperici* cannot pupate in hard, dry ground. **Egg laying:** Up to 2000 reddish, 1.2 × 0.5 mm eggs are laid singly or in small clusters on winter and spring foliage.

Duration of development: *Chrysolina hyperici* are pupae for about 12 days in the late spring. The adults feed for a few weeks in the spring and then they return to the soil to rest in the summer. When it begins to rain in the autumn, the beetles emerge, mate and lay eggs. If it does not rain in the autumn, they will mate and lay eggs in the spring. **Efficacy:** The larvae eat the roots of the weed, which stops its growth almost completely in the spring. The adults and the larvae eat the buds, preventing leaf growth and flower development. Larval consumption of early spring foliage causes St John's wort cover to decline from 100% to 1% within 3 years of beetle increase, but it can be up to 10 years before they increase. Summer defoliation by the adults is striking, but less effective. **Key reference(s):** N Rees *et al.* (eds). 1996. *Biological Control of Weeds in the West*, Western Society of Weed Science in cooperation with USDA-ARS, Montana Department of Agriculture and Montana State University, Color World Printers, Bozeman, Montana, USA.

COMMERCIALISATION: Formulation: Sold as fecund adults. **Tradenames:** 'Klamath Weed Beetle' (Integrated Weed Control).

APPLICATION: The adults should be released in groups of 250 or more to each infested area.

PRODUCT SPECIFICATIONS: Purity: Supplied as fecund adults, with no contaminants. **Storage conditions:** *Chrysolina hyperici* adults can be kept for 2 weeks in a cool place such as a refrigerator. They can be transported over several days (in a cooler with an ice pack) without any adverse effects. **Shelf-life:** Adults remain alive for several weeks, if stored as directed.

COMPATIBILITY: It is not usual to apply *Chrysolina hyperici* with chemical herbicides.

MAMMALIAN TOXICITY: There have been no reports of allergic or other adverse reactions from research workers, manufacturing staff or from the field release of *Chrysolina hyperici*.

ENVIRONMENTAL IMPACT AND NON-TARGET TOXICITY: *Chrysolina hyperici* occurs widely in Nature and is not expected to have any adverse effects on non-target organisms or on the environment. It has not been shown to attack plants other than St John's wort.

3:287 *Chrysolina quadrigemina*

Phytophagous beetle

Phytophagous beetle: Coleoptera: Chrysomelidae

NOMENCLATURE: Approved name: *Chrysolina quadrigemina* (Suffr.). **Other names:** St John's wort beetle; klamath weed beetle.

SOURCE: *Chrysolina quadrigemina* occurs in Europe and has been developed in Australia for control of St John's wort. It was introduced into the USA from Australia in 1946.

PRODUCTION: *Chrysolina quadrigemina* is reared on living St John's wort (*Hypericum perforatum* L.).

TARGET PESTS: *Chrysolina quadrigemina* is used to control St John's wort.

TARGET CROPS: Pastures and grassland.

BIOLOGICAL ACTIVITY: Biology: *Chrysolina quadrigemina* has a somewhat confusing life-cycle. Females may lay eggs in autumn or spring. It is possible to find eggs, larvae or adults year round. Any of the three stages may overwinter. Sunlight kills the larvae, but the adults avoid shade. Adults enter the soil and become dormant during summer. Autumn and spring rain is thought to stimulate them to emerge, mate and lay eggs. **Predation:** When the eggs hatch, the larvae start feeding on younger leaves. Larvae hide from the sun during the day under leaves or in the soil. When the larvae are mature, they construct cells in the soil and pupate. Adult beetles are metallic green, blue, bronze or black in colour and about 5 to 7 mm long. **Egg laying:** Eggs are deposited singly or in clusters of 2 to 4 on the underside of leaves in the autumn or spring. A female can lay hundreds of eggs in her lifetime. Eggs are oval in shape and orange in colour. **Efficacy:** Both adults and larvae consume the leaves of St John's wort. Larvae feed nocturnally, adults feed during the day.

COMMERCIALISATION: Formulation: Sold as fecund adults in summer.
Tradenames: 'Klamath Weed Beetle' (Biological Control of Weeds).

APPLICATION: Release adults during June and July.

PRODUCT SPECIFICATIONS: Purity: Supplied as fecund adults, with no contaminants.
Storage conditions: *Chrysolina quadrigemina* adults can be kept for 2 weeks in a cool place such as a refrigerator. They can be transported over several days (in a cooler with an ice pack) without any adverse effects. **Shelf-life:** Adults remain alive for several weeks, if stored as directed.

COMPATIBILITY: It is not usual to use *Chrysolina quadrigemina* with chemical herbicides.

3. Macro-organisms

MAMMALIAN TOXICITY: There have been no reports of allergic or other adverse reactions from research workers, manufacturing staff or from the field release of *Chrysolina quadrigemina*.

ENVIRONMENTAL IMPACT AND NON-TARGET TOXICITY: *Chrysolina quadrigemina* occurs widely in Nature and is not expected to have any adverse effects on non-target organisms or on the environment. It has not been shown to attack plants other than St John's wort.

3:288 *Chrysoperla carnea*   *Predator of aphids*

Lacewing: Neuroptera: Chrysopidae

NOMENCLATURE: Approved name: *Chrysoperla carnea* (Stephens). **Other names:** Pearly green lacewing; aphid lion; golden eye. Formerly known as: *Chrysopa carnea* Stephens.

SOURCE: Widespread general insect predator.

PRODUCTION: Reared on aphids or on an artificial diet and sold as eggs or, more usually, as larvae.

TARGET PESTS: All species of aphid. *Chrysoperla carnea* is a voracious predator and will also consume a variety of other slow-moving, soft-bodied arthropods, including whitefly, scale, thrips, mites, beetles and lepidopteran eggs.

TARGET CROPS: Particularly useful in interiorscapes. Can be used in protected and outdoor crops, especially strawberries, hops and top fruit (apples and pears).

BIOLOGICAL ACTIVITY: Biology: Eggs are laid on slender stalks on the underside of leaves. The larvae are pale brown or grey, grow to about 1 cm in length and feed on aphids immediately after emergence. The larvae camouflage themselves by covering their bodies in prey debris. After pupation, they emerge as adults that feed only on honey, pollen and nectar, which they need to reproduce. **Predation:** Larvae will consume over 400 aphids during development. Older larvae can consume between 30 and 50 aphids per day.
Egg laying: The adult female lays more than 100 eggs in her lifetime.
Duration of development: Larvae feed for 3 to 4 weeks, undergoing 3 moults before pupation. Adults emerge after about 7 days, begin to lay eggs after a further 6 days and live for about 14 days, depending upon temperature. At 26 °C, it takes between 20 and 25 days from egg to adult, but *Chrysoperla carnea* shows a good tolerance of temperature fluctuations and, even if the temperature reaches very low levels for few hours, the larva

can survive and start feeding again as soon as suitable conditions return. **Efficacy:** Very active and aggressive predator of aphids and other insects. Will consume beneficial, as well as phytophagous, insects and may even be cannibalistic, although most beneficials move too quickly for *C. carnea* to catch them. The common name, aphid lion, relates to its voracious appetite. **Key reference(s):** R A Sundby. 1966. A comparative study of the efficiency of three predatory insects *Coccinella septempunctata* (Coleoptera: Coccinelidae), *Chrysopa carnea* St (Neuroptera: Chrysopidae) and *Syrphus ribesii* L. (Diptera: Syrphidae) at two different temperatures, *Entomophaga*, **11**, 395–404.

COMMERCIALISATION: Formulation: Sold as eggs in bran, rice hulls or other material and with moth eggs present as a source of food. More usually sold as larvae in the same carrier or as individual adults in cells of cardboard trays. 'CrisoPAK' shaker bottles contain 1000 2nd instar larvae in vermiculite and buckwheat glumes. **Tradenames:** 'Chrysoline c' (Syngenta Bioline), 'Chrysopa-System' (Biobest), 'Chrysoperla' (Sautter & Stepper), 'Chrysoperla carnea' (Rincon-Vitova), (Neudorff), (IPM Laboratories) and (Praxis), 'Chrysosure (c)' (Biological Crop Protection), 'CrisoPAK' (Bioplanet), 'Green Lacewing' (Arbico) and (Biocontrol Network), 'Lacewing' (Kunafin).

APPLICATION: Release one lacewing for 10 aphids or use 10 eggs per plant or 5 larvae per square metre every other week. Ten to 30 larvae per square metre are recommended for inundative release. Repeat applications after 14 days, if the population has not been checked. More frequent application may lead to the older larvae consuming the younger ones. Eggs and larvae can be applied in the field by aircraft, using equipment designed for pollen application. Adults tend to migrate away from release sites. Often used to control hot spots.

PRODUCT SPECIFICATIONS: Purity: No contaminants. **Storage conditions:** Adults can be stored in the refrigerator for up to 2 days. **Shelf-life:** Pupae hatch within 5 days. Adults should be released within 2 days of receipt.

COMPATIBILITY: Lacewing eggs are eaten by ants. Incompatible with many beneficial insects, as the larvae will eat these, as well as pest species. Incompatible with persistent insecticides.

MAMMALIAN TOXICITY: No allergic or other adverse reactions have been noted following use in the field. Occasionally, the larvae will bite humans.

ENVIRONMENTAL IMPACT AND NON-TARGET TOXICITY: Lacewings are common in Nature and are voracious predators of a wide range of small arthropods. Although the favoured diet is aphids, it is possible, in the absence of sufficient prey, that other insect species, including beneficials, will be predated. **Approved for use in organic farming:** Yes.

3. Macro-organisms

3:289 *Chrysoperla rufilabris*

Predator of aphids

Lacewing: Neuroptera: Chrysopidae

NOMENCLATURE: Approved name: *Chrysoperla rufilabris* (Burmeister).
Other names: Green lacewing; aphid lion; rufi. Formerly known as: *Chrysopa rufilabris* Burmeister.

SOURCE: Widespread general insect predator.

PRODUCTION: Reared on aphids or on an artificial diet and sold as eggs or, more usually, as larvae. Also supplied as pupae or adults.

TARGET PESTS: Used against a wide range of soft-bodied insects, such as aphids, mealybugs, spider mites, leafhopper nymphs, caterpillar eggs, scales, thrips, whiteflies and phytophagous insect eggs.

TARGET CROPS: Can be used in protected and outdoor crops, especially strawberries, hops and top fruit (apples and pears).

BIOLOGICAL ACTIVITY: Biology: Eggs are laid on slender stalks on the underside of leaves. The larvae grow to about 1 cm in length and feed on aphids and other soft-bodied insects. The larvae camouflage themselves by covering their bodies in prey debris. After pupation, they emerge as adults that feed only on nectar and pollen. **Predation:** Larvae will consume over 400 aphids during development. Older larvae can consume between 30 and 50 aphids per day. **Egg laying:** The fertilised adult female will lay more than 100 eggs in her lifetime. **Duration of development:** Larvae feed for 3 to 4 weeks, undergoing 3 moults before pupation. Adults emerge after about 7 days, begin to lay eggs after a further 6 days and live for about 14 days, depending upon temperature. **Efficacy:** Very active and aggressive predator of aphids and other insects. Will consume beneficial, as well as phytophagous, insects and may even be cannibalistic. The common name, aphid lion, relates to its voracious appetite.

COMMERCIALISATION: Formulation: Sold as eggs in bran, rice hulls or other material and with moth eggs present as a source of food. More usually sold as larvae in the same carrier or as individual adults or larvae in cells of cardboard trays. **Tradenames:** 'Chrysoperla rufilabris' (Rincon-Vitova).

APPLICATION: Release one lacewing for 10 aphids or use 10 eggs per plant or 5 larvae per square metre every other week. Repeat applications after 14 days if the population has not been checked. More frequent application may lead to the older larvae consuming the younger ones. Eggs and larvae can be applied in the field by aircraft, using equipment designed for pollen application. Adults tend to migrate away from release sites. Often used to control hot spots.

PRODUCT SPECIFICATIONS: Storage conditions: Adults can be stored in the refrigerator for up to 2 days. **Shelf-life:** Pupae hatch within 5 days. Adults should be released within 2 days of receipt.

COMPATIBILITY: Lacewing eggs are eaten by ants. Incompatible with many beneficial insects, as the larvae will eat these, as well as pest species. Incompatible with persistent insecticides.

MAMMALIAN TOXICITY: No allergic or other adverse reactions have been noted following use in the field. Occasionally, the larvae will bite humans.

ENVIRONMENTAL IMPACT AND NON-TARGET TOXICITY: Lacewings are common in Nature and are voracious predators of a wide range of small arthropods. Although the favoured diet is aphids, it is possible, in the absence of sufficient prey, that other insect species, including beneficials, will be predated. **Approved for use in organic farming:** Yes.

3:290 *Coleomegilla maculata*

Predator of insects

Predatory beetle: Coleoptera: Coccinellidae

NOMENCLATURE: Approved name: *Coleomegilla maculata* De Geer. **Other names:** Pink spotted lady beetle; twelve spotted lady beetle; pink lady beetle.

SOURCE: A North American native that can be very abundant in north-eastern, southern and central states.

PRODUCTION: *Coleomegilla maculata* is a polyphagous insect that can be bred on aphids or collected in the wild.

TARGET PESTS: *Coleomegilla maculata* is an important aphid predator, but also preys on mites, insect eggs and small larvae.

TARGET CROPS: Wheat, sorghum, alfalfa (lucerne), soybeans, cotton, potatoes, sweet corn, vegetables, tomatoes, asparagus, apples and many other crops attacked by aphids.

BIOLOGICAL ACTIVITY: Biology: *Coleomegilla maculata* is a pink to red, oval beetle with six spots on each forewing. The area behind the head is often pink or yellowish with two large triangular black marks. The adult is about 5 to 6 mm long. The dark and alligator-like larvae have three pairs of prominent legs and grow to 5 to 6 mm in length. Eggs are spindle shaped and small, about 1 mm long. Adults overwinter in large aggregations beneath leaf litter and stones along hedgerows or in protected sites along crop borders, especially those of fields planted to maize (corn) in the previous season. They emerge from early to mid-spring and disperse, often by walking along the ground, to seek prey and egg laying sites in nearby crops. Eggs are usually deposited near prey such as aphids, often in small clusters in protected sites on leaves and stems. Larvae grow from about 1 mm to 5 to 6 mm in length and may wander up to 12 metres in search of prey. The larva attaches itself by the abdomen to a leaf or other surface to pupate. The pupal stage may last from 3 to 12 days depending on the temperature. There are from 2 to 5 generations per year. *Coleomegilla maculata* is unusual for a ladybird in that plant pollen may constitute up to 50% of its diet and the planting or preservation of refuges or interplanting early-flowering species with a high pollen load may be beneficial, providing a food source during late spring before the build-up of aphids. **Egg laying:** Female *Coleomegilla maculata* may lay from 200 to more than 1000 eggs over a 1 to 3 month period commencing in spring or early summer. **Efficacy:** *Coleomegilla maculata* is a voracious predator of a very wide range of insects. It prefers aphids (of all types), but will consume European corn borer (*Ostrinia nubilalis* (Hübner)), cabbage white butterfly (*Pieris brassicae* (L.) and *P. rapae* (L.)), fall armyworm (*Spodoptera frugiperda* (J. E. Smith)) and corn earworm (*Helicoverpa zea* Boddie) eggs; and Colorado potato beetle (*Leptinotarsa decemlineata* (Say)) eggs and larvae. **Key reference(s):** 1) M P Hoffmann & A C Frodsham. 1993. *Natural Enemies of Vegetable Insect Pests*, Cooperative Extension, Cornell University, Ithaca, NY, USA, 63 pp. 2) R D Gordon. 1985. The Coccinellidae (Coleoptera) of America North of Mexico, *J. N.Y. Entomol. Soc.*, **93**, 1–912.

COMMERCIALISATION: Formulation: Sold as adults. **Tradenames:** 'Coleomegilla maculata' (Entocare), (Integrated BioControl Systems) and (Rincon-Vitova).

APPLICATION: Distribute amongst aphid colonies, preferably when there is an abundance of pollen.

PRODUCT SPECIFICATIONS: Storage conditions: May be stored for short periods, in a cool, dark place.

COMPATIBILITY: Overwintering adults are more susceptible to residual, broad-spectrum insecticides than those that are actively feeding. In the absence of prey, *Coleomegilla maculata* may consume the eggs and larave of other predatory species.

MAMMALIAN TOXICITY: There have been no reports of allergic or other adverse toxicological effects arising from contact with *Coleomegilla maculata* from research staff, producers or users.

ENVIRONMENTAL IMPACT AND NON-TARGET TOXICITY: *Coleomegilla maculata* has a wide distribution in Nature and is a voracious predator of a wide range of small arthropods. Although the favoured diet is aphids, it is possible, in the absence of sufficient prey, that other insect species, including beneficials, will be predated. **Approved for use in organic farming:** Yes.

3:291 *Comperia merceti*
Parasitoid of cockroaches

Parasitoid wasp: Hymenoptera: Encyrtidae

NOMENCLATURE: Approved name: *Comperia merceti* (Compère).

SOURCE: A commonly occurring wasp; present where cockroaches are found.

PRODUCTION: *Comperia merceti* is reared on the egg cases of the brown-banded cockroach (*Supella longipalpa* (Fabricius)).

TARGET PESTS: The brown-banded cockroach (*Supella longipalpa*) and the German cockroach (*Blattella germanica* (L.))

TARGET CROPS: Used in warehouses, industrial properties and domestic houses.

BIOLOGICAL ACTIVITY: Biology: *Comperia merceti* attacks the egg capsules (oothecae) of only two domestic cockroach species, the brown-banded and the German cockroach. Females of *Comperia merceti* lay their eggs inside cockroach oothecae. The eggs hatch and the developing parasitoid larvae consume the developing cockroach nymphs. The larvae continue to mature and eventually pupate inside the cockroach oothecae. *Comperia merceti* reaches the adult stage inside the cockroach ootheca. Adults emerge by chewing a hole in the oothecal shell. Male and female *Comperia merceti* mate immediately upon emergence. After mating, *Comperia merceti* females search for more cockroach egg capsules. **Predation:** A female adult will attack 1 to 2 hosts in her lifetime. **Duration of development:** The developmental time from egg to adult is 32 to 36 days, dependent on temperature and number of individuals developing in the host. On average, there are 13 female progeny per host. The female lives for between 3 and 5 days. **Key reference(s):** 1) R R Coler, R G Van Driesche & J S Elkinton. 1984. Effect of an oothecal parastoid, *Comperia merceti* (Compère) (Hymenoptera: Encyrtidae), on a population of the brownbanded cockroach (Orthoptera: Blattellidae), *Environ. Entomol.*, **13**, 603–6. 2) L Lebeck. 1985. Host-parasite relationships between *Comperia merceti* (Compère) (Hymenoptera: Encyrtidae) and *Supella longipalpa* (F.) (Orthoptera: Blattelidae), Ph.D. Dissertation, 104 pp. 3) L M Lebeck. 1991. A review of the hymenopterous natural enemies of cockroaches with emphasis on biological control, *Entomophaga*, **36**(3), 335–52. 4) F Agudelo-Silva. 1995. Biological control of roaches, *IPM Practitioner*, **17**(5/6), 12–3. 5) A Hechmer & R G Van Driesche. 1996. Ten year persistence of a non-augmented population of the brownbanded cockroach (Orthoptera: Blattellidae) parasitoid, *Comperia merceti* (Hymenoptera: Encyrtidae), *Florida Entomologist*, **79**(1), 77–9.

COMMERCIALISATION: Formulation: Sold as adults as part of a cockroach control programme. **Tradenames:** 'Comperia merceti' (Rincon-Vitova).

3. Macro-organisms

APPLICATION: The wasps should be released where there is evidence of cockroach activity. *Comperia merceti* is more effective when used in combination with other cockroach control systems, such as baits and traps.

PRODUCT SPECIFICATIONS: **Storage conditions:** The wasps should be released immediately upon receipt.

COMPATIBILITY: *Comperia merceti* is incompatible with residual insecticides.

MAMMALIAN TOXICITY: There have been no reports of allergic or other adverse toxicological effects arising from contact with *Comperia merceti* from research staff, producers or users.

ENVIRONMENTAL IMPACT AND NON-TARGET TOXICITY: *Comperia merceti* has a wide distribution in Nature and there is no evidence that it has any adverse effects on non-target organisms or on the environment. It is a specific cockroach parasitoid.

3:292 *Cotesia* species   *Parasitoid of Lepidoptera*

Parasitoid wasp: Hymenoptera: Aphidiidae

NOMENCLATURE: **Approved name:** *Cotesia* species. **Other names:** Lepidopteran parasitic wasp. Formerly known as: *Apanates* species.

SOURCE: Natural parasitoids of the larvae of Lepidoptera. A number of species have been commercialised, including *C. plutellae* Kurdj., a parasitoid of diamondback moth (*Plutella xylostella* (L.)) larvae and *C. marginiventris* (Cresson), a parasitoid of several caterpillars, including *Trichoplusia ni* (Hübner), *Spodoptera* spp. and *Helicoverpa* spp.

PRODUCTION: Reared under controlled conditions on host caterpillars.

TARGET PESTS: Various caterpillar pests.

TARGET CROPS: Brassicae and other vegetables in glasshouses and outside.

BIOLOGICAL ACTIVITY: **Predation:** It is usual for the adult female to be attracted to the host by chemical stimuli produced when the phytophagous insect host damages the crop by feeding. There is also evidence that the frass of host insects attracts the female wasps. **Duration of development:** The level of parasitoids within a crop is enhanced by leaving old plants in or near new plantings, so the pupae can hatch and enter the establishing crop. **Efficacy:** An effective parasitoid of Lepidoptera, with good mobility and ability to find larvae.

Can be parasitised by hyperparasitoids. **Key reference(s)**: R G van Driesche & T S Bellows (eds). 1996. *Biological Control*, Chapman & Hall, London, UK.

COMMERCIALISATION: Formulation: Sold as adult wasps. **Tradenames:** 'Cotesia plutella' (Arbico), (Biofac), (CalTec), (Praxis) and (Rincon-Vitova), 'Cotesia marginiventris' (Arbico), (Biofac), (CalTec), (Praxis) and (Rincon-Vitova), 'Cotesia melanoscela' (Praxis).

APPLICATION: Release 400 adults per hectare per week during periods of infestation. Release on the upwind side of the area to be treated.

PRODUCT SPECIFICATIONS: Storage conditions: Store at temperatures between 10 and 20 °C. **Shelf-life:** Use as soon as possible after delivery.

COMPATIBILITY: Compatible with *Bacillus thuringiensis* sprays, but do not use with persistent chemical insecticides.

MAMMALIAN TOXICITY: There is no record of allergic or other adverse toxicological effects in research workers or production or field staff. Considered to be non-toxic to mammals.

ENVIRONMENTAL IMPACT AND NON-TARGET TOXICITY: *Cotesia* species occur naturally and are not expected to have any adverse effects on non-target organisms or on the environment. They will prey on many different species of caterpillar.
Approved for use in organic farming: Yes.

3:293 *Cryptolaemus montrouzieri*
Predator of mealybugs

Predatory beetle: Coleoptera: Coccinellidae

NOMENCLATURE: Approved name: *Cryptolaemus montrouzieri* Mulsant.
Other names: Australian ladybird beetle; mealybug predator; mealybug destroyer; crypts.

SOURCE: *Cryptolaemus montrouzieri* is a native of Australia.

PRODUCTION: Reared on mealybugs and shipped as live beetles or larvae with a food source.

TARGET PESTS: Will consume all species of mealybug, but its primary host is *Planococcus citri* (Risso). Mealybugs such as *Pseudococcus longispinus* (Targioni) are viviparous and there are no eggs for the adult predators to eat, and this restricts their establishment. *Cryptolaemus* *montrouzieri* has been known to consume other phytophagous insects, such as aphids and the young stages of soft scale.

TARGET CROPS: Widely used to control mealybugs in Californian citrus plantations and in US grape orchards. Effective against mealybug infestations in a wide range of situations, including orchards and plantations, glasshouse-grown crops and interiorscapes. Not effective on tomatoes.

BIOLOGICAL ACTIVITY: Biology: Larvae can reach a length of 13 mm and their bodies are covered with wax-like projections. They pupate in sheltered regions on stems or on the underside of leaves. The adult insect is dark brown, with an orange head, thorax and abdomen. It is about 4 mm long. **Predation:** All active stages of *Cryptolaemus montrouzieri* consume mealybugs. Young larvae and adult beetles prefer host eggs and young nymphs, but older *Cryptolaemus montrouzieri* larvae will consume all stages. They will also consume insects in families related to mealybugs, such as aphids, and it has been reported that they will eat their own kind. **Egg laying:** Females copulate shortly after emergence and begin egg laying within 5 days. Eggs are laid within the egg masses of the mealybug. The number of eggs laid is very dependent upon the female's diet. Food shortages cause a reduction in the number of eggs laid. The egg-laying capacity of a female varies between 200 and 700 eggs in her lifetime, with between 7 and 11 eggs being laid each day. **Duration of development:** The duration of development is dependent upon the temperature. Eggs hatch in 8 to 9 days at 21 °C and in 5 to 6 days at 27 °C. At the same temperatures, full larval development takes 19 to 26 and 12 to 17 days, respectively. The pupae take 14 to 20 and 7 to 10 days to hatch and the full cycle from egg laying to adult takes between 28 and 47 days. The adults live for between 27 and 70 days. **Efficacy:** Because the adults can fly, large areas can be covered in the search for food. At 21 °C, a larva will consume more than 250 mealybug larvae through its development to an adult. All stages are most active in sunlight and between temperatures of 21 and 29 °C, with humidities between 70% and 80%. Mealybugs are completely devoured. *Cryptolaemus montrouzieri* is not effective at temperatures below 21 °C.

COMMERCIALISATION: Formulation: Sold with a food and water source as adults (sometimes with a few added larvae) in bottles, vials or punnets containing at least 40 insects. **Tradenames:** 'Cryptobug' (Koppert), 'Cryptolaemus montrouzieri' (Arbico), (Biowise), (Harmony), (Hydro-Gardens), (International Technology), (IPM Laboratories), (Praxis), (Rincon-Vitova) and (Neudorff), 'Cryptolaemus montrouzieri mealy bug predator' (M&R Durango), 'Cryptolaemus-System' (Biobest), 'Cryptoline m' (Syngenta Bioline), 'CriptoPAK 1000' (Bioplanet), 'Cryptosure (m)' (Biological Crop Protection), 'Mealybug Destroyer' (Biocontrol Network).

APPLICATION: Make the first application in the spring when mealybug populations are low, at a rate of 5 beetles per infested plant or 2 to 5 beetles per square metre or at 1200 to

12000 beetles per hectare. Repeat as necessary, usually twice a week, if populations are high and twice a year, as a prophylactic treatment.

PRODUCT SPECIFICATIONS: Purity: Should contain adult beetles only, although larvae are sometimes introduced to aid identification. **Storage conditions:** Store under cool conditions, but do not refrigerate. Keep storage temperature above 15 °C. **Shelf-life:** Release as soon as possible.

COMPATIBILITY: Control ants, as these protect mealybugs for their honeydew. Do not release following application of residual insecticides. *Cryptolaemus montrouzieri* is particularly sensitive to diazinon. Attracted to light colours, so yellow sticky traps should be avoided. Often used in combination with *Leptomastix dactylopii* Howard.

MAMMALIAN TOXICITY: There have been no reports of allergic or other adverse reactions following use of *Cryptolaemus montrouzieri* in enclosed or field situations.

ENVIRONMENTAL IMPACT AND NON-TARGET TOXICITY: *Cryptolaemus montrouzieri* occurs widely in Nature and is not expected to have any adverse effects on non-target organisms or on the environment. Although the primary target is mealybugs, it will consume other species if the mealybug population is low. **Approved for use in organic farming:** Yes.

3:294 *Cybocephalus nipponicus*
Predator of euonymus scale

Predatory beetle: Coleoptera: Cybocephalidae

NOMENCLATURE: Approved name: *Cybocephalus nipponicus* Endrody-Younga.
Other names: Scale picnic beetle.

SOURCE: Occurs throughout North America.

PRODUCTION: Reared in insectaries on euonymus scale (*Unaspis euonymi* (Comstock)).

TARGET PESTS: Euonymus scale (*Unaspis euonymi*) and Siebold scale (*Lepidosaphes yangicola* Kuwana), but will also consume San José scale (*Quadraspidiotus perniciosus* (Comstock)).

TARGET CROPS: Euonymus shrubs (*Euonymus* spp.).

BIOLOGICAL ACTIVITY: Biology: The hemispherical adults are about 1 mm long. Female *Cybocephalus nipponicus* are black, while the males have a large beige to orange-coloured head and pronotum. **Predation:** The yellowish larvae bite into scale to feed on them, resulting in distinctive broken areas on the scale covers. The larvae hide under scale covers as they consume the liquid content of the scale body. They eat an average of 20 scale insects during their developmental period. They are cannibalistic when scale are not available, eating other beetle larvae or eggs. After about 12 days, the larvae complete their development and build a pupal chamber from the armour coverings of three or more adjacent scale. Adults also attack scale, chewing a hole in the scale cover and then feeding on the body contents. The number of scale eaten increases with scale density and some scale are only partially consumed, but are still killed. The adults prefer older, larger scale. Each pair of adult beetles can kill an average of 92 scale in a 2-day period. **Egg laying:** Four days after mating, the females lay an average of 3 eggs a day under the scale covers, laying more than 500 eggs if they live long enough. The eggs are transparent ovoids when laid and turn greyish or purple before hatching.

Efficacy: Like other beetles, *Cybocephalus nipponicus* is very mobile and seeks out prey actively. Both larvae and adults are aggressive predators. The small size of the beetles restricts the number and variety of prey consumed, but there is evidence that *C. nipponicus* will consume other scale species, including San José scale (*Quadraspidiotus perniciosus*).

Key reference(s): J M Alvarez & R van Driesche. 1998. Biology of *Cybocephalus* sp. nr. *nipponicus* (Coleoptera: Cybocephalidae), a natural enemy of euonymus scale (Homoptera: Diaspididae), *Environ. Entomol.*, **27**(1), 130–6.

COMMERCIALISATION: Formulation: Sold as adult beetles. **Tradenames:** 'Cybocephalus nipponicus adults' (Green Spot) and (Rincon-Vitova).

APPLICATION: Release between 20 and 30 beetles per cubic metre of foliage, with higher numbers for very heavy infestations.

PRODUCT SPECIFICATIONS: Storage conditions: Keep in a cool, dark location. **Shelf-life:** Use as soon as possible after receipt, but may be kept for 3 to 4 days.

COMPATIBILITY: Incompatible with persistent insecticides.

MAMMALIAN TOXICITY: *Cybocephalus nipponicus* has shown no allergic or other adverse effects on research workers, manufacturing or field staff. It is considered to be safe to mammals.

ENVIRONMENTAL IMPACT AND NON-TARGET TOXICITY: *Cybocephalus nipponicus* is not expected to have any adverse effects on non-target organisms or on the environment. It is specific to scale insects.

3:295 *Dacnusa sibirica*

Parasitoid of leaf miners

Parasitoid wasp: Hymenoptera: Braconidae

NOMENCLATURE: Approved name: *Dacnusa sibirica* Telenga. **Other names:** Leaf miner parasitoid; leaf miner wasp.

SOURCE: Common in temperate climates.

PRODUCTION: Produced on leaf miners and supplied as adult wasps.

TARGET PESTS: Leaf miners, such as *Liriomyza bryoniae* (Kaltenbach), *Liriomyza huidobrensis* (Blanchard), *Liriomyza trifolii* (Burgess) and *Phytomyza syngenesiae* (Hardy).

TARGET CROPS: Glasshouse-grown vegetables and ornamentals. Also effective in interiorscapes.

BIOLOGICAL ACTIVITY: Biology: Eggs are laid in leaf miner larvae by adult female wasps. Females are able to distinguish between parasitised and non-parasitised larvae. The wasp larva develops within the leaf miner larva, which is not killed until it pupates. The wasp pupa is formed within the pupated leaf miner and can hibernate therein. **Predation:** Females lay between 50 and 225 eggs within leaf miner larvae in their lifetime, depending on temperature. The eggs hatch and the larvae feed on the leaf miner larvae. The parasite has a preference for leaf miner larvae at the first or second larval stage. **Egg laying:** Eggs are laid throughout the year in heated glasshouses. Optimal temperatures are between 15 and 20°C. **Duration of development:** Development time from egg hatch to adult emergence is dependent upon temperature, taking about 16 days at 22°C. Adults live for between 8 and 20 days. **Efficacy:** The wasps are very mobile and, once an infested plant is located, the leaf miner larvae are rapidly parasitised. *Dacnusa sibirica* will parasitise leaf miners at low temperatures.

COMMERCIALISATION: Formulation: Live adults supplied with a food source in bottles. **Tradenames:** 'Minusa' and 'Diminex' (50% *Dacnusa* and 50% *Diglyphus isaea*) (Koppert), 'Dacline s' (Syngenta Bioline), 'Dacdigline si' (90% *Dacnusa* and 10% *Diglyphus*) (Syngenta Bioline), 'Dacnusa-System' and 'Dacnusa-Mix-System' (225 *Dacnusa* adults and 25 *Diglyphus* adults) (Biobest), 'Dacsure (si)' (in combination with *Diglyphus isaea*) (Biological Crop Protection), 'Dacnusa sibirica – Schlupfwespen' and 'Dacnusa/Diglyphus (225:25)' (Neudorff), 'Dacnusa' and 'Diglyphus/Dacnusa' (Sautter & Stepper), 'Dacnusa sibirica' (Arbico), (International Technology), (Praxis) and (Rincon-Vitova).

APPLICATION: Release in winter and spring at a rate of 1200 to 5000 wasps per hectare weekly, or at a rate of one wasp per 10 leaf miners.

3. Macro-organisms

PRODUCT SPECIFICATIONS: Purity: No contaminating insects. **Storage conditions:** Keep cool, but do not refrigerate. Keep out of direct sunlight. **Shelf-life:** Release immediately.

COMPATIBILITY: Avoid the use of residual insecticides for 4 weeks prior to release. Often used in combination with *Diglyphus isaea* (Walker).

MAMMALIAN TOXICITY: There have been no reports of allergic or other adverse reactions following use in glasshouses or interiorscapes.

ENVIRONMENTAL IMPACT AND NON-TARGET TOXICITY: *Dacnusa sibirica* occurs widely in Nature and is not expected to have any adverse effects on non-target organisms or on the environment. It is specific to leaf miners. **Approved for use in organic farming:** Yes.

3:296 *Delphastus catalinae*
Predator of whiteflies

Predatory beetle: Coleoptera: Coccinellidae

NOMENCLATURE: Approved name: *Delphastus catalinae* (Horn). **Other names:** Whitefly ladybeetle.

SOURCE: Widely distributed across central and southern USA and Central and South America.

PRODUCTION: Reared on glasshouse-grown whiteflies (*Trialeurodes vaporariorum* (Westwood) or *Bemisia tabaci* (Gennadius)) under controlled conditions.

TARGET PESTS: Glasshouse whitefly (*Trialeurodes vaporariorum*) and tobacco whitefly (*Bemisia tabaci*).

TARGET CROPS: Glasshouse cucumber, pepper, tomato, eggplant, tropical flowering and foliage plants. *Delphastus catalinae* is often used by commercial avocado and citrus growers.

BIOLOGICAL ACTIVITY: Biology: Adult *Delphastus catalinae* feeds on whitefly eggs and pupae and can consume over 200 whitefly eggs per day. The larval stage feeds on all stages of whitefly, but avoids feeding on whitefly larvae parasitised by *Encarsia formosa*, making them excellent partners in whitefly control. Unlike *Encarsia formosa*, which is hampered by high densities of whitefly, *Delphastus catalinae* prefers whitefly hot spots and the adults fly and are attracted to these areas. **Predation:** *Delphastus* feeds on all species of whitefly and continues to work at low temperatures and light levels that limit the effectiveness of other whitefly parasitoids. The optimal conditions for the predator are 19 to 30 °C (65 to 90 °F) and 20% to 80% r.h.

COMMERCIALISATION: Formulation: Sold as adult beetles (with no diapause) in bottles. A DEFRA licence is required for release in the UK. **Tradenames:** 'Delsure (c)' (Biological Crop Protection), 'Delphastus catalinae' (American Insectaries), (Praxis), (Harmony), (Beneficial Insectary) and (Rincon-Vitova), 'Whitefly Ladybeetle' (Applied Bio-Nomics), (Rincon-Vitova), (IPM Laboratories) and (Green Spot).

APPLICATION: Release 100 adults in each whitefly hot spot area of 10 plants, and repeat weekly, 5 times.

PRODUCT SPECIFICATIONS: Purity: Adults only, with no host material present. **Storage conditions:** Use as soon as possible after receipt. **Shelf-life:** May be stored for a few days at 10 °C. Longer periods reduce egg laying and predator viability.

COMPATIBILITY: Incompatible with residual insecticides. May be used with other whitefly parasitoids. Most effective at high populations of whitefly.

MAMMALIAN TOXICITY: There have been no reports of any adverse or allergic reactions from laboratory, manufacturing or field trial staff.

ENVIRONMENTAL IMPACT AND NON-TARGET TOXICITY: *Delphastus catalinae* occurs widely in Nature and there is no evidence of adverse environmental effects nor of effects on non-target organisms. It is specific to whiteflies. **Approved for use in organic farming:** Yes.

3:297 *Delphastus pusillus* *Predator of whiteflies*

Predatory beetle: Coleoptera: Coccinellidae

NOMENCLATURE: Approved name: *Delphastus pusillus* Leconte. **Other names:** Whitefly predatory beetle; black lady beetle.

SOURCE: Widely distributed across central and southern USA and Central and South America.

PRODUCTION: Reared on glasshouse-grown whiteflies (*Trialeurodes vaporariorum* (Westwood) or *Bemisia tabaci* (Gennadius)) under controlled conditions.

TARGET PESTS: Glasshouse whitefly (*Trialeurodes vaporariorum*) and tobacco whitefly (*Bemisia tabaci*).

TARGET CROPS: Cucumbers, peppers and other glasshouse vegetable crops, ornamentals and interiorscapes.

BIOLOGICAL ACTIVITY: Biology: Adult and larval stages feed on whitefly eggs and pupae, with the adults preferring eggs. Older larvae migrate down the plant and pupate on the underside of older leaves. Adults avoid feeding on whitefly larvae parasitised by *Encarsia formosa* Gahan. **Predation:** Whitefly eggs and nymphs are eaten by adult and larval stages. Adults also feed on honeydew produced by their prey, but females must feed on whitefly eggs for maximum egg production. **Egg laying:** Females lay an average of 3 eggs per day, to give a total of about 180 eggs in their life span. Eggs are laid amongst whitefly eggs and larvae. **Duration of development:** The development time from egg to adult is 21 days at 23 °C. Adult beetles can live for up to 60 days. **Efficacy:** Adults and larvae search actively for prey. Consumption of 100 to 150 whitefly eggs per day is required to maintain adult oviposition. Individual beetles can consume as many as 10 000 whitefly eggs or 700 whitefly scale during their lifetime.

COMMERCIALISATION: Formulation: Sold as adult beetles in bottles. A DEFRA licence is required for release in the UK. **Tradenames:** 'Delphastus-A' (Applied Bio-Nomics), 'Delphastus pusillus' (IPM Laboratories), (Rincon-Vitova), (Arbico), (Biocontrol Network) and (Praxis), 'Delphaline p' (Syngenta Bioline), 'Whitefly Destroyer' (Nature's Alternative Insectary), 'Delphastus-System' (Biobest), 'Delsure (p)' (Biological Crop Protection), 'Delphastus pusillus predator beetle' (M&R Durango).

APPLICATION: Adult beetles are released within infested crops at a rate of one beetle per 10 square metres as a preventive treatment and 5 beetles per 10 square metres as a curative treatment.

PRODUCT SPECIFICATIONS: Purity: Adults only, with no host material present. **Storage conditions:** Use as soon as possible after receipt. **Shelf-life:** May be stored for a few days at 10 °C. Longer periods reduce egg laying and predator viability.

COMPATIBILITY: Incompatible with residual insecticides. May be used with other whitefly parasitoids, such as *Encarsia formosa* and *Chrysoperla pusillus*. Most effective at high populations of whitefly.

MAMMALIAN TOXICITY: There have been no reports of any adverse or allergic reactions from laboratory, manufacturing or field trial staff.

ENVIRONMENTAL IMPACT AND NON-TARGET TOXICITY: *Delphastus pusillus* occurs widely in Nature and there is no evidence of adverse environmental effects nor of effects on non-target organisms. It is specific to whiteflies. **Approved for use in organic farming:** Yes.

3:298 *Deraeocoris brevis*
Predator of aphids and whiteflies

Predatory bug: Hemiptera: Anthocoridae and Miridae

NOMENCLATURE: Approved name: *Deraeocoris brevis* Knight. **Other names:** Mirid plant bug.

SOURCE: *Deraeocoris brevis* occurs widely in Nature.

PRODUCTION: Reared in insectaries on aphids or whiteflies.

TARGET PESTS: *Deraeocoris brevis* predates a wide range of insects. Although it favours aphids and whiteflies, it will consume thrips, small lepidopteran larvae, tarnished bugs, pear psyllids and the eggs of many different types of insect.

TARGET CROPS: Recommended for use in glasshouse vegetables and ornamentals, as well as in fields, interiorscapes, orchards and gardens.

BIOLOGICAL ACTIVITY: Biology: The female adult is about 5 mm long and lays up to 200 eggs in plant tissue in her lifetime. The eggs hatch and the second instar nymphs will predate phytophagous insects. Like all true bugs, *Deraeocoris brevis* attacks its prey by inserting its proboscis and sucking the prey dry. **Predation:** *Deraeocoris brevis* performs best at temperatures between 18 and 30 °C and a relative humidity between 30% and 60%. It will still give control of insect pests at lower temperatures, although it will reproduce less successfully and its growth cycle will be extended. In winter, temperatures must be kept above 21 °C and supplementary lighting must be provided to prevent the bugs from entering diapause. **Duration of development:** The predator spends about 30 days in its immature nymphal stages and about 21 days as an adult. **Efficacy:** *Deraeocoris brevis* requires the presence of insect prey and pollen, on which it will also feed, to become established, but a high inoculum is necessary for successful establishment of the bug. It is a very mobile predator and will attack a wide variety of plant insect pests.

COMMERCIALISATION: Formulation: It is sold as adults with some nymphs present. **Tradenames:** 'Deraeocoris brevis adults' (Green Spot) and (Applied Bio-Nomics).

APPLICATION: Open box carefully at the site of application and introduce the packing material plus bugs to the infested plants. Adult bugs may fly away from the place of introduction. It is recommended that at least 300 bugs are introduced in any given area.

PRODUCT SPECIFICATIONS: Purity: Contains mainly adults, plus a few nymphs. **Storage conditions:** Store at temperatures above 21 °C, but use as soon as possible after receipt.

COMPATIBILITY: Sensitive to most conventional insecticides.

3. Macro-organisms

MAMMALIAN TOXICITY: There have been no reports of allergic or other adverse toxicological effects arising from contact with *Deraeocoris brevis* from research staff, producers or users.

ENVIRONMENTAL IMPACT AND NON-TARGET TOXICITY: *Deraeocoris brevis* has a wide distribution in Nature and there is no evidence that it has any adverse effects on non-target organisms or on the environment. However, it is polyphagous and, when the population of its preferred hosts is low, it will prey on many insect species, including other beneficials. **Approved for use in organic farming:** Yes.

3:299 *Dicyphus hesperus*   *Predator of whiteflies*

Predatory bug: Heteroptera: Miridae

NOMENCLATURE: Approved name: *Dicyphus hesperus* Knight. **Other names:** Predatory mirid; whitefly predatory bug

SOURCE: Widely spread.

PRODUCTION: Bred under controlled conditions on whiteflies on tomatoes.

TARGET PESTS: All species of whitefly (glasshouse whitefly (*Trialeurodes vaporariorum* (Westwood)), tobacco whitefly (*Bemisia tabaci* (Gennadius)) and silverleaf whitefly (*Bemisia argentifolii* Bellows & Perring)). *Dicyphus hesperus* will also feed on aphids and, to a lesser extent, on two-spotted spider mites, insect eggs, leaf miner larvae and thrips, but will not control these pests. *Dicyphus* should not be used on its own to replace other biological control agents. It is best used along with other biological control agents in glasshouse crops.

TARGET CROPS: Protected ornamentals and vegetables under glass or plastic. *Dicyphus hesperus* is also a plant feeder and it should not be used on crops such as Gerbera which can be damaged.

BIOLOGICAL ACTIVITY: Biology: Adults live for 30 to 40 days and are 6 mm long, black with greenish abdomen and red eyes. The nymphs are green with red eyes, with a red stripe behind the eyes. They feed for 14 days before they moult into adults. **Predation:** Adult predatory bugs and nymphs search actively for their prey, insert their sucking mouthparts and suck out the contents. When whitefly eggs, larvae or pupae are eaten by *Dicyphus hesperus*, only the skin remains, usually in its original form, with a tiny hole where the mouthparts of the predatory bug were inserted. **Egg laying:** Eggs are laid in the tissue of plant stems and leaf veins and hatch within 2 weeks. The female lays from 90 to 100 eggs in leaves and stems in her lifetime. **Duration of development:** Development from egg to adult takes 5 weeks at 25 °C and 8 weeks at 20 °C. **Efficacy:** *Dicyphus hesperus* is a voracious predator of insects, but it also eats plants when prey is scarce. Therefore, it can survive in the absence of insect prey, but needs a large number of prey to reproduce. **Key reference(s)**: 1) R R McGregor, D R Gillespie, D M J Quiring & M R J Foisy. 1999. Potential use of *Dicyphus hesperus* Knight (Heteroptera: Miridae) for biological control of pests of greenhouse tomatoes, *Biol. Control*, **16**, 104–10. 2) R R McGregor, D R Gillespie, C G Park, D M J Quiring & M R J Foisy. 2000. Leaves or fruit? The potential for damage to tomato fruits by the omnivorous predator, *Dicyphus hesperus* Knight (Heteroptera: Miridae), *Entomol. Exp. Appl.*, **95**, 325–8.

COMMERCIALISATION: Formulation: Sold as adults and nymphs on leaf pieces. **Tradenames:** 'Dicyphus h' (Syngenta Bioline), 'Dicyphus hesperus' (IPM of Alaska), (Rincon-Vitova) and (Natural Insect Control), 'Dicyphus System' (Biobest).

APPLICATION: Turn and gently shake the bottle before use. Sprinkle the material directly onto the leaves where prey insects are present if possible. Introduce in clusters of at least 75 predatory bugs to ensure mating. Release in the morning or in the evening out of direct sunlight. Release in combination with *Ephestia* eggs early in the crop or when prey numbers are low to help establish the bugs.

PRODUCT SPECIFICATIONS: Purity: Adults and nymphs, but often with caterpillar eggs (*Ephestia*) as a source of food. **Storage conditions:** *Dicyphus hesperus* may be stored in the dark at 10 to 15 °C (50 to 59 °F), keeping the bottle horizontal. **Shelf-life:** If kept under recommended conditions, can be stored for 1 to 2 days.

COMPATIBILITY: Incompatible with residual broad-spectrum insecticides. Crops may be damaged in the absence of an adequate food source.

MAMMALIAN TOXICITY: There have been no reports of any adverse or allergic reactions from laboratory, production or field trial staff.

ENVIRONMENTAL IMPACT AND NON-TARGET TOXICITY: *Dicyphus hesperus* occurs widely in Nature and it is unlikely that any adverse environmental effects will result from its use. Although its preferred prey are whiteflies, it will consume many other insect species when populations are low, including beneficials. **Approved for use in organic farming:** Yes.

3:300 *Diglyphus isaea*

Parasitoid of leaf miners

Parasitoid wasp: Hymenoptera: Eulophidae

NOMENCLATURE: Approved name: *Diglyphus isaea* (Walker). **Other names:** Eulophid wasp; leaf miner parasitoid.

SOURCE: Widely distributed in Europe, North Africa and Japan. Introduced throughout the world.

PRODUCTION: Reared on leaf miners under controlled conditions, and supplied as adult wasps.

TARGET PESTS: Leaf miner larvae, such as *Liriomyza bryoniae* (Kaltenbach), *L. trifolii* (Burgess), *L. huidobrensis* (Blanchard) and *Phytomyza syngenesiae* (Hardy).

TARGET CROPS: Vegetables and ornamentals in glasshouses; in interiorscapes; and may control leaf miners on outdoor crops.

BIOLOGICAL ACTIVITY: Biology: Female wasps are about 2 mm long with short antennae. They paralyse the leaf miner larva and then lay one or more eggs next to the host, usually of late second or third larval stage. After egg hatch, the wasp larva lies next to the leaf miner and feeds on it. The leaf miner larva stops feeding after it is paralysed. The wasp larva moves away from the parasitised leaf miner to pupate using the mine of the host. The adult wasp leaves the leaf through a round hole in the upper surface of the leaf. **Predation:** Wasps sting leaf miner larvae to lay eggs and as a food source for the wasp and its progeny. Larvae of the second or third larval stages are preferred. When conditions are optimal for the wasp, a female will kill about 360 leaf miner larvae, of which 290 are used for egg laying and 70 as food. The higher the population of leaf miners, the greater the predation. It is possible that *Diglyphus isaea* can select areas with high leaf miner populations. **Egg laying:** Female *Diglyphus isaea* will lay between 200 and 300 eggs in their lifetime. **Duration of development:** The duration of the life-cycle is dependent upon temperature, with *Diglyphus isaea* being particularly active at higher temperatures. At 15 °C, the time taken from egg hatch to adult emergence is about 26 days, whilst, at 25 °C, it is about 10.5 days. Eggs hatch after about 2 days and the larvae moult three times. The pupal stage is about 6 days. **Efficacy:** *Diglyphus isaea* is a very effective ectoparasitoid of leaf miners, particularly at temperatures above 22 °C. The leaf miners cease feeding upon paralysis by the female wasp, which is a very efficient locator of its host.

COMMERCIALISATION: Formulation: Sold as adults in shaker bottles.
Tradenames: 'Miglyphus', 'Minex' (in combination with *Dacnusa sibirica*) and 'Diminex' (50% *Dacnusa* and 50% *Diglyphus isaea*) (Koppert), 'Digline i' (Syngenta Bioline), 'Dacdigline si'

(90% *Dacnusa sibirica* and 10% *Diglyphus*) (Syngenta Bioline), 'Digsure (i)' and 'Dacsure (si)' (in combination with *Dacnusa sibirica*) (Biological Crop Protection), 'Diglyphus/Dacnusa' (Sautter & Stepper), 'Leafminer Parasite' (Arbico) and (Biocontrol Network), 'Diglyphus-System' and 'Dacnusa-Mix-System' (225 adult *Dacnusa* and 25 adult *Diglyphus*) (Biobest), 'Diglyphus isaea' (Praxis) and (Rincon-Vitova), 'DigliPAK 100' (Bioplanet).

APPLICATION: Release when tunnels are first seen in the crop, at a rate of 1200 to 2400 wasps per hectare, evenly distributed throughout the crop. Repeat applications on a weekly basis for 3 weeks. Control should be seen within 15 days of application.

PRODUCT SPECIFICATIONS: Purity: Product contains only *Diglyphus isaea* adults. **Storage conditions:** Keep cool, but do not refrigerate. Do not expose to direct sunlight. **Shelf-life:** Release immediately.

COMPATIBILITY: Do not use residual insecticides within 4 weeks of release. Often used in conjunction with *Dacnusa sibirica* Telenga. Most effective at high temperatures.

MAMMALIAN TOXICITY: No allergic or other adverse reactions have been reported following use in glasshouses, interiorscapes or outdoors.

ENVIRONMENTAL IMPACT AND NON-TARGET TOXICITY: *Diglyphus isaea* occurs widely in Nature and is not expected to show any adverse effects on non-target organisms or on the environment. It is specific to leaf miners. **Approved for use in organic farming:** Yes.

3:301 *Encarsia deserti*   *Parasitoid of whiteflies*

Parasitoid wasp: Hymenoptera: Aphelinidae

NOMENCLATURE: Approved name: *Encarsia deserti* Rivnay & Gerling. **Other names:** Whitefly parasitoid; previously known as *Encarsia luteola* Howard.

SOURCE: Thought to have evolved in the same location as its host in tropical or subtropical regions. This species was found in California and has since been exported to Israel as a biocontrol agent.

PRODUCTION: Reared on whitefly nymphs under controlled glasshouse conditions.

TARGET PESTS: Glasshouse whitefly (*Trialeurodes vaporariorum* (Westwood) and *Bemisia tabaci* (Gennadius)). *E. derseti* is much more effective at controlling *B. tabaci* than *E. formosa*.

TARGET CROPS: Tomatoes, cucumbers, peppers and other vegetable crops and ornamentals in glasshouses. It is also used in interiorscapes.

BIOLOGICAL ACTIVITY: Biology: *Encarsia deserti* resembles *E. formosa* Gahan, but has a light brown head and thorax, is smaller, and is bi-parental. Adult female wasps can lay eggs in all nymphal stages of the whitefly, but it is usual to select third and fourth nymphal stages. The parasitoid develops within the whitefly nymph, passing through six developmental stages (egg, four larval stages and the pupal stage). When the wasp pupates within the whitefly, the host 'pupa' turns black. The adult wasp emerges from the parasitised 'pupa' and feeds on honeydew and the body fluids of whitefly nymphs; some hosts are killed by this feeding. **Predation:** *Encarsia deserti* adult females attack young whitefly nymphs by stinging and laying eggs within them. Adult wasps also feed directly on the scales. **Duration of development:** The life-cycle takes between 2 and 4 weeks, depending upon the temperature. At temperatures above 30 °C, the female lives only for a few days. **Efficacy:** The parasitoid wasp searches actively for a host. It can cover distances of 10 to 30 metres and is very effective in locating whitefly. The presence of honeydew restricts the movement of the adult and, consequently, large infestations are more difficult to control. **Key reference(s)**: D Gerling & T Rivnay. 1984. A new species of *Encarsia* [Hym.: Aphelinidae] parasitizing *Bemisia tabaci* [Hom.: Aleyrodidae], *BioControl*, **29**(4), 439–44.

COMMERCIALISATION: Formulation: Sold as pupae or on small cards either attached to the surface or protected within a well in the card. **Tradenames:** 'Encarsia deserti' (Applied BioPest) and (BioBest).

APPLICATION: Cards are hung within the crop and the adults emerge into the infested crop. Preventive applications are made in tomatoes at 0.5 to 1 cards per square metre from planting. Increase to 1 to 5 per square metre when whiteflies are seen and repeat every other week. Continue until 90% parasitism is achieved. Removal of lower leaves bearing parasitised scales may reduce the level of control.

PRODUCT SPECIFICATIONS: Storage conditions: Use as soon as possible after receipt. May be stored for a few days at 6 to 8 °C within a sealed container. Longer periods of storage reduce viability. Emergence occurs at room temperature.

COMPATIBILITY: Incompatible with residual insecticides. Do not release if methomyl or a synthetic pyrethroid has been used. Less effective at high populations of whitefly, because of high levels of honeydew. Can be used in conjunction with *Delphastus pusillus* LeConte and *Chrysoperla carnea* (Stephens).

MAMMALIAN TOXICITY: No allergic or other adverse reactions have been reported by producers or formulators, nor from the use of *Encarsia deserti* in glasshouses or interiorscapes.

ENVIRONMENTAL IMPACT AND NON-TARGET TOXICITY: *Encarsia deserti* occurs widely in Nature and it is unlikely that any adverse environmental effects will result from its use. It is specific to whiteflies and will not survive winters in temperate climates. **Approved for use in organic farming:** Yes.

3:302 *Encarsia formosa*   *Parasitoid of whiteflies*

The Pesticide Manual **Fifteenth Edition entry number:** 317

Parasitoid wasp: Hymenoptera: Aphelinidae

NOMENCLATURE: Approved name: *Encarsia formosa* Gahan. **Other names:** Glasshouse whitefly parasitoid; TAB-1 (Tomono).

SOURCE: Thought to have evolved in the same location as its host in tropical or subtropical regions. Now can be found in Europe, Australia, New Zealand and North America.

PRODUCTION: Reared on whitefly nymphs under controlled glasshouse conditions.

TARGET PESTS: Glasshouse whitefly (*Trialeurodes vaporariorum* (Westwood)). *Bemisia tabaci* (Gennadius) is also parasitised, but is a poor host and higher numbers of *Encarsia formosa* are required.

TARGET CROPS: Tomatoes, cucumbers, peppers and other vegetable crops and ornamentals in glasshouses. It is also used in interiorscapes.

BIOLOGICAL ACTIVITY: Biology: Adult female wasps can lay eggs in all nymphal stages of the whitefly, but it is usual to select third and fourth nymphal stages. The parasitoid develops within the whitefly nymph, passing through six developmental stages (egg, four larval stages and the pupal stage). When the wasp pupates within the whitefly, the host 'pupa' turns black. The adult wasp emerges from the parasitised 'pupa' and feeds on honeydew and the body fluids of whitefly nymphs; some hosts are killed by this feeding. **Predation:** *Encarsia formosa* adult females attack young whitefly nymphs by stinging and laying eggs within them. Adult wasps also feed directly on the scales. **Egg laying:** The adult wasp will not fly at temperatures below 15 °C. Activity is low below 18 °C, but adults continue to search by walking on the leaves. It is normal for the adult female wasp to lay between 60 and 100 eggs. **Duration of development:** The life-cycle takes between 2 and 4 weeks, depending upon the temperature. At temperatures above 30 °C, the female lives only for a few days. **Efficacy:** The parasitoid wasp searches actively for a host. It can cover distances of 10 to 30 metres and is very effective in locating whitefly. The presence of honeydew restricts the movement of the adult and, consequently, large infestations are more difficult to control. **Key reference(s):** 1) R G van Driesche & T S Bellows (eds). 1996. *Biological Control*, Chapman & Hall, London, UK. 2) M Malais & W J Ravensberg (eds). 1992. *Knowing and Recognising: the Biology of Glasshouse Pests and their Natural Enemies*, Koppert Biological Systems, Berkel en Rodenrijs, The Netherlands. 3) P Stary. 1988. Aphelinidae. In *Aphids: their Biology, Natural Enemies and Control*, A K Minks & P Harrewijn (eds), Vol. B, pp. 185–8, Elsevier, Amsterdam, The Netherlands.

3. Macro-organisms

COMMERCIALISATION: Formulation: Sold as pupae or on small cards either attached to the surface or protected within a well in the card. **Tradenames:** 'Encarline f' and 'Enc/Eretline' (mixture of *Encarsia formosa* and *Eretmocerus eremicus*) (Syngenta Bioline), 'En-Strip' and 'Enermix' (mixture of *Encarsia formosa* and *Eretmocerus eremicus*) (Koppert), 'Para-bulk' and 'Para-strip' (Applied Bio-Nomics), 'Encarsia formosa Glasshouse Whitefly Parasite' (M&R Durango), 'Sweet Potato Whitefly Predator' (Arbico), 'Encarsia formosa Whitefly Parasite' (Biofac), 'Encarsia Whitefly Parasites' (Biowise), 'Encarsia formosa Wasps' and 'Encarsia/Eretmocerus' (mixture of *Encarsia formosa* and *Eretmocerus eremicus*) (Rincon-Vitova), 'Encarsia-System' and 'Eretmix-System' (mixture of *Encarsia formosa* and *Eretmocerus eremicus*) (Biobest), 'Encsure (f)' and 'Encsure (cf)' (Biological Crop Protection), 'Encarsia formosa' (Biological Services), (Biocontrol Network), (Neudorff) and (Praxis), 'EnPAK 1000' (Bioplanet), 'EnerPAK' (mixture of *Encarsia formosa* and *Eretmocerus eremicus*) (Bioplanet).

APPLICATION: Cards are hung within the crop and the adults emerge into the infested crop. Preventive applications are made in tomatoes at 0.5 to 1 cards per square metre from planting. Increase to 1 to 5 per square metre when whiteflies are seen and repeat every other week. Continue until 90% parasitism is achieved. Removal of lower leaves bearing parasitised scales may reduce the level of control. Double these rates for cucumbers. For poinsettia, use 0.3 *E. formosa* per plant each week.

PRODUCT SPECIFICATIONS: Purity: Pupae of *Encarsia formosa*, with no live whitefly and no debris. **Storage conditions:** Use as soon as possible after receipt. May be stored for a few days at 6 to 8 °C within a sealed container. Longer periods of storage reduce viability. Emergence occurs at room temperature. **Shelf-life:** Three to four days maximum at 6 to 8 °C.

COMPATIBILITY: Incompatible with residual insecticides. Do not release if methomyl or a synthetic pyrethroid has been used. Less effective at high populations of whitefly, because of high levels of honeydew. Can be used in conjunction with *Delphastus pusillus* and *Chrysoperla carnea*.

MAMMALIAN TOXICITY: No allergic or other adverse reactions have been reported by producers or formulators nor from the use of *Encarsia formosa* in glasshouses or interiorscapes.

ENVIRONMENTAL IMPACT AND NON-TARGET TOXICITY: *Encarsia formosa* occurs widely in Nature and it is unlikely that any adverse environmental effects will result from its use. It is specific to whiteflies and will not survive winters in temperate climates. **Approved for use in organic farming:** Yes.

3:303 *Episyrphus balteatus* *Predator of aphids*

Predatory fly: Diptera: Syrphidae

NOMENCLATURE: Approved name: *Episyrphus balteatus* (De Geer).
Other names: Marmalade hoverfly.

SOURCE: Widely distributed.

PRODUCTION: Reared under controlled conditions on aphids.

TARGET PESTS: Aphids, all genera and species.

TARGET CROPS: A wide range of field and glasshouse-grown crops, including vegetables and ornamentals.

BIOLOGICAL ACTIVITY: Biology: *Episyrphus balteatus* adults are medium-sized insects. The eggs are white and surprisingly long. As larvae emerge from the eggs, they immediately begin to feed on any aphids that are present. At this stage, the larvae are transparent, are between 10 and 20 mm long and have no legs. Each larva, depending upon temperature, will consume 400 aphids in the 1 to 2 weeks before pupation. Pupae are orange-brown in colour and are pear-shaped. Adult hoverflies are 10 to 20 mm long, with yellow abdomens that have alternating wide and narrow black stripes. They hover above the crop, feeding on pollen and nectar. A source of pollen or nectar is essential for the growth of the adults, and the number of eggs laid by a female is dependent upon this energy source. **Egg laying:** A female can lay up to 100 eggs in her lifetime.

COMMERCIALISATION: Formulation: *Episyrphus balteatus* is supplied in packs of 500 larvae per tube. **Tradenames:** 'Episyrphus-System' (Biobest), 'Epiline B' (Syngenta Bioline).

APPLICATION: Introduce 500 tubes per hectare for at least 6 weeks.

PRODUCT SPECIFICATIONS: Purity: Tubes contain only *Episyrphus balteatus* larvae. **Storage conditions:** May be stored for a short period in cool, dry conditions away from direct sunlight. **Shelf-life:** Release as soon as possible after receipt.

COMPATIBILITY: Do not use broad-spectrum, residual insecticides when releasing *Episyrphus balteatus*.

MAMMALIAN TOXICITY: There have been no reports of allergic or other adverse toxicological effects arising from contact with *Episyrphus balteatus* from research staff, producers or users.

ENVIRONMENTAL IMPACT AND NON-TARGET TOXICITY: *Episyrphus balteatus* has a wide distribution in Nature and there is no evidence that it has any adverse effects on non-target organisms or on the environment. It is specific to aphids.
Approved for use in organic farming: Yes.

3:304 *Eretmocerus eremicus*
Parasitoid of whiteflies

Parasitoid wasp: Hymenoptera: Aphelinidae.

NOMENCLATURE: Approved name: *Eretmocerus eremicus* Rose and Zolnerowich.
Other names: Whitefly parasitoid. Formerly known as: *Eretmocerus* sp. nr. *californicus* Howard.

SOURCE: Species identification is difficult and the species definition, *Eretmocerus* near *californicus*, was recognised from 1980. Recently, the species has been redefined as *Eretmocerus eremicus*. It is the dominant species in the south-western USA, where it occurs in the desert regions of California and Arizona.

PRODUCTION: *Eretmocerus eremicus* is an obligate parasitoid of whiteflies and the species is reared on *Bemisia tabaci* (Gennadius) or *Trialeurodes vaporariorum* (Westwood) under controlled conditions.

TARGET PESTS: For control of tobacco whitefly (*Bemisia tabaci*). It is also capable of parasitising glasshouse whitefly (*Trialeurodes vaporariorum*).

TARGET CROPS: Recommended for use in vegetables, ornamentals and interiorscapes.

BIOLOGICAL ACTIVITY: Biology: All *Eretmocerus* species are obligate parasitoids of whiteflies. When the egg hatches, the first instar parasitoid larva burrows into the host, where it completes its development. **Egg laying:** Adult females lay single eggs beneath the immobile second or third instar nymphs of the host. **Efficacy:** The female wasp is very mobile and actively seeks whitefly to parasitise. The adults also feed directly on the scale.
Key reference(s): M Rose & G Zolnerowich. 1997. *Eretmocerus* Haldeman (Hymenoptera: Aphelinidae) in the United States, with descriptions of new species attacking *Bemisia* (*tabaci* complex) (Homoptera: Aleyrodidae), *Proc. Entomol. Soc. Wash.*, **99**(1), 1–27.

COMMERCIALISATION: Formulation: Sold as parasitised whitefly pupae mixed with carrier. DEFRA licence required for release in the UK. **Tradenames:** 'Ercal' and 'Enermix' (mixture of *Eretmocerus eremicus* and *Encarsia formosa*) (Koppert), 'Eretline e' (Syngenta Bioline), 'Eretmocerus californicus Small Parasitic Wasp' (M&R Durango), 'Eretmocerus californicus' (Beneficial Insectary), (IPM Laboratories), (Praxis) and (Rincon-Vitova), 'Encarsia/ Eretmocerus' (*Encarsia formosa* and *Eretmocerus eremicus*) (Rincon-Vitova), 'Eretsure (e)' (Biological Crop Protection), 'Eretmocerus-System' and 'Eretmix-System' (mixture of *Encarsia formosa* and *Eretmocerus eremicus*) (Biobest), 'Eretmocerus nr. californicus' (Biocontrol Network).

APPLICATION: The parasitoid is sold as parasitised *Bemisia tabaci* or *Trialeurodes vaporariorum* pupae that are sometimes attached to the surface of small cards. These should be placed on the infested plants and the adult wasps allowed to move into the area to be treated.

PRODUCT SPECIFICATIONS: Purity: Containers contain parasitised pupae in bran and no impurity.

COMPATIBILITY: Incompatible with residual insecticides.

MAMMALIAN TOXICITY: *Eretmocerus eremicus* has not demonstrated evidence of toxicity, infectivity, irritation or hypersensitivity to mammals. No allergic responses or other adverse health problems have been observed by research workers, manufacturing staff or users.

ENVIRONMENTAL IMPACT AND NON-TARGET TOXICITY: *Eretmocerus eremicus* occurs in Nature and, as such, is not expected to show any adverse effects on non-target organisms or on the environment. It is an obligate parasitoid of whiteflies.

Approved for use in organic farming: Yes.

3:305 *Eretmocerus hayati* *Parasitoid of whiteflies*

Parasitoid wasp: Hymenoptera: Aphelinidae

NOMENCLATURE: Approved name: *Eretmocerus hayati* Zolnerowich & Rose.
Other names: Whitefly parasitoid.

SOURCE: *Eretmocerus hayati* was originally from Pakistan and has been introduced into the USA and Australia as a biocontrol agent.

PRODUCTION: *Eretmocerus hayati* is an obligate parasitoid of whiteflies and is reared on *Bemisia tabaci* (Gennadius) under controlled conditions.

TARGET PESTS: Silverleaf whitefly (*Bemisia tabaci* Biotype B, also known as *Bemisia argentifolia* Perring & Bellows).

TARGET CROPS: Broccoli, cabbage, cauliflower, cotton, cucumber, eggplant, melons, pumpkins, soybeans, squash, tomatoes and zucchini.

BIOLOGICAL ACTIVITY: Biology: *Eretmocerus hayati* adults are less than 1mm long. The female lays her eggs under a host nymph. The wasp larva, when it hatches, bores into the nymph slowly developing along with the whitefly. Once the whitefly enters the final development stage, the parasitoid kills the whitefly and completes its development and the adult finally emerges through a hole it chews in the surface of the whitefly. **Egg laying:** Adult females lay single eggs beneath the immobile second or third instar nymphs of the host. **Efficacy:** The female wasp is very mobile and actively seeks whitefly to parasitise. The adults also feed directly on the scale. **Key reference(s):** L Lawrence. 2005. A minute wasp to tackle a big job – control of silverleaf whitefly, *Outlooks on Pest Management*, **16**(1), 35–6.

COMMERCIALISATION: Formulation: After extensive testing in quarantine in Brisbane, the Australian Government Departments of Environment and Heritage, and Agriculture, Fisheries and Forestry granted permission for the release of *Eretmocerus hayati* in Australia to evaluate its potential as a biological control agent. It is not sold commercially.

APPLICATION: Parasitised whitefly are distributed throughout the area to be treated.

COMPATIBILITY: Incompatible with residual insecticides.

MAMMALIAN TOXICITY: *Eretmocerus hayati* has not demonstrated evidence of toxicity, infectivity, irritation or hypersensitivity to mammals. No allergic responses or other adverse health problems have been observed by research workers, manufacturing staff or users.

ENVIRONMENTAL IMPACT AND NON-TARGET TOXICITY: *Eretmocerus hayati* occurs in Nature and, as such, is not expected to show any adverse effects on non-target organisms or on the environment. It is an obligate parasitoid of whiteflies.
Approved for use in organic farming: Yes.

3:306 *Eretmocerus mundus*
Parasitoid of whiteflies

Parasitoid wasp: Hymenoptera: Aphelinidae

NOMENCLATURE: Approved name: *Eretmocerus mundus* Mercet. **Other names:** Whitefly parasitoid.

SOURCE: *Eretmocerus mundus* was isolated from crops growing in the Mediterranean region.

PRODUCTION: *Eretmocerus mundus* is an obligate parasitoid of whiteflies and the species is reared on *Bemisia tabaci* (Gennadius) under controlled conditions.

TARGET PESTS: *Eretmocerus mundus* is recommended for the control of several biotypes of whitefly (*Bemisia* spp.).

TARGET CROPS: Recommended for use in protected vegetables, ornamentals and interiorscapes.

BIOLOGICAL ACTIVITY: Biology: *Eretmocerus mundus* is a wasp that parasitises several species of *Bemisia*. The adult is yellow with green eyes and three red ocelli. It can develop in any nymphal stage of *Bemisia*, but prefers the second and the beginning of the third nymphal stage. *Eretmocerus mundus* lays its eggs under the nymphs of *Bemisia*. After 3 days, the egg turns a brownish colour. If the eggs are laid under the first nymphal stage, the larva of *Eretmocerus mundus* does not develop before the *Bemisia* reaches the second nymphal stage. The life-cycle is completed in approximately 14 days, depending on the temperature and the stage of *Bemisia* when it was parasitised. During the winter, the life-cycle might be longer. In non-heated glasshouses, the life-cycle can take more than one month to complete. A recently parasitised *Bemisia* nymph does not acquire any distinct colouration. After 2 weeks of parasitism, the *Bemisia* nymphs swell, are shinier and their colour turns golden-yellow. At the moment of eclosion, *Eretmocerus mundus* makes a round hole in its host, just as *Eretmocerus eremicus* does. The presence of this hole is a good indication of parasitism. The *E. mundus* adult resembles that of *E. eremicus*. It is only possible to distinguish them under a microscope. *E. mundus* has 4 hairs on the prothorax, while *E. eremicus* has 6. **Duration of development:** At 25 to 30 °C, adult *E. mundus* can live about 10 days laying more than 50 eggs. At these temperatures, the life cycle lasts from 15 to 18 days, but, in unheated glasshouses, this can be more than 35 days. **Efficacy:** The large temperature fluctuations that affect the crops of southern Spain stimulated the search for new alternative solutions. *Eretmocerus mundus*, a native of the Mediterranean, was present and is very well adapted to the climatic conditions of the region. It is much more active in lower winter temperatures, is able to withstand high summer temperatures and is more tolerant of most insecticides than *E. eremicus*.

COMMERCIALISATION: Formulation: *Eretmocerus mundus* is supplied on cards with at least 75 pupae per card. One package contains 10 strips of 10 cards each. **Tradenames:** 'Mundus-System' and 'Mundus-Mix-System' (Biobest), 'MunduPAK 1000' (Bioplanet), 'Erotmocerus mundus' (Rincon-Vitova), 'Eretline m' (Syngenta Bioline).

APPLICATION: The cards should be hung on the plants to ensure a good distribution in the crop. Introduce the cards when *Bemisia tabaci* is the whitefly present. If the species of whitefly present is *Trialeurodes vaporariorum* (Westwood), introduce *Eretmocerus mundus* in combination with *Encarsia formosa* Gahan. As a curative treatment, begin introductions of 6 *Eretmocerus* per square metre for several weeks. As a preventive treatment, it is necessary to begin introductions from the moment the first *Bemisia* are observed. Introduce 2 to 3 *Eretmocerus* per square metre until parasitism reaches 80%.

PRODUCT SPECIFICATIONS: Purity: Contains only *Eretmocerus mundus*.
Storage conditions: May be stored for short periods in cool, dry conditions, out of direct sunlight.

COMPATIBILITY: Generally *Eretmocerus mundus* is considered to be more resistant than *Encarsia* to pesticides.

3. Macro-organisms

MAMMALIAN TOXICITY: *Eretmocerus mundus* has not demonstrated evidence of toxicity, infectivity, irritation or hypersensitivity to mammals. No allergic responses or other adverse health problems have been observed by research workers, manufacturing staff or users.

ENVIRONMENTAL IMPACT AND NON-TARGET TOXICITY: *Eretmocerus mundus* occurs in Nature and, as such, is not expected to show any adverse effects on non-target organisms or on the environment. It is an obligate parasitoid of whiteflies.

Approved for use in organic farming: Yes.

3:307 *Feltiella acarisuga*

Predator of spider mites

Predatory gall-midge: Diptera: Cecidomyiidae

NOMENCLATURE: Approved name: *Feltiella acarisuga* (Vallot). **Other names:** Red spider mite gall midge predator. Formerly known as: *Therodiplosis persicae* Kieffer.

SOURCE: Originated in western Europe and the Meditteranean and now widely occurring throughout Europe.

PRODUCTION: Reared under controlled conditions on red spider mites, *Tetranychus urticae* Koch.

TARGET PESTS: Spider mites, including *Tetranychus urticae* and *T. cinnabarinus* (Boisduval).

TARGET CROPS: Glasshouse-grown vegetable and ornamental crops.

BIOLOGICAL ACTIVITY: Biology: Adult *Feltiella acarisuga* lay yellowish eggs (about 0.25 mm long) in colonies of red spider mites. The creamy-brown larvae have four stages and each predates spider mites. Pupae appear as white fluff near the veins of the leaves. **Predation:** All larval stages of the midge consume eggs, nymphs and adult spider mites. Overwintered mites are also controlled. The predator consumes about five times as many mites per individual as the predatory mite *Phytoseiulus persimilis* Athios-Henriot.
Egg laying: Eggs are laid in spider mite colonies. **Duration of development:** Under normal temperature conditions, eggs hatch within 2 days and the four larval stages take approximately 7 days. The total life-cycle from egg to egg is between 2 and 4 weeks.
Efficacy: *Feltiella acarisuga* is a very mobile predator that is able to locate its prey whilst it is

in flight. It is effective under cold and dark conditions in the spring and autumn and is able to locate its prey in crops that are difficult to monitor, such as ornamentals. It is easy to see within the crop and provides long-lasting protection from spider mite damage.

COMMERCIALISATION: Formulation: Sold as larvae or pupae in tubs.
Tradenames: 'Feltiella-System' and 'Feltiella LV-System' (Biobest), 'Felsure (a)' (Biological Crop Protection), 'Feltiline a' (Syngenta Bioline), 'Feltiella acarisuga' (Rincon-Vitova) and (Biocontrol Network).

APPLICATION: Tubs containing pupae or cocoons are opened in the shade of the crop and the emerging adults are allowed to escape. In the spring, release 500 to 750 adults per hectare, weekly. It is best to place the tubs near spider mite colonies. As temperatures increase, release the midges less frequently.

PRODUCT SPECIFICATIONS: Purity: It is essential that the product contains only *Feltiella acarisuga* pupae and not the parasitoid *Aphanogmus parvulus* Roberti.
Storage conditions: Keep tubs under cool, dark conditions. Do not expose to direct sunlight. **Shelf-life:** Use as soon as possible after receipt.

COMPATIBILITY: Most fungicides have no effect on the midge, but acaricides delay the population build-up by removing its prey.

MAMMALIAN TOXICITY: There has been no evidence of adverse allergic or other effects following the rearing or release of *Feltiella acarisuga* under glasshouse conditions.

ENVIRONMENTAL IMPACT AND NON-TARGET TOXICITY: *Feltiella acarisuga* occurs widely in Nature and is not expected to have any adverse effects on non-target organisms or on the environment. It is specific to spider mites. **Approved for use in organic farming:** Yes.

3:308 *Franklinothrips orizabensis*
Predator of thrips

Predatory thrips: Thysanoptera: Aeolothripidae

NOMENCLATURE: Approved name: *Franklinothrips orizabensis* Johansen.
Other names: Avocado thrips predator; scirtothrips predator.

SOURCE: Originally found in Latin and Central America, but now introduced into North America.

PRODUCTION: Reared on *Ephestia* eggs in the presence of bean plants (*Phaseolus* sp.), under controlled conditions.

TARGET PESTS: Avocado thrips (*Scirtothrips persea* Nakahara).

TARGET CROPS: Citrus and avocado orchards.

BIOLOGICAL ACTIVITY: Biology: Female *Franklinothrips orizabensis* lay eggs directly into plant tissue. Developing larvae pass through two instars, with the second instar being distinguished from the first by the presence of red pigmentation. The larva falls from the tree and pupates in selected sites by spinning a cocoon made from secretions from its anal area. **Predation:** *Franklinothrips orizabensis* is a voracious predator of avocado thrips and, when given a choice between first and second instar larvae, females preferentially attack and feed on second instar larvae. When presented with second instar glasshouse-reared thrips larvae with protective faecal droplets carried on the tip of the abdomen or pro-pupae lacking protective droplets, *Franklinothrips orizabensis* females only attack pro-pupae and exhibit violent avoidance reactions when they contact a protective faecal droplet during unsuccessful attacks on second instar glasshouse-reared thrips larvae. **Egg laying:** *Franklinothrips orizabensis* exhibits haplodiploidy, with unfertilised eggs developing into males and fertilised eggs developing into females. **Duration of development:** On average, females live longer than males, with longevity being associated with the availability of food. **Efficacy:** *Franklinothrips orizabensis* adults will consume between 14 and 19 avocado thrips larvae a day, depending upon the temperature. **Key reference(s):** M S Hoddle, L Robinson, R Drescher & J Jones. 2000. Developmental and reproductive biology of a predatory *Franklinothrips orizabensis* (Thysanoptera: Aeolothripidae), *Biol. Control*, **18**, 27–38.

COMMERCIALISATION: Formulation: Sold as adults. **Tradenames:** 'Franklinothrips orizabensis' (Rincon-Vitova).

APPLICATION: Release adults into trees at the first sign of thrips damage.

PRODUCT SPECIFICATIONS: Purity: Product contains *Franklinothrips orizabensis* and no impurities. **Shelf-life:** Use as soon as possible after receipt.

MAMMALIAN TOXICITY: There have been no reports of adverse allergic or other reactions from research workers, manufacturing staff or from the field release of *Franklinothrips orizabensis*.

ENVIRONMENTAL IMPACT AND NON-TARGET TOXICITY: *Franklinothrips orizabensis* occurs widely in Nature and it is unlikely that any adverse environmental effects will result from its use. It is specific to thrips. **Approved for use in organic farming:** Yes.

3:309 *Galendromus annectens*

Predator of persea mite

Predatory mite: Acari: Phytoseiidae

NOMENCLATURE: Approved name: *Galendromus annectens* (DeLeon).
Other names: Persea mite predator.

SOURCE: A mite that is thought to have originated in Mexico. Well adapted to predate the avocado or persea mite (*Oligonychus perseae* Tuttle, Baker & Abbatiello).

PRODUCTION: Reared on phytophagous mites under controlled conditions.

TARGET PESTS: Spider mites, including *Tetranychus* species and *Oligonychus perseae*.

TARGET CROPS: Recommended for use in a wide range of protected and outdoor crops. Particularly useful for control of the avocado or persea mite, *Oligonychus perseae*, on crops such as avocados, citrus and fruit trees.

BIOLOGICAL ACTIVITY: Biology: Eggs are laid near a food source on the surface of leaves. These hatch into larvae with three pairs of legs; these larvae do not eat. The protonymph emerges from the larva and begins to feed. The adults emerge from the nymphal stage, mate within a few hours and the females then begin to lay eggs. *Galendromus* species are slower-acting predators than *Phytoseiulus persimilis* Athios-Henriot, but all stages are able to withstand wider ranges of temperature and are more tolerant of starvation. The adults are able to survive, but not breed, on pollen. *G. annectens* is well adapted to penetrate the thick web laid down by *O. perseae* to protect its eggs and can, thereby, lay its eggs very close to the phytophagous mite eggs. **Efficacy:** *G. annectens* is a well-adapted predator that feeds on all stages of the pest mites. It is particularly tolerant of high temperatures and will survive for relatively long periods in the absence of mites as a food source. *G. annectens* is particularly useful for the control of 'difficult' mites, such as *O. perseae*.

COMMERCIALISATION: Formulation: Usually supplied as active predators (adults and protonymphs) plus some eggs. The product will usually contain some corn meal cob or bran plus some host mite eggs as a food source. **Tradenames:** 'Galendromus annectens' (American Insectaries), (Applied BioPest), (Arbico), (BioSmith Pest Management), (Biotactics), (Buena Biosystems), (BugLogical) and (Rincon-Vitova).

APPLICATION: Release at a rate of about 10 predators per square metre every 2 weeks in glasshouses. Two or three applications should be sufficient. Outside, rates of between 12 000 and 50 000 adults per hectare every 2 weeks should be used. Again, two to three applications

3. Macro-organisms

should be sufficient. It is suggested that at least 100 predators should be applied to each tree in an orchard or grove that is attacked by *O. perseae*. Application can be made by placing the adult/egg/bran product mixture in a bag near a few active mite colonies. As soon as the predators emerge from the bag, they will begin to attack the mites. It usually takes about 3 months for the predatory mite to move over the entire tree. It is common in protected crops to apply in conjunction with a faster-acting predator, such as *Phytoseiulus persimilis*, particularly if mite infestations are high.

PRODUCT SPECIFICATIONS: Storage conditions: Do not expose to extremes of temperature or to direct sunlight. Keep cool, but do not freeze. **Shelf-life:** As *Galendromus annectens* can survive in the absence of a food source, it may be kept for several days before release.

COMPATIBILITY: Avoid the application of chemical pesticides for one week before and one week after release of the predators. *G. annectens* will survive applications of acaricides such as propargite, once it is established.

MAMMALIAN TOXICITY: There are no records of allergic or other adverse effects from *Galendromus annectens* in research workers, production or field staff. It is regarded as being non-toxic to mammals.

ENVIRONMENTAL IMPACT AND NON-TARGET TOXICITY: *Galendromus annectens* occurs in Nature and is not expected to have any adverse effects on non-target organisms or on the environment. It is specific to spider mites. **Approved for use in organic farming:** Yes.

3:310 *Galendromus helveolus*
Predator of spider mites

Predatory mite: Acari: Phytoseiidae

NOMENCLATURE: Approved name: *Galendromus helveolus* (Chant). **Other names:** Formerly known as: *Typhlodromus helveolus* Chant.

SOURCE: *Galendromus helveolus* is found throughout Florida, Texas, Mexico and Central America, and has been widely introduced into other states of the USA.

PRODUCTION: Reared on persea mite (*Oligonychus perseae* Tuttle, Baker & Abbatiello) under controlled conditions.

TARGET PESTS: The primary target pest is the persea mite (*Oligonychus perseae*), but *G. helveolus* will also prey upon the avocado brown mite (*Oligonychus punicae* (Hirst)) and the six-spotted mite (*Eotetranychus sexmaculatus* (Riley)).

TARGET CROPS: Avocado and citrus orchards.

BIOLOGICAL ACTIVITY: Biology: *Galendromus helveolus* has five life stages: egg, larva, protonymph, deutonymph and adult, with the rate of development being dependent upon the temperature. It completes a generation in 1 to 2 weeks depending on the average temperature (12.6 days at 18 °C (64 °F), 4.8 days at 33 °C (90 °F)). Females lay about 2 eggs per day for 2 weeks and live for about 30 days. They are well adapted to the persea mite and can penetrate the dense protective webbing formed by the persea mite to protect its eggs and can, thereby, lay their eggs very close to the phytophagous mite eggs. *Galendromus helveolus* will feed on all stages of the mites, but prefers eggs and protonymphs and, in the absence of food, females will feed on their own eggs. **Predation:** The nymphal stages consume eggs, larvae and protonymphs of the mite and adults eat all stages of the prey.

COMMERCIALISATION: Formulation: *Galendromus helveolus* is usually shipped in plastic vials containing a carrier such as corn cob meal, a small number of host mite eggs as a food source and active predators plus predator eggs. **Tradenames:** 'Galendromus helveolus' (Applied BioPest), (Arbico), (Biotactics), (BioSmith Pest Management), (Buena Biosystems), and (Rincon-Vitova).

APPLICATION: Releases of *Galendromus helveolus* can be made when persea mites are present and laying eggs at a rate of 12 000 per hectare (5000 per acre). Approximately 100 predators should be placed on each tree in the grove. Place a small paper bag around a cluster of several leaves at the end of a branch and staple the bag to the leaves. Place the predators in the bag. The predators crawl onto the leaves and begin feeding on the persea mites. They take about 3 months to spread around an avocado tree from a single release site.

PRODUCT SPECIFICATIONS: Storage conditions: May be stored in cool, dark conditions. **Shelf-life:** May be kept for 2 or 3 days, if stored according to the manufacturers' recommendations.

COMPATIBILITY: Use of pesticides one week prior to and one week after release of the predators should be avoided. *Galendromus helveolus* will tolerate aerial treatments of propargite reasonably well, once they become established. Ground applications of pesticides that completely cover all the leaves on the tree will probably cause substantial mortality of the predator.

MAMMALIAN TOXICITY: There are no records of allergic or other adverse effects from *Galendromus helveolus* in research workers, production or field staff. It is regarded as being non-toxic to mammals.

ENVIRONMENTAL IMPACT AND NON-TARGET TOXICITY: *Galendromus helveolus* occurs in Nature and is not expected to have any adverse effects on non-target organisms or on the environment. It is specific to spider mites. **Approved for use in organic farming:** Yes.

3:311 *Galendromus occidentalis*
Predator of spider mites

Predatory mite: Acari: Phytoseiidae

NOMENCLATURE: Approved name: *Galendromus occidentalis* (Nesbitt).
Other names: Mite predator. Formerly known as: *Typhlodromus occidentalis* Nesbitt; *Metaseiulus occidentalis* (Nesbitt).

SOURCE: Originally a Nearctic species, but now a widespread mite predator found primarily in regions of high temperature and humidity.

PRODUCTION: Reared on phytophagous mites under controlled conditions.

TARGET PESTS: *Tetranychus* species of red spider mite, particularly *T. urticae* Koch and *T. cinnabarinus* (Boisduval).

TARGET CROPS: Glasshouse and outdoor vegetables and ornamentals, grape vines and nut crops.

BIOLOGICAL ACTIVITY: Biology: Eggs are laid near a food source on the surface of leaves. These hatch into larvae with three pairs of legs; these larvae do not eat. The protonymph emerges from the larva and begins to feed. The adults emerge from the nymphal stage, mate within a few hours and the females then begin to lay eggs. *Galendromus occidentalis* is a slower-acting predator than *Phytoseiulus persimilis* Athios-Henriot, but all stages are able to withstand wider ranges of temperature and are more tolerant of starvation. **Predation:** The nymphal stages consume eggs, larvae and protonymphs of the spider mite and adults eat all stages of the prey. Adult *G. occidentalis* consume between 1 and 3 adult mites or 6 eggs per day. **Efficacy:** *Galendromus occidentalis* is a very versatile predator of spider mites and is well adapted to high temperature and humidity. It can survive in the absence of mites as a food source.

COMMERCIALISATION: Formulation: Sold as adults. **Tradenames:** 'Galendromus occidentalis' (Arbico), (Biocontrol Network), (IPM Laboratories), (Praxis) and (Rincon-Vitova).

APPLICATION: Release at a rate of about 10 predators per square metre every 2 weeks in glasshouses. Between two and three applications should be sufficient. Outside, rates of between 12 000 and 50 000 adults per hectare every 2 weeks should be used. Again, two to three applications should be sufficient. It is common to apply in conjunction with a faster-acting predator, such as *Phytoseiulus persimilis*, particularly if mite infestations are high.

PRODUCT SPECIFICATIONS: Storage conditions: Do not expose to extremes of temperature or to direct sunlight. Keep cool, but do not freeze. **Shelf-life:** As *Galendromus occidentalis* can survive in the absence of a food source, it may be kept for several days before release.

COMPATIBILITY: Used in conjunction with other mite predators. Do not use with persistent insecticides or acaricides.

MAMMALIAN TOXICITY: There are no records of allergic or other adverse effects from *Galendromus occidentalis* in research workers, production or field staff. It is regarded as being non-toxic to mammals.

ENVIRONMENTAL IMPACT AND NON-TARGET TOXICITY: *Galendromus occidentalis* occurs in Nature and is not expected to have any adverse effects on non-target organisms or on the environment. It is specific to spider mites. **Approved for use in organic farming:** Yes.

3:312 *Geocoris punctipes*   *Predator of insects*

Predatory bug: Heteroptera: Geocoridae

NOMENCLATURE: Approved name: *Geocoris punctipes* (Say). **Other names:** Big-eyed bug.

SOURCE: Occurs widely in North America.

PRODUCTION: Reared on an artificial diet of ground beef and beef liver, under controlled conditions.

TARGET PESTS: *Geocoris punctipes* is polyphagous and will prey upon aphids, mites, thrips, whitefly, flea beetles, spider mites, insect eggs and small caterpillars.

TARGET CROPS: *Geocoris punctipes* can be used in a wide range of crops, including cotton, soybean, peanut and managed turf.

BIOLOGICAL ACTIVITY: Biology: The adults are small (3 to 4 mm long) insects ranging in colour from black and white to tan. Their eyes are widely separated which gives them a wide field of vision for spotting their prey. The nymphs resemble miniature greyish adults.

Each nymph may consume up to 1600 spider mites during its immature stages, and as many as 80 mites a day as an adult. Other laboratory studies have shown that each *G. punctipes* nymph consumes about 250 soybean looper eggs before it reaches the adult stage. In another laboratory study, adult *G. punctipes* consumed up to four *Lygus* bug eggs per day. In laboratory and field studies, they have been observed to feed on dead insects, although, when given a choice, they prefer live prey. *G. punctipes* also feeds on plants, and maximal survival and reproduction is said to occur when they have a mix of plant and insect food. There is no evidence that this plant feeding causes significant injury to the plant. Like all true bugs, they go through incomplete metamorphosis. Nymphs are oval and somewhat flattened and they lack functional wings. Nymphal stages have similar behaviour and feeding habits to adults, but tend to feed on smaller prey. **Predation:** *Geocoris punctipes* has piercing, sucking mouthparts and usually attacks prey simply by approaching it, extending its beak and quickly inserting it into the prey. *Geocoris punctipes* may lift prey into the air, preventing it from attempting escape by running. The adult may drop to the ground if disturbed. **Egg laying:** *Geocoris punctipes* eggs are laid singly, usually on the leaf surface. Eggs are oblong and pale-coloured and develop a reddish eyespot shortly after being laid. **Duration of development:** *Geocoris punctipes* goes through five nymphal stages, with nymphal development taking about 30 days at 25 °C and 60 days at 20 °C. It is reported that *Geocoris punctipes* overwinter as adults or as eggs depending on the location. **Efficacy:** A very effective predator of many insect pest sepcies.

COMMERCIALISATION: Formulation: *Geocoris punctipes* adults and nymphs are sold in vials. **Tradenames:** 'Geocoris punctipes' (Arbico), (Biotactics) and (Rincon-Vitova).

APPLICATION: Release 100 adults or nymphs into a maximum area of 4000 sq. m. (44 000 sq. ft.) as prey population increases.

PRODUCT SPECIFICATIONS: Purity: Contains adults and nymphs of *Geocoris punctipes* and no impurities. **Storage conditions:** Use as soon as possible after receipt.

COMPATIBILITY: The need for *Geocoris punctipes* to feed on plant tissue renders it susceptible to systemic insecticides. When prey species are scarce, *Geocoris punctipes* may feed on other beneficial species.

MAMMALIAN TOXICITY: No allergic or other adverse reactions have been recorded from the production of *Geocoris punctipes* or its use under field conditions.

ENVIRONMENTAL IMPACT AND NON-TARGET TOXICITY: *Geocoris punctipes* occurs widely in Nature and has not been shown to have any adverse effects on non-target organisms or on the environment. It is polyphagous and will prey upon a wide range of insect species, including other beneficials. **Approved for use in organic farming:** Yes.

3:313 *Glyptapanteles liparidis*

Parasitoid of gypsy moth

Parasitoid wasp: Hymenoptera: Braconidae

NOMENCLATURE: Approved name: *Glyptapanteles liparidis* (Marsh).

SOURCE: Imported into the USA from India.

PRODUCTION: Reared on gypsy moth (*Lymantria dispar* (L.)) larvae.

TARGET PESTS: Gypsy moth (*Lymantria dispar*).

TARGET CROPS: Ornamental and amenity trees.

BIOLOGICAL ACTIVITY: Biology: Female *Glyptapanteles liparidis* actively seek larvae of the gypsy moth and lay eggs within the larval body. The preferred stages of larvae are third and fourth instar, although other stages may be parasitised if caterpillar populations are low. Sex determination is arrhenotokous, a condition in which fertilised (diploid) eggs give rise to female progeny and unfertilised (haploid) eggs male progeny.

COMMERCIALISATION: Formulation: Sold as parasitised larvae.
Tradenames: 'Glyptapanteles liparidis' (Praxis).

APPLICATION: Release in winter and spring when gypsy moth adults are first seen.

PRODUCT SPECIFICATIONS: Purity: No contaminating insects. **Storage conditions:** Keep cool, but do not refrigerate. Keep out of direct sunlight. **Shelf-life:** Release immediately.

COMPATIBILITY: Avoid the use of residual insecticides for 4 weeks prior to release.

MAMMALIAN TOXICITY: There have been no reports of allergic or other adverse reactions by research workers or field staff following use.

ENVIRONMENTAL IMPACT AND NON-TARGET TOXICITY: *Glyptapanteles liparidis* occurs widely in Nature and is not expected to have any adverse effects on non-target organisms or on the environment. Its preferred host is gypsy moth larvae, although it may prey on related caterpillars.

3. Macro-organisms

3:314 *Goniozus legneri*
Parasitoid of navel orangeworm

Parasitoid wasp: Hymenoptera: Bethylidae

NOMENCLATURE: Approved name: *Goniozus legneri* Gordh. **Other names:** Navel orangeworm parasitoid.

SOURCE: *Goniozus legneri* was introduced to California from South America for the biological control of navel orangeworm.

PRODUCTION: Reared on navel orangeworm (*Amylois transitella* (Walker)) larvae.

TARGET PESTS: Navel orangeworm (*Amylois transitella*).

TARGET CROPS: Soft-shelled almond, walnut and pistachio nut orchards.

BIOLOGICAL ACTIVITY: Biology: *Goniozus legneri* adults are streamlined to enable them to crawl into small spaces to locate hidden navel orangeworm larvae. They will attack all stages of larvae; the females sting and then lay eggs on the larva, which serves as food for the developing parasitoids. Navel orangeworm larvae that are stung by *Goniozus legneri* are immediately paralysed and no longer feed. After the *Goniozus legneri* eggs hatch, the externally developing parasitoid larvae feed on the paralysed navel orangeworm larva. As the parasitoid larvae grow, the entire parasitised larva is consumed, with the exception of the head capsule. Fully developed *Goniozus legneri* larvae spin white, loosely woven cocoons that are usually found in flocculent groups. New adult parasitoids emerge from the pupae inside the cocoons. The rate of development by the parasitoid is governed by temperature. Development is most rapid during the summer generations. **Egg laying:** *Goniozus legneri* females attach between 5 and 15 eggs, depending on the size of the host, on the outside of each larva that is stung. Each female parasitoid can produce about 150 eggs over a lifetime of some 3 months. Because the parasitoid develops about twice as quickly as the navel orangeworm, two generations of parasitoids can be produced on a single generation of navel orangeworms during the warm season. **Duration of development:** Parasitoid pupae are formed inside the cocoons and adult *Goniozus legneri* emerge from the cocoons about 5 to 10 days later during the warm season. Male *Goniozus legneri* emerge from the cocoons about 12 hours before the female parasitoids. The male parasitoids chew into the cocoons and mate with the unemerged females. Recent studies have shown that few *Goniozus legneri* successfully overwinter in almond mummies. Careful sanitation to remove mummified nuts may reduce overwintering *Goniozus legneri* populations. **Efficacy:** Field studies have shown that *Goniozus legneri* readily locates, attacks and reproduces on navel orangeworm larvae in orchards, and efficacy studies have shown that the parasitoid has a regulatory impact on these populations.

Key reference(s): 1) E F Legner & G Gordh. 1992. Low navel orangeworm (Lepidoptera: Phycitidae) population densities following establishment of *Goniozus legneri* (Hymenoptera: Bethylidae) in California, *J. Econ. Entomol.*, **85**(6), 2153–60. 2) BIOS for Almonds. *A Practical Guide to Biologically Integrated Orchard Systems Management*, Community Alliance with Family Farmers Foundation and the Almond Board of California, USA, 104 pp. 3) *Almond Pest Management Guidelines*, University of California Pest Management Guidelines, USA Publication 1, 33 pp.

COMMERCIALISATION: Formulation: *Goniozus legneri* pupae inside cocoons with honey as food for emerging adult parasitoids. **Tradenames:** 'Goniozus legneri' (M&R Durango), (Rincon-Vitova) and (Praxis).

APPLICATION: Releases of at least 2400 *Goniozus* per hectare (1000 per acre) each year are recommended, regardless of sanitation practices. Adult *Goniozus* are preferred for release, because ants and earwigs have proved to be voracious predators of *Goniozus* when released as pupae in cocoons. The first *Goniozus* releases should be initiated from late February and March, before the first moth flight, and followed up with additional releases during May and June. The objective is to increase navel orangeworm mortality prior to the third moth flight. A minimum of 1200 *Goniozus legneri* per hectare (500 per acre) should be released during the February–March period. The subsequent releases during May and June should result in a minimum total of 2400 *Goniozus* per hectare (1000 per acre) released prior to harvest. These release periods can vary with meteorological and environmental conditions. Packets containing adult *Goniozus* can be opened and placed directly in release trees when, or just before, temperatures are conducive to adult parasitoid activity (about 21 °C (70 °F)). Newly emerged *Goniozus legneri* will sting and paralyse navel orangeworm larvae, but the parasitoids do not lay eggs until they are about 3 days old. Releases during temperature extremes should be avoided.

PRODUCT SPECIFICATIONS: Purity: Contains only *Goniozus legneri* pupae inside cocoons, with honey as food. **Storage conditions:** Packets should be held at about 25 °C (80 °F) and 50% r.h. until all adult *Goniozus* have emerged from the cocoons. Male parasitoids emerge about 12 hours before the females. Adult male parasitoids are good indicators of when to expect all adult *Goniozus* to emerge. The parasitoids should emerge within 1 to 3 days of receipt, when held under the recommended conditions. **Shelf-life:** Use as soon as adults have emerged from cocoons.

COMPATIBILITY: Incompatible with persistent, broad-spectrum insecticides.

MAMMALIAN TOXICITY: There have been no reports of allergic or other adverse reactions by research workers or field staff following use.

ENVIRONMENTAL IMPACT AND NON-TARGET TOXICITY: *Goniozus legneri* occurs widely in Nature and is not expected to have any adverse effects on non-target organisms or on the environment. It is specific to navel orangeworm larvae.

3:315 *Harmonia axyridis*   *Predator of aphids*

Predatory ladybird: Coleoptera: Coccinellidae

NOMENCLATURE: Approved name: *Harmonia axyridis* Pallas. **Other names:** Predatory ladybird; Chinese ladybird; harlequin ladybird, multicoloured Asian ladybeetle.

SOURCE: Originated from Asia (probably China), but now introduced into North America and Europe.

PRODUCTION: Reared in insectaries under controlled conditions on eggs of moths, such as *Ephestia kuehniella* Zeller, or on aphids.

TARGET PESTS: Many species of aphid.

TARGET CROPS: A wide variety of glasshouse crops and ornamentals.

BIOLOGICAL ACTIVITY: Biology: As soon as larvae emerge from eggs, they begin to consume aphids and they continue until pupation. Released larvae are active at temperatures above 11 to 12 °C. In the absence of food, the larvae can become cannibalistic, but adults can survive some days without food. **Predation:** Both adults and larvae consume aphids, but they may also prey on other insects, such as scale and lepidopteran eggs. **Egg laying:** Adult females will lay approximately 20 eggs a day throughout a lifetime of 2 to 3 months. **Duration of development:** At 25 °C, it takes 15 to 20 days from egg hatch to adult. Adults live for several months. **Efficacy:** Larvae are very mobile and prospect widely for prey from their release point. They reduce aphid populations very quickly, being comparable to chemical treatments.

COMMERCIALISATION: Formulation: Sold as larvae in boxes containing a food source (usually *Ephestia* eggs) to allow development during transportation. **Tradenames:** 'Harmonia' (Biotop), 'Harmonia-System' (Biobest), 'Harmoline a' (Syngenta Bioline), 'Harmonia axyridis' (Applied Bio-Nomics) and (Rincon-Vitova).

APPLICATION: Release second or third instar larvae on infested plants. Rate of release depends upon plant size and the extent of the aphid infestation.

PRODUCT SPECIFICATIONS: Purity: Only larvae of *Harmonia axyridis*, with no contaminants. **Storage conditions:** Store in the dark, under cool conditions. Do not expose to bright sunlight. **Shelf-life:** Can be kept for several days, if stored under recommended conditions.

COMPATIBILITY: Incompatible with foliar insecticides.

MAMMALIAN TOXICITY: No allergic or other adverse reactions have been reported from the use of *Harmonia axyridis* in glasshouses or outside conditions.

ENVIRONMENTAL IMPACT AND NON-TARGET TOXICITY: *Harmonia axyridis* occurs widely in Nature and would not be expected to have any significant effects on non-target organisms or on the environment. However, where introduced, this species is considered to be an undesirable alien invader as it out-competes native ladybirds and is considered to be a threat to them. **Approved for use in organic farming:** *Harmonia axyridis* was initially sold in Europe and North America as a commercial biological control agent of aphids and coccids and, although effective in controlling a range of pest insects in a variety of crops, it is now classified as an invasive alien species. *H. axyridis* is a voracious polyphagous predator out of its native range, which displaces native aphidophagous species through direct competition resulting in biodiversity declines. It has been present in the United Kingdom since 2004 and, with populations increasing rapidly, a national survey website (www.harlequin-survey.org) has been launched to monitor its spread and the possible impacts on native ladybird species.

3:316 *Hippodamia convergens*
Predator of insects

Predatory ladybird: Coleoptera: Coccinellidae

NOMENCLATURE: Approved name: *Hippodamia convergens* Guérin. **Other names:** Ladybird; ladybug beetle; ladybird beetle.

SOURCE: Native to North America.

PRODUCTION: Usually collected in the wild, rather than insectary-reared.

TARGET PESTS: Polyphagous, eating a wide variety of insects, including aphids, beetles, chinch bugs, whiteflies and red spider mites, as well as many other soft-bodied insects and eggs.

TARGET CROPS: Many outdoor and glasshouse crops.

BIOLOGICAL ACTIVITY: Biology: Adults are about 5 mm in length, orange-brown in colour, with black spots and white stripes on the head shield. They mate and lay large numbers of oval, orange eggs in clusters on the underside of leaves. Eggs hatch into black larvae with orange spots. Larvae grow from about 1 mm to 5 to 6 mm in length and may wander up to 12 metres in search of prey. After 21 days, the larva attaches itself by the abdomen to a leaf or other surface to pupate and adults emerge in 2 to 8 days, depending on the temperature. **Predation:** Adults and larvae eat insects. *Hippodamia convergens* will consume about 400 aphids in its larval stages and over 5000 as an adult. **Egg laying:** Eggs are usually deposited near prey such as aphids, often in small clusters in protected sites under leaves, shortly after mating. **Duration of development:** At optimum temperatures, the life-cycle takes about 30 days. The adults can live for 3 months. **Efficacy:** Very efficient, polyphagous insect. It requires a source of nectar or pollen to mature to the adult stage and this must be provided if not available within the crop. Most effective under cool conditions. Mobile insects that often migrate from the site of application.

COMMERCIALISATION: Formulation: Adults collected in the wild and sold packed in a suitable substrate such as wood-wool. **Tradenames:** 'Ladybugs' (Kunafin), 'Hippodamia convergens Lady Beetle' (M&R Durango), 'Hippodamia System' (Biobest), 'Hippodamia convergens (Ladybugs) Aphid Destroyer' (Nature's Alternative Insectary), 'Ladybird Beetle' (Arbico), 'Ladybug' (BioPac), 'Lady Beetle' (Biocontrol Network), 'Aphidamia' (Koppert), 'Hippodamia convergens' (Rincon-Vitova), (Harmony), (Hydro-Gardens), (Praxis) and (IPM Laboratories).

APPLICATION: Release at a rate of between 180000 and 500000 beetles per hectare. An early evening release is preferred. If used in glasshouses, cover openings to prevent escape. If used in the field, the wings can be sealed together by sugar solutions as a temporary measure to encourage feeding and egg laying and reduce migration. Ensure there is moisture in the release environment.

PRODUCT SPECIFICATIONS: Purity: Dependent upon harvesting site.
Storage conditions: Store in a refrigerator for up to 3 months. Do not expose to direct sunlight. **Shelf-life:** Remain active for up to 3 months, if stored under suitable conditions. Release within 2 days, if stored at room temperature.

COMPATIBILITY: Do not use residual foliar insecticides within 1 month of release.

MAMMALIAN TOXICITY: No allergic or other adverse reactions have been reported following their release in protected crops or outdoors.

ENVIRONMENTAL IMPACT AND NON-TARGET TOXICITY: *Hippodamia convergens* occurs widely in Nature and is unlikely to show any adverse effects on non-target organisms or on the environment. It is polyphagous and will consume a wide variety of different insects, including other beneficials. It is able to overwinter in cool climates, if shelter is provided.
Approved for use in organic farming: Yes.

3:317 *Hyles euphorbiae*
Phytophagous caterpillar

Phytophagous caterpillar: Lepidoptera: Sphingidae

NOMENCLATURE: Approved name: *Hyles euphorbiae* (L.). **Other names:** Leafy spurge hawk moth.

SOURCE: Common in Europe, particularly in leafy spurge infested pastures. Imported into the USA in 1965.

PRODUCTION: Reared on *Euphorbia* plants in the subgenus *Esula.*

TARGET PESTS: In the USA, the target pests are the leafy spurges, *Euphorbia esula* L. and *E. cyparissias* L., but, in Europe, several species of *Euphorbia* are attacked.

TARGET CROPS: Rangelands.

BIOLOGICAL ACTIVITY: Biology: *Hyles euphorbiae* larvae are found feeding on leafy spurge leaves. The caterpillars are conspicuously coloured, with a pronounced tail or 'horn' near the rear end. Young larvae are variously patterned with green, yellow and black; older larvae have a distinctive red, black, yellow and white colour pattern. Mature larvae may approach 10 cm in length; when disturbed, they regurgitate a slimy green liquid. Pupae are 3.5 to 5 cm long and dark brown, and are found in the soil. Adult moths are large (length: 2 to 3 cm, wingspan: 5 to 7 cm) day-flying moths that often exhibit a hummingbird-like flight while visiting flowers. The body is light brown with various white and dark brown markings, while the wings have a conspicuous tan, brown and pink or red colour pattern. **Predation:** Adult moths are present beginning in early to mid-summer. After mating, females lay small clusters of eggs on leafy spurge foliage. After hatching, larvae consume leafy spurge leaves and flowers. Mature larvae enter the soil to pupate. There are one or two generations per year, with soil-inhabiting pupae as the overwintering stage. **Key reference(s):** 1) S W T Batra. 1983. Establishment of *Hyles euphorbiae* (L.) (Lepidoptera: Sphingidae) in the United States for control of two weedy spurges, *Euphorbia esula* L. and *E. cyparissias* L., *J. N.Y. Entomol. Soc.,* **91**, 304–11. 2) N E Rees & P K Fay. 1989. Survival of leafy spurge hawk moths (*Hyles euphorbiae*) when exposed to 2,4-D or picloram, *Weed Technol.,* **3**, 429–31.

COMMERCIALISATION: Formulation: Supplied as pupae. **Tradenames:** 'Hyles euphorbiae' (Praxis).

APPLICATION: Pupae should be distributed amongst the spurge plants in late spring to early summer.

PRODUCT SPECIFICATIONS: Purity: Supplied as pupae, with no contaminants. **Storage conditions:** Distribute immediately after receipt.

COMPATIBILITY: Compatible with broad-leaved herbicides such as 2,4-D.

MAMMALIAN TOXICITY: There have been no reports of allergic or other adverse reactions from research workers, manufacturing staff or from the field release of *Hyles euphorbiae.*

ENVIRONMENTAL IMPACT AND NON-TARGET TOXICITY: *Hyles euphorbiae* occurs widely in Nature and is not expected to have any adverse effects on non-target organisms or on the environment. It will not feed on poinsettia (*Euphorbia pulcherrima* Willd.). Crop species and native plants outside the genus *Euphorbia* will not be attacked.

3. Macro-organisms

3:318 *Hypoaspis aculeifer*

Predator of sciarid flies

Predatory mite: Mesostigmata: Laelapidae

NOMENCLATURE: Approved name: *Hypoaspis aculeifer* (Canestrini). **Other names:** Sciarid fly predator; fungus gnat predator; thrips predator. Formerly known as: *Stratiolaelaps miles* (Berlese).

SOURCE: *Hypoaspis aculeifer* is cosmopolitan and occurs widely in Nature in Europe and North America. It is a soil-dwelling predatory mite and is often found in association with decaying plants.

PRODUCTION: Reared commercially on sciarid flies in insectaries.

TARGET PESTS: Main prey are sciarid flies (fungus gnats) (*Bradysia* spp.) and bulb mites (*Rhizoglyphus robini* Claperède), although other soil-inhabiting insects, such as thrips (*Frankliniella occidentalis* Pergande) pupae, collembola and nematodes may also be preyed upon.

TARGET CROPS: Glasshouse vegetable crops and ornamentals including bulbs.

BIOLOGICAL ACTIVITY: Biology: A predatory mite that feeds on a wide range of insect, mite and nematode species in the soil. *Hypoaspis aculeifer* lays its eggs in the soil and these hatch into motile larval stages. There are three immature stages, reddish-brown in colour, and these will feed upon soil arthropods, but eat less than the adults. The adult mite reaches a length of about 1 mm. The mite does not enter diapause. **Egg laying:** An adult female will lay up to 87 eggs in her lifetime at optimal temperatures (22 °C).
Duration of development: The ideal temperature range for the mite is between 17 and 26 °C and development stops when the temperature falls below 7 or 8 °C.
Efficacy: *Hypoaspis aculeifer* is very effective at controlling a range of soil insects and mites. It will reduce thrips populations by predating the pupae when they fall to the ground.

COMMERCIALISATION: Formulation: Sold as a vermiculite formulation with 10 000 predatory mites (all stages) in a one-litre bottle. **Tradenames:** 'Entomite' (Koppert), 'Aculeifer-System' (Biobest), 'Hypoaspis aculifer' (Kuida).

APPLICATION: Turn and shake the bottle gently before use. Spread material carefully and evenly on to the soil or rockwood blocks. Recommended rates of use are 100 mites per square metre as a preventive treatment and 200 to 250 mites per square metre as a curative treatment. Apply once per season.

PRODUCT SPECIFICATIONS: Purity: No phytophagous mites, mould or other contaminants. **Storage conditions:** Store at 10 to 15 °C in the dark. **Shelf-life:** Use within 3 days of receipt, if stored under recommended conditions.

COMPATIBILITY: *Hypoaspis aculeifer* is sensitive to most conventional pesticides.

MAMMALIAN TOXICITY: No allergic or other adverse reactions have been reported following the use of *Hypoaspis aculeifer* in glasshouse crops by producers, research workers or growers.

ENVIRONMENTAL IMPACT AND NON-TARGET TOXICITY: *Hypoaspis aculeifer* has a wide geographical distribution and there is no evidence of adverse effects on non-target organisms or on the environment. It feeds primarily on sciarid flies, although other soil-dwelling insects and mites may be consumed. It is a soil-inhabiting species and, therefore, will not move far from its point of introduction. It is able to overwinter in temperate climates. **Approved for use in organic farming:** Yes.

3:319 *Hypoaspis miles*   *Predator of sciarid flies*

Predatory mite: Mesostigmata: Laelapidae

NOMENCLATURE: Approved name: *Hypoaspis miles* (Berlese). **Other names:** Fungus gnat predator; fungus fly predator; sciarid fly predator. Formerly known as: *Geolaelaps miles* Berlese; and *Stratiolaelaps miles* (Berlese).

SOURCE: Originally found in a decaying oat spillage at Leith, Scotland. Recorded in the literature from a variety of habitats in Europe, the former Soviet Union and the USA (Palaearctic). It is a soil-dwelling predatory mite and is often found in association with decaying plants.

PRODUCTION: Produced in mixtures of peat and vermiculite and fed on grain mites, such as *Tyrophagus putrescentriae* (Schrank). Supplied as mixed stages in moist carrier.

TARGET PESTS: Sciarid flies (fungus gants) (*Bradysia* spp.), springtails, mites, thrips pupae and nematodes.

TARGET CROPS: Glasshouse-grown vegetables and ornamentals and, in some cases, in interiorscapes.

BIOLOGICAL ACTIVITY: Biology: A predatory mite that feeds on a wide range of insect and mite species in the soil. *Hypoaspis miles* lays its eggs in the soil and these hatch into motile larval stages. These immature stages will feed upon soil arthropods, but eat less than the adults. The adult mite reaches a length of about 1 mm. **Predation:** One adult *Hypoaspis miles* can kill up to 7 sciarid fly larvae a day. The juvenile stages also predate, but consume fewer larvae. Adults can survive for long periods without feeding, but will not reproduce in the absence of food. They can survive on algae and plant debris when insect prey is not available. **Egg laying:** Female mites lay eggs in the soil and these hatch within 1 to 2 days at temperatures of 25 °C. The nymphs become adults within 5 to 6 days and inhabit the upper 1 to 2 cm of the soil surface. Development is slower at lower temperatures and no development occurs at all at 10 °C and lower. Fertilised eggs develop into females and unfertilised eggs develop into males. **Duration of development:** The ideal temperature range for *Hypoaspis miles* is 17 to 26 °C. Development stops at temperatures below 7 to 8 °C. *Hypoaspis miles* completes its life-cycle in 7 to 11 days. **Efficacy:** *Hypoaspis miles* is very effective at controlling a range of soil insects and mites. It will reduce thrips populations by predating the pupae when they fall to the ground.

COMMERCIALISATION: Formulation: Mixed populations of growth stages supplied in peat and/or vermiculite. One-litre bottles contain 10 000 predatory mites in the carrier. **Tradenames:** 'Hypoline m' (Syngenta Bioline), 'Hypoaspis Sciarid Fly Predators' (Biowise), 'Hyposure (m)' (Biological Crop Protection), 'Geolaelaps sp. (= Hypoaspis sp.)-Fungus Gnat Destroyer' (Nature's Alternative Insectary), 'Entomite' (Koppert), 'Hypex' (Svenska Predator), 'Hypoaspis-System' (Biobest), 'Hypoaspis miles' (Applied Bio-Nomics), (Arbico), (Buena Biosystems), (Biological Services), (BugLogical), (IPM Laboratories), (Neudorff) and (Rincon-Vitova), 'Hypoaspis sp.' (Biocontrol Network).

APPLICATION: Ensure that the soil is wet before *Hypoaspis miles* is added. Turn and shake the bottle gently before applying at the rate of 25 *Hypoaspis miles* for preventive treatment and 55 for curative treatment per square metre of growing medium or bench area. Rates of 150 to 250 per square metre are recommended for high populations of pests. Apply evenly to the soil of glasshouse-grown crops. A single application should be sufficient to establish a population for the season.

PRODUCT SPECIFICATIONS: Purity: Mixed population of different growth stages plus some soil insects as a food source. No phytophagous mites, mould or other contaminants. **Storage conditions:** Hold at room temperature and do not cool. Store in the dark. **Shelf-life:** Use as soon as possible after delivery, but may be kept for 3 days, if stored as recommended.

COMPATIBILITY: Do not use soil insecticides. Killed by freezing conditions, but well adapted to moist soil. *Hypoaspis miles* can survive in dry soil. Can be used in conjuction with other biologicals, including *Orius* spp., *Chrysoperla carnea* (Stephens), *Amblyseius cucumeris* (Oudemans), beneficial nematodes and *Bacillus thuringiensis subsp. israeliensis* Berliner.

MAMMALIAN TOXICITY: Allergic reactions have been recorded following use, although this is thought to be caused by the prey rather than *Hypoaspis miles*.

ENVIRONMENTAL IMPACT AND NON-TARGET TOXICITY: *Hypoaspis miles* occurs widely in Nature and is not expected to have any adverse effects on non-target organisms or on the environment. It feeds primarily on sciarid flies, although other soil-dwelling insects and mites may be consumed. It is a soil-inhabiting species and, therefore, will not move far from its point of introduction. It is able to overwinter in temperate climates.

Approved for use in organic farming: Yes.

3:320 *Leptomastix dactylopii*
Parasitoid of mealybugs

Parasitoid wasp: Hymenoptera: Encyrtidae

NOMENCLATURE: Approved name: *Leptomastix dactylopii* Howard. **Other names:** Chalcid mealybug parasitoid.

SOURCE: Native to South America, probably Brazil, and introduced into California in 1934. It has since spread around the world.

PRODUCTION: Reared on its only known host, the citrus mealybug (*Planococcus citri* (Risso)).

TARGET PESTS: Citrus mealybug (*Planococcus citri*).

TARGET CROPS: Ornamentals and vegetables in glasshouses and in interiorscapes. Also released to protect citrus plantations from attack.

BIOLOGICAL ACTIVITY: Biology: Females lay eggs in third instar mealybug nymphs, as their ovipositors often pass through younger instars. The egg hatches and the larva goes through four larval stages before pupating within the mummified mealybug. The adult emerges through a round hole in the mummy. Unfertilised females lay eggs which develop only into males. **Predation:** The mealybug into which an egg is laid is consumed by the developing wasp larva. **Egg laying:** Under ideal conditions, a female wasp will lay between 60 and 100 eggs in 10 to 14 days. It is unusual for more than a single egg to be laid within any one mealybug. **Duration of development:** The duration of development is dependent upon temperature, varying from about 45 days at 17 °C to 12 days at 30 °C. **Efficacy:** The adult parasitoid wasp is a very good flyer and has excellent searching ability. Even low wasp densities can control mealybug populations.

COMMERCIALISATION: Formulation: The parasitoid is supplied as both adults and pupae within the mummified mealybug. **Tradenames:** 'Leptopar' (Koppert), 'Mealybug Parasite' (Arbico) and (Biocontrol Network), 'Leptomastix' (Sautter & Stepper), 'Leptosure (d)' (Biological Crop Protection), 'Leptomastix-System' (Biobest), 'Leptomastix dactylopii' (Bugs for Bugs), (IPM Laboratories), (Neudorff), (Praxis) and (Rincon-Vitova), 'Leptoline d' (Syngenta Bioline).

APPLICATION: Release between 1 and 2 wasps per square metre or up to 5 wasps per plant if pest populations are high. Release weekly for 4 to 6 weeks, twice a year.

PRODUCT SPECIFICATIONS: Purity: No contaminating insects should be present. **Storage conditions:** Do not expose to direct sunlight and do not freeze. Keep cool if not used immediately. **Shelf-life:** Use as soon as possible after delivery.

COMPATIBILITY: Do not use residual insecticides. Often used in combination with *Cryptolaemus montrouzieri* Mulsant.

MAMMALIAN TOXICITY: No allergic or other adverse reactions have been reported from the use of *Leptomastix dactylopii* in glasshouse or field conditions.

ENVIRONMENTAL IMPACT AND NON-TARGET TOXICITY: *Leptomastix dactylopii* occurs widely in Nature and is not thought to pose any threat to non-target organisms or to have an adverse effect on the environment. Its only known host is the citrus mealybug. **Approved for use in organic farming:** Yes.

3:321 *Leucoptera spartifoliella*
Phytophagous caterpillar

Phytophagous caterpillar: Lepidoptera: Lyonetiidae

NOMENCLATURE: Approved name: *Leucoptera spartifoliella* (Hübner). **Other names:** Twig mining moth. Formerly known as: *Tinea punctaurella* Haworth.

SOURCE: Common in Europe. Introduced into the USA and Australia.

PRODUCTION: The moths are reared on scotchbroom (*Sarothamnus scoparius* Wimmer) plants.

TARGET PESTS: Scotchbroom (*Sarothamnus scoparius* (*Cytisus scoparius* Link.)).

TARGET CROPS: Pastures and rangeland.

BIOLOGICAL ACTIVITY: Biology: The adult moth is satin white, with yellow marks near the tips of the forewings. It has a wingspan of about 0.8 cm. The moths emerge in late spring and are most abundant in midsummer. There is one generation per year. There are six larval instars. Development to the third instar normally occurs by early winter with the final instar being found the following spring. The larvae mine up and down the stems of the previous season's growth in the spring. Fully developed mines often reach 50 to 80 cm in length. First and second instar larvae are extremely small. The mines created by these larvae are up to 10 mm long and less than 0.5 mm wide and are almost invisible to the naked eye. At 30 mm long, the mine is 0.5 to 1 mm wide and more readily visible. Most damage to the plant is caused by the fifth and sixth instar larvae, the mines at this stage being raised and approximately 2 mm wide. When larvae have fully developed, they form a cocoon from which the adult moths emerge in the spring. **Efficacy:** Heavy attack by the moth can result in stunted plant growth and death of stems. *Leucoptera spartifoliella* cannot control scotchbroom on its own.

COMMERCIALISATION: Formulation: Sold as pupae. **Tradenames:** 'Leucoptera spartifoliella' (Praxis).

APPLICATION: Releases of adult moths or cut stems containing cocoons can be made. These are generally free released. No insecticides should be sprayed around the immediate release site.

PRODUCT SPECIFICATIONS: Purity: Sold as pupae, with no contaminants.
Storage conditions: May be stored for up to 3 days on leaves out of direct sunlight and under cool conditions.

COMPATIBILITY: Incompatible with persistent, broad-spectrum insecticides.

MAMMALIAN TOXICITY: There have been no reports of allergic or other adverse reactions from research workers, manufacturing staff or from the field release of *Leucoptera spartifoliella*.

ENVIRONMENTAL IMPACT AND NON-TARGET TOXICITY: *Leucoptera spartifoliella* occurs widely in Nature and is not expected to have any adverse effects on non-target organisms or on the environment. Trials in the USA and Australia have demonstrated that the caterpillars do not feed on any species other than scotchbroom.

3. Macro-organisms

3:322 *Longitarsus jacobaeae*

Phytophagous beetle

Root flea beetle: Coleoptera: Chrysomelidae

NOMENCLATURE: Approved name: *Longitarsus jacobaeae* (Waterhouse).
Other names: Tansy ragwort root beetle; ragwort flea beetle.

SOURCE: Occurs commonly in regions where ragwort is present. The flea beetle originated in Europe and has been introduced into the USA and Australia.

PRODUCTION: Reared on common or tansy ragwort (*Senecio jacobaea* L.).

TARGET PESTS: Common or tansy ragwort (*Senecio jacobaea*).

TARGET CROPS: Rangeland and pastures.

BIOLOGICAL ACTIVITY: Biology: The adult beetle is sandy-brown to yellowish in colour and is 2.5 to 3 mm in length. They emerge in summer and lay eggs from early autumn up until the soil surface freezes. The life span of the adults is not known. The beetle overwinters in the soil as both pupae and larvae. The larvae feed on the roots of the ragwort plants, pupating in early to late summer. Ovipositing adults also feed on the ragwort foliage. *Longitarsus jacobaeae* has a search and destroy strategy that eliminates the weed, except for buried seed. Reinfestations occur following ground disturbance and are usually rapidly discovered by the beetle. **Efficacy:** The flea beetle performs well on well-drained soils where there is a dense stand of the weed. Mild winters and low elevations are preferred.
Key reference(s): 1) K E Frick. 1970. *Longitarsus jacobaeae* (Coleoptera: Chrysomelidae), a flea beetle for the biological control of Tansy Ragwort. 1. Host plant specificity studies, *Ann. Entomol. Soc. Am.*, **63**(1), 284–96. 2) K E Frick & G R Johnson. 1973. *Longitarsus jacobaeae* (Coleoptera: Chrysomelidae), a flea beetle for the biological control of Tansy Ragwort. 4. Life history and adult aestivation of an Italian biotype, *Ann. Entomol. Soc. Am.*, **66**(2), 358–67.

COMMERCIALISATION: Formulation: Sold as pupae. **Tradenames:** 'Longitarsus jacobaeae' (Praxis).

APPLICATION: Release pupae in summer.

PRODUCT SPECIFICATIONS: Purity: Contains only *Longitarsus jacobaeae* pupae.
Storage conditions: Release immediately.

COMPATIBILITY: Compatible with herbicides. May be used in conjunction with *Tyria jacobaeae* (L.) for better control of ragwort.

MAMMALIAN TOXICITY: There have been no reports of allergic or other adverse reactions from research workers, manufacturing staff or from the field release of *Longitarsus jacobaeae*.

ENVIRONMENTAL IMPACT AND NON-TARGET TOXICITY: *Longitarsus jacobaeae* occurs widely in Nature and is not expected to have any adverse effects on non-target organisms or on the environment. Trials in the USA and Australia have demonstrated that the beetles do not feed on any commercially valuable or environmentally sensitive species.

3:323 *Lydella thompsoni*
Parasitoid of European corn borer

Tachinid fly: Diptera: Tachinidae

NOMENCLATURE: Approved name: *Lydella thompsoni* G Herting.

SOURCE: Widely distributed in Europe and the Orient. Introduced into the USA between 1920 and 1938.

PRODUCTION: Reared on corn borer larvae or collected from the field.

TARGET PESTS: European corn borer (*Ostrinia nubilalis* (Hübner)).

TARGET CROPS: Maize (corn).

BIOLOGICAL ACTIVITY: Biology: *Lydella thompsoni* is a solitary internal parasitoid of European corn borer larvae. The adult fly resembles a large, very bristly housefly. The female retains her eggs inside her body until each larva is ready to hatch. Each female is capable of producing up to 1 000 eggs, although far fewer are deposited or ever find a host borer. After the eggs have incubated for about 5 days, the female fly finds a potential host. She runs along a stalk and appears to be searching from side to side, being attracted to volatiles produced from host frass pushed out of the borer tunnel. Hatching usually takes place just as the female is depositing her offspring. Living larvae are deposited at the entrance to the host tunnel. The female stands over the burrow, bends her abdomen under until the ovipositor is pointing downwards and then a larva wriggles out or is brushed off with a quick movement of the tip of the abdomen. This first instar must then move into the tunnel and find its way to the borer. It prefers to attack fourth instar borers. The maggot has sharp mouthparts with a sawtooth edge to penetrate the host body. The borer may squirm or wiggle, or bite at the area being attacked in an attempt to prevent parasitism, but rarely succeeds. Once in the host, the maggot feeds first on the body fluids, then on the fatty tissues and internal organs. Upon the death of the borer, the maggot forces an opening in the skin, but continues to feed until its development is complete. It then leaves the host remains and pupates nearby in the tunnel. Larval development takes about 8 days and another 8 days is spent in the pupal stage. **Predation:** The seasonal cycle of this fly is poorly synchronised with that of the European corn borer in the Midwest, generally emerging too early in the spring. It will, however, parasitise the common stalk borer, which allows for survival of the population until the next generation of the corn borer is available to parasitise. There are usually two generations a year, but three generations often occur when another host is utilised for the first generation. The winter is passed as second instar maggots in the hibernating host larvae. A small proportion of first generation larvae may go into diapause and delay adult emergence until the following season. **Efficacy:** For many years after its introduction, *Lydella thompsoni* was the most important parasitoid of European corn borer in many areas of the United States. Parasitism of up to 75% of the second borer generation was recorded in the early years and it was considered a major controlling factor of borer populations.

COMMERCIALISATION: Formulation: Sold as pupae. **Tradenames:** 'Lydella thompsoni' (Praxis).

APPLICATION: Distribute pupae within the crop when *Ostrinia* adults are seen.

PRODUCT SPECIFICATIONS: Purity: The product contains only *Lydella thompsoni* pupae. **Storage conditions:** May be stored for several days in a cool, dry situation. Keep away from sunlight. **Shelf-life:** The pupae should be released within 7 days.

COMPATIBILITY: Incompatible with broad-spectrum, residual insecticides.

MAMMALIAN TOXICITY: There have been no reports of allergic or other adverse reactions from research workers, manufacturing staff or from the field release of *Lydella thompsoni*.

ENVIRONMENTAL IMPACT AND NON-TARGET TOXICITY: *Lydella thompsoni* occurs widely in Nature and is not expected to have any adverse effects on non-target organisms or on the environment. Its preferred host is the corn borer larva, although other small, related species may be preyed upon.

3:324 *Lysiphlebus testaceipes*
Parasitoid of aphids

Parasitoid wasp: Hymenoptera: Aphidiinae

NOMENCLATURE: Approved name: *Lysiphlebus testaceipes* Cresson.
Other names: Greenbug parasitoid; aphid parasitoid.

SOURCE: *Lysiphlebus testaceipes* is a native of Central American climates. It has been successfully introduced in many areas of the Mediterranean region and has spread in other countries in which it is established, especially on citrus crops.

PRODUCTION: Reared on aphids, particularly *Aphis gossypii* Glover, under controlled conditions.

TARGET PESTS: Preys on all aphids that infest cereal crops and is very effective in controlling *Aphis gossypii*.

TARGET CROPS: Wheat and grain sorghum and protected vegetable crops.

BIOLOGICAL ACTIVITY: Biology: *Lysiphlebus testaceipes* is a tiny (<3 mm long) black wasp and is not easy to see. However, the distinctive aphid mummies which remain on leaves after the parasitoid has killed the aphid can easily be detected. The mummy consists of the outer skin of the aphid, which becomes modified into a tough protective shell after the developing wasp kills the aphid by its internal feeding. Aphids parasitised by *Lysiphlebus testaceipes* are beige or tan in colour and are round and swollen compared with healthy aphids. *Lysiphlebus testaceipes* overwinters as a grub or pupa inside a parasitised aphid. The newly emerged wasp mates and then begins to search for new aphids to attack. The female wasp inserts an egg into the aphid and, in about 2 days, a tiny wasp grub hatches and feeds internally on the living aphid. The wasp grub completes feedings in about 6 to 8 days, resulting in the death of the aphid. Movement of the wasp grub inside the aphid expands the aphid, giving it a swollen appearance. The larva cuts a hole in the bottom of the aphid, attaches the aphid to a leaf with silk and a glue, and the dead aphid changes colour from green to a brown mummy. Then the wasp grub moults to the pupal stage and, after 4 to 5 days, a wasp emerges by cutting a circular hole in the top of the mummy. At 22 °C (70 °F), development from egg to adult takes about 14 days. Wasps disperse by flying, or by being carried inside winged aphids which may undergo long migration flights. **Predation:** Temperature is an important factor influencing the efficacy of wasps as biological control agents of aphids. Wasps develop most rapidly when temperatures are above 19 °C (65 °F) and adults are not active if temperatures are below 13 °C (56 °F). However, aphids are much more tolerant of cool temperatures and continue to reproduce until temperatures drop to 5 °C (40 °F). Thus, wasps may not be effective in controlling aphids in wheat in the autumn and spring, due to cool weather. **Efficacy:** Wasp parasitoids contribute to aphid suppression in two ways. There is direct mortality caused by the wasp parasitism, but, in addition, parasitised aphids have reduced reproductive rates. Parasitised aphids stop reproducing within 1 to 5 days, while healthy aphids give birth to 3 to 4 live young a day for 25 to 30 days. Thus, the activity of these wasps can greatly reduce the rate of aphid increase. Hyperparasite attack may reduce the effectiveness of *Lysiphlebus testaceipes* if the hyperparasite is abundant. **Key reference(s):** 1) A E Knutson, P Boring III, G J Michaels Jr & F Gilstrap. 1993. *Biological Control of Insect Pests in Wheat*, Texas Agric. Ext. Service Publication B-5044, 8 pp. 2) R Wright. 1995. Know your friends: wasp parasites of greenbugs, *Midwest Biological Control News Online*, **II**, 9. 3) P Stary, J P Lyon & F Leclant. 1988. Biocontrol of aphids by the introduced *Lysiphlebus testaceipes* (Cress.) in Mediterranean France, *J. Appl. Entomol.*, **105**, 74–87.

COMMERCIALISATION: Formulation: Sold as parasitised mummies.
Tradenames: 'Lysiphlebus testaceipes' (Praxis), 'LisiPAK 200' (Bioplanet).

APPLICATION: Distribute mummies within the field in late spring when the temperature is above 13 °C. In protected vegetables, release near infested areas early in the season. Less effective at high aphid population. Release throughout the infested area by allowing parasites to escape from the shipping vial. Treat every other week, at a rate of between 2 and 3 wasps per square metre. *Lysiphlebus testaceipes* can be used at lower rates as a preventive treatment.

PRODUCT SPECIFICATIONS: Purity: Products contain only mummies of *Lysiphlebus testaceipes*. **Storage conditions:** Use as soon as possible after delivery, but may be stored for up to 7 days, if kept below 13 °C and away from direct sunlight. **Shelf-life:** Up to 7 days, if stored according to the manufacturer's instructions.

COMPATIBILITY: Pesticide use in wheat or grain sorghum may decrease activity of these parasitoid wasps. Insecticides applied as sprays will kill adult wasps, as well as immature wasps developing inside aphids killed by insecticides. Research in Texas has shown that parathion-methyl and chlorpyrifos are more toxic to adult wasps and to immature wasps inside aphids than systemic insecticides, such as dimethoate or disulfoton, especially at lower rates. However, the shorter residual activity of parathion-methyl allows parasitoids to recolonise a field sooner after treatment. Triadimefon, used to control leaf rust in wheat, is also very toxic to adult wasps.

MAMMALIAN TOXICITY: There have been no reports of allergic or other adverse toxicological effects arising from contact with *Lysiphlebus testaceipes* from research staff, producers or users.

ENVIRONMENTAL IMPACT AND NON-TARGET TOXICITY: *Lysiphlebus testaceipes* occurs widely in Nature and would not be expected to have any significant effects on non-target organisms or on the environment. It is specific to aphids.
Approved for use in organic farming: Yes.

3:325 *Macrocentrus ancylivorus*
Parasitoid of oriental fruit moth

Parasitoid wasp: Hymenoptera: Braconidae

NOMENCLATURE: Approved name: *Macrocentrus ancylivorus* Rohwer.

SOURCE: *Macrocentrus ancylivorus* is a parasitoid native to north-eastern portions of the USA. This braconid wasp was described in 1921 as a parasitoid of strawberry leafroller (*Ancylis comptana* (Froe.)), but will attack many other caterpillars, such as fruitworms, stem borers and other leafroller species. It rapidly adapted to the oriental fruit moth after that pest was introduced from Japan in 1913.

PRODUCTION: Reared on oriental fruit moth larvae feeding on apple slices.

TARGET PESTS: Oriental fruit moth (*Grapholitha molesta* (Busck)) and strawberry leafroller (*Ancylis comptana*).

TARGET CROPS: Peach and late-season apple orchards. It can also be used in strawberry plantations.

BIOLOGICAL ACTIVITY: Biology: The 3 to 5 mm long adults are amber-yellow to reddish-brown in colour. The antennae and ovipositor of the female are as long as its body. They are active only when temperatures are between 18 and 25 °C (65 and 80 °F); in the summer, they often become active only at twilight and will oviposit throughout the night. *Macrocentrus ancylivorus* also becomes inactive in low relative humidity (<40%). The female deposits eggs singly in the bodies of host larvae during its life span of less than 2 weeks. Up to 50 hosts may be parasitised. The wasps find their hosts by homing in on frass, rather than on the caterpillar itself, and will even try to oviposit in fresh frass, but not in free-crawling larvae. They prefer to oviposit in second and third instars. *Macrocentrus ancylivorus* overwinters as first instar larvae in the hibernating host caterpillars. The larvae feed internally during the first three instars and then emerge to feed externally for the final instar. The wasp cocoon is spun in the cocoon of its host. **Predation:** The seasonal cycle of the wasp is correlated with that of its host, with two or more generations per year, depending on the host species. *Macrocentrus ancylivorus* can provide good control of oriental fruit moth in peaches and apples. It is most effective against larvae infesting twigs rather than those in large fruits later in the season. **Efficacy:** *Macrocentrus ancylivorus* is most effective in controlling oriental fruit moth in those orchards with a varied weed ground flora. Caterpillars of various species, feeding on the weeds, are an excellent reservoir for the parasitoid at times when its preferred host is unavailable. The ragweed borer (*Epiblema strenuana* (Walker)), which bores in the stems of ragweed species, is a widely distributed alternate host.

3. Macro-organisms

COMMERCIALISATION: Formulation: Sold as cocoons. **Tradenames:** 'Macrocentrus ancylivorus' (Praxis).

APPLICATION: Releases of 3 to 6 *Macrocentrus ancylivorus* females per tree for the first and second broods of oriental fruit moth effectively reduced damage by this pest on peaches. Releases work best if made as soon as the first wilted twigs are observed. This parasitoid could be used to enhance control of oriental fruit moth in orchards using mating disruption or in combination with a reduced insecticide treatment schedule.

PRODUCT SPECIFICATIONS: Purity: Products contain only *Macrocentrus ancylivorus*. **Storage conditions:** Keep cool and out of direct sunlight.

COMPATIBILITY: Incompatible with broad-spectrum, residual insecticides.

MAMMALIAN TOXICITY: There have been no reports of allergic or other adverse reactions from research workers, manufacturing staff or from the field release of *Macrocentrus ancylivorus*.

ENVIRONMENTAL IMPACT AND NON-TARGET TOXICITY: *Macrocentrus ancylivorus* occurs widely in Nature and is not expected to have any adverse effects on non-target organisms or on the environment. Its preferred host is the larva of the oriental fruit moth, although it will lay eggs in other small caterpillars. It can overwinter in temperate conditions.

3:326 *Macrolophus caliginosus*
Predator of whiteflies and spider mites

Predatory bug: Hemiptera: Miridae

NOMENCLATURE: Approved name: *Macrolophus caliginosus* Wagner. **Other names:** Capsid bug.

SOURCE: Palaearctic species, native to southern Europe. Common in the Mediterranean region, where it regularly occurs on various vegetable crops, both in open fields and in glasshouses. Identified in the 1980s as a candidate biological control agent for use in glasshouses.

PRODUCTION: Reared in insectaries, most often on tobacco plants, although tomato plants have also been used. Whiteflies and eggs of the flour moth, *Ephestia kuehniella* Zeller, are used as food.

TARGET PESTS: Main targets are the glasshouse whitefly (*Trialeurodes vaporariorum* (Westwood)) and tobacco whitefly (*Bemisia tabaci* (Gennadius)), but it will also prey on spider mites, aphids, thrips and moth eggs.

TARGET CROPS: Recommended for use on glasshouse vegetable crops and ornamentals. Use on *Gerbera* spp. is not recommended.

BIOLOGICAL ACTIVITY: Biology: A long-legged, bright green bug, 3 to 6 mm long, with five nymphal instars. The adults are pale green with a whitish hairiness. They have red eyes and the base of the antennae is black. Young nymphal stages are paler, yellowish-green in colour. Although it is mainly predatory, it will also feed on plant sap, which is thought to be necessary for successful development, but rarely causes economic damage to crop plants. However, damage to *Gerbera* spp. has been recorded and there have been occasional reports of problems on cherry tomatoes when bug numbers are very high with few prey available. **Predation:** Adults and nymphs search plants actively for prey, which they pierce with their needle-like mouthparts and suck dry. *Macrolophus caliginosus* will eat eggs, larvae and pupae of prey such as whitefly. **Egg laying:** Eggs are laid singly in plant tissue, usually in the leaf veins, petioles and stems. The number of eggs laid is greatly influenced by diet, but averages 100 to 250 per female. Fecundity seems to be greatest on diets of flour moth eggs and/or whitefly. **Duration of development:** Development is very dependent upon temperature. Eggs hatch after about 11 days at 25 °C, but hatching is delayed at lower temperatures and can take more than a month at 15 °C. Nymphal development is quite slow, taking about 19 days at 25 °C and almost 2 months at 15 °C. **Efficacy:** Long legs allow it to move rapidly, even on plants with glandular hairs.

COMMERCIALISATION: Formulation: Sold as adults. **Tradenames:** 'Macrolophus-System' and 'Macrolophus-N-System' (Biobest), 'Macroline c' (Syngenta Bioline), 'MiriPAK 250' (Bioplanet).

APPLICATION: Release into crops in the vicinity of the pest insects. *Macrolophus caliginosus* performs most effectively at temperatures above 25 °C.

PRODUCT SPECIFICATIONS: Purity: Contains mostly adults, with some eggs of *Ephestia* as a food source. **Storage conditions:** Will remain viable for several days, if stored in cool conditions.

COMPATIBILITY: Incompatible with persistent insecticides.

MAMMALIAN TOXICITY: There have been no reports of adverse allergic or other reactions from research workers, manufacturing staff or from the field release of *Macrolophus caliginosus*.

ENVIRONMENTAL IMPACT AND NON-TARGET TOXICITY: *Macrolophus caliginosus* occurs widely in Nature and is not expected to have any adverse effects on non-target organisms or on the environment. Its preferred hosts are whiteflies, although it will prey upon other insect species. It rarely survives winters under temperate conditions.

Approved for use in organic farming: Yes.

3:327 *Mesoseiulus longipes*
Predator of two-spotted spider mite

Predatory mite: Mesostigmata: Phytoseiidae

NOMENCLATURE: Approved name: *Mesoseiulus longipes* Evans. **Other names:** Mite predator. Formerly known as: *Phytoseiulus longipes* Evans.

SOURCE: Native of South Africa.

PRODUCTION: Reared in insectaries on *Tetranychus urticae* Koch on beans.

TARGET PESTS: Two-spotted spider mites (*Tetranychus urticae*).

TARGET CROPS: Effective in glasshouses and interiorscapes with artificial lighting. Used on almonds, grapes, strawberries and ornamentals.

BIOLOGICAL ACTIVITY: Biology: Eggs are laid near a food source on the surfaces of leaves. These hatch into six-legged larvae that do not feed. The protonymph emerges from the larva and immediately starts to feed. At the second nymphal stage, the predator searches for food constantly. The adults emerge from the nymphal stage, mate within a few hours and the females then begin to lay eggs. In the absence of food, the predator can survive for some time on water and honey, but reproduction then ceases. **Predation:** Larvae do not feed; the nymphal stages consume eggs, larvae and protonymphs of the spider mite; and adults eat all stages of the prey. *Mesoseiulus longipes* is almost completely dependent on the two-spotted spider mite. **Efficacy:** *Mesoseiulus longipes* is very similar to *Phytoseiulus persimilis*, but can tolerate lower relative humidity (40% at 21 °C (70 °F)). The adults are active at higher temperatures up to 38 °C (100 °F), but higher relative humidities are needed at these temperatures. The life span of the adults is about 34 days.

COMMERCIALISATION: Formulation: Sold as adults. **Tradenames:** 'Mesoseiulus longipes' (Arbico), (Harmony), (IPM Laboratories), (Praxis) and (Rincon-Vitova).

APPLICATION: Release 30 to 35 adults per square metre every 2 weeks, 3 to 5 times a season in glasshouses and interiorscapes; and 12 000 to 48 000 per hectare every 2 weeks, 3 to 4 times a season outdoors.

PRODUCT SPECIFICATIONS: Purity: The product contains only *Mesoseiulus longipes* adults. **Storage conditions:** If storage is absolutely necessary, refrigerate at 6 to 10 °C (40 to 50 °F) for no longer than 5 days.

COMPATIBILITY: Susceptible to pesticides. Field tolerance will vary with spray timing, application methods, weather and crop. Avoid spraying crop one week before or after releasing predators.

MAMMALIAN TOXICITY: No allergic or other adverse reactions have been reported from its use in glasshouse or outdoor crops.

ENVIRONMENTAL IMPACT AND NON-TARGET TOXICITY: *Mesoseiulus longipes* is widespread in Nature and is not thought to be damaging to non-target species or to the environment. It is specific to spider mites. **Approved for use in organic farming:** Yes.

3:328 *Metaphycus bartletti*
Parasitoid of scale insects

Parasitoid wasp: Hymenoptera: Encyrtidae

NOMENCLATURE: Approved name: *Metaphycus bartletti* Annecke and Mynhardt. **Other names:** Soft scale parasite; black scale parasite; scale parasitoid.

SOURCE: Widely occurring scale parasite.

PRODUCTION: Reared in insectaries under controlled conditions on scale insects.

TARGET PESTS: Mediterranean black scale (*Saissetia oleae* (Bernhard)).

TARGET CROPS: Olive plantations and ornamentals such as laurel.

BIOLOGICAL ACTIVITY: Biology: The larvae consume scale insects from the egg to the adult stages and the adult wasps feed on honey. The life-cycle of the wasp is between 25 and 40 days, depending on the temperature. **Predation:** The larvae eat the inside of parasitised scale. Females will parasitise only third instar nymphs and very young scale. **Egg laying:** The female wasp is very long-lived and can lay over 100 eggs in her lifetime. **Duration of development:** Development time is very dependent on the temperature. The egg can develop into an adult in 11 days and the adult wasp can survive for over 50 days. **Efficacy:** The adult female wasp is very mobile and very effective at seeking and parasitising its prey.

COMMERCIALISATION: Formulation: Sold as adult wasps or as parasitised scale. **Tradenames:** 'Metaphycus bartletti' (Biotop).

APPLICATION: Apply to infested olive trees at a rate of 5 to 10 wasps per tree in the spring and autumn. Introduce as soon as scale reach the third nymphal instar.

PRODUCT SPECIFICATIONS: Purity: Adults consist of fecund females. All scales supplied are parasitised. **Storage conditions:** Adults may be kept under normal conditions in the presence of a suitable food source. Very active in direct sunlight. **Shelf-life:** Adults may be kept for a few days, if fed on honey solution.

COMPATIBILITY: Incompatible with foliar insecticides.

MAMMALIAN TOXICITY: No allergic or other adverse reactions have been reported following the release of *Metaphycus bartletti* in glasshouses or outdoors.

ENVIRONMENTAL IMPACT AND NON-TARGET TOXICITY: *Metaphycus bartletti* occurs widely in Nature and is not expected to have any adverse effects on non-target organisms or on the environment. It is specific to scale insects. **Approved for use in organic farming:** Yes.

3:329 *Metaphycus flavus*
Parasitoid of scale insects

Parasitoid wasp: Hymenoptera: Encyrtidae

NOMENCLATURE: Approved name: *Metaphycus flavus* (Howard). **Other names:** Black scale parasite; soft scale parasite; scale parasitoid.

SOURCE: Originated in North Africa, but now widely distributed.

PRODUCTION: Reared under controlled conditions on scale insects.

TARGET PESTS: Soft scales, such as *Saissetia coffeae* (Walker), *S. oleae* (Bernhard) and, particularly, *Coccus hesperidum* L. Not effective against hard scale insects.

TARGET CROPS: Primary use is in citrus plantations.

BIOLOGICAL ACTIVITY: Biology: *Metaphycus flavus* is a small, parasitoid wasp, with metallic gloss over its wings. Females are yellow, with antennae transversally striped; males are also yellow, but with a dark abdomen. Host size and quality strongly influence offspring sex ratios and brood sizes; larger hosts yield more female offspring and larger broods. Larger hosts lead to a larger size in the emerging wasps and larger wasps have greater egg loads and live longer than smaller wasps. However, wasp longevity and the influence of wasp size on longevity are mediated by the wasp's diet. *Metaphycus flavus* females live the longest when they have access to hosts, nectar and water. **Predation:** *Metaphycus flavus* parasitises all stages of the soft scale, with several eggs being laid in larger stages. Adults also consume scale. Ants will attack wasps, so avoid use when ants are present and do not use when the temperature exceeds 30 °C. **Efficacy:** *Metaphycus flavus* is very mobile and very effective at seeking and parasitising its prey.

COMMERCIALISATION: Formulation: Sold as adults in capsules containing 25 or 100 wasps. **Tradenames:** 'Metaphycus flavus' (Entocare) and (Sautter & Stepper).

APPLICATION: Remove lids from capsules within the crop canopy and leave for 2 days to allow all the adults to emerge, protected from direct sunlight and rain. Repeat after 3 weeks if the infestation is heavy. Ideal conditions are when the temperature is between 15 and 30 °C and the r.h. is above 40%.

PRODUCT SPECIFICATIONS: Purity: Fecund female adults. **Storage conditions:** May be stored for short periods at 8 to 12 °C, away from direct sunlight.

COMPATIBILITY: Extremely low humidity and temperature reduce the wasp's activity. Control ants, as they attack *Metaphycus flavus*. Do not use in areas of dull light.

MAMMALIAN TOXICITY: No allergic or other adverse reactions have been reported following release of *Metaphycus flavus* in glasshouses or outdoors.

ENVIRONMENTAL IMPACT AND NON-TARGET TOXICITY: *Metaphycus flavus* occurs widely in Nature and there is no evidence that it affects non-target organisms or that it has any adverse effects on the environment. It is specific to soft scale insects.
Approved for use in organic farming: Yes.

3:330 *Metaphycus helvolus*
Parasitoid of scale insects

Parasitoid wasp: Hymenoptera: Encyrtidae

NOMENCLATURE: Approved name: *Metaphycus helvolus* (Compère). **Other names:** Black scale parasite; soft scale parasite; scale parasitoid.

SOURCE: Originated in South Africa, but now widely introduced.

PRODUCTION: Reared in insectaries on scale insects.

TARGET PESTS: Citrus black scale (*Parlatoria ziziphi* (Lucas)) and soft scale, such as *Saissetia coffeae* (Walker), *S. oleae* (Bernhard) and *Coccus hesperidum* L. Not effective against hard scale insects.

TARGET CROPS: Citrus plantations, fruit orchards, ornamentals outside and under glass.

BIOLOGICAL ACTIVITY: Biology: The larvae consume scale insects from the egg to the adult stages and the adults feed on older scale. The adults can survive on nectar. The life-cycle of the wasp is very rapid, particularly at temperatures above 25 °C, varying from 11 to 33 days. **Predation:** Females parasitise only second and third instar scale nymphs.
Egg laying: The female wasp can lay over 400 eggs in her lifetime. Eggs are laid in young nymphal instars. **Duration of development:** Very dependent on the temperature. Adults can live for over 50 days. **Efficacy:** *Metaphycus helvolus* is very mobile and very effective at seeking and parasitising its prey.

COMMERCIALISATION: Formulation: Sold as adult wasps. **Tradenames:** 'Metasure (h)' (Biological Crop Protection), 'Metaphycus helvolus wasps' (Rincon-Vitova), 'Metaphycus helvolus' (IPM Laboratories), (Natural Pest Control) and (Praxis).

APPLICATION: Apply to infested citrus trees at a rate of 1000 to 3000 wasps per tree in the spring. Under glasshouse conditions, release about 5 wasps per square metre, repeating 3 times at 2-week intervals. Introduce as soon as scale is seen.

PRODUCT SPECIFICATIONS: Purity: Fecund female adults. **Storage conditions:** Adults may be kept under warm conditions in the presence of a food source. Very active in direct sunlight. **Shelf-life:** Use as soon as possible, as quality declines with storage.

COMPATIBILITY: Extremely low humidity and temperature reduce the wasp's activity. Control ants, as they attack *Metaphycus helvolus*. Do not use in areas of dull light.

MAMMALIAN TOXICITY: No allergic or other adverse reactions have been reported following release of *Metaphycus helvolus* in glasshouses or outdoors.

ENVIRONMENTAL IMPACT AND NON-TARGET TOXICITY: *Metaphycus helvolus* occurs widely in Nature and there is no evidence that it affects non-target organisms or that it has any adverse effects on the environment. It is specific to soft scale insects.
Approved for use in organic farming: Yes.

3:331 *Meteorus* species

Parasitoid of Lepidoptera

Parasitoid wasp: Hymenoptera: Braconidae

NOMENCLATURE: Approved name: *Meteorus autographae* Muesebeck, *Meteorus trachynotus* Viereck and *Meteorus laphygmae* Viereck.

SOURCE: *Meteorus* species occur widely in North America, for example *Meteorus autographae* is found as far north as Newfoundland, as far south as Florida and as far west as Wisconsin and Louisiana.

PRODUCTION: *Meteorus* species attack a wide range of caterpillars and are bred on live larvae under controlled conditions.

TARGET PESTS: A wide range of caterpillars are controlled; for example, typical hosts of *Meteorus autographae* include eastern blackheaded budworm (*Acleris variana* (Fern.)), black cutworm (*Agrotis ipsilon* (Hufn.)), velvet bean caterpillar (*Anticarsia gemmatalis* Hbn.), American bollworm/corn earworm/tomato fruitworm (*Helicoverpa zea* (Boddie)), whitemarked tussock moth (*Orgyia leucostigma* (Sm.)), southern armyworm (*Spodoptera eridania* (Cram.)), beet armyworm (*S. exigua* (Hbn.)), fall armyworm (*S. frugiperda* (Sm.)) and cabbage looper (*Trichoplusia ni* (Hbn.)).

TARGET CROPS: Effective at controlling caterpillars in a wide range of field crops, including soybeans and brassicae.

BIOLOGICAL ACTIVITY: Biology: Eggs of *Meteorus* species are clear and thin-walled. Soon after oviposition, the folded larva can be observed inside. The larvae are translucent, long and slender, with a pronounced sclerotised head. Often more than one egg is laid, but the first larva to emerge kills its siblings. Pupae are brown-coloured, with the 5 mm-long cocoon usually suspended from the edge of the leaf on a silk thread. The adult wasps are orange, with black eyes and antennae. The body does not exceed 6 mm in length. The female has a well-defined black ovipositor. **Egg laying:** During its life span in the laboratory, a female lays, on average, 240 eggs, though sometimes the number of progeny can reach 350.
Duration of development: *Meteorus* species develop from egg to pupa in 8 days at 27 °C. Six days later, the adult hatches. This time triples when the temperature drops to 16 °C. The adult lives, on average, 40 days. Wasps also develop and survive better at cooler temperatures. **Key reference(s):** 1) J F Grant & M Shepard. 1984. Laboratory biology of *Meteorus autographae* (Hymenoptera: Braconidae), an indigenous parasitoid of soybean looper (Lepidoptera: Noctuidae) larvae, *Environ. Entomol.*, **13**, 838–42. 2) J F Grant & M Shepard. 1986. Seasonal incidence of *Meteorus autographae* on soybean looper larvae on soybean in South Carolina, and the influence of the host density on parasitization, *J. Entomol.*

3. Macro-organisms

Sci., **21**, 338–45. 3) E R Mitchell *et al*. February 2000. Stage by stage comparison of parasitoids important in biocontrol of cabbage pests. USDA, www.usda.ufl.edu/biocontrol/guide (March 2000).

COMMERCIALISATION: Formulation: Sold as adults. **Tradenames:** 'Meteorus species' (Praxis).

APPLICATION: Release adults when pest moths are seen in the crop.

PRODUCT SPECIFICATIONS: Storage conditions: Use immediately upon receipt.

COMPATIBILITY: Incompatible with persistent insecticides.

MAMMALIAN TOXICITY: There have been no reports of allergic or other adverse reactions from research workers, manufacturing staff or from the field release of *Meteorus* species.

ENVIRONMENTAL IMPACT AND NON-TARGET TOXICITY: *Meteorus* species occur widely in Nature and are not expected to have any adverse effects on non-target organisms or on the environment. It will predate most lepidopteran larvae and is able to overwinter, even under harsh weather conditions.

3:332 *Microctonus aethiopoides*
Parasitoid of weevils

Parasitoid wasp: Hymenoptera: Braconidae

NOMENCLATURE: Approved name: *Microctonus aethiopoides* Loan. **Other names:** Alfalfa weevil parasite.

SOURCE: The wasp is widespread in Europe and it was first imported into the United States for use against the sweetclover weevil.

PRODUCTION: *Microctonus aethiopoides* is raised on alfalfa weevil.

TARGET PESTS: Alfalfa weevil (*Hypera postica* (Gyllenhal)).

TARGET CROPS: Alfalfa (lucerne).

BIOLOGICAL ACTIVITY: Biology: Unlike most other parasitoids of weevils, *Microctonus aethiopoides* attacks the adult stage. There are several strains of the wasp that are very specific for different weevil species and not all strains attack the alfalfa weevil. Selection of the appropriate strain has ensured that *Microctonus aethiopoides* is considered one of the most important parasitoids responsible for maintaining alfalfa weevil at or below economic levels in the USA. The stingless *Microctonus aethiopoides* wasps are about 30 mm long. Females are reddish-brown and males are black in colour. **Predation:** Female *Microctonus aethiopoides* will lay eggs only in moving adult weevils. The parasitoid takes a position behind the weevil and inserts her ovipositor into the membranous area at the tip of the weevil's abdomen, which is exposed only while the weevil is in motion. *M. aethiopoides* lays a single egg directly inside the body cavity of the weevil. The larva that hatches from the egg then feeds and completes its development internally in 22 to 26 days. There are no external signs of parasitism until the wasp larva is fully grown and it forces its way out of the back end of the weevil. The wasp larva spins a whitish silken cocoon and pupates in the soil or under debris. An adult wasp emerges 6 to 9 days later. *M. aethiopoides* overwinters as first instar larvae inside hibernating weevils. *M. aethiopoides* has two generations per year that are well synchronised with the adult weevil's activity. The first generation females attack and lay eggs in the surviving overwintered weevil adults (spring population) that are actively laying eggs in late spring. The second generation wasps attack newly emerged weevil adults in the summer population. Parasitism rates of alfalfa weevil adults vary from year to year, averaging 40% and 52% for the spring and summer weevil populations, respectively. **Efficacy:** The effect of *Microctonus aethiopoides* on the alfalfa weevil population results not only from death of the adult weevils themselves, but also by reducing the number of eggs a female weevil lays. Parasitised female alfalfa weevils cease oviposition within 3 days; male weevils are castrated shortly after parasitisation. The impact of *M. aethiopoides* as a biological control agent of alfalfa weevil is significant because of this sterilisation. Although the parasitised summer generation weevils will survive until the following spring, they are unable to reproduce and, hence, contribute nothing to the alfalfa weevil population.

COMMERCIALISATION: Formulation: Sold as parasitised weevils. **Tradenames:** 'Microctonus aethiopoides' (Praxis).

APPLICATION: Release the parasitised weevils into the crop at the time of weevil activity. It is important to conserve *Microctonus aethiopoides* by using agricultural practices that are not detrimental to the parasitoids. Effective conservation can be achieved by using degree days for timing sprays to avoid insecticide applications during the periods when the adult parasitoids are active. Leaving uncut refuge areas of alfalfa allows parasitised weevils to survive.

PRODUCT SPECIFICATIONS: Purity: Products contain only parasitised weevils. **Storage conditions:** *Microctonus aethiopoides* should be kept in cool, dry conditions, out of direct sunlight. **Shelf-life:** *Microctonus aethiopoides* should be used as soon as possible after receipt, but may be stored for up to 10 days.

COMPATIBILITY: *Microctonus aethiopoides* is incompatible with broad-spectrum, residual insecticides.

MAMMALIAN TOXICITY: *Microctonus aethiopoides* has shown no allergic or other adverse reactions to workers in the field. There is no evidence of acute or chronic toxicity, eye or skin irritation or hypersensitivity to mammals.

3. Macro-organisms

ENVIRONMENTAL IMPACT AND NON-TARGET TOXICITY: *Microctonus aethiopoides* occurs widely in Nature and it is not expected to have any effect on non-target organisms or to have any adverse effects on the environment. It is specific to weevils, such as the alfalfa weevil.

3:333 *Microlarinus* species

Phytophagous beetle

Phytophagous beetle: Coleoptera: Curculionidae

NOMENCLATURE: Approved name: *Microlarinus lypriformis* (Wollaston) and *Microlarinus lareynii* (Jacquilin du Val). **Other names:** Puncture vine seed weevil.

SOURCE: Both species of *Microlarinus* originated in Europe, but both have been released in the USA from weevils collected in Italy.

PRODUCTION: *Microlarinus* species are reared on puncture vine seed pods.

TARGET PESTS: Puncture vine (*Tribulus terrestris* L.).

TARGET CROPS: Pasture and rangeland.

BIOLOGICAL ACTIVITY: Biology: Female weevils chew into the side of a young puncture vine bur, deposit eggs into the seed and seal it with faecal material. Weevil grubs develop inside the seed and pupate therein. Each seed may produce 1 to 3 weevils. The life-cycle from egg to adult requires about 25 days. Adult weevils may feed on the plant, but do not cause appreciable damage to the plant. The number of generations per year depends on the climate. Adults overwinter. **Egg laying:** A female may deposit up to 324 eggs in her lifetime. **Efficacy:** Weevil establishment is favoured by warm temperature areas associated with mild winters. Sufficient puncture vine density is another factor necessary to support substantial weevil populations. **Key reference(s):** 1) R D Goeden & R L Kirkland. 1978. An insecticidal-check study of the biological control of puncturevine (*Tribulus terrestris* L.) by imported weevils, *Microlarinus lareynii* and *M. lypriformis* (Col.: Curculionidae), *Environ. Entomol.*, **7**(3), 349–54. 2) R L Kirkland & R D Goeden. 1978. Biology of *Microlarinus lareynii* (Col.: Curculionidae) on puncturevine in Southern California, *Ann. Entomol. Soc. Amer.*, **71**, 13–18. 3) D M Maddox. 1976. History of weevils on puncturevine in and near the United States, *Weed Sci.*, **24**(4), 414–9.

COMMERCIALISATION: Formulation: Sold as adult weevils. **Tradenames:** 'Microlarinus' (Praxis).

APPLICATION: It is usual to release between 250 and 1000 weevils in a population of puncture vine.

PRODUCT SPECIFICATIONS: Purity: Products contain only adult *Microlarinus*. **Storage conditions:** Release as soon as possible after delivery.

COMPATIBILITY: Incompatible with persistent insecticides.

MAMMALIAN TOXICITY: There have been no reports of allergic or other adverse reactions from research workers, manufacturing staff or from the field release of *Microlarinus*.

ENVIRONMENTAL IMPACT AND NON-TARGET TOXICITY: *Microlarinus* species occur widely in Nature and are not expected to have any adverse effects on non-target organisms or on the environment. Tests in the USA have shown that *Microlarinus lareynii* feeds on puncture vine (*Tribulus terrestris*), Jamaica feverplant (*Tribulus cistoides* L.) and *Kallstroemia* spp., but *Microlarinus lypriformis* is specific to puncture vine.

3:334 *Microplitis plutellae*
Parasitoid of Lepidoptera

Parasitoid wasp: Hymenoptera: Braconidae

NOMENCLATURE: Approved name: *Microplitis plutellae* (Muesebeck).
Other names: Diamond-back moth parasitoid.

SOURCE: Native North American species recorded from temperate regions of the United States and Canada, but is also found in subtropical regions. Abundant.

PRODUCTION: *Microplitis plutellae* will only parasitise the diamondback moth (*Plutella xylostella* (L.)) and the cabbage looper (*Trichoplusia ni* (Hübner)) and it is reared on one of these species under controlled conditions.

TARGET PESTS: Diamond-back moth (*Plutella xylostella*).

TARGET CROPS: Cruciferous crops, such as cabbages.

BIOLOGICAL ACTIVITY: Biology: Adult wasps are brownish-black. Females of *Microplitis plutellae* attack the early larval instars of the diamondback moth. Females locate their host with their antennae and may lay one or more eggs in the body cavity of the diamondback moth caterpillar. The first instar searches for and attempts to kill any other parasitoids within the caterpillar, including those of its own species. Following completion of its larval development, third instar larvae leave the host and spin a cocoon nearby. The new adult cuts its way out through the cocoon at one end. A portion of the parasite population does not enter diapause and continues its development to attack the second generation of diamondback moth caterpillars. **Egg laying:** Each female will lay up to 230 eggs in her lifetime. **Duration of development:** The service life of this parasitoid, evaluated during four generations, was of 30.7 days. *Efficacy: Microplitis plutellae* occurs in both univoltine and multivoltine populations of *Plutella xylostella*, and parasitism varies between 20% and 77% in the first generation and between 7% and 27% in the second generation. Because of this high parasitism incidence, *Microplitis plutellae* is a significant biocontrol agent of the diamondback moth. **Key reference(s):** L G Putnam. 1973. Effects of the larval parasites *Diadegma insularis* and *Microplitis plutellae* on the abundance of the diamondback moth in Saskatchewan rape and mustard crops, *Can. J. Plant Sci.,* **53**(4), 911–4.

COMMERCIALISATION: Formulation: *Microplitis plutellae* are sold as adults. **Tradenames:** 'Microplitis plutellae' (Praxis).

APPLICATION: Release adults into the crop as diamondback moth adults are first noticed.

PRODUCT SPECIFICATIONS: Purity: The product contains only *Microplitis plutellae* adults. **Shelf-life:** Release immediately upon receipt.

COMPATIBILITY: Incompatible with broad-spectrum, residual insecticides.

MAMMALIAN TOXICITY: No allergic or other adverse reactions have been reported from the use of *Microplitis plutellae* in glasshouse or field conditions. Allergies associated with host Lepidoptera are known in production.

ENVIRONMENTAL IMPACT AND NON-TARGET TOXICITY: *Microplitis plutellae* is widespread in Nature and it is not expected to show any adverse effects on the environment. The only two lepidopteran pests that act as a host are *Plutella xylostella* and *Trichoplusia ni*. **Approved for use in organic farming:** Yes.

3:335 *Microterys nietneri*
Parasite of soft scales

Parasitic wasp: Hymenoptera: Aphelinidae.

NOMENCLATURE: Approved name: *Microterys nietneri* (Motschulsky) **Other names:** *Microterys flavus* (Howard); *Encyrtus nietneri* Motschlsky; *Encyrtus flavus* Howard; parasitic wasp.

SOURCE: A widespread parasitic wasp found in many countries including Bangladesh, India, Pakistan and South Africa. It has been recorded in North America, Hawaii, New Zealand, Australia, Japan, Malaya, India, Ceylon and South Africa. Originally isolated from *Pulvinaria psidii* Maskell on guava in Bangladesh.

PRODUCTION: Bred on the soft brown scale (*Coccus hesperidum* L.).

TARGET PESTS: Soft scale, especially soft brown scale (*Coccus hesperidum*). It is less effective at controlling other soft scale species.

TARGET CROPS: Deciduous flowering fruit trees, ornamental trees and shrubs.

BIOLOGICAL ACTIVITY: Biology: *Microterys nietneri* grows up from egg to adult inside its host. At 25 °C this development takes about 2 weeks. The females of *Microterys flavus* are yellow. Their wings and antennae are marked with a pattern of black and white bands. Males are black. Both sexes are 2 mm in length. Parasitised soft scales can be recognised by one or more dark rings under their scale-cover in the centre of which the larva of the parasitic wasp grows. The young parasitic wasp leaves the scale through a hole in the scale-cover at the end of the pupa, clearly visible from above. **Predation:** In addition to parasitising adult soft scale, *Microterys nietneri* also feeds on young scales. This behaviour increases the life span of the adult wasp. **Egg laying:** A female can lay about 70 eggs during her lifetime. The number of eggs deposited in one scale depends on the size of the scale. Up to seven new adult parasitoids can emerge from one scale. **Duration of development:** Adults can live for a maximum of 2 months. The temperature range in which the wasp is active is between 18 and 32 °C. **Key reference(s):** 1) D P Annecke. 1964. The encyrtid and aphelinid parasites (Hymenoptera: Chalcidoidea) of soft brown scale, *Coccus hesperidum* L. (Hemiptera: Coccidae) in South Africa, *Entomol. Mem. Dept. Agric. Tech. Servs Rep. Sth Africa* **7**, 1–74. 2) B R Bartlett & C F Lagace. 1961. A new biological race of *Microterys flavus* introduced into California for the control of lecanine coccids, with an analysis of its behaviour in host selection. *Annals of the Entomological Society of America*, **54**(2), 222–7.

COMMERCIALISATION: Formulation: Supplied as parasitised soft brown scale. **Tradenames:** 'Microterys flavus' (Neudorff), (NIJHOF) and (Flora Nützlinge).

APPLICATION: Distribute throughout the crop at first sign of scale attack.

PRODUCT SPECIFICATIONS: Purity: Products contain only brown scale parasitised by *Microterys nietneri*, and no contaminants. **Storage conditions:** May be stored for short periods, under cool, dry conditions.

MAMMALIAN TOXICITY: No allergic or other adverse reactions have been reported from the use of *Microterys nietneri* in glasshouse or field conditions.

ENVIRONMENTAL IMPACT AND NON-TARGET TOXICITY: *Microterys nietneri* occurs widely in Nature and is not expected to have any adverse effects on non-target organisms or on the environment. **Approved for use in organic farming:** Yes.

3:336 *Muscidifurax* species   *Parasitoid of flies*

Parasitoid wasp: Hymenoptera: Pteromalidae

NOMENCLATURE: Approved name: *Muscidifurax raptor* Girault & Sanders and *Muscidifurax raptorellus* Kogan & Legner. **Other names:** Filth fly parasitoid.

SOURCE: Widely dispersed.

PRODUCTION: Reared under controlled conditions on houseflies (*Musca domestica* L.).

TARGET PESTS: The housefly (*Musca domestica*), the stable fly (*Stomoxys calcitrans* (L.)) and other fly species.

TARGET CROPS: Dairy farms and poultry ranches, as well as suburban and urban localities where flies exist.

BIOLOGICAL ACTIVITY: Biology: The adult parasitoid stings the fly pupa, killing the pupa and then lays eggs in the pupal case. *Muscidifurax raptorellus* can lay 6 to 8 eggs in each fly pupa, but *M. raptor* only lays one egg per pupa. When the eggs hatch, the larvae feed on the dead fly pupa. In 19 to 21 days, the adults emerge from the fly pupal case and begin to search for fly pupae on which to feed and deposit eggs. Adults feed on fly pupae body fluids. The female may sting the pupa, killing it, without depositing an egg. A droplet of fluid appears at the sting site and the female feeds on the fluid and may share with a nearby male. **Efficacy:** Flies develop into adults in half the time and lay more eggs than the parasitoids and are thus more numerous. For control of flies, it is helpful to augment parasitoid populations in early summer. **Key reference(s):** 1) L Smith & D A Rutz. 1991. Relationship of microhabitat to incidence of housefly (Diptera: Muscidae) immatures and their parasitoids at dairy farms in Central New York, *Environ. Entomol.*, **20**, 669–74. 2) D W Watson, J K Waldron & D A Rutz. 1994. Pest Management Fact Sheet, Cooperative Extension, Cornell University, IPM Dairy Management 102DMFS450.00, 4 pp.

COMMERCIALISATION: Formulation: Sold as adult wasps. **Tradenames:** 'Muscidifurax' (Arbico), (Beneficial Insectary), (IPM Laboratories), (Natural Pest Control) and (Rincon-Vitova).

APPLICATION: The optimal time to begin parasitoid releases is mid- to late May, continuing on a weekly basis until mid August.

PRODUCT SPECIFICATIONS: Purity: All parasitoids are free from microsporidosis. **Storage conditions:** *Muscidifurax* species should be released as soon as they are received.

COMPATIBILITY: *Muscidifurax* species have a longer life-cycle than the flies and have not developed resistance to commonly used pesticides to which flies have developed resistance. In many cases, a pesticide will scarcely affect the fly population but will destroy its parasitoids.

Research has shown that, in general, the parasitoids are more susceptible than flies to insecticides. If insecticides must be used, space sprays and baits are most compatible with the parasitoids.

MAMMALIAN TOXICITY: No allergic or other adverse reactions have been reported following use of *Muscidifurax* species in animal houses, stables, barns or outdoors.

ENVIRONMENTAL IMPACT AND NON-TARGET TOXICITY: *Muscidifurax* species occur widely in Nature and are not expected to show any adverse effects on non-target organisms or on the environment. They are specific to fly species, but reproduce more slowly and are, therefore, unlikely to eliminate them once introduced.

3:337 *Neodryinus typhlocybae*
Parasitoid of planthoppers

Parasitoid wasp: Hymenoptera: Dryinidae

NOMENCLATURE: Approved name: *Neodryinus typhlocybae* Ashmead.
Other names: Planthopper parasitoid.

SOURCE: Originated in North America, but now introduced widely.

PRODUCTION: Bred under controlled conditions on planthoppers.

TARGET PESTS: Planthoppers, such as the citrus planthopper (*Metcalfa pruinosa* Say).

TARGET CROPS: Many crops that are susceptible to attack by planthoppers, but particularly apple and pear orchards and ornamental trees.

BIOLOGICAL ACTIVITY: Biology: *Neodryinus typhlocybae* females search for the young nymphs of *Metcalfa pruinosa* and lay an egg on its body. The wasp larva develops as a 'cyst' outside the victim. Finally, the larva constructs a cocoon below the empty host body. The adults emerge from the pupae either in the summer a few months after parasitisation or emergence may be delayed over winter, with the adults emerging in the spring. Adult wasps also prey on young planthopper larvae, contributing to control. **Efficacy:** The females are mobile and are very effective at seeking out host nymphs.

COMMERCIALISATION: Formulation: Sold as pupae. **Tradenames:** 'Neodryinus typhlocybae' (Bioplanet).

APPLICATION: *Neodryinus typhlocybae* pupae are distributed in infested areas and the progressive spread of the parasitoid leads to a stable control of planthoppers.

PRODUCT SPECIFICATIONS: Storage conditions: Keep cool, out of direct sunlight. Do not allow to freeze. **Shelf-life:** Use immediately upon receipt.

COMPATIBILITY: *Neodryinus typhlocybae* should not be used with residual insecticides.

MAMMALIAN TOXICITY: *Neodryinus typhlocybae* has not demonstrated evidence of toxicity, infectivity, irritation or hypersensitivity in mammals. No allergic responses or other adverse health problems have been observed by research workers, manufacturing staff or users.

ENVIRONMENTAL IMPACT AND NON-TARGET TOXICITY: *Neodryinus typhlocybae* occurs in Nature and, as such, is not expected to cause any adverse effects on non-target organisms. It is specific to planthoppers. **Approved for use in organic farming:** Yes.

3:338 *Nesidiocoris tenuis*

Predator of whiteflies

Predatory bug: Heteroptera: Miridae

NOMENCLATURE: Approved name: *Nesidiocoris tenuis* (Reuter) **Other names:** Previously known as *Cyrtopeltis tenuis* Reuter; *Engytatus volucer* (Kirkaldy); *Gallobelicus crassicornis* Distant; *Dicyphus nocivus* Fulmek. Common names include tomato bug; tobacco capsid; tomato mirid.

SOURCE: Widely spread.

PRODUCTION: Bred under controlled conditions on whiteflies on tomatoes.

TARGET PESTS: All species of whitefly (glasshouse whitefly (*Trialeurodes vaporariorum* (Westwood)), tobacco whitefly (*Bemisia tabaci* (Gennadius)) and silverleaf whitefly (*Bemisia argentifolii* Bellows & Perring)).

TARGET CROPS: Protected ornamentals and vegetables under glass or plastic.

BIOLOGICAL ACTIVITY: Predation: Adult predatory bugs and nymphs search actively for their prey, insert their sucking mouthparts and suck out the contents. **Efficacy:** *Nesidiocoris tenuis* is a voracious predator of whitefly.

COMMERCIALISATION: Formulation: Under investigation by Biobest as an additional predatory bug for the control of whitefly.

PRODUCT SPECIFICATIONS: Purity: Adults and nymphs, but often with caterpillar eggs (*Ephestia*) as a source of food. **Storage conditions:** *Nesidiocoris tenuis* may be stored in the dark at 10 to 15 °C (50 to 59 °F), keeping the bottle horizontal. **Shelf-life:** If kept under recommended conditions, can be stored for 1 to 2 days.

COMPATIBILITY: Incompatible with residual broad-spectrum insecticides. Crops may be damaged in the absence of an adequate food source.

MAMMALIAN TOXICITY: There have been no reports of any adverse or allergic reactions from laboratory, production or field trial staff.

ENVIRONMENTAL IMPACT AND NON-TARGET TOXICITY: *Nesidiocoris tenuis* occurs widely in Nature and it is unlikely that any adverse environmental effects will result from its use. Although its preferred prey are whiteflies, it will consume many other insect species when populations are low, including beneficials. **Approved for use in organic farming:** Yes.

3:339 *Oberea erythrocephala*
Phytophagous beetle

Phytophagous beetle: Coleoptera: Cerambycidae

NOMENCLATURE: Approved name: *Oberea erythrocephala* (Schrank). **Other names:** Leafy spurge beetle.

SOURCE: *Oberea erythrocephala* is widely distributed in Europe and was introduced into the USA in 1980.

PRODUCTION: Reared under controlled conditions on leafy spurge (*Euphorbia esula* L.) plants.

TARGET PESTS: Leafy spurge (*Euphorbia esula*) plants.

TARGET CROPS: Rangelands and riparian areas.

BIOLOGICAL ACTIVITY: Biology: *Oberea erythrocephala* adults are slender beetles, 10 to 12 mm in length, with long, dark antennae. The beetles are slate-grey, with a reddish-orange head or light grey with reddish markings; the legs are yellowish-brown. They are active fliers and may be seen moving above the leafy spurge canopy. The legless larvae are found within leafy spurge roots and may reach a length of 20 mm. They are white, with a yellowish head and the body is obviously segmented. **Egg laying:** After mating, the female beetle chews a hole in the upper part of a leafy spurge stem, into which an egg is deposited. Generally, a single egg is laid per stem, but a female may lay a total of 40 eggs. **Duration of development:** Adult beetles are present in early to mid summer and feed on leafy spurge leaves and flowers. After hatching, young larvae tunnel down the leafy spurge stem until reaching the root crown area, just below the soil surface. Larvae do most of their feeding within the root crown and larger lateral roots. They overwinter in the larval stage and development resumes in the spring. Larvae construct a cell within the upper root crown in which pupation occurs. Newly eclosed adults chew through remaining root tissue and emerge from the soil. There is one generation per year in Europe, but it appears that two years may be required to complete the life-cycle in the northern USA. **Efficacy:** This species is potentially a valuable biological control agent, because of the extensive damage it can cause to leafy spurge root systems. **Key reference(s):** D Schroeder. 1980. Investigations on *Oberea erythrocephala* (Schrank) (Col.: Cerambycidae), a possible biocontrol agent of leafy spurge, *Euphorbia* spp. (Euphorbiaceae) in Canada, *Z. Angew. Entomol.*, **90**, 237–54.

COMMERCIALISATION: Formulation: Sold as adult beetles. **Tradenames:** 'Oberea erythrocephala' (Praxis).

APPLICATION: Release beetles into leafy spurge infested rangeland in early summer.

PRODUCT SPECIFICATIONS: Storage conditions: Use as soon as possible after receipt.

COMPATIBILITY: It is unusual to use *Oberea erythrocephala* with chemical pesticides.

MAMMALIAN TOXICITY: There have been no reports of allergic or other adverse reactions from research workers, manufacturing staff or from the field release of *Oberea erythrocephala*.

ENVIRONMENTAL IMPACT AND NON-TARGET TOXICITY: *Oberea erythrocephala* occurs widely in Nature and is not expected to have any adverse effects on non-target organisms or on the environment. Trials in North America have shown that the host range of *O. erythrocephala* is restricted to plants in the subgenus *Esula* of the genus *Euphorbia*. In Europe, this beetle feeds on leafy spurge and several other closely related spurge species. It will not feed on poinsettia (*Euphorbia pulcherrima* Willd.).

3:340 *Oecophylla longinoda*   *Predatory ant*

Predatory ant: Hymenoptera: Formicidae

NOMENCLATURE: Approved name: *Oecophylla longinoda* Latreille. **Other names:** African weaver ant.

SOURCE: Native to Africa.

PRODUCTION: *Oecophylla longinoda* is not bred commercially.

TARGET PESTS: A wide variety of pest insects including weevils (*Pantorhytes* spp.), coreid bugs (*Amblypelta* and *Pseudotheraptus* spp.) and mirids (*Creontiades*, *Lygus* and *Adelphocoris* spp.).

TARGET CROPS: Tropical tree fruit crops, such as coconut, cocoa, citrus, cashew and mango.

BIOLOGICAL ACTIVITY: Biology: The African weaver ant is a tree inhabiting species which protects the tree crop from insect attack by moving through the tree canopy, where it preys upon or deters a wide range of potential pests. Ropes hung between trees can be used to allow colonies to move from tree to tree. **Predation:** Weaver ants build their nests in trees and forage for pest insects. **Efficacy:** The African weaver ant is territorial and forages in a 3-dimensional system when attending honeydew secreted by insect colonies. When moving off the nest, territorial faecal marking by the major workers is not randomly placed over the home-range area. The chemical trail leading from the nest to a given food site is reinforced by faecal materials on both the horizontal and the vertical planes. In addition to visual cues and chemical trails laid by the major workers, foragers use the territorial marking both to localise the food site and to return to the nest. Thus, anal-drop deposition in *O. longinoda* also has a clear dual-purpose function: territorial and orientational marking. **Key reference(s):** P van Mele. 2008. Biological control with the weaver ant *Oecophylla longinoda* in Africa, *Outlooks on Pest Management*, **19**(4), 180–3.

COMMERCIALISATION: Formulation: *Oecophylla longinoda* is not sold commercially.

COMPATIBILITY: Sprays of azadirachtin can be used on trees containing ant nests.

MAMMALIAN TOXICITY: *Oecophylla longinoda* can inflict painful bites on humans collecting fruit from infested trees.

ENVIRONMENTAL IMPACT AND NON-TARGET TOXICITY: Naturally occurring and unlikley to cause any adverse environmental effects. **Approved for use in organic farming:** Yes.

3:341 *Orius albidipennis*   *Predator of thrips*

Predatory bug: Hemiptera: Anthocoridae

NOMENCLATURE: Approved name: *Orius albidipennis* (Reuter). **Other names:** Minute pirate bug.

SOURCE: Palaearctic species, found mainly in the southern Mediterranean from North Africa to the Near East, Iran and Russian Asia Minor. It is also found in the Canary Islands, Cape

Verde and Spain. *Orius* spp. were first reported as predators of thrips in 1914. Subsequent work in the USA in the late 1970s and early 1980s led to the introduction of *Orius* spp. as commercial products in Canada and Europe in the 1990s.

PRODUCTION: Bred in insectaries on soft-bodied phytophagous adult, larval, nymphal and egg-stage insects. Lepidopteran eggs are a common food source.

TARGET PESTS: Thrips are the main prey, but *Orius albidipennis* will also consume a wide range of insects, including aphids and the eggs of some lepidopteran (caterpillar) species, in addition to mites.

TARGET CROPS: Glasshouse-grown vegetables and ornamentals. 'Oristar' is only recommended for glasshouse-grown bell peppers.

BIOLOGICAL ACTIVITY: Biology: There are seven development stages: egg, five nymphal stages and the adult. Eggs are laid within the plant tissue, usually in the leaf stem or in the main vein on the underside of the leaf. Eggs are usually laid separately. All the nymphal stages have characteristic red eyes. Adults vary in size from 2 to 3 mm, with the females being larger than the males. Males are clearly asymmetrical. **Predation:** All growth stages catch and kill small insects, holding them motionless with their forelegs and sucking them dry. If insect numbers are high, the bugs will kill more insects than they need to eat, and they also kill other *Orius* species and other beneficial insects. Prey is located by touch rather than by sight. The predatory nymphs and adults are relatively fast-moving. **Egg laying:** Eggs are laid, usually singly, with only their tops protruding above the plant surface, 2 to 3 days after mating. Temperature and food supply have a significant effect on the number of eggs laid by a female, with the normal being between 125 and 160 eggs in her lifetime. **Duration of development:** Development is very dependent upon temperature and, at 28 °C, total pre-imaginal development takes 14.5 days, whilst, at 20 °C, it takes 23 days. A poor food supply slows this development time-scale significantly. Females live for 20 to 23 days. **Efficacy:** The adults can fly reasonably well and can locate new infested plants very easily. *Orius albidipennis* is able to survive in the absence of prey.

Key reference(s): 1) R Chyzik, M Klein & Y Ben-Dov. 1995. Reproduction and survival of the predatory bug *Orius albidipennis* on various arthropod prey, *Entomologia Experimentalis et Applicata*, **75**, 27–31. 2) M Salim, S A Masud & H M Khan. 1987. *Orius albidipennis* (Reut.) (Hemiptera: Anthocoridae) – a predator of cotton pests, *Philippine Entomologist*, **7**, 37–42. 3) J Riudavets. 1995. Predators of *Frankliniella occidentalis* (Perg.) and *Thrips tabaci* Lind.: a review, *Wageningen Agricultural University Papers*, **95–1**, 43–87.

COMMERCIALISATION: Formulation: Sold as adults in packaging material with a food supply. **Tradenames:** 'Minute Pirate Bug' (species not stated) (Arbico), 'Orius' (species not stated) (Sautter & Stepper), 'Oristar' (species not stated) (Sumitomo), (Agros) and (Sumika Technoservice).

APPLICATION: Spread carrier material onto plants or place in release boxes. Apply at a rate of one adult per two infested plants. Release close to site of infestation in early morning or late evening, under low light conditions and when the glasshouse vents are closed.

PRODUCT SPECIFICATIONS: Purity: Containers include adult bugs in a carrier, plus a food source and no contaminants. **Storage conditions:** Store at 5 to 10 °C. Do not expose to direct sunlight. **Shelf-life:** Use as soon as possible after receipt.

COMPATIBILITY: Do not use residual insecticides. Adults are easily disturbed. Egg laying is optimal in the presence of a high-quality food source and at temperatures around 22 °C.

MAMMALIAN TOXICITY: No allergic or other adverse reactions have been reported following its use under glasshouse conditions.

ENVIRONMENTAL IMPACT AND NON-TARGET TOXICITY: *Orius albidipennis* occurs widely in Nature and has not shown any adverse effects on the environment, but it will consume other insects. *Orius* species' main prey are thrips, although they will consume other insect species when thrips populations are low. They are native to Mediterranean climates, suggesting that they are unlikely to survive winters in temperate climates.
Approved for use in organic farming: Yes.

3:342 *Orius insidiosus*   *Predator of thrips*

Predatory bug: Hemiptera: Anthocoridae

NOMENCLATURE: Approved name: *Orius insidiosus* (Say). **Other names:** Minute pirate bug.

SOURCE: Nearctic, very widely spread in Nature.

PRODUCTION: Bred in insectaries on soft-bodied phytophagous adult, larval, nymphal and egg-stage insects.

TARGET PESTS: Many thrips, including *Frankliniella occidentalis* (Pergande), *Sericothrips variabilis* (Beach), *Caliothrips fasciatus* (Pergande) and *Thrips tabaci* Lindeman, but will also consume a wide range of insects, including aphids and the eggs of some lepidopteran (caterpillar) species, as well as mites.

TARGET CROPS: Glasshouse-grown vegetables and ornamentals. Widely used in glasshouses in Canada. 'Oristar' is only recommended for glasshouse-grown bell peppers.

BIOLOGICAL ACTIVITY: Biology: There are seven development stages: egg, five nymphal stages and the adult. Eggs are laid within the plant tissue, usually in the leaf stem or in the main vein on the underside of the leaf. Eggs are usually laid separately. All the nymphal stages have characteristic red eyes. Adults vary in size from 2 to 3 mm, with the females being larger than the males. In the field, *Orius insidiosus* are slightly aggregated, with the crowding of nymphs being slightly greater than of adults. **Predation:** All growth stages catch and kill small insects, holding them motionless with their forelegs and sucking them dry, with little difference in the numbers of prey consumed by adults and nymphs. If insect numbers are high, the bugs will kill more insects than they need to eat and they also kill other *Orius* species and other beneficial insects. Prey is located by touch rather than by sight. The predatory nymphs and adults move rapidly. **Egg laying:** Eggs are laid, usually singly, with only their tops protruding above the plant surface, 2 to 3 days after mating. Eggs are laid in or adjacent to the growing tips, with very few eggs laid in flowers and fruits. Temperature and food supply have a significant effect on the number of eggs laid by a female, with the normal being about 1 to 3 eggs per day, giving a total of 30 to 40 eggs in her lifetime. **Duration of development:** Development is very dependent upon temperature, with development time between 28 and 32 °C being 12 days and at 24 °C being 20 days. Development ceases at temperatures below 10 °C. Adult longevity and female fecundity at 26 °C vary with diet, being, respectively, 42 days and 144 eggs on *Ephestia kuehniella* Zeller eggs and 17 days and 66 eggs on *Frankliniella occidentalis* (Pergande) adults. Female longevity is significantly longer on diets of pollen alone, whilst fecundity is higher on diets of *Heliothis virescens* (Fabricius) eggs than pollen. Nymphs can complete their development on diets of pollen. **Efficacy:** The adults can fly reasonably well and can locate new infested plants easily. **Key reference(s):** 1) R J Jacobson. 1997. Integrated pest management (IPM) in greenhouses. In *Thrips as Crop Pests*, T Lewis (ed.), pp. 639–66, CABI, Wallingford, UK. 2) M W Sabelis & P J C van Rijn. 1997. Predation by insects and mites. In *Thrips as Crop Pests*, T Lewis (ed.), pp. 259–354, CABI, Wallingford, UK. 3) J Riudavets. 1995. Predators of *Frankliniella occidentalis* (Perg.) and *Thrips tabaci* Lind.: a review, *Wageningen Agricultural University Papers*, **95–1**, 43–87.

COMMERCIALISATION: Formulation: Sold as adults in buckwheat or vermiculite, with a food supply of moth eggs. **Tradenames:** 'Oriline i' (Syngenta Bioline), 'Predatory Minute Pirate Bug' (M&R Durango), 'Orius insidiosus Pirate Bug' (Nature's Alternative Insectary), 'Orius insidiosus' (IPM Laboratories), (Harmony), (Hydro-Gardens), (Praxis) and (Rincon-Vitova), 'Minute Pirate Bug' (species not stated) (Arbico), 'Minute Pirate Bugs' (*Orius insidiosus*) (Biocontrol Network), 'Orius' (species not stated) (Sautter & Stepper), 'Orius-System' (contains *O. insidiosus*, *O. laevigatus* and *O. majusculus*) and 'Orius-N-System' (nymphs) (Biobest), 'Thripor' (general trade name for *Orius* species) (Koppert), 'Oristar' (species not stated) (Sumitomo), (Agros) and (Sumika Technoservice).

APPLICATION: Sprinkle carrier onto plants or use release boxes, allowing adults to escape on their own. Apply at a rate of one adult per two infested plants. Release close to site of infestation in early morning or late evening, under low light conditions and with the glasshouse vents closed. They can also be used outdoors.

PRODUCT SPECIFICATIONS: Purity: Containers include adult bugs in a carrier, plus a food source and no contaminants. **Storage conditions:** Store at 5 to 10 °C. Do not expose to direct sunlight. **Shelf-life:** Use as soon as possible after receipt.

COMPATIBILITY: Do not use residual insecticides. Adults are easily disturbed. Egg laying is optimal in the presence of a high-quality food source and at temperatures around 22 °C.

MAMMALIAN TOXICITY: No allergic or other adverse reactions have been reported following its use under glasshouse conditions.

ENVIRONMENTAL IMPACT AND NON-TARGET TOXICITY: *Orius insidiosus* occurs in Nature and is not expected to have any adverse effects on non-target organisms or the environment. *Orius* species' main prey are thrips, although they will consume other insect species when thrips populations are low. They are native to Mediterranean climates, suggesting that they are unlikely to survive winters in temperate climates.
Approved for use in organic farming: Yes.

3:343 *Orius laevigatus*   *Predator of thrips*

Predatory bug: Hemiptera: Anthocoridae

NOMENCLATURE: Approved name: *Orius laevigatus* (Fieber). **Other names:** Minute pirate bug.

SOURCE: West palaearctic in areas with marine influence. Widespread throughout the Mediterranean area, especially in the Iberian Peninsula and Atlantic coasts of western Europe. Very widely spread in Nature.

PRODUCTION: Bred in insectaries on soft-bodied phytophagous adult, larval, nymphal and egg-stage insects.

TARGET PESTS: Thrips, especially *Frankliniella occidentalis* (Pergande), *Thrips tabaci* Lindeman and *Caliothrips fasciatus* (Pergande), but will also consume a wide range of insects, including aphids and the eggs of some lepidopteran (caterpillar) species, in addition to mites.

TARGET CROPS: Glasshouse-grown vegetables and ornamentals. *Orius laevigatus* has also found success in strawberry tunnels. 'Oristar' is only recommended for glasshouse-grown bell peppers.

BIOLOGICAL ACTIVITY: Biology: There are seven development stages: egg, five nymphal stages and the adult. Eggs are laid within the plant tissue, usually in the leaf stem or in the main vein on the underside of the leaf. Eggs are usually laid separately. All the nymphal stages have characteristic red eyes. Adults vary in size, with the females being larger than the males. *Orius laevigatus* adapts very well to a protected environment and it can survive, even without thrips prey. **Predation:** All growth stages catch and kill small insects, holding them motionless with their forelegs and sucking their contents. Nymphs and adults consume about two *Frankliniella occidentalis* nymphs or adults per day at 20 °C. If insect numbers are high, the bugs will kill more insects than they need to eat and they also kill other *Orius* species and other beneficial insects. Prey is located by touch rather than by sight. The predatory nymphs and adults are relatively fast-moving. **Egg laying:** Eggs are laid, usually singly, with only their tops protruding above the plant surface, 2 to 3 days after mating. Temperature and food supply have a significant effect on the number of eggs laid by a female. Oviposition periods range from 18 to 33 days. A female will lay about 150 eggs in her lifetime. **Duration of development:** Development is very dependent upon temperature and, at 27 °C, the egg will hatch in about 4 days and the five nymphal stages take about 13 days. Males do not live as long as females. Lower temperatures and a poor food supply slow this development time-scale significantly. *Orius laevigatus* enters reproductive diapause in day-lengths of less than 16 hours and this limits its use in more northerly countries during the early part of the season. It hibernates as adults in European climates. **Efficacy:** The adults can fly reasonably well and can locate new infested plants very easily.

Key reference(s): 1) R J Jacobson. 1997. Integrated pest management (IPM) in greenhouses. In *Thrips as Crop Pests*, T Lewis (ed.), pp. 639–66, CABI, Wallingford, UK. 2) J Riudavets. 1995. Predators of *Frankliniella occidentalis* (Perg.) and *Thrips tabaci* Lind.: a review, *Wageningen Agricultural University Papers*, **95–1**, 43–87.

COMMERCIALISATION: Formulation: Sold as adults in packaging material with a food supply. **Tradenames:** 'Oriline I' (Syngenta Bioline), 'Orisure (I)' (Biological Crop Protection), 'Minute Pirate Bug' (species not stated) (Arbico), 'Orius' (species not stated) (Sautter & Stepper), 'Orius-System' (contains *O. insidiosus*, *O. laevigatus* and *O. majusculus*) and 'Orius-N-System' (nymphs) (Biobest), 'Thripor' (general trade name for *Orius* species) (Koppert), 'Oristar' (species not stated) (Sumitomo), (Agros) and (Sumika Technoservice), 'LeviPAK 500' (Bioplanet).

APPLICATION: Sprinkle carrier over crops or place in a release box and allow adults to escape from packaging on their own. Apply at a rate of one adult per two infested plants or 2 to 4 adults per square metre. Release close to site of infestation in early morning or late evening, when light intensity is low and the glasshouse vents are closed.

PRODUCT SPECIFICATIONS: Purity: Containers include adult bugs in a carrier, plus a food source and no contaminants. **Storage conditions:** Store at 10 to 15 °C. Do not expose to direct sunlight. **Shelf-life:** Use as soon as possible after receipt.

COMPATIBILITY: Do not use residual insecticides. Adults are easily disturbed. Egg laying is optimal in the presence of a high-quality food source and at temperatures around 22 °C.

MAMMALIAN TOXICITY: No allergic or other adverse reactions have been reported following its use under glasshouse conditions.

ENVIRONMENTAL IMPACT AND NON-TARGET TOXICITY: *Orius laevigatus* occurs widely in Nature and would not be expected to show any adverse effects on non-target organisms or on the environment. *Orius* species' main prey are thrips, although they will consume other insect species when thrips populations are low. They are native to Mediterranean climates, suggesting that they are unlikely to survive winters in temperate climates. **Approved for use in organic farming:** Yes.

3:344 *Orius majusculus*   *Predator of thrips*

Predatory bug: Hemiptera: Anthocoridae

NOMENCLATURE: Approved name: *Orius majusculus* Reuter. **Other names:** Minute pirate bug.

SOURCE: Palaearctic. Common throughout central Europe, parts of the British Isles and southern Scandinavia. Also found in southern Europe, but not in North Africa.

PRODUCTION: Bred in insectaries on soft-bodied phytophagous adult, larval, nymphal and egg-stage insects.

TARGET PESTS: Thrips, including *Frankliniella occidentalis* (Pergande), *Thrips tabaci* Lindeman and *Caliothrips fasciatus* (Pergande), but will also consume a wide range of insects, including aphids and the eggs of some lepidopteran (caterpillar) species, in addition to mites.

TARGET CROPS: Glasshouse-grown vegetables and ornamentals. 'Oristar' is only recommended for glasshouse-grown bell peppers.

BIOLOGICAL ACTIVITY: Biology: There are seven development stages: egg, five nymphal stages and the adult. Eggs are laid within the plant tissue, usually in the leaf stem or in the main vein on the underside of the leaf. Eggs are usually laid separately. All the nymphal stages have characteristic red eyes. Adults vary in size, with the females being larger than the males. Winter populations in the wild consist mainly of females that become active in March to April. Oviposition begins in May. **Predation:** All growth stages catch and kill small insects, holding them motionless with their forelegs and sucking them dry. The immature stages of *Orius majusculus* will consume about 130 *Frankliniella occidentalis* nymphs at 25 °C. If prey are abundant, the bugs will kill more insects than they need to eat and they also kill other *Orius* species and other beneficial insects. Prey is located by touch rather than by sight. The predatory nymphs and adults are relatively fast-moving. **Egg laying:** Eggs are laid, 2 to 3 days after mating, usually singly, with only their tops protruding above the plant surface. Temperature and food supply have a significant effect on the number of eggs laid by a female. At 15 °C, 20 °C and 25 °C, females fed on caterpillar eggs laid 195, 158 and 237 eggs, respectively. **Duration of development:** Development is temperature-dependent and, at 25 °C, the egg will hatch in about 4 days, whilst, at 15 °C, it takes 9 to 10 days. The duration of the five nymphal stages is between 14 and 15 days. Adult longevity is dependent on food supply and temperature, but males do not live as long as females. At 26 °C, with caterpillar eggs as food, *Orius majusculus* lives for about 50 days, but with *Frankliniella occidentalis* adults as food, it lives for about 20 days. **Efficacy:** The adults can fly reasonably well and can locate new infested plants very easily. **Key reference(s):** 1) R J Jacobson. 1997. Integrated pest management (IPM) in greenhouses. In *Thrips as Crop Pests*, T Lewis (ed.), pp. 639–66, CABI, Wallingford, UK. 2) J Riudavets. 1995. Predators of *Frankliniella occidentalis* (Perg.) and *Thrips tabaci* Lind.: a review, *Wageningen Agricultural University Papers*, **95–1**, 43–87.

COMMERCIALISATION: Formulation: Sold as adults in packaging material with a food supply. **Tradenames:** 'Oriline m' (Syngenta Bioline), 'Minute Pirate Bug' (species not stated) (Arbico), 'Orius' (species not stated) (Sautter & Stepper), 'Orius-System' (contains *O. insidiosus*, *O. laevigatus* and *O. majusculus*) and 'Orius-N-System' (nymphs) (Biobest), 'Orius-Raubwanzen' (Neudorff), 'Thripor' (general trade name for *Orius* species) (Koppert), 'Oristar' (species not stated) (Sumitomo), (Agros) and (Sumika Technoservice).

APPLICATION: Sprinkle carrier onto crops or place in release boxes and allow adults to escape on their own. Apply at a rate of one adult per two infested plants. Release close to site of infestation in early morning or late evening, when the light intensity is low and the glasshouse vents are closed. Best results in glasshouses were obtained with one predator per 100 thrips when starting with infestations of 10 thrips per leaf. Do not release in early crops (before late March in northern Europe), as the bugs often do not establish.

PRODUCT SPECIFICATIONS: Purity: Containers include adult bugs in a carrier, plus a food source and no contaminants. **Storage conditions:** Store at 10 to 15 °C. Do not expose to direct sunlight. **Shelf-life:** Use as soon as possible after receipt.

COMPATIBILITY: Do not use residual insecticides. Adults are easily disturbed. Egg laying is optimal in the presence of a high-quality food source and at temperatures around 22 °C. *Orius majusculus* has occasionally been observed to be phytophagous on chrysanthemums in The Netherlands.

MAMMALIAN TOXICITY: No allergic or other adverse reactions have been reported following its use under glasshouse conditions.

ENVIRONMENTAL IMPACT AND NON-TARGET TOXICITY: *Orius majusculus* occurs widely in Nature and has not been shown to have any adverse effects on non-target organisms or on the environment. *Orius* species' main prey are thrips, although they will consume other insect species when thrips populations are low.
Approved for use in organic farming: Yes.

3:345 *Pediobius foveolatus*
Parasitoid of Mexican bean beetle

Parasitoid wasp: Hymenoptera: Eulophidae

NOMENCLATURE: Approved name: *Pediobius foveolatus* (Crawford). **Other names:** Mexican bean beetle parasitoid.

SOURCE: *Pediobius foveolatus* was introduced from India and now occurs widely in North America.

PRODUCTION: Reared in insectaries on Mexican bean beetles (*Epilachna varivestis* Mulsant) and supplied as mummies.

TARGET PESTS: Mexican bean beetle (*Epilachna varivestis*) and squash bug (*Anasa tristis* (De Geer)) nymphs.

TARGET CROPS: Beans and other legumes and squash and other cucurbit crops.

BIOLOGICAL ACTIVITY: Biology: *Pediobius foveolatus* adults lay eggs in the larvae of the Mexican bean beetle or nymphs of the squash bug. The parasites develop within these juvenile stages, eventually turning them into mummies. The reproductive cycle is very rapid and there are usually several generations within a season, depending upon the temperature and food source. *P. foveolatus* is intolerant of cold conditions and will not survive the winter in colder climates. **Predation:** The fertile female wasps are very mobile and seek out host larvae actively. **Key reference(s):** E M Barrows & M E Hooker. 1981. Parasitization of the Mexican bean beetle by *Pediobius foveolatus* in urban vegetable gardens, *Environ. Entomol.*, **10**, 782–6.

COMMERCIALISATION: Formulation: Sold as mummies in Mexican bean beetle larvae. **Tradenames:** 'Pediobius foveolatus mummies' (Arbico), (Biocontrol Network), (Green Spot), (Praxis) and (Rincon-Vitova).

APPLICATION: The mummies should be introduced into the field when eggs of the host insects are beginning to hatch or, at the latest, when the first generation Mexican bean beetle larvae are about half-grown. Two applications are recommended at 7-day intervals, at a rate of 250 to 750 mummies per hectare.

PRODUCT SPECIFICATIONS: Storage conditions: Use as soon as possible after receipt. Store in cool, dry conditions and do not expose to direct sunlight. **Shelf-life:** The mummies hatch very quickly and should be used as soon as possible.

COMPATIBILITY: Incompatible with persistent insecticides.

MAMMALIAN TOXICITY: *Pediobius foveolatus* has caused no allergic or other adverse effects on research workers, manufacturing or field staff. It is considered to be non-toxic to mammals.

ENVIRONMENTAL IMPACT AND NON-TARGET TOXICITY: *Pediobius foveolatus* is not expected to have any adverse effects on non-target organisms or on the environment. Its prey are bean beetles and cucurbit bugs. A native of India, it is unlikley to survive in cold winter conditions. **Approved for use in organic farming:** Yes.

3:346 *Pentalitomastix plethorica*
Parasitoid of navel orangeworm

Parasitoid wasp: Hymenoptera: Encyrtidae

NOMENCLATURE: Approved name: *Pentalitomastix plethorica* Caltagirone.
Other names: Navel orangeworm parasitoid.

SOURCE: *Pentalitomastix plethorica* is native to Mexico.

PRODUCTION: Bred under controlled conditions on navel orangeworm (*Amyelois transitella* (Walker)).

TARGET PESTS: Navel orangeworm (*Amyelois transitella*).

TARGET CROPS: Orchard nut crops (walnuts, almonds, pistachios).

BIOLOGICAL ACTIVITY: Biology: The parasitoid lays eggs in the egg of the navel orangeworm. It is polyembryonic and develops up to 1250 offspring from one navel orangeworm mummy. **Duration of development:** At 27 °C (80 °F), it takes 45 days from egg lay until adult emergence.

COMMERCIALISATION: Formulation: *Pentalitomastix plethorica* is sold by mummy case, 750 to 1250 wasps per mummy case. **Tradenames:** 'Pentalitomastix plethorica' (Arbico), (Rincon-Vitova) and (Praxis).

APPLICATION: Release rate varies according to the weather conditions and the level of navel orangeworm infestation.

PRODUCT SPECIFICATIONS: Storage conditions: Store at low temperature (10 °C (50 °F)) and at a r.h. of around 75%. **Shelf-life:** If stored under the manufacturer's recommended conditions, *Pentalitomastix plethorica* may be kept for 7 days before release.

COMPATIBILITY: It is advisable not to release *Pentalitomastix plethorica* following applications of broad-spectrum, residual insecticides.

MAMMALIAN TOXICITY: *Pentalitomastix plethorica* has caused no allergic or other adverse effects on research workers, manufacturing or field staff. It is considered to be non-toxic to mammals.

ENVIRONMENTAL IMPACT AND NON-TARGET TOXICITY: *Pentalitomastix plethorica* is not expected to have any adverse effects on non-target organisms or on the environment. Its prey is the navel orange worm and it is a native of Mexico. It does not survive temperate winters.

3:347 *Phytoseiulus macropilis*
Predator of spider mites

Predatory mite: Mesostigmata: Phytoseiidae

NOMENCLATURE: Approved name: *Phytoseiulus macropilis* (Banks). **Other names:** Red spider mite predator; formerly known as *Phytoseiulus speyeri* Evans; *Phytoseiulus chanti* Ehara; and *Typhlodromus macrosetis* Hirschmann.

SOURCE: *Phytoseiulus macropilis* is found in the United States, Brazil, the Canary Islands and Africa as well as on numerous islands in the Pacific Ocean.

PRODUCTION: Reared in insectaries on *Tetranychus urticae* Koch on beans under controlled conditions.

TARGET PESTS: Spider mites, especially the two-spotted spider mite (*Tetranychus urticae*).

TARGET CROPS: Glasshouse-grown vegetables and ornamentals and outdoor crops such as strawberries, cotton, vegetables and ornamentals.

BIOLOGICAL ACTIVITY: Biology: *Phytoseiulus macropilis* is about 0.5 mm in length, reddish orange in colour and has a pear shaped body. It is very similar to *Phytoseiulus persimilis*. It has five life stages, egg, larva, protonymph, deutonymph and adult. The sex ratio is typically four females to one male. **Predation:** It does not reproduce on pollen, insects or plants. It is well adapted to mild temperatures and high humidity (60–80% RH). It does not do well in hot glasshouses or dry interior areas (>35 °C). Under ideal conditions (20 to 25 °C, 70 to 80% r.h.), *Phytoseiulus macropilis* can consume 7 adult spider mites or 20 immatures or 25 eggs/day. It is capable of multiplying its numbers 44 times in 17 days. This high rate of reproduction enables it to overcome its slower reproducing host in a relatively short time. **Egg laying:** The female lays an average of 2.4 eggs per day for about 30 days, depending on temperature. **Duration of development:** *Phytoseiulus macropilis* can complete egg to egg development in as little as 5 five days (30 °C). At 15 °C it takes 25 days. **Efficacy:** Under ideal conditions of temperature and humidity, the predator can destroy an infestation and will subsequently starve to death.

COMMERCIALISATION: Formulation: Sold as a mixture of eggs, larvae, protonymphs, deutonymphs and adults in an inert carrier. **Tradenames:** 'Macropolis' (Olive Trees).

APPLICATION: Introduce *Phytoseiulus macropilis* at first sign of spider mite activity or when spider mites infest less than 10% of the leaves. If the spider mites average more than one per leaf, treat with a miticide before using *Phytoseiulus macropilis*. Apply at the rate of 2 per square metre of planted area or 2 per damaged leaf or 2 per plant, if plants are small.

PRODUCT SPECIFICATIONS: Purity: No contaminants. **Storage conditions:** The bottles of predators should be examined for mite activity upon arrival. Release the predators on infested plants as soon as possible. The predators are packed in an inert medium (vermiculite, corn cob grits); this material aids in the distribution of the predators on the plants.

COMPATIBILITY: Avoid using pesticides one week prior to or one week after release of *Phytoseiulus macropilis*. Pyrethroids and organophosphates are highly toxic to most predators. Some materials such as methomyl may be toxic for up to four weeks.

MAMMALIAN TOXICITY: No allergic or other adverse reactions have been reported from its use in glasshouse or outdoor crops.

ENVIRONMENTAL IMPACT AND NON-TARGET TOXICITY: *Phytoseiulus macropilis* is widespread in Nature and is not thought to be damaging to non-target species or to the environment. It is specific to spider mites. **Approved for use in organic farming:** Yes

3:348 *Phytoseiulus persimilis*

Predator of spider mites

Predatory mite: Mesostigmata: Phytoseiidae

NOMENCLATURE: Approved name: *Phytoseiulus persimilis* Athios-Henriot.
Other names: Red spider mite predator; Chilean mite. Formerly known as: *Phytoseiulus riegeli* Dosse; and *Phytoseiulus tardi* (Lombardini).

SOURCE: Claimed to have been first identified on roots of orchids imported into Germany from Chile in 1958, although the species is now said to be Mediterranean in origin, with type specimens from Algeria.

PRODUCTION: Reared in insectaries on *Tetranychus urticae* Koch on beans under controlled conditions.

TARGET PESTS: Two-spotted spider mite (*Tetranychus urticae*).

TARGET CROPS: Glasshouse-grown vegetables and ornamentals, interiorscapes and outdoor crops such as strawberries, cotton, vegetables and ornamentals.

BIOLOGICAL ACTIVITY: Biology: Eggs hatch into six-legged larvae that do not feed. The protonymph emerges from the larva and immediately starts to feed. At the second nymphal stage, the predator searches for food constantly. The adults emerge from the nymphal stage, mate within a few hours and the females then begin to lay eggs. Adult females have pear-shaped bodies, are shiny orange, very mobile and slightly bigger than a red spider mite. In the absence of food, the predator can survive for some time on water and honey, but reproduction then ceases. **Predation:** Larvae do not feed, the nymphal stages consume eggs, larvae and protonymphs of the spider mite. Adults eat all stages of the prey. *Phytoseiulus persimilis* is almost completely dependent on the spider mite. **Egg laying:** Eggs are laid near a food source on the surfaces of leaves. A female will lay eggs for the duration of her life, following mating. No eggs are laid by unfertilised females. A female can deposit up to 5 eggs per day under ideal conditions of temperature, humidity and food supply and will lay a maximum of 60 eggs in her lifetime. **Duration of development:** Rate of development is very dependent upon the temperature. At 15 °C, the period of time from egg to egg is about 25 days, but, at 30 °C, this is reduced to about 5 days. **Efficacy:** The rapid breeding rate of *Phytoseiulus persimilis* means that it is a very effective predator of spider mites. It reproduces much faster than the pest mite and the nymphal stages remain on the leaf on which they hatch. The adults move from plant to plant through physical contact of adjacent plants and along spider mite webs. An adult female predator will consume about 5 adult spider mites or 20 young larvae and eggs in a day. Under ideal conditions of temperature (15 to 25 °C) and humidity (60% to 70%), the predator can destroy an infestation and will subsequently starve to death.

3. Macro-organisms

COMMERCIALISATION: Formulation: Sold as adults in bags, paper sachets, bottles and vials with a suitable carrier (such as vermiculite) and *Tetranychus urticae* eggs as a food source. Also supplied on bean leaves that can be distributed throughout the crop. **Tradenames:** 'Phytoline p' (Syngenta Bioline), 'Spidex' (Koppert), 'Phytoseiulus-System' and 'Pré-Phytoseiulus-System' (Biobest), 'Phytosure (p)' and 'Phytosure (pt)' (Biological Crop Protection), 'Phytoseiulus persimilis' (Applied Bio-Nomics), (Applied BioPest), (Arbico), (Beneficial Insectary), (Biocontrol Network), (Biowise), (Neudorff) and (Rincon-Vitova), 'TriPak' (*Phytoseiulus persimilis* plus *Amblyseius fallacis* plus *Stethorus punctillum*) (Rincon-Vitova), 'FitoPAK 2000' (Bioplanet).

APPLICATION: Under glasshouse conditions, release one predatory mite for every 15 to 25 spider mites or 5 predatory mites per square metre as soon as the first sign of infestation is noticed. Shake adults carefully and evenly throughout the crop. Outdoors, release between 2500 and 50 000 per hectare.

PRODUCT SPECIFICATIONS: Purity: Fecund adults with carrier and a few red spider mite eggs. No contaminants. **Storage conditions:** Store at 12 to 15 °C and protect from freezing and high temperatures. **Shelf-life:** Use as soon as possible, as the prey mites on leaf pieces will soon be consumed and the predators may turn to cannibalism.

COMPATIBILITY: Incompatible with benzimidazole fungicides and residual insecticides. Humidities below 60% slow life-cycle and inhibit egg hatch. Mites go dormant below 10 °C. May be used in conjunction with *Amblyseius californicus* (McGregor).

MAMMALIAN TOXICITY: No allergic or other adverse reactions have been reported from its use in glasshouse or outdoor crops.

ENVIRONMENTAL IMPACT AND NON-TARGET TOXICITY: *Phytoseiulus persimilis* is widespread in Nature and is not thought to be damaging to non-target species or to the environment. It is specific to spider mites. **Approved for use in organic farming:** Yes.

3:349 *Podisus maculiventris*
Predator of caterpillars

Predatory bug: Hemiptera: Pentatomidae

NOMENCLATURE: Approved name: *Podisus maculiventris* (Say). **Other names:** Caterpillar predator; spined soldier bug; predatory stink bug.

SOURCE: *Podisus maculiventris* occurs widely in Nature in North and South America. It is a plant-dwelling, predatory bug.

PRODUCTION: Reared under controlled conditions on lepidopteran larvae.

TARGET PESTS: Lepidoptera (caterpillars) and Coleoptera (beetles).

TARGET CROPS: Vegetable, agronomic and ornamental crops in glasshouses and in open fields.

BIOLOGICAL ACTIVITY: Biology: Young nymphs are red, with a black pattern on the abdomen. Older nymphs have a pattern of black, white, orange and yellow on the abdomen. The development threshold for *Podisus maculiventris* is 11 to 12 °C, with a temperature optimum between 17 and 26 °C. A constant temperature of 33 °C is lethal. *Podisus maculiventris* does not enter diapause. **Predation:** Both nymphs and adults feed on caterpillars and the larvae of beetles. They can eat caterpillars that are larger than other predators can manage. The prey is 'harpooned', injected with a substance that quickly paralyses it and is killed as body fluids are sucked out through the 'harpoon'. Nymphs begin feeding four or five days after hatching and continue to attack prey throughout the next three weeks of their development. **Egg laying:** Yellow eggs are deposited in clusters (20 to 25 eggs per cluster), with the cluster bearing a crop of spines. **Duration of development:** At temperatures between 21 and 27 °C, eggs take about 5 days to hatch. There are five nymphal stages, totalling between 20 and 28 days. **Efficacy:** A very mobile predator of Lepidoptera and Coleoptera. *Podisus maculiventris* will provide approximately 30 days of control from time of release, but under optimum conditions it can provide control for up to 85 days.

COMMERCIALISATION: Formulation: Sold as mixed instar nymphs (third and fourth instars) and adults mixed with paper scraps or vermiculite. Each box contains 100 predatory bugs. **Tradenames:** 'Podibug' (Koppert), 'Podisus maculiventris' (Arbico), (Praxis) and (Rincon-Vitova), 'Podiline m' (Syngenta Bioline), 'Spined Soldier Bug' (Biocontrol Network).

APPLICATION: Open box carefully at the site of application and introduce the packing material plus bugs to the infested plants. *Podisus maculiventris* is highly mobile: nymphs can walk quickly from plant to plant and adults will spread rapidly by flying and walking. Adult bugs may fly away from the place of introduction. Use one bug per square metre for light infestations and 5 per square metre for heavy infestations or hot spots. Both rates are curative.

PRODUCT SPECIFICATIONS: Purity: No phytophagous mites, moulds or other contaminants. **Storage conditions:** Store at 8 to 10 °C in the dark. **Shelf-life:** Use within 3 days, if stored as recommended.

COMPATIBILITY: Sensitive to most conventional pesticides.

MAMMALIAN TOXICITY: There have been no reports of allergic or other adverse toxicological effects arising from contact with *Podisus maculiventris* from research staff, producers or users.

ENVIRONMENTAL IMPACT AND NON-TARGET TOXICITY: *Podisus maculiventris* has a wide distribution in Nature and there is no evidence that it has any adverse effects on non-target organisms or on the environment. It preys upon lepidopteran larvae, but is unable to withstand elevated temperatures (above 33 °C) and its development stops at temperatures below 11 °C. **Approved for use in organic farming:** Yes.

3:350 *Psyllaephagus pilosus*

Parasitoid of psyllids

Parasitoid wasp: Hymenoptera: Encyrtidae

NOMENCLATURE: Approved name: *Psyllaephagus pilosus* Noyes. **Other names:** Eucalyptus psyllid parasitoid.

SOURCE: *Psyllaephagus pilosus* originated in Australia.

PRODUCTION: Reared under controlled conditions on the eucalyptus psyllid (*Ctenarytaina eucalypti* (Maskell)).

TARGET PESTS: Eucalyptus psyllid (*Ctenarytaina eucalypti*).

TARGET CROPS: Eucalyptus.

BIOLOGICAL ACTIVITY: Biology: The female *Psyllaephagus pilosus* wasp lays a single egg in a number of psyllid nymphs. These remain alive for some time as each resulting wasp larva feeds inside its host. The psyllid's metabolism is diverted to feed the young parasitoid larva, which grows at its host's expense. It is important for the parasitoid that its host survives as long as possible, so essential psyllid nymph organs are not destroyed immediately. Eventually, however, like all parasitoid Hymenoptera, the *Psyllaephagus* larva kills its host, creating a mummy, which remains stuck to the plant. Once its development is complete, the adult wasp emerges from the mummy through an exit hole cut with its mouthparts. This exit opening is always situated dorsally on the abdomen of the psyllid. Adult *Psyllaephagus pilosus* wasps also feed as predators of younger eucalyptus psyllid nymphs, biting into the psyllid's body and licking the haemolymph that leaks from the resulting wound.

COMMERCIALISATION: Formulation: Supplied as adult wasps or as mummies. **Tradenames:** 'Psyllaephagus pilosus'.

APPLICATION: The mummies are distributed within the plantation at the time of psyllid invasion. Adults are released when psyllids are noticed.

PRODUCT SPECIFICATIONS: Shelf-life: Use as soon as possible, as quality declines with storage.

COMPATIBILITY: It is advisable not to release *Psyllaephagus pilosus* following applications of broad-spectrum, residual insecticides.

MAMMALIAN TOXICITY: *Psyllaephagus pilosus* has not demonstrated evidence of toxicity, infectivity, irritation or hypersensitivity to mammals. No allergic responses or other adverse health problems have been observed by research workers, manufacturing staff or users.

ENVIRONMENTAL IMPACT AND NON-TARGET TOXICITY: *Psyllaephagus pilosus* is not expected to have any adverse effects on non-target organisms or on the environment. It preys upon psyllids and, being a native of Australia, is unable to over-winter under temperate conditions.

3:351 *Pyemotes tritici*　　　　　*Predator of insects*

Mite: Acari: Pyemotidae

NOMENCLATURE: Approved name: *Pyemotes tritici* (Lagrèze-Fossat & Montané). **Other names:** Straw itch mite; grain itch mite.

SOURCE: *Pyemotes tritici* was first described in 1959 from specimens collected by W.W. Boyle from the dry-wood termite, *Cryptotermes brevis* (Walker), and the koa haole seed weevil, *Araecerus levipennis* Jordan, in Honolulu, Hawaii.

PRODUCTION: Bred on a variety of insects, under controlled conditions.

TARGET PESTS: Grain insect pests, especially Coleoptera (beetles).

TARGET CROPS: Used in grain stores.

BIOLOGICAL ACTIVITY: Biology: *Pyemotes tritici* is known to parasitise a wide variety of insects. Some of the more common hosts are coleopterous or lepidopterous larvae which infest the seeds of plants. **Predation:** It has been found that at least 90% of one or more of the developmental stages of *Plodia interpunctella* (Hb.), *Lasioderma serricorne* (F.), *Tribolium castaneum* (Hbst.), *Oryzaephilus mercator* (Fauv.) and *Ephestia cautella* (Wlk.) were parasitised by *Pyemotes tritici*. The ease with which the mites can be reared and populations synchronised, and the short duration of a generation (about 4–7 days) ensure that sufficient numbers of the parasite can be made available to reduce a pest population effectively. **Egg laying:** Fertilised females give birth to male and female offspring, but unfertilised females give birth only to males. A female can produce over 200 offspring in her life-time.

Duration of development: Time of emergence of this ovoviviparous species ranged from 4 days at 30 °C to 17 days at 18 °C. Above or below this temperature range few offspring emerged. **Key reference(s)**: 1) P Vaivanijkul. 1979. Life history of the grain itch mite, *Pyemotes tritici* (Acarina: Pyemotidae), *J. Sci. Soc. Thailand*, **5**, 89–96. 2) W Hoschele & L K Tanigoshi. 1993. *Pyemotes tritici* (Acari: Pyemotidae), a potential biological control agent of *Anagasta kuehniella* (Lepidoptera: Pyralidae), *Experimental and Applied Acarology*, **17**(11).

COMMERCIALISATION: Tradenames: 'Pyemotes tritici' (BugLogical), (CropKing) and (Panhandle Biological).

APPLICATION: Introduce adults into grain stores shortly after filling.

PRODUCT SPECIFICATIONS: Purity: No contaminating insects. **Storage conditions:** Can be stored in a cooled, insulated container at 15 to 19 °C, away from direct sunlight.

COMPATIBILITY: Cannot be used with any broad-spectrum grain store fumigant or broad-spectrum insecticide.

MAMMALIAN TOXICITY: Man is sometimes attacked by *Pyemotes tritici*; however, they are not known to reproduce on him. Man is bitten by these pyemotid mites when he comes in close contact with products infested by insects which are parasitised by them.

ENVIRONMENTAL IMPACT AND NON-TARGET TOXICITY: *Pyemotes tritici* has a wide distribution in Nature and there is no evidence that it has any adverse effects on non-target organisms or on the environment. It is polyphagous, but is unlikely to leave the grain stores into which it was introduced. **Approved for use in organic farming:** Yes.

3:352 *Rhinocyllus conicus*   *Phytophagous weevil*

Phytophagous weevil: Coleoptera: Curculionidae

NOMENCLATURE: Approved name: *Rhinocyllus conicus* Froelich. **Other names:** Thistle-head weevil.

SOURCE: *Rhinocyllus conicus* is a native of southern and central Europe, North Africa and western Asia. It was the first insect introduced into North America for the biological control of musk or nodding thistle. It was collected in France and released in Canada in 1968. In 1969, it was introduced into California, Virginia and Montana. Subsequently, it was relocated from Virginia to other states in the USA.

PRODUCTION: Raised on musk or nodding thistle (*Carduus nutans* L.).

TARGET PESTS: Musk or nodding thistle (*Carduus nutans*).

TARGET CROPS: Rangeland and pasture.

BIOLOGICAL ACTIVITY: Biology: Adult *Rhinocyllus conicus* are dark brown in colour and 10 to 15 mm long. Each female lays between 100 and 200 eggs on the bracts of thistle heads. Eggs hatch in 6 to 9 days and newly hatched larvae feed through the bracts into the receptacle. Developing larvae feed on the receptacle and the young seeds, reducing or preventing the production of viable seeds. There are four larval growth stages, which last about 4 to 6 weeks. Pupation occurs in a black oval cell inside the thistle head and takes another 7 to 10 days. Adults emerge from seed heads in mid-summer. They seek shelter and pass the summer in hiding and then hibernate in winter. The following spring, adults emerge from winter shelters to lay eggs on the developing buds before dying. There is one adult generation per year. **Efficacy:** *Rhinocyllus conicus* is very effective in the control of musk or nodding thistle. The first success in biological control of musk thistle was documented in 1975, soon after the weevils were released in Virginia. Typical musk thistle stand reductions of 80% to 95% occurred in sites where the weevil became established. However, it only provides partial control of plumeless thistle (*Carduus acanthoides* L.), because the egg-laying period of the weevil only covers the development of the terminal thistle heads, but not the lateral heads. **Key reference(s):** 1) L T Kok & W W Surles. 1975. Successful biocontrol of musk thistle by an introduced weevil, *Rhinocyllus conicus*, *Environ. Entomol.*, **4**, 1025–7. 2) L M Smith, F W Ravlin, L T Kok & W T Mays. 1984. Seasonal model of the interaction between *Rhinocyllus conicus* (Coleoptera: Curculionidae) and its weed host, *Carduus thoermeri* (Campanulatae: Asteraceae), *Environ. Entomol.*, **13**, 1417–26. 3) W W Surles & L T Kok. 1977. Ovipositional preference and synchronization of *Rhinocyllus conicus* with *Carduus nutans* and *C. acanthoides*, *Environ. Entomol.*, **6**, 222–4.

COMMERCIALISATION: Formulation: *Rhinocyllus conicus* is available commercially and is shipped in the adult stage. **Tradenames:** 'Rhinocyllus conicus' (Praxis).

APPLICATION: Overwintered adults must be released in early spring without delay, to allow them to lay their eggs before dying. Adults of the new generation that are collected in the summer will not lay eggs until the following spring. Thus, they can be released any time after collection. These weevils will hibernate and the survivors will lay eggs in the spring.

PRODUCT SPECIFICATIONS: Purity: Products contain only overwintered adults. **Storage conditions:** Adults must be released immediately upon receipt.

COMPATIBILITY: *Rhinocyllus conicus* is susceptible to commonly used insecticides and should not be exposed to insecticidal sprays. It is not directly affected by commonly used herbicides and can be used in conjunction with herbicides for thistle control. Larval development is indirectly affected by herbicides if the thistle dies before larval feeding is completed.

3. Macro-organisms

Herbicides are best applied after the insect has completed development.

MAMMALIAN TOXICITY: There have been no reports of allergic or other adverse reactions from research workers, manufacturing staff or from the field release of *Rhinocyllus conicus*.

ENVIRONMENTAL IMPACT AND NON-TARGET TOXICITY: *Rhinocyllus conicus* occurs widely in Nature and is not expected to have any adverse effects on non-target organisms or on the environment. It is specific to the musk or nodding thistle.

3:353 *Rhizophagus grandis*
Predator of bark beetles

Predatory beetle: Coleoptera: Rhizophagidae

NOMENCLATURE: Approved name: *Rhizophagus grandis* Gyllenhal.

SOURCE: Originated in Eurasia, but now widely occurring.

PRODUCTION: Bred on *Dendroctonus micans* (Kugelann) larvae.

TARGET PESTS: Bark beetles (*Dendroctonus micans* and *Dendroctonus terebrans* (Oliver)).

TARGET CROPS: Forests, especially spruce.

BIOLOGICAL ACTIVITY: Biology: *Rhizophagus grandis* adults are brown in colour and are the size of a rice grain. They are attracted to volatile chemicals present in larval frass of *Dendroctonus* spp., a system that is very effective, with evidence that pest populations as low as 1 to 4 per hectare showed up to 90% predation. The females mate shortly after emergence and lay their eggs in the larval galleries of the bark beetles. Adults and larvae feed on the pests, but the larvae are more damaging. The *Dendroctonus* larvae are completely consumed, with only the cuticle left. At this stage, the *Rhizophagus grandis* larvae move on to attack nearby larvae, pupating within the pest colonies. **Key reference(s):** 1) M Baisier. 1990. Biologie des stades immatures du prédateur *Rhizophagus grandis* Gyll. (Coleoptera: Rhizophagidae), *Thèse de Doctorat en Sciences Zoologiques, Université Libre de Bruxelles*, Belgium, 195 pp. 2) J H Frank & J L Foltz. 1997. Classical Biological Control of Pest Insects of Trees in the Southern United States, A Review and Recommendations, USDA, FS, FHTET-96-20.

COMMERCIALISATION: Formulation: Sold or released as adults. **Tradenames:** 'Rhizophagus grandis'.

APPLICATION: Adults are released in forests where bark beetle infestations are known.

PRODUCT SPECIFICATIONS: Purity: Only fecund adults. **Storage conditions:** Use as soon as possible.

COMPATIBILITY: Should not be used with residual insecticides.

MAMMALIAN TOXICITY: There have been no reports of allergic or other adverse toxicological effects arising from contact with *Rhizophagus grandis* from research staff, producers or users.

ENVIRONMENTAL IMPACT AND NON-TARGET TOXICITY: *Rhizophagus grandis* has a wide distribution in Nature and there is no evidence that it has any adverse effects on non-target organisms or on the environment. Its habitat is forests, where its prey are widely dispersed and there is considerable competition for resources.

3:354 *Rhyzobius forestieri*
Predator of scale insects

Predatory beetle: Coleoptera: Coccinellidae

NOMENCLATURE: Approved name: *Rhyzobius forestieri* (Mulsant). **Other names:** Black ladybird.

SOURCE: Common in Mediterranean regions, but now found in North America and Australia.

PRODUCTION: *Rhyzobius forestieri* is reared on scale insects and supplied as adults with a food source available.

TARGET PESTS: Olive scale (*Saissetia oleae* Olivier).

TARGET CROPS: Olive and citrus groves.

BIOLOGICAL ACTIVITY: Predation: The adults consume scale insects. Eggs hatch under the scale cover and the larvae consume scale crawlers and eggs. **Egg laying:** Females lay a large number of eggs under scale insects. Each female is able to lay several hundred eggs. **Duration of development:** The larval stages last about 3 weeks, with final instar larvae reaching a length of about 2.5 mm. Adults are around 2.5 mm in length and live for 5 to 8 weeks, depending upon the temperature and the food source. The beetles will consume scale insects at temperatures around 5 °C, but are most effective at temperatures between 23 and 30 °C, with a r.h. of about 65%. Low temperatures delay or inhibit reproduction and egg laying. **Efficacy:** *Rhyzobius forestieri* is a very effective predator of scale insects and, being very mobile, will seek out new infestations actively. The adults will consume pollen and it is suggested that reproductive ability is improved in the presence of flowering, pollen-producing plants.

COMMERCIALISATION: Formulation: Sold as pre-mated and fed, insectary-reared adults, primarily for the citrus industry. **Tradenames:** 'Rhyzobius forestieri' (Applied BioPest), (Arbico) and (BugLogical).

APPLICATION: Open containers should be placed within the crop to allow the adult beetles to emerge. It is advisable to place the containers within the canopy of trees. Do not use yellow sticky insect traps. Control ants, as these will protect the scale from beetle attack. *Rhyzobius forestieri* cannot tolerate high levels of honeydew.

PRODUCT SPECIFICATIONS: Storage conditions: Store in a cool area, away from direct sunlight.

COMPATIBILITY: The predator can be released with other beneficial insects, but there may be predation of other beneficial species. Persistent insecticides should not be used.

MAMMALIAN TOXICITY: No allergic or other adverse reactions have been reported following release of *Rhyzobius forestieri* in glasshouses or outdoors.

ENVIRONMENTAL IMPACT AND NON-TARGET TOXICITY: *Rhyzobius forestieri* occurs widely in Nature and there is no evidence that it affects non-target organisms or that it has any adverse effects on the environment, although, in the absence of pest species, it may consume other insects, including beneficial species. **Approved for use in organic farming:** Yes.

3:355 *Rhyzobius lophantae*
Predator of scale insects

Predatory beetle: Coleoptera: Coccinellidae.

NOMENCLATURE: Approved name: *Rhyzobius lophantae* Blaisdell. **Other names:** Lady beetle scale destroyer; lindorus; black lady beetle. Formerly known as: *Lindorus lophanthae* (Blais.).

SOURCE: Found in North America.

PRODUCTION: *Rhyzobius lophantae* is reared on scale insects and supplied as adults with a food source available.

TARGET PESTS: All types of scale insect, but particularly effective against citrus black scale (*Parlatoria ziziphi* (Lucas)) and soft scale, such as *Saissetia coffeae* (Walker), *S. oleae*

(Bernhard) and *Coccus hesperidum* L. It is also effective against purple scale (*Lepidosaphes beckii* Newman), oriental red scale (*Aonidiella orientalis* (Newstead)) and red scale (*Aspidiotus nerii* (Bouché)).

TARGET CROPS: May be used in orchard crops, citrus, ornamentals and interiorscapes.

BIOLOGICAL ACTIVITY: Predation: The adults consume both soft and armoured scale insects. Eggs hatch under the scale cover and the larvae consume scale crawlers and eggs. **Egg laying:** Females lay a large number of eggs under scale insects. Each female is able to lay several hundred eggs. **Duration of development:** The larval stages last about 3 weeks, with final instar larvae reaching a length of about 3 mm. Adults are around 2.5 mm in length and live for 5 to 8 weeks, depending upon the temperature and the food source. The beetles will consume scale insects at temperatures around 5 °C, but are most effective at temperatures between 23 and 30 °C, with a r.h. of about 65%. Low temperatures delay or inhibit reproduction and egg laying. **Efficacy:** *Rhyzobius lophantae* is a very effective predator of scale insects and, being very mobile, will seek out new infestations actively. There are reports of *Rhyzobius lophantae* consuming other insects, including mealybugs and various other insect eggs, although its use for control of insects other than scale is not recommended. The adults will consume pollen and it is suggested that reproductive ability is improved in the presence of flowering, pollen-producing plants.

COMMERCIALISATION: Formulation: Sold as pre-mated and fed, insectary-reared adults. **Tradenames:** 'Rhyzobius lophantae adults' (Green Spot), 'Rhyzobius lophantae' (Rincon-Vitova), (Arbico), (IPM Technology) and (Praxis), 'Purple Scale Predator' (Biocontrol Network).

APPLICATION: Open containers should be placed within the crop to allow the adult beetles to emerge. It is advisable to place the containers within the canopy of trees. Do not use yellow sticky insect traps. Control ants, as these will protect the scale from beetle attack. *Rhyzobius lophantae* cannot tolerate high levels of honeydew.

PRODUCT SPECIFICATIONS: Storage conditions: Store in a cool area, away from direct sunlight. **Shelf-life:** It is advised that *Rhyzobius lophantae* is used as soon as possible after delivery, but adults should survive for 3 to 4 days.

COMPATIBILITY: The predator can be released with other beneficial insects, but there may be predation of other beneficial species. Persistent insecticides should not be used.

MAMMALIAN TOXICITY: No allergic or other adverse reactions have been reported following release of *Rhyzobius lophantae* in glasshouses or outdoors.

ENVIRONMENTAL IMPACT AND NON-TARGET TOXICITY: *Rhyzobius lophantae* occurs widely in Nature and there is no evidence that it affects non-target organisms or that it has any adverse effects on the environment, although, in the absence of pest species, it may consume other insects, including beneficial species. **Approved for use in organic farming:** Yes.

3. Macro-organisms

3:356 *Rodolia cardinalis* *Predator of scale insects*

Predatory beetle: Coleoptera: Coccinellidae

NOMENCLATURE: Approved name: *Rodolia cardinalis* (Mulsant). **Other names:** Vedalia beetle. Formerly known as: *Vedalia cardinalis* Mulsant.

SOURCE: Originally found in Australasia, but subsequently introduced into Argentina, Chile, Cyprus, Egypt, Hawaii, North America, the Soviet Union, Peru, the Philippines, Portugal, Puerto Rico, Taiwan, Uruguay and Venezuela.

PRODUCTION: *Rodolia cardinalis* is specific to cottony cushion scale (*Icerya purchasi* Maskell), on which it is reared.

TARGET PESTS: Cottony cushion scale (*Icerya purchasi*).

TARGET CROPS: Citrus plantations, olive groves and several ornamental species, including magnolia, boxwood and roses.

BIOLOGICAL ACTIVITY: Biology: Adults are very small, densely pubescent, red and black ladybirds (beetles), about 2.5 to 4 mm long. Eggs are red, young larvae are reddish and pupal cases are whitish. Adults and mature larvae feed on all scale stages; the young larvae feed on eggs. There are eight generations per year in cooler areas and twelve generations per year in hot, dry areas. **Egg laying:** Adult female beetles lay eggs underneath the scale or attached to scale egg sacs. A female will lay from 150 to 190 eggs during her lifetime. **Key reference(s)**: 1) L E Caltagirone & R L Doutt. 1989. The history of the vedalia beetle importation to California and its impact on the development of biological control, *Ann. Rev. Entomol.*, **34**, 1–16. 2) P DeBach. 1974. *Biological Control by Natural Enemies*, Cambridge University Press, New York, USA, 323 pp.

COMMERCIALISATION: Formulation: Introduced as adults. **Tradenames:** 'Rodolia cardinalis' (Rincon-Vitova).

APPLICATION: *Rodolia cardinalis* should be introduced throughout the orchard to be protected. The adults are very mobile and seek prey actively.

PRODUCT SPECIFICATIONS: Purity: Product contains only adults, and no contaminants.

COMPATIBILITY: *Rodolia cardinalis* is very susceptible to some broad-spectrum insecticides, including insect growth regulators.

MAMMALIAN TOXICITY: There have been no reports of any adverse or allergic reactions from laboratory, manufacturing or field trial staff with *Rodolia cardinalis*.

ENVIRONMENTAL IMPACT AND NON-TARGET TOXICITY: *Rodolia cardinalis* occurs widely in Nature and there is no evidence of adverse environmental effects or of effects on non-target organisms. It is specific to scale insects. **Approved for use in organic farming:** Yes.

3:357 *Rumina decollata*

Predator of slugs and snails

Decollate snail: Gastropoda: Subulinidae

NOMENCLATURE: Approved name: *Rumina decollata* (L.) **Other names:** Decollate snail.

SOURCE: The decollate snail is native to portions of southern Europe, northern Africa and western Asia that border the Mediterranean Sea.

PRODUCTION: *Rumina decollata* is reared on slugs and decaying organic matter under controlled conditions.

TARGET PESTS: Slugs and snails.

TARGET CROPS: A wide range of horticultural crops.

BIOLOGICAL ACTIVITY: Biology: The shell of *Rumina decollata* is narrowly rimate, cylindric or cylindric-tapering, truncate and closed by a spiral convex plug at the summit. It is thin, irregularly striate, usually showing some spiral lines or vertical punctures. Adult shells are pale yellow or whitish in colour, but are light brown when young. There are typically four to six whorls in adults. The aperture is ovate, the outer lip simple, more or less thickened within. The columnella is vertical and nearly straight. Decollate snails are small predaceous land snails, capable of tolerating several months without water. They may aestivate during periods of high temperatures and low relative humidity and are capable of tolerating cold weather by burying themselves, having been known to burrow into the soil to considerable depths. Decollate snails do not move great distances from their home range. They are active mainly at night or after rainfall. They are known as decollate snails because, at a length of about 10 mm, the top 3 to 3.5 whorls of the shell are broken off, with the shell becoming decollate. A calcareous septum, resembling the protoconch, is formed as a cover for the top of the broken shell. Decollation has been reported to be advantageous because it provides increased mobility, a reduction in shell weight and enhances resistance to desiccation. Decollate snails may grow to 40 mm in length. **Predation:** Decollate snails are omnivorous. However, they can be voracious predators, feeding on other small land snails, slugs and worms. They enter the shells of *Helix* spp., consuming the snails from the inside out. Decollate snails also feed on most species of snail and, in some cases, even consume the shells. Several species of worms and slugs are also used as food. Decollate snails will also eat decaying plant material, but not healthy, growing specimens. **Egg laying:** The decollate snail is hermaphroditic, with both self-fertilisation and cross-fertilisation having been reported. Snails begin reproductive behaviour at approximately 10 months of age. Eggs are typically produced in late winter to mid summer and from early autumn to early winter. Nests consist of shallow depressions in the soil, with eggs being deposited individually, but close enough that they form clusters of between 15 and 55 eggs.

The incubation period ranges from 10 to 30 days. Eggs measure approximately 2 mm in diameter. **Efficacy:** Decollate snails may negatively impact small native snails and slugs through direct predation. **Key reference(s):** 1) J H Batts. 1957. Anatomy and life cycle of the snail *Rumina decollata, Southwest Naturalist,* **2**, 74–82. 2) D S Dundee. 1986. Notes on the habits and anatomy of the introduced land snails, *Rumina* and *Lamellaxis* (Subulinidae), *The Nautilus,* **100**(1), 32–7. 3) R K Selander & R O Hudson. 1976. Animal population structure under close inbreeding: The land snail *Rumina* in southern France, *The American Naturalist,* **110**, 695–718. 4) R K Selander & D W Kaufman. 1973. Self-fertilization and genetic population structure in a colonizing land snail, *Proc. Nat. Acad. Sci. USA,* **70**(4), 1186–90.

COMMERCIALISATION: Formulation: Supplied as dormant adults. **Tradenames:** 'Decollate Snails' (Biocontrol Network) and (Rincon-Vitova).

APPLICATION: Place the snails in a pan or bucket and run cool water over them for 2 to 3 minutes. Drain off the water and place them in a moist, shady area where snails are present, at a rate of 2400 per hectare (1000 per acre), distributed in several locations where organic matter is plentiful.

PRODUCT SPECIFICATIONS: Storage conditions: Adults are dormant when despatched and they can be kept under cool conditions for a few days before being used.

COMPATIBILITY: Should not be used with mollusc-specific beneficial nematodes.

MAMMALIAN TOXICITY: *Rumina decollata* has caused no allergic or other adverse effects on research workers, manufacturing or field staff. It is considered to be non-toxic to mammals.

ENVIRONMENTAL IMPACT AND NON-TARGET TOXICITY: *Rumina decollata* is not expected to have any adverse effects on non-target organisms or on the environment. It is specific to molluscs. **Approved for use in organic farming:** Yes.

3:358 *Scolothrips sexmaculatus*
Predator of mites

Predatory thrips: Thysanoptera: Thripidae

NOMENCLATURE: Approved name: *Scolothrips sexmaculatus* (Pergande). **Other names:** Six-spotted thrips.

SOURCE: Common in North America.

PRODUCTION: Bred on mites such as two-spotted mite (*Tetranychus urticae* Koch), under controlled conditions.

TARGET PESTS: Pacific mite (*Tetranychus pacificus*), two-spotted mite (*Tetranychus urticae*), willamette mite (*Eotetranychus willamettei* (McGregor)) and strawberry mite (*Tetranychus atlanticus* McGregor), although it will prey on other mites.

TARGET CROPS: Effective on a wide range of crops including top fruit (apples and pears), vegetables and ornamentals.

BIOLOGICAL ACTIVITY: Biology: *Scolothrips sexmaculatus* is a small (2 mm long), tan, cigar-shaped insect, with long, narrow wings fringed with hairs. Thrips eggs are laid in the leaf where they hatch in 7 days. There are four larval stages, of which two are active. During the active stages the larvae eat 10 mite eggs per day over a period of 10 days. When adult thrips emerge, they feed on about 60 mite eggs per day over a 30 day life span. All stages will prey on all stages of mite, but early life stages are preferred. When prey becomes scarce, thrips will become cannibalistic. **Predation:** Prey are consumed following puncture of the host, with the contents being sucked out. Female thrips are more effective than males, with immature females consuming up to 52 prey per day, whilst male immatures eat 12 prey per day. **Egg laying:** Female *Scolothrips sexmaculatus* will lay about 220 eggs in her lifetime. Fertilised eggs develop into females and unfertilised eggs develop into males. **Duration of development:** The mean life span of adult *Scolothrips sexmaculatus* is 9 days at 27°C and 50% r.h.; for first instar larva, 28 days; second instar larva, 2.5 days; prepupa, 1.2 days; and pupal stage, about 2.5 days. **Efficacy:** *Scolothrips sexmaculatus* is a very effective predator of mites, searching efficiently and killing more prey than it needs to survive. It requires relatively high populations to sustain a colony. If mite populations fall, *Scolothrips sexmaculatus* adults will leave the location in search for more abundant prey, but larvae stay and feed on each other at low pest densities, so some predators survive to adulthood.

COMMERCIALISATION: Formulation: Products contain all stages of thrips life-cycle in paper bags on bean leaves. **Tradenames:** 'Scolothrips sexmaculatus' (Rincon-Vitova) and (Integrated Pest Management Services).

APPLICATION: Release 2500 to 10 000 per hectare (1000 to 4000 per acre) depending on type of crop and severity of mite infestation. Very effective for control of mite hot spots.

PRODUCT SPECIFICATIONS: Purity: Products contain all stages of thrips on bean leaves plus some prey mites. **Storage conditions:** *Scolothrips sexmaculatus* can survive for 2 to 3 days in the bag at cool temperatures (10 to 12°C) and can be stored for up to 4 days in a refrigerator. **Shelf-life:** If possible, use within 24 hours of receipt.

MAMMALIAN TOXICITY: There have been no reports of allergic or other reactions following the use of *Scolothrips sexmaculatus* from research workers, producers or users.

ENVIRONMENTAL IMPACT AND NON-TARGET TOXICITY: *Scolothrips sexmaculatus* occurs widely in Nature and is not expected to have any adverse effects on non-target organisms or the environment. It is specific to mites, although it may consume other predatory mites when prey mites are in short supply.

Approved for use in organic farming: Yes.

3:359 *Serangium parcesetosum*
Predator of whiteflies

Predatory beetle: Coleoptera: Coccinellidae

NOMENCLATURE: Approved name: *Serangium parcesetosum* Sicard. **Other names:** Whitefly predatory beetle.

SOURCE: *Serangium parcesetosum* originated in India, but is now very widespread, particularly in citrus-growing regions.

PRODUCTION: *Serangium parcesetosum* is specific to whiteflies, particularly citrus whitefly (*Dialeurodes citri* Ashmead), on which it is bred. It has also been reared on tobacco whitefly (*Bemisia tabaci* (Gennadius)).

TARGET PESTS: The main target pest is the citrus whitefly (*Dialeurodes citri*), although control of silverleaf whitefly (*Bemisia argentifolii* Bellows & Perring) has been claimed.

TARGET CROPS: Citrus orchards and some ornamentals, such as poinsettia.

BIOLOGICAL ACTIVITY: Biology: Both larvae and adults of *Serangium parcesetosum* are voracious feeders, capable of consuming large numbers of immature whitefly in short periods of time. Adults consume approximately 400 whitefly nymphs in a 24-hour period; larvae consume 25 to 50 whitefly eggs or nymphs in 24 hours, depending on the larval stage. The cumulative lifetime predation rate is approximately 5000 whitefly nymphs per adult *Serangium parcesetosum*, showing that the predator gives very effective control of whitefly at moderate to high levels. However, low whitefly infestation levels may not be adequate to sustain reproduction or even adult survival.

COMMERCIALISATION: Formulation: Sold as adults. **Tradenames:** 'Serangium parcesetosum'.

APPLICATION: Release into the crop to be protected as soon as whitefly activity is noticed.

PRODUCT SPECIFICATIONS: Shelf-life: Release as soon as possible after receipt.

COMPATIBILITY: *Serangium parcesetosum* is very susceptible to some broad-spectrum insecticides, including insect growth regulators.

MAMMALIAN TOXICITY: There have been no reports of any adverse or allergic reactions from laboratory, manufacturing or field trial staff with *Serangium parcesetosum*.

ENVIRONMENTAL IMPACT AND NON-TARGET TOXICITY: *Serangium parcesetosum* occurs widely in Nature and there is no evidence of adverse environmental effects or of effects on non-target organisms. It is specific to whiteflies and originated in India. It does not overwinter in temperate conditions. **Approved for use in organic farming:** Yes.

3:360 *Sphenoptera jugoslavica*
Phytophagous beetle

Phytophagous beetle: Coleoptera: Buprestidae

NOMENCLATURE: Approved name: *Sphenoptera jugoslavica* Obenberger.
Other names: Knapweed root beetle. Formerly known as: *Chilostetha jugoslavica* Obenberger.

SOURCE: *Sphenoptera jugoslavica* is native to eastern Europe and Asia. It was introduced into North America in 1979 as part of a programme to control alien knapweed populations.

PRODUCTION: It is reared on diffuse knapweed under controlled conditions.

TARGET PESTS: Loose or diffuse knapweed (*Centaurea diffusa* Lam.).

TARGET CROPS: Rangeland, pasture, grasslands and riparian sites.

BIOLOGICAL ACTIVITY: Biology: *Sphenoptera jugoslavica* is a flat, metallic, copper-coloured, elongated beetle, measuring 7 mm in length. The larvae have an enlarged head and a long, thin body. The larvae cause a gall-like swelling in the knapweed root near the crown. They overwinter as larvae in the root of the knapweed plant. The newly formed adults remain motionless inside knapweed roots in their pupal chambers for 2 to 5 days. Then they bite a hole in the gall wall and leave the root. Adult *S. jugoslavica* emerge in July. The new adults feed on diffuse knapweed leaves for 2 to 3 days before mating. Adult *S. jugoslavica* live up to 30 days. **Egg laying:** Egg laying commences 10 to 12 days after emergence, 7 to 10 days after mating. **Efficacy:** Small knapweed rosettes that have been attacked often die by the end of the next season. Diffuse knapweed plants that survive are stunted and produce fewer flowers. After the beetle population has built up for 5 to 6 years, the knapweed population may collapse, with only scattered knapweed plants remaining. This effect has been observed in drier areas where the beetles have been released. Stunted and unhealthy plants were observed in Europe, especially in drier areas. **Key reference(s):** 1) R D Powell & J H Myers. 1988. The effect of *Sphenoptera jugoslavica* Obenb. (Col., Burprestidae) on its host plant *Centaurea diffusa* Lam. (Compositae), *J. Appl. Entomol.*, **106**, 25–45. 2) H Zwolfer. 1976. Investigations on *Sphenoptera (Chilostetha) jugoslavica* Obenb. (Col., Burprestidae), a possible biocontrol agent of the weed *Centaurea diffusa* Lam. (Compositae) in Canada, *Z. Angew. Entomol.*, **80**, 170–90.

COMMERCIALISATION: Formulation: Sold as adult beetles. In some US states, *S. jugoslavica* adults may be obtained at no cost from state weed management agencies. **Tradenames:** 'Sphenoptera jugoslavica' (Praxis).

APPLICATION: *Sphenoptera jugoslavica* adults should be released in mid-summer. The numbers released depend upon the level of infestation. Releases should be made in sections of the diffuse knapweed infestation where the plants are at least one to two feet apart, with some bare ground showing. The beetle thrives in drier areas.

PRODUCT SPECIFICATIONS: Storage conditions: *Sphenoptera jugoslavica* may be stored for up to 3 days, out of direct sunlight and under cool conditions.

COMPATIBILITY: *Sphenoptera jugoslavica* is susceptible to commonly used insecticides and, thus, should not be exposed to insecticidal sprays. It is not directly affected by commonly used herbicides and can be used in conjunction with herbicides for knapweed control. This can be carried out as long as the insect has completed development before the knapweed dies from herbicide action.

MAMMALIAN TOXICITY: There have been no reports of allergic or other adverse reactions from research workers, manufacturing staff or from the field release of *Sphenoptera jugoslavica*.

ENVIRONMENTAL IMPACT AND NON-TARGET TOXICITY: *Sphenoptera jugoslavica* occurs widely in Nature and is not expected to have any adverse effects on non-target organisms or on the environment. Its only known hosts are diffuse knapweed and, rarely, spotted knapweed (*Centaurea maculosa* Lam.).

3:361 *Spurgia esulae*   *Phytophagous midge*

Phytophagous gall midge: Diptera: Cecidomyiidae

NOMENCLATURE: Approved name: *Spurgia esulae* Gagné. **Other names:** Leafy spurge gall midge. Formerly known as: *Bayeria capitigena* Bremi.

SOURCE: Widely distributed in Europe. Samples introduced into the USA in 1985 were collected in northern Italy.

PRODUCTION: *Spurgia esulae* is reared on leafy spurge (*Euphorbia esula* L.) plants, under controlled conditions.

TARGET PESTS: Leafy spurge (*Euphorbia esula*).

TARGET CROPS: Pasture, rangelands and riparian areas.

BIOLOGICAL ACTIVITY: Biology: *Spurgia esulae* overwinters in the soil as a mature larva. Overwintered larvae pupate in the spring and adults emerge from the soil from April to June, depending on location. *S. esulae* adults are small (<2 mm long) flies that probably live only 2 or 3 days under field conditions. However, adult females appear able, actively or passively (via wind dispersal), to move at least 10 m from the pupation site. After egg-hatch, first-instar larvae migrate to leafy spurge buds, where they begin feeding on the meristematic tissue. Larval feeding induces hypertrophy in the bud tissues, causing the formation of a gall. *S. esulae* induced galls are somewhat cabbage-like in appearance and consist of a tight cluster of abnormal warty and flattened leafy spurge leaves, among which are found the developing *S. esulae* larvae. Larvae are legless, orange in colour and pass through three instars. Larval development generally requires 2 to 4 weeks, depending on local weather conditions. Mature larvae of the final generation drop from the galls and burrow into the soil, where they spend the winter. During earlier generations, *Spurgia esulae* larvae construct white silken cocoons inside the gall, within which pupation occurs. The pupal stage lasts about a week, after which adult flies emerge from the galls. *Spurgia esulae* is multivoltine, but the number of generations completed per year varies with location. *S. esulae* was reported to complete four or five generations in northern Italy. In the United States, three generations occur per year in south-western Montana, while *S. esulae* completes four generations in eastern North Dakota. After gall formation, leafy spurge buds lose their meristematic activity, so development of flower buds and further stem elongation are prevented. In addition, gall formation appears to disrupt apical dominance, which stimulates the formation of lateral branches on galled spurge stems. Buds on these lateral branches provide attack sites for subsequent *Spurgia esulae* generations. **Egg laying:** After mating, adult females deposit groups of small (<1 mm in diameter), orange-coloured eggs on leafy spurge leaves near the apical buds. Generally, female flies produce 20 to 150 eggs, laid in groups of 8 to 20. **Efficacy:** *Spurgia esulae* will reduce flower and seed production by,

3. Macro-organisms

and perhaps reduce the vigour of, leafy spurge plants. However, attacks by this insect should rarely, if ever, directly kill the weed. Thus, the primary role of this agent in leafy spurge control lies in reducing the seed pool available for long-range dispersal.

Key reference(s): 1) R J Gagné. 1990. Gall midge complex (Diptera: Cecidomyiidae) in bud galls of Palearctic *Euphorbia* (Euphorbiaceae), *Ann. Entomol. Soc. Am.*, **83**, 335–45.

2) P Pecora, R W Pemberton, M Stazi & G R Johnson. 1991. Host specificity of *Spurgia esulae* Gagné (Diptera: Cecidomyiidae), a gall midge introduced into the United States for control of leafy spurge (*Euphorbia esula* L. 'complex'), *Environ. Entomol.*, **20**, 282–7.

COMMERCIALISATION: Formulation: Sold as adult flies. **Tradenames:** 'Spurgia esulae' (Praxis).

APPLICATION: *Spurgia esulae* adults should be released in grassland infested with leafy spurge in the spring, when weed growth is vigorous.

PRODUCT SPECIFICATIONS: Purity: Products contain only *Spurgia esulae* adults. **Storage conditions:** The adults may be stored for a few days at low temperatures (5 °C) and away from direct sunlight.

COMPATIBILITY: It is unusual to release *Spurgia esulae* with chemical pesticides.

MAMMALIAN TOXICITY: There have been no reports of allergic or other adverse reactions from research workers, manufacturing staff or from the field release of *Spurgia esulae*.

ENVIRONMENTAL IMPACT AND NON-TARGET TOXICITY: *Spurgia esulae* occurs widely in Nature and is not expected to have any adverse effects on non-target organisms or on the environment. Trials in the USA have shown the host range of *Spurgia esulae* to be limited to a subset of plant species in the subgenus *Esula* of the genus *Euphorbia*, including the target weed (leafy spurge). Under experimental conditions, *Spurgia esulae* has been shown to cause galls on several native *Euphorbia* species, although it is not yet known if these plants will be suitable hosts under field conditions. Two US federally protected native spurges (*Euphorbia garberi* Engelm. ex Chapm. and *E. deltiodes* Engelm. ex Chapm. subgenus Chamaesyce) and two rare native US spurges that are being considered for protection (*E. purpurea* (Raf.) Fernald and *E. telephiodes* Chapm.) are not suitable hosts for *Spurgia esulae*.

3:362 *Stethorus punctillum*

Predator of spider mites

Predatory beetle: Coleoptera: Coccinellidae

NOMENCLATURE: Approved name: *Stethorus punctillum* Weise. **Other names:** Spider mite destroyer.

SOURCE: Widespread, particularly in North America. First documented in 1936.

PRODUCTION: Reared on phytophagous mites, such as *Tetranychus urticae* Koch on bean plants, or on acarid grain mites on bran.

TARGET PESTS: Spider mites, including the European red mite (*Panonychus ulmi* (Koch)), the two-spotted spider mite (*Tetranychus urticae*), the spruce spider mite (*Oligonychus ununguis* (Jacobi)) and the southern red mite (*Oligonychus illicis* (McGregor)).

TARGET CROPS: May be used in orchards and indoors in glasshouses and interiorscapes.

BIOLOGICAL ACTIVITY: Biology: Emerging adults are reddish-orange for a few hours after emerging from the pupal case and then darken in colour. Adults are tiny (1.5 mm long), dark brown to black, oval beetles, with brownish-yellow antennae, mouthparts and legs. Eggs hatch into greyish, alligator-shaped larvae, about 2 mm in length. Larvae can be slow-moving, grey to blackish, with many long, branched hairs and black patches. They move from plant to plant on the leaves. As the larva matures, it becomes reddish, at first on the edges of the body and, just prior to pupation, the entire larva turns reddish. The larva has thirteen segments, plus the head. It passes through four larval stages in 12 to 14 days (depending on climate and food availability), feeding on all stages of spider mites. The pupae are black and flattened, teardrop-shaped, and the entire pupa is covered with yellow hairs. Pupae are often found along leaf veins on the underside of leaves. Pupation lasts about 5 to 7 days. **Predation:** The adults and larvae are aggressive predators, consuming large numbers of mites throughout their development. Predation is optimal at temperatures between 20 and 32 °C. Lower temperatures delay reproduction and subsequent development. **Egg laying:** Adult female beetles lay 8 eggs per day, singly, in or near mite colonies, usually on the underside of leaves, near the primary vein. Eggs adhere tightly to the leaf and are oval and yellowish in colour, turning blackish just before the larvae emerge (5 to 7 days). **Duration of development:** Egg hatch timing is dependent upon the temperature, but the larvae usually take 18 days to pupation (the pupae appear as grey balls on the veins of the underside of leaves). Adults live for between 4 and 5 weeks. **Efficacy:** *Stethorus punctillum* is able to fly and will distribute easily into trees. It is an active seeker of prey and is very mobile. Both larvae and adult beetles feed on all stages of spider mites. Adults can eat 50 mite eggs or 10 adults per day and can consume 240 spider mites during their developmental period. They will also feed on other small arthropod eggs, aphids, nectar and pollen, if food supplies are short, and will even cannibalise each other for survival. Optimum conditions are moderate to high temperatures of 16 to 35 °C. They will not fly at temperatures below 13 °C. *Stethorus punctillum* enters diapause under short daylight conditions and will stop reproducing unless provided with supplementary light (16 hours). Misting of the treated area to maintain a high relative humidity improves the establishment of the predator.

COMMERCIALISATION: Formulation: Sold as pre-mated adults packaged with filter paper to provide protection and hiding places during transport. **Tradenames:** 'Stethorus punctillum' (Applied Bio-Nomics), (Praxis) and (Rincon-Vitova), 'TriPak' (*Phytoseiulus persimilis* plus *Amblyseius fallacis* plus *Stethorus punctillum*) (Rincon-Vitova), 'Stethorus-System' (Biobest).

APPLICATION: It is recommended that the beetles are released at the rate of 1 to 2 adults per square metre a month for prevention, and between 3 and 4 beetles per square metre tri-weekly for control of low infestations to a maximum of 5 to 6 beetles per square metre bi-weekly for control of high infestations. Lower rates are required for outdoors wide-area use and higher rates are needed for protected tomatoes. Best control is achieved if *Stethorus* is used along with predatory mites, such as *Phytoseiulus persimilis* Athios-Henriot or *Amblyseius californicus* (McGregor).

PRODUCT SPECIFICATIONS: Purity: Products consist of newly mated adults only. **Storage conditions:** Keep away from direct sunlight. Store in cool conditions. **Shelf-life:** Use as soon after delivery as possible.

COMPATIBILITY: Incompatible with residual insecticides. May be used with other mite predators. Most effective at high populations of phytophagous mites.

MAMMALIAN TOXICITY: There have been no reports of any adverse or allergic reactions from laboratory, manufacturing or field trial staff.

ENVIRONMENTAL IMPACT AND NON-TARGET TOXICITY: *Stethorus punctillum* occurs widely in Nature and there is no evidence of adverse environmental effects or of effects on non-target organisms. It is specific to spider mites.
Approved for use in organic farming: Yes.

3:363 *Tenodera aridifolia sinensis*

Predatory insect

Praying mantis: Orthoptera: Mantidae

NOMENCLATURE: Approved name: *Tenodera aridifolia sinensis* Saussure.
Other names: Praying mantis; Chinese praying mantis.

SOURCE: Originated in China, but now widespread in warm climates.

PRODUCTION: Bred under controlled conditions on a variety of insect as food.

TARGET PESTS: Praying mantids consume a wide range of insects, mites and insect pest eggs.

TARGET CROPS: Used on a wide range of crops, but particularly popular in amateur gardens.

BIOLOGICAL ACTIVITY: Biology: Optimal conditions for hatching of the eggs from within the egg case are 40% to 95% r.h., with temperatures between 21 and 32 °C (70 and 90 °F). Praying mantids will eat most pest insects, mites and their eggs, but, if food is scarce, they will also eat other beneficial insects, with the exception of ladybirds, which are rejected because of their bitter taste. The Chinese mantis is brown/green in colour, 80 to 90 mm in length, and both male and female have the classical praying mantis appearance. The male has a blue band on its back.

COMMERCIALISATION: Formulation: *Tenodera aridifolia sinensis* are shipped as egg cases, each of which contains about 200 immature mantids. **Tradenames:** 'Praying Mantis' (Arbico), (Biocontrol Network), (Natural Pest Control), (Rincon-Vitova) and (Praxis).

APPLICATION: The egg cases should be distributed in early to mid summer, at a rate of 3 cases per 460 square metres (5000 square feet) or 25 to 250 per hectare (10 to 100 cases per acre) per year. The cases should be attached to twigs, leaves, fences or in the crotch of a bush or tree. Do not place on the ground, as they will be attacked by ants.

PRODUCT SPECIFICATIONS: Purity: The products contain only egg cases of *Tenodera aridifolia sinensis*. **Storage conditions:** May be stored for a few days, if kept out of direct sunlight in a cool, dry container.

COMPATIBILITY: *Tenodera aridifolia sinensis* can be used in conjunction with other beneficial species, but may consume these if prey becomes scarce. It is recommended that the insects are not used in crops where broad-spectrum, residual insecticides have been applied.

MAMMALIAN TOXICITY: There have been no reports of allergic or other adverse reactions from research workers, manufacturing staff or from the field release of *Tenodera aridifolia*

3. Macro-organisms

sinensis. Adults are often purchased as pets and there is no evidence of any toxicological problems associated with this practice.

ENVIRONMENTAL IMPACT AND NON-TARGET TOXICITY: *Tenodera aridifolia sinensis* occurs widely in Nature and is not expected to have any adverse effects on the environment, although its extensive range of prey may result in the loss of some beneficial and other non-pest species. **Approved for use in organic farming:** Yes.

3:364 *Tetrastichus* species   *Parasitoid of insects*

Parasitoid wasp: Hymenoptera: Eulophidae

NOMENCLATURE: Approved name: *Tetrastichus* species.

SOURCE: *Tetrastichus* species are found widely around the world. Some species are native to Europe (*Tetrastichus gallerucae* (Boy.)), whilst others are found in the USA (*Tetrastichus brevistigma* Gahan).

PRODUCTION: Reared under controlled conditions in beetles. There has been some success in breeding the insects *in vitro* on artificial diet.

TARGET PESTS: A wide range of insects are parasitised by *Tetrastichus* species, particularly beetles.

TARGET CROPS: Used in a wide range of crops, as diverse as elm trees and wheat.

BIOLOGICAL ACTIVITY: Biology: The adult *Tetrastichus gallerucae* is 3 mm long, with a dark, metallic blue thorax and a black abdomen with a faint metallic lustre. It is a solitary internal parasitoid, completing its life-cycle in about 17 days. There are several generations each season, usually twice the number of annual generations of its host. It probably overwinters as adults in sheltered places. The females also destroy many eggs by host feeding; this is considered an important factor in reducing the pest populations in Europe. *T. brevistigma* is a gregarious 3 mm-long internal parasitoid. An average of 12 individuals develop in each host, but up to 37 have been recorded from a single beetle pupa. Larval development can be completed in 9 to 15 days, but many individuals enter larval diapause until the following spring. The percentage of larvae that will hibernate increases from generation to generation and nearly all the full-grown larvae of the last generation hibernate. They overwinter as full-grown larvae in the pupal skin of their host. These dead beetle pupae are found in the grass and debris surrounding the bases of treated plants. The parasitoids emerge from the host pupae in June and July to attack the first generation of beetles as they begin to pupate. **Egg laying:** Each female *Tetrastichus gallerucae* can lay up to 100 eggs and female *T. brevistigma* lay from 1 to 8 eggs at a time in beetle pupae or prepupae.

COMMERCIALISATION: Formulation: Sold as pupae. **Tradenames:** 'Tetrastichus parasitoids' (Praxis).

APPLICATION: Distribute the pupae within the crop in early summer, as beetle populations are beginning to build up.

PRODUCT SPECIFICATIONS: Storage conditions: Store in a cool, dry place, away from direct sunlight.

COMPATIBILITY: Incompatible with broad-spectrum, residual insecticides.

MAMMALIAN TOXICITY: *Tetrastichus* species have caused no allergic or other adverse effects on research workers, manufacturing or field staff. They are considered to be non-toxic to mammals.

ENVIRONMENTAL IMPACT AND NON-TARGET TOXICITY: *Tetrastichus* species are not expected to have any adverse effects on non-target organisms or on the environment. They will consume a wide range of insect species. **Approved for use in organic farming:** Yes.

3:365 *Thripobius semiluteus* *Parasitoid of thrips*

Parasitoid wasp: Hymenoptera: Eulophidae

NOMENCLATURE: Approved name: *Thripobius semiluteus* Boucek. **Other names:** Glasshouse thrips parasitoid; greenhouse thrips parasitoid.

SOURCE: *Thripobius semiluteus* is native to South America, but is now found widely in North America, Europe and Australasia, following introduction for control of the glasshouse thrips.

PRODUCTION: Reared on thrips, under controlled conditions.

TARGET PESTS: Glasshouse thrips (*Heliothrips haemorrhoidalis* Bouché).

TARGET CROPS: Effective on a wide range of crops, including glasshouse-grown vegetables and ornamentals and orchard crops, such as citrus and avocado.

BIOLOGICAL ACTIVITY: Biology: *Thripobius semiluteus* adults are about 0.5 mm in length, with a black head and thorax and a yellow abdomen. The female wasp lays her eggs inside the glasshouse thrips, parasitising mainly small larvae. Larvae develop within the thrips larva. Parasitised larvae turn yellow initially, but later they turn black, becoming small cocoons.

The adult wasp emerges from these cocoons, destroying them completely. Immediately after emergence, the wasps mate and seek new thrips larvae to parasitise. *Thripobius semiluteus* is more effective at high temperatures, with high light intensities.

COMMERCIALISATION: Formulation: Sold as cards containing pupae.
Tradenames: 'Thripobius semiluteus' (Entocare) and (Arbico).

APPLICATION: Hang the cards holding the wasp pupae in trees in late spring/early summer.

PRODUCT SPECIFICATIONS: Purity: The product contains only *Thripobius semiluteus* pupae. **Storage conditions:** May be stored for short periods, under cool, dark conditions.

COMPATIBILITY: Incompatible with broad-spectrum, residual insecticides.

MAMMALIAN TOXICITY: No allergic or other adverse reactions have been recorded from the production of *Thripobius semiluteus* or its use in glasshouse or field conditions.

ENVIRONMENTAL IMPACT AND NON-TARGET TOXICITY: *Thripobius semiluteus* species are not expected to have any adverse effects on non-target organisms or on the environment. Studies in the laboratory and field have shown that the host range is restricted to pest thrips species only. **Approved for use in organic farming:** Yes.

3:366 *Trichogramma brassicae*
Parasitoid of lepidopteran eggs

Parasitoid wasp: Hymenoptera: Trichogrammatidae

NOMENCLATURE: Approved name: *Trichogramma brassicae* Westwood.
Other names: Moth egg predator; lepidopteran egg parasitoid; trichogramm. Formerly known as: *Trichogramma maidis* (Pint et Voeg).

SOURCE: Widely occurring parasitoid wasp.

PRODUCTION: Reared in insectaries under controlled conditions on the eggs of moths, such as *Ephestia kuehniella* Zeller.

TARGET PESTS: Eggs of lepidopteran caterpillar pests, particularly European corn borer (*Ostrinia nubilalis* (Hübner)), *Mamestra brassicae* (L.) and *Helicoverpa armigera* (Hübner).

TARGET CROPS: Maize (corn), peppers, tomatoes and ornamentals. The main use is in maize.

BIOLOGICAL ACTIVITY: Biology: Adult female wasps are yellowish in colour and about 0.5 mm long. They lay their eggs into a lepidopteran egg, preferring recently laid eggs. When the wasp egg hatches, the wasp larva consumes the developing caterpillar inside the egg. It pupates inside the egg and emerges as an adult wasp. Adults feed on nectar. **Predation:** Most lepidopteran eggs can be parasitised by *Trichogramma brassicae*. **Egg laying:** Adult wasps mate shortly after emergence and fertilised females lay eggs that develop into either males or females. Unfertilised females lay viable eggs that develop only into males. A female wasp will parasitise over 50 eggs in her life span of 5 to 14 days. **Duration of development:** The time from egg laying to wasp emergence varies with temperature, from as little as 7 to as long as 20 days. **Efficacy:** If the recommendations for use are followed, *Trichogramma brassicae* will destroy about 80% of pest eggs and will keep the lepidopteran population in check. The adults are very mobile and are expert at locating insect eggs.

COMMERCIALISATION: Formulation: Sold as parasitised eggs fixed on cardboard or in capsules. **Tradenames:** 'Trig' (UNCAA) and (Biotop), 'Trichocap' and 'Pyratyp' (BASF), 'Tricholine b' (Syngenta Bioline), 'TR 16' (UNCAA), 'Trichogramma-System' and 'Trichogramma-Mix-System' (contains both *T. brassicae* and *T. evanescens*) (Biobest), 'Trichogramma brassicae' (Rincon-Vitova), (IPM Laboratories) and (Beneficial Insectary), 'TrichoPAK' (Bioplanet).

APPLICATION: Spread cardboard carrier or capsules within the crop. The release rate should be between 12 000 and 500 000 parasitised eggs per hectare, depending on the stage of growth of the crop, the area to be treated, the type of pest targeted, the development stage of the pest and the climatic conditions.

PRODUCT SPECIFICATIONS: Purity: Only parasitised, stored-product lepidopteran eggs, with no contaminants. **Storage conditions:** Diapausing *Trichogramma brassicae* must be stored at 3 °C and 75% r.h. Always store following the manufacturer's instructions.
Shelf-life: Diapausing parasitoids can be kept under recommended storage conditions for several months. Reactivated *Trichogramma brassicae* must be used within a few days of delivery.

COMPATIBILITY: Incompatible with foliar insecticides.

MAMMALIAN TOXICITY: No allergic or other adverse reactions have been recorded from the production of *Trichogramma brassicae* or its use in glasshouse or field conditions. Allergies associated with host Lepidoptera are known in production.

ENVIRONMENTAL IMPACT AND NON-TARGET TOXICITY: *Trichogramma brassicae* occurs widely in Nature. It has such a wide host range that it is possible that some lepidopteran non-target species may be affected. It is not expected to show any adverse effects on the environment. **Approved for use in organic farming:** Yes.

3. Macro-organisms

3:367 *Trichogramma evanescens*
Parasitoid of lepidopteran eggs

Parasitoid wasp: Hymenoptera: Trichogrammatidae

NOMENCLATURE: Approved name: *Trichogramma evanescens* Westwood.
Other names: Moth egg parasitoid; trichogramm.

SOURCE: Widely occurring parasitoid wasp.

PRODUCTION: Reared in insectaries on the eggs of moths, such as *Ephestia kuehniella* Zeller.

TARGET PESTS: Eggs of lepidopteran caterpillar pests, such as bollworm, corn earworm and tomato fruitworm (*Helicoverpa zea* (Boddie)), tobacco budworm (*Heliothis virescens* (Fabricius)) and European corn borer (*Ostrinia nubilalis* (Hübner)).

TARGET CROPS: A wide range of field and glasshouse-grown crops, including vegetables and ornamentals.

BIOLOGICAL ACTIVITY: Biology: Adult female wasps are about 0.5 mm long. They lay their eggs into a lepidopteran egg, preferring freshly laid eggs. Upon hatching, the wasp larva consumes the developing caterpillar within the egg. It pupates inside the egg and emerges as an adult wasp. Adults feed on nectar. **Predation:** Most lepidopteran eggs can be parasitised by *Trichogramma evanescens*. **Egg laying:** Adult wasps mate shortly after emergence and fertilised females lay eggs that develop into either males or females. Unfertilised females lay eggs that develop only into males. A female wasp will parasitise over 50 eggs in her life span of 5 to 14 days. **Duration of development:** The time from egg laying to wasp emergence varies with temperature, from as little as 7 to as long as 20 days. **Efficacy:** The parasitoid can have 30 or more generations in a season and, if numbers are high, will keep a lepidopteran population in check. The adults are very mobile and are expert at locating insect eggs.

COMMERCIALISATION: Formulation: Sold as parasitised eggs fixed to a card.
Tradenames: 'Tricho-strip' (Koppert), 'Trichogramma evanescens' (Praxis), 'Tricho-line' (Syngenta Bioline), 'Trichogramma-Mix-System' (contains both *T. brassicae* and *T. evanescens*) (Biobest).

APPLICATION: Hang cards within the crop, evenly distributed. Release rates should be about 12 000 to 500 000 parasitised eggs per hectare, depending upon the incidence of caterpillars. Use pheromone traps to monitor the population of Lepidoptera within the crop.

PRODUCT SPECIFICATIONS: Purity: Only parasitised, stored-product lepidopteran eggs, with no contaminants. **Storage conditions:** Store in the dark at 10 to 15 °C. **Shelf-life:** Can be kept for 3 to 4 days, under recommended storage conditions.

COMPATIBILITY: Incompatible with foliar insecticides.

MAMMALIAN TOXICITY: No allergic or other adverse reactions have been reported from the use of *Trichogramma evanescens* in glasshouse or field conditions. Allergies associated with host Lepidoptera are known in production.

ENVIRONMENTAL IMPACT AND NON-TARGET TOXICITY: *Trichogramma evanescens* is widespread in Nature. It has such a wide host range that it is possible that some lepidopteran non-target species may be affected. It is not expected to show any adverse effects on the environment. **Approved for use in organic farming:** Yes.

3:368 *Trichogramma minutum*
Parasitoid of lepidopteran eggs

Parasitoid wasp: Hymenoptera: Trichogrammatidae

NOMENCLATURE: Approved name: *Trichogramma minutum* Riley. **Other names:** Minute egg parasitoid.

SOURCE: Widely occurring parasitoid wasp.

PRODUCTION: Reared in insectaries under controlled conditions on the eggs of moths, such as *Ephestia kuehniella* Zeller. Tests are under way to produce *Trichogramma minutum in vitro*.

TARGET PESTS: Tree crop Lepidoptera, such as codling moth (*Cydia pomonella* L.) and other fruit moths.

TARGET CROPS: Used primarily in orchards. *T. minutum* is recommended in the USA for ornamentals, orchards, and grapes east of the Rocky Mountains.

BIOLOGICAL ACTIVITY: Biology: Adult female wasps are less than 0.5 mm long. They lay their eggs into a lepidopteran egg, preferring freshly laid eggs. Upon hatching, the wasp larva consumes the developing caterpillar within the egg. It pupates inside the egg and emerges as an adult wasp. Adults feed on nectar. **Predation:** Most lepidopteran eggs can be parasitised by *Trichogramma minutum*, but it is most effective in tree fruit crops. **Egg laying:** Adult wasps mate shortly after emergence and fertilised females lay eggs that develop into either males or females. Unfertilised females lay eggs that develop only into males. A female wasp will parasitise over 50 eggs in her life span of 5 to 14 days. **Duration of development:** The time from egg laying to wasp emergence varies with temperature, from as little as 7 to as long as 20 days. **Efficacy:** The parasitoid can have many generations in a season and, if numbers are high, will keep a lepidopteran population in check. The adults are very mobile and are expert at locating insect eggs.

COMMERCIALISATION: Formulation: Sold as parasitised eggs fixed to a card or in capsules. **Tradenames:** 'Trichogramma minutum' (Arbico), (Beneficial Insectary), (Harmony), (IPM Laboratories) and (Rincon-Vitova), 'Moth Egg Parasite' (Biocontrol Network).

APPLICATION: Spread cardboard carrier or capsules within the crop. The release rate should be between 12 000 and 500 000 parasitised eggs per hectare, depending on the stage of growth of the crop, the area to be treated, the type of pest targeted, the development stage of the pest and the climatic conditions.

PRODUCT SPECIFICATIONS: Purity: Only parasitised, stored-product lepidopteran eggs, with no contaminants. **Storage conditions:** Store in the dark at 10 to 15 °C. **Shelf-life:** Can be kept for 3 to 4 days, under recommended storage conditions.

COMPATIBILITY: Incompatible with broad-spectrum, residual foliar insecticides.

MAMMALIAN TOXICITY: No allergic or other adverse reactions have been reported from the use of *Trichogramma minutum* in glasshouse or field conditions. Allergies associated with host Lepidoptera are known in production.

ENVIRONMENTAL IMPACT AND NON-TARGET TOXICITY: *Trichogramma minutum* is widespread in Nature. It has such a wide host range that it is possible that some lepidopteran non-target species may be affected. It is not expected to show any adverse effects on the environment. **Approved for use in organic farming:** Yes.

3:369 *Trichogramma ostriniae*
Parasitoid of lepidopteran eggs

Parasitoid wasp: Hymenoptera: Trichogrammatidae

NOMENCLATURE: Approved name: *Trichogramma ostriniae* Pang & Chen. **Other names:** European corn borer egg parasitoid; trichogramm.

SOURCE: *Trichogramma ostriniae* was imported from China in 1990 and has been shown to be effective at controlling European corn borer (*Ostrinia nubilalis* (Hübner)).

PRODUCTION: Reared in insectaries under controlled conditions on the eggs of moths, such as *Ephestia kuehniella* Zeller.

TARGET PESTS: European corn borer (*Ostrinia nubilalis*). It may also hold potential against other pests such as cabbage looper (*Trichoplusia ni* (Hübner)), diamondback moth (*Plutella xylostella* (L.)) and grape berry moth (*Eupoecilia ambiguella* (Hübner)).

TARGET CROPS: Maize (corn).

BIOLOGICAL ACTIVITY: Biology: Adult female wasps are about 0.5 mm long. They lay their eggs into a lepidopteran egg, preferring freshly laid eggs. Upon hatching, the wasp larva consumes the developing caterpillar within the egg. It pupates inside the egg and emerges as an adult wasp. Adults feed on nectar. **Predation:** When released into maize (corn), most *Trichogramma ostriniae* remain in the field, but a few may disperse into other habitats where they may parasitise non-pest species. This potential impact on 'non-target' species is real, but research to date indicates that the risk is minimal. Although *Trichogramma* wasps can parasitise the eggs of several species of moths and butterflies, in reality their host range is limited by their inability to find non-target eggs. While *T. ostriniae* is able to move into non-crop habitats, it appears to prefer maize when given a choice. It searches corn much more successfully than other plants. Recent studies suggest that *T. ostriniae* are attracted to European corn borer sex pheromones, but not by volatile chemicals coming from maize plants. So, although they may travel to habitats surrounding agricultural fields, the impact will be limited by their inability to find host eggs in non-agricultural habitats. In addition, *T. ostriniae* is unable to survive very cold winters. **Egg laying:** Adult wasps mate shortly after emergence and fertilised females lay eggs that develop into either males or females. Unfertilised females lay eggs that develop only into males. A female wasp will parasitise over 50 eggs in her life span of 5 to 14 days.

Duration of development: The time from egg laying to wasp emergence varies with temperature, from as little as 7 to as long as 20 days. **Efficacy:** In trials of inoculative releases in commercial fields of processing sweet corn, overall parasitism was about 50% over two years of trials. A single release of 75 000 wasps per hectare (30 000 wasps per acre) resulted in European corn borer control roughly equivalent to an insecticide application.

Key reference(s): T P Kuhar, M P Hoffmann & M G Wright. 2003. Controlling European corn borer in vegetables with a parasitic wasp, *Pesticide Outlook*, **14**(3), 99–101.

COMMERCIALISATION: Formulation: For commercial production, *Trichogramma ostriniae* wasps are reared on grain moth eggs. Once the grain moth eggs are parasitised, they are glued inside release packets that protect them from the elements and from predation. As wasps emerge from the grain moth eggs they leave the release packets and search out pest eggs. **Tradenames:** 'Trichogramma ostriniae' (IPM Laboratories).

APPLICATION: The recommended release method for *Trichogramma ostriniae* is a single inoculative release (75 000 per hectare (30 000/acre)) at an early stage of maize development.

PRODUCT SPECIFICATIONS: Purity: Only parasitised, stored-product lepidopteran eggs, with no contaminants. **Shelf-life:** *Trichogramma ostriniae* can be stored for one to two weeks at 10°C (48°F).

COMPATIBILITY: Incompatible with broad-spectrum, residual foliar insecticides.

MAMMALIAN TOXICITY: No allergic or other adverse reactions have been reported from the use of *Trichogramma ostriniae* in glasshouse or field conditions. Allergies associated with host Lepidoptera are known in production.

ENVIRONMENTAL IMPACT AND NON-TARGET TOXICITY: While *Trichogramma ostriniae* is able to move into non-crop habitats, it appears to prefer maize when given a choice. It searches maize much more successfully than other plants. Recent studies suggest that *Trichogramma ostriniae* are attracted to European corn borer sex pheromones, but not by volatile chemicals coming from maize or pepper plants. So, although they may travel to habitats surrounding agricultural fields, the impact will be limited by their inability to find host eggs in non-agricultural habitats. In addition, *Trichogramma ostriniae* is unable to survive very cold winters. **Approved for use in organic farming:** Yes.

3:370 *Trichogramma platneri*
Parasitoid of lepidopteran eggs

Parasitoid wasp: Hymenoptera: Trichogrammatidae

NOMENCLATURE: Approved name: *Trichogramma platneri* Nagarkatti (Fowler strain). **Other names:** Moth egg parasitoid; trichogramm.

SOURCE: Widely occurring parasitic wasp.

PRODUCTION: Reared in insectaries under controlled conditions on the eggs of moths, such as *Ephestia kuehniella* Zeller.

TARGET PESTS: Codling moth (*Cydia pomonella* L.).

TARGET CROPS: Apple and pear orchards and in avocados. In the USA, it is recommended for grapes west of the Rocky Mountains.

BIOLOGICAL ACTIVITY: Biology: Adult female wasps are about 0.5 mm long. They lay their eggs into a lepidopteran egg, preferring freshly laid eggs. Upon hatching, the wasp larva consumes the developing caterpillar within the egg. It pupates inside the egg and emerges as an adult wasp. Adults feed on nectar. **Predation:** Most lepidopteran eggs can be parasitised by *Trichogramma platneri*, but it is most effective in tree fruit crops. **Egg laying:** Adult wasps mate shortly after emergence and fertilised females lay eggs that develop into either males or females. Unfertilised females lay eggs that develop only into males. A female wasp will parasitise over 50 eggs in her life span of 5 to 14 days. **Duration of development:** The time from egg laying to wasp emergence varies with temperature, from as little as 7 to as long as 20 days. **Efficacy:** The parasitoid can have many generations in a season and, if numbers are high, will keep the codling moth population in check. The adults are very mobile and are expert at locating insect eggs.

COMMERCIALISATION: Formulation: Sold as parasitised eggs fixed to a card or in capsules. **Tradenames:** 'Trichogramma platneri' (Arbico), (Beneficial Insectary), (Rincon-Vitova), (Harmony), (IPM Laboratories) and (Praxis), 'Moth Egg Parasite' (Biocontrol Network).

APPLICATION: Spread cardboard carrier or capsules under the tree canopy. The release rate should be between 12000 and 500000 parasitised eggs per hectare, depending on the stage of growth of the crop, the area to be treated, the type of pest targeted, the development stage of the pest and the climatic conditions.

PRODUCT SPECIFICATIONS: Purity: Only parasitised, stored-product lepidopteran eggs, with no contaminants. **Storage conditions:** Store in the dark at 10 to 15 °C. **Shelf-life:** Can be kept for 3 to 4 days, under recommended storage conditions.

COMPATIBILITY: Should not be released after treatment with a broad-spectrum, residual insecticide. Can be used in combination with other beneficial insects.

MAMMALIAN TOXICITY: No allergic or other adverse reactions have been reported from the use of *Trichogramma platneri* in glasshouse or field conditions. Allergies associated with host Lepidoptera are known in production.

ENVIRONMENTAL IMPACT AND NON-TARGET TOXICITY: *Trichogramma platneri* is widespread in Nature. It has such a wide host range that it is possible that some lepidopteran non-target species may be affected. It is not expected to show any adverse effects on the environment. **Approved for use in organic farming:** Yes.

3. Macro-organisms

3:371 *Trichogramma pretiosum*
Parasitoid of lepidopteran eggs

Parasitoid wasp: Hymenoptera: Trichogrammatidae

NOMENCLATURE: Approved name: *Trichogramma pretiosum* Riley. **Other names:** Moth egg parasitoid; trichogramm.

SOURCE: Widely occurring parasitoid wasp.

PRODUCTION: Reared in insectaries under controlled conditions on the eggs of moths, such as *Ephestia kuehniella* Zeller.

TARGET PESTS: Noctuids, such as *Helicoverpa* and *Heliothis* spp., vegetable loopers, such as *Trichoplusia ni* (Hübner), cabbage moth (*Pieris* spp.) and other lepidopterans.

TARGET CROPS: Vegetables, sweet corn, field crops and fruit crops.

BIOLOGICAL ACTIVITY: Biology: *Trichogramma pretiosum* is a tiny wasp, less than 0.5 mm long, which lays its eggs into moth eggs. The wasp larva develops into a fully formed wasp inside the moth egg, in the process killing the developing caterpillar. A wasp emerges from the moth egg, instead of a caterpillar. Adults feed on nectar. **Predation:** Most lepidopteran eggs can be parasitised by *Trichogramma pretiosum*. **Egg laying:** Adult wasps mate shortly after emergence and fertilised females lay eggs that develop into either males or females. Unfertilised females lay eggs that develop only into males. A female wasp will parasitise over 50 eggs in her life span of 5 to 14 days. **Duration of development:** The time from egg laying to wasp emergence varies with temperature, from as little as 7 to as long as 20 days. **Efficacy:** The parasitoid can have many generations in a season and, if numbers are high, will keep the caterpillar population in check. The adults are very mobile and are expert at locating insect eggs.

COMMERCIALISATION: Formulation: Supplied as parasitised moth eggs in capsules (in sheets of 60), which are simply broken into individual capsules. **Tradenames:** 'Trichogramma pretiosum' (Bugs for Bugs), (Arbico), (Beneficial Insectary), (Harmony), (IPM Laboratories), (Praxis) and (Rincon-Vitova), 'Moth Egg Parasite' (Biocontrol Network).

APPLICATION: Place capsules evenly through the crop when moths are laying eggs. In sweet corn, place the capsules in the whorl. In tomatoes, staple the capsule to a leaf or post. In cotton, it is advantageous to have more advanced, low-insecticide crops, such as alfalfa (lucerne), sorghum and sunflowers, near the cotton into which *Trichogramma pretiosum* can be released. They can then build up and migrate into the cotton. Chickpeas are not suitable for this parasitoid. Releases can be made directly into cotton, but, in this case, insect-tolerant crops are preferred, as sprays for *Heliothis* in the early stages can be minimised. Release rates may vary from 30 to 90 capsules per hectare. Inoculative releases as low as 30 capsules per hectare will enable establishment of *Trichogramma pretiosum*, if there is a good egg lay at the time of release. If there is a steady egg lay, parasitism will increase with each generation of wasps (only 8 days in summer). It is not uncommon for parasitism to exceed 90%, even under heavy pressure. In sweet corn, two releases of 60 to 120 capsules per hectare, depending on pest pressure, should be made into each planting as soon as lepidopteran eggs are present in reasonable numbers (1 egg per 5 plants) or else weeks before silking. In tomatoes, two releases of 60 to 120 capsules per hectare, a week apart, should be made at the first sign of moth eggs. Regular releases may be the best option, where pressure is variable. *Trichogramma pretiosum* can be applied in water, with a special thickener added. The mixture can then be dosed onto foliage. A dosing device that is soft on the eggs is necessary. This method of application is fast and is suited to larger areas, but needs to be better timed than the capsules for best results.

PRODUCT SPECIFICATIONS: Purity: Only parasitised, stored-product lepidopteran eggs, with no contaminants. **Storage conditions:** Store in the dark at 10 to 15 °C.
Shelf-life: *Trichogramma pretiosum* adults will emerge from the parasitised eggs within 5 days of delivery.

COMPATIBILITY: *Trichogramma pretiosum* is best used in conjunction with other compatible and complementary control measures, such as biological insecticides, selective chemical insecticides and other biocontrol agents, including lacewings, local parasitoids and predatory beetles and bugs.

MAMMALIAN TOXICITY: No allergic or other adverse reactions have been reported from the use of *Trichogramma pretiosum* in glasshouse or field conditions. Allergies associated with host Lepidoptera are known in production.

ENVIRONMENTAL IMPACT AND NON-TARGET TOXICITY: *Trichogramma pretiosum* is widespread in Nature. It has such a wide host range that it is possible that some lepidopteran non-target species may be affected. It is not expected to show any adverse effects on the environment. **Approved for use in organic farming:** Yes.

3:372 *Trichogrammatoidea bactrae*
Parasitoid of lepidopteran eggs

Parasitoid wasp: Hymenoptera: Trichogrammatidae

NOMENCLATURE: Approved name: *Trichogrammatoidea bactrae* Nagaraja.

SOURCE: Native to Australia, but now widely spread following introduction for control of pest Lepidoptera.

PRODUCTION: Raised on eggs of Lepidoptera.

TARGET PESTS: Pink bollworm (*Pectinophora gossypiella* (Saunders)), cocoa pod borer (*Conopomorpha cramerella* Snellen), codling moth (*Cydia pomonella* L.) and other pest Lepidoptera.

TARGET CROPS: *Trichogrammatoidea bactrae* has been used in a wide range of crops, including cotton, cocoa, kiwifruit, avocado, grapes and apples.

BIOLOGICAL ACTIVITY: Biology: Adult female wasps are less than 0.5 mm long. They lay their eggs into a lepidopteran egg, preferring freshly laid eggs. Upon hatching, the wasp larva consumes the developing caterpillar within the egg. It pupates inside the egg and emerges as an adult wasp. Adults feed on nectar and will also eat lepidopteran eggs. **Predation:** *Trichogrammatoidea bactrae* is very mobile and actively seeks out newly laid pest eggs. Parasitised eggs turn black. Mated females lay eggs that develop into females whilst unmated females' eggs develop into males. **Efficacy:** Egg parasitism has been recorded to be as high as 95% in some studies.

COMMERCIALISATION: Formulation: Supplied as parasitised moth eggs in capsules or on cards. **Tradenames:** 'Trichogrammatoidea bactrae' (Rincon-Vitova) and (Praxis).

APPLICATION: Hang cards within the crop, evenly distributed. Release rates should be about 12000 to 500000 parasitised eggs per hectare, depending upon the incidence of caterpillars. Use pheromone traps to monitor the population of Lepidoptera within the crop.

PRODUCT SPECIFICATIONS: Purity: The products contain only eggs that have been fully parasitised by *Trichogrammatoidea bactrae*. **Storage conditions:** Eggs parasitised by *Trichogrammatoidea bactrae* may be stored under cool (<8 °C) conditions, out of direct sunlight for no more than 8 days.

COMPATIBILITY: *Trichogrammatoidea bactrae* can be used in conjunction with other beneficial insects and with *Bacillus thuringiensis*-based products. It is not compatible with broad-spectrum, residual insecticides.

MAMMALIAN TOXICITY: No allergic or other adverse reactions have been reported from the use of *Trichogrammatoidea bactrae* in glasshouse or field conditions. Allergies associated with host Lepidoptera are known in production.

ENVIRONMENTAL IMPACT AND NON-TARGET TOXICITY: *Trichogrammatoidea bactrae* is widespread in Nature. It has such a wide host range that it is possible that some lepidopteran non-target species may be affected. It is not expected to show any adverse effects on the environment. **Approved for use in organic farming:** Yes.

3:373 *Trichosirocalus horridus*

Phytophagous weevil

Phytophagous weevil: Coleoptera: Curculionidae

NOMENCLATURE: Approved name: *Trichosirocalus horridus* (Panzer). **Other names:** Thistle rosette weevil. Formerly known as: *Ceuthorhynchidius horridus* (Panzer).

SOURCE: *Trichosirocalus horridus* is a native of south and central Europe. It was imported from Italy into the USA in 1970–1972 for quarantine testing. Host specificity testing revealed that it would not be a threat to non-target plants and it was approved for field release in the USA in 1974.

PRODUCTION: Reared on thistle plants or collected from the wild as adults.

TARGET PESTS: *Trichosirocalus horridus* is used to control mainly *Carduus* and *Cirsium* species in the subtribe Carduinae of the Asteraceae family.

TARGET CROPS: Pasture and grassland.

BIOLOGICAL ACTIVITY: Biology: Newly emerged adult *Trichosirocalus horridus* are reddish in colour and about 10 mm long. They change to dark brown to black with time. The weevil overwinters in the adult, egg and larval stages. Adults emerge in late winter and early spring to lay eggs on young thistle rosettes. Each female lays an average of 500 to 800 eggs. Eggs are laid within the mid-rib of the underside of the leaves, in clusters of 2 or 3, and hatch in about 10 to 12 days. Newly hatched larvae feed within the mid-rib towards the centre growth point of the thistle rosette. The larvae reach the growing point in 7 days and continue feeding on the young tissues, causing the centre point to become dark in colour. They complete development in 6 to 8 weeks. The full-grown larva leaves the plant to pupate in the soil in pupation chambers. The new adult emerges in 12 to 20 days. The new generation of adults appears in May or June and goes into hiding until the autumn, when it emerges to feed. Some individuals may start laying eggs until the first frost before hibernation. There is usually one generation per year. **Efficacy:** The thistle rosette weevil is very effective in the control of musk thistle (*Carduus nutans* L.), as well as plumeless thistle (*Carduus acanthoides* L.). Biological control of musk thistle occurs after 5 to 6 years, but it takes longer (10 to 12 years) to control plumeless thistle. Typical thistle stand reductions of up to 95% have been obtained at sites where the weevil became established. **Key reference(s):** 1) L T Kok. 1975. Host specificity studies on *Ceuthorhynchidius horridus* (Panzer) (Coleoptera: Curculionidae) for the control of musk and plumeless thistle, *Weed Research*, **15**, 21–25. 2) L T Kok. 1986. Impact of *Trichosirocalus horridus* (Coleoptera: Curculionidae) on *Carduus* thistles in pastures, *Crop Protection*, **5**, 214–7. 3) L T Kok. 1992. Biological control of musk and plumeless thistles, *Virginia Coop. Ext. Pub.* 444-019, 1–8. 4) L T Kok & W T Mays. 1989. Comparison of the seasonal occurrence of *Trichosirocalus horridus* (Coleoptera: Curculionidae) in Virginia between 1981–83 and 1979, *J. Entomol. Sci.*, **24**, 465–71. 5) L T Kok & W T Mays. 1991. Successful biological control of plumeless thistle, *Carduus acanthoides* L. [Campanulatae: Asteraceae (=Compositae)], by *Trichosirocalus horridus* (Panzer) (Coleoptera: Curculionidae) in Virginia, *Biological Control*, **1**, 197–202. 6) J T Trumble & L T Kok. 1982. Integrated pest management techniques in thistle suppression in pastures of North America, *Weed Research*, **22**, 345–59.

COMMERCIALISATION: Formulation: *Trichosirocalus horridus* is available in the adult stage. **Tradenames:** 'Ceuthorhynchidius horridus' (Praxis).

APPLICATION: Overwintered adults must be released in early spring without delay to allow them to lay their eggs before they die. Adults of the new generation that are collected in the summer will not lay eggs until the following spring. Thus, they can be released any time after collection, as they will hibernate and the surviving overwintered weevils will lay eggs in the spring.

PRODUCT SPECIFICATIONS: Purity: Supplied as fecund adults, with no contaminants. **Storage conditions:** The adults should be released as soon as possible after delivery.

COMPATIBILITY: This weevil is susceptible to the commonly used insecticides and thus should not be exposed to insecticidal sprays. It is not directly affected by commonly used herbicides and thus can be used in conjunction with herbicides for thistle control. This can be carried out as long as the insect has completed development before the thistles die from herbicide action.

MAMMALIAN TOXICITY: There have been no reports of allergic or other adverse reactions from research workers, manufacturing staff or from the field release of *Trichosirocalus horridus*.

ENVIRONMENTAL IMPACT AND NON-TARGET TOXICITY: *Trichosirocalus horridus* occurs widely in Nature and is not expected to have any adverse effects on non-target organisms or on the environment. Host specificity testing carried out in the USA revealed that it would not be a threat to non-target plants.

Predatory mite: Mesostigmata: Phytoseiidae

NOMENCLATURE: Approved name: *Typhlodromus pyri* Scheuten (Mikulov strain).
Other names: Fruit tree predatory mite; *Typhlodromus* mite predator.
Synonymous with: *Galendromus pyri* Scheuten and *Metaseiulus pyri* (Scheuten).

SOURCE: Occurs widely, particularly in warm climates. Frequently found in fruit trees.

PRODUCTION: Reared, under controlled conditions, on two-spotted spider mites
(*Tetranychus urticae* Koch) on bean plants.

TARGET PESTS: Spider mites: *Tetranychus urticae*, *T. cinnabarinus* (Boisduval) and *Panonychus ulmi* (Koch); tarsonemid mites: *Polyphagotarsonemus latus* (Banks) and *Phytonemus pallidus* (Banks); and eriophyid mites: *Calepitrimerus vitis* (Nalepa), *Eriophyes vitis* (Prendergast) and *Aculus schlechtendali* (Nalepa).

TARGET CROPS: Fruit trees, especially top fruit, and vineyards. Also recommended for use in protected and outdoor edible and ornamental crops.

BIOLOGICAL ACTIVITY: Biology: Eggs are laid in the vicinity of mite eggs. These hatch into larvae. Nymphs develop from the larvae and consume mite eggs, larvae, nymphs and adults. Adults live for 17 to 20 days and predate mite eggs, larvae, nymphs and adults. Depending upon climatic conditions, there are between two and four generations per season. At low temperatures and in short days, adults will enter diapause, with the females overwintering in the bark of trees. If *Typhlodromus pyri* is to be used in the winter, supplementary lighting is needed, with the preferred lighting regime 16:8 light:dark. *Typhlodromus pyri* can survive on pollen and fungi. **Predation:** Predatory mites pierce their prey and suck the contents. Adults will consume about 8 mites or eggs each day. Most effective under conditions of low humidity and high temperature. **Egg laying:** A female *Typhlodromus pyri* will lay about 50 eggs in her lifetime, at the rate of 2 to 3 per day. **Duration of development:** Eggs hatch within a few days of laying and adults develop within 7 to 8 days, under ideal conditions. Optimum temperatures are between 27 and 32 °C and the minimum effective temperature is 10 °C, although the predator will survive at lower temperatures. **Efficacy:** *Typhlodromus pyri* is a very effective predator of mites, but control in orchards and vineyards takes time. Often good control is not achieved until the season following application. *T. pyri* will tolerate extremes of temperature and low relative humidity better than *Phytoseiulus persimilis* Athios-Henriot and is, therefore, particularly useful in interiorscapes and on outdoor nursery stock. *T. pyri* spreads more slowly through the crop, however, so an initially higher introduction rate is needed.

COMMERCIALISATION: Formulation: Sold as packs containing textile belts with 10 to 20 predatory female mites in diapause and as intact colonies on bean leaves. Sometimes

3. Macro-organisms

supplied in combination with *Phytoseiulus persimilis*. **Tradenames:** 'Typex' (Svenska Predator), 'Typhlodromus mite predators' (Biowise).

APPLICATION: Apply at a rate of 25 000 predators per hectare. Attach textile strips to the trunks or branches of the crop to be treated in the late winter or early spring. For intact colonies, the bean leaves should be placed within the infested foliage. Protected crops should be treated as soon as the phytophagous mites are noticed. Outdoor crops should be treated every 7 to 14 days through the spring and summer, while the pest mites are active.

PRODUCT SPECIFICATIONS: Purity: Supplied as adult female mites in diapause or on bean leaves that may contain some phytophagous mites, in addition to *T. pyri*.
Storage conditions: *Typhlodromus pyri* is supplied only in winter and should be stored in a cooled room, an unheated cellar or in the open air. Do not allow the storage temperature to rise above 10 °C. Do not expose to direct sunlight. Adults on bean leaves should be stored at temperatures between 5 and 10 °C. **Shelf-life:** Diapaused *Typhlodromus pyri* may be stored for 2 months, under recommended storage conditions. Adults on bean leaves should be used within 3 days of receipt.

COMPATIBILITY: *Typhlodromus pyri* is very sensitive to chemical pesticides, although the Mikulov strain is 200 times less sensitive than the wild type. Nevertheless, it is recommended that chemical pesticides are not applied following the release of the predator.

MAMMALIAN TOXICITY: There have been no reports of allergic or other adverse reactions from research workers, manufacturing staff or from the field release of *Typhlodromus pyri*.

ENVIRONMENTAL IMPACT AND NON-TARGET TOXICITY: *Typhlodromus pyri* occurs widely in Nature and is not expected to have any adverse effects on non-target organisms or on the environment. It is specific to mites. **Approved for use in organic farming:** Yes.

3:375 *Typhlodromus rickeri*   *Predator of mites*

Predatory mite: Mesostigmata: Phytoseiidae

NOMENCLATURE: Approved name: *Typhlodromus rickeri* Chant. **Other names:** Mite predator.

SOURCE: India.

PRODUCTION: Reared, under controlled conditions, on two-spotted spider mites (*Tetranychus urticae* Koch) on bean plants.

TARGET PESTS: Spider mites: *Tetranychus urticae*, *T. cinnabarinus* (Boisduval) and *Panonychus ulmi* (Koch); broad mites (*Polyphagotarsonemus latus* (Banks)); and avocado six spotted mite (*Eotetranychus sexmaculatus* (Riley)).

TARGET CROPS: Fruit trees, especially top fruit, and vineyards.

BIOLOGICAL ACTIVITY: Biology: Eggs are laid in the vicinity of mite eggs. These hatch into larvae. Nymphs develop from the larvae and consume mite eggs, larvae, nymphs and adults. Adults live for 17 to 20 days and predate mite eggs, larvae, nymphs and adults. Depending upon climatic conditions, there are between two and four generations per season. At low temperatures and in short days, adults will enter diapause, with the females overwintering in the bark of trees. *Typhlodromus rickeri* can survive on pollen and fungi. **Predation:** Predatory mites pierce their prey and suck the contents. Adults will consume about 8 mites or eggs each day. Most effective under conditions of low humidity and high temperature. **Egg laying:** A female *Typhlodromus rickeri* will lay about 50 eggs in her lifetime, at the rate of 2 to 3 per day. **Efficacy:** *Typhlodromus rickeri* is a very effective predator of mites, but control in orchards and vineyards takes time. Often good control is not achieved until the season following application. It spreads slowly through the crop, however, so a high introduction rate is needed. **Key reference(s):** J A McMurtry, M H Badii & H G Johnson. 1984. The broad mite, *Polyphagotarsonemus latus*, as a potential prey for phytoseiid mites in California, *BioControl*, **29**(1), 83–6.

COMMERCIALISATION: Formulation: Sold as packs containing textile belts with 10 to 20 predatory female mites in diapause and as intact colonies on bean leaves. **Tradenames:** 'Typhlodromus rickeri' (Biotactics) and (Rincon-Vitova).

APPLICATION: Apply at a rate of 25 000 predators per hectare. Attach textile strips to the trunks or branches of the crop to be treated in the late winter or early spring. For intact colonies, the bean leaves should be placed within the infested foliage. Protected crops should be treated as soon as the phytophagous mites are noticed. Outdoor crops should be treated every 7 to 14 days through the spring and summer, while the pest mites are active.

PRODUCT SPECIFICATIONS: Storage conditions: Do not allow the storage temperature to rise above 10 °C. Do not expose to direct sunlight.

COMPATIBILITY: *Typhlodromus rickeri* is very sensitive to chemical pesticides and it is recommended that chemical pesticides are not applied following the release of the predator.

MAMMALIAN TOXICITY: There have been no reports of allergic or other adverse reactions from research workers, manufacturing staff or from the field release of *Typhlodromus rickeri*.

ENVIRONMENTAL IMPACT AND NON-TARGET TOXICITY: *Typhlodromus rickeri* occurs widely in Nature and is not expected to have any adverse effects on non-target organisms or on the environment. It is specific to mites. **Approved for use in organic farming:** Yes.

3. Macro-organisms

Phytophagous caterpillar: Lepidoptera: Arctiidae

NOMENCLATURE: Approved name: *Tyria jacobaeae* (L.). **Other names:** Cinnabar moth. Formerly known as: *Callimorpha senecionis* Godart.

SOURCE: The moth is common in Europe where its host, the common or tansy ragwort (*Senecio jacobaea* L.), is widespread. It has been introduced into North America and Australasia.

PRODUCTION: *Tyria jacobaeae* is reared under controlled conditions on its host, the common or tansy ragwort.

TARGET PESTS: Common or tansy ragwort (*Senecio jacobaea*).

TARGET CROPS: Pasture, rangeland and prairie.

BIOLOGICAL ACTIVITY: Biology: Eggs, which are spherical and yellow when first laid, but becoming darker later, are laid in groups of about 50, usually on the underside of leaves of a host plant. Caterpillars are initially yellow, with a black head. Later instars are brightly coloured, with black and yellow bands. They feed initially on the leaves, but, as the host plant matures, they move onto the flowers. When fully grown, the caterpillars leave the host plant and pupate some distance away in a crevice or similarly sheltered spot. The adult moth is shiny black, with a red line along the costa and two red spots on each forewing. It has red hind-wings and a wingspan of about 3 cm. Nevertheless, the moths appear to be reluctant fliers, just moving on a metre or so when disturbed. **Efficacy:** The larvae of *Tyria jacobaeae* are voracious feeders and can completely defoliate the host plant within a few weeks. **Key reference(s):** D Carter. 1992. *Butterflies and Moths*, Collins Eyewitness Handbooks, Sydney, Australia, 282 pp.

COMMERCIALISATION: Formulation: Sold as pupae. **Tradenames:** 'Tyria jacobaeae' (Praxis).

APPLICATION: Distribute the pupae within the weed-infested plots.

PRODUCT SPECIFICATIONS: Purity: Products contain only *Tyria jacobaeae* pupae, and no contaminants. **Storage conditions:** May be stored for up to 7 days, in a cool, dry place out of direct sunlight.

COMPATIBILITY: Compatible with most herbicides that might be used for control of ragwort. May be used together with *Longitarsus jacobaeae* (Waterhouse) for more effective control of ragwort.

MAMMALIAN TOXICITY: There have been no reports of allergic or other adverse reactions from research workers, manufacturing staff or from the field release of *Tyria jacobaeae*.

ENVIRONMENTAL IMPACT AND NON-TARGET TOXICITY: *Tyria jacobaeae* occurs widely in Nature and is not expected to have any adverse effects on non-target organisms or on the environment. Host specificity testing carried out in Australia and the USA revealed that it would not be a threat to non-target plants.

3:377 *Urophora quadrifasciata*
Phytophagous seedhead fly

Phytophagous fly: Diptera: Tephritidae

NOMENCLATURE: Approved name: *Urophora quadrifasciata* (Meigen).
Other names: Knapweed seedhead fly.

SOURCE: *Urophora quadrifasciata* is native to Europe, but has been introduced in North America for control of knapweeds.

PRODUCTION: *Urophora quadrifasciata* is reared on knapweed plants.

TARGET PESTS: Spotted knapweed (*Centaurea maculosa* Lam.) and diffuse knapweed (*C. diffusa* Lam.).

TARGET CROPS: Pasture, grassland and prairie.

BIOLOGICAL ACTIVITY: Biology: *Urophora quadrifasciata* is a small (2 to 3 mm long), dark-bodied fly, with a distinctive V- or U-shaped mark on each wing. The female has a long, retractable ovipositor. The gall in the seedhead is paper-thin, with pappus on top. The larva is white. *U. quadrifasciata* produces two generations per year. The overwintering generation emerges as adults from galls in the mature seedheads in the spring, as the new flower buds are beginning to form. Eggs hatch in 3 to 4 days and the larva chews down through a floret into the ovary, causing the plant to start forming a gall about 8 days after the larva has hatched. The gall will have reached maximum size 15 days after hatching. It takes about 8 to 9 weeks from hatching for the new adult generation to appear. The next generation or late summer generation will overwinter as mature larvae in the seedheads, emerging the following spring as adults. *U. quadrifasciata* is a strong flier and disperses great distances. This fly is rapidly colonising the knapweed patches of North America. **Egg laying:** The female fly, after mating, uses her ovipositor to penetrate flower buds that are over half grown, to lay her eggs among the developing stamens. She may lay more than one egg per seedhead.

Efficacy: *Urophora quadrifasciata*, when used in conjunction with other beneficial insects, such as *Agapeta zoegana* (L.) and *Sphenoptera jugoslavica* Obenb., has reduced seed production in knapweed plants by as much as 95%. **Key reference(s):** 1) P Harris. 1980. Establishment of *Urophora affinis* Frfld. and *U. quadrifasciata* (Meig.) (Diptera: Tephritidae) in Canada for the biological control of diffuse and spotted knapweed, *Z. Angew. Entomol.*, **89**, 504–14. 2) E R Hoebeke. 1993. Establishment of *Urophora quadrifasciata* (Diptera: Tephritidae) and *Chrysolina quadrigemina* (Coleoptera: Chrysomelidae) in portions of the eastern United States, *Entomol. News*, **104**, 143–52. 3) N E Rees & J M Story. 1991. Host plant testing of *Urophora quadrifasciata* (Diptera: Tephritidae) against *Carthamus tinctorius* and two North American species of *Centaurea*, *Entomophaga*, **36**, 115–19.

COMMERCIALISATION: Formulation: Sold as adults. In some US states, adults may be obtained, at no cost, from state weed management agencies. **Tradenames:** 'Urophora quadrifasciata' (Praxis).

APPLICATION: Release adults in sites that will not be disturbed for at least 10 years by development or pesticide use. *U. quadrifasciata* prefers open, sunny sites to closed shady areas.

PRODUCT SPECIFICATIONS: Purity: Products contain only *Urophora quadrifasciata* adults, and no contaminants. **Storage conditions:** Adults may be stored under cool, dry conditions for up to 7 days.

COMPATIBILITY: High rates of 2,4-D may cause increased larval and pupal mortality, but spring applications of the herbicides 2,4-D and picloram (applied when knapweeds are primarily in the rosette stage) appear to have no significant negative impacts on *Urophora quadrifasciata* populations.

MAMMALIAN TOXICITY: There have been no reports of allergic or other adverse reactions from research workers, manufacturing staff or from the field release of *Urophora quadrifasciata*.

ENVIRONMENTAL IMPACT AND NON-TARGET TOXICITY: *Urophora quadrifasciata* occurs widely in Nature and is not expected to have any adverse effects on non-target organisms or on the environment. *U. quadrifasciata* attacks plants in the genus *Centaurea*. The fly was tested against safflower (*Carthamus tinctorius* L.), an economically important relative of knapweed, and two US native knapweeds. No oviposition or larval development was found on these three plant species.

3:378 *Xylocoris flavipes*   *Predator of insects*

Pirate bug: Hemiptera: Anthocoridae

NOMENCLATURE: Approved name: *Xylocoris flavipes* (Reuter). **Other names:** Warehouse pirate bug.

SOURCE: Widely occurring where grain is stored.

PRODUCTION: Bred on a variety of insects, under controlled conditions.

TARGET PESTS: A general predator, consuming all commonly occurring insect and mite pests of stored grain.

TARGET CROPS: Used in grain stores.

BIOLOGICAL ACTIVITY: Biology: *Xylocoris flavipes* is a generalist predator, whose adults and nymphs feed on any life stage of pests that can be found in stored grain. Pirate bugs have the greatest impact on external feeding species of beetles and moths. It does not feed on or adversely affect grain, and the adult natural enemies occur outside grain kernels and, hence, can be cleaned from grain before milling. **Efficacy:** Research during the 1970s and 1980s demonstrated the efficacy of the predator in controlling storage pests. Releases of the warehouse pirate bug, *Xylocoris flavipes*, resulted in a 79% to 100% suppression of moth populations in small storages of peanuts, up to 99% reduction of sawtoothed grain beetle populations in maize (corn) and a 90% to 98% suppression of red flour beetles in a simulated peanut warehouse.

COMMERCIALISATION: Formulation: Sold as adults. **Tradenames:** 'Xylocoris flavipes' (Praxis), (Rincon-Vitova) and (Arbico).

APPLICATION: Introduce adults into grain stores shortly after filling.

PRODUCT SPECIFICATIONS: Purity: Products contain only adult *Xylocoris flavipes*. **Storage conditions:** May be stored for short periods, under cool, dry conditions. The bug is an aggressive predator and will not live long without a food source.

COMPATIBILITY: Cannot be used with any broad-spectrum grain store fumigant or broad-spectrum insecticide.

MAMMALIAN TOXICITY: There have been no reports of allergic or other adverse toxicological effects arising from contact with *Xylocoris flavipes* from research staff, producers or users.

ENVIRONMENTAL IMPACT AND NON-TARGET TOXICITY: *Xylocoris flavipes* has a wide distribution in Nature and there is no evidence that it has any adverse effects on non-target organisms or on the environment. It is polyphagous, but is unlikely to leave the grain stores into which it was introduced.

3. Macro-organisms

4. Semiochemicals

4:379 (Z)-13-acetoxy-8-heptadecen-2-one
Apple leaf midge sex pheromone

NOMENCLATURE: Approved name: (Z)-13-acetoxy-8-heptadecen-2-one.
Other names: Apple leaf midge sex pheromone.

SOURCE: Identified from compounds emitted by virgin female apple leaf midges (*Dasineura mali* (Kieffer)).

PRODUCTION: Manufactured for use in lures.

TARGET PESTS: Apple leaf midge (*Dasineura mali*).

TARGET CROPS: Apple orchards.

BIOLOGICAL ACTIVITY: Mode of action: (Z)-13-acetoxy-8-heptadecen-2-one is the female sex pheromone of the apple leaf midge. Males are attracted to lures containing the pheromone where they are trapped. Counting the number of males caught allows a decision to be made on whether or not to spray the crop with an insecticide to protect the crop. Trials are currently under way to use the pheromone in mating disruption. **Efficacy:** The numbers of *Dasineura mali* caught in the pheromone traps give a reliable indication of the onset of flight activity in the vicinity of orchards and have a strong correlation with the severity of subsequent crop infestation levels. The high sensitivity of the traps outweighs the fact that pheromone traps catch male rather than female insects. Pheromone traps can be used to give a warning to growers of when midge levels are expected to be damagingly high. Insecticide should be applied soon after the detection of high levels of adult midge. **Key reference(s):** J J Heath, S O Gaul, D M Nash & R F Smith. 1998. Evidence for a female-produced sex pheromone in the apple leaf midge, *Dasineura mali* (Kieffer) (Diptera: Cecidomyiidae), *Canadian Entomologist*, **130**, 109–10.

COMMERCIALISATION: Formulation: Supplied as a trap that releases the pheromone to attract the males. Estimates of the size of the invading population of midges can then be calculated. **Patents:** International Patent Application Number: PCT/GB2005/002504.

4. Semiochemicals

APPLICATION: The traps are set before the first flight of the midges in Spring at the rate of one trap for each plantation/cropping situation towards the centre of the orchard.

PRODUCT SPECIFICATIONS: Manufactured for use in lures.

MAMMALIAN TOXICITY: The product has shown no adverse toxicological effects on research or manufacturing workers or on users.

ENVIRONMENTAL IMPACT AND NON-TARGET TOXICITY: There is no evidence that (Z)-13-acetoxy-8-heptadecen-2-one has caused significant adverse effects on non-target organisms nor had an adverse environmental impact.
Approved for use in organic farming: Yes.

4:380 2-acetoxy-5-undecanone
Raspberry cane midge sex pheromone

NOMENCLATURE: Approved name: 2-acetoxy-5-undecanone. **Other names:** Raspberry cane midge sex pheromone.

SOURCE: Isolated from the volatiles released by virgin females.

PRODUCTION: Manufactured for use in lures.

TARGET PESTS: Raspberry cane midge (*Resseliella theobaldi* (Barnes)).

TARGET CROPS: Raspberries.

BIOLOGICAL ACTIVITY: Mode of action: The pheromone is enclosed in a trap which is designed to catch adult raspberry cane midges. Male adults are attracted to the female sex pheromone contained in a rubber septum and become stuck on the glue surface. Trap catches can be used to time sprays in the open air as well as in protected crops, where attacks occur much earlier than outdoors. **Efficacy:** The numbers of *Resseliella theobald* caught in the pheromone traps give a reliable indication of the onset of flight activity in the vicinity of the crop and have a strong correlation with the severity of subsequent crop infestation levels.

The high sensitivity of the traps outweighs the fact that pheromone traps catch male rather than female insects. Pheromone traps can be used to give a warning to farmers of when midge levels are expected to be damagingly high. Insecticide then should be applied soon after the detection of high levels of midge.

COMMERCIALISATION: Formulation: Sold as a trap that releases the pheromone to attract the males. Estimates of the size of the invading population of midges can then be calculated. **Tradenames:** 'Raspberry Cane Midge Kit' (Suterra). **Patents:** International Patent Application Number: PCT/GB2005/002504.

APPLICATION: The traps are set before the first flight of the midges in Spring at the rate of one trap for each plantation/cropping situation towards the centre of the plantation. The trap base should be at a height of 0.5 m above the ground. Lures last 4 weeks and will need to be changed monthly in the course of the season.

PRODUCT SPECIFICATIONS: Manufactured for use in lures.

MAMMALIAN TOXICITY: The product has shown no adverse toxicological effects on research or manufacturing workers or on users.

ENVIRONMENTAL IMPACT AND NON-TARGET TOXICITY: There is no evidence that 2-acetoxy-5-undecanone has caused significant adverse effects on non-target organisms nor had an adverse environmental impact. **Approved for use in organic farming:** Yes.

4:381 4-allyl anisole
Insect repellent and anti-aggregation pheromone

NOMENCLATURE: Approved name: 4-allyl anisole. **Other names:** Estragole. **OPP Chemical Code:** 062150. **CAS RN:** *[140-67-0]*.

SOURCE: 4-Allyl anisole is produced naturally in conifer trees in response to infestation by beetles. Herbs such as basil, fennel and tarragon also contain high levels of 4-allyl anisole.

PRODUCTION: Manufactured for use in forestry management.

TARGET PESTS: Bark beetles, such as southern pine beetle (*Dendroctonus frontalis* Zimmerman).

TARGET CROPS: Conifer forests.

BIOLOGICAL ACTIVITY: Mode of action: The exact mode of action is unknown, but it is proposed that 4-allyl anisole either acts as a repellent when large amounts are released to signal bark beetles that too many beetles are present on a diseased tree, or that the release of the compound by an infested tree attracts parasites that attack the bark beetles. **Efficacy:** The naturally occurring chemical acts as a beetle repellent/anti-aggregant. When small amounts of 4-allyl anisole are applied to trees, beetles are prevented from aggregating and from large scale reproduction.

COMMERCIALISATION: Formulation: The product is sold as technical material containing 97.5% 4-allyl anisole, which is formulated into paint balls for application to infested or damaged trees. **Tradenames:** 'Beetleball Technical' (Taensa).

APPLICATION: The technical material is incorporated into paint balls, which are fired at infested or damaged trees.

PRODUCT SPECIFICATIONS: Manufactured for use in forestry management.

COMPATIBILITY: It is not usual to apply 4-allyl anisole with other crop protection agents.

MAMMALIAN TOXICITY: No harm to mammals is expected from use of pesticide products containing 4-allyl anisole. **Acute oral LD$_{50}$:** rats >1230 mg/kg (Toxicity Category III). **Acute dermal LD$_{50}$:** rabbits >2020 mg/kg (Toxicity Category III). **Inhalation:** Rats: LC$_{50}$ >2.64 mg/litre (Toxicity Category IV). **Skin and eye:** Primary eye irritation slightly irritating to rabbit eyes (Toxicity Category IV). Primary dermal irritation moderately irritating to rabbit skin (Toxicity Category III). **Other toxicological effects:** 4-Allyl anisole has been classified as GRAS (generally recognised as safe) by the US Food and Drug Administration for use in perfumes and as flavourings in foods and liqueurs.

ENVIRONMENTAL IMPACT AND NON-TARGET TOXICITY: No harm to non-target organisms or the environment is expected from use of pesticide products containing 4-allyl anisole.

4:382 codlemone

Codling moth sex pheromone

The *Pesticide Manual* **Fifteenth Edition entry number:** 178

$$\underset{CH_3}{H}C=C\overset{H}{\underset{H}{\diagdown}}C=C\overset{(CH_2)_7OH}{\underset{H}{\diagup}}$$

NOMENCLATURE: Approved name: codlemone ((*E,E*)-dodeca-8,10-dien-1-ol).
Other names: *Cydia pomonella* pheromone; codling moth pheromone; E8,10-12OH (IOBC); EE8,10-12OH. **Development code:** E101 (Exosect). **CAS RN:** [33956-49-9].

SOURCE: The sex pheromone of the codling moth (*Cydia pomonella* L.) was originally isolated from the terminal abdominal segments of virgin females.

PRODUCTION: Manufactured for use in crop protection.

TARGET PESTS: Codling moth (*Cydia pomonella*). Also effective against the hickory shuckworm (*C. caryana* (Fitch)).

TARGET CROPS: Recommended for use in pome fruit crops such as apples, pears and quinces; in peaches, nectarines, plums, and other stone fruit orchards; and in walnut, pecans and other tree nuts.

BIOLOGICAL ACTIVITY: Mode of action: Codlemone is the female sex pheromone of the codling moth. Male moths locate and subsequently mate with female moths by following the trail or pheromone plume emitted by virgin females. The indiscriminate application of artificially high levels of codlemone interferes with this process, as a constant exposure to high levels of pheromone makes trail following impossible (habituation/adaptation). Alternatively, the use of discrete sources of pheromone released over time presents the male with a false trail to follow (confusion). Control is subsequently achieved through the prevention of mating and the laying of fertile eggs. The natural female sex pheromone contains a number of components, some of which enhance the attractiveness of the main component, (*E,E*)-dodeca-8,10-dien-1-ol to male moths (dodecanol) and others that reduce the attraction of male moths ((*E,E*)-dodeca-8,10-dien-1-ol acetate). **Efficacy:** Low rates are effective in causing mating disruption. Codlemone is volatile and distributes throughout the crop easily. Codlemone has been shown to maintain populations of codling moth effectively at economically tolerable levels between fruiting and harvest. The pheromone can also be used as a means of monitoring the incidence of the moths, so that insecticidal sprays (such as *Cydia pomonella* granulosis virus) can be applied at the most susceptible stages of the insect larvae.

4. Semiochemicals

The use of the pheromone to attract the moths to a contact insecticide (lure and kill) is also successful in controlling the moths. **Key reference(s):** R J Bartell & T E Bellas. 1981. Evidence for naturally occurring, secondary compounds of the codling moth female sex pheromones, *J. Australian Entomological Society*, **20**, 197.

COMMERCIALISATION: Formulation: Sold as a coil or as polyethylene ampoules that release the pheromone slowly as a vapour. Consep also manufactures a sprayable formulation. **Tradenames:** 'Isomate-C' (mixture) (Shin-Etsu) and (Biocontrol), 'Codlemone' (plastic tubes), 'Isomate-C Special' (plus (*Z*)-tetradec-11-enyl acetate) and 'Isomate C LR' (plus dodecan-1-ol plus tetradecan-1-ol plus (*Z*)-9-tetradecenyl acetate plus (*Z*)-11-tetradecenyl acetate) (Shin-Etsu), 'NoMate CM' (Scentry), 'RAK 3' (BASF), 'Sirene-CM' (plus permethrin) (IPM Technologies), 'Hercon Disrupt CM' (laminated plastic) and 'Disrupt CM-Xtra' (Hercon), 'CheckMate CM', 'Checkmate CM-F', 'Checkmate CM-WS' and 'Checkmate CM-XL 1000' (Consep), 'Checkmate CM-WS' (hand-held dispenser), 'Puffer CM' (aerosol) and 'Paramount Aerosol NOW/CM' (plus (*Z,Z*)-11,13-hexadecadienal) (Suterra), 'MEC.CM' (Suterra), 'Ecopom' (mating disruption) (Isagro), 'A-9497b' (plus permethrin) (IPM Technologies), 'Appeal' (plus cyfluthrin) (Bayer CropScience), 'Isomate-CM/LR' (plus *n*-decanol plus *n*-dodecanol plus tetradec-11-enyl acetate) (Pacific Biocontrol), 'Isomate-CTT' (plus *n*-dodecanol plus *n*-tetradecanol) and 'Isomate-C Plus' (plus *n*-dodecanol plus *n*-tetradecanol) (Shin-Etsu) and (Pacific Biocontrol), 'RAK 3+4' (plus tetradec-11-enyl acetate) (BASF), 'SPLAT Cydia' (ISCA), 'Exosex CM' (Exosect).

APPLICATION: Slow-release containers are placed at intervals within the crop from bud burst to small fruit stage and prior to the flight of over-wintering adults. The pheromone is allowed to diffuse into the air and disperse throughout the orchard. The recommended number of lures per hectare varies with release rates, but 1000 per hectare with additional protection at the edges of the orchard is common. Lures should be replaced every 5 weeks. Lure and kill strategies use codlemone together with a contact insecticide such as permethrin. It is recommended that no more than 370 g a.i. be dispensed per hectare per year. 'SPLAT Cydia' (Specialized Pheromone and Lure Application Technology) is an amorphous polymer matrix for the sustained release of insect pheromones. It has been designed to dispense, optimise and modulate the semiochemical release in mating disruption pest management programmes. The product can be applied by hand, metered backpack sprayers, caulking guns, metered dosing guns, paintball guns, tractor or aircraft. For most orchards, one application of 'SPLAT Cydia' has been demonstrated to provide mating disruption protection for the whole season.

PRODUCT SPECIFICATIONS: Manufactured for use in crop protection. **Purity:** The mating pheromone of the codling moth (*Cydia pomonella*). Typical composition for mating disruption contains 62.5% codlemone, with 31% dodecan-1-ol and 6% *n*-tetradecan-1-ol.
Shelf-life: Stored in their original packaging at 5 °C, the products will keep for 2 years. When

the products are refrigerated for use the following season, store them at room temperature for one month prior to field application.

COMPATIBILITY: Compatible with all crop protection agents that do not repel codling moth adults.

MAMMALIAN TOXICITY: Codlemone has shown no adverse toxicological effects on research or manufacturing workers or on users. Considered to be non-toxic.
Acute oral LD$_{50}$: rats and mice >4000 mg/kg. **Inhalation:** LC$_{50}$ (4 h) 5 mg/litre.
Skin and eye: No toxicity was observed at the highest test level, but it can cause some irritation to eyes and skin.

ENVIRONMENTAL IMPACT AND NON-TARGET TOXICITY: Codlemone is a natural insect pheromone that is specific to the codling moth. There is no evidence that it has caused any adverse effects on non-target organisms nor had any deleterious effect on the environment. **Fish toxicity:** LC$_{50}$ (96 h) for rainbow trout >120 mg/litre.
Other aquatic toxicity: EC$_{50}$ (48 h) for *Daphnia* 2.8 mg/litre. EC$_{50}$ (72 h) for algae 0.83 mg/litre.
Behaviour in soil: Rapidly degraded in soil to carbon dioxide and water.
Approved for use in organic farming: Yes.

4:383 (*E*)-dec-5-enyl acetate plus (*E*)-dec-5-enol   *Peach tree borer sex pheromone*

The Pesticide Manual **Fifteenth Edition entry number:** 237

$$\underset{CH_3(CH_2)_3}{H} \diagdown \diagup \underset{H}{(CH_2)_4OCOCH_3}$$

(*E*)-dec-5-enyl acetate

$$\underset{CH_3(CH_2)_3}{H} \diagdown \diagup \underset{H}{(CH_2)_4OH}$$

(*E*)-dec-5-enol

NOMENCLATURE: Approved name: (*E*)-dec-5-enyl acetate plus (*E*)-dec-5-enol.
Other names: Peach tree borer sex pheromone; E5-10Ac; E5-10OH.
Development code: BAS 287 I. **CAS RN:** *[38421-90-8]* (*E*)-dec-5-enyl acetate; *[56578-18-8]* (*E*)-dec-5-enol.

SOURCE: The sex pheromone of the peach tree borer or peach twig borer (*Anarsia lineatella* Zeller) was isolated from virgin female pheromone glands.

PRODUCTION: Manufactured for use in crop protection.

TARGET PESTS: Peach tree borer (*Anarsia lineatella*).

TARGET CROPS: Peach tree orchards.

BIOLOGICAL ACTIVITY: Mode of action: The combination of (*E*)-dec-5-enyl acetate and (*E*)-dec-5-enol comprises the female sex pheromone of the peach tree borer. Male moths locate and subsequently mate with female moths by following the pheromone trail or pheromone plume emitted by virgin females. The application of high levels of this combination product makes the trail impossible for male *Anarsia lineatella* to follow (camouflage, competition between artificial and female plume, false trail following). Control is subsequently achieved through the prevention of mating and the laying of fertile eggs. **Efficacy:** Very low rates are required to cause mating disruption. (*E*)-dec-5-enyl acetate and (*E*)-dec-5-enol are volatile and distribute throughout the crop very easily. **Key reference(s):** W Roelofs, J Kochansky, E Anthon, R Rice & R Cardé. 1975. Sex pheromone of the peach twigborer moth (*Anarsia lineatella*), *Environ. Entomol.*, **4**, 580.

COMMERCIALISATION: Formulation: Sold as a slow-release formulation of polyethylene ampoules. Consep also offers 'Checkmate SF', a dual dispenser containing the peach tree borer and oriental fruit moth pheromones. **Tradenames:** 'RAK 5+6' (sold in mixture with (*Z*)-dodec-8-enyl acetate plus (*E*)-dodec-8-enyl acetate plus (*Z*)-dodec-8-enol) (BASF), 'Checkmate PTB', 'Checkmate SF' (plus dodec-8-enyl acetate as mixture of (*E*)- and (*Z*)-isomers with the (*Z*)- alcohol) and 'Paramount NOW/PTB' (plus (*Z*,*Z*)-hexadeca-11,13-dienal) (Suterra), 'Hercon Disrupt PTB' (Hercon), 'Isonet A' (plus dodec-8-enyl acetate) (Shin-Etsu).

APPLICATION: The slow-release dispensers are attached to individual trees in the orchard at a density of between 500 and 1000 per hectare, or 375 per hectare for 'Checkmate PTB'.

PRODUCT SPECIFICATIONS: Manufactured for use in crop protection. **Shelf-life:** Unopened packages can be stored at temperatures below 10° C. Dispensers must be kept at room temperature for at least one month prior to field application in the following season.

COMPATIBILITY: The pheromone is used alone for mating disruption without the addition of insecticides.

MAMMALIAN TOXICITY: (*E*)-dec-5-enyl acetate and (*E*)-dec-5-enol combinations have not shown any allergic or other adverse toxicological effects on manufacturers, formulators, field workers or farmers. **Acute oral LD$_{50}$:** rats >2000 mg/kg.

ENVIRONMENTAL IMPACT AND NON-TARGET TOXICITY: (*E*)-dec-5-enyl acetate plus (*E*)-dec-5-enol is a natural insect pheromone that is specific to the peach tree borer. There is no evidence that it has caused any adverse effects on any non-target organisms or on the environment. **Approved for use in organic farming:** Yes.

4:384 (R,Z)-5-(1-decenyl)dihydro-2(3H)-furanone

Japanese beetle sex pheromone

The Pesticide Manual **Fifteenth Edition** entry number: 512

$$CH_3(CH_2)_7$$

NOMENCLATURE: Approved name: (R,Z)-5-(1-decenyl)dihydro-2(3H)-furanone.
Other names: Japanese beetle sex pheromone; japonilure. **CAS RN:** *[64726-91-6]*, formerly *[133494-64-1]*, *[77518-55-9]* and *[111192-53-1]*; *[64726-93-8]* (S)- isomer.

SOURCE: Originally isolated from deposits such as faeces in flasks that had contained female Japanese beetles (*Popillia japonica* Newman).

PRODUCTION: Manufactured for use in beetle control programmes.

TARGET PESTS: Japanese beetles (*Popillia japonica*).

TARGET CROPS: Ornamental turf and horticultural crops.

BIOLOGICAL ACTIVITY: Mode of action: The lures for the Japanese beetle are sold as a dual scented trap. One scent is a floral lure (phenethyl propionate: eugenol: geraniol in the proportion 3:7:3) that attracts female beetles, whilst the other is the sex pheromone that attracts the males. The strategy is a lure and trap or a lure and kill, with the beetles being held within the traps and unable to move to the crops to be protected. **Efficacy:** The use of these baits in lure and trap or kill strategies has been shown to keep the beetle population at a level that is acceptable. **Key reference(s):** J H Tumlinson, M G Klein, R F Doolittle, T L Ladd & A T Proveaux. 1977. Identification of the female Japanese beetle sex pheromone: inhibition of male response by an enantiomer, *Science (Washington)*, **197**, 798.

COMMERCIALISATION: Formulation: Sold as a combination dual lure trap containing food plant extracts attractive to females and (R,Z)-5-(1-decenyl)dihydro-2(3H)-furanone attractive to males. **Tradenames:** 'BioLure Japanese Beetle Trap' (Arbico).

APPLICATION: The traps are hung in trees from a metal stand, ten to twenty metres downwind of the crops to be protected, before the adults migrate. The lures should be replaced after 4 to 6 weeks.

PRODUCT SPECIFICATIONS: Manufactured for use in beetle control programmes.
Purity: (R,Z)-5-(1-decenyl)dihydro-2(3H)-furanone is >95% chemically and isomerically pure.

COMPATIBILITY: It is unusual to use the Japanese beetle lures in combination with other crop protection chemicals.

4. Semiochemicals

MAMMALIAN TOXICITY: There have been no reports of allergic or other adverse toxicological effects from (R,Z)-5-(1-decenyl)dihydro-2(3H)-furanone by researchers, manufacturers, formulators or field workers. It is considered to be non-toxic. **Inhalation:** LC_{50} for rats >1.35 mg/litre (aerosol).

ENVIRONMENTAL IMPACT AND NON-TARGET TOXICITY: (R,Z)-5-(1-decenyl)dihydro-2(3H)-furanone is the naturally occurring female sex pheromone of the Japanese beetle and is specific to this insect. Consequently, it is not expected to have any adverse effects on non-target organisms or on the environment, although the floral attractants may attract some pollinators. **Approved for use in organic farming:** Yes.

4:385 2,4-dimethyl-5-ethyl-6,8-dioxa-bicyclo[3.2.1]octane (multistriatin) plus (−)-4-methyl-3-heptanol

European elm bark beetle pheromone

multistriatin

4-methylheptan-3-ol

NOMENCLATURE: Approved name: multistriatin (2,4-dimethyl-5-ethyl-6,8-dioxabicyclo[3.2.1]octane) plus (−)-4-methyl-3-heptanol. **Other names:** European elm bark beetle pheromone. **CAS RN:** *[54815-06-4]* multistriatin; *[59014-03-8]* (−)-α-multistriatin; *[14979-39-6]* (−)-4-methyl-3-heptanol (general CAS RN).

SOURCE: The two components were identified from the volatiles emanating from virgin female smaller European elm bark beetle in elm bark. When first reported, the pheromone was given a general CAS RN, then, after nomenclature/indexing changes at CAS and the characterisation of the individual isomers, additional/separate numbers were assigned for all components.

PRODUCTION: Manufactured for use in tree protection strategies.

TARGET PESTS: Smaller European elm bark beetle (*Scolytus multistriatus* (Marsham)).

TARGET CROPS: Elm trees.

BIOLOGICAL ACTIVITY: Mode of action: (+/−)-Multistriatin is relatively cheap and is the main component of commercial products, while (−)-α-multistriatin is the natural, most active component. The lures' likely components would be an isomeric mixture of (+/−)- and (−)-α-multistriatin plus 4-methyl-3-heptanol. Male adults are attracted to the virgin females by following a plume of sex pheromones. The blend of components that make up the pheromone is mediated by the presence of volatiles from the host elm and allows males to locate and mate with females. The combination product is used to attract and trap male adults, reduce their numbers and thereby lower the numbers of fecund adult females. Also used for insect monitoring. **Efficacy:** The use of the product in urban settings in the USA and Canada has been very successful in lowering the incidence of bark beetle attack to tolerable levels. **Key reference(s):** 1) G T Pearce, W E Gore, R M Silverstein, J W Peacock, R A Cuthbert, G N Lanier & J B Simeone. 1975. Chemical attractants for the smaller European elm bark beetle, *Scolytus multistriatus*, *J. Chem. Ecol.*, **1**, 115. 2) M M Blight, N C Henderson & L J Wadhams. 1983. The identification of 4-methyl-3-heptanone from *Scolytus scolytus* (F.) and *S. multistriatus* (Marsham). Absolute configuration, laboratory bioassay and electrophysiological studies on *S. scolytus*, *Insect Biochem.*, **13**, 27–38. 3) M M Blight, N C Henderson, L J Wadhams, N J Fielding & C J King. 1982. Field response of elm bark beetles to baits containing 4-methyl-3-heptanone, *Naturwissenschaften*, **69**, 554–5.

COMMERCIALISATION: Formulation: Sold as the attractant in male beetle traps. The beetles are trapped on sticky surfaces or killed by contact with an insecticide.
Tradenames: 'Multilure' (a patented slow-release trap).

APPLICATION: The traps are placed in urban areas near the host elm trees and males are attracted to them.

PRODUCT SPECIFICATIONS: Manufactured for use in tree protection strategies.
Purity: The pheromones used in traps are >95% chemically pure.

MAMMALIAN TOXICITY: There is no evidence of any allergic or other adverse toxicological effects in research workers, manufacturers, formulators or field staff from use. Considered to be non-toxic.

ENVIRONMENTAL IMPACT AND NON-TARGET TOXICITY: The components that make up the European elm bark beetle pheromone are naturally occurring and are specific for their target insect. Consequently, it is unlikely that they will have any adverse effects on non-target organisms or on the environment. In addition, before evaporation, the pheromone blend is contained within the traps and is not applied to the crop.
Approved for use in organic farming: Yes.

4. Semiochemicals

4:386 1,7-dioxaspiro[5.5]undecane

Olive fly sex pheromone

The Pesticide Manual Fifteenth Edition entry number: 295

NOMENCLATURE: Approved name: 1,7-dioxaspiro[5.5]undecane. **Other names:** Olive fly sex pheromone. **CAS RN:** *[180-84-7]*.

SOURCE: Produced by virgin female olive flies (*Bactrocera oleae* Gmel.). The major component of pheromones isolated from the abdomen of virgin females.

PRODUCTION: Manufactured for use in control strategies.

TARGET PESTS: Olive fly (*Bactrocera oleae*).

TARGET CROPS: Olives.

BIOLOGICAL ACTIVITY: Mode of action: Male flies are attracted to females by following the pheromone trail released by virgin females. The use of the pheromone in slow-release devices will attract the flies to the traps, where they can be collected as a means of monitoring the level of flies in the olive grove. Monitoring in this way ensures the application of insecticide to the trees at the best time for fly control. In addition, the traps can be used as a lure and kill device in which an insecticide is incorporated so that the attracted flies come into contact with the toxicant and, thereby, are eliminated from the olive grove.
Key reference(s): 1) R Baker, R H Herbert, P E Howse, O T Jones, W Franke & W Reith. 1980. Identification and synthesis of the major sex pheromone of the olive fruit fly (*Dacus oleae*), *J. Chem. Soc., Chemical Communications*, 52–3. 2) O T Jones, J C Lisk, A W Mitchell, R Baker & P Ramos. 1985. A sex pheromone baited trap that catches the olive fruit fly (*Dacus oleae*) with a measurable degree of selectivity. In *Integrated Pest Control in Olive Groves*, R Cavalloro & A Crovetti (eds.), *Proc. CEC/IOBC Int. Joint Meeting*, Pisa, Italy, 3–6 April, pp. 104–12.

COMMERCIALISATION: Formulation: Sold as technical material for use in lures for monitoring. Formulated as a sprayable formulation entrapped in polyurea microcapsules, 5 to 10 μm in size, or polymer-stabilised emulsions for tank-mixing with an insecticide. Also sold as a lure and kill trap containing 1,7-dioxaspiro[5.5]undecane plus lambda-cyhalothrin. **Tradenames:** 'Polycore SKL' and 'Magnet Oli' (attract and kill) (Suterra).

APPLICATION: Slow-release lures containing either 1,7-dioxaspiro[5.5]undecane and insecticide or baited with ammonium salt dispensers as a food attractant plus insecticide, in the ratio 1:2–4 (sex pheromone:food source), are dispersed throughout the olive grove. The sprayable formulations are polymer entrapped beads that contain 20 g of pheromone per litre together with malathion or dimethoate. This mixed, slow-release formulation is sprayed onto the olive trees either by air, treating 20 metre swathes every 100 metres of grove, or using ground equipment, treating only the south side of each row or tree. Organic farms can use these sprayable formulations if the insecticide used is natural pyrethrum or rotenone.

PRODUCT SPECIFICATIONS: Manufactured for use in control strategies. **Purity:** The pheromone is >95% chemically pure.

COMPATIBILITY: Compatible with all compounds that are recommended for use in olive groves.

MAMMALIAN TOXICITY: There have been no reports of allergic or other adverse toxicological effects from the pheromone by researchers, manufacturers, formulators or field workers. Considered to be non-toxic.

ENVIRONMENTAL IMPACT AND NON-TARGET TOXICITY: 1,7-dioxaspiro[5.5]undecane is a naturally occurring pheromone that is specific to the olive fly. It is not expected to have any adverse effects on non-target organisms or on the environment.
Approved for use in organic farming: Yes.

4:387 disparlure

Gypsy moth sex pheromone

The Pesticide Manual **Fifteenth Edition entry number:** 300

$$CH_3(CH_2)_9\text{''''}\overset{O}{\underset{H\quad H}{\triangle}}\text{'''}(CH_2)_4CH(CH_3)_2$$

NOMENCLATURE: Approved name: disparlure ((7R,8S)-7,8-epoxy-2-methyloctadecane). **Development code:** SHA 114301; ENT 34886. **CAS RN:** *[29804-22-6]* disparlure (racemate); *[54910-51-9]* (+)-disparlure; *[54910-52-0]* (–)-disparlure.

SOURCE: Female gypsy moth adults (*Lymantria dispar* L.) emit a volatile component shown to be attractive to males. The main component was identified from extracts of the abdominal tips of virgin females.

PRODUCTION: Manufactured for use in protection of temperate fruit tree crops and forests.

TARGET PESTS: Gypsy moth (*Lymantria dispar*).

TARGET CROPS: Gypsy moth is a major forest pest and serious threat to urban shade trees and ornamentals. It has 300 known hosts, including temperate fruit tree crops such as apples, pears and peaches, and many forest species. Disparlure has been widely tested in forests as a selective gypsy moth control agent and is still used extensively.

BIOLOGICAL ACTIVITY: Mode of action: Disparlure is a synthetic pheromone that was developed as a male moth attractant when the natural pheromones, gyplure ((Z)-7,10-acetoxy-7-hexadecen-1-ol) and gyptol ((Z)-O-10-acetoxy-7-hexadecen-l-ol), were found to be inconsistent or inactive in the field. This was explained as inhibition effects or deactivations caused by the presence of traces of the geometrically isomeric E-compounds, as well as of completely different substances such as traces of solvent or impurities originating from the synthesis. (7R,8S)-Disparlure is the synthetic sex pheromone of the gypsy moth. Males locate females by following a plume of air rich in the odour of the pheromone. Evaporation of enantiomerically pure (7R,8S)-disparlure from traps attracts male moths to traps. Permeation of the canopy of an orchard with vapours of the racemic disparlure is used to disrupt location of females by males and decreases mating. It has been shown that the (+)- isomer is more effective at trapping and confusing males than the (–)- isomer. **Efficacy:** Several commercial formulations of racemic disparlure have been shown to maintain populations of gypsy moth effectively at economically tolerable levels during the time between fruiting and harvest. **Key reference(s):** 1) J R Plimmer, C P Schwalbe, E C Paszek, B A Bierl, R E Webb, S Marumo & S Iwaki. 1977. Contrasting effectiveness of (+) and (–) enantiomers of disparlure for trapping native populations of the gypsy moth in Massachusetts, *Environ. Entomol.*, **6**, 518. 2) R T Cardé, C C Doane, T C Baker, S Iwaki & S Marumo. 1977. Attractancy of optically active pheromone for male gypsy moths, *Environ. Entomol.*, **6**, 768.

COMMERCIALISATION: Formulation: Slow-release formulations of pheromone from laminated plastics, plastic tubes or plastic barrier film. **Tradenames:** 'Disparlure' (laminated plastic) (Hercon), 'Disparlure' (plastic tubes) (Shin-Etsu), 'Disparlure' (Cyclo International), (International Speciality Group) and (IPM Technologies), 'Luretape Disparlure' and 'Hercon Disrupt II' (Hercon), 'SPLAT GM' (ISCA).

APPLICATION: Pheromone dispensers are attached to lower branches of trees at heights recommended by formulators. The recommended number of lures per hectare varies with release rates. The lures are effective for about 12 weeks and should be applied before the adults fly. 'SPLAT GM' (Specialized Pheromone and Lure Application Technology) is an amorphous polymer matrix for the sustained release of insect pheromones. It has been designed to dispense, optimise and modulate the semiochemical release in mating disruption pest management programmes. The product can be applied by hand, metered backpack sprayers, caulking guns, metered dosing guns or tractor.

PRODUCT SPECIFICATIONS: Manufactured for use in protection of temperate fruit tree crops and forests. **Purity:** Pheromone that is >90% chemically and isomerically pure is used in lures.

COMPATIBILITY: Pheromone disruption of mating does not require insecticide, but disparlure can be used with the application of insecticides to control developing larvae.

MAMMALIAN TOXICITY: Disparlure has shown no adverse toxicological effects on manufacturers, formulators or field workers. Considered to be non-toxic. **Acute oral LD$_{50}$:** rats >34 600 mg/kg. **Inhalation:** LC$_{50}$ (1 h) for rats >5.0 mg/litre. **Skin and eye:** Acute percutaneous LD$_{50}$ for rabbits >2025 mg/kg. It is a skin irritant and a mild eye irritant.

ENVIRONMENTAL IMPACT AND NON-TARGET TOXICITY: Disparlure is a pheromone that is specific to the gypsy moth. There is no evidence that it has caused any adverse effects on non-target organisms nor had an adverse environmental impact. **Bird toxicity:** LC$_{50}$ (8 d) for mallard ducks and quail >5000 ppm. **Fish toxicity:** LC$_{50}$ for trout and bluegill sunfish >100 ppm.

4:388 (Z)-dodec-7-enyl acetate
Tomato looper moth sex pheromone

The Pesticide Manual **Fifteenth Edition entry number:** 309

$$CH_3(CH_2)_3 \quad (CH_2)_6OCOCH_3$$
$$C=C$$
$$H \qquad H$$

NOMENCLATURE: Approved name: (Z)-dodec-7-enyl acetate. **Other names:** Z7-12Ac. **CAS RN:** [14959-86-5].

SOURCE: Originally isolated from virgin female tomato looper moths (*Chrysodeixis chalcites* Esper).

PRODUCTION: The pheromone is now synthesised for commercial use in crop protection.

TARGET PESTS: Tomato looper (*Chrysodeixis chalcites*). Other common names include green garden looper, golden twin spot, groundnut semilooper and Turkish moth.

TARGET CROPS: Tomatoes and other vegetable crops.

BIOLOGICAL ACTIVITY: Mode of action: Males locate females by following a plume of air rich in the odour of a complete pheromone blend emitted by the females. Permeation of

(Z)-dodec-7-enyl acetate 597

the canopy of the target crop with vapours of this pheromone is used to disrupt location of females by males and decreases mating. **Efficacy:** Commercial formulations of the pheromone have been shown to be very effective at maintaining populations of tomato looper at economically tolerable levels between flowering and harvest.

COMMERCIALISATION: Formulation: Sold as slow-release formulations within release devices such as plastic tubes, laminated plastic or plastic barrier film. Manufactured by Bedoukian and Shin-Etsu. **Tradenames:** 'Fersex CCh' (SEDQ), 'Disrupt FAW' (plus (*Z*)-tetradec-9-enyl acetate) (Hercon), 'Isomate FAW' (plus (*Z*)-tetradec-9-enyl acetate) (Shin-Etsu), 'No-Mate FAW' (plus (*Z*)-tetradec-9-enyl acetate) (Scentry).

APPLICATION: Pheromone dispensers are attached to plants at heights recommended by formulators. The recommended number of lures per hectare varies with release rates.

PRODUCT SPECIFICATIONS: The pheromone is now synthesised for commercial use in crop protection. **Purity:** The pheromone in the product is >95% chemically pure.

COMPATIBILITY: Pheromone disruption of mating does not require insecticide, but can be used with application of insecticides.

MAMMALIAN TOXICITY: There are no reports of tomato looper moth sex pheromone causing allergic or other adverse toxicological effects to research, manufacturing or field workers. Considered to be non-toxic.

ENVIRONMENTAL IMPACT AND NON-TARGET TOXICITY: Tomato looper moth sex pheromone is a natural insect pheromone that is specific to the tomato looper moth. There is no evidence that it has caused any adverse effects on non-target organisms or had an adverse environmental impact. **Approved for use in organic farming:** Yes.

4:389 (*Z*)-dodec-8-enyl acetate
Plum fruit moth sex pheromone

The Pesticide Manual **Fifteenth Edition entry number:** 310

$$CH_3(CH_2)_2 \quad (CH_2)_7OCOCH_3$$

with the structure showing H and H on a (Z)-configured double bond.

NOMENCLATURE: Approved name: (*Z*)-dodec-8-enyl acetate. **Other names:** Z8-12Ac; Z8DDA. **CAS RN:** *[28079-04-1]*.

SOURCE: Originally isolated from virgin female plum fruit moth (*Cydia funebrana* (Treitschke)) adults.

PRODUCTION: Manufactured by SEDQ for use in crop protection.

TARGET PESTS: Plum fruit moth (*Cydia funebrana*).

TARGET CROPS: Plum, cherry, sour cherry and apricot orchards.

BIOLOGICAL ACTIVITY: Mode of action: Males locate females by following a plume of air rich in the odour of a complete pheromone blend emitted by virgin females. Evaporation of the pheromone from traps attracts male moths. Permeation of the canopy of an orchard with vapours of the same pheromone is sometimes used to disrupt location of females by males and decrease mating, but (Z)-dodec-8-enyl acetate is often used to attract males to sticky traps.

COMMERCIALISATION: Formulation: Sold as a slow-release formulation associated with sticky traps. **Tradenames:** 'Fersex CF' ((Z)- isomer) (SEDQ), 'Checkmate OFM' (mixture of (E)- and (Z)- isomers plus (Z)- alcohol) (Suterra), 'Disrupt OFM' (Hercon), 'Ecodian' (disorientation) (Isagro), 'Isomate C/OFM' (plus codlemone), 'Confuser A' (plus tetradec-11-enyl acetate plus (E,Z)-tetradeca-4,10-dienyl acetate plus (Z)-icos-13-en-10-one), 'Confuser P' (plus tetradec-11-enyl acetate plus 14-methyloctadec-1-ene plus (Z)-icos-13-en-10-one) and 'Confuser Z' (plus dodec-9-enyl acetate plus tetradec-11-enyl acetate plus (Z)-icos-13-en-10-one plus (Z)-tetradec-9-enyl acetate) (Shin-Etsu), 'Isomate-M 100' (Shin-Etsu) and (Pacific Biocontrol), 'OPM Disrupt' (ChemTica), 'RAK 5' (BASF).

APPLICATION: Traps are placed within the canopy of individual trees in an orchard, at a height of about 3 metres and at a rate of about one trap per 5 trees (more if insect pressure is significant).

PRODUCT SPECIFICATIONS: Manufactured by SEDQ for use in crop protection.

COMPATIBILITY: It is unusual to apply (Z)-dodec-8-enyl acetate with any other crop protection chemicals.

MAMMALIAN TOXICITY: There are no reports of allergic or other adverse toxicological effects from researchers, manufacturers, formulators or field workers. It is considered to be non-toxic. **Acute oral LD$_{50}$:** rats >5000 mg/kg. **Inhalation:** No toxicity seen on rats at the highest test level. **Skin and eye:** There were no signs of toxicity at the highest test levels.

ENVIRONMENTAL IMPACT AND NON-TARGET TOXICITY: (Z)-dodec-8-enyl acetate is the sex pheromone of the plum fruit moth. It is not expected that it will have any adverse effects on non-target organisms or on the environment. **Behaviour in soil:** It is broken down to carbon dioxide and water very rapidly in soil. **Approved for use in organic farming:** Yes.

4:390 (Z)-dodec-8-enyl acetate plus (E)-dodec-8-enyl acetate plus (Z)-dodec-8-enol

Oriental fruit moth sex pheromone

The Pesticide Manual **Fifteenth Edition entry numbers:** 308 and 310

$$CH_3(CH_2)_2 \quad (CH_2)_7OCOCH_3$$
$$C=C$$
$$H \qquad H$$

(Z)- isomer

$$H \qquad (CH_2)_7OCOCH_3$$
$$C=C$$
$$CH_3(CH_2)_2 \qquad H$$

(E)- isomer

$$CH_3(CH_2)_2 \quad (CH_2)_7OH$$
$$C=C$$
$$H \qquad H$$

(Z)-dodec-8-enol

NOMENCLATURE: Approved name: (Z)-dodec-8-enyl acetate plus (E)-dodec-8-enyl acetate plus (Z)-dodec-8-enol. **Other names:** *Grapholitha molesta* sex pheromone; oriental fruit moth sex pheromone; Z8-12Ac; E8-12Ac; Z8DDA. **Development code:** BAS 284 I.
CAS RN: *[28079-04-1]* (Z)-dodec-8-enyl acetate; *[38363-29-0]* (E)-dodec-8-enyl acetate; *[40642-40-8]* (Z)-dodec-8-enol.

SOURCE: Female *Grapholitha molesta* (Busck) emit several volatile components shown to be attractive to males. Originally extracted from the terminal segments of virgin females.

PRODUCTION: Manufactured for use in crop protection.

TARGET PESTS: Oriental fruit moth (*Grapholitha [Cydia] molesta*). Also effective against the macadamia nut borer (*Cryptophlebia ombrodelta* (Lower)), the koa seedworm (*C. illepida* Butler) and the plum fruit moth (*Cydia funebrana* (Treitschke))).

TARGET CROPS: Orchard fruit, such as apples, pears, quinces and other pome fruit; peaches, nectarines, plums, prunes, cherries, apricots and other stone fruit; almonds, macadamias and other tree nut crops.

BIOLOGICAL ACTIVITY: Mode of action: Males locate females by following a plume of air rich in the odour of a complete pheromone blend emitted by the females. Evaporation of the pheromone blend given above from traps attracts male moths to traps. Permeation of the canopy of an orchard with vapours of the pheromone is used to disrupt location of females by males and decrease mating. **Efficacy:** The combination of the three components has given more effective disruption than the use of the (Z)-dodec-8-enyl acetate alone. Some products contain only one component. Very low rates are required to cause mating disruption. The pheromone blend is volatile and distributes throughout the crop easily.
Key reference(s): T C Baker, W Meyer & W L Roelofs. 1981. Sex pheromone dosage and blend specificity of response by oriental fruit moth, *Environ. Entomol.*, **30**, 269.

COMMERCIALISATION: Formulation: Sold in slow-release devices. Consep sells a sprayable formulation and offers 'Checkmate SF', a dual dispenser containing the oriental fruit moth and peach tree borer pheromones. **Tradenames:** 'Isomate-M' (Shin-Etsu) and (Biocontrol), 'Isomate OFM rosso' (Shin-Etsu), 'Isomate OFM Plus' (long life formulation) (Biocontrol), 'Confusalin' (Bayer CropSciences) and (Calliope), 'RAK 5' and 'Quant G.m.' (BASF), 'Checkmate OFM' and 'Checkmate OFM-F' (Consep), 'Hercon Disrupt OFM' (Hercon), 'MEC.OFM' (Suterra), 'SPLAT OFM' (ISCA).

APPLICATION: The controlled-release dispensers contain synthetic pheromones that are chemically similar to the natural pheromone produced by *Grapholitha molesta*. The dispenser is composed of two parallel polymer tubes, one of which contains an aluminum wire and enables the positioning of the dispenser in the field and the other filled with the specific pheromone mixture. The slow-release devices are placed within the orchard at a height specified by the manufacturer prior to the flight of over-wintering adults. It is usual to allow 500 dispensers per hectare ('Checkmate OFM'). For 'Isomate OFM rosso', use 600 dispensers per hectare, plus additional dispensers to reinforce protection at the orchard edges. The pheromone diffuses out of the dispenser and disperses throughout the orchard. The products should be reapplied as necessary based on insect monitoring and field scouting programmes, although the maximum applied per hectare should not exceed 150 g per year. Apply spray formulations directly to the foliage in sufficient water to cover the crop uniformly, but not to run-off. 'SPLAT OFM' (Specialized Pheromone & Lure Application Technology) is a formulation that has been designed to dispense, optimise and modulate the semiochemical release in mating disruption pest management programmes. The product can be applied by hand, caulking gun, metered backpack sprayer, tractor or aircraft.

PRODUCT SPECIFICATIONS: Manufactured for use in crop protection. **Purity:** The pheromone components are >90% chemically pure. **Shelf-life:** Stored in their original packaging at 5 °C, the products will normally keep for 2 years. When the products are refrigerated for use the following season, store them at room temperature for one month prior to field application.

COMPATIBILITY: The products can be used with all crop protection agents that do not repel the oriental fruit moth.

MAMMALIAN TOXICITY: Oriental fruit moth pheromone blend has shown no adverse toxicological effects on manufacturers, formulators or field workers. Considered to be non-toxic. **Acute oral LD$_{50}$:** rats >15 000 mg/kg ((Z)- and (E)-dodec-8-enyl acetate). **Acute dermal LD$_{50}$:** rabbits >15 000 mg/kg ((Z)- and (E)-dodec-8-enyl acetate). **Skin and eye:** Irritating to skin.

ENVIRONMENTAL IMPACT AND NON-TARGET TOXICITY: Oriental fruit moth pheromone blend is a natural insect pheromone that is specific to the oriental fruit moth. There is no evidence that it has caused any adverse effects on non-target organisms or had any adverse environmental impact. **Approved for use in organic farming:** Yes.

4. Semiochemicals

4:391 (E)-dodec-9-enyl acetate

European pine shoot moth and
Eastern pine shoot borer sex pheromone

The Pesticide Manual **Fifteenth Edition entry number:** 311

$$CH_3(CH_2)_2 \quad (CH_2)_7OCOCH_3$$

(Z)- isomer

$$H \qquad (CH_2)_7OCOCH_3$$
$$CH_3(CH_2)_2 \qquad H$$

(E)- isomer

NOMENCLATURE: Approved name: (E)-dodec-9-enyl acetate. **Other names:** European pine shoot moth sex pheromone. **CAS RN:** [35148-19-7].

SOURCE: Sex pheromone of the Eastern pine shoot borer (*Eucosma gloriola* Heinrich) and European pine shoot moth (*Rhyacionia buoliana* (Denis and Schiffermüller)). (This chemical is part of the Pheromone Joint Review Pilot Program currently under way between Canada Pest Management Regulatory Agency (PMRA) and the United States EPA). Originally isolated from the abdominal segments of virgin female European pine shoot moths (*Rhyacionia buoliana*).

PRODUCTION: Manufactured for use in forestry management.

TARGET PESTS: Lepidopteran insects (moths), such as the Eastern pine shoot borer (*Eucosma gloriola*) and the European pine shoot moth (*Rhyacionia buoliana*).

TARGET CROPS: Recommended for use in pine forests and other woodland applications.

BIOLOGICAL ACTIVITY: Mode of action: The male moths locate virgin females for mating by following a plume of volatile components emitted by the females. This location process can be disrupted by saturating the air of the crop to be protected with synthetic pheromone. This confuses the males and means that the females cannot be found and mating does not take place. **Efficacy:** The use of the pheromone in Chile has been very effective in keeping the population of moths below the damage threshold and data generated in the Pheromone Joint Review Pilot Program led to its registration in North America. **Key reference(s):** 1) R G Smith, G E Daterman, G D Daves Jr, K D McMurtrey & W L Roelofs. 1974. Sex pheromone of the European pine shoot moth: chemical identification and field tests, *J. Insect Physiol.*, **20**, 661. 2) L Sukovata, A M Kolk & Cieslak. 2004. Effect of attract and kill formulations and application rates on trap catches of European pine shoot moth (Lepidoptera: Tortricidae) and shoot damage in Scots pine saplings, *J. Econ. Entomol.*, **97**(5), 1619–23.

COMMERCIALISATION: Formulation: Sold as a slow-release formulation. Formulated as a spray for application to trees and as a slow-release formulation for long-term release. 'MEC Eastern Pine Shoot Borer' is a timed-release microencapsulated pheromone concentrate. **Tradenames:** 'Selibate EPSM' and 'MEC Eastern Pine Shoot Borer' (Suterra), '9-Dodecenyl Acetate Technical Pheromone' (Bedoukian Research).

APPLICATION: The pheromone can be used as a spray, in traps and in dispensers. The slow-release devices are deployed throughout the pine forest at the time of flight of the moths.

PRODUCT SPECIFICATIONS: Manufactured for use in forestry management. **Purity:** The product is >95% chemically pure.

COMPATIBILITY: It is unusual to use the product with other crop protection agents.

MAMMALIAN TOXICITY: There are no reports of allergic or other adverse toxicological effects from the use of (E)-dodec-9-enyl acetate by researchers, manufacturers, formulators or field workers. It is considered to be non-toxic. **Acute oral LD$_{50}$:** rats >15 000 mg/kg. **Acute dermal LD$_{50}$:** rabbits >3000 mg/kg.

ENVIRONMENTAL IMPACT AND NON-TARGET TOXICITY: (E)-dodec-9-enyl acetate is the sex pheromone of the European pine shoot moth (*Rhyacionia buoliana*) and Eastern pine shoot borer (*Eucosma gloriola*) and is specific to these species. It is not expected that its use will have any adverse effects on non-target organisms or on the environment.

4. Semiochemicals

4:392 (Z)-dodec-9-enyl acetate
European grape berry moth sex pheromone

The Pesticide Manual **Fifteenth Edition entry number:** 308

$$CH_3CH_2 \qquad (CH_2)_8OCOCH_3$$
$$C=C$$
$$H \qquad H$$

NOMENCLATURE: Approved name: (Z)-dodec-9-enyl acetate. **Other names:** European grape berry moth sex pheromone; Z9-12Ac. **Development code:** BAS 281 I. **CAS RN:** [16974-11-1].

SOURCE: The sex pheromone of the European grape berry moth (*Eupoecilia ambiguella*

(Hübner)) that was originally isolated from the terminal segments of virgin females.

PRODUCTION: Manufactured for use in crop protection.

TARGET PESTS: European grape berry moth (*Eupoecilia ambiguella*).

TARGET CROPS: Vineyards.

BIOLOGICAL ACTIVITY: Mode of action: (*Z*)-dodec-9-enyl acetate is a component of the female sex pheromone of the grape berry moth. Males locate and subsequently mate with female moths by following the pheromone trail or pheromone plume emitted by virgin females. The application of (*Z*)-dodec-9-enyl acetate makes trail following impossible (competition between applied and natural pheromone plume, false trail following). Control is eventually achieved by preventing mating of the moths and the subsequent lack of fertile egg production. **Efficacy:** Very low rates are required to cause mating disruption. (*Z*)-dodec-9-enyl acetate is volatile and is rapidly dispersed throughout the crop. It has been shown that, of the constituents of the female sex pheromone, (*Z*)-dodec-9-enyl acetate is the best male attractant. The pheromone products are often applied in combination with other compounds that are not components of the natural pheromone, notably (*Z*)-icos-13-en-10-one, tetradec-11-enyl acetate and dodeca-7,9-dienyl acetate. **Key reference(s):** 1) H Arn, R Roehrich, C Descoins & S Rauscher. 1979. Performance of five sex attractant formulations for the grape moth, *Eupoecilia ambiguella* Hb. in European vineyards, *Mitt. Schweiz. Entomol. Ges.*, **52**, 45. 2) H Arn, S Rauscher & A Schmid. 1979. Sex attractant formulations and traps for the grape moth, *Eupoecilia ambiguella* Hb., *Mitt. Schweiz. Entomol. Ges.*, **52**, 49.

COMMERCIALISATION: Formulation: Sold as a slow-release formulation of polyethylene ampoules. **Tradenames:** 'RAK 1+2' (plus dodeca-7,9-dienyl acetate), 'RAK 1 Plus' and 'RAK 1 Neu' (BASF), 'Decoy EPSM' and 'Selibate EPSM' (Suterra), 'Disrupt EPSM' and 'Disrupt WPSB' (Hercon), 'Isonet E' ((*Z*)- isomer), 'Confuser Z' (plus tetradec-11-enyl acetate plus (*Z*)-icos-13-en-10-one plus (*Z*)-tetradec-9-enyl acetate plus (*Z*)-dodec-8-enyl acetate), 'Hamaki-con N' (plus tetradec-11-enyl acetate plus (*Z*)-tetradec-9-enyl acetate), 'Isonet L Plus' (plus dodeca-7,9-dienyl acetate), 'Isonet LE' (plus dodeca-7,9-dienyl acetate) and 'Isomate-GBM' (plus tetradec-11-enyl acetate) (Shin-Etsu), 'Gbm Rope' (plus tetradec-11-enyl acetate) (both as (*Z*)- ester) (Nihon Nohyaku), 'Mstrs BHFW' (plus tetradec-11-enyl acetate) (both as (*Z*)- ester), 'Mstrs BHFW-2' (plus tetradec-11-enyl acetate) (as E9-12Ac and Z11-14Ac) and 'Mstrs SS' (plus tetradec-11-enyl acetate) (both as (*E*)- ester) (Mstrs).

APPLICATION: The slow-release dispensers are attached to individual vines in the vineyard at a rate of 500 per hectare.

PRODUCT SPECIFICATIONS: Manufactured for use in crop protection. **Purity:** The pheromone in the product is >95% chemically pure.

COMPATIBILITY: (*Z*)-dodec-9-enyl acetate is used alone without the use of chemical insecticides.

MAMMALIAN TOXICITY: (Z)-dodec-9-enyl acetate has shown no allergic or other adverse toxicological effects on manufacturers, formulators, research workers or farmers.
Acute oral LD$_{50}$: rats >5000 mg/kg. **Acute dermal LD$_{50}$:** rabbits >3000 mg/kg. **Skin and eye:** It is not a skin or eye irritant.

ENVIRONMENTAL IMPACT AND NON-TARGET TOXICITY: (Z)-dodec-9-enyl acetate is a natural insect pheromone that is specific to the grape berry moth, although the single compound is released by other species and it is the blend that provides species-specificity. There is no evidence that it has had any adverse effects on non-target organisms or on the environment. **Fish toxicity:** LC$_{50}$ (96 h) for zebra fish (*Brachydanio rerio*) 10–22 mg/litre. **Other aquatic toxicity:** EC$_{50}$/LC$_{50}$ (48 h) for *Daphnia* 2.6 mg/litre; EC$_{50}$/LC$_{50}$ (72 h) for algae 0.3 mg/litre. **Effects on beneficial insects:** Not toxic to bees.
Approved for use in organic farming: Yes.

4:393 (E,Z)-dodeca-3,5-dienyl acetate plus (Z)-hexadec-9-enyl acetate
Brazilian apple leaf roller sex pheromone

(E,Z)-3,5-dodecadienyl acetate

(Z)-9-hexadecenyl acetate

NOMENCLATURE: Approved name: (E,Z)-dodeca-3,5-dienyl acetate plus (Z)-hexadec-9-enyl acetate. **Other names:** Brazilian apple leaf roller sex pheromone; E3,Z5D12Ac plus Z9D16Ac.

SOURCE: Isolated from the pheromone glands of virgin female Brazilian apple leaf roller (*Bonagota cranaodes* Meyrick).

TARGET PESTS: Brazilian apple leaf roller (*Bonagota cranaodes*).

TARGET CROPS: Apple orchards.

BIOLOGICAL ACTIVITY: Mode of action: Used in mating disruption and attract and kill programmes. **Efficacy:** Very low rates are required to cause mating disruption.
Key reference(s): A E Eiras, A Kovaleski, E F Vilela, J P Chambon, C R Unelius, A-K Borg-Karlson, I Liblikas, R Mozuraitis, M Bengtsson & P Witzgall. 1999. Sex pheromone of the Brazilian apple leaf roller, *Bonagota cranaodes* Meyrick (Lepidoptera: Tortricidae), *Z. Naturforsch.*, **54c**, 595–601.

COMMERCIALISATION: Formulation: 'SPLAT BAL' (Specialized Pheromone and Lure Application Technology) is an amorphous polymer matrix for the sustained release of insect pheromones. It has been designed to dispense, optimise and modulate the semiochemical release in mating disruption pest management programmes. **Tradenames:** 'SPLAT BAL' (ISCA).

APPLICATION: The product can be applied by hand, metered backpack sprayers, caulking guns, metered dosing guns or tractor.

COMPATIBILITY: The pheromone trap does not require insecticide applications, but traps containing insecticide can be used for control within orchards.

MAMMALIAN TOXICITY: The pheromone blend has shown no adverse toxicological effects on manufacturers, formulators, research workers or farmers. It is considered to be non-toxic.

ENVIRONMENTAL IMPACT AND NON-TARGET TOXICITY: Brazilian apple leaf roller pheromone blend is a natural insect pheromone that is specific to the Brazilian apple leaf roller. There is no evidence that it has caused any adverse effects on non-target organisms or had any adverse environmental impact.

4:394 dodeca-7,9-dienyl acetate
European grapevine moth sex pheromone

The Pesticide Manual **Fifteenth Edition entry number:** 306

$$CH_3CH_2 \quad H \quad (CH_2)_6OCOCH_3$$

(7E, 9Z)- isomer

NOMENCLATURE: Approved name: (E,9)-dodeca-7,9-dienyl acetate plus (E,9)-dodeca-7,9-dienyl acetate. **Other names:** European grapevine moth pheromone; *Lobesia botrana* pheromone; E7Z9-12Ac (IOBC). **Development code:** BAS 288 I. **CAS RN:** *[54364-62-4]* (*E,Z*)- isomer; *[54364-63-5]* (*E,E*)- isomer.

SOURCE: The female sex pheromone of the European grapevine moth (*Lobesia botrana* (Denis and Schiffermüller)). It was originally isolated from the female pheromone glands.

PRODUCTION: Manufactured for use in crop protection.

TARGET PESTS: European grapevine moth (*Lobesia botrana*) and the European grape berry moth (*Eupoecilia ambiguella* Hübner).

TARGET CROPS: Vineyards.

BIOLOGICAL ACTIVITY: Mode of action: Dodeca-7,9-dienyl acetate is the sex pheromone of the European grapevine moth. Male moths locate and subsequently mate with female moths by following the trail or pheromone plume emitted by the virgin females. The application of high levels of dodeca-7,9-dienyl acetate makes trail following impossible (camouflage, competition between artificial and female plume, false trail following). Control is subsequently achieved through the prevention of mating and the laying of fertile eggs. **Efficacy:** Very low rates are required to cause mating disruption. Dodeca-7,9-dienyl acetate is volatile and distributes throughout the crop easily. Acts as an attractant and by disruption of mating in the disorientation mode. **Key reference(s):** 1) R Roehrich, J-P Carles, Y Darrioumerie, P Pargade & B Lalanne-Cassou. 1976. Essais en vignoble de phéromones de synthèse pour le capture des males de l'eudemis (*Lobesia botrana* Schiff.), *Ann. Zool. Ecol. Anim.*, **8**, 473. 2) U Neumann, V Harries, A Gasser, W Waldner & W K Kast. 1992. Recent advances with the mating disruption technique in apples and grapes – factors influencing the success of pheromones. *Brighton Crop Protection Conference – Pests and Diseases*, **8** C-10, 1045–50.

COMMERCIALISATION: Formulation: Sold as a slow-release vapour-releasing dispenser. Consep also manufactures a sprayable formulation. **Tradenames:** 'RAK 2', 'Quant LB' and 'RAK 1+2' (plus dodec-9-enyl acetate) (BASF), 'Checkmate LB' and 'Checkmate LB-F' (Consep), 'Rimilure GVM' (Rimi), 'Isonet L' (($7E,9Z$)- isomer), 'Isonet L Plus' (plus dodec-9-enyl acetate) and 'Isonet LE' (plus dodec-9-enyl acetate) ((Z)- and ($7E,9Z$)- isomers) (Shin-Etsu).

APPLICATION: Dispensers are distributed in the vineyard being attached to individual vines to give a density of 500 dispensers per hectare for trellis trained vines (plus additional numbers for reinforcement of the vineyard borders) and 750 dispensers per hectare for overhead training systems (plus additional numbers for reinforcement of the vineyard borders) or 300 units per hectare for 'Checkmate LB'. The pheromone diffuses out of the dispenser and is dispersed throughout the vineyard.

PRODUCT SPECIFICATIONS: Manufactured for use in crop protection. **Purity:** The pheromone components in the product are >95% chemically pure.

COMPATIBILITY: Can be used with all crop protection agents that do not repel the European grapevine moth.

MAMMALIAN TOXICITY: The product has shown no adverse toxicological effects on research or manufacturing workers or on users. **Acute oral LD$_{50}$:** rats >5000 mg/kg. **Skin and eye:** Not a skin or eye irritant.

ENVIRONMENTAL IMPACT AND NON-TARGET TOXICITY: The product is a natural insect pheromone that is specific to the European grapevine moth. There is no evidence that it has caused any adverse effects on non-target organisms nor had any deleterious effect on the environment. **Fish toxicity:** LC$_{50}$ (96 h) for rainbow trout 10 mg/litre.
Other aquatic toxicity: EC$_{50}$/LC$_{50}$ (48 h) for *Daphnia* 1.1 mg/litre; EC$_{50}$/LC$_{50}$ (72 h) for algae 0.45 mg/litre. **Approved for use in organic farming:** Yes.

4:395 (E,E)-dodeca-8,10-dienyl acetate
Pea moth sex pheromone

The Pesticide Manual Fifteenth Edition entry number: 307

$$\begin{array}{c} H \qquad (CH_2)_7OCOCH_3 \\ \diagdown \qquad \diagup \\ C=C \\ \diagup \qquad \diagdown \\ H \qquad H \\ \diagdown C=C \diagup \\ \diagup \qquad \diagdown \\ CH_3 \qquad H \end{array}$$

NOMENCLATURE: **Approved name:** (E,E)-dodeca-8,10-dienyl acetate. **Other names:** *Cydia [Laspeyresia] nigricana* pheromone; pea moth pheromone. **CAS RN:** [53880-51-6].

SOURCE: The female sex pheromone of the pea moth (*Cydia [Laspeyresia] nigricana* (Fabricius)), solvent extracted from the pheromone gland of virgin females.

PRODUCTION: Manufactured for use in crop protection.

TARGET PESTS: Pea moth (*Cydia [Laspeyresia] nigricana*).

TARGET CROPS: Peas.

BIOLOGICAL ACTIVITY: **Mode of action:** (E,E)-dodeca-8,10-dienyl acetate is the major component of the sex pheromone of the pea moth. Male moths locate and subsequently mate with female moths by following the trail or pheromone plume emitted by virgin females. The indiscriminate application of high levels of the product interferes with this process, as a constant exposure to high levels of pheromone makes trail following impossible (habituation/adaptation). Alternatively, the use of discrete sources of pheromone released over time presents the male with a false trail to follow (confusion). The pea moth pheromone is of special interest, as it has served as a model for a series of investigations on pheromone dispersal, orientation to a pheromone, trap design, trap interactions and population monitoring. The pea moth sex pheromone is usually used to monitor the presence of moths so that insecticide applications can be made at the time when the most vulnerable larval stages are present, shortly after egg lay. **Efficacy:** Very low rates are required to cause mating disruption. (E,E)-dodeca-8,10-dienyl acetate is volatile and distributes throughout the crop easily, attracting male moths to traps. **Key reference(s):** 1) A R Greenaway & C Wall. 1981. Attractant lures for males of the pea moth, *Cydia nigricana* (F.), containing (E)-10-dodecen-1-yl acetate and (E,E)-8,10-dodecadien-1-yl acetate, *J. Chem. Ecol.*, **7**, 563. 2) J N Perry & C Wall. 1985. Orientation of male pea moth, *Cydia nigricana*, to pheromone traps in a wheat crop, *Entomol. Exp. Appl.*, **37**, 161.

COMMERCIALISATION: **Formulation:** Sold in controlled-release lures and traps.
Tradenames: 'Pea Moth Pheromone' (Agralan), 'Pea Moth Monitoring System' (Oecos).

4. Semiochemicals

APPLICATION: The slow-release lures are distributed throughout the crop. The pheromone diffuses out of the dispensers and males follow the plume to the traps.

PRODUCT SPECIFICATIONS: Manufactured for use in crop protection. **Purity:** (E,E)-dodeca-8,10-dienyl acetate is >95% chemically pure.

COMPATIBILITY: Can be used with all crop protection agents that do not repel the pea moth.

MAMMALIAN TOXICITY: The product has shown no adverse toxicological effects on research or manufacturing workers or on users.

ENVIRONMENTAL IMPACT AND NON-TARGET TOXICITY: The product is a natural insect pheromone that is specific to the pea moth. There is no evidence that it has caused any adverse effects on non-target organisms nor had any deleterious effect on the environment. **Approved for use in organic farming:** Yes.

4:396 (Z)-3-dodecen-1-ol-(E)-2-butenoate
Sweet potato weevil sex pheromone

NOMENCLATURE: Approved name: (Z)-3-dodecen-1-ol-(E)-2-butenoate.
Other names: Sweet potato weevil sex pheromone.

SOURCE: Isolated from virgin female sweet potato weevil (*Cylas formicarius* (Fabricius)) adults.

PRODUCTION: Manufactured for commercial use.

TARGET PESTS: Sweet potato weevil (*Cylas formicarius*).

TARGET CROPS: Sweet potatoes.

BIOLOGICAL ACTIVITY: Mode of action: The pheromone is used in survey and trapping programmes in sweet potato. Males are attracted to the pheromone and can be mass trapped, thereby reducing the population of weevils in the crop, or the numbers caught can be monitored, thereby identifying the best timing for the application of insecticide sprays. The commercial baits are also used in a lure and kill strategy, whereby the adults are attracted to the lures where they are exposed to a fast-acting, contact insecticide.
Efficacy: Pheromone traps show great promise for monitoring of adult population density. Weevils respond to low concentrations of pheromone, and apparently will move up to 280 m to a pheromone source. The sex pheromone also shows great potential for mating disruption and mass trapping. **Key reference(s):** 1) R R Heath, J A Coffelt, P E Sonnet, F I Proshold, B Dueben & J H Tumlinson. 1986. Identification of sex pheromone produced by female sweet potato weevil, *Cylas formicarius elegantulus* (Summers), *J. Chem. Ecol.*, **12**, 1489–503. 2) R K Jansson, F I Proshold, L J Mason, R R Heath & S H Lecrone. 1990. Monitoring sweetpotato weevil (Coleoptera: Curculionidae) with sex pheromone: effects of dosage and age of septa, *J. Tropical Pest Management*, **36**, 263–9.

COMMERCIALISATION: Formulation: Sold as a monitoring and attract and kill unit. **Tradenames:** '*PHERO*BANK sweet potato weevil pheromone' (Plant Research).

APPLICATION: Place the pheromone lure above the sticky surface of a suitable trap or near any situation where the pest is to be retained, with at least 10 metres between traps. Check the trap and remove caught insects at least twice a week. Lures are attractive for at least 8 weeks.

PRODUCT SPECIFICATIONS: Manufactured for commercial use. **Shelf-life:** Keep the product in the refrigerator or freezer in the tightly closed original packaging. Lures are guaranteed for two years when kept at −25 °C and for one year when kept in the refrigerator. Before use after storage, allow the product to reach room temperature before opening to prevent condensation.

MAMMALIAN TOXICITY: There is no evidence of allergic or other adverse toxicological effects from (Z)-3-dodecen-1-o1(E)-2-butenoate on researchers, manufacturers, formulators or field workers. Considered to be non-toxic.

ENVIRONMENTAL IMPACT AND NON-TARGET TOXICITY: There is no evidence that the product has caused significant adverse effects on non-target organisms nor had an adverse environmental impact. Not expected to be detected in the environment under normal low rate of use conditions.

4. Semiochemicals

4:397 (2R',5R')-ethyl-1,6-dioxaspiro [4.4]nonane plus (2R',5S')-ethyl-1,6-dioxaspiro[4.4]nonane

Six-toothed spruce bark beetle aggregation pheromone

The Pesticide Manual Fifteenth Edition entry number: 133

CH₃CH₂

NOMENCLATURE: **Approved name:** chalcogran ((2R',5R')-ethyl-1,6-dioxaspiro[4.4]nonane plus (2R',5S')-ethyl-1,6-dioxaspiro[4.4]nonane). **Other names:** Six-toothed spruce bark beetle aggregation pheromone. **CAS RN:** *[38401-84-2]* chalcogran; *[69744-43-0]* (2R)-*trans*- isomer; *[69744-44-1]* (2R)-*cis*- isomer.

SOURCE: The volatiles were isolated from adult male beetles and were shown to be the aggregation pheromone of the six-toothed spruce bark beetle (*Pityogenes chalcographus* (L.)).

PRODUCTION: Manufactured for use in forestry management.

TARGET PESTS: Six-toothed spruce bark beetle (*Pityogenes chalcographus*).

TARGET CROPS: Spruce forests.

BIOLOGICAL ACTIVITY: **Mode of action:** (2R',5R')-ethyl-1,6-dioxaspiro[4.4]nonane and (2R',5S')-ethyl-1,6-dioxaspiro[4.4]nonane are the components of the aggregation pheromone of the six-toothed spruce bark beetle. Males and females locate host trees by following a plume of air enriched with the odour of the host tree and the aggregation pheromone. Evaporation of pheromone vapours from lures attached to host trees attracts both male and female bark beetles to baited trees and establishes conditions for mass attack of the baited trees. Also used for insect monitoring. **Efficacy:** The baiting of selected areas of the forest has reduced the number of attacks in the main forest area. The baited trees are felled before the brood emerges from the infested trees. In 1988, it was shown that methyl (*E,Z*)-deca-2,4-dienoate, although inactive on its own, was synergistic to chalcogran.
Key reference(s): W Francke, V Heeman, B Gerken, J A A Renwick & J P Vité. 1977. 2-Ethyl-1,6-dioxaspiro[4.4]nonane, principal aggregation pheromone of *Pityogenes chalcographus* (L.), *Naturwissenschaften*, **64**, 590.

COMMERCIALISATION: Formulation: Sold as a slow-release dispenser containing the pheromone. **Tradenames:** 'Chalcoprax' (BASF) and (Bayer).

APPLICATION: The idea is to attract the bark beetle to specific areas where baited trees are attacked. These heavily infested trees are then felled and the population of insects is destroyed before the brood emerges. To accomplish this, the pheromone dispensers are attached to trees, at chest height, at a distance of 50 metres apart in the baiting area. Baiting four trees per hectare per year is usually sufficient to concentrate the beetles.

PRODUCT SPECIFICATIONS: Manufactured for use in forestry management. **Purity:** The pheromones are >95% chemically pure.

COMPATIBILITY: It is unusual to apply the pheromones with other crop protection agents, although insecticides can be used to kill the insects in the baited trees.

MAMMALIAN TOXICITY: There are no reports of allergic or other adverse toxicological effects from the use, manufacture or formulation of (2R',5R')-ethyl-1,6-dioxaspiro[4.4]nonane plus (2R',5S')-ethyl-1,6-dioxaspiro[4.4]nonane. It is considered to be non-toxic.

ENVIRONMENTAL IMPACT AND NON-TARGET TOXICITY: (2R',5R')-Ethyl-1,6-dioxaspiro[4.4]nonane plus (2R',5S')-ethyl-1,6-dioxaspiro[4.4]nonane is the aggregation pheromone of the six-toothed spruce bark beetle and it is specific to this species. It is not expected that it will have any adverse effects on non-target organisms or on the environment.

4:398 1-ethyl 3,5,7 trimethyl-2,8-dioxabi-cyclo[3,2,1]octane (sordidin)
Banana weevil aggregation pheromone

The Pesticide Manual **Fifteenth Edition entry number:** 782

A   B   C   D

NOMENCLATURE: Approved name: 1-ethyl 3,5,7 trimethyl-2,8-dioxabicyclo[3,2,1]octane (sordidin). **Other names:** Banana weevil aggregation pheromone; sordidin.
Development code: Qlure-COS. **CAS RN:** *[178152-25-5].*

SOURCE: Isolated from the volatile compounds emitted by adult weevils.

PRODUCTION: Manufactured for use in crop protection.

TARGET PESTS: Banana weevil (*Cosmopolites sordidus* (Germar)). (See also entry 4:437 – banana corm weevil aggregation pheromone).

TARGET CROPS: All varieties of cultivated and wild banana, as well as plants belonging to the genus *Musa*.

BIOLOGICAL ACTIVITY: Mode of action: 'COS Lure' can be used to attract adult weevils to a trap where the population can be monitored to allow an accurate prediction of pesticide application timing to be made or to a trap infused with an insecticide which kills the adults. **Key reference(s):** W Tinzaara, C S Gold, G H Kagezi, M Dicke, A Van Huis, C M Nankinga, W Tushemereirwe & P E Ragama. 2005. Effects of two pheromone trap densities against banana weevil, *Cosmopolites sordidus*, populations and their impact on plant damage in Uganda, *J. Applied Entomol.*, **129**(5), 265–71.

COMMERCIALISATION: Formulation: Sold as the aggregation pheromone within an adult insect trap. **Tradenames:** 'COS Lure' (Russell IPM).

APPLICATION: Use two traps per hectare for smallholdings and in fields of uneven topography and one trap for every two hectares in large scale unicrop fields. Place the traps near the highest point of the plant using supporting posts approximately 1 metre high (higher if the crop is higher). Lures should be changed every 4–6 weeks for the most accurate population density results. Collect data weekly from the start of the flight of the over-wintering generation. At the population peak, more frequent readings may be needed to allow decisions on pesticide application to be made.

PRODUCT SPECIFICATIONS: Manufactured for use in crop protection. **Shelf-life:** Store in a cool, dry place. Shelf-life can vary from 3–36 months depending on the storage temperature.

MAMMALIAN TOXICITY: Not considered to be toxic to mammals. **Skin and eye:** May cause skin and eye irritation.

ENVIRONMENTAL IMPACT AND NON-TARGET TOXICITY: There is no evidence that the product has caused significant adverse effects on non-target organisms nor had an adverse environmental impact. Not expected to be detected in the environment under normal low rate of use conditions.

4:399 *exo*-brevicomin plus *endo*-brevicomin
Western balsam bark beetle aggregation pheromone

exo-brevicomin   endo-brevicomin

NOMENCLATURE: Approved name: *exo*-brevicomin (*exo*-7-ethyl-5-methyl-6,8-dioxabicyclo[3.2.1]octane) plus *endo*-brevicomin (*endo*-7-ethyl-5-methyl-6,8-dioxabicyclo[3.2.1]octane). **Other names:** Western balsam bark beetle aggregation pheromone. **CAS RN:** *[60018-04-4]* *exo*-brevicomin; *[62532-53-0]* *endo*-brevicomin.

SOURCE: Male *Dryocoetes confusus* Swaine volatile components that have been shown to attract adult beetles. Originally isolated from extracts of males.

PRODUCTION: Manufactured for use in crop protection.

TARGET PESTS: Western balsam bark beetle (*Dryocoetes confusus*).

TARGET CROPS: Fir forests.

BIOLOGICAL ACTIVITY: Mode of action: Western balsam bark beetle tree bait is the aggregation pheromone of the western balsam bark beetle. Males and females locate host trees by following a plume of air enriched with the odour of the host tree and the aggregation pheromone. Evaporation of pheromone vapours from lures attached to host trees attracts both male and female western balsam bark beetles to baited trees and establishes conditions for mass attack of baited trees. **Efficacy:** The baiting of selected areas of the forest has reduced the number of attacks in the main forest area. The baited trees are felled before the brood emerges from the infested trees. **Key reference(s):** J H Borden, A M Pierce Jr, L J Chong, A J Stock & A C Oehlschlager. 1987. Semiochemicals produced by western balsam bark beetle, *Dryocoetes confusus* Swaine (Coleoptera: Scolytidae), *J. Chem. Ecol.*, **13**, 823.

COMMERCIALISATION: Formulation: Sold as a slow-release formulation of the mixed products from polymer plugs or plastic bags. **Tradenames:** 'Western Balsam Bark Beetle Tree Bait' (PheroTech) and (ChemTica).

APPLICATION: Pheromone dispensers are attached to trees, at chest height, at a distance of 50 metres apart in the baiting area where the beetle population is to be concentrated. Four baited trees per hectare per year are usually sufficient for effective concentration of the beetles.

4. Semiochemicals

PRODUCT SPECIFICATIONS: Manufactured for use in crop protection. **Purity:** The chemicals are >95% pure.

COMPATIBILITY: Pheromone-baited trap trees do not require insecticide to kill the beetles as they arrive. Baited trees and surrounding trees are felled and removed before the brood emerges.

MAMMALIAN TOXICITY: 'Western Balsam Bark Beetle Tree Bait' has shown no allergic or other adverse toxicological effects on manufacturers, formulators and field workers.

ENVIRONMENTAL IMPACT AND NON-TARGET TOXICITY: 'Western Balsam Bark Beetle Tree Bait' contains only naturally occurring insect pheromones that are specific to the western balsam bark beetle. There is no evidence that it has had any adverse effects on non-target organisms or on the environment.

4:400 exo-brevicomin plus *trans*-verbenol
Mountain pine beetle aggregation pheromone

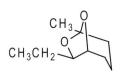

exo-brevicomin

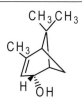

trans-verbenol

NOMENCLATURE: Approved name: exo-brevicomin (exo-7-ethyl-5-methyl-6,8-dioxabicyclo[3.2.1]octane) plus *trans*-verbenol (*trans*-4,6,6-trimethyl(bicyclo[3.3.1]hept-3-en-2-ol). **Other names:** Mountain pine beetle aggregation pheromone. **CAS RN:** *[60018-04-4]* exo-brevicomin; *[1845-30-3] trans*-verbenol.

SOURCE: Male *Dendroctonus ponderosae* Hopkins volatile components that have been shown to attract adult beetles. Isolated from the extracts of the hindgut of emergent and feeding males and females.

PRODUCTION: Manufactured for use in crop protection.

TARGET PESTS: Mountain pine beetle (*Dendroctonus ponderosae*).

TARGET CROPS: Pine forests.

BIOLOGICAL ACTIVITY: Mode of action: Mountain pine beetle tree bait is the aggregation pheromone of the mountain pine beetle. Male and female adults locate host trees by following a plume of air enriched with the odour of the host pine tree and containing the aggregation pheromone. The pheromone is used by *Dendroctonus ponderosae* to recruit others to participate in mass attacks of trees. The beetles have a preference for stressed trees, usually at the edge of plantations and relatively old trees, where the production of sap has reduced, making them more susceptible to attack. The attachment of lures to host trees attracts both male and female mountain pine beetles to the baited trees and establishes conditions that are appropriate for mass attack. **Efficacy:** The attraction of beetles to baited trees and the subsequent felling and destruction of these trees before the insect brood emerges has been very successful in reducing attack in non-baited areas and reducing the population of the mountain pine beetle. **Key reference(s):** J H Borden, L C Ryker, L J Chong, H D Pierce Jr, B D Johnston & A C Oehlschlager. 1983. Semiochemicals for the mountain pine beetle, *Dendroctonus ponderosae*, in British Columbia: baited tree studies, *Can. J. For. Res.*, **13**, 325.

COMMERCIALISATION: Formulation: The pheromones are formulated separately. *Exo*-brevicomin is produced as a slow-release formulation in a polymer plug or plastic sachet and *trans*-verbenol is a slow-release formulation in a plastic sachet. **Tradenames:** 'Mountain Pine Beetle Tree Bait' (PheroTech) and (ChemTica).

APPLICATION: The idea is to attract the mountain pine beetle to specific areas where baited trees are attacked. These heavily infested trees are then felled and the population of insects is destroyed before the brood emerges. To accomplish this, the pheromone dispensers are attached to trees, at chest height, at a distance of 50 metres apart in the baiting area. Baiting four trees per hectare is usually sufficient to concentrate the beetles.

PRODUCT SPECIFICATIONS: Manufactured for use in crop protection. **Purity:** *Exo*-brevicomin is >95% chemically pure and *trans*-verbenol is >65% chemically pure.

COMPATIBILITY: Pheromone-baited trees do not require insecticide to kill the arriving beetles. Baited trees and surrounding trees are felled and removed before the brood emerges.

MAMMALIAN TOXICITY: 'Mountain Pine Beetle Tree Bait' has shown no allergic or other adverse toxicological effects on manufacturers, formulators or field workers.

ENVIRONMENTAL IMPACT AND NON-TARGET TOXICITY: 'Mountain Pine Beetle Tree Bait' contains only naturally occurring insect pheromones that are specific to the mountain pine beetle. There is no evidence that its use has had any adverse effects on non-target organisms or on the environment.

4:401 farnesol with nerolidol
Spider mite alarm pheromone

The *Pesticide Manual* **Fifteenth Edition** entry number: 349

farnesol (2Z, 6E)- isomer

nerolidol *trans*- isomer

NOMENCLATURE: Approved name: (Z,E)-3,7,11-trimethyl-2,6,10-dodecatrien-1-ol (farnesol) plus 3,7,11-trimethyl-1,6,10-dodecatrien-3-ol (nerolidol). **Other names:** Spider mite alarm pheromone. **CAS RN:** *[4602-84-0]* farnesol; *[7212-44-4]* nerolidol.

SOURCE: The alarm pheromone of the two-spotted spider mite (*Tetranychus urticae* Koch), which was isolated from the crude extracts of homogenised female deutonymphs.

TARGET PESTS: Two-spotted spider mite (*Tetranychus urticae*).

TARGET CROPS: All crops infested with spider mites.

BIOLOGICAL ACTIVITY: Mode of action: The product is the alarm pheromone of *Tetranychus urticae*. It is released under natural conditions when the population is threatened or is being attacked by a mite predator. The result is an increase in the activity of mites, with consequent greater exposure to a co-applied acaricide. **Efficacy:** *Tetranychus urticae* is very sensitive to the alarm pheromone and it has been shown that mixtures with conventional acaricides show significant increases in mortality over those found with acaricides used alone. In addition, the alarmed spider mites feed less than undisturbed mites.
Key reference(s): S Regev & W W Cone. 1975. Evidence of farnesol as a male sex attractant for the two-spotted spider mite, *Tetranychus urticae* Koch, *Environ. Entomol.*, **4**, 307.

COMMERCIALISATION: Formulation: Formulated as a controlled-release liquid concentrate. **Tradenames:** 'Stirrup M' and 'Stirrup Mylox' (plus sulfur) (Troy).

APPLICATION: Applied in combination with conventional acaricides. Used at a rate of 150 to 425 ml of product per hectare.

PRODUCT SPECIFICATIONS: Purity: Mixture of structural isomers.

COMPATIBILITY: Compatible with most crop protection agents, particularly acaricides.

MAMMALIAN TOXICITY: The product has shown no adverse toxicological effects on research or manufacturing workers or on users. **Acute oral LD_{50}:** rats >5000, rabbits >2000 mg/kg. **Toxicity class:** EPA (formulation) IV.

ENVIRONMENTAL IMPACT AND NON-TARGET TOXICITY: The product is a natural insect pheromone that is specific to the two-spotted spider mite. There is no evidence that it has caused any adverse effects on non-target organisms nor had any deleterious effect on the environment. **Approved for use in organic farming:** Yes (when used alone).

4:402 ferrolure+
Red palm weevil aggregation pheromone

The Pesticide Manual **Fifteenth Edition entry number:** 582

4-methylnonan-5-ol   4-methylnonan-5-one

NOMENCLATURE: Approved name: ferrolure+; 4-methylnonan-5-ol plus 4-methylnonan-5-one. **Other names:** Red palm weevil aggregation pheromone. **CAS RN:** *[35900-26-6]* 4-methylnonan-5-one ('general' number); *[152045-16-4]* 4-methylnonan-5-ol (also called ferrugineol); *[152203-43-5]* 4-methylnonan-5-one (or ferruginone); *[152783-91-0]* isomeric mixture (also known as rhynchophorol III).

SOURCE: Ferrolure+ is the male *Rhynchophorus ferrugineus* (Olivier) volatile component that has been shown to attract adult weevils.

TARGET PESTS: Red palm weevil (*Rhynchophorus ferrugineus*). It has also been shown to be effective in attracting *Rhynchophorus palmarum* (L.), the American or black palm weevil.

TARGET CROPS: Mature coconut and date palm plantations.

BIOLOGICAL ACTIVITY: Mode of action: Ferrolure+ is the aggregation pheromone of the red palm weevil. Adults locate mates by following a plume of air enriched with the odour of the pheromone. Evaporation of pheromone vapours from traps containing palm, palm fruit, apples or sugarcane stalks attracts male and female weevils to the traps.
Key reference(s): A L Perez, R H Hallett, R Gries, G Gries, A C Oehlschlager & J H Borden. 1996. Pheromone chirality of Asian palm weevils, *Rhynchophorus ferrugineus* (Oliv.) and *R. vulneratus* (Panz.) (Coleoptera: Curculionidae), *J. Chem. Ecol.*, **22**, 357–68.

COMMERCIALISATION: Formulation: Sold as a slow-release formulation from plastic sachets containing liquid pheromone. SEDQ manufactures the active ingredient.
Tradenames: 'Ferrolure+' (ChemTica), 'Red Date Palm Weevil Attract and Kill Dispensers' (Calliope), 'Ferag RF' (SEDQ), 'Rhyncho Kill' (NPP).

APPLICATION: The pheromone dispensers are attached inside bucket traps containing food pieces. These traps are placed on the ground near palms in infested areas, at a density of one trap per 3 hectares. It has been shown that the addition of ethyl acetate increases the number of weevils captured.

PRODUCT SPECIFICATIONS: Purity: The product is ≥95% chemically pure.

COMPATIBILITY: Pheromone use does not require the use of insecticides, but trapping can be used in addition to the chemical treatment of infested palms. Non-repellent insecticides such as deltamethrin or chlorpyrifos can be added to the trap.

MAMMALIAN TOXICITY: Ferrolure+ has shown no allergic or other adverse toxicological effects on manufacturers, formulators or field workers.

ENVIRONMENTAL IMPACT AND NON-TARGET TOXICITY: Ferrolure+ is a naturally occurring insect pheromone that is specific to the red palm weevil. There is no evidence that it has caused any adverse effects on any non-target organisms or on the environment.

4:403 frontalin plus camphene
Douglas fir beetle aggregation pheromone

CH₃

O

O

CH₃

frontalin

CH₃

CH₃

CH₂

camphene

NOMENCLATURE: Approved name: frontalin (1,5-dimethyl-6,8-dioxabicyclo[3.2.1]octane) plus camphene. **Other names:** Douglas fir beetle aggregation pheromone. **CAS RN:** [60478-96-8] frontalin; [279-92-5] camphene.

SOURCE: Frontalin and camphene are two of several components isolated from the adult female Douglas fir beetle (*Dendroctonus pseudotsuga* Hopkins). These compounds are key components of the aggregation pheromone to which male and female beetles respond.

PRODUCTION: The chemicals are manufactured for use in forest management.

TARGET PESTS: Douglas fir beetle (*Dendroctonus pseudotsuga*).

TARGET CROPS: Douglas fir (*Pseudotsuga menziesii* Franco).

BIOLOGICAL ACTIVITY: Mode of action: Douglas fir beetle tree bait contains the aggregation pheromone of the Douglas fir beetle. Males and females locate host trees by following a plume of air enriched with the odour of the host tree and containing the aggregation pheromone. Evaporation of the pheromone from lures attached to host trees attracts both male and female beetles to the baited trees and establishes conditions for mass attack of baited trees. **Efficacy:** The use of four baited trees per hectare is successful in concentrating the beetle population in the baited areas and reduces the population in non-baited areas. Baited trees and those close to baited trees are felled before the brood emerges. **Key reference(s):** G B Pitman & J P Vité. 1970. Field response of *Dendroctonus pseudotsugae* to synthetic frontalin, *Ann. Entomol. Soc. Am.*, **63**, 661.

COMMERCIALISATION: Formulation: Formulated as a slow-release product releasing frontalin and camphene. **Tradenames:** 'Douglas Fir Beetle Aggregation Pheromone' (PheroTech).

APPLICATION: The pheromone dispensers are attached to trees, at chest height, 50 metres apart in the baited area where the beetles are to be concentrated. Four baited trees per hectare have been used successfully to reduce populations in non-baited areas.

4. Semiochemicals

PRODUCT SPECIFICATIONS: The chemicals are manufactured for use in forest management. **Purity:** The pheromones are >95% chemically pure.

COMPATIBILITY: Pheromone-baited trap trees do not require insecticide to kill arriving beetles. Baited trees and surrounding trees are felled and removed before the brood emerges.

MAMMALIAN TOXICITY: There are no reports of allergic or other adverse toxicological effects from the use of the aggregation pheromone of the Douglas fir beetle. Considered to be non-toxic.

ENVIRONMENTAL IMPACT AND NON-TARGET TOXICITY: The aggregation pheromone is specific to the Douglas fir beetle and is not expected to have any adverse effects on non-target organisms or on the environment.

4:404 frontalin plus *endo*-brevicomin plus 3-carene plus α-pinene

Southern pine beetle aggregation pheromone

frontalin   *endo*-brevicomin   3-carene   α-pinene

NOMENCLATURE: Approved name: frontalin (1,5-dimethyl-6,8-dioxabicyclo[3.2.1]octane) plus *endo*-brevicomin (*endo*-7-ethyl-5-methyl-6,8-dioxabicyclo[3.2.1]octane) plus 3-carene plus α-pinene. **Other names:** Southern pine beetle aggregation pheromone. **CAS RN:** *[60478-96-8]* frontalin; *[62532-53-0]* endo-brevicomin; *[13466-78-9]* 3-carene; *[80-56-8]* α-pinene.

SOURCE: Frontalin and *endo*-brevicomin are insect-produced aggregation pheromones. 3-Carene and α-pinene are host-produced synergists of the pheromone components.

PRODUCTION: Manufactured for use in forest protection strategies.

TARGET PESTS: Southern pine beetle (*Dendroctonus frontalis* Zimmerman).

TARGET CROPS: Pine forests.

BIOLOGICAL ACTIVITY: Mode of action: Both male and female beetles are attracted by the aggregation pheromone and follow its trail in order to locate host trees. The attachment of lures to selected trees establishes conditions for mass insect attack of these baited trees. **Efficacy:** The use of four baited trees per hectare is successful in concentrating the beetle population in the baited areas and reduces the population in non-baited areas.
Key reference(s): T L Payne, J E Coster, J V Richerson, L J Edson & E R Hart. 1978. Field response of the Southern pine beetle to behavioural chemicals, *Environ. Entomol.*, **7**, 578.

COMMERCIALISATION: Formulation: Sold as slow-release formulations from devices such as polymer plugs or plastic bags. **Tradenames:** 'Southern Pine Beetle Aggregation Pheromone' (PheroTech).

APPLICATION: The slow-release devices are attached to the trees, at chest height, at a distance of 50 metres apart in the baiting area where the beetles are to be concentrated. The baited trees and those surrounding the baited trees are felled and removed before the brood emerges.

PRODUCT SPECIFICATIONS: Manufactured for use in forest protection strategies. **Purity:** All components are >95% chemically pure.

COMPATIBILITY: It is unusual to use these pheromone lures in conjunction with other crop protection agents.

MAMMALIAN TOXICITY: There are no records of allergic or other adverse toxicological effects from research staff, manufacturers, formulators or field workers from the product. It is considered to be non-toxic. **Acute oral LD$_{50}$:** α-Pinene: rats 3.7 g/kg. **Skin and eye:** α-Pinene is an irritant to rabbits at concentrations of 364–572 mg/litre.

ENVIRONMENTAL IMPACT AND NON-TARGET TOXICITY: Southern pine beetle aggregation pheromone is composed of natural *Dendroctonus frontalis* pheromone and host tree volatiles, all of which occur widely in Nature. It is not expected that the product will have any adverse effects on non-target organisms or on the environment at the rates used for control.

4. Semiochemicals

4:405 frontalin plus α-pinene
Spruce beetle aggregation pheromone and host kairomone

frontalin   α-pinene

NOMENCLATURE: **Approved name:** frontalin (1,5-dimethyl-6,8-dioxabicyclo[3.2.1]octane) plus α-pinene. **Other names:** Spruce beetle aggregation pheromone and host kairomone. **CAS RN:** *[60478-96-8]* frontalin; *[80-56-8]* α-pinene.

SOURCE: Male *Dendroctonus rufipennis* (Kirby) volatile components and spruce-produced α-pinene have been shown to attract male and female adult beetles.

PRODUCTION: Manufactured for use in forest management.

TARGET PESTS: Spruce beetle (*Dendroctonus rufipennis*).

TARGET CROPS: Spruce forests.

BIOLOGICAL ACTIVITY: **Mode of action:** 'Spruce Beetle Tree Bait' contains the aggregation pheromone of the spruce beetle. Males and females locate host trees by following a plume of air enriched with the odour of the host tree and containing the aggregation pheromone. Evaporation of the pheromone from lures attached to host trees attracts both male and female spruce beetles to the baited trees and establishes conditions for mass attack of baited trees. **Efficacy:** The use of four baited trees per hectare is successful in concentrating the beetle population in the baited areas and reduces the population in non-baited areas. **Key reference(s):** E D A Dyer & P M Hall. 1980. Effect of living host tree (*Picea*) on the responses of *Dendroctonus rufipennis* and a predator *Thanasimus undulatus* to frontalin and suedenol, *Can. Entomol.*, **107**, 979.

COMMERCIALISATION: **Formulation:** Formulated as a slow-release product releasing frontalin and α-pinene from a polymer plug or plastic sachet. **Tradenames:** 'Spruce Beetle Tree Bait' (PheroTech) and (ChemTica).

APPLICATION: The pheromone dispensers are attached to trees, at chest height, 50 metres apart, in the baiting area where the beetles are to be concentrated. Four baited trees per hectare have been used successfully to reduce populations in non-baited areas.

PRODUCT SPECIFICATIONS: Manufactured for use in forest management. **Purity:** Frontalin is >95% chemically pure.

COMPATIBILITY: Pheromone-baited trap trees do not require insecticide to kill arriving beetles. Baited trees and surrounding trees are felled and removed before the brood emerges.

MAMMALIAN TOXICITY: 'Spruce Beetle Tree Bait' has not shown any allergic or other adverse toxicological effects on manufacturers, formulators or field workers.

ENVIRONMENTAL IMPACT AND NON-TARGET TOXICITY: 'Spruce Beetle Tree Bait' contains only natural insect pheromones that are specific to the spruce beetle and host volatile components. There is no evidence of any adverse effects on non-target organisms or on the environment.

4:406 German cockroach pheromone
German cockroach pheromone

NOMENCLATURE: Approved name: German cockroach pheromone. (A mixture of about 150 chemicals.)

SOURCE: Extracted from German cockroach (*Blatta germanica* (L.)) droppings.

TARGET PESTS: German cockroach (*Blatta germanica*).

TARGET CROPS: Used in houses, restaurants, health care facilities, schools, factories, offices and garages.

BIOLOGICAL ACTIVITY: Efficacy: The product is a mixture of about 150 chemicals which attract the cockroaches to the bait station. It is not known which of the chemicals in the product are the most active.

COMMERCIALISATION: Tradenames: 'Cockroach Bait Station' (Woodstream).

APPLICATION: The pheromone mixture is part of a bait station that can be placed anywhere indoors.

MAMMALIAN TOXICITY: No adverse human health effects are expected from use of German cockroach pheromone mixture in bait stations. There is no evidence of mammalian toxicity from amounts much greater than people would encounter from these bait stations. Furthermore, both the pheromone and the accompanying insecticide are attached to the inside of the plastic bait station. As a result, no exposure to either the pheromone mixture or the insecticide is expected from approved uses.

4. Semiochemicals

ENVIRONMENTAL IMPACT AND NON-TARGET TOXICITY: No harmful environmental effects are expected because: the bait stations are approved only for use indoors; there is virtually no exposure to any species except the German cockroach; and the pheromone is not known to be toxic.

4:407 gossyplure   *Pink bollworm sex pheromone*

The Pesticide Manual **Fifteenth Edition entry number:** 446

(Z,Z)-   (Z,E)-

NOMENCLATURE: Approved name: gossyplure (name in common use); (Z,Z)- plus (Z,E)-hexadeca-7,11-dien-1-yl acetate (50:50 blend of isomers). **Other names:** Pink bollworm sex pheromone; PBW sex pheromone; hexadecadienyl acetate; Z7Z11-16Ac; Z7E11-16Ac. **Development code:** PP761 (Zeneca (now Syngenta)). **CAS RN:** *[53042-79-8]* (7Z,11E)-isomer; *[52207-99-5]* (Z,Z)- isomer; *[122616-64-2]* (7Z,11-unspecified stereochemistry)-isomer; *[50933-33-0]* unspecified stereochemistry.

SOURCE: The sex pheromone of the pink bollworm (*Pectinophora gossypiella* (Saunders)). Isolated from the abdominal tips of virgin females.

PRODUCTION: Manufactured for use in the protection of cotton.

TARGET PESTS: Pink bollworm (*Pectinophora gossypiella*).

TARGET CROPS: Cotton.

BIOLOGICAL ACTIVITY: Mode of action: Gossyplure is the sex pheromone of the pink bollworm. Male moths locate and subsequently mate with female moths by following the trail or pheromone plume emitted by the virgin females. The indiscriminate application of high levels of gossyplure interferes with this process, as a constant exposure to high levels of pheromone makes trail following impossible (habituation/adaptation). Alternatively, the use of discrete sources of pheromone released over time presents the male with a false trail to follow (confusion). Control is subsequently achieved through the prevention of mating and laying of fertile eggs. **Efficacy:** Very low rates are required to cause mating disruption.

Gossyplure is volatile and distributes throughout the crop easily. It was used on over 2 million acres in the mid-1990s for control of PBW by mating disruption, principally in the USA and Egypt. However, this technology has largely been replaced by transgenic cotton developed to control the related bollworms *Heliothis virescens* (Fabricius) and, latterly, *Helicoverpa armigera* (Hübner). **Key reference(s):** J R Merkl & H M Flint. 1981. Responses of male pink bollworms to various mixtures of the (Z,Z)- and (Z,E)- isomers of gossyplure, *Environ. Entomol.*, **6**, 114.

COMMERCIALISATION: Formulation: Slow-release formulations of hollow fibres of polyacrylate resin containing gossyplure, laminated flakes covering a porous layer impregnated with gossyplure, polyamide micro-capsules containing gossyplure, twist tie dispensers and polymer bands are all commercially available. Consep has introduced a sprayable controlled-release granule formulation. **Tradenames:** 'Nomate PBW' (Scentry), 'Pectone' and 'Sirene PBW' (plus permethrin) (Syngenta), 'PBW Rope-L' and 'PB Rope' (Shin-Etsu), 'Frustrate PBW' and 'DeCoy PBW Band' (Suterra), 'Selibate PBW' (Monterey), 'Last Flight PBW' and 'Lost Dream' (plus methomyl plus chlorpyrifos) (Troy), 'Checkmate PBW' and 'Checkmate PBW-F' (Consep), 'Disrupt PBW' (Hercon), 'Dismate' (Russell), 'Pherocon' (Trece), 'Sirene-PBW' (plus permethrin) (IPM Technologies), 'SPLAT PBW' (ISCA).

APPLICATION: Many slow-release formulations are applied by aerial spraying with the slow-release plastic device adhering to the crop's foliage and the pheromone diffusing into the field, but the product is also suitable for Ultra Low Volume (ULV) application or it may be applied using a back pack or trailer-mounted sprayer. Other dispensers are attached by hand to individual plants within the cotton field. Apply at the pinhead square stage of cotton crop growth at a rate of 10 to 30 g a.i. per hectare. Do not exceed 150 g per hectare per year. Diluted mixtures should be sprayed within 12 hours of preparation. 'SPLAT PBW' (Specialized Pheromone and Lure Application Technology) is an amorphous polymer matrix for the sustained release of insect pheromones. It has been designed to dispense, optimise and modulate the semiochemical release in mating disruption pest management programmes. The product can be applied by hand, tractor or aircraft.

PRODUCT SPECIFICATIONS: Manufactured for use in the protection of cotton. **Purity:** The isomers are >95% chemically pure. **Storage conditions:** It should be stored in a cool place until used and freezing extends the shelf-life indefinitely except for the water-based micro-encapsulated formulation.

COMPATIBILITY: Gossyplure can be used alone or in combination with other chemicals, such as foliage fertilisers or with insecticides as a lure and kill strategy. When used in conjunction with a monitoring programme, the results from the monitoring traps may be used to determine the timing of applications. Typically, three spray applications will be needed between May and August, but hand-applied formulations are applied once to give season long control. Under optimum conditions, gossyplure should suppress trap catch for two weeks. Increase in catch above tolerance level indicates the need for re-application.

4. Semiochemicals

MAMMALIAN TOXICITY: Gossyplure has shown no adverse toxicological effects on manufacturers, formulators, research workers or farmers. It is considered to be non-toxic. **Acute oral LD$_{50}$:** rats >5000 mg/kg. **Acute dermal LD$_{50}$:** rabbits >2000 mg/kg. **Inhalation:** LC$_{50}$ (4 h) for rats >2000 mg/litre. **Skin and eye:** Acute percutaneous LD$_{50}$ for rats >2000 mg/kg. Slight erythema (rats). **Toxicity class:** EPA (formulation) III.

ENVIRONMENTAL IMPACT AND NON-TARGET TOXICITY: Gossyplure is a natural insect pheromone that is specific to the pink bollworm. There is no evidence that it has caused any adverse effects on any non-target organisms nor had any adverse environmental impact. **Bird toxicity:** Acute oral LD$_{50}$ for bobwhite quail >2000 mg/kg. Dietary LC$_{50}$ for bobwhite quail >5620 ppm. **Fish toxicity:** LC$_{50}$ (96 h) for rainbow trout >120 ppm. **Other aquatic toxicity:** LC$_{50}$ (48 h) for *Daphnia* 0.70 ppm. **Effects on beneficial insects:** Practically non-toxic to honeybees. It has been shown that the use of gossyplure in place of insecticides increased honey production and improved boll size. **Behaviour in soil:** DT$_{50}$ in soil 1 day; in water 7 days. **Approved for use in organic farming:** Yes.

4:408 grandlure   *Boll weevil sex pheromone*

The Pesticide Manual Fifteenth Edition entry number: 447

grandlure I   grandlure II   grandlure III   grandlure IV

NOMENCLATURE: Approved name: grandlure; (+)-*cis*-2-isopropenyl-1-methylcyclobutaneethanol (grandlure I) plus *cis*-3,3-dimethyl-Δ^{β}-cyclohexaneethanol (grandlure II) plus *cis*-3,3-dimethyl-$\Delta^{1\alpha}$-cyclohexaneacetaldehyde (grandlure III) plus *trans*-3,3-dimethyl-$\Delta^{1\alpha}$-cyclohexaneacetaldehyde (grandlure IV). **Other names:** Boll weevil sex pheromone; grandisol (for grandlure I); (+)-grandisol (for grandlure I). **Development code:** SHA 112401. **CAS RN:** *[11104-05-5]* mixture of grandlure I, II and IV; *[26532-22-9]* grandlure I; *[30820-22-5]* racemic grandlure I; *[26532-23-0]* grandlure II; *[26532-24-1]* grandlure III; *[26532-25-2]* grandlure IV.

SOURCE: The compounds were identified from both male and female boll weevils (*Anthonomus grandis* Boheman) and their frass.

PRODUCTION: Manufactured for use in cotton.

TARGET PESTS: Boll weevil (*Anthonomus grandis*).

TARGET CROPS: Cotton.

BIOLOGICAL ACTIVITY: Mode of action: The pheromones are used in survey and trapping programmes in cotton. Both males and females are attracted to the volatiles and can be mass trapped, thereby reducing the population of boll weevils in the crop, or the numbers caught can be monitored, thereby identifying the best timing for the application of insecticide sprays. The grandlure baits are also used in a lure and kill strategy, whereby the boll weevil adults are attracted to the lures where they are exposed to a fast-acting, contact insecticide.

Efficacy: Grandlure has been widely used in the southern USA, with over 12.5 million lures having been deployed. The use of the three pest control strategies has reduced the damage caused by boll weevils in cotton. The major use of the 'Tubo' products is in South America, notably Paraguay. The range of *Anthomonus grandis* is increasing in South America and, as a result, the opportunity for employing the pheromone lure and kill technology for control is increasing. Over one million 'Boll Weevil Attract and Control Tubes (BWACT)' are produced each year. The use of one per hectare has reduced boll weevil populations by >85%.

Key reference(s): 1) G H McKibben, D D Hardee, T B Davich, R C Gueldner and P A Hedin. 1971. Slow-release formulations of Grandlure, the synthetic pheromone of the boll weevil, *J. Econ. Entomol.*, **64**, 317. 2) D D Hardee, G H McKibben, D R Rummel, P M Huddleston and J R Coppedge. 1972. Boll weevils in Nature respond to Grandlure, a synthetic pheromone, *J. Econ. Entomol.*, **65**, 97.

COMMERCIALISATION: Formulation: Sold in slow-release dispensers for use in surveys and mass trapping programmes. Sold in combination with insecticides in lure and kill strategies. **Tradenames:** 'Grandlure' (International Speciality Products) and (IPM Technologies), 'Sirene-BW' (plus profenfos) (IPM Technologies), 'Tubo Mata Bicudo', 'Tubo Mata Picudo', 'Boll Weevil Attract and Control Tubes (BWACT)' (plus malathion) and 'TM' (plus malathion) (Plato).

APPLICATION: The slow-release dispensers are distributed throughout the crop to attract the adults to the traps.

PRODUCT SPECIFICATIONS: Manufactured for use in cotton. **Purity:** The chemicals are >95% pure.

COMPATIBILITY: Compatible with crop protection agents that do not repel the boll weevil.

MAMMALIAN TOXICITY: There is no evidence of allergic or other adverse toxicological effects from grandlure on researchers, manufacturers, formulators or field workers. Considered to be non-toxic.

4. Semiochemicals

ENVIRONMENTAL IMPACT AND NON-TARGET TOXICITY: Grandlure is a synthetic pheromone that is specific to the boll weevil (*Anthonomus grandis*). As such, it is not expected to have any adverse effects on non-target organisms or on the environment.

4:409 (*Z*)-hexadec-11-enal
Artichoke plume moth sex pheromone

The Pesticide Manual **Fifteenth Edition entry number:** 462

$$\underset{CH_3(CH_2)_3 \quad (CH_2)_9CHO}{\overset{\displaystyle H \qquad H}{\underset{\displaystyle }{C=C}}}$$

NOMENCLATURE: Approved name: (*Z*)-hexadec-11-enal. **Other names:** Artichoke plume moth sex pheromone; Z11-16Al; Z11HDAL. **Development code:** E103 (Exosect). **CAS RN:** *[53939-28-9]* (*Z*)-hexadec-11-enal; *[72194-83-3]* unspecified stereochemistry.

SOURCE: Isolated from the ovipositors of virgin females of the artichoke plume moth (*Platyptilia carduidactyla* (Riley)).

PRODUCTION: Manufactured for use in crop protection.

TARGET PESTS: Artichoke plume moth (*Platyptilia carduidactyla*).

TARGET CROPS: Vegetables.

BIOLOGICAL ACTIVITY: Mode of action: The product is the sex pheromone of the artichoke plume moth. Male moths locate and subsequently mate with female moths by following the trail or pheromone plume emitted by the virgin females. The indiscriminate application of high levels of the product interferes with this process, as a constant exposure to high levels of pheromone makes trail following impossible (habituation/adaptation). Alternatively, the use of discrete sources of pheromone released over time presents the male with a false trail to follow (confusion). Control is subsequently achieved through the prevention of mating and laying of fertile eggs. **Efficacy:** Very low rates are required to cause mating disruption. (*Z*)-hexadec-11-enal is volatile and distributes throughout the crop easily.

Acts as an attractant and by disruption of mating in the disorientation mode.

Key reference(s): J A Klun, K F Haynes, B A Bierl-Leonhardt, M C Birch & J R Plimmer. 1981. Sex pheromone of the female artichoke plume moth, *Platyptilia carduidactyla*, *Environ. Entomol.*, **10**, 763.

COMMERCIALISATION: Formulation: Sold as the pheromone contained in slow-release sprayable, plastic tubes. 'Apm-Rope' is not sprayable, but a hand-applied twist tie formulation, and 'Selibate' is a hand-applied PVC resin formulation. **Tradenames:** 'Disrupt APM' (Hercon), 'Isomate APM', 'Isomate DBM' (plus (Z)-hexadec-11-enyl acetate), 'Konaga-con' (plus (Z)-hexadec-11-enyl acetate) and 'Isomate RSB' (plus (Z)-octadec-9-enal plus (Z)-octadec-13-enal) (Shin-Etsu), 'Apm-Rope' (Shin-Etsu) and (Nihon Nohyaku), 'Fersex ChS' (plus (Z)-hexadec-9-enal plus (Z)-octadec-13-enal) (SEDQ), 'Checkmate DBM' (plus (Z)-hexadec-11-enyl acetate) and 'Selibate CS' (plus (Z)-octadec-9-enal plus (Z)-octadec-13-enal) (Suterra), 'NoMate DBM' (plus (Z)-hexadec-11-enyl acetate) (Scentry), 'Exosex APM' (Exosect).

APPLICATION: The slow-release carriers are sprayed onto or hand-applied within the crop to be protected at the time of adult flight.

PRODUCT SPECIFICATIONS: Manufactured for use in crop protection. **Purity:** (Z)-hexadec-11-enal is >95% chemically pure.

COMPATIBILITY: It is not usual to apply the pheromone with other crop protection chemicals.

MAMMALIAN TOXICITY: There have been no reports of allergic or other adverse toxicological effects from the use of (Z)-hexadec-11-enal by researchers, manufacturers, formulators or field workers. It is considered to be non-toxic. **Acute oral LD$_{50}$:** rats >5000 mg/kg. **Inhalation:** LC$_{50}$: for rats >5 mg/litre. **Skin and eye:** Acute percutaneous LD$_{50}$ for rats >2000 mg/kg. The pheromone is not a skin or eye irritant.

ENVIRONMENTAL IMPACT AND NON-TARGET TOXICITY: (Z)-hexadec-11-enal is a component of several lepidopteran sex pheromones, but it is considered unlikely to have any adverse effects on non-target organisms or on the environment at the dose rates applied. **Approved for use in organic farming:** Yes.

4:410 (Z)-hexadec-11-enal plus (Z)-hexadec-9-enal

Yellow stem borer sex pheromone

The Pesticide Manual **Fifteenth Edition entry number:** 462

$$\begin{array}{cc} H \quad\quad H \\ \diagdown \quad\diagup \\ C=C \\ \diagup \qu_\quad\diagdown \\ CH_3(CH_2)_3 \quad (CH_2)_9CHO \end{array}$$

(Z)-hexadec-11-enal

$$\begin{array}{cc} H \quad\quad H \\ \diagdown \quad\diagup \\ C=C \\ \diagup \quad\quad\diagdown \\ CH_3(CH_2)_5 \quad (CH_2)_7CHO \end{array}$$

(Z)-hexadec-9-enal

NOMENCLATURE: Approved name: (Z)-hexadec-11-enal plus (Z)-hexadec-9-enal. **Other names:** Yellow stem borer sex pheromone. **Development code:** E208 (Exosect). **CAS RN:** *[53939-28-9]* (Z)-hexadec-11-enal; *[56219-04-6]* (Z)-hexadec-9-enal.

SOURCE: The female sex pheromone of the yellow rice stem borer (*Scirpophaga incertulas* (Walker)) was originally isolated from the terminal segments of virgin females.

PRODUCTION: Manufactured for use in crop protection.

TARGET PESTS: Yellow rice stem borer (*Scirpophaga incertulas*).

TARGET CROPS: Rice.

BIOLOGICAL ACTIVITY: Efficacy: The compounds (Z)-hexadec-11-enal plus (Z)-hexadec-9-enal are the sex pheromone of the yellow rice stem borer. Male moths locate and subsequently mate with female moths by following the trail or pheromone plume emitted by the virgin females. The application of (Z)-hexadec-11-enal plus (Z)-hexadec-9-enal makes trail following impossible (camouflage, competition between artificial and female plume, false trail following). Control is subsequently achieved through the prevention of mating and laying of fertile eggs. In some cases, the pheromone is used for mass trapping of males, thereby reducing the incidence of egg laying.

COMMERCIALISATION: Formulation: Sold as a mixture of the volatiles in slow-release devices such as plastic tubes or plastic laminated film. **Tradenames:** 'Exosex YSBTab' (Exosect).

APPLICATION: The devices are distributed throughout the crop about 14 days after transplanting, which is after the first adult flight, but before the economically important second and, perhaps, third flights.

PRODUCT SPECIFICATIONS: Manufactured for use in crop protection. **Purity:** All components present in the slow-release devices are >95% chemically pure.

MAMMALIAN TOXICITY: There are no reports of the product causing allergic or other adverse toxicological effects to researchers, manufacturers, formulators or field workers. Considered to be non-toxic. **Acute oral LD$_{50}$:** rats >5000 mg/kg (both components).

ENVIRONMENTAL IMPACT AND NON-TARGET TOXICITY: The product is the sex pheromone of the yellow rice stem borer (*Scirpophaga incertulas*) and is specific to that species. It is not expected that it will have any adverse effects on non-target organisms or on the environment. **Approved for use in organic farming:** Yes.

4:411 (Z)-hexadec-11-enal plus (Z)-hexadec-9-enal plus (Z)-octadec-13-enal
Striped rice stem borer sex pheromone

The Pesticide Manual Fifteenth Edition entry number: 462

(Z)-hexadec-11-enal   (Z)-hexadec-9-enal   (Z)-octadec-13-enal

NOMENCLATURE: Approved name: (Z)-hexadec-11-enal plus (Z)-hexadec-9-enal plus (Z)-octadec-13-enal. **Other names:** Striped rice stem borer sex pheromone.
CAS RN: [53939-28-9] (Z)-hexadec-11-enal; [56219-04-6] (Z)-hexadec-9-enal; [58594-45-9] (Z)-octadec-13-enal; [106049-29-0] mixture of (Z)-hexadec-11-enal with (Z)-octadec-13-enal and (Z)-hexadec-9-enal.

SOURCE: The female sex pheromone of the striped rice stem borer (*Chilo suppressalis* (Walker)) is a combination of three components. Originally isolated from the terminal segments of virgin females.

PRODUCTION: Manufactured for use in crop protection.

TARGET PESTS: Striped rice stem borer (*Chilo suppressalis*).

TARGET CROPS: Rice.

BIOLOGICAL ACTIVITY: Mode of action: The compounds (Z)-hexadec-11-enal plus (Z)-hexadec-9-enal plus (Z)-octadec-13-enal are the sex pheromone of the rice stem borer. Male moths locate and subsequently mate with female moths by following the trail or pheromone plume emitted by the virgin females. The application of (Z)-hexadec-11-enal plus (Z)-hexadec-9-enal plus (Z)-octadec-13-enal makes trail following impossible (camouflage, competition between artificial and female plume, false trail following). Control is subsequently achieved through the prevention of mating and laying of fertile eggs. In some cases, the pheromone is used for mass trapping of males, thereby reducing the incidence of egg laying. **Efficacy:** Very low rates are required to cause mating disruption, typically 40 g a.i. per ha per season. The pheromone is volatile and distributes throughout the crop easily.
Key reference(s): O Mochida, G S Arida, S Tatsuki & J Fukami. 1984. A field test on a third component of the female sex pheromone of the rice striped stem borer, *Chilo suppressalis*, in the Philippines, *Entomol. Exp. Appl.*, **36**, 295.

COMMERCIALISATION: Formulation: Sold as a mixture of the volatiles in slow-release devices such as plastic tubes or plastic laminated film. Hercon has developed a sprayable formulation. **Tradenames:** 'Isomate RSB' and 'Konaga-con' (Shin-Etsu), 'Fersex ChS' (SEDQ), 'Selibate CS' (Suterra), 'Disrupt RSB' (Hercon).

APPLICATION: The devices are distributed throughout the crop about 14 days after transplanting, which is after the first adult flight, but before the economically important second and, perhaps, third flights.

PRODUCT SPECIFICATIONS: Manufactured for use in crop protection. **Purity:** All components present in the slow-release devices are >95% chemically pure.

COMPATIBILITY: The pheromones are compatible with all crop protection agents that do not repel rice stem borers.

MAMMALIAN TOXICITY: There are no reports of the product causing allergic or other adverse toxicological effects to researchers, manufacturers, formulators or field workers. Considered to be non-toxic. **Acute oral LD_{50}:** rats >5000 mg/kg (all components). **Inhalation:** LC_{50} for rats >5 mg/litre ((Z)-hexadec-11-enal).

ENVIRONMENTAL IMPACT AND NON-TARGET TOXICITY: The product is the sex pheromone of the striped rice stem borer (*Chilo suppressalis*) and is specific to that species. It is not expected that it will have any adverse effects on non-target organisms or on the environment. **Approved for use in organic farming:** Yes.

4:412 (Z)-hexadec-11-enal plus (Z)-hexadec-11-enyl acetate

Diamondback moth pheromone

The Pesticide Manual Fifteenth Edition entry numbers: 462 and 463

$$CH_3(CH_2)_3 \quad (CH_2)_9CHO$$

(Z)-hexadec-11-enal

$$CH_3(CH_2)_3 \quad (CH_2)_{10}OCOCH_3$$

(Z)-hexadec-11-enyl acetate

NOMENCLATURE: Approved name: (Z)-hexadec-11-enal plus (Z)-hexadec-11-enyl acetate. **Other names:** *Plutella xylostella* sex pheromone; diamondback moth sex pheromone. **CAS RN:** *[53939-28-9]* (Z)-hexadec-11-enal; *[34010-21-4]* (Z)-hexadec-11-enyl acetate.

SOURCE: Female *Plutella xylostella* (L.) emit several volatile components shown to be attractive to males. These components were originally extracted from the whole bodies of virgin females.

TARGET PESTS: Diamondback moth (*Plutella xylostella*).

TARGET CROPS: Vegetables such as cabbage and other cruciferous crops.

BIOLOGICAL ACTIVITY: Mode of action: Males locate females by following a sex pheromone trail containing the complete pheromone blend emitted by virgin female moths during specific periods in the scotophase. Permeation of the canopy of a vegetable crop with vapours of the two-component blend above is used to disrupt location of females by males and decreases mating. **Efficacy:** Several commercial formulations of the two-component blend above have been shown to be effective in maintaining populations of diamondback moths at economically tolerable levels in vegetables until harvest. A slow-release formulation containing a blend of Z-hexadec-11-enyl acetate (27%), Z-hexadec-11-en-1-ol (1%), Z-tetradec-11-en-1-ol (9%), Z-hexadec-11-enal (63%) and the insecticide permethrin was shown to be highly attractive to male moths for at least four weeks. The permethrin insecticide had no influence on response of moths to lure baited traps. The formulation suppressed captures of male diamondback moths >90% for up to 3 weeks post treatment.
Key reference(s): 1) Y S Chow, Y M Lin & C L Hsu. 1977. Sex pheromone of the diamondback moth (Lepidoptera: Plutellidae), *Bull. Inst. Zool. Acad. Sinica*, **16**, 99.
2) G V P Reddy & A Guerrero. 2000. Pheromone-based integrated pest management to control the diamondback moth, *Plutella xylostella*, in cabbage fields, *Pest Man. Sci.*, **56**, 882–8.
3) G V P Reddy & A Guerrero. 2001. Optimum timing of insecticide applied against the diamondback moth, *Plutella xylostella*, in cole crops using threshold catches in sex pheromone

4. Semiochemicals

traps, *Pest Man. Sci.*, **57**, 90–4. 4) E R Mitchell. 2002. Promising new technology for managing diamondback moth (Lepidoptera: Plutellidae) in cabbage with pheromone, *J. Environmental Science and Health*, Part B, **37**(3), 277–90.

COMMERCIALISATION: Formulation: Sold as slow-release formulations within slow-release devices such as laminated plastic, plastic barrier film and plastic tubes. Consep also offers a sprayable formulation. The pheromone is produced by Hercon, Shin-Etsu, Konaga-con and Scentry Biologicals (acquired from Ecogen). **Tradenames:** 'NoMate DBM' (Scentry), 'Isomate DBM' (Shin-Etsu), 'Checkmate DBM-F' (Consep).

APPLICATION: Pheromone dispensers are attached to sticks about 20 cm above ground. The recommended number of lures per hectare varies with release rates.

PRODUCT SPECIFICATIONS: Purity: The pheromones used in mating disruption devices are >95% chemically pure.

COMPATIBILITY: Pheromone disruption of mating does not require insecticide but can be used with application of insecticides.

MAMMALIAN TOXICITY: Diamondback moth pheromone blend has shown no adverse toxicological effects on manufacturers, formulators or field workers. Considered to be non-toxic. **Acute oral LD$_{50}$:** rats >5000 mg/kg ((Z)-hexadec-11-enal). **Inhalation:** LC$_{50}$: for rats >5 mg/litre ((Z)-hexadec-11-enal).

ENVIRONMENTAL IMPACT AND NON-TARGET TOXICITY: Diamondback moth pheromone blend is a natural insect pheromone that is specific to the diamondback moth. There is no evidence that it has caused any adverse effects on non-target organisms nor had an adverse environmental impact. **Approved for use in organic farming:** Yes.

4:413 (Z)-hexadec-11-enal plus (Z)-tetradec-9-enal *Tobacco budworm sex pheromone*

The Pesticide Manual **Fifteenth Edition entry number:** 462

$$
\begin{array}{cc}
\text{H} \qquad \text{H} & \qquad \text{H} \qquad \text{H} \\
\text{C} = \text{C} & \qquad \text{C} = \text{C} \\
CH_3(CH_2)_3 \; (CH_2)_9CHO & \qquad CH_3(CH_2)_3 \; (CH_2)_7CHO
\end{array}
$$

(Z)-hexadec-11-enal   (Z)-tetradec-9-enal

NOMENCLATURE: Approved name: (Z)-hexadec-11-enal plus (Z)-tetradec-9-enal.

Other names: *Heliothis virescens* sex pheromone; tobacco budworm sex pheromone.
CAS RN: *[53939-28-9]* (Z)-hexadec-11-enal; *[53939-27-8]* (Z)-tetradec-9-enal.

SOURCE: Both components were first isolated from the terminal segments of virgin female *Heliothis virescens* (Fabricius) moths.

PRODUCTION: Manufactured for use in crop protection.

TARGET PESTS: Tobacco budworm (*Heliothis virescens*).

TARGET CROPS: Cotton, maize (corn), tobacco and tomatoes.

BIOLOGICAL ACTIVITY: Mode of action: The male moth locates a virgin female by following her pheromone trail. If this plume of volatiles is masked with the release of additional and random pheromone components, the trail is disrupted and the male will be unable to locate a mate or it will follow a false trail, again failing to find and mate with a female. **Efficacy:** It has been shown that the use of the sex pheromone of the tobacco budworm has been effective in maintaining the insect population at economically acceptable levels. **Key reference(s):** J H Tumlinson, D E Hendricks, E R Mitchell, R E Doolittle & M M Brennan. 1975. Isolation, identification and synthesis of the sex pheromone of the tobacco budworm, *J. Chem. Ecol.*, **2**, 1535.

COMMERCIALISATION: Formulation: Sold as a combination of the two components in slow-release devices such as plastic tubes, laminated plastic or plastic film.
Tradenames: 'Isomate TBW' (Shin-Etsu).

APPLICATION: Applied throughout the field to be treated by placing the dispensers within the crop canopy.

PRODUCT SPECIFICATIONS: Manufactured for use in crop protection. **Purity:** The components of the product are >95% chemically pure.

COMPATIBILITY: The pheromone can be used with other crop protection agents that do not repel the moths.

MAMMALIAN TOXICITY: There are no reported cases of allergic or other adverse toxicological effects from the product by researchers, manufacturers, formulators or field workers. It is considered to be non-toxic. **Acute oral LD$_{50}$:** rats >5000 mg/kg (both components). **Inhalation:** LC$_{50}$: for rats >5 mg/litre (both components).

ENVIRONMENTAL IMPACT AND NON-TARGET TOXICITY: The pheromone occurs in Nature and is specific to *Heliothis virescens*. As such, it is not expected to have any adverse effects on non-target organisms or on the environment.
Approved for use in organic farming: Yes.

4. Semiochemicals

4:414 (Z)-hexadec-11-enyl acetate
European corn borer sex pheromone

The Pesticide Manual Fifteenth Edition entry number: 463

$$\underset{CH_3(CH_2)_3 \quad (CH_2)_{10}OCOCH_3}{\overset{\overset{\displaystyle H \qquad H}{\underset{\displaystyle C=C}{\diagdown \quad \diagup}}}{}}$$

NOMENCLATURE: Approved name: (Z)-hexadec-11-enyl acetate. **Other names:** Z11-16Ac; Z11HDA; European corn borer sex pheromone; corn stem borer sex pheromone. **CAS RN:** *[34010-21-4].*

SOURCE: Originally isolated from virgin female European corn borers (*Ostrinia nubilalis* (Hübner)).

PRODUCTION: Manufactured by SEDQ and Shin-Etsu for use in crop protection.

TARGET PESTS: European corn borer (*Ostrinia nubilalis*), pink stalk borer (*Sesamia nonagrioides* (Lefèbvre)) and other lepidopteran pests in maize. When used in combination with (Z)-hexadec-11-enal, it has been shown to be attractive to the diamondback moth (*Plutella xylostella* (L.)).

TARGET CROPS: Maize (corn).

BIOLOGICAL ACTIVITY: Mode of action: Male moths locate virgin female moths for mating by following a plume of the sex pheromone emitted by the females. The European corn borer sex pheromone is more usually used, however, as a means of monitoring the flight of adults so that insecticides can be applied at the most effective timing for the control of the larvae. **Efficacy:** The pheromone is very effective at attracting the European corn borer and related moths, thereby allowing the appropriate application of insecticide at the time of maximum movement of adult moths into the crop. **Key reference(s):** J A Klun & J F Robinson. 1972. Olfactory discrimination in the European corn borer and several pheromonally analogous moths, *Ann. Entomol. Soc. Am.*, **65**, 1337.

COMMERCIALISATION: Formulation: Sold as slow-release lures for trapping male moths. **Tradenames:** 'Fersex SN' (plus (Z)-hexadec-11-enol) (SEDQ), 'Checkmate DBM' (plus (Z)-hexadec-11-enal) (Suterra), 'Konaga-con' (plus (Z)-hexadec-11-enal) and 'Isomate DBM' (plus (Z)-hexadec-11-enal) (Shin-Etsu), 'Disrupt BAW' (plus (Z)-tetradec-9-enol plus (Z,E)-tetradeca-9,12-dienyl acetate) (Hercon), 'NoMate DBM' (plus (Z)-hexadec-11-enal) (Scentry).

APPLICATION: Traps are placed around the field and the numbers of male moths captured give an indication of the intensity of the population, allowing insecticide spray application to be made when, and if, necessary.

PRODUCT SPECIFICATIONS: Manufactured by SEDQ and Shin-Etsu for use in crop protection.

MAMMALIAN TOXICITY: There have been no reports of allergic or other adverse toxicological effects from the use, manufacture or formulation of pheromone traps containing (Z)-hexadec-11-enyl acetate. It is considered to be non-toxic.

ENVIRONMENTAL IMPACT AND NON-TARGET TOXICITY: It is not expected that (Z)-hexadec-11-enyl acetate will have any adverse effects on non-target organisms or on the environment.

4:415 (E)-hexadec-11-enyl acetate plus (E)-hexadec-11-en-1-ol
Brinjal shoot and fruit borer sex pheromone

The Pesticide Manual **Fifteenth Edition entry number:** 462

$$\underset{CH_3(CH_2)_3}{\overset{H}{\diagdown}}C=C\underset{(CH_2)_{10}OCOCH_3}{\overset{H}{\diagup}}$$

(E)-hexadec-11-enyl acetate

$$\underset{CH_3(CH_2)_3}{\overset{H}{\diagdown}}C=C\underset{(CH_2)_{10}OH}{\overset{H}{\diagup}}$$

(E)-hexadec-11-en-1-ol

NOMENCLATURE: Approved name: (E)-hexadec-11-enyl acetate plus (E)-hexadec-11-en-1-ol. Other names: Brinjal shoot and fruit borer sex pheromone; *Leucinodes orbonalis* sex pheromone; E11-16Ac plus E11-16OH.

SOURCE: Isolated from the pheromone glands of virgin female brinjal shoot and fruit borer (*Leucinodes orbonalis* Guenée).

PRODUCTION: Manufactured for use in crop protection.

TARGET PESTS: Brinjal shoot and fruit borer (*Leucinodes orbonalis*).

TARGET CROPS: Egg plant (aubergine or brinjal).

BIOLOGICAL ACTIVITY: Mode of action: Used in attract and kill strategies.
Key reference(s): 1) A Cork, S N Alam, A Das, C S Das, G C Ghosh, S Phythian, D I Farman,

D R Hall, N R Maslen, K Vedham, F M A Rouf & K Srinivasan. 2001. Female sex pheromone of Brinjal fruit and shoot borer, *Leucinodes orbonalis* (Lepidoptera: Pyralidae): Blend optimization, *J. Chem. Ecol.*, **27**, 1867–77. 2) A Cork, S N Alam, F M A Rouf & N S Talekar. 2003. Female sex pheromone of brinjal fruit and shoot borer, *Leucinodes orbonalis* (Lepidoptera: Pyralidae): trap optimization and application in IPM trials, *Bull. Entomol. Res.*, **93**, 107–13.

COMMERCIALISATION: Tradenames: 'Pheromone trap for Brinjal shoot and fruit borer' (ChemTica and Phero Asia).

APPLICATION: The traps are placed in the crop near the top of the canopy at the beginning of flowering at a rate of 100 per hectare at 10 metre intervals. Control is enhanced by the removal of infested shoots by hand. The lures are effective for 6 weeks.

PRODUCT SPECIFICATIONS: Manufactured for use in crop protection.

COMPATIBILITY: Pheromone traps should use a non-repellent insecticides such as deltamethrin or chlorpyrifos.

MAMMALIAN TOXICITY: Not considered to be toxic to mammals.

ENVIRONMENTAL IMPACT AND NON-TARGET TOXICITY: The blend (*E*)-11-hexadecenyl acetate plus (*E*)-11-hexadecen-1-ol is the sex pheromone of the brinjal shoot and fruit borer. It is not expected that it will have any adverse effects on non-target organisms or on the environment.

4:416 (Z,Z)-hexadeca-11,13-dien-1-al
Apple leaf miner sex pheromone

The Pesticide Manual Fifteenth Edition entry number: 461

NOMENCLATURE: Approved name: (*Z,Z*)-hexadeca-11,13-dien-1-al. **CAS RN:** *[35153-16-3]*.

SOURCE: The sex pheromone gland was located at the intersegmental membrane between the eighth and ninth abdominal segments of the female. (*Z,Z*)-hexadeca-11,13-dienal was originally isolated from this gland. (*E,Z*)-tetradeca-4,10-dienyl acetate was also isolated from the pheromone gland.

PRODUCTION: Manufactured for use in crop protection.

TARGET PESTS: Apple leaf miner moth (*Phyllonorycter ringoniella* (Matsumura)). (*Z,Z*)-hexadeca-11,13-dien-1-al is also the sex pheromone for the navel orangeworm (*Amyelois transitella* (Walker)) (entry 4:392).

TARGET CROPS: Apple orchards.

BIOLOGICAL ACTIVITY: Mode of action: Males locate females by following a plume of air rich in the odour of the pheromone emitted by females. Permeation of the canopy of an orchard with vapours of the pheromone (*Z,Z*)-hexadeca-11,13-dienal is used to disrupt location of females by males and decrease mating. **Efficacy:** Commercial formulations of this pheromone have been shown to be effective at maintaining populations of apple leaf miner moth at economically tolerable levels until harvest.

COMMERCIALISATION: Tradenames: 'Confuser A' (plus (*E,Z*)-tetradeca-4,10-dienyl acetate plus (*Z*)-icos-13-en-10-one plus dodec-8-enyl acetate plus tetradec-11-enyl acetate) (Shin-Etsu).

APPLICATION: Pheromone dispensers are attached to trees at heights recommended by the formulators. The recommended number of lures per hectare varies with release rates.

PRODUCT SPECIFICATIONS: Manufactured for use in crop protection. **Purity:** The pheromone products are >95% chemically pure.

COMPATIBILITY: Pheromone disruption of mating does not require insecticide, but can be used with application of insecticides and other crop protection agents.

MAMMALIAN TOXICITY: Apple leaf miner moth sex pheromone has shown no adverse toxicological effects on manufacturers, formulators or field workers. Considered to be non-toxic.

ENVIRONMENTAL IMPACT AND NON-TARGET TOXICITY: Apple leaf miner moth sex pheromone is a natural insect pheromone blend that is specific to the apple leaf miner moth. There is no evidence that any component has caused any adverse effects on non-target organisms or had an adverse environmental impact.

Approved for use in organic farming: Yes.

4. Semiochemicals

4:417 (Z,Z)-hexadeca-11,13-dien-1-al
Navel orangeworm sex pheromone

The Pesticide Manual Fifteenth Edition entry number: 461

$$CH_3CH_2 \quad \overset{H}{\underset{}{C}} = \overset{H}{\underset{}{C}} \quad \overset{}{\underset{H}{C}} = \overset{}{\underset{H}{C}} \quad (CH_2)_9CHO$$

NOMENCLATURE: **Approved name:** (Z,Z)-hexadeca-11,13-dien-1-al **CAS RN:** [7131-73-2]

SOURCE: Isolated from the pheromone gland of virgin female navel orangeworms (*Amyelois transitella* (Walker)). (Z,Z)-hexadeca-11,13-dien-1-al is also the sex pheromone for the apple leaf miner moth (*Phyllonorycter ringoniella* (Matsumura)).

TARGET PESTS: Navel orange worm (*Amyelois transitella*).

TARGET CROPS: Nut and fig plantations.

BIOLOGICAL ACTIVITY: **Mode of action:** (Z,Z)-hexadeca-11,13-dien-1-al is the sex pheromone of the navel orangeworm moth. Male moths locate and subsequently mate with female moths by following the pheromone trail or plume emitted by the virgin females. The application of high levels of the pheromone makes trail following impossible (camouflage, competition between artificially applied and the natural female plume, false trail following). Control is achieved by the prevention of mating and laying of fertile eggs. **Efficacy:** Very low rates are required to cause mating disruption. The pheromone is volatile and disperses through the orchard very easily. **Key reference(s):** C S Burks & D G Brandl. 2004. Seasonal abundance of the navel orangeworm, *Amyelois transitella*, in figs and the effect of peripheral aerosol dispensers on sexual communication, *J. Insect Sci.*, 2004, **4**, 40.

COMMERCIALISATION: **Formulation:** Sold as an aerosol or powder formulation. **Tradenames:** 'Puffer NOW' (Suterra), 'Paramount Aerosol NOW/CM' (plus codlemone) and 'Paramount Aerosol NOW/PTB' (plus (E)-dec-5-enyl acetate and (E)-dec-5-enol) (Suterra).

APPLICATION: The pheromone is distributed throughout the orchard.

PRODUCT SPECIFICATIONS: **Purity:** The chemical is >95% pure.

MAMMALIAN TOXICITY: There have been no reports of allergic or other adverse toxicological effects from the pheromone by researchers, manufacturers, formulators or field workers. Considered to be non-toxic. **Acute oral LD$_{50}$:** rats >2000–5000 mg/kg. **Inhalation:** LC$_{50}$ (4 h) rats 2–20 mg/litre. **Skin and eye:** Acute percutaneous LD$_{50}$ for rabbits >2000 mg/kg.

ENVIRONMENTAL IMPACT AND NON-TARGET TOXICITY: (Z,Z)-hexadeca-11,13-dien-1-al is a naturally occurring insect pheromone that is specific to the navel orangeworm moth. There is no evidence that it has caused any adverse effects on any non-target organisms or on the environment. **Approved for use in organic farming:** Yes.

4:418 (Z)-hexadec-13-en-11-ynyl acetate
Pine processionary moth sex pheromone

The Pesticide Manual **Fifteenth Edition entry number:** 461

$$\begin{array}{c} CH_2CH_3 \\ H-C \\ \diagdown C-C\equiv C-(CH_2)_{10}OCOCH_3 \\ H \end{array}$$

(Z)- isomer

NOMENCLATURE: Approved name: (Z)-hexadec-13-en-11-ynyl acetate. **Other names:** Pine processionary moth sex pheromone; Y11Z13-16Ac; Z13Y11-16Ac. **CAS RN:** *[78617-58-0]* (Z)-hexadec-13-en-11-ynyl acetate.

SOURCE: Isolated from the abdomens of virgin female pine processionary moths (*Thaumetopoea pityocampa* (Denis and Schiffermüller)).

PRODUCTION: Manufactured for use in forestry management.

TARGET PESTS: Pine processionary moths (*Thaumetopoea pityocampa*).

TARGET CROPS: Pine forests.

BIOLOGICAL ACTIVITY: Mode of action: The presence of high levels of the processionary moth is monitored by the use of traps baited with the sex pheromone. These traps emit a plume of the volatile components, attracting males in search of virgin females for mating. The number of male moths captured is an indication of the size of the moth population and indicates when control measures should be taken. In addition, the pheromone can be used as a mass trapping system within the forest, thereby lowering the population of adults and the numbers of viable eggs laid. **Efficacy:** The attractants are very effective at capturing males. **Key reference(s):** J Einhorn, P Menassieu, D Michelot & J Riom. 1983. Piégeage sexuel de la processionaire du pin, *Thaumetopoea pityocampa* Schiff. par des attractifs de synthèse. Premiers essais dans le Sud-Ouest de la France, *Agronomie*, **3**, 499.

4. Semiochemicals

COMMERCIALISATION: **Formulation:** Sold as slow-release lures for trapping the male moths and as monitoring dispensers. **Tradenames:** 'Fersex TP' and 'Pityolure' (SEDQ).

APPLICATION: The traps are placed within the pine plantation, at chest height. A rate of one dispenser per hectare is recommended.

PRODUCT SPECIFICATIONS: Manufactured for use in forestry management. **Purity:** The components are >95% chemically pure.

MAMMALIAN TOXICITY: There is no record of allergic or other adverse toxicological effects from researchers, manufacturers, formulators or field workers from the pine processionary moth sex pheromone. It is considered to be non-toxic. **Acute oral LD$_{50}$:** rats >5000 mg/kg. **Acute dermal LD$_{50}$:** rats >2000 mg/kg. **Skin and eye:** Acute percutaneous LD$_{50}$ for rats >2000 mg/kg. It is not a skin or eye irritant.

ENVIRONMENTAL IMPACT AND NON-TARGET TOXICITY: The product is the sex pheromone of the pine processionary moth and it is specific to that species. It is not expected that it will have any adverse effects on non-target organisms or on the environment.

4:419 (Z)-icos-13-en-10-one
Peach fruit moth sex pheromone

The Pesticide Manual **Fifteenth Edition entry number:** 474

$$CH_3(CH_2)_8CO(CH_2)_2\underset{}{\overset{H\ \ \ \ H}{C=C}}(CH_2)_5CH_3$$

NOMENCLATURE: **Approved name:** (Z)-icos-13-en-10-one. **Other names:** *Carposina niponensis* sex pheromone; peach fruit moth sex pheromone; Z7-20-11Kt; Z13-20-10Kt. **CAS RN:** [63408-44-6].

SOURCE: Female *Carposina niponensis* Walsingham emit several volatile components shown to be attractive to male moths. The major sex attractant, however, is (Z)-icos-13-en-10-one.

PRODUCTION: Manufactured for use in crop protection.

TARGET PESTS: Peach fruit moth (*Carposina niponensis*).

TARGET CROPS: Orchard tree crops such as peaches, apricots and apples.

BIOLOGICAL ACTIVITY: Mode of action: Males locate females by following a plume of air rich in the odour of a complete pheromone blend emitted by females. Permeation of the canopy of an orchard with vapours of the single component of the pheromone (*Z*)-icos-13-en-10-one is used to disrupt location of females by males and decreases mating.

Efficacy: Several commercial formulations of this pheromone have been shown to be effective at maintaining populations of peach fruit moth below economic threshold levels until harvest.

Key reference(s): K Honma, K Kawasaki & Y Tamaki. 1978. Sex pheromone of the peach fruit moth, *Carposina niponensis* Walsingham, *Jpn. J. Appl. Entomol. Zool.*, **22**, 87.

COMMERCIALISATION: Formulation: Sold as a variety of slow-release formulations in different dispensing devices. Manufactured by Shin-Etsu and Shinkui-con. **Tradenames:** 'Isomate PFM', 'Shinkui-con', 'Confuser A' (plus dodec-8-enyl acetate plus tetradec-11-enyl acetate plus (*E,Z*)-tetradeca-4,10-dienyl acetate with (*Z*)-tetradec-10-enyl acetate), 'Confuser P' (plus 14-methyloctadecene plus (*Z*)-dodec-8-enyl acetate plus tetradec-11-enyl acetate), 'Confuser Z' (plus (*Z*)-tetradec-9-enyl acetate plus dodec-8-enyl acetate plus (*Z*)-dodec-9-enyl acetate plus (*Z*)-tetradec-11-enyl acetate) 'Confuser N' and 'Confuser R' (plus (*Z*)-tetradec-9-enyl acetate plus (*Z*)-dodec-8-enyl acetate plus tetradec-11-enyl acetate) (Shin-Etsu).

APPLICATION: Pheromone dispensers are attached to trees at heights recommended by the formulators. The number of lures used per hectare varies with release rates.

PRODUCT SPECIFICATIONS: Manufactured for use in crop protection. **Purity:** The pheromone products are >95% chemically pure.

COMPATIBILITY: Pheromone disruption of mating does not require insecticide, but can be used with application of insecticides and other crop protection agents.

MAMMALIAN TOXICITY: Peach fruit moth sex pheromone has shown no adverse toxicological effects on manufacturers, formulators or field workers. Considered to be non-toxic.

ENVIRONMENTAL IMPACT AND NON-TARGET TOXICITY: Peach fruit moth sex pheromone is a natural insect pheromone that is specific to the peach fruit moth. There is no evidence that (*Z*)-icos-13-en-10-one has caused any adverse effects on non-target organisms or had an adverse environmental impact. **Approved for use in organic farming:** Yes.

4. Semiochemicals

4:420 ipsdienol plus ipsenol plus (Z)-verbenol plus (E)-verbenol

Spruce bark beetle aggregation pheromone-2

The Pesticide Manual **Fifteenth Edition entry number:** 500

ipsdienol　　　　　　　ipsenol　　　　　　　verbenol

NOMENCLATURE: Approved name: ipsdienol ((+)-2-methyl-6-methylene-2,7-octadien-4-ol) plus ipsenol ((−)-2-methyl-6-methylene-7-octen-4-ol) plus (Z)-verbenol ((Z)-(+)-2,6,6-trimethylbicyclo[3.1.1]hept-2-en-4-ol) plus (E)-verbenol ((E)-(+)-2,6,6-trimethylbicyclo[3.1.1]hept-2-en-4-ol). **Other names:** Spruce bark beetle aggregation pheromone. **CAS RN:** *[35628-00-3]* ipsdienol; *[35628-05-8]* ipsenol; *[473-67-6]* (E)-verbenol.

SOURCE: All four compounds have been isolated from the hindgut of male *Ips typographus* (L.) beetles and are behaviourally active components of the aggregation pheromones.

PRODUCTION: Manufactured for use in forestry management strategies.

TARGET PESTS: Spruce bark beetle (*Ips typographus*).

TARGET CROPS: Spruce forests.

BIOLOGICAL ACTIVITY: Mode of action: The spruce bark beetle aggregation pheromone is vital for the insects to locate a suitable food source. The male and female adults follow a trail of the pheromone to locate suitable host trees. The release of the volatiles from a lure attached to baited trees concentrates the beetles in the baited trees and reduces their population in non-baited areas by establishing conditions for mass attack of the baited trees. More recently, the forests have been protected by the mass trapping of the beetles in bucket traps. However, in Sweden, there has been and continues to be much debate about whether mass trapping works for this species. High capacity pheromone traps are also now used to catch emerging insects from infested plantations to reduce the possibility of them successfully attacking neighbouring trees. However, catch is not necessarily higher than 70%, so whether this will be a successful strategy remains to be seen. Experiments are ongoing to test out combinations of repellents and attractants in order to develop a push-pull strategy for control of *Ips typographus*. **Efficacy:** At a rate of four baited trees per hectare, tree baits have been very successful in reducing insect populations and subsequent tree damage in the non-baited areas. Both this combination of components and ipsdienol plus (*E*)-verbenol plus α-pinene (entry 4:) are effective aggregation pheromones for *Ips typographus*.

Key reference(s): A Bakke. 1976. Spruce bark beetle, *Ips typographus*: pheromone production and field response to synthetic pheromones, *Naturwissenschaften*, **63**, 92.

COMMERCIALISATION: Formulation: Sold as a mixture of the volatile components in a slow-release device such as a polymer plug or a plastic bag. **Tradenames:** 'Spruce Bark Beetle Aggregation Pheromone' (PheroTech).

APPLICATION: The baits are strapped to trees, at chest height, about 50 metres apart, in the area where the beetles are to be concentrated. Baited trees and surrounding trees are felled and removed before the brood emerges.

PRODUCT SPECIFICATIONS: Manufactured for use in forestry management strategies. **Purity:** All components of the lures are >95% chemically pure.

MAMMALIAN TOXICITY: There have been no reports of allergic or other adverse toxicological effects from the volatiles on researchers, manufacturers, formulators or field workers. The product is considered to be non-toxic.

ENVIRONMENTAL IMPACT AND NON-TARGET TOXICITY: All components of the spruce bark beetle aggregation pheromone occur in Nature. It is not expected that they will have any adverse effects on non-target organisms or on the environment.

4. Semiochemicals

4:421 ipsdienol plus (*E*)-verbenol plus α-pinene

Spruce bark beetle aggregation pheromone-1

The Pesticide Manual Fifteenth Edition entry number: 500

CH₂ / C / CH₂ CH / H–C CH₂ / HO''''' CH / C / CH₃ CH₃

CH₃CH₃ / CH₃ / HO H

CH₃CH₃ / CH₃

ipsdienol   cis-verbenol   α-pinene

NOMENCLATURE: **Approved name:** ipsdienol ((+)-2-methyl-6-methylene-2,7-octadien-4-ol) plus (*E*)-verbenol ((*E*)-(+)-2,6,6-trimethylbicyclo[3.1.1]hept-2-en-4-ol) plus α-pinene. **Other names:** Spruce bark beetle aggregation pheromone. **CAS RN:** *[35628-00-3]* ipsdienol; *[473-67-6]* (*E*)-verbenol; *[80-56-8]* α-pinene.

SOURCE: The components of the aggregation pheromone were isolated from spruce bark beetle (or European spruce bark beetle) (*Ips typographus* (L.)) adults. Isolated from dissected hindguts of tree-reared beetles.

PRODUCTION: Manufactured for use in forest control strategies.

TARGET PESTS: The eight-spined spruce bark beetle (*Ips typographus*).

TARGET CROPS: Coniferous forests.

BIOLOGICAL ACTIVITY: **Mode of action:** *Ips typographus* locates new trees by following the trail of the species-specific aggregation pheromone, often with some host tree volatiles present as well. These aggregation pheromones lead the adult beetles to new food sources and to other adults so that mating can take place. Because of the copious amounts of resin produced by these trees, the beetles have to 'mass attack' them in order to overcome this defence mechanism. Pioneer beetles that begin galleries and start the pheromone trail identify the weaker trees that are under stress. Typically, older trees on the edges of newly cut plantations are most susceptible. If these pheromones are attached to specific trees within a forest, then beetles are attracted to these baited trees and the invasion of the majority of the forest is prevented. The baited trees and those surrounding the baited trees are felled before the brood emerges, thereby reducing the beetle population and protecting most of the trees. More recently, the forests have been protected by the mass trapping of the beetles in bucket traps. However, in Sweden, there has been and continues to be much debate about whether mass trapping works for this species. High capacity pheromone traps are also now used to catch emerging insects from infested plantations to reduce the possibility of them successfully attacking neighbouring trees. However, catch is not necessarily higher than 70%, so whether this will be a successful strategy remains to be seen. Experiments are on-going to test out combinations of repellents and attractants in order to develop a push-pull strategy for control of *Ips typographus*. **Efficacy:** It has been shown that the baiting of trees significantly reduces the invasion of the other forest trees. Both this combination of components and ipsdienol plus ipsenol plus (*Z*)-verbenol plus (*E*)-verbenol (entry 4:) are effective aggregation pheromones for *Ips typographus*. **Key reference(s):** C Chararas. 1980. Attraction primaire et secondaire chez trois espèces de scolytidae (*Ips*) et mécanisme de colonisation, *C. R. Acad. Sci. (Paris)*, **290**, 375.

COMMERCIALISATION: Formulation: Sold as slow-release dispensers containing the pheromone. **Tradenames:** 'Pheropax' (BASF).

APPLICATION: The dispensers are attached to bait trees within the forest, at chest height. It is usual to bait four trees per hectare. These baited trees and surrounding trees are felled before the brood emerges.

PRODUCT SPECIFICATIONS: Manufactured for use in forest control strategies. **Purity:** The biologically active compounds in the product are >95% chemically pure.

COMPATIBILITY: It is unusual to use the pheromone with other crop protection agents.

MAMMALIAN TOXICITY: There are no reports of allergic or other adverse toxicological effects from the products by researchers, manufacturers, formulators or field workers. Considered to be non-toxic.

ENVIRONMENTAL IMPACT AND NON-TARGET TOXICITY: *Ips typographus* aggregation pheromone is a naturally occurring mixture of volatile components that have a specific effect on *Ips typographus*. It is not expected that the use of the product will have any adverse effects on non-target organisms or on the environment.

4:422 lineatin

Ambrosia beetle aggregation pheromone

The Pesticide Manual Fifteenth Edition entry number: 523

NOMENCLATURE: **Approved name:** lineatin ((1*R*,4*S*,5*R*,7*R*)-(+)-3,3,7-trimethyl-2,9-dioxa-tricyclo[3.3.1.0^{4,7}]nonane). **Other names:** Ambrosia beetle aggregation pheromone. **CAS RN:** *[71899-16-6]*.

SOURCE: Originally extracted from the frass of the ambrosia beetle (*Trypodendron lineatum* (Olivier)).

PRODUCTION: Manufactured for use in forestry management.

TARGET PESTS: Ambrosia beetle (*Trypodendron lineatum*).

TARGET CROPS: Harvested trees in pine and fir forests.

BIOLOGICAL ACTIVITY: **Mode of action:** Lineatin is the aggregation pheromone of this ambrosia beetle. It attracts *Trypodendron lineatum* to traps. Male and female beetles locate suitable harvested host trees by following a plume of air enriched with the odour of harvested trees and containing the pheromone. Evaporation of the aggregation pheromone and vapours, from host trees from lures attached to traps, attracts beetles to traps and diverts attack from harvested timber in the vicinity. The mixture of lineatin, sulcatol, *S*-(+)-sulcatol and α-pinene is also used to attract other species of ambrosia beetle. **Efficacy:** The use of this lure has been shown to reduce attacks by the beetles on harvested timber in commercial forests. **Key reference(s):** T L Shore & J A McLean. 1983. A further evaluation of the interactions between the pheromones and two host kairomones of the ambrosia beetles, *Trypodendron lineatum* and *Gnathotrichus sulcatus*, *Can. Entomol.*, **115**, 1.

COMMERCIALISATION: **Formulation:** Sold as a slow-release formulation of lineatin as polymer plugs or from individual plastic bags and vials with plastic lids. **Tradenames:** 'Linoprax' (Biosystemes) and (BASF).

APPLICATION: Pheromone dispensers are attached to barrier traps, at chest height, 50 metres apart, along the perimeter of areas containing stacked harvested pine and fir.

PRODUCT SPECIFICATIONS: Manufactured for use in forestry management. **Purity:** The lineatin in the product is >95% chemically pure.

COMPATIBILITY: It is unusual to use the product in combination with other crop protection chemicals.

MAMMALIAN TOXICITY: There are no reports of allergic or other adverse toxicological effects from lineatin by researchers, manufacturers, formulators or field workers. It is considered to be non-toxic.

ENVIRONMENTAL IMPACT AND NON-TARGET TOXICITY: Lineatin is a naturally occurring insect aggregation pheromone that is specific to ambrosia beetles. It is not expected that it will have any adverse effects on non-target organisms or on the environment.

4:423 lineatin plus sulcatol plus S-(+)-sulcatol plus α-pinene plus ethanol
Ambrosia beetle aggregation pheromone and host kairomones

The Pesticide Manual Fifteenth Edition entry number: 523

lineatin   (S)-sulcatol   α-pinene

NOMENCLATURE: Approved name: lineatin ((1R,4S,5R,7R)-(+)-3,3,7-trimethyl-2,9-dioxatricyclo[3.3.1.0^{4,7}]nonane) plus sulcatol ((±)-5-methylhepten-2-ol) plus S-(+)-sulcatol (S-(+)-5-methylhepten-2-ol) plus α-pinene plus ethanol. **Other names:** Ambrosia beetle aggregation pheromone and host kairomones. **CAS RN:** *[71899-16-6]* lineatin; *[1569-60-4]* sulcatol; *[4630-06-2]* (+/-)-sulcatol; *[58917-26-3]* (S)-(+)-sulcatol; *[80-56-8]* α-pinene.

SOURCE: Male *Trypodendron lineatum* (Olivier), *Gnathotrichus sulcatus* (LeConte) and *G. retusus* (LeConte) volatile components have been shown to attract male and female adult beetles.

PRODUCTION: Manufactured for use in forestry management.

TARGET PESTS: Ambrosia beetles (*Trypodendron lineatum*, *Gnathotrichus sulcatus* and *G. retusus*).

TARGET CROPS: Harvested trees in pine and fir forests.

BIOLOGICAL ACTIVITY: Mode of action: These components are the aggregation pheromones of ambrosia beetles. Lineatin, ethanol and α-pinene are used together to attract *Trypodendron lineatum* to traps; sulcatol, ethanol and α-pinene are used to attract *Gnathotricus sulcatus* to traps; and S-(+)-sulcatol, ethanol and α-pinene are used to attract *G. retusus* to traps. Male and female beetles of each species locate suitable harvested host trees by following a plume of air enriched with the odour of harvested trees and containing the pheromone. Evaporation of the aggregation pheromone and vapours from host trees from lures attached to traps, attracts beetles to traps and diverts attack from harvested timber in the vicinity. **Efficacy:** The use of these lures has been shown to reduce attacks by the beetles on harvested timber in commercial forests. **Key reference(s):** J H Borden, L Chong, K N Slessor, A C Oehlschlager, H D Pierce Jr & S Lindgren. 1981. Allelochemical activity of aggregation pheromones between three sympatric species of ambrosia beetles, *Can. Entomol.*, **113**, 557.

COMMERCIALISATION: Formulation: Sold as a slow-release formulation of lineatin, sulcatol, S-(+)-sulcatol, ethanol and α-pinene as polymer plugs or from individual plastic bags and vials with plastic lids. **Tradenames:** 'Ambrosia Beetle Trap Lures' (PheroTech) and (ChemTica), 'T. lineatum Lure' (Boehringer).

APPLICATION: Pheromone dispensers are attached to barrier traps, at chest height, 50 metres apart, along the perimeter of areas containing stacked harvested pine and fir.

PRODUCT SPECIFICATIONS: Manufactured for use in forestry management. **Purity:** The insect pheromone components are >95% chemically pure.

COMPATIBILITY: Pheromone-baited trap trees do not require insecticide to kill the beetles as they arrive. The same traps cannot be used to attract both *Gnathotrichus sulcatus* and *G. retusus*, because of pheromone inhibition. Different traps must be baited with either sulcatol or S-(+)-sulcatol.

MAMMALIAN TOXICITY: Ambrosia beetle lures have shown no allergic or other adverse toxicological effects on manufacturers, formulators or field workers.

ENVIRONMENTAL IMPACT AND NON-TARGET TOXICITY: Ambrosia beetle lures contain natural insect pheromones and host volatile components. There is no evidence that their use has had any adverse effects on non-target organisms or on the environment.

4:424 3-methylcyclohex-2-en-1-one
Beetle repellent

The Pesticide Manual **Fifteenth Edition entry number:** 576

NOMENCLATURE: Approved name: 3-methylcyclohex-2-en-1-one. **Other names:** MCH; methylcyclohexenone.

SOURCE: 3-Methylcyclohex-2-en-1-one is produced by beetles when the population on a particular tree becomes critical for reproduction and survival. This compound repels additional beetles from the tree.

TARGET PESTS: Douglas fir and spruce beetles.

TARGET CROPS: Douglas fir (*Pseudotsuga menziesii* Franco) and spruce (*Picea abies* Karst.) trees.

BIOLOGICAL ACTIVITY: Mode of action: Dead and dying Douglas fir and spruce trees release the chemical seudenol (3-methylcyclohex-2-en-1-ol) that is used by the beetles to locate the trees (it is a kairomone). As the insects gather on dead trees in large numbers, they are stimulated to reproduce. When the number of beetles at a tree reaches a critical density, the beetles produce 3-methylcyclohex-2-en-1-one, a pheromone that repels additional beetles and, thereby, protects the food supply needed by the initial beetles and their offspring. The beetles modify the seudenol to produce 3-methylcyclohex-2-en-1-one, changing the attractant kairomone into the repellent semiochemical.
Key reference(s): 1) D W Ross, G E Daterman & K E Gibson. 2002. Elution rate and spacing of anti-aggregation pheromone dispensers for protecting live trees from *Dendroctonus pseudotsugae* (Coleoptera: Scolytidae), *J. Econ. Entomol.*, **95**(4), 778–81. 2) E H Holsten, P J Shea & R R Borys. 2003. MCH released in a novel pheromone dispenser prevents spruce beetle, *Dendroctonus rufipennis* (Coleoptera: Scolytidae), attacks in south-central Alaska, *J. Econ. Entomol.*, **96**(1), 31–4.

COMMERCIALISATION: Formulation: 3-Methylcyclohex-2-en-1-one is contained in polyethylene, 'bubble-capsule', slow-release containers. **Tradenames:** 'MCH' (PheroTech).

APPLICATION: The polyethylene slow-release containers are stapled to dead or dying trees, 2 to 4 metres (6 to 12 feet) above the ground. The number of containers used per tree and frequency of application depend on the level of beetle infestation.

4. Semiochemicals

MAMMALIAN TOXICITY: Use of 3-methylcyclohex-2-en-1-one in polyethylene containers is not expected to harm humans. Toxicity tests show that 3-methylcyclohex-2-en-1-one has very low toxicity. Exposure of humans should be minimal to non-existent, based on the approved use of 3-methylcyclohex-2-en-1-one only in containers that are placed on trees in forests, 2 to 4 metres above the ground. 3-Methylcyclohex-2-en-1-one has been used as a beetle repellent for more than 20 years, with no reports of adverse effects.

ENVIRONMENTAL IMPACT AND NON-TARGET TOXICITY: Risks to non-target species are expected to be minimal because 3-methylcyclohex-2-en-1-one shows no adverse effects or very minor effects on non-target species, and exposure of non-target species is not expected to occur to any large extent because of the specific, localised method of application. Furthermore, the amount of 3-methylcyclohex-2-en-1-one released from the product is less than would be released naturally from heavily infested trees. No adverse effects have been reported during more than 20 years of use.

4:425 14-methyloctadec-1-ene
Peach miner sex pheromone

The Pesticide Manual **Fifteenth Edition entry number:** 583

$$CH_3(CH_2)_3 \diagdown_{CH} \diagup (CH_2)_{11}CH=CH_2$$
$$| $$
$$CH_3$$

NOMENCLATURE: Approved name: (S)-14-methyloctadec-1-ene. **Other names:** Peach miner sex pheromone; *Lyonetia clerkella* sex pheromone. **CAS RN:** [93091-95-3] racemate, formerly [97424-51-6].

SOURCE: The sex pheromone of the peach leaf miner (*Lyonetia clerkella* (L.)) was isolated from virgin female terminal segments.

PRODUCTION: Manufactured for use in crop protection.

TARGET PESTS: Peach leaf miner (*Lyonetia clerkella*) (known as apple leaf miner in Japan).

TARGET CROPS: Peach orchards.

BIOLOGICAL ACTIVITY: Mode of action: Males locate females by following a plume of air rich in the odour of the pheromone emitted by females. Permeation of the canopy of an orchard with vapours of the pheromone (sold as the racemate 14-methyloctadec-1-ene) is used to disrupt location of females by males and decrease mating. **Efficacy:** Several commercial formulations of this pheromone have been shown to be effective at maintaining populations of peach miner moth at economically tolerable levels until harvest.

Key reference(s): 1) H Sugie, Y Tamaki, R Sato, F Jizaka & P E Sonnet. 1986. Biological activity of the stereoisomers of the sex pheromone of the peach leafminer moth, *Lyonetia clerkella* (L.), *J. Appl. Entomol. Zool. (Japan)*, **20**, 411–5. 2) R Sato, N Abe, H Sugie, M Kato, K Mori & Y Tamaki. 1986. Biological activity of the chiral sex pheromone of the peach leafminer moth, *Lyonetia clerkella* Linné (Lepidoptera: Lyonetiidae), *Appl. Entomol. Zool.*, **21**, 478–80.

COMMERCIALISATION: Formulation: Sold as a variety of slow-release formulations in different dispensing devices. **Tradenames:** 'Peach Leafminer Pheromone' and 'Confuser P' (plus (*Z*)-dodec-8-enyl acetate plus tetradec-11-enyl acetate plus (*Z*)-icos-13-en-10-one) (Shin-Etsu).

APPLICATION: Pheromone dispensers are attached to trees at heights recommended by the formulators. The recommended number of lures per hectare varies with release rates.

PRODUCT SPECIFICATIONS: Manufactured for use in crop protection. **Purity:** The pheromone products are >95% chemically pure.

COMPATIBILITY: Pheromone disruption of mating does not require insecticide, but can be used with application of insecticides and other crop protection agents.

MAMMALIAN TOXICITY: Peach miner moth sex pheromone has shown no adverse toxicological effects on manufacturers, formulators or field workers. Considered to be non-toxic. **Acute oral LD_{50}:** rats and mice 4000 mg/kg.

ENVIRONMENTAL IMPACT AND NON-TARGET TOXICITY: Peach miner moth sex pheromone is a natural insect pheromone that is specific to the peach miner moth. There is no evidence that 14-methyloctadec-1-ene has caused any adverse effects on non-target organisms or had an adverse environmental impact. **Fish toxicity:** LC_{50} (96 h) for carp >10 ppm. **Other aquatic toxicity:** LC_{50} (48 h) for *Daphnia* >10 ppm.
Approved for use in organic farming: Yes.

4. Semiochemicals

The Pesticide Manual **Fifteenth Edition entry number:** 601

$$CH_3(CH_2)_{12} \quad (CH_2)_7CH_3$$
$$\underset{H}{C} = \underset{H}{C}$$

NOMENCLATURE: Approved name: muscalure ((Z)-9-tricosene). **CAS RN:** *[27519-02-4]* (Z)- isomer; *[35857-62-6]* (E)- isomer.

SOURCE: The sex pheromone of the housefly was isolated from female flies in 1971 by Carlson *et al.*

PRODUCTION: The product is now manufactured for use.

TARGET PESTS: House flies (*Musca domestica* L.).

TARGET CROPS: Used in animal houses.

BIOLOGICAL ACTIVITY: Mode of action: The product is the major component of the sex pheromone of the house fly. Male flies locate female flies by following the trail or pheromone plume emitted by the virgin females and subsequently mate. The indiscriminate application of high levels of the product interferes with this process, as a constant exposure to high levels of pheromone makes trail following impossible (habituation/adaptation). Muscalure is used primarily as an attractant in attract and kill strategies where males are attracted to the bait, which is spiked with a rapid-knockdown insecticide. Yellow sticky fly papers are also used in control strategies to increase the catch. The removal of adult males from the animal houses reduces the level of female fertilisation and hence the incidence of fly egg laying.
Key reference(s): D A Carlson, M S Mayer, D L Silhacek, J D James, M Beroza & B A Beirl. 1971. Sex attractant pheromone from the housefly: isolation, identification and synthesis, *Science,* **174,** 76–8.

COMMERCIALISATION: Tradenames: 'Muscamone' (Cyclo), 'Flybait' (plus methomyl) (Denka), 'Goldben' (plus methomyl) (public health) (Vapco), 'Golden Malrin Fly Bait' (plus methomyl) (Syngenta A H), 'SFB' (plus cypermethrin) (Trithin), 'Stimukil' (plus methomyl) (Troy Biosciences).

PRODUCT SPECIFICATIONS: The product is now manufactured for use. **Purity:** Primarily the (Z)- isomer, with the (E)- isomer present as a minor component.

MAMMALIAN TOXICITY: Generally considered to be non-toxic. **Acute oral LD$_{50}$:** rats >10000 mg/kg. **Inhalation:** LC$_{50}$ (4 h) for rats >5.71 g/m^3 (technical, containing 85% muscalure). **Skin and eye:** Acute percutaneous LD$_{50}$ for rabbits >2000 mg/kg. Not irritating to eyes or skin of rabbits. Moderate skin sensitiser in guinea pigs. **Toxicity class:** WHO (active ingredient) III; EPA (formulation) IV ('Flybait'). **Other toxicological effects:** Not mutagenic in the Ames test; not teratogenic to rats at >5 g/kg.

ENVIRONMENTAL IMPACT AND NON-TARGET TOXICITY: At the rates of use employed, it is considered unlikely that muscalure will have any adverse effects on non-target organisms or on the environment. **Bird toxicity:** Acute oral LD_{50} for mallard ducks >4640 mg/kg. LC_{50} for bobwhite quail >4640 mg/kg. NOEL in one-generation reproductive study for bobwhite quail >20 ppm, for mallard ducks 0.1 ppm; reproduction hazard at 2 ppm. **Fish toxicity:** LC_{50} (96 h) for rainbow trout and bluegill sunfish >1000 mg/litre (not toxic within water solubility). **Other aquatic toxicity:** LC_{50} (48 h) for *Daphnia* 265.7 µg/litre.

4:427 (2S,7S)-nonadiyl dibutyrate
Wheat orange blossom midge sex pheromone

NOMENCLATURE: Approved name: (2S,7S)-nonadiyl dibutyrate.

SOURCE: Isolated from the pheromone gland of virgin female wheat orange blossom midges.

PRODUCTION: Manufactured for use as the racemate.

TARGET PESTS: The wheat orange blossom midge (*Sitodiplosis mosellana* (Gehin)).

TARGET CROPS: Wheat, barley and rye.

BIOLOGICAL ACTIVITY: Mode of action: (2S,7S)-Nonadiyl dibutyrate is the female sex pheromone of the wheat orange blossom midge. The racemate has been shown to be as effective as the SS-enantiomer and is used commercially. Males are attracted to lures containing the pheromone where they are trapped. Counting the number of males caught each night allows a decision to be made on whether or not to spray the crop with an insecticide to protect the crop. **Efficacy:** The numbers of *Sitodiplosis mosellana* caught in the pheromone traps give a reliable indication of the onset of flight activity in the vicinity of wheat fields and have a strong correlation with the severity of subsequent crop infestation levels. The high sensitivity of the traps outweighs the fact that pheromone traps catch male rather than female insects. Pheromone traps can be used to give a warning to farmers of when midge levels are expected to be damagingly high. Insecticide should be applied soon after the detection of high levels of midge, because insecticide only targets adults and larvae before they reach the grain. **Key reference(s):** 1) T J A Bruce, M Hooper, L Ireland, O T Jones, J L Martin, L E Smart, J Oakley & L J Wadhams. 2007. Development of a pheromone trap monitoring system for orange wheat blossom midge, *Sitodiplosis mosellana*, in the UK, *Pest Manag. Sci.*, **63**, 49–56. 2) R Gries, G Gries, G Khaskin, S King, O Olfert, L A Kaminski, *et al.* 2000. Sex pheromone of orange wheat blossom midge, *Sitodiplosis mosellana*, *Naturwissenschaften*, **87**, 450–4.

COMMERCIALISATION: Formulation: Sold as a trap that releases the pheromone to attract the males. Estimates of the size of the invading population of midges can then be calculated. **Tradenames:** 'Wheat Blossom Midge Monitoring Kit' (Suterra).

APPLICATION: The traps should be positioned a week before the first ears emerge. This should be at GS45, when the boots are swollen. Traps should then be left in position until the latest crop in the vicinity reaches flowering (GS61). Traps should be attached to stakes at crop height. A minimum of two traps should be used in each field being monitored. The lures have a field life of at least six weeks and do not need to be changed during the monitoring season.

PRODUCT SPECIFICATIONS: Manufactured for use as the racemate.

MAMMALIAN TOXICITY: The product has shown no adverse toxicological effects on research or manufacturing workers or on users.

ENVIRONMENTAL IMPACT AND NON-TARGET TOXICITY: There is no evidence that (2S,7S)-nonadiyl dibutyrate has caused significant adverse effects on non-target organisms nor had an adverse environmental impact.

4:428 (E,Z)-2,13-octadecadienal plus (E,Z)-2,13-octadecadienol

Webbing clothes moth sex pheromone

(E,Z)-2,13-octadecadienal

+

(E,Z)-2,13-octadecadienol

NOMENCLATURE: Approved name: (E,Z)-2,13-octadecadienal plus (E,Z)-2,13-octadecadienol. **Other names:** Webbing clothes moth sex pheromone.

SOURCE: Isolated from the pheromone gland of virgin female webbing clothes moths (*Tineola bisselliella* Hummel).

PRODUCTION: Manufactured for use commercially.

TARGET PESTS: Webbing clothes moth (*Tineola bisselliella*)

TARGET CROPS: Used in fabric stores.

BIOLOGICAL ACTIVITY: Mode of action: (E,Z)-2,13-Octadecadienal plus (E,Z)-2,13-octadecadienol are the sex pheromones of the webbing clothes moth. Male moths locate and subsequently mate with female moths by following the trail or pheromone plume emitted by the virgin females. The indiscriminate application of high levels of the product interferes with this process, as a constant exposure to high levels of pheromone makes trail following impossible (habituation/adaptation). Alternatively, the use of discrete sources of pheromone released over time presents the male with a false trail to follow (confusion). Control is subsequently achieved through the prevention of mating and the laying of fertile

4. Semiochemicals

eggs. The pheromone blend is also used within a trap that allows the population levels to be determined. **Efficacy:** Very low rates are required to cause mating disruption. (E,Z)-2,13-Octadecadienal plus (E,Z)-2,13-octadecadienol are volatile and distribute throughout the area to be treated easily. They act as attractants and by disruption of mating in the disorientation mode. **Key reference(s):** 1) P Trematerra & F Fontana. 1996. Monitoring of webbing clothes moth, *Tineola bisselliella* (Hummel), by sex pheromone. *J. Pest Sci.*, **69**(5), 119–21. 2) S Takacs, G Gries & R Gries. 2001. Communication ecology of webbing clothes moth: 4. Identification of male- and female-produced pheromones, *Chemoecology*, **11**, 153–9.

COMMERCIALISATION: Formulation: 'Exosex CLTab' is a pesticide-free, pheromone-based monitoring trap enhancer for clothes moth. Entostat powder, a natural food grade product, is combined with the pheromone specific to the female clothes moth to create an intelligent and pesticide free system for use on clothes moth. **Tradenames:** 'Exosex CLTab' (Exosect).

APPLICATION: 'Exosex CLTab' should be placed in a 5m grid pattern throughout the affected area. It contains Entostat powder formulated with synthetic female pheromone. The increased presence of pheromone attracts male moths into the open. Male moths are attracted to either a monitoring trap, where they provide an idea of the overall infestation level or to 'Exosex CLTab', where they pick up the powder and female pheromone. A male carrying Entostat powder will form a mobile pheromone dispenser, producing 'false' pheromone trails, which additionally attracts males into the open. Contact between the males ensures that Entostat powder and the mating confusion effect is continued.

PRODUCT SPECIFICATIONS: Manufactured for use commercially. **Purity:** The insect pheromone components are >95% chemically pure.

COMPATIBILITY: It is not recommended that 'Exosex CLTab' be used with any other product.

MAMMALIAN TOXICITY: The product has shown no adverse toxicological effects on research or manufacturing workers or on users.

ENVIRONMENTAL IMPACT AND NON-TARGET TOXICITY: There is no evidence that 'Exosex CLTab' has caused significant adverse effects on non-target organisms nor had an adverse environmental impact. Its method of use will restrict any contact with other insect species.

4:429 (*E,Z*)-octadeca-2,13-dienyl acetate

Leopard moth sex pheromone (1)

The Pesticide Manual **Fifteenth Edition entry number:** 623

$$CH_3(CH_2)_3 \underset{H}{\overset{H}{C}}=\underset{(CH_2)_9}{\overset{H}{C}} \quad \underset{(CH_2)_9}{\overset{H}{C}}=\underset{H}{\overset{CH_2OCOCH_3}{C}}$$

(2*E*,13*Z*)- isomer

NOMENCLATURE: Approved name: (*E,Z*)-octadeca-2,13-dienyl acetate.
Other names: E2Z13-18Ac. **CAS RN:** *[86252-74-6]* (2*E*,13*Z*)- isomer.

SOURCE: Originally isolated from adult virgin female leopard moths (*Zeuzera pyrina* (L.)).

PRODUCTION: Manufactured for commercial use.

TARGET PESTS: Leopard moths (*Zeuzera pyrina*).

TARGET CROPS: Apple, pear and other pome fruit orchards.

BIOLOGICAL ACTIVITY: Mode of action: The (2*E*,13*Z*)- isomer is the main component of the sex pheromone the leopard moth (*Zeuzera pyrina*) and the currant clearwing moth (*Synanthedon tipuliformis* (Clerk)). Male moths locate and subsequently mate with female moths by following the trail or pheromone plume emitted by the virgin females. The indiscriminate application of high levels of the product interferes with this process, as a constant exposure to high levels of pheromone makes trail following impossible (habituation/ adaptation). Alternatively, the use of discrete sources of pheromone released over time presents the male with a false trail to follow (confusion). Control is subsequently achieved through both the prevention of mating and laying of fertile eggs.

COMMERCIALISATION: Formulation: Under development by SEDQ as an attractant for use in orchards either to monitor the presence of the moth or to disrupt mating. **Tradenames:** '(*E,Z*)-2,13-Octadecadienyl Acetate'and 'Fersex ZP' (plus (*E,Z*)-octadeca-3,13-dienyl acetate) (SEDQ), 'Isonet Z' (plus (*E,Z*)-octadeca-3,13-dienyl acetate) and 'Isomate CCM' (Shin-Etsu).

APPLICATION: Slow-release dispensers are placed within the tree canopy, at a height of about 3 metres.

PRODUCT SPECIFICATIONS: Manufactured for commercial use. **Shelf-life:** If stored in its original packaging at 5 °C temperature, the product will keep for 2 years. When the product is refrigerated, keep it at room temperature for one month prior to field application the following season.

4. Semiochemicals

COMPATIBILITY: It is unusual to use the product with any other crop protection agent.

MAMMALIAN TOXICITY: (E,Z)-octadeca-2,13-dienyl acetate has not shown any adverse or allergic effects on researchers, formulators, manufacturers or users. It is considered to be non-toxic.

ENVIRONMENTAL IMPACT AND NON-TARGET TOXICITY: (E,Z)-octadeca-2,13-dienyl acetate is the naturally occurring sex pheromone of the leopard moth and is not expected to have any harmful effects on non-target organisms or on the environment.
Other aquatic toxicity: Harmful to aquatic organisms, may cause long term adverse effects in aquatic environments. Approved for use in organic farming: Yes.

4:430 (E,Z)-octadeca-3,13-dienyl acetate
Leopard moth sex pheromone (2)

The Pesticide Manual Fifteenth Edition entry number: 625

$$CH_3(CH_2)_3 \overset{H}{\underset{}{}} C = C \overset{H}{\underset{(CH_2)_8}{}} C = C \overset{(CH_2)_2OCOCH_3}{\underset{H}{}}$$

NOMENCLATURE: Approved name: (E,Z)-octadeca-3,13-dienyl acetate.
Other names: Z3Z13-18Ac. CAS RN: [53120-26-6] (3E,13Z)- isomer; [53120-27-7] (3Z,13Z)- isomer.

SOURCE: Originally isolated from adult virgin female leopard moths (Zeuzera pyrina (L.)). The compound is also the main component of the sex pheromone of many Synanthedon spp.; that of the peach tree borer (S. exitiosa (Say)) and the apple clearwing moth (S. myopaeformis (Borkhausen)) is mainly the (3Z,13Z)- isomer. The sex pheromone of the lesser peach tree borer (S. pictipes (Grote & Robinson)) is the (3E,13Z)- isomer, whilst that of the cherry tree borer (S. hector Butler) is a 1:1 mixture of these isomers.

TARGET PESTS: Leopard moths (Zeuzera pyrina).

TARGET CROPS: Apple, pear and other pome fruit orchards.

BIOLOGICAL ACTIVITY: Mode of action: The product is the sex pheromone of the leopard moth. Male moths locate and subsequently mate with female moths by following the trail or pheromone plume emitted by the virgin females. The indiscriminate application of high levels of the product interferes with this process, as a constant exposure to high levels of pheromone makes trail following impossible (habituation/adaptation). Alternatively, the use of discrete sources of pheromone released over time presents the male with a false trail to follow (confusion). Control is subsequently achieved through both the prevention of mating and laying of fertile eggs.

COMMERCIALISATION: Formulation: Under development by SEDQ as an attractant for use in orchards either to monitor the presence of the moth or to disrupt mating. **Tradenames:** '(*E,Z*)-Octadeca-3,13-dienyl acetate' and 'Fersex ZP' (plus (*E,Z*)-octadeca-2,13-dienyl acetate) (SEDQ), 'Sukasiba-con' (mixture of (*E,Z*)- and (*Z,Z*)- isomers) (Shin-Etsu), 'Isonet Z' (plus (*E,Z*)-octadeca-2,13-dienyl acetate) (Shin-Etsu), 'Isomate LPTB' (mixture of (*E,Z*)- and (*Z,Z*)- isomers) and 'Isomate-P' (mixture of (*Z,Z*)- and (*E,Z*)- isomers) (Shin-Etsu) and (Pacific Biocontrol).

APPLICATION: Slow-release dispensers are placed within the tree canopy, at a height of about 3 metres.

PRODUCT SPECIFICATIONS: Shelf-life: If stored in its original packaging at 5 °C temperature, the product will keep for 2 years. When the product is refrigerated, keep it at room temperature for one month prior to field application the following season.

COMPATIBILITY: It is unusual to use the product with any other crop protection agent.

MAMMALIAN TOXICITY: (*E,Z*)-octadeca-3,13-dienyl acetate has not shown any adverse or allergic effects on researchers, formulators, manufacturers or users. It is considered to be non-toxic. **Acute oral LD$_{50}$:** rats >5000 mg/kg.

ENVIRONMENTAL IMPACT AND NON-TARGET TOXICITY: (*E,Z*)-octadeca-3,13-dienyl acetate is the naturally occurring sex pheromone of the leopard moth and is not expected to have any harmful effects on non-target organisms or on the environment.
Other aquatic toxicity: Harmful to aquatic organisms, may cause long term adverse effects in aquatic environments. **Approved for use in organic farming:** Yes.

4. Semiochemicals

4:431 (Z,Z)-octadeca-3,13-dienyl acetate
Apple clearwing moth sex pheromone

The Pesticide Manual Fifteenth Edition entry number: 625

NOMENCLATURE: Approved name: (Z,Z)-octadeca-3,13-dienyl acetate.
Other names: Apple clearwing moth sex pheromone; Z3Z13-18Ac.
Development code: BAS 286 I. **CAS RN:** [53120-27-7].

SOURCE: Originally isolated from adult virgin female apple clearwing moth (*Synanthedon myopaeformis* (Borkhausen)). Female adults emit this volatile component which has been shown to be attractive to males. The compound is also the main component of the sex pheromone of many *Synanthedon* spp.; that of the peach tree borer (*S. exitiosa* (Say)) and the apple clearwing moth (*S. myopaeformis*) is mainly the (3Z,13Z)- isomer. The sex pheromone of the lesser peach tree borer (*S. pictipes* (Grote & Robinson)) is the (3E,13Z)- isomer, whilst that of the cherry tree borer (*S. hector* Butler) is a 1:1 mixture of these isomers.

PRODUCTION: Manufactured for use in crop protection.

TARGET PESTS: Apple clearwing moth (*Synanthedon myopaeformis*).

TARGET CROPS: Apple orchards.

BIOLOGICAL ACTIVITY: Mode of action: (Z,Z)-octadeca-3,13-dienyl acetate is the sex pheromone of the apple clearwing moth. Male moths locate and subsequently mate with female moths by following the pheromone trail or plume emitted by the virgin females. The application of high levels of (Z,Z)-octadeca-3,13-dienyl acetate makes trail following impossible (camouflage, competition between artificially applied and the natural female plume, false trail following). Control is achieved by the prevention of mating and laying of fertile eggs. **Efficacy:** Very low rates are required to cause mating disruption. The pheromone is volatile and disperses through the orchard very easily. **Key reference(s):** S Voerman, A K Minks, G Vanwetswinkel & J H Tumlinson. 1978. Attractivity of the 3,13-octadecadiene-1-ol acetates to the male clearwing moth *Synanthedon myopaeformis* (Borkhausen) (Lepidoptera: Sesiidae), *Entomol. Exp. Appl.*, **23**, 301.

COMMERCIALISATION: Formulation: Sold as a slow-release formulation of polyethylene ampoules. **Tradenames:** 'RAK 7' (BASF).

APPLICATION: The slow-release dispensers are attached to individual trees within the orchard, at a rate of 500 per hectare.

PRODUCT SPECIFICATIONS: Manufactured for use in crop protection. **Purity:** The chemical is >95% pure.

COMPATIBILITY: (Z,Z)-octadeca-3,13-dienyl acetate is used alone, without the need for chemical insecticides.

MAMMALIAN TOXICITY: (Z,Z)-octadeca-3,13-dienyl acetate has shown no allergic or other adverse toxicological effects on manufacturers, formulators, research workers or farmers. **Acute oral LD$_{50}$:** rats >5000 mg/kg.

ENVIRONMENTAL IMPACT AND NON-TARGET TOXICITY: (Z,Z)-octadeca-3,13-dienyl acetate is a naturally occurring insect pheromone that is specific to the apple clearwing moth. There is no evidence that it has caused any adverse effects on any non-target organisms or on the environment. **Approved for use in organic farming:** Yes.

4:432 (E,Z)-octadeca-3,13-dienyl acetate plus (Z,Z)-octadeca-3,13-dienyl acetate
Lesser peach tree borer sex pheromone

The Pesticide Manual Fifteenth Edition entry number: 625

(Z, Z)- isomer   (E, Z)- isomer

NOMENCLATURE: Approved name: (E,Z)-octadeca-3,13-dienyl acetate plus (Z,Z)-octadeca-3,13-dienyl acetate. **Other names:** *Synanthedon pictipes* sex pheromone; lesser peach tree borer sex pheromone. **CAS RN:** *[53120-27-7]* (3Z,13Z)- isomer; *[53120-26-6]* (3E,13Z)- isomer.

SOURCE: Originally isolated from adult virgin female moths of the lesser peach tree borer (*Synanthedon pictipes* (Grote & Robinson)). Adult females emit several volatile components shown to be attractive to males. These pheromones were extracted from the ovipositors of unmated females.

PRODUCTION: Manufactured for use in crop protection.

TARGET PESTS: Lesser peach tree borer (*Synanthedon pictipes*), peach twig borer (*Sanninoidea exitiosa* (Say.)), cherry tree borer (*Synanthedon hector* Butler) and currant clearwing moth (*Synanthedon tipuliformis* (Clerk)).

TARGET CROPS: Stone fruit orchards such as peaches and apricots.

BIOLOGICAL ACTIVITY: Mode of action: Males locate females by following a plume of air rich in the odour of a complete pheromone blend emitted by the females. Permeation of the canopy of an orchard with vapours of the two-component blend of pheromone is used to disrupt location of females by males and decrease mating. **Efficacy:** Several commercial formulations of this two-component blend have been shown to be effective at maintaining populations of lesser peach tree borer at economically tolerable levels until harvest. This blend also disrupts mating of peach twig borer, cherry tree borer and currant clearwing moth. **Key reference(s):** J H Tumlinson, C Yonce, R E Doolittle, R R Heath, C R Gentry & E R Mitchell. 1974. Sex pheromones and reproductive isolation of the lesser peachtree borer and peachtree borer, *Science (Washington)*, **185**, 614.

COMMERCIALISATION: Formulation: Sold as a slow-release formulation in slow-release devices. Manufactured by Scentry Biologicals, Shin-Etsu and Consep. **Tradenames:** 'Isomate-L' (Shin-Etsu), 'No-Mate PTB' (Scentry).

APPLICATION: The pheromone dispensers are attached to trees at heights recommended by formulators. The recommended number of lures per hectare varies with release rates.

PRODUCT SPECIFICATIONS: Manufactured for use in crop protection. **Purity:** The pheromones are >95% chemically pure.

COMPATIBILITY: Pheromone disruption of mating does not require insecticide, but can be used with application of insecticides.

MAMMALIAN TOXICITY: The lesser peach tree borer blend has shown no adverse toxicological effects on manufacturers, formulators or field workers. Considered to be non-toxic. **Acute oral LD$_{50}$:** rats >5000 mg/kg.

ENVIRONMENTAL IMPACT AND NON-TARGET TOXICITY: The lesser peach tree borer pheromone blend is a natural insect pheromone that is specific to lesser peach tree borer and related moths in the genus *Synanthedon*. There is no evidence that it has caused an adverse effect on non-target organisms nor had an adverse environmental impact. **Approved for use in organic farming:** Yes.

4:433 1-octen-3-ol plus 4-methylphenol plus 3-propylphenol

Tse-tse fly kairomone

1-octen-3-ol   3-propylphenol   4-methylphenol

NOMENCLATURE: Approved name: 1-octen-3-ol plus 4-methylphenol plus 3-propylphenol. **Other names:** Tse-tse fly kairomone.

SOURCE: Identified as the chemical attractants produced by the hosts of tse-tse flies (*Glossina pallidipes* Austen and *Glossina morsitans* Westwood).

TARGET PESTS: Tse-tse flies (*Glossina* spp.).

BIOLOGICAL ACTIVITY: Mode of action: Kairomones are odours emitted by species that benefit the receptor organism (in this case the tse-tse fly) and are disadvantageous to the emitter (the host), in interspecific interactions. Kairomones are used by haematophagous insects in the process of searching and locating the host for blood feeding. 1-Octen-3-ol is attractive for the tse-tse fly and was identified from volatile compounds emanated from bovines. (The attraction of the tse-tse fly and other haematophagous insects by 1-octen-3-ol baited traps was also reported for mosquitoes, screwworm, tabanides and culicoides and is thought to occur quite commonly). Blends of 1-octen-3-ol plus 4-methylphenol plus 3-propylphenol in the ratio 1:4:2 with added acetone or ethylmethyl ketone are placed behind a wire mesh or on a black and blue target that has been impregnated with insecticides (the insecticide is only placed on the black portion of the target, allowing the quantity of insecticide used to be reduced). The tse-tse flies pass through the mesh or land on the target in search of a blood meal and are killed by the insecticide. Thousands of lures per year are used in Africa to control tse-tse fly by this lure-and-kill strategy. Originally it was used to eradicate species from areas, but is now used as a barrier to prevent re-invasion.
Key reference(s): D R Hall, A Beevor, B F Cork & G A Vale. 1984. 1-Octen-3-ol: a potent olfactory stimulant and attractant for tse-tse isolated from cattle odours, *Insect. Sci. Appl.*, **5**, 335–9.

COMMERCIALISATION: Formulation: Introduced by UN agencies and local governments for the control of tse-tse fly.

4. Semiochemicals

APPLICATION: Distributed in areas where tse-tse flies are active.

MAMMALIAN TOXICITY: As the compounds are impregnated into a wire mesh or target together with an insecticide, the kairomone blend is unlikely to come into contact with individuals. The product has shown no adverse toxicological effects on research or manufacturing workers or on users.

ENVIRONMENTAL IMPACT AND NON-TARGET TOXICITY: The product is a natural kairomone that is attractive to haematophagous insects. There is no evidence that it has caused any adverse effects on non-target organisms nor had any deleterious effect on the environment.

4:434 oryctalure

Rhinoceros beetle aggregation pheromone

The Pesticide Manual Fifteenth Edition entry number: 340

$$CH_3(CH_2)_3CH(CH_3)(CH_2)_2CO_2CH_2CH_3$$

NOMENCLATURE: Approved name: oryctalure (ethyl 4-methyloctanoate).
Other names: Rhinoceros beetle aggregation pheromone. CAS RN: *[56196-53-3]*.

SOURCE: The volatile component from male coconut rhinoceros beetles (*Oryctes rhinoceros* (L.)) has been shown to attract both male and female adult beetles.

PRODUCTION: Manufactured for commercial use by SEDQ.

TARGET PESTS: Coconut rhinoceros beetle (*Oryctes rhinoceros*).

TARGET CROPS: Young oil palm plantations.

BIOLOGICAL ACTIVITY: Mode of action: Oryctalure is the aggregation pheromone of the coconut rhinoceros beetle. Males are located by following a plume of air rich in the odour of the pheromone. Evaporation of the pheromone from traps attracts both male and female rhinoceros beetles to the traps. Efficacy: Use of pheromone-primed traps in young oil palm plantations reduces damage caused by rhinoceros beetle attack.
Key reference(s): R H Hallett, A L Perez, G Gries, R Gries, H D Pierce Jr, J Yue, A C Oehlschlager, L M Gonzalez & J H Borden. 1995. Aggregation pheromone of coconut rhinoceros beetle, *Oryctes rhinoceros* (L.) (Coleoptera: Scarabaeidae), *J. Chem. Ecol.*, **21**, 1549–70.

COMMERCIALISATION: **Formulation:** Sold as a slow-release formulation of the pheromone from plastic bags. **Tradenames:** 'Oryctalure' (ChemTica), 'Sime RB Pheromone' (Sime Darby), 'Ferag OR' (SEDQ), 'SPLAT RB Lure' (ISCA). **Patents:** Malaysian patent MY-113487-A (30 March 2002).

APPLICATION: Pheromone dispensers are attached to vanes of a vaned bucket that is elevated two metres above the ground, at a density of one trap per two hectares. Dispensers should be replaced every 12 to 20 weeks depending on the environmental conditions.

PRODUCT SPECIFICATIONS: Manufactured for commercial use by SEDQ. **Purity:** The lures contain >95% chemically pure pheromone. **Shelf-life:** If stored in the original container under cold, dry conditions, the lures will last for 18 to 24 months. They should not be frozen.

COMPATIBILITY: The pheromone trap does not require insecticide applications, but traps containing insecticide can be used for control within any affected palms.

MAMMALIAN TOXICITY: Oryctalure has shown no adverse toxicological effects on manufacturers, formulators or field workers.

ENVIRONMENTAL IMPACT AND NON-TARGET TOXICITY: Oryctalure is a natural insect pheromone that is specific to the coconut rhinoceros beetle. There is no evidence that it has caused any adverse effects on any non-target organisms or had any adverse environmental impact.

4:435 4-(3-oxobutyl)phenyl acetate (cuelure)
Melon fly pheromone

The Pesticide Manual **Fifteenth Edition entry number:** 191

$$CH_3CO(CH_2)_2 - \langle \rangle - OCOCH_3$$

NOMENCLATURE: **Approved name:** 4-(3-oxobutyl)phenyl acetate (cuelure).
Other names: *Dacus cucurbitae* pheromone; melon fly attractant. **CAS RN:** *[3572-06-3]*.

SOURCE: Synthetic component similar in action to the natural secretions of the male melon fly (*Dacus cucurbitae* Coquillet) that attracts females.

PRODUCTION: Manufactured for use in US federal and state insect control strategies.

TARGET PESTS: Melon fly (*Dacus cucurbitae*).

TARGET CROPS: Melons and other cucurbits.

BIOLOGICAL ACTIVITY: Mode of action: 'Cuelure' is used in US-based federal and state control strategies to attract the adult melon fly to traps in a lure-and-kill programme. **Efficacy:** Lure-and-kill programmes have been shown to keep the populations of the melon fly at acceptable levels. **Key reference(s):** M Jacobson, L Keiser, E J Harris & D H Miyshita. 1976. Impurities in Cue-Lure attractive to female Tephritidae, *J. Agric. Food Chem.*, **24**, 782.

COMMERCIALISATION: Formulation: Distributed in US federal and state control strategies as traps. **Tradenames:** 'Cuelure' (Cyclo International) and (International Speciality Products).

APPLICATION: First introduced commercially by Cyclo International in 1995 as a lure and has since been used for control. Lure-and-kill strategies require placing the traps containing the pheromone in a slow-release formulation within the flight path of the adult female moths. It is usual to use sticky traps to capture the adults or to use rapid knockdown contact insecticides within the lures.

PRODUCT SPECIFICATIONS: Manufactured for use in US federal and state insect control strategies. **Purity:** The chemical is >95% pure.

COMPATIBILITY: The pheromone is not applied to the crop but is used in a lure containing an entrapment device or an insecticide.

MAMMALIAN TOXICITY: There are no reports of allergic or other adverse toxicological effects from researchers, manufacturers, formulators or field workers from the use of 'Cuelure'. It is considered to be non-toxic. **Acute oral LD_{50}:** rats 3038 mg/kg. **Acute dermal LD_{50}:** rabbits >2025 mg/kg.

ENVIRONMENTAL IMPACT AND NON-TARGET TOXICITY: 'Cuelure' is a synthetic pheromone that is specific to the melon fly (*Dacus cucurbitae*). It is not expected that it will have any adverse effects on non-target organisms or on the environment. **Fish toxicity:** LC_{50} (24 h) for rainbow trout 21 mg/litre, for bluegill sunfish 18 mg/litre. **Approved for use in organic farming:** Yes.

4:436 rhyncolure
American palm weevil aggregation pheromone

The Pesticide Manual **Fifteenth Edition entry number:** 579

$$(CH_3)_2CHCH_2-\overset{\overset{\displaystyle OH}{|}}{CH} \quad \underset{\underset{\displaystyle CH_3}{}}{\overset{\displaystyle H}{C}}$$

(E)- isomer

NOMENCLATURE: Approved name: rhyncolure (6-methylhept-2-en-4-ol).
Other names: American palm weevil aggregation pheromone; *Rhynchophorus palmarum* aggregation pheromone. **CAS RN:** *[83212-30-0]* racemic (E)- isomer of rhyncolure; *[66900-48-9]* (E,Z)- isomer of rhyncolure.

SOURCE: Rhyncolure is the male *Rhynchophorus palmarum* (L.) volatile component that attracts male and female adult weevils. The pheromone can be obtained either by commercial synthesis or by filling the trap with males, activating them with ethyl acetate odours to induce production of the pheromone. Captured insects can then be killed with insecticide or by other means.

PRODUCTION: Manufactured for use in crop protection.

TARGET PESTS: American or black palm weevil (*Rhynchophorus palmarum*).

TARGET CROPS: Mature oil palm plantations in which red ring disease is a problem.

BIOLOGICAL ACTIVITY: Mode of action: Rhyncolure is the aggregation pheromone of the American palm weevil. Adults locate males by following a plume of air rich in the pheromone. Evaporation of the pheromone from traps containing palm or sugar cane stalks attracts both male and female weevils to the traps. **Efficacy:** The use of rhyncolure in traps in mature oil palm plantations reduces the weevil population and lowers the incidence of the associated red ring disease by 80% in a single year. This low level of disease is maintained as long as trapping is continued. **Key reference(s):** 1) C Chinchilla & A C Oehlschlager. 1992. Captures of *Rhynchophorus palmarum* in traps baited with the male-produced aggregation pheromone, *ASD Oil Palm Papers*, **5**, 1–8. 2) J Moura, E Vilela, R Sgrillo, M Aguilar & M Resende. 1989. A behavioural olfactory study of *Rhynchophorus palmarum* in the field, *Anais da Sociedade Entomológica do Brasil*, **18**, 267–74. 3) A C Oehlschlager, C M Chinchilla, L González, L Jirón, R Mexón & B Morgan. 1993. Development of pheromone-based trapping system for *Rhynchophorus palmarum*, *J. Econ. Entomol.*, **86**, 1381–2. 4) D Rochat, A Gonzalez, D Marian, A Villanueva & P Sagatti. 1991. Evidence for male-produced aggregation pheromone in

4. Semiochemicals

Rhynchophorus palmarum, *J. Chem. Ecol.*, **17**, 1221–30. 5) D Rochat, C Malosse, M Lettere, P H Ducrot, P Zagatti, M Renou & C Descoins. 1991. Male-produced aggregation pheromone of *Rhynchophorus palmarum*: collection, identification, electrophysiological activity, and laboratory bioassay. *J. Chem. Ecol.*, **17**, 2127–41. 6) A C Oehlschlager, R S McDonald, C M Chinchilla & S N Patschke. 1995. Influence of a pheromone-based mass trapping system on the distribution of *Rhynchophorus palmarum* (L.) and the incidence of red ring disease in oil palm, *Environ. Entomol.*, **224**, 1004–12.

COMMERCIALISATION: Formulation: Rhyncolure is sold as a slow-release formulation from plastic bags containing liquid pheromone. **Tradenames:** 'Rhynkolure' and 'Rhyncopherol' (ChemTica).

APPLICATION: The pheromone dispenser is attached to the inside of bucket traps that contain pieces of palm or sugarcane. These are strapped, at chest height, to palms in infested areas, at a density of one trap per five hectares. The most modern versions of the trap use natural or synthetic aggregation pheromones to help attract the insects. Various efficient trapping methods have been proposed, all of which are based on containers which attract the insect with odours produced by plant tissue (mostly sugarcane) and the aggregation pheromone.

PRODUCT SPECIFICATIONS: Manufactured for use in crop protection. **Purity:** The pheromone used in the traps is >95% chemically pure.

COMPATIBILITY: The pheromone trap does not require the additional use of insecticides, but the traps can be used in combination with application of insecticides to affected palm trees.

MAMMALIAN TOXICITY: Rhyncolure has not shown any allergic or other adverse toxicological effects on manufacturers, formulators or field workers.

ENVIRONMENTAL IMPACT AND NON-TARGET TOXICITY: Rhyncolure is a natural insect pheromone that is specific to the American palm weevil. There is no evidence that it has caused any adverse effects on any non-target organisms or had any adverse effects on the environment.

4:437 sordidin
Banana corm weevil aggregation pheromone

The Pesticide Manual Fifteenth Edition entry number: 782

A   B   C   D

NOMENCLATURE: Approved name: sordidin isomers (A,B,C,D). **Other names:** Banana corm weevil aggregation pheromone. **CAS RN:** *[162490-88-2]* isomer A; *[162428-76-4]* isomer B; *[162334-33-0]* isomer C; *[162428-75-3]* isomer D.

SOURCE: Originally isolated from adult *Cosmopolites sordidus* (Germar), sordidin A, B, C and D are the aggregation pheromones of the banana corm weevil.

PRODUCTION: Manufactured for use in crop protection.

TARGET PESTS: Banana corm weevil (*Cosmopolites sordidus*) (see also entry ?? – banana weevil aggregation pheromone).

TARGET CROPS: Banana and plantain.

BIOLOGICAL ACTIVITY: Mode of action: Male *Cosmopolites sordidus* produce an aggregation pheromone (sordidin isomers) that attracts male and female *C. sordidus.* The pheromone (sordidin isomers) attracts adult male and female *C. sordidus* to ground-level traps, in which they are killed in water. **Efficacy:** Four traps per hectare, each containing a pheromone lure, capture sufficient adult *C. sordidus* to lower populations and corm damage by 65%–85% over one crop cycle. **Key reference(s):** 1) D Alpizar, M Fallas, A C Oehlschlager, L M Gonzalez & S Jayaraman. 1998. Pheromone-based mass trapping of the banana weevil, *Cosmopolites sordidus* (Germar) and the West Indian sugarcane weevil, *Metamasius hemipterus* L. in plantain and banana, *Association for Cooperation in Banana Research in the Caribbean and Tropical America Bi-Annual Conference*, Guayaquil, Ecuador, November 23–27. 2) D Alpizar, M Fallas, A C Oehlschlager, L M Gonzalez & S Jayaraman. 1999. Pheromone-based mass trapping of the banana weevil and the West Indian sugarcane weevil in plantain and banana, *5th International Conference on Plant Protection in the Tropics (ACORBAT)*, Malaysian Plant Protection Society, Kuala Lumpur, Malaysia, March 15–18. 3) D Alpizar, M Fallas, A C Oehlschlager, L M Gonzalez & S Jayaraman. 2000. Pheromone-based mass trapping of the banana weevil, *Cosmopolites sordidus* (Germar) in plantain and banana, *Association for*

4. Semiochemicals

Cooperation in Banana Research in the Caribbean and Tropical America (ACORBAT) Bi-Annual Conference, San Juan, Puerto Rico, July 25–31.

COMMERCIALISATION: Formulation: Sold as a mixture of sordidin isomers in a slow-release formulation. **Tradenames:** 'Cosmolure' (ChemTica).

APPLICATION: One pheromone lure is used per trap. Four traps are used per hectare. Pheromone lures each emit 0.5 to 3 mg/day of pheromone.

PRODUCT SPECIFICATIONS: Manufactured for use in crop protection. **Purity:** The pheromone is >90% chemically pure. It can be diluted in non-repellent solvent. **Shelf-life:** The product is stable for 1 to 2 years at room temperature and can be stored at this temperature in impermeable packaging available from the manufacturer.

COMPATIBILITY: Pheromone trapping is compatible with herbicide and nematicide applications normally conducted in banana and plantain. Most commercial nematicides are repellent to *Cosmopolites sordidus*. Capture rates in pheromone-baited traps decrease immediately after application of nematicide, but increase about two weeks later.

MAMMALIAN TOXICITY: Not considered to be toxic to mammals. **Acute oral LD$_{50}$:** rats 2538 mg/kg.

ENVIRONMENTAL IMPACT AND NON-TARGET TOXICITY: There is no evidence that 'Cosmolure' has caused significant adverse effects on non-target organisms nor had an adverse environmental impact. Not expected to be detected in the environment under normal use conditions of 2 to 9 mg per hectare per day. Expected to be rapidly degraded in the environment. **Bird toxicity:** Not expected to be toxic to birds under normal use conditions of 2 to 9 mg per hectare per day. **Fish toxicity:** Not expected to be toxic to fish under normal use conditions of 2 to 9 mg per hectare per day. **Other aquatic toxicity:** Not expected to be detectable under normal use conditions of 2 to 9 mg per hectare per day. **Effects on beneficial insects:** No beneficial organisms are attracted to pheromone (sordidin isomers) traps. **Behaviour in soil:** Not known. Not expected to be detected in soil at trap sites under normal use conditions of 2 to 9 mg per hectare per day.

4:438 (E,Z)-tetradeca-4,10-dienyl acetate
Apple leaf miner sex pheromone

The Pesticide Manual Fifteenth Edition entry number: 826

$$\begin{array}{c} \text{H} \quad (\text{CH}_2)_4 \quad (\text{CH}_2)_2\text{CH}_3 \\ \text{C}=\text{C} \quad \text{C}=\text{C} \\ \text{CH}_3\text{COO}(\text{CH}_2)_3 \quad \text{H H} \quad \text{H} \end{array}$$

NOMENCLATURE: Approved name: (E,Z)-tetradeca-4,10-dienyl acetate.
Other names: E4Z10-14Ac. **CAS RN:** *[105700-87-6]*, formerly *[149745-51-7]*.

SOURCE: The sex pheromone gland was located at the intersegmental membrane between the eighth and ninth abdominal segments of the female. (E,Z)-tetradeca-4,10-dienyl acetate was originally isolated from this gland. (Z,Z)-hexadeca-11,13-dienal was also isolated from this gland.

PRODUCTION: Manufactured for use in crop protection.

TARGET PESTS: Apple leaf miner moth (*Phyllonorycter ringoniella* (Matsumura)).

TARGET CROPS: Apple orchards.

BIOLOGICAL ACTIVITY: Mode of action: Males locate females by following a plume of air rich in the odour of the pheromone emitted by females. Permeation of the canopy of an orchard with vapours of the pheromone, (E,Z)-tetradeca-4,10-dienyl acetate, is used to disrupt location of females by males and decreases mating. **Efficacy:** Several commercial formulations of this pheromone have been shown to be effective at maintaining populations of apple leaf miner moth at economically tolerable levels until harvest.
Key reference(s): 1) K S Boo & C H Jung. 1998. Field tests of synthetic sex pheromone of the apple leafminer moth, *Phyllonorycter ringoniella*, *J. Chem. Ecol.*, **24**, 1939–47. 2) H Sugie, Y Tamaki, K Kawasaki, M Wakou, T Oku, C Hirano & M Horiike. 1986. Sex pheromone of the apple leafminer moth, *Phyllonorycter ringoniella* (Matsumura) (Lepidoptera: Gracillariidae): activity of geometrical isomers of tetradecadienyl acetates, *Appl. Entomol. Zool.*, **21**, 578–81.

COMMERCIALISATION: Formulation: Sold as a variety of slow-release formulations in different dispensing devices. **Tradenames:** 'Confuser A' (plus (Z)-icos-13-en-10-one plus dodec-8-enyl acetate plus tetradec-11-enyl acetate plus (Z,Z)-hexadeca-11,13-dienal) (Shin-Etsu).

APPLICATION: Pheromone dispensers are attached to trees at heights recommended by the formulators. The recommended number of lures per hectare varies with release rates.

4. Semiochemicals

PRODUCT SPECIFICATIONS: Manufactured for use in crop protection. **Purity:** The pheromone products are >95% chemically pure.

COMPATIBILITY: Pheromone disruption of mating does not require insecticide, but can be used with application of insecticides and other crop protection agents.

MAMMALIAN TOXICITY: Apple leaf miner moth sex pheromone has shown no adverse toxicological effects on manufacturers, formulators or field workers. Considered to be non-toxic. **Acute oral LD$_{50}$:** rats and mice >5000 mg/kg.

ENVIRONMENTAL IMPACT AND NON-TARGET TOXICITY: Apple leaf miner moth sex pheromone is a natural insect pheromone that is specific to the apple leaf miner moth. There is no evidence that (E,Z)-tetradeca-4,10-dienyl acetate has caused any adverse effects on non-target organisms or had an adverse environmental impact.
Approved for use in organic farming: Yes.

4:439 (Z,E)-tetradeca-9,11-dienyl acetate
Egyptian cotton leafworm sex pheromone

The Pesticide Manual **Fifteenth Edition** entry number: 827

NOMENCLATURE: Approved name: (Z,E)-tetradeca-9,11-dienyl acetate
Other names: Z9E11-14Ac. **CAS RN:** [50767-79-8].

SOURCE: Several volatile components originally isolated from the terminal segments of virgin female Egyptian cotton leafworms (*Spodoptera littoralis* (Boisch.)) have been shown to be attractive to males.

TARGET PESTS: Egyptian cotton leafworm (*Spodoptera littoralis*).

TARGET CROPS: Cotton.

BIOLOGICAL ACTIVITY: Mode of action: Males locate females by following a plume of air rich in the odour of the pheromone emitted by the females. Permeation of the canopy of cotton with vapours is used to disrupt location of females by males and decrease mating. Often sold in combination with the tobacco cutworm (*Spodoptera litura* Fabricius) pheromone – (*Z,E*)-tetradeca-9,12-dienyl acetate (see entry 4:441). **Efficacy:** A commercial formulation of the two-component pheromone blend has been shown to be effective at maintaining populations of Egyptian cotton leafworm at economically tolerable levels.

COMMERCIALISATION: Formulation: Sold as a slow-release formulation.
Tradenames: 'Yotoh-con H' (Shin-Etsu).

APPLICATION: Pheromone dispensers are attached to plants at heights recommended by formulators.

PRODUCT SPECIFICATIONS: Purity: The pheromones in the product are >95% chemically pure.

COMPATIBILITY: Pheromone disruption of mating does not require insecticide.

MAMMALIAN TOXICITY: There are no reports of Egyptian cotton leafworm sex pheromone causing allergic or other adverse toxicological effects to research, manufacturing or field workers. Considered to be non-toxic. **Acute oral LD$_{50}$:** rats and mice >5000 mg/kg.

ENVIRONMENTAL IMPACT AND NON-TARGET TOXICITY: Egyptian cotton leafworm pheromone is a natural insect pheromone. There is no evidence that it has caused any adverse effects on non-target organisms or had an adverse environmental impact.
Approved for use in organic farming: Yes.

4. Semiochemicals

4:440 (9Z,12E)-tetradecadien-l-ol plus (9Z,12E)-tetradecadien-l-ol acetate

Flour moth sex pheromone

$$\text{(9Z,12E)-tetradecadien-l-ol acetate}$$

(9Z,12E)-tetradecadien-l-ol

NOMENCLATURE: Approved name: (9Z,12E)-tetradecadien-l-ol plus (9Z,12E)-tetradecadien-l-ol acetate. **Other names:** Flour moth sex pheromone. **Development code:** E290 (Exosect).

SOURCE: Isolated from the pheromone glands of virgin female flour moths *(Ephestia kuehniella* Zeller).

TARGET PESTS: Grain moth *(Ephestia elutella* (Hübner)), Mediterranean flour moth *(Ephestia kuehniella* Zeller), raisin moth *(Ephestia figulilella* Gregson), almond moth *(Cardra cautella* Walker) and Indian meal moth *(Plodia interpunctella* (Hübner)).

TARGET CROPS: Used in grain stores and flour mills.

BIOLOGICAL ACTIVITY: Mode of action: Male moths are attracted to the 'Exosex SPTab' tablet, which contains Entostat powder formulated with synthetic female pheromone. Male moths pick up the Entostat powder and with it the female pheromone. The male's pheromone receptors become overloaded and so the male is unable to locate females. A male carrying Entostat powder will form a moving pheromone dispenser, producing 'false' pheromone trails, which attract additional males. Contact between the males ensures that the Entostat powder and the confusion effect is passed from male moth to male moth, a process known as auto-confusion. The result is a state of sexual confusion amongst the male moth population, leading to effective mating disruption. The auto-confusion effect dramatically reduces the chances of the female moth to mate. In the unlikely event that mating does occur, delays beyond the optimum period of fertility of the female moth will reduce the number and viability of her eggs. **Efficacy:** 'Exosex SPTab' provides continuous activity. Should raw material enter the building with an undetected infestation, the system will already be in place and active. There are no restrictions on entering the building whilst 'Exosex SPTab' is in place. It is very effective in giving control even where resistance to chemicals already exists. **Key reference(s):** P Trematerra. 1994. Control of *Ephestia kuehniella* Zell. by sex pheromones in the flour mills, *J. Pest Sci.*, **67**(4), 74–7.

COMMERCIALISATION: Formulation: 'Exosex CLTab' is a pesticide-free, pheromone-based monitoring trap enhancer for the clothes moth. Entostat powder, a natural food grade product, is combined with the pheromone specific to the female clothes moth to create an intelligent and pesticide free system for use on the clothes moth. **Tradenames:** 'Exosex SPTab' (Exosect).

APPLICATION: Entostat powder is impregnated with minute quantities of the female sex pheromone common to five of the major warehouse moth pests – the grain moth (*Ephestia elutella*), Mediterranean flour moth (*Ephestia kuehniella*), raisin moth (*Ephestia figulilella*), almond moth (*Cardra cautella*) and Indian meal moth (*Plodia interpunctella*). The powder is placed inside dispensers in a tablet format, which are located throughout the facility at 5 metre intervals.

PRODUCT SPECIFICATIONS: Purity: The insect pheromone components are >95% chemically pure.

COMPATIBILITY: It is not recommended that 'Exosex CLTab' be used with any other product.

MAMMALIAN TOXICITY: The product has shown no adverse toxicological effects on research or manufacturing workers or on users.

ENVIRONMENTAL IMPACT AND NON-TARGET TOXICITY: There is no evidence that 'Exosex CLTab' has caused significant adverse effects on non-target organisms nor had an adverse environmental impact. Its method of use will limit contact with other insect species. **Approved for use in organic farming:** Yes.

4. Semiochemicals

4:441 (Z,E)-tetradeca-9,12-dienyl acetate plus (Z)-tetradec-9-en-1-yl acetate

Beet armyworm sex pheromone

The Pesticide Manual Fifteenth Edition entry numbers: 828 and 833

$$CH_3COO(CH_2)_8\ CH_2-C\overset{\overset{H}{\underset{}{|}}C-CH_3}{\underset{\underset{C=C}{\underset{H\quad H}{}}}{}}$$

(Z,E)-tetradeca-9,12-dienyl acetate

$$\overset{H\qquad H}{\underset{CH_3(CH_2)_3\quad (CH_2)_8OCOCH_3}{C=C}}$$

(Z)-tetradec-9-en-1-yl acetate

NOMENCLATURE: **Approved name:** (Z,E)-tetradeca-9,12-dienyl acetate plus (Z)-tetradec-9-en-1-yl acetate. **Other names:** *Spodoptera exigua* sex pheromone; beet armyworm sex pheromone. **CAS RN:** [31654-77-0] (Z,E)-tetradeca-9,12-dienyl acetate; [16725-53-4] (Z)-tetradec-9-en-1-yl acetate.

SOURCE: Female beet armyworm (*Spodoptera exigua* (Hübner)) emit several volatile components shown to be attractive to males. Whole body extracts of virgin females revealed the presence of many components, but (Z,E)-tetradeca-9,12-dienyl acetate has been shown to be essential for male moth attraction.

PRODUCTION: Manufactured for use in crop protection.

TARGET PESTS: Beet armyworm (*Spodoptera exigua*).

TARGET CROPS: A wide variety of crops including alfalfa (lucerne), asparagus, beans, beets, cabbage, celery, cole crops, cotton, cucumbers, peanuts, lettuce, onions, peas, peppers, soybeans, strawberries, sweet potatoes, tomatoes and tobacco.

BIOLOGICAL ACTIVITY: **Mode of action:** Males locate females by following a plume of air rich in the odour of a complete pheromone blend emitted by the females. Evaporation of the pheromone blend given above from traps attracts male moths to traps. Permeation of the canopy of a crop with vapours of this two-component blend of pheromones is used to disrupt location of females by males and thereby decrease mating. **Efficacy:** Several commercial formulations of the two-component pheromone blend have been shown to be effective at maintaining populations of beet armyworm at economically tolerable levels until harvest. **Key reference(s):** C J Persoons, C van der Kraan, W J Nooijen, F J Ritter, S Voerman

& T C Baker. 1981. Sex pheromone of the beet armyworm, *Spodoptera exigua*: isolation, identification and preliminary field evaluation, *Entomol. Exp. Appl.*, **30**, 98.

COMMERCIALISATION: Formulation: Sold as slow-release formulations within slow-release devices. Manufactured by Hercon, Scentry Biologicals and Shin-Etsu. **Tradenames:** 'Isomate BAW' (Shin-Etsu) and (Biocontrol), 'Hamaki-con N' (plus (*Z*)-dodec-9-enyl acetate plus tetradec-11-enyl acetate), 'Isomate FAW' (plus (*Z*)-dodec-7-enyl acetate), 'Confuser N' (plus (*Z*)-dodec-8-enyl acetate plus tetradec-11-enyl acetate plus (*Z*)-icos-13-en-10-one), 'Confuser R' (plus (*Z*)-dodec-8-enyl acetate plus tetradec-11-enyl acetate plus (*Z*)-icos-13-en-10-one) and 'Confuser Z' (plus (*Z*)-dodec-8-enyl acetate plus (*Z*)-dodec-9-enyl acetate plus tetradec-11-enyl acetate plus (*Z*)-icos-13-en-10-one) (Shin-Etsu), 'No-Mate BAW' (plus (*Z*)-tetradec-9-enol) and 'No-Mate FAW' (plus (*Z*)-dodec-7-enyl acetate) (Scentry), 'Disrupt FAW' (plus (*Z*)-dodec-7-enyl acetate) (Hercon).

APPLICATION: Pheromone dispensers are attached to plants or sticks within the canopy of the crop as recommended by formulators. The number of lures recommended to be used per hectare varies with release rates.

PRODUCT SPECIFICATIONS: Manufactured for use in crop protection. **Purity:** The pheromones are >95% chemically pure.

COMPATIBILITY: Pheromone disruption of mating does not require insecticide, but can be used with application of insecticides.

MAMMALIAN TOXICITY: The beet armyworm pheromone blend has shown no allergic or other adverse toxicological effects on manufacturers, formulators or field workers. Considered to be non-toxic. **Acute oral LD$_{50}$:** rats >5000 mg/kg.

ENVIRONMENTAL IMPACT AND NON-TARGET TOXICITY: The beet armyworm pheromone blend is a natural insect pheromone that is specific to the beet armyworm. There is no evidence that it has caused an adverse effect on non-target organisms nor had an adverse environmental impact. **Approved for use in organic farming:** Yes.

4. Semiochemicals

4:442 (3E,8Z,11Z)-tetradecatrien-1-yl acetate plus (3E,8Z)-tetradecadien-1-yl acetate

Tomato moth sex pheromone

(3E,8Z,11Z)-tetradecatrien-1-yl acetate

(3E,8Z)-tetradecadien-1-yl acetate

NOMENCLATURE: Approved name: (3E,8Z,11Z)-tetradecatrien-1-yl acetate plus (3E,8Z)-tetradecadien-1-yl acetate. **Other names:** Tomato moth sex pheromone.

SOURCE: Isolated from the pheromone glands of virgin female tomato moths (*Tuta absoluta* (Meyrick)).

PRODUCTION: Manufactured for use in crop protection.

TARGET PESTS: Tomato moths (*Tuta absoluta*).

TARGET CROPS: Tomatoes and potatoes.

BIOLOGICAL ACTIVITY: Mode of action: The product is the sex pheromone of the tomato moth. Male moths locate and subsequently mate with female moths by following the trail or pheromone plume emitted by the virgin females. The indiscriminate application of high levels of the product interferes with this process, as a constant exposure to high levels of pheromone makes trail following impossible (habituation/adaptation). Alternatively, the use of discrete sources of pheromone released over time presents the male with a false trail to follow (confusion). Control is subsequently achieved through the prevention of mating and the laying of fertile eggs. The pheromone can also be used to monitor populations prior to insecticide applications. **Efficacy:** Each pheromone trap, baited with the synthetic sex pheromone caught, on average, 1200 males per trap per night, whilst those baited with virgin females caught only 201 males. The high biological activity of the synthetic pheromone suggests that it could be useful for pest monitoring and in mating disruption. Initial trapping tests in glasshouse tomato showed that very large numbers of males were caught and, at the same time, the level of infestation was reduced to an acceptable level.

Key reference(s): 1) M M Filho, E F Vilela, A B Attygalle, J Meinwald, A Svatos & G N Jham. 2000. Field trapping of tomato moth, *Tuta absoluta* with pheromone traps, *J. Chemical Ecology*, **26**(4), 875–81. 2) F A A Ferrara, E F Vilela, G N Jham, A E Eiras, M C Picanço, A B Attygalle, A Svatos, R T S Frighetto & J Meinwald. 2001. Evaluation of the synthetic major component of the sex pheromone of *Tuta absoluta* (Meyrick) (Lepidoptera: Gelechiidae), *J. Chemical Ecology*, **27**(5), 907–17.

COMMERCIALISATION: Tradenames: 'PHEROBANK tomato moth pheromone' (Plant Research).

APPLICATION: Pheromone traps should be used from before the emergence of the first generation and until the end of the season. Water traps are most suitable for this insect. The trap should be placed near the highest point of the plant using supporting posts approximately 1 metre high. Trap density should be one trap every hectare, but no less than two traps per farm. Replace pheromone dispenser every 4–6 weeks.

PRODUCT SPECIFICATIONS: Manufactured for use in crop protection.

MAMMALIAN TOXICITY: The product has shown no adverse toxicological effects on research or manufacturing workers or on users.

ENVIRONMENTAL IMPACT AND NON-TARGET TOXICITY: The product is a natural insect pheromone that is specific to the tomato moth. There is no evidence that it has caused any adverse effects on non-target organisms nor had any deleterious effect on the environment.

4. Semiochemicals

4:443 (Z)-tetradec-7-enal
Olive moth sex pheromone

The Pesticide Manual Fifteenth Edition entry number: 829

$$\underset{CH_3(CH_2)_5}{\overset{H}{\underset{}{\diagup}}}C=C\underset{(CH_2)_5CHO}{\overset{H}{\diagdown}}$$

NOMENCLATURE: Approved name: (Z)-tetradec-7-enal. **Other names:** *Prays oleae* sex pheromone; olive moth sex pheromone; Z7-14Al. **CAS RN:** [65128-96-3].

SOURCE: Originally isolated from the female glands of the olive moth (*Prays oleae* (Bernard)).

PRODUCTION: Manufactured for use in crop protection.

TARGET PESTS: Olive moth (*Prays oleae*) and citrus flower moth (*Prays citri* Millière).

TARGET CROPS: Olives and citrus.

BIOLOGICAL ACTIVITY: Mode of action: Males locate females by following a plume of air rich in the odour of a complete pheromone blend emitted by the females. Evaporation of the pheromone from traps attracts male moths to the traps. Permeation of the canopy of an orchard with vapours of the same pheromone is used to disrupt location of females by males and decrease mating. **Efficacy:** Very low concentrations confuse male moths and thereby disrupt mating. **Key reference(s):** D G Campion, L J McVeigh, J Polyrakis, S Michaelakis, G N Stravarakis, P S Beevor, D R Hall & B F Nesbitt. 1979. Laboratory and field studies of the female sex pheromone of the olive moth, *Prays oleae*, *Experientia*, **35**, 1146.

COMMERCIALISATION: Formulation: Sold as the sex pheromone in a slow-release device. **Tradenames:** 'Olive Moth Pheromone' (Suterra), 'Fersex PO' (SEDQ), 'Rimilure CFM' (Rimi).

APPLICATION: The slow-release devices are distributed throughout the olive grove or citrus plantation and the pheromone is allowed to permeate through the area.

PRODUCT SPECIFICATIONS: Manufactured for use in crop protection. **Purity:** (*Z*)-tetradec-7-enal is >95% chemically pure.

COMPATIBILITY: It is unusual to use the pheromone with other crop protection agents.

MAMMALIAN TOXICITY: There are no reports of allergic or other adverse toxicological effects from researchers, manufacturers, formulators or field workers. It is considered to be non-toxic. **Acute oral LD$_{50}$:** rats >5000 mg/kg.

ENVIRONMENTAL IMPACT AND NON-TARGET TOXICITY: (*Z*)-tetradec-7-enal is the sex pheromone of the olive moth and the citrus flower moth. It is not expected that it will have any adverse effects on non-target organisms or on the environment.
Approved for use in organic farming: Yes.

4:444 (Z)-tetradec-9-enol
Beet armyworm sex pheromone

The Pesticide Manual Fifteenth Edition entry number: 830

$$\begin{array}{cc} H & H \\ \diagdown & \diagup \\ C=C \\ \diagup & \diagdown \\ CH_3(CH_2)_3 & (CH_2)_8OH \end{array}$$

NOMENCLATURE: Approved name: (Z)-tetradec-9-enol. **Other names:** Beet armyworm sex pheromone. **CAS RN:** [35153-15-2].

SOURCE: Female beet armyworm (*Spodoptera exigua* (Hübner)) emit several volatile components shown to be attractive to males. Whole body extracts of virgin females revealed the presence of many different components. (Z)-tetradec-9-enol has been shown to act synergistically with these components.

PRODUCTION: Manufactured for use in crop protection.

TARGET PESTS: Beet armyworm (*Spodoptera exigua*).

TARGET CROPS: A wide variety of crops, including alfalfa (lucerne), asparagus, beans, beets, cabbage, celery, cole crops, cotton, cucumbers, peanuts, lettuce, onions, peas, peppers, soybeans, strawberries, sweet potatoes, tomatoes and tobacco.

BIOLOGICAL ACTIVITY: Mode of action: Males locate females by following a plume of air rich in the odour of the pheromone emitted by the females. Evaporation of the pheromone blend given above from traps attracts male moths to the traps. Permeation of the canopy of a crop with vapours of this two component blend of pheromones is used to disrupt location of females by males and thereby decrease mating. **Efficacy:** (Z)-tetradec-9-enol has been shown to be an effective enhancer of the effects of beet armyworm pheromones.

COMMERCIALISATION: Formulation: Sold as a component of slow-release formulations within slow-release devices. Manufactured by Hercon, Scentry Biologicals and Shin-Etsu. **Tradenames:** 'Disrupt BAW' (plus (Z,E)-tetradeca-9,12-dienyl acetate plus (Z)-hexadec-11-enyl acetate) (Hercon), 'Isomate BAW' (plus (Z,E)-tetradeca-9,12-dienyl acetate plus (Z)-tetradec-9-enyl acetate) and 'Yotoh-con-S' (plus (Z,E)-tetradeca-9,12-dienyl acetate) (Shin-Etsu), 'No-Mate BAW' (plus (Z,E)-tetradeca-9,12-dienyl acetate plus (Z)-tetradec-9-enyl acetate) (Scentry).

APPLICATION: Pheromone dispensers are attached to plants or sticks within the canopy of the crop as recommended by formulators. The recommended number of lures per hectare varies with release rates.

4. Semiochemicals

PRODUCT SPECIFICATIONS: Manufactured for use in crop protection. **Purity:** The pheromones are >95% chemically pure.

COMPATIBILITY: Pheromone disruption of mating does not require insecticide, but can be used with application of insecticides.

MAMMALIAN TOXICITY: The beet armyworm pheromone blend has shown no allergic or other adverse toxicological effects on manufacturers, formulators or field workers. Considered to be non-toxic.

ENVIRONMENTAL IMPACT AND NON-TARGET TOXICITY: The beet armyworm pheromone blend is a natural insect pheromone that is specific to the beet armyworm. There is no evidence that it has caused an adverse effect on non-target organisms nor had an adverse environmental impact. **Approved for use in organic farming:** Yes.

4:445 (Z)-tetradec-9-enyl acetate plus (Z)-dodec-7-enyl acetate

Fall armyworm sex pheromone

The Pesticide Manual **Fifteenth Edition entry numbers:** 309 and 833

$$
\underset{CH_3(CH_2)_3 \quad (CH_2)_8OCOCH_3}{\overset{H \qquad H}{C=C}}
\qquad
\underset{H \qquad H}{\overset{CH_3(CH_2)_3 \quad (CH_2)_6OCOCH_3}{C=C}}
$$

NOMENCLATURE: Approved name: (Z)-tetradec-9-enyl acetate plus (Z)-dodec-7-enyl acetate. **Other names:** *Spodoptera frugiperda* sex pheromone; fall armyworm sex pheromone. **CAS RN:** *[16725-53-4]* (Z)-tetradec-9-enyl acetate; *[14959-86-5]* (Z)-dodec-7-enyl acetate.

SOURCE: Several volatile components originally isolated from the terminal segments of virgin female fall armyworms (*Spodoptera frugiperda* (J. E. Smith)) have been shown to be attractive to males.

PRODUCTION: Manufactured for use in crop protection.

TARGET PESTS: Fall armyworm (*Spodoptera frugiperda*).

TARGET CROPS: Maize (corn) and cotton.

BIOLOGICAL ACTIVITY: Mode of action: Males locate females by following a plume of air rich in the odour of a complete pheromone blend emitted by the females. Permeation of the canopy of maize or cotton with vapours of this two-component pheromone blend is used to disrupt location of females by males and decrease mating. **Efficacy:** Several commercial formulations of the two-component blend have been shown to be very effective at maintaining populations of fall armyworm at economically tolerable levels between flowering and harvest. **Key reference(s):** J H Tumlinson, E R Mitchell, P E A Teal, R R Heath & L J Mengelkoch. 1986. Sex pheromone of fall armyworm, *Spodoptera frugiperda* (Smith). Identification of components critical to attraction in the field, *J. Chem. Ecol.*, **12**, 1909.

COMMERCIALISATION: Formulation: Sold as slow-release formulations within release devices such as plastic tubes, laminated plastic or plastic barrier film. Manufactured by several companies, including Hercon, Scentry (formerly Ecogen), Shin-Etsu and Consep. **Tradenames:** 'Isomate FAW' (Shin-Etsu), 'No-Mate FAW' (Scentry), 'Hercon Disrupt FAW' (Hercon).

APPLICATION: Pheromone dispensers are attached to plants at heights recommended by formulators. The recommended number of lures per hectare varies with release rates.

PRODUCT SPECIFICATIONS: Manufactured for use in crop protection. **Purity:** The pheromones in the product are >95% chemically pure.

COMPATIBILITY: Pheromone disruption of mating does not require insecticide, but can be used with application of insecticides.

MAMMALIAN TOXICITY: There are no reports of fall armyworm sex pheromone causing allergic or other adverse toxicological effects to research, manufacturing or field workers. Considered to be non-toxic. **Acute oral LD$_{50}$:** rats >12 000 mg/kg (both components). **Acute dermal LD$_{50}$:** rabbits >2000 mg/kg (both components).

ENVIRONMENTAL IMPACT AND NON-TARGET TOXICITY: Fall armyworm pheromone blend is a natural insect pheromone that is specific to the fall armyworm. There is no evidence that it has caused any adverse effects on non-target organisms or had an adverse environmental impact. **Approved for use in organic farming:** Yes.

4. Semiochemicals

4:446 tetradec-11-enyl acetate

Tea tortrix sex pheromone

The Pesticide Manual Fifteenth Edition entry number: 832

(E)- isomer

(Z)- isomer

NOMENCLATURE: Approved name: tetradec-11-enyl acetate. **Other names:** *Homona magnanima* sex pheromone; tea tortrix sex pheromone. **CAS RN:** *[20711-10-8]* (Z)- isomer; *[33189-72-9]* (E)- isomer.

SOURCE: The product is the sex pheromone of the tea tortrix (*Homona magnanima* Diakonoff). Originally isolated from the terminal abdominal segments of virgin females.

PRODUCTION: Manufactured for use in crop protection.

TARGET PESTS: Tea tortrix (*Homona magnanima*), smaller tea tortrix (*Adoxophyes* sp.) and leaf rollers (*Adoxophyes orana* (F. von R.) and *Platynota stultana* (Walsingham)).

TARGET CROPS: Tea; peaches, nectarines, apricots, plums, prunes, cherries and other stone fruit; apples, quinces and other pome fruit; almonds and other tree nut crops; and grapes.

BIOLOGICAL ACTIVITY: Mode of action: The product is the sex pheromone of the tea tortrix. Male moths locate and subsequently mate with female moths by following the trail or pheromone plume emitted by the virgin females. The indiscriminate application of high levels of the product interferes with this process, as a constant exposure to high levels of pheromone makes trail following impossible (habituation/adaptation). Alternatively, the use of discrete sources of pheromone released over time presents the male with a false trail to follow (confusion). Control is subsequently achieved through the prevention of mating and the laying of fertile eggs. **Efficacy:** Very low rates are required to cause mating disruption. Tetradec-11-enyl acetate is volatile and distributes throughout the crop easily. It acts as an attractant and by disruption of mating in the disorientation mode.

Key reference(s): H Noguchi, Y Tamaki, S Arai, M Shimoda & I Ishikawa. 1981. Field evaluation of synthetic sex pheromone of the oriental tea tortrix moth, *Homona magnanima* Diakonoff, *Jpn. J. Appl. Entomol. Zool.*, **25**, 170.

COMMERCIALISATION: Formulation: The pheromone is sold in slow-release tubes. Consep has produced a sprayable formulation, 'Checkmate OLR-F'. **Tradenames:** 'Hamaki-con' ((Z)-

isomer) and 'Isomate-C Special' ((Z)- isomer plus codlemone) (Shin-Etsu), 'NoMate OLR' (mainly (E)- isomer) (Scentry), 'Checkmate OLR' and 'Checkmate OLR-F' (Consep).

APPLICATION: The slow-release tubes are distributed throughout the tea plantation or orchard at a rate of 250 to 375 dispensers per hectare (100 to 150 per acre). The active ingredient diffuses out of the tubes and is dispersed around the plantation. The first application should be made early in the season after the first moth is detected in a pheromone trap. Apply 'Checkmate OLR-F', at 6 to 24 g per hectare, directly to the foliage, ensuring good leaf coverage, but not to run-off. The product should be reapplied as necessary, but rates per season should not exceed 350 g per hectare.

PRODUCT SPECIFICATIONS: Manufactured for use in crop protection.

COMPATIBILITY: Can be used with any crop protection agent that does not repel the tortrix or leaf roller moths.

MAMMALIAN TOXICITY: The product has shown no adverse toxicological effects on research or manufacturing workers or on users.

ENVIRONMENTAL IMPACT AND NON-TARGET TOXICITY: The product is a natural insect pheromone that is specific to the tea tortrix. There is no evidence that it has caused any adverse effects on non-target organisms nor had any deleterious effect on the environment. **Approved for use in organic farming:** Yes.

4:447 (Z)-tetradec-11-enyl acetate
Tortrix moth sex pheromone

The Pesticide Manual **Fifteenth Edition** entry number: 832

$$CH_3CH_2 \quad (CH_2)_{10}OCOCH_3$$
$$C=C$$
$$H \qquad H$$

NOMENCLATURE: Approved name: (Z)-tetradec-11-enyl acetate. **Other names:** Tortrix moth sex pheromone; apple leaf roller sex pheromone; Z11-14Ac. **Development code:** BAS 280 I. **CAS RN:** *[20711-10-8]* (Z)- isomer; *[26532-95-6]* unspecified stereochemistry.

SOURCE: A component of the sex pheromones of tortrix moths (leaf rollers) (*Adoxophyes orana* (Fischer von Rösslerstamm), *Pandemis heparana* (Denis and Schiffermüller) and *Archips podanus* (Scopoli)).

PRODUCTION: Manufactured for use in crop protection.

TARGET PESTS: Tortrix moth complex; leaf roller complex (*Adoxophyes orana*, *Pandemis heparana* and *Archips podanus*).

TARGET CROPS: Pome fruit (apples and pears).

BIOLOGICAL ACTIVITY: Mode of action: (Z)-tetradec-11-enyl acetate is a component of the sex pheromones of several tortrix moths. Male moths locate and subsequently mate with female moths by following the pheromone trail or pheromone plume emitted by the virgin females. The application of high levels of (Z)-tetradec-11-enyl acetate makes trail following impossible, as the elevated levels of this component confuse the male moths and camouflage the female pheromone plume. Control is achieved by preventing the mating of moths and laying of fertile eggs. **Efficacy:** Very low rates are required to cause mating disruption. (Z)-tetradec-11-enyl acetate is volatile and is easily distributed throughout the orchard. The (E)- isomer reduces the attraction shown by the pheromone. The presence of (Z)-tetradec-9-enyl acetate, in the ratio 95:5; (Z)-11:(Z)-9, increases the activity of the pheromone to *P. heparana*, whilst ratios of 25:75; (Z)-11:(Z)-9, were preferred by *A. orana* males.

Key reference(s): M A El-Adl & P J Charmillot. 1982. Laboratory studies with the sex pheromones of the summerfruit tortrix, *Adoxophyes orana* F. v. R. and its components, *Acta Phytopathol. Acad. Sci. Hung.*, **17**, 133.

COMMERCIALISATION: Formulation: Sold as a slow-release formulation of polyethylene ampoules. **Tradenames:** 'RAK 4' (BASF), 'Isomate' (Shin-Etsu), 'Isomate C LR' (plus (E,E)-8,10-dodecadien-1-ol plus dodecan-1-ol plus tetradecan-1-ol plus (Z)-9-tetradecenyl acetate) (Shin-Etsu), 'MEC.LRX' ('Selibate CS' (plus (Z)-octadec-9-enal plus (Z)-octadec-13-enal)) (Suterra).

APPLICATION: Dispensers are attached to individual trees within the orchard at a rate of 500 per hectare.

PRODUCT SPECIFICATIONS: Manufactured for use in crop protection. **Purity:** The pheromone is >95% chemically pure.

COMPATIBILITY: (Z)-tetradec-11-enyl acetate is used without the need for added chemical insecticides.

MAMMALIAN TOXICITY: (Z)-tetradec-11-enyl acetate has not shown any allergic or other adverse toxicological effects to manufacturers, formulators, research workers or farmers. **Acute oral LD$_{50}$:** rats >5000 mg/kg.

ENVIRONMENTAL IMPACT AND NON-TARGET TOXICITY: (Z)-tetradec-11-enyl acetate is a component of natural insect pheromone systems that are specific to some Tortricid species. There is no evidence that it has had any adverse effects on non-target organisms or on the environment. **Approved for use in organic farming:** Yes.

4:448 (*Z*)-tetradec-11-enyl acetate plus (*E*)-tetradec-11-enyl acetate

European corn borer sex pheromone

The Pesticide Manual Fifteenth Edition entry number: 832

$$CH_3CH_2 \quad H$$
$$C=C$$
$$H \quad (CH_2)_{10}OCOCH_3$$

$$CH_3CH_2 \quad (CH_2)_{10}OCOCH_3$$
$$C=C$$
$$H \quad H$$

(*E*)- isomer   (*Z*)- isomer

NOMENCLATURE: Approved name: (*Z*)-tetradec-11-enyl acetate plus (*E*)-tetradec-11-enyl acetate. **Other names:** European corn borer sex pheromone; *Ostrinia nubilalis* sex pheromone. **CAS RN:** *[20711-10-8]* (*Z*)- isomer; *[33189-72-9]* (*E*)- isomer.

SOURCE: The sex pheromone of female European corn borer (*Ostrinia nubilalis* (Hübner)) moths, originally extracted from the whole, homogenised bodies of virgin females.

PRODUCTION: Manufactured for use in crop protection and insect monitoring.

TARGET PESTS: European corn borer (*Ostrinia nubilalis*).

TARGET CROPS: Maize (corn).

BIOLOGICAL ACTIVITY: Mode of action: Male moths locate virgin female moths for mating by following a plume of the sex pheromone emitted by the females. If synthetic pheromones are released into the field, the high concentration of the volatile components may confuse the males and thereby prevent them from locating the females. The European corn borer sex pheromone is more usually used, however, as a means of monitoring the flight of adults, so that insecticides can be applied at the most effective timing for the control of the larvae. **Efficacy:** The monitoring of the moths is very effective, allowing growers to identify the time when the population has exceeded threshold levels and spray applications are necessary. **Key reference(s):** J A Klun & J F Robinson. 1972. Olfactory discrimination in the European corn borer and several pheromonally analogous moths, *Ann. Entomol. Soc. Am.*, **65**, 1337.

COMMERCIALISATION: Formulation: Sold as slow-release lures for trapping the male moths. **Tradenames:** 'European Corn Borer Lure' (Scentry).

APPLICATION: Traps are placed around the field and the numbers of male moths captured gives an indication of the intensity of the population, allowing insecticide spray application to be made when, and if, necessary.

4. Semiochemicals

PRODUCT SPECIFICATIONS: Manufactured for use in crop protection and insect monitoring. **Purity:** Both isomers are >95% chemically pure.

COMPATIBILITY: It is unusual to use the lures in conjunction with other crop protection chemicals.

MAMMALIAN TOXICITY: There have been no reports of allergic or other adverse toxicological effects from the use, manufacture or formulation of pheromone traps containing (Z)-tetradec-11-enyl acetate plus (E)-tetradec-11-enyl acetate. It is considered to be non-toxic. **Acute oral LD$_{50}$:** rats >5000 mg/kg.

ENVIRONMENTAL IMPACT AND NON-TARGET TOXICITY: (Z)-tetradec-11-enyl acetate plus (E)-tetradec-11-enyl acetate is the sex pheromone of the European corn borer and is specific to that species. It is not expected that it will have any adverse effects on non-target organisms or on the environment.

4:449 (E)-tetradec-11-enol plus (E)-tetradec-11-enyl acetate
Tufted apple bud moth sex pheromone

The Pesticide Manual **Fifteenth Edition entry number:** 832

(E)-tetradec-11-enol   (E)-tetradec-11-enyl acetate

NOMENCLATURE: Approved name: (E)-tetradec-11-enol plus (E)-tetradec-11-enyl acetate. **Other names:** Tufted apple bud moth sex pheromone. **CAS RN:** *[35153-18-5]* (E)-tetradec-11-enol; *[33189-72-9]* (E)-tetradec-11-enyl acetate.

SOURCE: Both compounds were isolated from the abdomens of female tufted apple bud moths (*Platynota idaeusalis* (Walker)).

PRODUCTION: Manufactured for use in crop protection.

TARGET PESTS: Tufted apple bud moth (*Platynota idaeusalis*).

TARGET CROPS: Orchards, especially apples.

BIOLOGICAL ACTIVITY: Mode of action: Male moths locate virgin females by following a plume of the sex pheromone. The use of the pheromone in orchards either confuses the males by masking the emission from the females or causes the male to follow a false trail. In addition, the pheromone can be used as a means of trapping the males. The acetate is unattractive on its own and the (*Z*)- isomer of tetradec-11-enol is inhibitory to the (*E*)-isomer in the field. **Efficacy:** It has been shown that the use of the product in apple orchards has kept the population of moths at a level that is acceptable from flowering until harvest. **Key reference(s):** A S Hill, R T Cardé, A Comeau, W Bode & W L Roelofs. 1974. Sex pheromones of the tufted apple bud moth (*Platynota idaeusalis*), *Environ. Entomol.*, **3**, 249.

COMMERCIALISATION: Formulation: Sold as a combination product in slow-release devices such as plastic tubes, laminated plastic and plastic barrier film. **Tradenames:** 'NoMate TABM' 'NoMate OLR' (mainly (*E*)- isomer of tetradec-11-enyl acetate alone) 'Nomate LRX' ((*Z*)-tetradec-11-enyl acetate alone) (Scentry), 'Checkmate OLR' ((*E*)- and (*Z*)- isomers of tetradec-11-enyl acetate alone) (Suterra), 'RAK 4' ((*Z*)-tetradec-11-enyl acetate alone) and 'RAK 3+4' ((*Z*)-tetradec-11-enyl acetate plus codlemone) (BASF), 'Ecb' ((*Z*)-tetradec-11-enyl acetate alone), 'Mstrs BHFW' ((*Z*)-tetradec-11-enyl acetate plus (*Z*)-dodec-9-enyl acetate), 'Mstrs BHFW-2' ((*Z*)-tetradec-11-enyl acetate plus (*Z*)-dodec-9-enyl acetate) and 'Mstrs SS' ((*E*)-tetradec-11-enyl acetate plus (*E*)-dodec-9-enyl acetate) (Mstrs), 'Hamaki-con N' ((*Z*)-tetradec-11-enyl acetate plus (*Z*)-tetradec-9-enyl acetate plus dodec-9-enyl acetate), 'Isomate' ((*Z*)-tetradec-11-enyl acetate alone), 'Isomate LBAM' ((*Z*)-tetradec-11-enyl acetate alone), 'Isomate-GBM' ((*Z*)-tetradec-11-enyl acetate plus dodec-9-enyl acetate), 'Isomate-C Special' ((*Z*)-tetradec-11-enyl acetate plus codlemone), 'Confuser A' ((*Z*)-tetradec-11-enyl acetate plus (*E,Z*)-tetradeca-4,10-dienyl acetate plus (*Z*)-icos-13-en-10-one plus dodec-8-enyl acetate plus (*Z*)-tetradec-10-enyl acetate) (Shin-Etsu), 'Confuser N' ((*Z*)-tetradec-11-enyl acetate plus (*Z*)-icos-13-en-10-one plus (*Z*)-tetradec-9-enyl acetate plus dodec-8-enyl acetate), 'Confuser P' ((*Z*)-tetradec-11-enyl acetate plus (*Z*)-icos-13-en-10-one plus 14-methyloctadec-1-ene plus dodec-8-enyl acetate) 'Confuser R' ((*Z*)-tetradec-11-enyl acetate plus (*Z*)-icos-13-en-10-one plus (*Z*)-tetradec-9-enyl acetate plus dodec-8-enyl acetate) and 'Confuser Z' ((*Z*)-tetradec-11-enyl acetate plus (*Z*)-icos-13-en-10-one plus (*Z*)-tetradec-9-enyl acetate plus dodec-8-enyl acetate plus dodec-9-enyl acetate) (Shin-Etsu), 'Isomate-OBLR/PLR' (Shin-Etsu) and (Pacific Biocontrol), 'Isomate-CM/LR' ((*Z*)-tetradec-11-enyl acetate plus codlemone plus *n*-decanol plus *n*-dodecanol) (Pacific Biocontrol), 'Gbm Rope' ((*Z*)-tetradec-11-enyl acetate plus (*Z*)-dodec-9-enyl acetate) (Nihon Nohyaku).

APPLICATION: Dispensers are attached to trees throughout the orchards at a height of about 3 metres. The number of dispensers used depends upon the release rate of the pheromone.

4. Semiochemicals

PRODUCT SPECIFICATIONS: Manufactured for use in crop protection. **Purity:** All pheromone constituents are >95% chemically pure.

COMPATIBILITY: The products can be used with any crop protection agent that is not inhibitory to the tufted apple bud moth.

MAMMALIAN TOXICITY: There have been no reports of allergic or other adverse toxicological effects from the products on researchers, manufacturers, formulators and field workers. It is considered to be non-toxic. **Acute oral LD$_{50}$:** rats and mice >5000 mg/kg ((Z)-tetradec-11-enyl acetate). **Inhalation:** LC$_{50}$ >5 mg/litre ((Z)-tetradec-11-enyl acetate). **Skin and eye:** Acute percutaneous LD$_{50}$ >2000 mg/kg. Non-irritating to skin and eyes (rabbits); not a skin sensitiser (guinea pigs) ((Z)-tetradec-11-enyl acetate).

ENVIRONMENTAL IMPACT AND NON-TARGET TOXICITY: The tufted apple bud moth sex pheromone occurs in Nature and it is specific to the tufted apple bud moth. Consequently, it is not expected that the pheromone will have any adverse effects on non-target organisms or on the environment. **Fish toxicity:** LC$_{50}$ (96 h) for rainbow trout >10 mg/litre ((Z)-tetradec-11-enyl acetate). **Other aquatic toxicity:** EC/LC$_{50}$ (48 h) for *Daphnia magna* >10 mg/litre ((Z)-tetradec-11-enyl acetate). **Approved for use in organic farming:** Yes.

4:450 tridec-4-enyl acetate
Tomato pinworm sex pheromone

The Pesticide Manual **Fifteenth Edition entry number:** 880

$$CH_3(CH_2)_7 \quad H$$
$$C=C$$
$$H \quad (CH_2)_3OCOCH_3$$

(E)- isomer

NOMENCLATURE: **Approved name:** tridec-4-enyl acetate. **Other names:** *Keiferia lycopersicella* pheromone; tomato pinworm pheromone. **CAS RN:** *[72269-48-8]* (E)- isomer; *[65954-19-0]* (Z)- isomer.

SOURCE: The sex pheromone of the tomato pinworm (*Keiferia lycopersicella* (Walsingham)).

PRODUCTION: Manufactured for use in crop protection.

TARGET PESTS: Tomato pinworm (*Keiferia lycopersicella*).

TARGET CROPS: Tomato and aubergine (eggplant).

BIOLOGICAL ACTIVITY: Mode of action: The product is the sex pheromone of the tomato pinworm moth. Male moths locate and subsequently mate with female moths by following the trail or pheromone plume emitted by the virgin females. The indiscriminate application of high levels of the product interferes with this process, as a constant exposure to high levels of pheromone makes trail following impossible (habituation/adaptation). Alternatively, the use of discrete sources of pheromone released over time presents the male with a false trail to follow (confusion). Control is subsequently achieved through the prevention of mating and the laying of fertile eggs. Tridec-4-enyl acetate is used extensively in Mexico for tomato pinworm control in conjunction with natural enemies. **Efficacy:** Very low rates are required to cause mating disruption. Tridec-4-enyl acetate is volatile and distributes throughout the crop easily. It acts as an attractant and by disruption of mating in the disorientation mode.

COMMERCIALISATION: Formulation: Sold as a microencapsulated slow-release formulation on a polymer matrix. Consep also offers a sprayable formulation. **Tradenames:** 'NoMate TPW' (Scentry), 'Checkmate TPW' (hand-held dispenser), 'CheckMate TPW' and 'CheckMate TPW-F' (sprayable formulations) (Suterra), 'Isomate TPW' (Shin-Etsu), 'Frustrate TPW' (Suterra), 'Disrupt TPW' (Hercon).

APPLICATION: Applied to the foliage of the tomato crop as adult flight commences. The pheromone diffuses out of the microencapsulated product and is dispersed throughout the field. Applied at a rate of 250 dispensers per hectare. Spray 'Checkmate TPW-F' in sufficient water for uniform coverage, but not to the point of run-off, at a rate of between 20 and 100 g per hectare, depending upon intensity of infestation. Do not apply more than 350 g per hectare per year.

PRODUCT SPECIFICATIONS: Manufactured for use in crop protection. **Purity:** The compound is >95% chemically pure. **Storage conditions:** The product should be stored in a cool place and protected from freezing.

COMPATIBILITY: Tridec-4-enyl acetate can be used in combination with other crop protection agents that do not repel the tomato pinworm.

MAMMALIAN TOXICITY: The product has shown no adverse toxicological effects on research or manufacturing workers or on users. **Acute oral LD$_{50}$:** rats >5000 mg/kg. **Acute dermal LD$_{50}$:** rabbits >2000 mg/kg. **Inhalation:** LC$_{50}$ (4 h) for rats >5 mg/litre. **Toxicity class:** EPA (formulation) III.

ENVIRONMENTAL IMPACT AND NON-TARGET TOXICITY: The product is a natural insect pheromone that is specific to the tomato pinworm. There is no evidence that it has caused any adverse effects on non-target organisms nor had any deleterious effect on the environment. **Approved for use in organic farming:** Yes.

4. Semiochemicals

4:451 trimedlure

Mediterranean fruit fly attractant

The Pesticide Manual Fifteenth Edition entry number: 890

A

B_1

B_2

C

NOMENCLATURE: **Approved name:** trimedlure (*tert*-butyl-2-methyl-4(or 5)-chlorocyclohexanecarboxylate). **Other names:** Mediterranean fruit fly attractant; *Ceratitis capitata* attractant. **CAS RN:** *[12002-53-8]* trimedlure (for the mixture of the 16 isomers); *[5748-22-1]* trimedlure A; *[5748-20-9]* trimedlure B_1; *[5748-21-0]* trimedlure B_2; *[5748-23-2]* trimedlure C.

SOURCE: Synthetic attractant for male *Ceratitis capitata* (Wiedemann), found to be attractive by screening tests.

PRODUCTION: Manufactured for use in protection of citrus.

TARGET PESTS: Mediterranean fruit fly (*Ceratitis capitata*).

TARGET CROPS: Citrus.

BIOLOGICAL ACTIVITY: **Mode of action:** Trimedlure 4/5-chloro- isomers are strong attractants of male Mediterranean fruit fly. Males are attracted to traps baited with trimedlure containing these isomers, where they are retained by adhesive or killed by insecticide. **Efficacy:** Monitoring helps time management decisions on insecticide application. **Key reference(s):** 1) M Jacobsen, K Ohinata, D L Chambers, W A Jones & M J Fujimoto. 1973. Insect sex attractants, XIII. Isolation, identification, and synthesis of sex pheromones of the Mediterranean fruit fly, *J. Med. Chem.*, **16**, 248. 2) T P McGovern & M Beroza. 1966. Structure of the four isomers of the insect attractant trimedlure, *J. Org. Chem.*, **31**, 1472–7. 3) M Beroza, N Green, S I Gertler, L F Steiner & D H Miyashita. 1961. New attractants for the Mediterranean fruit fly, *J. Agric. Food Chem.*, **9**, 361–5. 4) R E Doolittle, R T Cunningham, T P McGovern & P E Sonnet. 1991. Trimedlure enantiomers: differences in attraction for the Mediterranean fruit fly, *Ceratitis capitata* (Wied.) (Diptera: Tephritidae), *J. Chem. Ecol.*, **17**, 475–84.

COMMERCIALISATION: Formulation: Slow-release formulations of attractant from polymer plugs or from plastic bags containing about 2 ml of liquid trimedlure isomers.
Tradenames: 'Trimedlure' (liquid) (Suterra), 'Trimedlure' (plugs) (Farma-Tech) and (Suterra), 'Trimedlure' (sachets) (ChemTica).

APPLICATION: Pheromone dispensers are attached to Jackson traps (triangular) with sticky inserts. Traps are hung in citrus trees in the area to be monitored. Alternatively, 2 ml of trimedlure are applied to a dental wick, which is placed in a Jackson trap (triangular). Plugs and bags emit trimedlure for about 8 weeks under field conditions.

PRODUCT SPECIFICATIONS: Manufactured for use in protection of citrus. **Purity:** The product is a >90% chemically pure mixture of the 4- and 5-chloro isomers.

COMPATIBILITY: The trap does not require insecticide, but trapping can be used with application of insecticides.

MAMMALIAN TOXICITY: Some skin irritation has been reported in California by field workers using trimedlure. Not considered to be toxic to mammals.

ENVIRONMENTAL IMPACT AND NON-TARGET TOXICITY: There is no evidence that trimedlure has caused significant adverse effects on non-target organisms nor had an adverse environmental impact. **Approved for use in organic farming:** Yes.

4:452 verbenone
Anti-aggregation pheromone of the southern pine beetle

The Pesticide Manual **Fifteenth Edition entry number:** 899

NOMENCLATURE: Approved name: verbenone ((1R,5R)-4,6,6-trimethylbicyclo[3.1.1]hept-3-en-2-one). **Other names:** pin-2-en-2-one; (R)-(+)-verbenone; d-verbenone.
CAS RN: *[18309-32–5]*; *[1196-01-6]* (1S,3S)-verbenone; *[80-57-9]* racemic mixture.

SOURCE: Verbenone is produced predominantly by male southern pine beetles (*Dendroctonus frontalis* Zimmermann). It is derived from α-pinene upon the oxidation of trans-verbenol and is also produced outside the beetle by the autoxidation of trans-verbenol in the presence of air or from the symbiotic fungi introduced into the host tree by beetles.

TARGET PESTS: Southern pine beetles (*Dendroctonus frontalis*).

TARGET CROPS: Coniferous forests, especially pine forests.

BIOLOGICAL ACTIVITY: Mode of action: During mass attack, verbenone is released primarily by males in low concentrations where it acts to balance the sex ratio of beetles attracted to the host by enhancing the attractiveness of the aggregation pheromones to females. At the height of mass attack (3 to 5 days after initial attack) large numbers of males release high concentrations of verbenone, which inhibit the response of both sexes to frontalin and cause a significant drop in the number of arriving beetles. Treatment of freshly attacked trees and uninfested trees with verbenone has shown considerable success in slowing or halting the growth of small to moderate-sized infestations.

Key reference(s): R F Billings, C W Berisford, S M Salom & T L Payne. 1995. Application of semiochemicals in the management of southern pine beetle infestations: current status of research. pp. 30–8, in S M Salom and K R Hobson (eds.). *Application of Semiochemicals for Management of Bark Beetle Infestations.* Proceedings of an Informal Conference. December 12–16, 1994, Dallas, TX, USA. USDA, Forest Service, Intermountain Res. Stat., Gen. Tech. Rep. No. 318.

COMMERCIALISATION: Tradenames: 'Southern Pine Beetle Repellent' (Phero Tech).

APPLICATION: Verbenone is applied to susceptible trees enclosed in a pouch in order to repel beetle attack.

MAMMALIAN TOXICITY: The product has shown no adverse toxicological effects on research or manufacturing workers or on users. **Acute oral LD_{50}:** male rats 3400 mg/kg; female rats 1800 mg/kg. **Skin and eye:** Acute percutaneous LD_{50} for rabbits >2000 mg/kg. A mild skin irritant and a moderate eye irritant.

ENVIRONMENTAL IMPACT AND NON-TARGET TOXICITY: Bird toxicity: Acute oral LD_{50} for bobwhite quail >300 mg/kg. **Fish toxicity:** LC_{50} for rainbow trout 130 and bluegill sunfish 210 mg/litre. **Other aquatic toxicity:** LC_{50} (48 h) *Daphnia* 200 mg/litre.

Appendix: Plant and animal extracts

The products that the International Federation of Organic Agricultural Movements (IFOAM) will allow to be used in organic farming are listed on pages xl–xliii of this manual. IFOAM does point out that the items in this list can be used only if they are registered in the country in question and if they are approved by the national accreditation body. There are different criteria applied to the registration of 'natural' products in different parts of the world. The European Union, for example, requires a full toxicological package for chemicals, regardless of origin, if their use is associated with a claim for pest, disease or weed control. If a natural extract is marketed with no such crop protection claim, the regulatory requirements are much less stringent and commercial introductions are much easier (and cheaper), although product safety is still a fundamental requirement.

In the USA, by contrast, the regulatory authorities are much less demanding of toxicological data for products that are derived from natural sources. This does not mean that compounds can be registered and sold just because they are natural, but there are tiers of testing that natural compounds (or biochemicals) must pass before they can be sold. Failure of any aspect of the first tier means that further testing has to be undertaken. The argument is that if preliminary but thorough testing suggests a product is safe, then it can be sold. There is also no necessity for a product to demonstrate efficacy, and it may well be that a claim by the manufacturer of a response that has not been shown to be reliable in the field is the claimed, and registered, use. This leads to inconsistencies between different regions regarding which products are registered, the uses for which the registration applies, and the acceptance of these products in organic farming systems. An example is that of azadirachtin. There are a very large number of neem-based products registered in North America (many of which are listed in entry 2:156), but there are no European registrations. Neem-based products are also sold in countries such as India and, more recently, China has developed an industry based on the extraction of azadirachtin from the seeds of the neem tree grown in plantations specifically for that purpose. It is important to remember that all pesticides sold or distributed in the United States must be registered by the Environmental Protection Agency (EPA), based on scientific studies showing that they can be used without posing unreasonable risks to people or the environment. Because of advances in scientific knowledge, the law requires that pesticides that were first registered years ago be re-registered to ensure that they meet today's more stringent standards.

In evaluating pesticides for re-registration, EPA obtains and reviews a complete set of studies from pesticide producers, describing the human health and environmental effects of each pesticide. The Agency imposes any regulatory controls that are needed to manage each pesticide's risks effectively. EPA then re-registers pesticides that can be used without posing unreasonable risk to human health or the environment. When a pesticide is eligible for re-registration, EPA announces this and explains why in a Re-registration Eligibility Decision document. The re-registration case for Flower and Vegetable Oils is composed of a group of compounds that are natural components of plants. These oils are active ingredients in pesticide products registered for use as animal repellents, feeding depressants, insecticides and miticides. Some of the essential oils also are included as active ingredients in antimicrobial pesticide products (disinfectants, sanitisers, bacteriostats, microbiocides and fungicides). However, since the essential oils have no independent pesticidal activity in antimicrobial products, these uses are not eligible for re-registration. Many of the flower and vegetable oils have other, more significant, non-pesticidal uses as food additives, flavourings and components of cosmetics, soaps, perfumes, plastics, resins and other products.

Essential oils are defined by EPA as any volatile oil that gives distinctive odour or flavour to a plant, flower or fruit and were first registered as pesticide active ingredients in 1947.

It should be emphasised here that the products listed in the Main Entries of this Manual are included because they have a verified chemical structure or because they are extracts with a confirmed, reliable and registered use.

For the purposes of this review of the type of products that are available and how they relate to organic farming systems, the list of products that may be used, which was shown in the Introductory pages, has been reproduced, but reduced to those that refer to crop protection/crop health. Entries that are recommended for fertilisation or soil conditioning have been ignored, as have minerals, such as copper salts, and sulfur. Within this reduced list is a brief explanation of what is claimed with each preparation. Such an exercise will necessarily be incomplete, as the number of such 'growth-enhancing' products on the market is very large and the claims, in many cases, hard to verify. Some typical claims for products that have been released very recently include 'A future sprout suppressant, the product is a natural product, which as a volatile liquid atomises in the store', 'An inoculant/growth promoter that is said to be effective for cooler and early season planting soybean varieties' and 'Foliar products, which are claimed to provide essential micronutrients for alfalfa, beans and cotton'. Whilst it may well be the case that such products do have a beneficial effect on the growth of the target crops, it is difficult to report these effects with any confidence and it may well be that the only real effect is one of a foliar fertiliser or a general toxicant. For these reasons, and because of the huge number of products that are marketed with general claims for a beneficial effect, these products have been omitted from this overview.

IFOAM Regulations – products that may be used for plant pest and disease control, weed management and growth regulation

I Plant and animal origin

- Algal preparations – Section 2: Natural Products contains one entry which claims a crop protection effect (Main Entry 2:190 laminarine). It is included in this section because the fungicidal effect claimed falls within the remit of the manual, it is natural and it is accepted within organic farming systems. There are a number of seaweed-based products that do not have a specific crop protection claim and these are not included in the Main Entries. The claims made against extracts of this type include such things as improved plant vigour with the added suggestion that a healthy plant is better able to withstand the attacks of pathogens and insects and to compete for space with weeds. Without a doubt, algal products do contain a range of micronutrients which will benefit all plants, particularly those in nutrient deficient situations, and this nutritional benefit will lead to more vigorous growth from the treated plants. There is also an argument that foliar applications of such preparations increase the natural resistance to attack within the treated plant (they activate systemic acquired resistance), although such claims are often not made on the label (particularly in Europe). It is also true that algal extracts often contain natural plant cytokinins and it is well known that these compounds delay the onset of senescence and are involved in the process of cell division in higher plants (see Main Entry 2:238 zeatin). The incorporation of algae into soil has been claimed, by those who are sceptical of the claims of organic farming supporters, to lead to the release of methyl bromide, a well-known broad-spectrum pesticide that is also linked to ozone depletion. This claim is based on the fact that rotting seaweeds do, indeed, release quantities of methyl bromide into the atmosphere. The debate on this issue continues.

- Animal preparations and oils – animal-based preparations and oils are not represented within the Main Entries, because their use is limited and claims are not well substantiated, although components such as fish oil are sometimes added to products (see Main Entry 2:223 sesame oil).

- Beeswax – beeswax is also not listed in the Main Entries as a crop protection natural product, primarily because its use is more as a physical barrier to invasion rather than a protectant/eradicant in the true sense of the science. It is clear, therefore, that any substance which is naturally occurring and coats and, thereby, protects a plant, could be considered as suitable for organic culture.

- Chitin nematicides (natural origin) – chitin-based products are listed in Section 2, as Main Entry 2:208, under the entry name 'poly-D-glucosamine' and as Main Entry 2:210, under the entry name 'poly-N-acetyl-D-glucosamine'. The claims for such products include fungicidal activity as well as nematicidal effects. It is claimed that some products are elicitors that stimulate the natural defence response system in treated plants. Poly-D-glucosamine binds to fungal receptor sites, mimicking an attack by fungal spores. This, in turn, results in signals being sent to the nuclei of the plant cells. These signals elicit multiple genetic and biological responses, including the production of phytoalexins (anti-microbial compounds produced in plants), aimed at inhibiting fungal infections. The mode of action against nematodes is less well understood, but it, too, may be associated with the natural defence response.

- Coffee grounds – there are claims that items such as coffee grounds have a dramatic effect on the survival of slugs. It is thought that this is due to the high concentration of caffeine in the preparation. Whilst the use of coffee grounds has been shown to be effective against slugs and snails, the rate of use is very high and the product has not been commercialised.

- Corn gluten meal (weed control) – the effect of products such as corn gluten meal for control of weeds is to smother the weeds and, thereby, prevent the interception of light by the weed leading to significant growth inhibition. This allows the untreated crop to continue to grow in the absence of weed competition. The best form of herbicide has often been described as a healthy crop, and the inhibition of weed growth as a consequence of coverage by the gluten leads to good crop growth and the overtopping of any weeds that grow away from the treatment.

- Dairy products (such as milk, casein) – are allowed from non-synthetic sources without synthetic substances added. The role of such products in crop protection in agriculture is unclear, although some components of milk and milk products do have an anti-viral/anti-fungal effect on plant pathogens.

- Gelatine – as with products such as beeswax, gelatine has an effect that involves smothering the target pest or disease.

- Lecithin – natural lecithin occurs in mammals being produced in the liver. It is a fatty substance that is also an emulsifier of fats. In this role, it has been claimed to have crop protection properties.

- Natural acids (such as acetic acid) – some natural acids are included in the Main Entries in Section 2. Compounds such as vinegar (acetic acid) are effective because they are mildly corrosive and they have found use in organic farming systems as weak herbicides, often as components of mixtures of other naturally occurring components (Main Entry 2:151). Citric acid is also sold as an organic insecticide (Main Entries 2:166 and 2:167), as are mixed plant-derived fatty acids (Main Entries 2:176 and 2:203), although there are also claims that such products are also effective as fungicides and herbicides. Many plant-derived fatty acids are sold as sodium or potassium salts for use as insecticides (Main Entry 2:226). Formic acid (Main Entry 2:177) is sold as a miticide.

- Neem (*Azadirachta indica* extracts where registered for use) – extracts of the neem tree are accepted for use in organic systems where the products are registered for use. These products are described in Main Entry 2:156.

- Plant oils, Plant preparations and Plant-based repellents – have been grouped together for the purposes of this overview. Many such products are to be found in the Main Entries in Section 2: Natural Products and it could be considered surprising that other, more specific, plant-based products are listed separately within this IFOAM classification. Several plant oils can be found in the Manual, including 4-allyl-2-methoxyphenol, eugenol or clove oil (Main Entry 2:152); azadirachtin could be described as a plant oil extract (Main Entry 2:156); jojoba oil is sold for control of sucking insects and some fungal pathogens (Main Entry 2:186); canola oil is finding a wide acceptance as an insecticide in home gardening (Main Entries 2:162 and 2:207); and extract of *Chenopodium ambrosioides* near *ambrosioides* has been commercialised as an insecticide (Main Entry 2:175). Extracts of pepper, alone or in combination with other oils, are used as repellents or general insecticides (Main Entries 2:159 and 2:163), as is trans-α-ionone (Main Entry 2:185); cinnamaldehyde occurs as the major component of oil found in cassia plants and is used to control various fungal diseases of many different crops (Main Entry 2:165); and (*E*)-3,7-dimethyl-2,6-octadien-1-ol (geraniol) has found a use as an insect repellent (Main Entry 2:172). Also included in this classification is garlic extract, used as a repellent for insects, molluscs and some small mammals alone (Main Entry 2:178) or in combination with other plant oils (Main Entry 2:169); karanjin (3-methoxy-2-phenyl-4*H*-furo[2,3-*H*]-1-benzopyran-4-one), an extract of pongam or Indian beech (*Derris indica*) is used as an insecticide/acaricide and fungicide (Main Entry 2:187); and DMDP, isolated from tropical legumes in the Genera *Lonchocarpus* and *Derris*, is under development as a systemic nematicide (Main Entry 2:173). *Macleaya* extract (Main Entry 2:192) is sold for control of foliar diseases, as is milsana (*Reynoutria sachalinensis* extract) (Main Entry 2:198).

- Pyrethrum (*Tanacetum* (formerly *Chrysanthemum*) *cinerariaefolium* extracts), but without the synergist piperonyl butoxide after 2006 – these entries are to be found as the Main Entries 2:213, 2:214 and 2:215. More details on the chemistry of these products can be found in Entry 737 of *The Pesticide Manual* Fifteenth Edition.

- Quassia (*Quassia amara* extracts where registered for use), rotenone (extracts of *Derris elliptica*, *Lonchocarpus* spp. and *Thephrosia* spp.) (Main Entry 2:219), ryania (*Ryania speciosa* extracts where registered for use) (Main Entry 2:220), sabadilla (where registered for use) (Main Entry 2:221) and tobacco tea (pure nicotine is forbidden, but the extract of the tobacco leaf is allowed by some authorisation bodies) (Main Entry 2:200) fall into the category of compounds of known chemical composition, and the mode of action of the biologically active agents is well known. Often these compounds were used in traditional medicine or hunting strategies and their uses have been passed down through local folklore. It is not uncommon for them to have a number of biological effects.

Appendix

II Organisms used for biological pest control

- Fungal preparations, bacterial preparations (*Bacillus thuringiensis*), release of parasites, predators and other natural enemies, and sterilised insects and viral preparations (polyhedro- and granuloviruses) are widely accepted in organic farming and are gaining increasing use in protected cropping systems. These products can be found within the Sections 1 and 3.

III Others

- Soft soap – most of the entries within this section accepted by IFOAM relate to natural mineral-based products rather than living systems. Soft soap is the exception and the properties of soft soap as an insecticide are described in Main Entry 2:226.

IV Traps, barriers, repellents

- Pheromones – in traps and dispensers only – this restriction on use clearly applies to lure and kill strategies and would seem to prohibit wide application by conventional machinery. Many semiochemicals can be found in Section 4.

The use of synthetic growth regulators is prohibited, but the definition of synthetic is unclear. It is often the case that if a compound such as zeatin is isolated from a natural source, then its use is permitted. If, however, it is manufactured, it is not. Synthetic dyes may not be used for cosmetic alteration of organic products and the use of genetically engineered organisms or products thereof is prohibited.

Glossary: Latin–English

Names include species referred to in the Main Entries and other agriculturally important organisms, some of which are not specifically mentioned within a Main Entry. For each name that is identified in the first column at the Genus level (i.e. names in italics), the third column gives: for fungi, bacteria, insects and vertebrates, the Order then Family; for plants, the Family. The first column also includes some Families and Orders, with a corresponding higher level indicated in the third column. Wherever possible, the authorities for these Latin names are also given.

Latin	English	Order and/or Family
Abutilon theophrasti Medic.	Velvetleaf	Malvaceae
Acantholyda erythrocephala (L.)	Pine false webworms	Hymenoptera: Pamphiliidae
Acari	Mites	
Acaridae	Acarid mites	Acari
Acarina (see Acari)		
Acarus siro L.	Storage mite	Acarina: Acaridae
Aceria guerreronis Keifer	Coconut mite	Acari: Eriophyidae
Acrididae	Grasshoppers and locusts	Saltatoria
Actinomycetales	Filamentous bacteria	
Aculops spp.	Mites	Acari: Eriophyidae
Aculops pelekassi (Keifer)	Pink citrus rust mite	Acari: Eriophyidae
Aculus spp.	Rust mites	Acari: Eriophyidae
Aculus schlechtendali (Nalepa)	Rust mite, apple	Acari: Eriophyidae
Acyrthosiphon kondoi Shinji	Alfalfa aphid	Hemiptera: Aphididae
Acyrthosiphon pisum (Harris)	Pea aphid	Hemiptera: Aphididae
Adoryphorus couloni (Burmeister)	Redheaded cockchafer	Coleoptera: Scarabaeidae
Adoxophyes spp.	Tortrix moths and leaf rollers	Lepidoptera: Tortricidae
Adoxophyes orana Fischer von Roeslerstamm	Summer fruit tortrix moth	Lepidoptera: Tortricidae
Aedes aegypti (L.)	Yellow fever mosquito	Diptera: Culicidae
Aeschynomene spp.	Joint vetches	Fabaceae
Aeschynomene virginica L.	Northern joint vetch	Fabaceae
Agaricales	Mushrooms, etc.	
Agriotes spp.	Wireworms	Coleoptera: Elateridae
Agrobacterium radiobacter (Beijerink & van Delden)	Beneficial bacterium	Eubacteriales: Rhizobiaceae
Agrobacterium tumefasciens Conn.	Crown gall	Eubacteriales: Rhizobiaceae
Agromyza spp.	Leaf miners	Diptera: Agromyzidae
Agropyron repens Beauv. (see *Elytrigia repens*)		

Latin	English	Order and/or Family
Agrostis gigantea Roth	Black bent	Poaceae
Agrostis stolonifera L.	Creeping bent	Poaceae
Agrotis spp.	Cutworms	Lepidoptera: Noctuidae
Agrotis ipsilon (Hufnagel)	Black cutworm	Lepidoptera: Noctuidae
Agrotis segetum (Schiffermüller)	Turnip moth	Lepidoptera: Noctuidae
Alabama argillacea (Hübner)	Cotton leaf worm	Lepidoptera: Noctuidae
Albugo candida Kuntze	White blister	Chromista: Oomycetes: Peronosporales
Aleurothrixus floccosus (Mask.)	Whitefly	Hemiptera: Aleyrodidae
Aleyrodidae	Whiteflies	Hemiptera
Alopecurus myosuroides Huds.	Black-grass	Poaceae
Alphitobius spp.	Mealworms	Coleoptera: Tenebrionidae
Alternanthera philoxeroides (Mart.) (Griseb.)	Alligatorweed	Amarantheaceae
Alternaria spp.	Leaf spots, various	Deuteromycetes: Moniliales
Alternaria alternata	Leaf spot	Deuteromycetes: Moniliales
Alternaria brassicae Sacc.	Dark leaf spot, brassicas	Deuteromycetes: Moniliales
Alternaria brassicicola Wiltsh.	Dark leaf spot, brassicas	Deuteromycetes: Moniliales
Alternaria dauci Groves & Skolko	Carrot leaf blight	Deuteromycetes: Moniliales
Alternaria solani Jones & Grout	Potato early blight	Deuteromycetes: Moniliales
Amaranthus spp.	Amaranths	Amaranthaceae
Amaranthus retroflexus L.	Redroot pigweed; common amaranth	Amaranthaceae
Amblyseius barkeri (Hughes)	Predatory mite	Mesostigmata: Phytoseiidae
Amblyseius californicus (McGregor)	Predatory mite	Mesostigmata: Phytoseiidae
Amblyseius cucumeris (Oudemans)	Predatory mite	Mesostigmata: Phytoseiidae
Amblyseius degenerans (Berlese)	Fruit tree red spider mite predator	Mesostigmata: Phytoseiidae
Amblyseius fallacis (Garman)	Spider mite predator	Mesostigmata: Phytoseiidae
Ambrosia artemisifolia L.	Ragweed, common	Asteraceae
Ampelomyces quisqualis	Hyperparasite of fungi	Deuteromycetes: Sphaeropsidales
Amylois transitella (Walker)	Navel orangeworm	Lepidoptera: Pyralidae
Anagrapha falcifera (Kirby)	Celery looper; alfalfa looper	Lepidoptera: Noctuidae
Anagrus atomus (L.)	Leafhopper egg parasitoid	Hymenoptera: Mymaridae
Anarsia lineatella Zeller	Peach tree borer	Lepidoptera: Gelechiidae
Anopheles spp.	Mosquitoes	Diptera: Culicideae
Anthomyiidae, *Delia* spp. (= some *Hylemya* spp.) and others	Root flies or maggots	Diptera: Anthomyiidae
Anthonomus grandis Boheman	Cotton boll weevil	Coleoptera: Curculionidae
Anticarsia gemmatalis Hübner	Soybean looper; velvet bean caterpillar	Lepidoptera: Noctuidae
Aonidiella aurantii (Maskell)	Californian red scale	Hemiptera: Diaspididae
Aonidiella orientalis (Newstead)	Oriental red scale	Hemiptera: Diaspididae
Apera spica-venti Beauv.	Loose silky-bent	Poaceae

Latin	English	Order and/or Family
Aphanomyces spp.	Foot rot; root rot (various hosts)	Chromista: Oomycetes: Saprolegniales
Aphanomyces cochlioides Drechs.	Blackleg, beet	Chromista: Oomycetes: Saprolegniales
Aphelenchoides xylophilus see *Bursaphelenchus xylophilus* (Steiner & Buhrer) Nickle		
Aphelinus spp.	Aphid parasitoid wasps	Hymenoptera: Aphelinidae
Aphididae	Aphids	Hemiptera
Aphidius spp.	Aphid parasitoid wasps	Hymenoptera: Aphelinidae
Aphidoletes aphidimyza Rondani	Aphid gall midge	Diptera: Cecidomyiidae
Aphis citricida (Kirkaldy)	Black citrus aphid	Hemiptera: Aphididae
Aphis fabae Scopoli	Black bean aphid	Hemiptera: Aphididae
Aphis gossypii Glover	Melon and cotton aphid	Hemiptera: Aphididae
Aphyllophorales		Basidiomycotina
Aphytis lignanensis Compère	Red scale parasite	Hymenoptera: Aphelinidae
Aphytis melinus DeBach	Golden chalcid	Hymenoptera: Aphelinidae
Archips podanus (Scopoli)	Leaf roller	Lepidoptera: Tortricidae
Armillaria mellea Kumm.	Honey fungus	Basidiomycetes: Agaricales
Arrhenatherum elatius Beauv.	False oat-grass	Poaceae
Arrhenatherum elatius var. *bulbosum* Spenn.	Onion couch	Poaceae
Artemisia vulgaris L.	Mugwort; wormwood	Asteraceae
Ascochyta spp.	Leaf spots, various hosts	Deuteromycetes: Sphaeropsidales
Ascochyta chrysanthemi Stevens (see *Didymella ligulicola*)		
Ascochyta fabae Speg.	Leaf spot, beans	Deuteromycetes: Sphaeropsidales
Ascochyta pinodes Jones	Leaf and pod spot, peas	Deuteromycetes: Sphaeropsidales
Ascochyta pisi Lib.	Leaf and pod spot, peas	Deuteromycetes: Sphaeropsidales
Ascomycotina	Fungi, sexually produced spores in sacs	
Aspergillus spp.	Storage fungi	Deuteromycetes: Moniliales
Aspidiotus nerii (Bouché)	Red scale	Hemiptera: Diaspididae
Athous spp.	Garden wireworms	Coleoptera: Elateridae
Atomaria linearis Stephens	Pygmy mangold beetle	Coleoptera: Cryptophagidae
Atriplex patula L.	Common orache	Chenopodiaceae
Aulacorthum solani (Kaltenbach)	Glasshouse potato aphid	Hemiptera: Aphididae
Autographa californica (Speyer)	Alfalfa looper	Lepidoptera: Noctuidae
Autographa nigrisgna (Walker)	Beet semi-looper	Lepidoptera: Noctuidae
Avena spp.	Oats (wild and cultivated)	Poaceae
Avena barbata Brot.	Bearded oat	Poaceae
Avena fatua L.	Wild oat	Poaceae

Latin	English	Order and/or Family
Avena sterilis L.	Sterile oat	Poaceae
Avena sterilis L. ssp. *Ludoviciana* (= *A. ludoviciana* Durieu)	Wild oat, winter	Poaceae
Azadirachta indica A. Juss.	Neem tree	Meliaceae
Bacillus sphaericus Meyer & Neide		Schizomycetes: Eubacteriales
Bacillus subtilis Cohn.	Hay bacillus	Schizomycetes: Eubacteriales
Bacillus thuringiensis Berliner	Bt	Schizomycetes: Eubacteriales
Bactrocera oleae Gml. (= *Dacus oleae* Gml.)	Olive fruit fly	Diptera: Tephritidae
Basidiomycotina	Fungi, spores produced exogenously in basidia	
Beauvaria bassiana Balsamo	White muscardine	Deuteromycetes: Moniliales
Begonia elatior Hort.	Begonia	Begoniaceae
Belonolaimus longicausatus Rau	Sting nematode	Nematoda
Bemisia spp.	Whiteflies	Hemiptera: Aleyrodidae
Bemisia argentifolii Bellows & Perring	Silverleaf whitefly	Hemiptera: Aleyrodidae
Bemisia tabaci (Gennadius)	Tobacco whitefly	Hemiptera: Aleyrodidae
Betula lutea Michx.	Yellow birch	Betulaceae
Bilderdykia convolvulus Dum. (see *Fallopia convolvulus*)		
Bipolaris stenospila Shoemaker	Brown stripe, sugar cane	Ascomycetes: Sphaeriales
Blatella germanica (L.)	German cockroach	Dictyoptera: Blattidae
Blissus leucopterus (Say)	Chinch bug	Hemiptera: Lygaeidae
Blumeria graminis (DC.) Speer	Powdery mildew, cereals, grasses	Ascomycetes: Erysiphales
Blumeriella jaapii Arx.	Coccomycosis, cherry leaf spot	Ascomycetes: Helotiales
Botryosphaeria dothidea Ces. & De Not	Apple fruit rot, leaf spot and stem canker	Ascomycetes: Sphaeriales
Botryosphaeria obtusa Shoemaker (= *Physalospora obtusa* (Schw.) Cke.)	Leaf spot and black rot, apple	Ascomycetes: Sphaeriales
Botrytis allii Munn	Neck rot, onions	Deuteromycetes: Moniliales
Botrytis cinerea Pers.	Fruit rot, various hosts	Deuteromycetes: Moniliales
Brachiaria mutica (Forsk.) Stapf. (see *Urochloa mutica*)		
Bradysia spp.	Sciarid flies	Diptera: Sciaridae
Brassica napus L.	Rape	Brassicaceae
Bremia lactucae Regel	Downy mildew, lettuce	Chromista: Oomycetes: Peronosporales
Brevipalpus phoenicis (Geijskes)	Red crevice tea mite	Acari: Tenuipalpidae
Bromus sterilis L.	Barren brome	Poaceae
Bryobia praetiosa Koch	Clover bryobia mite	Acari: Tetranychidae
Bryobia ribis Thom.	Gooseberry bryobia mite	Acari: Tetranychidae
Bryophyta	Mosses and liverworts	Bryophyta
Bucculatrix thurberiella Busck	Cotton leaf perforator	Lepidoptera: Lyonetiidae
Bursaphelenchus lignicolus (see *Bursaphelenchus xylophilus* (Steiner & Buhrer) Nickle)		

Latin	English	Order and/or Family
Bursaphelenchus xylophilus (Steiner & Buhrer) Nickle	Pine wood nematode; pinewood nematode	Nematoda: Aphelenchoididae
Butomus umbellatus L.	Rush, flowering	Butomaceae
Cacopsylla pyri (L.)	Pear psylla	Hemiptera: Psyllidae
Cacopsylla pyricola (Förster)	Pear psylla	Hemiptera: Psyllidae
Calepitrimerus spp.	Mite	Acari: Eriophyidae
Caloptilia theivora (Wlsm.)	Tea leaf roller	Lepidoptera: Gracillariidae
Calystegia sepium R. Br. ssp. *sepium*	Bindweed, large	Convolvulaceae
Camponotus spp.	Carpenter ants	Hymenoptera: Formicidae
Candida spp.	Parasitic yeasts	Endomycetales
Capsella bursa-pastoris Medic.	Shepherd's purse	Brassicaceae
Carduus spp.	Thistles	Asteraceae
Carex spp.	Sedges	Cyperaceae
Carposina niponensis Walsingham	Oriental fruit tree moth	Lepidoptera: Tortricidae
Cassia obtusifolia L.	Sickle pod	Fabaceae
Cecidomyiidae	Gall midges and predacious midges	Diptera
Cecidophyopsis ribis (Westwood)	Blackcurrant gall-mite	Acari: Eriophyidae
Centaurea spp.	Knapweeds	Asteraceae
Centaurea calcitrapa L.	Purple star-thistle	Asteraceae
Centaurea diffusa Monnet De La Marck	Diffuse knapweed	Asteraceae
Centaurea maculosa Monnet De La Marck	Spotted knapweed	Asteraceae
Centaurea solstitialis L.	Yellow star-thistle	Asteraceae
Ceratitis capitata L.	Mediterranean fruit fly	Diptera: Tephritidae
Ceratobasidium cereale	Sharp eyespot, cereals	Basidiomycetes: Tulasnellales
Ceratocystis ulmi Moreau	Dutch elm disease	Ascomycetes: Sphaeriales
Ceratodon purpureus Brid.	Moss	Bryophyta
Ceratophyllum demersum L.	Hornweed, common	Ceratophyllaceae
Cercospora spp.	Leaf spots, various	Deuteromycetes: Moniliales
Cercospora beticola Sacc.	Leaf spot, beet	Deuteromycetes: Moniliales
Cercospora zonata	Cercospora leaf spot, beans	Deuteromycetes: Moniliales
Cercosporella herpotrichoides Fron. (see *Pseudocercosporella herpotrichoides* Deighton)		
Cercosporidium spp. (includes *C. sojinum* = *Cercospora sojina*)	Frog eye; leaf spot, soybean	Deuteromycetes: Moniliales
Ceutorhynchus spp.	Brassica gall and stem weevils	Coleoptera: Curculionidae
Ceutorhynchus assimilis (Paykull)	Cabbage seed weevil	Coleoptera: Curculionidae
Ceutorhynchus pleurostigmata (Marsh.)	Turnip gall weevil	Coleoptera: Curculionidae
Ceutorhynchus quadridens (Panz.)	Cabbage stem weevil	Coleoptera: Curculionidae
Chaetocnema spp.	Flea beetles	Coleoptera: Chrysomelidae
Chaetocnema concinna (Marsh.)	Mangold flea beetle	Coleoptera: Chrysomelidae

Latin	English	Order and/or Family
Chamomilla spp.	Mayweeds (some)	Asteraceae
Chenopodium album L.	Fat hen	Chenopodiaceae
Chilo spp.	Stem borers	Lepidoptera: Pyralidae
Chilo plejadellus Zk.	Rice stem borer	Lepidoptera: Pyralidae
Chilo suppressalis (Walker)	Rice stalk borer; rice stem borer	Lepidoptera: Pyralidae
Chondrilla juncea L.	Rush skeleton weed	Asteraceae
Chondrostereum purpureum Pouzar	Silver leaf	Agaricales: Agaricaceae
Chorioptes spp.	Mange mites	Acari: Psoroptidae
Chromatomyia syngenesia (see *Phytomyza syngenesiae*)		
Chromolaena odorata King & Rob.	Siam weed	Asteraceae
Chrysanthemum segetum L.	Corn marigold	Asteraceae
Chrysodeixis chalcites Esper	Tomato looper	Lepidoptera: Noctuidae
Chrysomelidae	Chrysomelid beetles	Coleoptera
Chrysomphalus aonidum L.	Citrus red scale	Hemiptera: Diaspididae
Chrysomphalus dictyospermi Morgan	Palm scale	Hemiptera: Diaspididae
Chrysopa carnea Stephens (see *Chrysoperla carnea*)		
Chrysoperla carnea (Stephens)	Pearly green lacewing	Neuroptera: Chrysopidae
Chrysoteuchia caliginosellus (= *Crambus caliginosellus*)	Grass moth	Lepidoptera: Pyralidae
Cicadellidae	Leafhoppers	Hemiptera
Cirsium arvense Scop.	Creeping thistle	Asteraceae
Cladosporium spp.	Black mould; sooty mould	Deuteromycetes: Moniliales
Cladosporium carpophilum Lev. (see *Stigmina carpophila*)		
Cladosporium fulvum Cke. (see *Fulvia fulva*)		
Clasterosporium carpophilum Aderh. (see *Stigmina carpophila*)		
Clavibacter michiganensis David *et al.*	Tomato canker	Eubacteriales
Cnaphalocrocis medinalis Gn.	Rice leaf roller	Lepidoptera: Pyralidae
Cnemidocoptes spp.	Bird skin mites	Acari: Sarcoptidae
Coccidae	Scale insects	Hemiptera
Coccomyces hiemalis Higgins (see *Blumeriella jaapii*)		
Coccus spp.	Scale insects	Hemiptera: Coccidae
Coccus hesperidum L.	Brown soft scale	Hemiptera: Coccidae
Cochliobolus miyabeanus Drechs.	Brown spot, rice	Ascomycetes: Sphaeriales
Cochliobolus sativus Drechs.	Foot rot; root rot, cereals and grasses	Ascomycetes: Sphaeriales
Coleoptera	Beetles	Insecta
Colletotrichum spp.	Anthracnose, various root rot and leaf curl diseases	Ascomycetes: Melanconiales

Latin	English	Order and/or Family
Colletotrichum atramentarium Taub. (see *Colletotrichum coccodes*)		
Colletotrichum coccodes Hughes	Root rot, tomato	Ascomycetes: Melanconiales
Colletotrichum coffeanum Noack (see *Glomerella cingulata*)		
Colletotrichum gloeosporoides Penz. (see *Glomerella cingulata*)		
Colletotrichum lagenarium (Pass.) Ell. & Halst.	Anthracnose, cucurbits	Ascomycetes: Melanconiales
Colletotrichum lindemuthianum Briosi & Cavara	Anthracnose, french beans	Ascomycetes: Melanconiales
Commelina spp.	Dayflower; wandering Jew	Commelinaceae
Comstockaspis perniciosus Comstock (see *Quadraspidiotus perniciosus*)		
Conopomorpha cramerella Snellen	Cocoa pod borer	Lepidoptera: Gracillariidae
Convolvulus arvensis L.	Field bindweed	Convolvulaceae
Coptotermes spp.	Termites	Isoptera: Rhinotermitidae
Coptotermes formosanus Shiraki	Formosan termite	Isoptera: Rhinotermitidae
Coquillettidea spp.	Mosquitoes	Diptera: Culicideae
Corticium cerealis (see *Ceratobasidium cereale*)		
Corticium fuciforme Wakef. (see *Laetisaria fuciformis*)		
Corticium sasakii Matsu. (see *Pellicularia sasakii*)		
Corynebacterium michiganense Jens. (see *Clavibacter michiganensis*)		
Corynespora melonis Lindau	Leaf spot, melon	Deuteromycetes: Moniliales
Cosmopolites sordidus (Germ.)	Banana root borer; banana weevil	Coleoptera: Curculionidae
Costelytra zealandica (White)	New Zealand grass grub	Coleoptera: Scarabaeidae
Cotesia spp.	Lepidopteran parasitic wasp	Hymenoptera: Aphidiidae
Crambus caliginosellus	Grass moth	Lepidoptera: Pyralidae
Cricetus spp.	Crickets	Saltatoria: Gryllidae
Crinipellis perniciosa (Stahel) Sing	Witches' broom	Basidiomycetes: Agaricales
Cronartium ribicola Fisch.	Blackcurrant rust	Basidiomycetes: Uredinales
Cryphonectria parasitica (Murrill) Barr.	Chestnut blight	Ascomycetes: Nectriaceae
Cryptolaemus montrouzieri Mulsant	Mealybug predator	Coleoptera: Coccinelidae
Cryptolestes spp.	Grain beetles	Coleoptera: Cucujidae
Cryptophlebia ombrodelta (Lower)	Macadamia nut borer	Lepidoptera: Tortricidae
Cryptophlebia illepida (Butler)	Koa seed worm	Lepidoptera: Tortricidae
Ctenarytaina eucalypti Maskell	Eucalyptus psyllid	Hemiptera: Psyllidae
Cucujidae	Flour beetles	Coleoptera
Culex spp.	Mosquitoes	Diptera: Culicidae

Latin	English	Order and/or Family
Culex fatigans Wd. (*Culex quinquefasciatus* Say)	House mosquito	Diptera: Culicidae
Culex quinquefasciatus Say (see *Culex fatigans* Wd.)		
Culicidae	Mosquitoes	Diptera
Culiseta spp.	Mosquitoes	Diptera: Culicidae
Curculionidae	Weevils	Coleoptera
Curvularia spp.	Leaf spot	Moniliales: Dematiaceae
Cuscuta spp.	Dodder	Convolvulaceae
Cuscuta australis R.Br.	Australian dodder	Convolvulaceae
Cuscuta campestris Yuncker	Field dodder	Convolvulaceae
Cuscuta chinensis Semen	Chinese dodder	Convolvulaceae
Cuscuta europaea L.	Large dodder; European dodder; greater dodder	Convolvulaceae
Cuscuta gronovii Willd. ex Schult.	Swamp dodder	Convolvulaceae
Cuscuta indecora Choisy	Large seed dodder	Convolvulaceae
Cuscuta planiflora Ten.	Small seed dodder	Convolvulaceae
Cydia caryana Fitch	Hickory shuckworm	Lepidoptera: Tortricidae
Cydia funebrana (Treitschke)	Plum fruit moth	Lepidoptera: Tortricidae
Cydia molesta (see *Grapholitha molesta*)		
Cydia nigicana Fabricius (= *Laspeyresia nigricana*)	Pea moth	Lepidoptera: Tortricidae
Cydia pomonella L.	Codling moth	Lepidoptera: Tortricidae
Cynodon spp.	Bermuda grass; star grasses	Poaceae
Cynodon dactylon Pers.	Bermuda grass	Poaceae
Cyperus spp.	Nutsedges	Cyperaceae
Cyperus brevifolius Hassk.	Kyllinga, green	Cyperaceae
Cyperus difformis L.	Umbrella plant	Cyperaceae
Cyperus esculentus L.	Yellow nutsedge	Cyperaceae
Cyperus rotundus L.	Nutgrass	Cyperaceae
Cyperus serotinus Rottb.	Late-flowering cyperus	Cyperaceae
Dacnusa sibirica Telenga	Chrysanthemum leaf miner parasitoid	Hymenoptera: Braconidae
Dacus spp.	Fruit flies	Diptera: Tephritidae
Dacus cucurbitae Coquillet	Melon fly	Diptera: Tephritidae
Dacus oleae (Gml.) (see *Bactrocera oleae* Gml.)		
Datura stramonium L.	Jimson weed; thorn apple	Solanaceae
Decoceras spp.	e.g. field slug	Mollusca: Gastropoda
Delia spp. (= some *Hylemya* spp.)	Root flies	Diptera: Anthomyiidae
Delia brassicae (Bouché) (see *D. radicum*)		
Delia coarctata (Fallen)	Wheat bulb fly	Diptera: Anthomyiidae

Latin	English	Order and/or Family
Delia radicum (L.)	Cabbage root fly	Diptera: Anthomyiidae
Delphasus pusillus Leconte	Whitefly predatory beetle	Coleoptera: Coccinellidae
Dendroctonus frontalis Zimmerman	Southern pine beetle	Coleoptera: Scolytidae
Dendroctonus ponderosae Hopkins	Mountain pine beetle	Coleoptera: Scolytidae
Dendroctonus pseudotsuga Hopkins	Douglas fir beetle	Coleoptera: Scolytidae
Dendroctonus rufipennis (Kirby)	Spruce beetle	Coleoptera: Scolytidae
Dermolepida albohirtum (Waterhouse)	Greyback canegrub	Coleoptera: Scarabaeidae
Deuteromycetes (= Fungi Imperfecti; mitosporitic fungi)	Fungi with no known sexual stage, or asexual stages of other fungi	
Diabrotica spp.	Corn rootworms	Coleoptera: Chrysomelidae
Diabrotica undecimpunctata Barber	Corn rootworm	Coleoptera: Chrysomelidae
Dialeurodes citri (Riley & Howard)	Citrus whitefly	Hemiptera: Aleyrodidae
Diaporthales		Ascomycetes
Diaporthe spp.	Includes stem canker fungi, various hosts	Ascomycetes: Sphaeriales
Diaporthe citri Wolf	Melanosis, citrus	Ascomycetes: Sphaeriales
Diaporthe helianthi Muntanola-Cvetkovic Mihaljcevic	Leaf spot and stem canker, sunflowers	Ascomycetes: Sphaeriales
Diaprepes abbreviatus (L.)	Sugar cane rootstalk borer	Coleoptera: Curculionidae
Diaspidae (and others)	Scale insects	Hemiptera
Diatraea saccharalis (Fabricius)	Maize stalk borer; sugar cane borer	Lepidoptera: Crambidae
Didesmococcus brevipes	Scale insect	Hemiptera: Coccidae
Didymella applanata Sacc.	Spur blight, cane fruit	Ascomycetes: Sphaeriales
Didymella chrysanthemi (Tassi) Garibaldi & Gullino (see *Didymella ligulicola*)		
Didymella ligulicola Arx.	Ray blight, chrysanthemum	Ascomycetes: Sphaeriales
Digitaria spp.	Crabgrasses	Poaceae
Digitaria adscendens Henr. (= *D. ciliaris*)	Crabgrass, tropical	Poaceae
Digitaria ciliaris Koeler (see *Digitaria adscendens*)		
Digitaria sanguinalis Scop.	Crabgrass	Poaceae
Diplocarpon earliana Wolf	Leaf scorch, strawberry	Ascomycetes: Helotiales
Diplocarpon rosae Wolf	Blackspot, roses	Ascomycetes: Helotiales
Diplodia spp.	Stalk rots, various hosts	Deuteromycetes: Sphaeropsidales
Diplodia pseudodiplodia Fckl. (perfect stage of *Nectria galligena*)	Apple and pear canker	Deuteromycetes: Sphaeropsidales
Diplopoda	Millepedes	Myriapoda
Diprion spp.	Sawflies	Hymenoptera: Diprionidae
Diptera	Flies	Insecta

Latin	English	Order and/or Family
Distantiella theobroma (Dist.)	Cocoa capsid	Heteroptera: Miridae
Ditylenchus dipsaci (Kuehn)	Stem nematode	Nematoda: Tylenchidae
Dothidiales		Ascomycetes
Drechslera graminea (Rabenh. ex Schltdl.) (see *Pyrenophora graminea*)		
Drepanopeziza ribis Hoehn. (see *Pseudopeziza ribis*)		
Drosophila spp.	Fruit flies	Diptera: Drosophilidae
Drosophilidae	Fruit flies	Diptera
Dryocoetes confusus Swaine	Western balsam bark beetle	Coleoptera: Scolytidae
Earias spp.	Spiny bollworms	Lepidoptera: Noctuidae
Echinochloa spp.	Barnyard grasses	Poaceae
Echinochloa colonum Link	Barnyard grass, awnless	Poaceae
Echinochloa crus-galli Beauv.	Barnyard grass	Poaceae
Echinochloa oryzicola (Vasinger) Vasinger (= *E. oryzoides*)	Cockspur, rice	Poaceae
Echinochloa oryzoides (Ard.) Fritsch (see *E. oryzicola*)		
Eichhornia crassipes Solms	Water hyacinth	Pontederiaceae
Elateridae	Click beetles; wireworms	Coleoptera
Eleocharis acicularis Roem. & Schult.	Spike rush	Cyperaceae
Eleusine indica Gaertn.	Goosegrass	Poaceae
Elodea canadensis Michx.	Water weed; Canadian pondweed	Hydrocharitaceae
Elsinoe fawcettii Bitanc. & Jenkins	Scab, citrus	Ascomycetes: Myriangiales
Elymus repens (L.) Gould (see *Elytrigia repens*)		
Elytrigia repens (L.) Nevski	Common couch; quackgrass	Poaceae
Empoasca spp.	Cotton leafhoppers	Hemiptera: Cicadellidae
Empoasca decipiens Poali	Leafhopper	Hemiptera: Cicadellidae
Empoasca fabae (Harris)	Green leafhopper	Hemiptera: Cicadellidae
Encarsia formosa Gahan	Glasshouse whitefly parasitoid	Hymenoptera: Aphelinidae
Endomycetales	Yeasts	Ascomycetes
Endothia parasitica Anders. & Anders.	Chestnut blight	Sphaeriales: Diaporthaceae
Eotetranychus spp.	Tetranychid mites	Acari: Tetranychidae
Eotetranychus sexmaculatus (Riley)	Six-spotted mite	Acari: Tetranychidae
Eotetranychus willamettei (McGregor)	Willamette mite	Acari: Tetranychidae
Ephestia elutella (Hübner)	Warehouse moth	Lepidoptera: Pyralidae
Epilachna spp.	Bean beetles	Coleoptera: Coccinellidae
Epilachna varivestis (Muls.)	Mexican bean beetle	Coleoptera: Coccinellidae

Latin	English	Order and/or Family
Epitrimerus pyri Nalepa	Pear rust mite	Acari: Eriophyidae
Epitrix hirtipennis (Marsh)	Tobacco flea beetle	Coleoptera: Chrysomelidae
Eretmocerus sp. nr. *californicus* Howard (see *Eretmocerus eremicus*)		
Eretmocerus eremicus Rose & Zolnerowich	Whitefly parasite	Hymenoptera: Aphelinidae
Eriophyes spp.	Mites	Acari: Eriophyidae
Eriophyidae	Eriophyid mites	Acari
Eriosoma lanigerum (Hausmann)	Woolly aphid	Hemiptera: Pemphigidae
Erwinia amylovora Winsl.	Fire blight of pome fruit	Eubacteriales: Enterobacteriaceae
Erwinia carotovora Holl.	Bacterial rot, celery; basal stem rot, cucurbits; blackleg, potatoes	Eubacteriales: Enterobacteriaceae
Erysiphaceae		Erysiphales
Erysiphales	Powdery mildews	Ascomycetes
Erysiphe spp.	Powdery mildew, various hosts	Ascomycetes: Erysiphales
Erysiphe betae (Vanha) Weltzien	Powdery mildew, beet	Ascomycetes: Erysiphales
Erysiphe cichoracearum DC.	Powdery mildew, cucurbits	Ascomycetes: Erysiphales
Erysiphe graminis DC. (see *Blumeria graminis*)		
Erythroneura elegantula (Osborn)	Grape leafhopper	Hemiptera: Cicadellidae
Eubacteriales	Cellular bacteria	
Eucosma gloriola Heinrich	Eastern pine shoot borer	Lepidoptera: Tortricidae
Eupatorium odoratum L. (= *Chromolaena odorata*)	Siam weed	Asteraceae
Euphorbia esula L.	Leafy spurge	Euphorbiaceae
Euphorbia maculata L.	Spotted spurge	Euphorbiaceae
Eupoecilia ambiguella (Hübner)	Grape berry moth	Lepidoptera: Cochylidae
Eupterycyba jucunda	Potato leafhopper	Hemiptera: Cicadellidae
Eutetranychus spp.	Tetranychid mites	Acari: Tetranychidae
Eutetranychus banksi (McGregor)	Texas citrus mite	Acari: Tetranychidae
Euxoa spp.	Cutworms; dart moths	Lepidoptera: Noctuidae
Exobasidium vexans Mass.	Blister blight, tea	Basidiomycetes: Exobasidiales
Fallopia convolvulus Adans.	Black bindweed	Polygonaceae
Feltiella acarisuga (Vallot)	Mite predator	Diptera: Cecidomyiidae
Fimbristylis spp.	Fringe rushes	Cyperaceae
Fomes annosus Cke.	Butt rot, conifers	Basidiomycetes: Agaricales
Formicidae	Ants	Hymenoptera
Frankliniella intonsa (Trybom)	Flower thrips	Thysanoptera: Thripidae
Frankliniella occidentalis (Pergande)	Western flower thrips	Thysanoptera: Thripidae
Fuchsia hybrida Voss	Fuchsia	Onagraceae
Fulvia spp.	Leaf moulds	Deuteromycetes: Hyphales
Fulvia fulva Cif.	Leaf mould, tomato	Deuteromycetes: Hyphales

Latin	English	Order and/or Family
Fusarium spp.	Rots, ear blights and wilts, various hosts (Imperfect fungi with perfect stages in various genera)	Deuteromycetes: Moniliales
Fusarium coeruleum Sacc.	Dry rot, post-harvest rot	Deuteromycetes: Moniliales
Fusarium culmorum Sacc.	Fusarium foot and root rots, various hosts	Deuteromycetes: Moniliales
Fusarium graminearum Schwabe (see *Gibberella zeae*)		
Fusarium moniliforme Sheldon (see *Gibberella fujikuroi* Wr.)		
Fusarium moniliforme var. *subglutinans* Wr. & Reinking	Pitch canker disease	Deuteromycetes: Moniliales
Fusarium nivale Ces. (see *Microdochium nivalis*)		
Fusarium oxysporum Schlect.	Fusarium wilt, various hosts	Deuteromycetes: Moniliales
Galendromus occidentalis (Nesbitt)	Mite predator	Acarina: Phytoseiidae
Galium aparine L.	Cleavers	Rubiaceae
Ganoderma spp.	White rot, timber	Basidiomycetes: Agaricales
Gastropoda	Slugs and snails	Mollusca
Geotrichum candidum Ferr.	Rubbery rot, potatoes	Deuteromycetes: Moniliales
Geranium spp.	Crane's bills	Geraniaceae
Gibberella spp. (= various *Fusarium* spp.)	Scab, cereals; brown foot rot and ear blight and other cereal diseases	Deuteromycetes: Hypocreales
Gibberella fujikuroi Wr.	Banana black heart; cotton boll rot; maize stalk rot	Deuteromycetes: Hypocreales
Gibberella zeae Petch	Scab, cereals	Deuteromycetes: Hypocreales
Globodera spp.	Potato cyst nematodes	Nematoda: Heteroderidae
Gloeodes pomigena Colby	Sooty blotch, apple pear and citrus	Deuteromycetes: Sphaeropsidales
Gloeosporium spp.	Gloeosporium rot, apples	Deuteromycetes: Sphaeriales
Gloeosporium fructigenum Berk. (see *Glomerella cingulata*)		
Glomerella cingulata Spauld. & Schrenk	Gloeosporium rot, apples	Deuteromycetes: Sphaeriales
Gnathotrichus retusus (LeConte)	Ambrosia beetle	Coleoptera: Scolytidae
Gnathotrichus sulcatus (LeConte)	Ambrosia beetle	Coleoptera: Scolytidae
Gonipterus gibberus Boisduval	Eucalyptus snout beetle	Coleoptera: Curculionidae
Grapholitha molesta (Busck)	Oriental fruit moth	Lepidoptera: Tortricidae
Gryllidae	True crickets	Orthoptera: Gryllotalpidae
Gryllotalpa spp.	Mole crickets	Orthoptera: Gryllotalpidae
Gryllotalpa gryllotalpa (L.)	Mole cricket	Orthoptera: Gryllotalpidae
Guignardia bidwellii Viala & Rivas	Black rot, grapevines	Ascomycetes: Sphaeriales
Gymnosporangium spp.	Leaf scorch, apples; rust, various hosts	Basidiomycetes: Uredinales

Latin	English	Order and/or Family
Gymnosporangium fuscum DC.	Pear rust	Basidiomycetes: Uredinales
Harmonia axyridis Pallas	Ladybird; ladybug	Coleoptera: Coccinellidae
Hauptidia maraccana Melichar	Leafhopper	Hemiptera: Cicadellidae
Hedera helix L.	Ivy	Araliaceae
Helianthus annuus L.	Sunflower	Asteraceae
Helicotylenchus spp.	Spiral nematodes	Nematoda: Tylenchidae
Helicoverpa armigera (Hübner)	Old World bollworm	Lepidoptera: Noctuidae
Helicoverpa assulta (Guen.)	Oriental tobacco budworm	Lepidoptera: Noctuidae
Helicoverpa zea Boddie	American bollworm; tomato fruitworm; corn earworm	Lepidoptera: Noctuidae
Heliothis armigera (Hübner) (see *Helicoverpa armigera*)		
Heliothis assulta (Guen.) (see *Helicoverpa assulta*)		
Heliothis virescens (Fabricius)	Tobacco budworm	Lepidoptera: Noctuidae
Heliothis zea (Boddie) (see *Helicoverpa zea*)		
Helminthosporium oryzae B. de Haan	Helminthosporium blight, rice	Deuteromycetes: Moniliales
Helminthosporium solani Dur. & Mont.	Silver scurf, potatoes	Deuteromycetes: Moniliales
Helminthosporium turcicum Pass.	Northern leaf blight, maize	Deuteromycetes: Moniliales
Helotiales		Ascomycotina
Hemileia vastatrix Berk. & Br.	Coffee rust	Basidiomycetes: Uredinales
Hemitarsonemus latus (Banks) (see *Polyphagotarsonemus latus*)		
Heterobasidion annosum Bref. (see *Fomes annosus*)		
Heterodera spp.	Lemon-shaped cyst nematodes	Nematoda: Heteroderidae
Heterodera cruciſereae Fran.	Brassica cyst nematode	Nematoda: Heteroderidae
Heterodera glycines Ichinohe	Soybean cyst nematode	Nematoda: Heteroderidae
Heterodera goettingiana Liebs.	Pea cyst nematode	Nematoda: Heteroderidae
Heterodera schachtii Schm.	Beet cyst nematode	Nematoda: Heteroderidae
Heteroderidae	Cyst nematodes	Nematoda
Heteropeza pygmaea Winn.	Mushroom cecid	Diptera: Cecidomyiidae
Heteroptera	Bugs	Hemiptera
Hippodamia convergens Guerin	Ladybird; ladybug	Coleoptera: Coccinellidae
Homona spp.	Tortrix moths and leaf rollers	Lepidoptera: Tortricidae
Homona magnanima Diakonoff	Tea tortrix	Lepidoptera: Tortricidae
Homoptera	Aphids, hoppers, etc.	Hemiptera
Hoplia philanthus (Foerster)	Welsh chafer	Coleoptera: Scarabaeidae
Hoplochelis marginalis (Fairmaire)	White grub	Coleoptera: Scarabaeidae
Hydrilla verticillata (L.F.) Royle	Florida elodea; waterthyme	Hydrocharitaceae
Hylemya spp. (see *Delia* spp.)		
Hymenoptera	Ants; bees; wasps; sawflies	Insecta

Latin	English	Order and/or Family
Hypera postica (Gyllenhal)	Alfalfa weevil	Coleoptera: Curculionidae
Hypericum montanum L.	Pale St John's Wort	Guttifereae
Hypericum perforatum L.	St John's Wort	Guttifereae
Hyphales		Deuteromycotina
Hypoaspis aculeifer (Canestrini)	Fungus gnat predator	Mesostigmata: Laelapidae
Hypoaspis miles (Berlese)	Sciarid fly predator	Mesostigmata: Laelapidae
Hypocreales		Ascomycotina
Icerya purchasi Maskell	Cottony cushion scale	Hemiptera: Margarodidae
Ipomoea hederacea Jacq.	Morning glory, ivyleaf	Convolvulaceae
Ipomoea purpurea Roth	Morning glory, tall	Convolvulaceae
Ips sexdentatus (Borner)	Six-spined ips	Coleoptera: Ipidae
Ips typographus (L.)	Douglas fir beetle	Coleoptera: Ipidae
Ischaemum rugosum Salisb.	Saramatta grass	Poaceae
Juncus maritimus Lam.	Sea-rush	Juncaceae
Jussiaea spp.	Water primroses	Onagraceae
Jussiaea diffusa Forsk.		
(see *Ludwigia peploides*)		
Keiferia lycopersicella (Walsingham)	Tomato pinworm	Lepidoptera: Gelechiidae
Kochia scoparia Roth.	Mock cypress	Chenopodiaceae
Laetisaria fuciformis (McAlpine) Burds.	Red thread	Basidiomycetes: Aphyllophorales
Lamium purpureum L.	Red dead-nettle	Labiateae
Laodelphax striatella (Fallen)	Small brown planthopper	Hemiptera: Delphacidae
Lapsana communis L.	Nipplewort	Asteraceae
Laspeyresia nigricana (see *Cydia nigricana*)		
Lepidoglyphus (*Glycyphagus*) *destructor* (Schrank)	Storage mite	Acarina: Acaridae
Lepidoptera	Butterflies; moths	Insecta
Lepidosaphes beckii (Newman) (*Cornuaspis* beckii (Newman))	Citrus mussel scale; citrus purple scale	Hemiptera: Coccidae
Lepidosaphes yangicola (Kuwana)	*Euonymus alatus* scale	Hemiptera: Coccidae
Leptinotarsa decemlineata (Say)	Colorado beetle	Coleoptera: Chrysomelidae
Leptochloa spp.	Sprangletop grasses	Poaceae
Leptochloa chinensis Nees	Sprangletop, red	Poaceae
Leptochloa fascicularis Gray (= *Diplachne fascicularis* (Lam.) P. Beauv.)	Sprangletop, bearded	Poaceae
Leptomastix dactylopii Howard	Mealybug parasite	Hymenoptera: Encrytidae
Leptosphaeria nodorum Muell. (= *Septoria nodorum*)	Glume blotch, wheat	Ascomycetes: Sphaeriales

Latin	English	Order and/or Family
Leucoptera spp.	Leaf-mining moths	Lepidoptera: Lyonetiidae
Leucoptera malifoliella (Costa)	Pear leaf blister moth	Lepidoptera: Lyonetiidae
Leucoptera scitella Zeller (see *L. malifoliella*)		
Leveillula spp.	Powdery mildew	Ascomycetes: Erysiphales
Leveillula taurica Arn.	Powdery mildew, pepper and tomato	Ascomycetes: Erysiphales
Lindernia procumbens Philcox	Pimpernel, false	Scrophulariaceae
Liriomyza spp.	Leaf miners	Diptera: Agromyzidae
Liriomyza bryoniae (Kalt.)	Tomato leaf miner	Diptera: Agromyzidae
Liriomyza huidobrensis (Blan.)	South American leaf miner	Diptera: Agromyzidae
Liriomyza trifolii (Burgess)	American serpentine leaf miner	Diptera: Agromyzidae
Lissorhoptrus oryzophilus Kusch	Rice water weevil	Coleoptera: Curculionidae
Lithocolletis spp. (see *Phyllonorycter* spp.)		
Lobesia botrana (Denis & Schiffermueller)	European grapevine moth	Lepidoptera: Tortricidae
Lolium spp.	Ryegrasses	Poaceae
Lolium multiflorum Lam.	Ryegrass, Italian	Poaceae
Lolium perenne L.	Ryegrass, perennial	Poaceae
Lolium rigidum Gaud.	Wimmera ryegrass; annual ryegrass	Poaceae
Longidorus spp.	Needle nematodes	Nematoda
Ludwigia palustris Ell.	Water purslane	Onagraceae
Ludwigia peploides (Kunth) Raven	Water primrose	Onagraceae
Lycoriella auripila (Fitch)	Mushroom sciarid	Diptera: Sciaridae
Lygocoris pabulinus L.	Common green capsid	Hemiptera: Miridae
Lygus herperus Knight	Western tarnished plant bug	Hemiptera: Miridae
Lygus lineolaris (Palisot de Beauvois)	Southern tarnished plant bug	Hemiptera: Miridae
Lygus pabulinus L.	Tarnished plant bug	Hemiptera: Miridae
Lymantria dispar (L.)	Gypsy moth	Lepidoptera: Lymantriidae
Lyonetia clerkella (L.)	Apple leaf miner	Lepidoptera: Lyonetiidae
Macrosiphum euphorbiae (Thomas)	Potato aphid	Hemiptera: Aphididae
Macrosiphum rosae (L.)	Rose aphid	Hemiptera: Aphididae
Macrotermes species	Macrotermites	Uniramia: Macrotermitinae
Magnaporthe grisea (T.T. Hebert) M.E. Barr (see also *Pyricularia oryzae*)	Rice blast (perfect stage)	Ascomycetes: Magnaporthaceae
Mahanarva postica (Stål)	Sugarcane spittle bug	Hemiptera: Cercopidae
Malva spp.	Mallow	Malvaceae
Mamestra brassicae (L.)	Cabbage moth	Lepidoptera: Noctuidae
Mamestra configurata Walker	Bertha armyworm	Lepidoptera: Noctuidae

Latin	English	Order and/or Family
Marasmius oreades (Bolton) Fr. and other species	Fairy rings	Agaricales
Margarodidae	Scale insects	Hemiptera
Marsilea spp.	Four-leaved water clover	Marsileaceae
Marssonina spp.	Leaf blotches etc., various hosts	Melanconiales
Marssonina potentillae (Desm.) Magnus ssp. *fragariae* (Lib.) Ohl (see *Diplocarpon earliana*)		
Mastigomycotina	Primitive fungi (Phycomycetes) producing motile spores	
Matricaria spp.	Mayweeds (some)	Asteraceae
Matricaria perforata Mérat. (= *M. inodora* L. and *Tripleurospermum perforatum* (Mérat))	Scentless mayweed	Asteraceae
Megalurothrips sjostedti (Trybom)	Legume flower thrips	Thysanoptera: Thripidae
Megaselia spp.	Scuttle flies	Diptera: Phoridae
Melanconiales		Deuteromycotina
Meligethes spp.	Blossom beetles; pollen beetles	Coleoptera: Nitidulidae
Meligethes aeneus (Fabricius)	Pollen beetle	Coleoptera: Nitidulidae
Meloidogyne spp.	Root-knot nematodes	Nematoda
Meloidogyne incognita (K. & W.)	Southern root-knot nematode	Nematoda
Melolontha spp.	Cockchafers	Coleoptera: Scarabaeidae
Melolontha hippocastani (L.)	Cockchafer	Coleoptera: Scarabaeidae
Melolontha melolontha (L.)	Cockchafer	Coleoptera: Scarabaeidae
Metaphycus bartletti Annecke & Mynhardt	Scale insect parasite	Hymenoptera: Encrytidae
Metaphycus helvolus (Compère)	Scale insect parasite	Hymenoptera: Encrytidae
Metcalfa pruinosa Say	Citrus planthopper	Hemiptera: Fulgoroidea
Microdochium nivalis (Fr.) Samuel & IC Hallett	Snow mould, grasses, cereals	Deuteromycetes: Moniliales
Microtermes spp.	Macrotermites	Uniramia: Macrotermitinae
Microthyriella rubi Petr. (see *Schizothyrium pomi*)		
Miridae	Capsid bugs	Heteroptera
Mollusca	Slugs and snails	
Monilia spp.	Various rots	Deuteromycetes: Moniliales
Monilia laxa Sacc. (see *Sclerotinia fructigena*)		
Monilia roreri Cif.	Pod rot, cocoa	Deuteromycetes: Moniliales
Monilinia spp. (see *Sclerotinia* spp.)		
Monilinia laxa Honey (see *Sclerotinia fructigena*)		
Monilinia mali (Takahashi) Whetzel	Monilinia leaf blight, apple	Deuteromycetes: Moniliales
Monochoria vaginalis Presl.	Pickerel weed	Pontederiaceae
Monographella nivalis (Schaffnit) E. Mull.	Rice leaf scald	Ascomycetes: Sphaeriales

Latin	English	Order and/or Family
Monomorium spp.	Seed-eating ants	Hymenoptera: Formicidae
Monomorium pharaonis (L.)	Pharaoh's ant	Hymenoptera: Formicidae
Morrenia odorata Lindl.	Strangler vine; milkweed vine	Asclepiadaceae
Mucor spp.	Fruit rot, strawberries	Mucorales
Mucorales		Zygomycotina
Musca domestica L.	Housefly	Diptera: Muscidae
Mycogone perniciosa Magn.	White mould, mushrooms	Deuteromycetes: Moniliales
Mycosphaerella spp.	Leaf spot diseases, various hosts	Ascomycetes: Sphaeriales
Mycosphaerella arachidis Deighton	Brown spot, peanut	Ascomycetes: Sphaeriales
Mycosphaerella brassicicola Dud.	Ring-spot, brassicas	Ascomycetes: Sphaeriales
Mycosphaerella fijiensis Deighton	Black leaf streak, banana	Ascomycetes: Sphaeriales
Mycosphaerella fragariae Lindau	White leaf spot, strawberry	Ascomycetes: Sphaeriales
Mycosphaerella graminicola (Fuckel) J. Schröt.	Septoria leaf spot, wheat	Ascomycetes: Sphaeriales
Mycosphaerella musicola Leach	Banana leaf spot, sigatoka	Ascomycetes: Sphaeriales
Mycosphaerella pinodes Stone (see *Ascochyta pinodes*)		
Mycosphaerella pomi Lindau	Brooks spot, apple	Ascomycetes: Sphaeriales
Myzus nicotianae Blackman	Tobacco aphid	Hemiptera: Aphididae
Myzus persicae (Sulzer)	Peach-potato aphid	Hemiptera: Aphididae
Nectria galligena Bres. (imperfect stage of *Diplodia pseudodiplodia* Fckl.)	Canker, apple, pear	Hypocreales
Nematoda	Nematodes	
Neodiprion abietis (Harris)	Balsam whitefly	Hymenoptera: Diprionidae
Neodiprion lecontei Fitch	Pine sawfly	Hymenoptera: Diprionidae
Neodiprion sertifer Geoffrey	European pine sawfly	Hymenoptera: Diprionidae
Nephotettix spp.	Green leafhoppers	Hemiptera: Cicadellidae
Nephotettix cincticeps (Uhl.)	Green rice leafhopper	Hemiptera: Cicadellidae
Nephotettix nigropictus (Stål.)	Tropical green rice leafhopper	Hemiptera: Cicadellidae
Nicotiana spp.	Tobacco	Solanaceae
Nicotiana rustica L.	Tobacco	Solanaceae
Nilaparvata spp.	Planthoppers	Hemiptera: Delphacidae
Nilaparvata lugens (Stal.)	Rice brown planthopper	Hemiptera: Delphacidae
Nitrosomonas spp.	N-fixing bacteria	Bacterium
Noctua spp.	Cutworms	Lepidoptera: Noctuidae
Noctua pronuba (L.)	Yellow underwing moth; cutworm	Lepidoptera: Noctuidae
Noctuidae	Noctuid moths	Lepidoptera
Oculimacula acuformis (Boerema, R. Pieters & Hamers) Crous & W. Gams	Eye-spot, cereals	Deuteromycetes: Moniliales
Oculimacula yallundae (Wallwork & Spooner) Crous & W. Gams	Eye-spot, cereals	Deuteromycetes: Moniliales
Odontotermes species	Macrotermites	Uniramia: Macrotermitinae

Latin	English	Order and/or Family
Oidiopsis taurica Salm.	Tomato powdery mildew	Deuteromycetes: Moniliales
Oidium hevea Steinm.	Powdery mildew	Ascomycetes: Erysiphales
Oligonychus illicis (McGregor)	Southern red mite	Acari: Tetranychidae
Oligonychus perseae Tuttle, Baker & Abbatiello	Persea mite	Acari: Tetranychidae
Oligonychus punicae (Hirst)	Avocado brown mite	Acari: Tetranychidae
Oligonychus ununguis (Jacobi)	Spruce spider mite	Acari: Tetranychidae
Chromista: Oomycetes:		Mastigomycotina
Oospora lactis Sacc. (see *Geotrichum candidum*)		
Oospora pustulans Owen & Wakef. (see *Polyscytalum pustulans*)		
Opomyza spp.	Grass flies; cereal flies	Diptera: Opomyzidae
Opomyza florum (Fabricius)	Yellow cereal fly	Diptera: Opomyzidae
Opuntia lindheimerii Engelmann	Prickly pear cactus	Cactaceae
Orgyia leucostigma (Sm.)	Whitemarked tussock moth	Lepidoptera: Lymantriidae
Orgyia pseudotsugata (McDunnough)	Douglas fir tussock moth	Lepidoptera: Lymantriidae
Orius spp.	Minute pirate bugs	Hemiptera: Anthocoridae
Oryctes rhinoceros (L.)	Coconut rhinoceros beetle	Coleoptera: Scarabaeidae
Oscinella frit (L.)	Frit fly	Diptera: Chloropidae
Ostrinia furnacalis Guenee	Asiatic corn borer	Lepidoptera: Pyralidae
Ostrinia nubilalis (Hübner)	European corn borer	Lepidoptera: Pyralidae
Otiorhynchus sulcatus (Fabricius)	Vine weevil	Coleoptera: Curculionidae
Oulema melanopus (L.)	Cereal leaf beetle	Coleoptera: Chrysomelidae
Oulema oryzae (Kuway.)	Rice leaf beetle	Coleoptera: Chrysomelidae
Pachnaeus litus Germar	Citrus weevil	Coleoptera: Curculionidae
Paecilomyces spp.	Saprophytic fungi	Hyphales
Pandemis heparana (Denis & Schiffermueller)	Leaf roller	Lepidoptera: Totricidae
Panicum spp.	Panic grasses	Poaceae
Panicum dichotomiflorum Michx.	Fall panicum; smooth witchgrass	Poaceae
Panicum purpurascens Raddi (see *Urochloa mutica*)		
Panicum texanum Buckl.	Millet, Texas	Poaceae
Panonychus spp.	Red spider mites	Acari: Tetranychidae
Panonychus citri (McGregor)	Citrus red mite	Acari: Tetranychidae
Panonychus ulmi (Koch)	Fruit tree red spider mite	Acari: Tetranychidae
Papaver spp.	Poppies	Papaveraceae
Parapediasia teterrella (Zincken)	Bluegrass webworm	Lepidoptera: Crambidae
Parlatoria ziziphi (Lucas)	Citrus black scale	Hemiptera: Diaspididae
Parthenothrips dracaenae (Hegeer)	Palm thrips	Thysanoptera: Thripidae
Pectinophora gossypiella (Saunders)	Pink bollworm	Lepidoptera: Gelechiidae
Pegomya betae (Curtis) (see *P. hyoscamni*)		
Pegomya hyoscamni (Panzer)	Beet leaf-miner; mangold fly	Diptera: Anthomyiidae

Latin	English	Order and/or Family
Pellicularia spp.	Rots, damping off, etc., various hosts	Basidiomycetes: Agaricales
Pellicularia sasakii Ito	Rice sheath blight	Basidiomycetes: Agaricales
Penicillium spp.	Penicillium rots	Deuteromycetes: Moniliales
Penicillium digitatum Sacc.	Green mould, citrus	Deuteromycetes: Moniliales
Penicillium expansum Link	Blue mould; blue rot, apples and pears	Deuteromycetes: Moniliales
Penicillium italicum Wehm.	Blue mould, citrus	Deuteromycetes: Moniliales
Periplaneta americana (L.)	American cockroach	Dictyoptera: Blattidae
Peronosclerospora spp.	Downy mildew, sorghum	Chromista: Oomycetes: Peronosporales
Peronospora spp.	Downy mildews	Chromista: Oomycetes: Peronosporales
Peronospora parasitica Fr.	Downy mildew, brassicae	Chromista: Oomycetes: Peronosporales
Peronospora tabacina Adam (= *Plasmopara tabacina*)	Blue mould, tobacco	Chromista: Oomycetes: Peronosporales
Peronosporales	Downy mildews, etc.	Chromista: Oomycetes
Petunia spp.	Petunia	Solanaceae
Phakopsora pachyrhizi Syd.	Rust, soybean	Basidiomycetes: Uredinales
Phalaris spp.	Canary grasses	Poaceae
Phalaris paradoxa L.	Canary grass, awned	Poaceae
Phoma spp.	Root and stem rots, various	Deuteromycetes: Sphaeropsidales
Phoma exigua Desh. var. *foveata*	Gangrene, potatoes	Deuteromycetes: Sphaeropsidales
Phomopsis spp. (see *Diaporthe* spp.)		
Phomopsis citri Fawc. (see *Diaporthe citri*)		
Phomopsis helianthi Munt.-Cvetk., Mihaljč. & M. Petrov, (see *Diaporthe helianthi*)		
Phomopsis viticola Sacc.	Dead arm, grape vines	Deuteromycetes: Sphaeropsidales
Phoridae	Scuttle flies	Diptera
Phorodon humuli (Schrank)	Damson-hop aphid	Hemiptera: Aphididae
Phragmidium mucronatum Schlect.	Rust, roses	Basidiomycetes: Uredinales
Phragmidium violaceum (Schultz) G. Winter	Rust, blackberry	Basidiomycetes: Uredinales
Phthorimaea operculella Zeller	Potato moth	Lepidoptera: Gelechiidae
Phycomycetes	Primitive fungi with coenocytic mycelium; includes the Divisions Mastigomycotina and Zygomycotina	
Phyllactinia spp.	Powdery mildew, various hosts	Erysiphales
Phyllocoptes spp.	Mites	Acari: Eriophyidae
Phyllocoptruta spp.	Rust mites	Acari: Eriophyidae

723

Latin	English	Order and/or Family
Phyllocoptruta oleivora (Ashmead)	Citrus rust mite	Acari: Eriophyidae
Phyllonorycter spp.	Leaf-mining moths	Lepidoptera: Gracillariidae
Phyllonorycter blancardella (Fabricius)	Apple leaf miner	Lepidoptera: Gracillariidae
Phyllopertha horticola (L.)	Garden chafer; bracket chafer	Coleoptera: Scarabaeidae
Phyllotreta spp.	Flea beetles	Coleoptera: Chrysomelidae
Phyllotreta striolata (Fabricius)	Flea beetle	Coleoptera: Chrysomelidae
Physalospora obtusa Cke. (see *Botryosphaeria obtusa*)		
Phytomyza spp.	Leaf miners	Diptera: Agromyzidae
Phytomyza syngenesiae (Hardy)	Chrysanthemum leaf miner	Diptera: Agromyzidae
Phytophthora spp.	Blight, damping off, foot-rot, various hosts	Chromista: Oomycetes: Peronosporales
Phytophthora cactorum Schroet.	Collar rot, crown rot, apple	Chromista: Oomycetes: Peronosporales
Phytophthora capsici Leonian	Blight, capsicums	Chromista: Oomycetes: Peronosporales
Phytophthora fragariae Hickman	Red core, strawberry	Chromista: Oomycetes: Peronosporales
Phytophthora infestans De Bary	Late blight, potato, tomato	Chromista: Oomycetes: Peronosporales
Phytophthora megasperma Drechs.	Root rot, brassicas	Chromista: Oomycetes: Peronosporales
Phytophthora palmivora (Butl.) Butl.	Rot, various crops	Chromista: Oomycetes: Peronosporales
Phytoseiulus persimilis Anthios-Henriot	Two-spotted spider mite predator	Mesostigmata: Phytoseiidae
Pieris spp.	Cabbage white butterflies	Lepidoptera: Pieridae
Pieris brassicae (L.)	Large white butterfly	Lepidoptera: Pieridae
Pieris rapae (L.)	Small white butterfly	Lepidoptera: Pieridae
Pikonema alaskensis (Rohwer)	Yellow-headed spruce sawfly	Hymenoptera: Tenthredinidae
Pistia stratiotes L.	Water duckweed	Araceae
Pityogenes chalcographus (L.)	Six-toothed spruce bark beetle	Coleoptera: Scolytidae
Planococcus citri (Risso)	Citrus mealybug	Hemiptera: Pseudococcidae
Plantago spp.	Plantains	Plantaginaceae
Plasmodiophora brassicae Woron.	Clubroot, brassicas	Chromista: Phycomycetes: Plasmodiophorales
Plasmodiophoromycetes	Parasitic members of the Myxomycota - a group with affinities with both primitive fungi and primitive animals	
Plasmopara spp.	Downy mildews, various hosts	Chromista: Oomycetes: Peronosporales
Plasmopara tabacina (see *Peronospora tabacina*)		
Plasmopara viticola Berl. & de T.	Downy mildew, grapevine	Chromista: Oomycetes: Peronosporales

Latin	English	Order and/or Family
Platynota idaeusalis (Walker)	Tufted apple moth	Lepidoptera: Tortricidae
Platynota stultana (Walsingham)	Leaf roller	Lepidoptera: Tortricidae
Platyptilia carduidactyla (Riley)	Artichoke plume moth	Lepidoptera: Pterophoridae
Plodia interpunctella (Hübner)	Indian meal moth	Lepidoptera: Pyralidae
Plusia spp.	e.g. silvery moth	Lepidoptera: Noctuidae
Plutella xylostella (L.)	Diamondback moth	Lepidoptera: Yponomeutidae
Poa spp.	Meadow grasses	Poaceae
Poa annua L.	Meadow grass, annual	Poaceae
Poa trivialis L.	Meadow grass, rough	Poaceae
Podisus maculiventris (Say)	Caterpillar predator	Heteroptera: Pentatomidae
Podosphaera spp.	Powdery mildew, various hosts	Ascomycetes: Erysiphales
Podosphaera leucotricha (Ellis & Everhart) Salmon	Powdery mildew, apple	Ascomycetes: Erysiphales
Polychrosis botrana Schiffermueller (see *Lobesia botrana* (Denis & Schiffermueller))		
Polygonum spp.	Knotweeds	Polygonaceae
Polygonum aviculare L.	Knot grass	Polygonaceae
Polygonum convolvulus L. (see *Fallopia convolvulus* Adans.)		
Polygonum cuspidatum Sieb. & Zucc. (see *Reynoutria japonica*)		
Polygonum lapathifolium L.	Pale persicaria	Polygonaceae
Polygonum persicaria L.	Redshank; persicaria; smartweed	Polygonaceae
Polygonum sachalinense F. Schmidt ex Maxim (see *Reynoutria sachalinensis*)		
Polymyxa betae Keskin	Fungal vector of rhizomania virus	Phycomycetes: Plasmodiophorales
Polyphagotarsonemus latus (Banks)	Broad mite	Acari: Tarsonemidae
Polyscytalum pustulans	Skin spot, potatoes	Deuteromycetes: Hyphales
Polystigmatales		Ascomycetales
Popillia japonica Newman	Japanese beetle	Coleoptera: Scarabaeidae
Populus spp.	Poplars	Salicaceae
Portulaca spp.	Purslanes	Portulacaceae
Portulaca oleracea L.	Purslane	Portulacaceae
Potamogeton spp.	Pondweeds	Potamogetonaceae
Potamogeton distinctus Benn.	Pondweed, American	Potamogetonaceae
Pratylenchus spp.	Root-lesion nematodes	Nematoda: Hoplolaimidae
Prays citri Millière	Citrus flower moth	Lepidoptera: Yponomeutidae
Prays oleae (Bernard)	Olive moth	Lepidoptera: Yponomeutidae
Prunus serotina Ehrh.	American black cherry	Rosaceae
Pseudocercosporella capsellae	White leaf spot, oilseed rape	Deuteromycetes: Hyphales
Pseudococcidae	Mealybugs	Hemiptera
Pseudococcus spp.	Mealybugs	Hemiptera: Pseudococcidae
Pseudococcus affinis (Maskell) Hemiptera: Pseudococcidae	Obscure mealybug	Hemiptera: Pseudococcidae

Latin	English	Order and/or Family
Pseudococcus longispinus (Targioni-Tozzetti)	Long-tailed mealybug	Hemiptera: Pseudococcidae
Pseudomonas spp.	Bacterial blights and leaf spots, various hosts	Pseudomonadales: Pseudomonadaceae
Pseudomonas glumae	Bacterial grain rot, rice	Pseudomonadales: Pseudomonadaceae
Pseudomonas lachrymans Carsner	Angular leaf spot, cucurbits	Pseudomonadales: Pseudomonadaceae
Pseudomonas mors-prunorum Wormould	Bacterial canker, prunus	Pseudomonadales: Pseudomonadaceae
Pseudomonas phaseolicola Dows	Halo blight, beans	Pseudomonadales: Pseudomonadaceae
Pseudomonas syringae Van Hall pv. *lachrymans* (see *P. lachrymans*)		
Pseudomonas syringae Van Hall pv. *mors-prunorum* (see *P. mors-prunorum*)		
Pseudomonas tabaci Stevens	Wild fire of tobacco and soybean	Pseudomonadales: Pseudomonadaceae
Pseudomonas tolaasii Paine	Brown fleck of mushrooms	Pseudomonadales: Pseudomonadaceae
Pseudoperonospora cubensis Rostow	Downy mildew, cucurbits	Chromista: Oomycetes: Peronosporaceae
Pseudoperonospora humuli Wils.	Downy mildew, hops	Chromista: Oomycetes: Peronosporaceae
Pseudopeziza ribis Kleb.	Leaf spot, currants and gooseberry	Ascomycetes: Helotiales
Pseudotsuga menziesii Franco	Douglas fir	Pinaceae
Psila rosae (Fabricius)	Carrot fly	Diptera: Psilidae
Psylla spp.	Psyllids	Hemiptera: Psyllidae
Psyllidae	Psyllids	Hemiptera
Pteridium aquilinum Kuhn	Bracken	Filicales
Puccinia spp.	Rust, various hosts	Basidiomycetes: Uredinales
Puccinia chrysanthemi Roze	Brown rust, chrysanthemum	Basidiomycetes: Uredinales
Puccinia dispersa f. sp. *tritici* Eriks. & Henn. (see *Puccinia triticana*)		
Puccinia graminis Pers.	Black stem rust, grasses	Basidiomycetes: Uredinales
Puccinia hordei Otth	Brown rust, barley	Basidiomycetes: Uredinales
Puccinia recondita f. sp. *tritici* (Eriks. & Henn.) Henderson (see *Puccinia triticana*)		
Puccinia striiformis West	Yellow rust, cereals	Basidiomycetes: Uredinales
Puccinia triticana Eriks.	Brown rust, wheat	Basidiomycetes: Uredinales
Pyralidae	Pyralid moths	Lepidoptera
Pyrenopeziza brassicae Sutton & Rawl.	Light leaf spot, brassicas	Ascomycetes: Helotiales
Pyrenophora graminea Ito & Kuribay	Leaf stripe, barley	Ascomycetes: Sphaeriales

Latin	English	Order and/or Family
Pyrenophora teres Drechs.	Net blotch, barley	Ascomycetes: Sphaeriales
Pyrenophora tritici-vulgaris Dicks	Tan spot, wheat	Ascomycetes: Sphaeriales
Pyricularia oryzae Cavara (see also *Magnaporthe grisea*)	Rice blast (imperfect stage)	Deuteromycetes: Moniliales
Pythium spp.	Root rots, various	Chromista: Oomycetes: Peronosporales
Quadraspidiotus perniciosus (Comstock)	San José scale	Hemiptera: Coccidae
Radopholus similis (Cobb) Thorne	Burrowing nematode	Nematoda: Tylenchidae
Ramularia spp.	Leaf spots, various	Deuteromycetes: Moniliales
Ramularia beticola Fautr. & Lambotte	Leaf spot, beet	Deuteromycetes: Moniliales
Ranunculus spp.	Buttercups	Ranunculaceae
Raphanus raphanistrum L.	Wild radish; runch	Brassicaceae
Reynoutria japonica Houtt. (= *Polygonum cuspidatum* Sieb. & Zucc.)	Japanese knotweed	Polygonaceae
Reynoutria sachalinensis (F. Schmidt) Nakai (= *Polygonum sachalinense*)	Giant knotweed	Polygonaceae
Rhagoletis pomonella (Walsh)	Apple maggot	Diptera: Tephritidae
Rhinotermitidae	Termites	Isoptera
Rhizoctonia spp.	Foot rot; root rot, various hosts	Deuteromycetes: Agonomycetiales
Rhizoctonia solani Kuehn (= *Thanetophorus cucumeris*)	Damping off; root rots, various	Deuteromycetes: Agonomycetiales
Rhizoglyphus callae Oudemans	Bulb mites	Acari: Acaridae
Rhizoglyphus echinopus (Fumouze & Robin) (see *R. callae* and *R. robini*)		
Rhizoglyphus robini Claperède	Bulb mites	Acari: Acaridae
Rhizopertha dominica (Fabricius) (see *Rhyzopertha dominica*)		
Rhizophora mangle L.	Mangrove	Rhizophoraceae
Rhizopus spp.	Post-harvest rots	Phycomycetes: Mucorales
Rhododendron ponticum L.	Rhododendron	Ericaceae
Rhopalosiphum padi (L.)	Bird-cherry aphid; apple grain aphid	Hemiptera: Aphididae
Rhus aromatica Aiton	Sumac	Anacardiaceae
Rhyacionia buoliana (Denis & Schiffermueller)	European pine shoot moth	Lepidoptera: Tortricidae
Rhynchophorus ferrugineus (Olivier)	Red palm weevil	Coleoptera: Curculionidae
Rhynchophorus palmarum (L.)	American palm beetle	Coleoptera: Curculionidae
Rhynchosporium spp.	Leaf spots, grasses	Deuteromycetes: Moniliales
Rhynchosporium secalis Davis	Leaf blotch, barley and rye	Deuteromycetes: Moniliales
Rhyzopertha dominica (Fabricius)	Lesser grain borer	Coleoptora: Bostrichidae

Latin	English	Order and/or Family
Rotylenchulus reniformis Linford & Oliveira	Reniform nematode	Nematoda
Rubus spp.	Brambles, blackberries	Rosaceae
Rubus fruticosus L.	Brambles, blackberries	Rosaceae
Rumex spp.	Docks and sorrels	Polygonaceae
Sagittaria sagittifolia L.	Arrowhead	Alismataceae
Sahlbergella singularis Hagl.	Cocoa capsid	Heteroptera: Miridae
Saissetia coffeae (Walker)	Hemispherical scale; helmet scale	Hemiptera: Coccidae
Saissetia oleae (Bernard)	Mediterranean black scale; black olive scale	Hemiptera: Coccidae
Salsola kali L.	Russian thistle	Chenopodiaceae
Saltatoria	Crickets, grasshoppers, etc.	Insecta
Sanninoidea exitiosa (Say.)	Peach tree borer	Lepidoptera: Aegeriidae
Saprolegniales		Chromista: Oomycetes
Sarothamnus scoparius Wimmer	Scotchbroom	Fabaceae
Scapteriscus vicinus Scudder	Mole cricket	Orthoptera: Gryllotapidae
Schizothyrium pomi Arx.	Fly speck disease, apple	Ascomycetes: Hemisphaeriales
Sciara spp.	Sciarid flies	Diptera: Sciaridae
Sciaridae	Fungus gnats, sciarid flies	Diptera
Scirpus spp.	Club-rushes	Cyperaceae
Scirpus juncoides	Japanese bullrush	Cyperaceae
Scirpus maritimus L.	Sea club-rush	Cyperaceae
Scirpus mucronatus L.	Roughseed bullrush	Cyperaceae
Scirtothrips persea Nakahara	Avocado thrips	Thysanoptera: Thripidae
Sclerophthora spp.	Downy mildew, wheat	Chromista: Oomycetes: Peronosporales
Sclerophthora macrospora (Sacc.) Thirum.	Downy mildew, cereals	Chromista: Oomycetes: Peronosporales
Sclerospora spp.	Downy mildews, e.g. on pearl millet	Chromista: Oomycetes: Peronosporales
Sclerotinia spp.	Sclerotinia rots, various hosts	Ascomycetes: Helotiales
Sclerotinia fructicola Rehm.	Brown rot, top fruit	Ascomycetes: Helotiales
Sclerotinia fructigena Anderh. & Ruhl.	Brown rot, apple, pear and plum	Ascomycetes: Helotiales
Sclerotinia homeocarpa Bennett	Dollar spot, turf	Ascomycetes: Helotiales
Sclerotinia laxa Anderh. & Ruhl.	Blossom wilt, apple and plum	Ascomycetes: Helotiales
Sclerotinia minor Jagger	Lettuce neck rot	Ascomycetes: Helotiales
Sclerotinia sclerotiorum De Bary	Rots of stems, storage organs, etc., various crops	Ascomycetes: Helotiales
Sclerotium spp.	Post-harvest rots, various hosts	Deuteromycetes: Agonomycetales
Sclerotium cepivorum Berk.	White rot, onion	Deuteromycetes: Agonomycetales
Sclerotium rolfsii Sacc.	Rots, various hosts	Deuteromycetes: Agonomycetales
Scolytus multistratus (Marsham)	Smaller European elm bark beetle	Coleoptera: Scolytidae

Latin	English	Order and/or Family
Senecio jacobaea L.	Common or tansy ragwort	Asteraceae
Septoria spp.	Leaf and glume spots, various hosts	Ascomycetes: Sphaeriales
Septoria glycines Hemmi	Soybean brown spot	Ascomycetes: Sphaeriales
Septoria nodorum Berk. (see *Leptosphaeria nodorum*)		
Septoria tritici Rob. (see *Mycosphaerella graminicola*)		
Sesamia cretica Lederer	Pink corn borer	Lepidoptera: Pyralidae
Sesamia nonagroides Lefebvre	Mediterranean corn borer	Lepidoptera: Pyralidae
Sesbania exaltata Cory	Hemp sesbania	Fabaceae
Setaria spp.	Foxtail grasses	Poaceae
Setaria faberi Herrm.	Foxtail, giant	Poaceae
Setaria glauca Beauv. (= *S. lutescens* Hurb.)	Foxtail, yellow	Poaceae
Setaria viridis Beav.	Foxtail, green	Poaceae
Sida spinosa L.	Spiny sida; prickly sida	Malvaceae
Simuliidae	Blackflies	Diptera
Simulium spp.	Blackflies	Diptera: Simuliidae
Sinapis alba L.	White mustard	Brassicaceae
Sinapis arvensis L.	Charlock	Brassicaceae
Sitobium avenae (Fabricius)	Grain aphid	Hemiptera: Aphididae
Sitona spp.	Pea weevils; bean weevils	Coleoptera: Curculionidae
Sitophilus oryzae (L.)	Rice weevil	Coleoptera: Curculionidae
Sitophilus zeamais Mutsch.	Maize weevil; rice weevil	Coleoptera: Curculionidae
Sitotroga cerealella (Oliver)	Angoumois grain moth	Lepidoptera: Gelechiidae
Sogatella furcifera (Howorth)	White-backed planthopper	Hemiptera: Delphacidae
Solanum nigrum L.	Black nightshade	Solanaceae
Solanum sarrachoides (Sendtner)	Hairy nightshade	Solanaceae
Solenopsis spp.	Fire ants	Hymenoptera: Formicidae
Sonchus oleraceus L.	Smooth sowthistle	Asteraceae
Sorghum spp.	Sorghum grasses	Poaceae
Sorghum almum Parodi	Columbus grass	Poaceae
Sorghum bicolor Moench	Shattercane	Poaceae
Sorghum halepense Pers.	Johnson grass	Poaceae
Sparganium erectum L.	Branched bur-reed	Sparganiaceae
Spergula arvensis L.	Corn spurrey	Caryophyllaceae
Sphacelotheca reiliana Clint.	Head smut, maize	Basidiomycetes: Ustilaginales
Sphaeriales		Ascomycotina
Sphaeropsidales		Deuteromycotina
Sphaerotheca spp.	Powdery mildew, various hosts	Ascomycetes: Erysiphales
Sphaerotheca fuliginea Poll.	Powdery mildew, cucurbits	Ascomycetes: Erysiphales
Sphaerotheca pannosa Lev.	Powdery mildew, rose	Ascomycetes: Erysiphales
Spodoptera spp.	Armyworms	Lepidoptera: Noctuidae
Spodoptera eridania (Cram.)	Southern armyworm	Lepidoptera: Noctuidae
Spodoptera exigua (Hübner)	Beet armyworm; lesser armyworm	Lepidoptera: Noctuidae
Spodoptera frugiperda (J. E. Smith)	Fall armyworm	Lepidoptera: Noctuidae

Latin	English	Order and/or Family
Spodoptera littoralis (Boisch.)	Egyptian cotton leafworm	Lepidoptera: Noctuidae
Spodoptera litura (Hübner)	Beet armyworm	Lepidoptera: Noctuidae
Stellaria media Vill.	Common chickweed	Caryophyllaceae
Steneotarsonemus laticeps (Halb.)	Bulb scale mite	Acari: Tarsonemidae
Stethorus punctillum Weise	Minute black ladybird	Coleoptera: Coccinellidae
Stigmina carpophila Ell.	Shothole, prunus	Deuteromycetes: Moniliales
Streptomyces scabies Walk.	Common scab of crops such as potato and beet	Schizomycetes: Actinomycetales
Symphyla spp.	Symphilids	Myriapoda
Synanthedon hector (Butler)	Cherry tree borer	Lepidoptera: Aegeriidae
Synanthedon myopaeformis (Borkhausen)	Apple clearwing moth	Lepidoptera: Aegeriidae
Synanthedon pictipes (Grote & Robinson)	Lesser peach tree borer	Lepidoptera: Aegeriidae
Synanthedon tipuliformis (Clerck)	Currant clearwing moth	Lepidoptera: Aegeriidae
Syngrapha falcifera (Kirby)	Celery looper	Lepidoptera: Noctuidae
Tanacetum cinerariaefolium (Trev.) Schultz-Bip.	Pyrethrum daisy	Asteraceae
Tanymecus palliatus Fabricius	Beet leaf weevil	Coleoptera: Curculionidae
Tapesia acuformis (Boerema, R. Pieters & Hamers) Crous (*see* *Oculimacula acuformis* (Boerema, R. Pieters & Hamers) Crous & W. Gams)		
Tapesia yallundae var. *yallundae* Wallwork & Spooner (*see* *Oculimacula yallundae* (Wallwork & Spooner) Crous & W. Gams)		
Taphrina deformans Tul.	Peach leaf-curl	Ascomycetes: Taphrinales
Tarsonemus spp. (= *Phytonemus*, in part)	Tarsonemid mites	Acari: Tarsonemidae
Taxus baccata L.	Yew	Taxaceae
Tephritidae	Large fruit flies	Diptera
Tetanops myopaeformis (Roeder)	Sugar beet root maggot	Diptera: Otitidae
Tetranychidae	Spider mites	Acari
Tetranychus spp.	Spider mites	Acari: Tetranychidae
Tetranychus atlanticus McGregor	Strawberry mite	Acari: Tetranychidae
Tetranychus cinnabarinus (Boisduval)	Carmine spider mite	Acari: Tetranychidae
Tetranychus kanzawai Kishida	Kanzawa spider mite	Acari: Tetranychidae
Tetranychus mcdanieli McG.	McDaniel's spider mite	Acari: Tetranychidae
Tetranychus urticae Koch	Two-spotted spider mite	Acari: Tetranychidae
Thanetophorus cucumeris Donk. (= *Rhizoctonia solani* Kuehn)	Damping-off disease	Basidiomycetes: Stereales
Thaumetopoea pityocampa (Denis & Schiffermueller)	Pine processionary caterpillar	Lepidoptera: Thaumetopoeidae

Latin	English	Order and/or Family
Thielaviopsis basicola Ferr.	Black root rot, tobacco	Deuteromycetes: Moniliales
Thripidae	Thrips	Thysanoptera
Thrips spp.	Thrips	Thysanoptera: Thripidae
Thrips fuscipennis Haliday	Rose thrips	Thysanoptera: Thripidae
Thrips obscuratus Crawford	New Zealand flower thrips	Thysanoptera: Thripidae
Thrips palmi Karny	Thrips	Thysanoptera: Thripidae
Thrips tabaci Lindeman	Onion thrips	Thysanoptera: Thripidae
Thysanoptera	Thrips	Insecta
Tilletia spp.	Smut, various hosts	Basidiomycetes: Ustilaginales
Tilletia caries Tul.	Bunt; stinking smut	Basidiomycetes: Ustilaginales
Tipula spp.	Crane flies; leatherjackets	Diptera: Tipulidae
Tortricidae	Tortrix moths	Lepidoptera
Tortrix spp.	Tortrix moths	Lepidoptera: Tortricidae
Tranzschelia discolor Tranz. & Litw. (see *Tranzschelia pruni-spinosae*)		
Tranzschelia pruni-spinosae Diet.	Plum rust	Basidiomycetes: Uredinales
Trialeurodes vaporariorum (Westwood)	Glasshouse whitefly	Hemiptera: Aleyrodidae
Tribulus terrestris L.	Puncture vine	Zygophyllaceae
Trichodorus spp.	Stubby-root nematodes	Nematoda
Trichogramma spp.	Lepidopteran egg parasites	Hymenoptera: Trichogrammatidae
Trichoplusia ni (Hübner)	Cabbage looper	Lepidoptera: Noctuidae
Tripleurospermum maritimum Koch (= *Matricaria inodora* L.)	Mayweed, scentless	Asteraceae
Trypodendron lineatum (Olivier)	Ambrosia beetle	Coleoptera: Scolytidae
Tulasnellales		Basidiomycotina
Typha spp.	Bullrushes	Typhaceae
Typhlodromus occidentalis Nesbitt (synonymous with *Galendromus occidentalis* and *Metaseiulus occidentalis*)	Predatory mite	Mesostigmata: Phytoseiidae
Typhlodromus pyri Scheuten	Fruit tree red spider mite predator	Mesostigmata: Phytoseiidae
Typhula incarnata Lasch	Snow rot, cereals	Basidiomycetes: Agaricales
Ulex europaeus L.	Gorse	Leguminaceae
Unaspis euonymi (Comstock)	Euonymus scale	Hemiptera: Coccidae
Uncinula necator Burr	Powdery mildew, grapevines	Erysiphales: Erysiphaceae
Uredinales	Rust fungi	Basidiomycetes
Urochloa mutica (Forsk.) Nguyen	Para grass, buffalo grass, California grass	Poaceae
Urocystis spp.	Leaf smuts, various hosts	Basidiomycetes: Ustilaginales
Uromyces spp.	Rusts, various crops	Basidiomycetes: Uredinales
Uromyces betae Lev.	Rust, beet crops	Basidiomycetes: Uredinales
Urtica spp.	Nettles	Urticaceae
Urtica dioica L.	Nettle, common	Urticaceae

References

Latin	English	Order and/or Family
Urtica urens L.	Nettle, small	Urticaceae
Ustilaginales	Smut fungi	Basidiomycetes
Ustilago spp.	Smut diseases, various hosts	Ustilaginales
Ustilago nuda Rostr.	Loose smut, barley and wheat	Basidiomycetes: Ustilaginales
Valsa ceratosperma (Tode ex Fr.) Maire	Valsa canker of apple	Ascomycetes: Sphaeriales
Vasates spp.	Mites	Acari: Eriophyidae
Venturia inaequalis Wint.	Scab, apples	Ascomycetes: Sphaeriales
Venturia pirina Aderh.	Scab, pears	Ascomycetes: Sphaeriales
Veronica spp.	Speedwells	Scrophulariaceae
Veronica filiformis Sm.	Speedwell, slender	Scrophulariaceae
Veronica hederifolia L.	Speedwell, ivy-leaved	Scrophulariaceae
Veronica persica Poir.	Speedwell, common field	Scrophulariaceae
Verticillium spp.	Verticillium wilt, various hosts	Deuteromycetes: Moniliales
Verticillium albo-atrum Reinke & Berth.	Wilt	Deuteromycetes: Moniliales
Verticillium dahliae Kleb.	Black heart, apricots	Deuteromycetes: Moniliales
Verticillium fungicola (Preuss) Hassebrauk	Dry bubble, mushrooms	Deuteromycetes: Moniliales
Viola spp.	Wild pansies	Violaceae
Viola arvensis Murr.	Field pansy	Violaceae
Xanthium pennsylvanicum Wallr.	Cocklebur	Asteraceae
Xanthium strumarium L.	Rough Cocklebur	Asteraceae
Xanthomonas spp.	Bacterial leaf spots, various hosts	Pseudomonadales: Pseudomonadaceae
Xanthomonas campestris Dows. pv. *citri* (see *X. citri*)		
Xanthomonas campestris Dows. pv. *malvacearum* (see *X. malvacearum*)		
Xanthomonas campestris Dows. pv. *oryzae* (see *X. oryzae*)		
Xanthomonas citri Dows.	Citrus canker	Pseudomonadales: Pseudomonadaceae
Xanthomonas malvacearum Dows.	Bacteriosis, cotton	Pseudomonadales: Pseudomonadaceae
Xanthomonas oryzae Dows.	Leaf blight, rice	Pseudomonadales: Pseudomonadaceae
Zeuzera pyrina (L.)	Leopard moth	Lepidoptera: Cossidae
Zygomycotina	Primitive fungi (Phycomycetes) which do not produce motile spores; sexually produced spores are non-motile zygospores	
Zygophiala jamaicensis Mason	Greasy blotch, carnation	Deuteromycetes: Moniliales

Glossary: English–Latin

This glossary is intended to help the reader to identify the Latin name in cases where the English name may be unfamiliar. The Latin name is presented as either genus and species or, if the English name is represented by a genus, the specific name is excluded. For example, the causal organisms for anthracnose diseases are fungal pathogens from the genus *Colletotrichum*, but the peach-potato aphid is *Myzus persicae*.

This glossary has been produced simply by inverting the Latin–English glossary, with limited subsequent editing. The reader should recognise that English-name and Latin-name groups of species are not congruent. For example, all *Peronospora* are downy mildews, but not all downy mildews are *Peronospora*; consequently, there are entries for various downy mildews, giving different Latin names.

In searching for English names, alternative forms of the name should be considered, especially with names containing an adjectival component.

Where these two problems occur together, this glossary needs to be used with particular care.

English	Latin	Order and/or Family
Alfalfa aphid	*Acyrthosiphon kondoi*	Hemiptera: Aphididae
Alfalfa looper	*Autographa californica*	Lepidoptera: Noctuidae
Alfalfa weevil	*Hypera postica*	Coleoptera: Curculionidae
Alligatorweed	*Alternanthera philoxeroides*	Amaranthaceae
Amaranths	*Amaranthus*	Amaranthaceae
Ambrosia beetle	*Gnathotrichus retusus*	Coleoptera: Scolytidae
Ambrosia beetle	*Gnathotrichus sulcatus*	Coleoptera: Scolytidae
Ambrosia beetle	*Trypodendron lineatum*	Coleoptera: Scolytidae
American black cherry	*Prunus serotina*	Rosaceae
American bollworm	*Helicoverpa zea*	Lepidoptera: Noctuidae
American cockroach	*Periplaneta americana*	Dictyoptera: Blattidae
American palm beetle	*Rhyncophorus palmatum*	Coleoptera: Curculionidae
American serpentine leaf miner	*Liriomyza trifolii*	Diptera: Agromyzidae
Angoumois grain moth	*Sitotroga cerealella*	Lepidoptera: Gelechiidae
Angular leaf spot, cucurbits	*Pseudomonas lachrymans*	Pseudomonadales: Pseudomonadaceae
Annual ryegrass	*Lolium rigidum*	Poaceae
Anthracnose, cucurbits	*Colletotrichum lagenarium*	Ascomycetes: Melanconiales
Anthracnose, french beans	*Colletotrichum lindemuthianum*	Ascomycetes: Melanconiales
Anthracnose, various root rot and leaf curl diseases	*Colletotrichum*	Ascomycetes: Melanconiales

English	Latin	Order and/or Family
Aphid, tobacco	*Myzus nicotianae*	Hemiptera: Aphididae
Aphid gall midge	*Aphidoletes aphidimyza*	Diptera: Cecidomyiidae
Aphid parasitoid wasps	*Aphelinus*	Hymenoptera Aphelinidae
Aphid parasitoid wasps	*Aphidius*	Hymenoptera: Aphelinidae
Apple and pear canker	*Diplodia pseudodiplodia*	Deuteromycetes: Sphaeropsidales
Apple clearwing moth	*Synanthedon myopaeformis*	Lepidoptera: Aegeriidae
Apple leaf miner	*Lyonetia clerkella*	Lepidoptera: Lyonetiidae
Apple leaf miner	*Phyllonorycter blancardella*	Lepidoptera: Gracillariidae
Apple grain aphid, bird-cherry aphid	*Rhopalosiphum padi*	Hemiptera: Aphididae
Apple maggot	*Rhagoletis pomonella*	Diptera: Tephritidae
Armyworm, beet	*Spodoptera exigua; S. litura*	Lepidoptera: Noctuidae
Armyworm, bertha	*Mamestra configurata*	Lepidoptera: Noctuidae
Armyworm, fall	*Spodoptera frugiperda*	Lepidoptera: Noctuidae
Armyworm, southern	*Spodoptera eridania*	Lepidoptera: Noctuidae
Armyworms	*Spodoptera*	Lepidoptera: Noctuidae
Arrowhead	*Sagittaria sagittifolia*	Alismataceae
Artichoke plume moth	*Platyptilia carduidactyla*	Lepidoptera
Asiatic corn borer	*Ostrinia furnacalis*	Lepidoptera: Pyralidae
Avocado thrips	*Scirtothrips persea*	Thysanoptera: Thripidae
Bacterial blights and leaf spots, various hosts	*Pseudomonas*	Pseudomonadales: Pseudomonadaceae
Bacterial canker, prunus	*Pseudomonas mors-prunorum*	Pseudomonadales: Pseudomonadaceae
Bacterial grain rot, rice	*Pseudomonas glumae*	Pseudomonadales: Pseudomonadaceae
Bacterial leaf spots, various hosts	*Xanthomonas*	Pseudomonadales: Pseudomonadaceae
Bacterial rot, celery	*Erwinia carotovora*	Eubacteriales: Enterobacteriaceae
Bacteriosis, cotton	*Xanthomonas malvacearum*	Pseudomonadales: Pseudomonadaceae
Balsam whitefly	*Neodiprion abietis*	Hymenoptera: Diprionidae
Banana black heart	*Gibberella fujikuroi*	Deuteromycetes: Hypocreales
Banana corm weevil	*Cosmopolites sordidus*	Coleoptera: Curculionidae
Banana leaf spot	*Mycosphaerella musicola*	Ascomycetes: Sphaeriales
Banana root borer	*Cosmopolites sordidus*	Coleoptera: Curculionidae
Barnyard grass	*Echinochloa crus-galli*	Poaceae
Barnyard grass, awnless	*Echinochloa colonum*	Poaceae
Barnyard grasses	*Echinochloa*	Poaceae
Barren brome	*Bromus sterilis*	Poaceae
Basal stem rot, cucurbits	*Erwinia carotovora*	Eubacteriales: Enterobacteriaceae
Bean beetles	*Epilachna*	Coleoptera: Coccinellidae
Bearded oat	*Avena barbata*	Poaceae

English	Latin	Order and/or Family
Beet armyworm	*Spodoptera exigua*	Lepidoptera: Noctuidae
Beet armyworm	*Spodoptera litura*	Lepidoptera: Noctuidae
Beet cyst nematode	*Heterodera schachtii*	Nematoda: Heteroderidae
Beet leaf-miner	*Pegomya hyoscamni*	Diptera: Anthomyiidae
Beet leaf weevil	*Tanymecus pallidus*	Coleoptera: Curculionidae
Beet semi-looper	*Autographa nigrisgna*	Lepidoptera: Noctuidae
Begonia	*Begonia elatior*	Begoniaceae
Beneficial bacterium	*Agrobacterium radiobacter*	Eubacteriales: Rhizobiaceae
Bermuda grass	*Cynodon*	Poaceae
Bermuda grass	*Cynodon dactylon*	Poaceae
Bindweed, large	*Calystegia sepium*	Convolvulaceae
Bird-cherry aphid	*Rhopalosiphum padi*	Hemiptera: Aphididae
Black bean aphid	*Aphis fabae*	Hemiptera: Aphididae
Black bent	*Agrostis gigantea*	Poaceae
Blackberries	*Rubus fruticosus*	Rosaceae
Blackberry rust	*Phragmidium violaceum*	Basidiomycetes: Uredinales
Black bindweed	*Fallopia convolvulus*	Polygonaceae
Blackcurrant gall-mite	*Cecidophyopsis ribis*	Acari: Eriophyidae
Blackcurrant rust	*Cronartium ribicola*	Basidiomycetes: Uredinales
Black cutworm	*Agrotis ipsilon*	Lepidoptera: Noctuidae
Blackflies	*Simulium*	Diptera: Simuliidae
Black-grass	*Alopecurus myosuroides*	Poaceae
Black heart, apricots	*Verticillium dahliae*	Deuteromycetes: Moniliales
Black leaf streak, banana	*Mycosphaerella fijiensis*	Ascomycetes: Sphaeriales
Blackleg, beet crops	*Aphanomyces cochlioides*	Saprolegniales: Saprolegniacea
Blackleg, potatoes	*Erwinia carotvora*	Eubacteriales: Enterobacteriaceae
Black mould	*Cladosporium*	Deuteromycetes: Moniliales
Black nightshade	*Solanum nigrum*	Solanaceae
Black olive scale	*Saissetia oleae*	Hemiptera: Coccidae
Black root rot, tobacco	*Thielaviopsis basicola*	Deuteromycetes: Moniliales
Black rot, grapevines	*Guignardia bidwellii*	Ascomycetes: Sphaeriales
Blackspot, roses	*Diplocarpon rosae*	Ascomycetes: Helotiales
Black stem rust, grasses	*Puccinia graminis*	Basidiomycetes: Uredinales
Blight, capsicums	*Phytophthora capsici*	Chromista: Oomycetes: Peronosporales
Blight, chestnut	*Cryphonectria parasitica*	Ascomycetes: Nectriaceae
Blight, damping off, foot-rot, various hosts	*Phytophthora*	Chromista: Oomycetes: Peronosporales
Blister blight, tea	*Exobasidium vexans*	Basidiomycetes: Exobasidiales
Blossom beetles	*Meligethes spp.*	Coleoptera: Nitidulidae
Blossom wilt, apple and plum	*Sclerotinia laxa*	Ascomycetes: Helotiales
Bluegrass webworm	*Parapediasia teterrella*	Lepidoptera: Crambidae
Blue mould, apples and pears	*Penicillium expansum*	Deuteromycetes: Moniliales

English	Latin	Order and/or Family
Blue mould, citrus	*Penicillium italicum*	Deuteromycetes: Moniliales
Blue mould, tobacco	*Peronospora tabacina* (= *Plasmopara tabacina*)	Chromista: Oomycetes: Peronosporales
Blue rot, apples and pears	*Penicillium expansum*	Deuteromycetes: Moniliales
Bollworms, spiny	*Earias*	Lepidoptera: Noctuidae
Bracken	*Pteridium aquilinum*	Filicales
Bracket chafer	*Phyllopertha horticola*	Coleoptera: Scarabaeidae
Bramble rust	*Phragmidium violaceum*	Basidiomycetes: Uredinales
Brambles	*Rubus*	Rosaceae
Branched bur-reed	*Sparganium erectum*	Sparganiaceae
Brassica cyst nematode	*Heterodera crucifereae*	Nematoda: Heteroderidae
Brassica gall and stem weevils	*Ceutorhynchus*	Coleoptera: Curculionidae
Broad mite	*Polyphagotarsonemus latus*	Acari: Tarsonemidae
Brooks spot, apple	*Mycosphaerella pomi*	Ascomycetes: Sphaeriales
Brown fleck of mushrooms	*Pseudomonas tolaasii*	Pseudomonadales: Pseudomonadaceae
Brown foot rot and ear blight and other cereal diseases	*Gibberella*	Deuteromycetes: Hypocreales
Brown rot, apple, pear and plum	*Sclerotinia fructigena*	Ascomycetes: Helotiales
Brown rot, top fruit	*Sclerotinia fructicola*	Ascomycetes: Helotiales
Brown rust, barley	*Puccinia hordei*	Basidiomycetes: Uredinales
Brown rust, chrysanthemum	*Puccinia chrysanthemi*	Basidiomycetes: Uredinales
Brown rust, wheat	*Puccinia triticana*	Basidiomycetes: Uredinales
Brown soft scale	*Coccus hesperidum*	Hemiptera: Coccidae
Brown spot, peanut	*Mycosphaerella arachidis*	Ascomycetes: Sphaeriales
Brown spot, rice	*Cochliobolus miyabeanus*	Ascomycetes: Sphaeriales
Brown spot, soybean	*Septoria glycines*	Ascomycetes: Sphaeriales
Brown stripe, sugar cane	*Bipolaris stenospila*	Ascomycetes: Sphaeriales
Bt	*Bacillus thuringiensis*	Schizomycetes: Eubacteriales
Bulb mites	*Rhizoglyphus callae, R. robini*	Acari: Acaridae
Bulb scale mite	*Steneotarsonemus laticeps*	Acari: Tarsonemidae
Bullrushes	*Typha*	Typhaceae
Bunt	*Tilletia caries*	Basidiomycetes: Ustilaginales
Burrowing nematode	*Radopholus similis*	Nematoda: Tylenchidae
Buttercups	*Ranunculus*	Ranunculaceae
Butt rot, conifers	*Fomes annosus*	Basidiomycetes: Agaricales
Cabbage looper	*Trichoplusia ni*	Lepidoptera: Noctuidae
Cabbage moth	*Mamestra brassicae*	Lepidoptera: Noctuidae
Cabbage root fly	*Delia radicum*	Diptera: Anthomyiidae
Cabbage seed weevil	*Ceutorhynchus assimilis*	Coleoptera: Curculionidae
Cabbage stem weevil	*Ceutorhynchus quadridens*	Coleoptera: Curculionidae
Cabbage white butterflies	*Pieris*	Lepidoptera: Pieridae
Californian red scale	*Aonidiella aurantii*	Hemiptera: Diaspididae
Canadian pondweed	*Elodea canadensis*	Hydrocharitaceae
Canary grass, awned	*Phalaris paradoxa*	Poaceae

English	Latin	Order and/or Family
Canary grasses	*Phalaris*	Poaceae
Canegrub, greyback	*Dermolepida albohirtum*	Coleoptera: Scarabaeidae
Canker, apple, pear	*Nectria galligena*	Hypocreales
Carmine spider mite	*Tetranychus cinnabarinus*	Acari: Tetranychidae
Carpenter ants	*Camponotus*	Hymenoptera: Formicidae
Carrot fly	*Psila rosae*	Diptera: Psilidae
Carrot leaf blight	*Alternaria dauci*	Deuteromycetes: Moniliales
Caterpillar predator	*Podisus maculiventris*	Heteroptera: Pentatomidae
Celery looper	*Syngrapha falcifera*	Lepidoptera: Noctuidae
Cercospora leaf spot, beans	*Cercospora zonata*	Deuteromycetes: Moniliales
Cereal leaf beetle	*Oulema melanopus*	Coleoptera: Chrysomelidae
Charlock	*Sinapis arvensis*	Brassicaceae
Cherry tree borer	*Synanthedon hector*	Lepidoptera: Aegeriidae
Chestnut blight	*Endothia parasitica*	Sphaeriales: Diaporthaceae
Chinch bug	*Blissus leucopterus*	Lygaeidae
Chrysanthemum leaf miner	*Phytomyza syngenesiae*	Diptera: Agromyzidae
Chrysanthemum leaf miner parasitoid	*Dacnusa sibirica*	Hymenoptera: Braconidae
Chrysomelid beetles	*Chrysomelidae*	Coleoptera
Citrus aphid	*Aphis citricida*	Hemiptera: Aphididae
Citrus black scale	*Parlatoria ziziphi*	Hemiptera: Diaspididae
Citrus canker	*Xanthomonas citri*	Pseudomonadales: Pseudomonadaceae
Citrus flower moth	*Prays citri*	Lepidoptera: Yponomeutidae
Citrus mealybug	*Planococcus citri*	Hemiptera: Pseudococcidae
Citrus mussel scale	*Lepidosaphes beckii* (*Cornuaspis beckii*)	Hemiptera: Coccidae
Citrus purple scale	*Lepidosaphes beckii* (*Cornuaspis beckii*)	Hemiptera: Coccidae
Citrus red mite	*Panonychus citri*	Acari: Tetranychidae
Citrus red scale	*Chrysomphalus aonidum*	Hemiptera: Diaspididae
Citrus rust mite	*Phyllocoptruta oleivora*	Acari: Eriophyidae
Citrus weevil	*Pachnaeus litus*	Coleoptera: Curculionidae
Citrus whitefly	*Dialeurodes citri*	Hemiptera: Aleyrodidae
Cleavers	*Galium aparine*	Rubiaceae
Clover bryobia mite	*Bryobia praetiosa*	Acari: Tetranychidae
Clubroot, brassicas	*Plasmodiophora brassicae*	Chromista: Oomycetes: Plasmodiophorales
Club-rushes	*Scirpus*	Cyperaceae
Coccomycosis, cherry leaf spot	*Blumeriella jaapii*	Ascomycetes: Helotiales
Cockchafer	*Melolontha melolontha and Melolontha hippocastani*	Coleoptera: Scarabaeidae
Cockchafer, redheaded	*Adoryphorus couloni*	Coleoptera: Scarabaeidae
Cockchafers	*Melolontha*	Coleoptera: Scarabaeidae
Cocklebur	*Xanthium pennsylvanicum*	Asteraceae

English	Latin	Order and/or Family
Cockroach, American	*Periplaneta americana*	Dictyoptera: Blattidae
Cockroach, German	*Blatella germanica*	Dictyoptera: Blattidae
Cockspur, rice	*Echinochloa oryzicola* (= *E. oryzoides*)	Poaceae
Cocoa capsid	*Distantiella theobroma*	Heteroptera: Miridae
Cocoa capsid	*Sahlbergella singularis*	Heteroptera: Miridae
Cocoa pod borer	*Conopomorpha cramerella*	Lepidoptera: Gracillariidae
Coconut mite	*Aceria guerreronis*	Acari: Eriophyidae
Coconut rhinoceros beetle	*Oryctes rhinoceros*	Coleoptera: Scarabaeidae
Codling moth	*Cydia pomonella*	Lepidoptera: Tortricidae
Coffee rust	*Hemileia vastatrix*	Basidiomycetes: Uredinales
Collar rot, apple	*Phytophthora cactorum*	Chromista: Oomycetes: Peronosporales
Colorado beetle	*Leptinotarsa decemlineata*	Coleoptera: Chrysomelidae
Columbus grass	*Sorghum almum*	Poaceae
Common amaranth	*Amaranthus retroflexus*	Amaranthaceae
Common chickweed	*Stellaria media*	Caryophyllaceae
Common couch	*Elytrigia repens* (= *Elymus repens*)	Poaceae
Common orache	*Atriplex patula*	Chenopodiaceae
Common scab of crops such as potato and beet	*Streptomyces scabies*	Schizomycetes: Actinomycetales
Corn earworm	*Helicoverpa zea*	Lepidoptera: Noctuidae
Corn marigold	*Chrysanthemum segetum*	Asteraceae
Corn rootworm	*Diabrotica undecimpunctata*	Coleoptera: Chrysomelidae
Corn rootworms	*Diabrotica*	Coleoptera: Chrysomelidae
Corn spurrey	*Spergula arvensis*	Caryophyllaceae
Cotton boll rot	*Gibberella fujikuroi*	Deuteromycetes: Hypocreales
Cotton boll weevil	*Anthonomus grandis*	Coleoptera: Curculionidae
Cotton leafhoppers	*Empoasca*	Hemiptera: Cicadellidae
Cotton leaf perforator	*Bucculatrix thurberiella*	Lepidoptera: Lyonetiidae
Cotton leaf worm	*Alabama argillacea*	Lepidoptera: Noctuidae
Crabgrass	*Digitaria sanguinalis*	Poaceae
Crabgrass, tropical	*Digitaria adscendens* (= *D. ciliaris*)	Poaceae
Crabgrasses	*Digitaria*	Poaceae
Crane flies	*Tipula*	Diptera: Tipulidae
Crane's bills	*Geranium*	Geraniaceae
Creeping bent	*Agrostis stolonifera*	Poaceae
Crickets	*Cricetus*	Saltatoria: Gryllidae
Crickets, grasshoppers, etc.	*Saltatoria*	Insecta
Crown gall	*Agrobacterium tumefasciens*	Eubacteriales: Rhizobiaceae
Crown rot, apple	*Phytophthora cactorum*	Chromista: Oomycetes: Peronsporales
Cucurbit anthracnose	*Colletotrichum lagenarium*	Ascomycetes: Melanconiales
Currant clearwing moth	*Synanthedon tipuliformis*	Lepidoptera: Aegeriidae
Cutworm	*Noctua pronuba*	Lepidoptera: Noctuidae

English	Latin	Order and/or Family
Cutworms	*Agrotis*	Lepidoptera: Noctuidae
Cutworms	*Euxoa*	Lepidoptera: Noctuidae
Cutworms	*Noctua*	Lepidoptera: Noctuidae
Cyst nematodes	*Heteroderidae*	Nematoda
Cyst nematode, soybean	*Heterodera glycines*	Nematoda: Heteroderidae
Damping-off	*Rhizoctonia solani* (= *Thanetophorus cucumeris*)	Deuteromycetes: Agonomycetiales
Damping-off disease	*Thanetophorus cucumeris* (= *Rhizoctonia solani*)	Basidiomycetes: Stereales
Damson-hop aphid	*Phorodon humuli*	Hemiptera: Aphididae
Dark leaf spot, brassicas	*Alternaria brassicae*	Deuteromycetes: Moniliales
Dark leaf spot, brassicas	*Alternaria brassicicola*	Deuteromycetes: Moniliales
Dart moths	*Euxoa*	Lepidoptera: Noctuidae
Dayflower	*Commelina*	Commelinaceae
Dead arm, grape vines	*Phomopsis viticola*	Deuteromycetes: Sphaeropsidales
Diamondback moth	*Plutella xylostella*	Lepidoptera: Yponomeutidae
Diffuse knapweed	*Centaurea diffusa*	Asteraceae
Docks and sorrels	*Rumex*	Polygonaceae
Dodder	*Cuscuta spp.*	Convolvulaceae
Dodder, Australian	*Cuscuta australis*	Convolvulaceae
Dodder, Chinese	*Cuscuta chinensis*	Convolvulaceae
Dodder, European	*Cuscuta europaea*	Convolvulaceae
Dodder, field	*Cuscuta campestris*	Convolvulaceae
Dodder, greater	*Cuscuta europaea*	Convolvulaceae
Dodder, large	*Cuscuta europaea*	Convolvulaceae
Dodder, large seed	*Cuscuta indecora*	Convolvulaceae
Dodder, small seed	*Cuscuta planiflora*	Convolvulaceae
Dodder, swamp	*Cuscuta gronovii*	Convolvulaceae
Dollar spot, turf	*Sclerotinia homeocarpa*	Ascomycetes: Helotiales
Douglas fir	*Pseudotsuga menziesii*	Pinaceae
Douglas fir beetle	*Dendroctonus pseudotsuga*	Coleoptera: Scolytidae
Douglas fir beetle	*Ips typographus*	Coleoptera: Ipidae
Douglas fir tussock moth	*Orgyia pseudotsugata*	Lepidoptera: Lymantriidae
Downy mildew, brassicae	*Peronospora parasitica*	Chromista: Oomycetes: Peronosporales
Downy mildew, cereals	*Scerophthora macrospora*	Chromista: Oomycetes: Peronosporales
Downy mildew, cucurbits	*Pseudoperonospora cubensis*	Chromista: Oomycetes: Peronosporales
Downy mildew, grapevine	*Plasmopara viticola*	Chromista: Oomycetes: Peronosporales
Downy mildew, hops	*Pseudoperonospora humuli*	Chromista: Oomycetes: Peronosporales
Downy mildew, lettuce	*Bremia lactucae*	Chromista: Oomycetes: Peronosporales

English	Latin	Order and/or Family
Downy mildew, sorghum	*Peronosclerospora*	Chromista: Oomycetes: Peronosporales
Downy mildew, wheat	*Sclerophthora*	Chromista: Oomycetes: Peronosporales
Downy mildews	*Peronospora*	Chromista: Oomycetes: Peronosporales
Downy mildews, e.g. on pearl millet	*Sclerospora*	Chromista: Oomycetes: Peronosporales
Downy mildews, various hosts	*Plasmopara*	Chromista: Oomycetes: Peronosporales
Dry bubble, mushrooms	*Verticillium fungicola*	Deuteromycetes: Moniliales
Dry rot, post-harvest rot	*Fusarium coeruleum*	Deuteromycetes: Moniliales
Dutch elm disease	*Ceratocystis ulmi*	Ascomycetes: Sphaeriales
Early blight, potato	*Alternaria solani*	Deuteromycetes: Moniliales
Earworm, corn	*Helicoverpa zea*	Lepidoptera: Noctuidae
Egyptian cotton leafworm	*Spodoptera littoralis*	Lepidoptera: Noctuidae
Elodea, Florida	*Hydrilla verticillata*	Hydrocharitaceae
Eucalyptus snout beetle	*Gonipterus gibberus*	Coleoptera: Curculionidae
European corn borer	*Ostrinia nubilalis*	Lepidoptera: Pyralidae
European grapevine moth	*Lobesia botrana*	Lepidoptera: Tortricidae
European pine sawfly	*Neodiprion sertifer*	Hymenoptera: Diprionidae
European pine shoot moth	*Rhyacionia buoliana*	Lepidoptera: Tortricidae
Eye-spot, cereals	*Pseudocercosporella herpotrichoides*	Deuteromycetes: Moniliales
Fairy rings	*Marasmius oreades*	Basidiomycetes: Agaricales
Fall armyworm	*Spodoptera frugiperda*	Lepidoptera: Noctuidae
Fall panicum	*Panicum dichotomiflorum*	Poaceae
False oat-grass	*Arrhenatherum elatius*	Poaceae
Fat hen	*Chenopodium album*	Chenopodiaceae
Field bindweed	*Convolvulus arvensis*	Convolvulaceae
Field pansy	*Viola arvensis*	Violaceae
Field slug	*Decoceras*	Mollusca: Gastropoda
Fire ants	*Solenopsis*	Hymenoptera: Formicidae
Fire blight of pome fruit	*Erwinia amylovora*	Eubacteriales: Enterobacteriaceae
Flea beetle	*Phyllotreta striolata*	Coleoptera: Chrysomelidae
Flea beetles	*Chaetocnema*	Coleoptera: Chrysomelidae
Flea beetles	*Phyllotreta*	Coleoptera: Chrysomelidae
Fly speck disease, apple	*Schizothyrium pomi*	Ascomycetes: Hemisphaeriales
Foot rot; root rot, cereals, grasses	*Cochliobolus sativus*	Ascomycetes: Sphaeriales
Foot rot, root rot, various hosts	*Aphanomyces*	Chromista: Oomycetes: Saprolegniales
Foot rot, root rot, various hosts	*Rhizoctonia*	Deuteromycetes: Agonomycetiales

English	Latin	Order and/or Family
Formosan termite	*Coptotermes formosanus*	Isoptera: Rhinotermitidae
Four-leaved water clover	*Marsilea*	Marsileaceae
Foxtail, giant	*Setaria faberi*	Poaceae
Foxtail, green	*Setaria viridis*	Poaceae
Foxtail, yellow	*Setaria glauca (= S. lutescens)*	Poaceae
Foxtail grasses	*Setaria*	Poaceae
Fringe rushes	*Fimbristylis*	Cyperaceae
Frit fly	*Oscinella frit*	Diptera: Chloropidae
Frog eye, soybean	*Cercosporidium (C. sojinum = Cercospora sojina)*	Deuteromycetes: Moniliales
Fruit flies	*Dacus*	Diptera: Tephritidae
Fruit flies	*Drosophila*	Diptera: Drosophilidae
Fruit flies	*Drosophilidae*	Diptera
Fruit moth, plum	*Cydia funebrana*	Lepidoptera: Tortricidae
Fruit rot, strawberries	*Mucor*	Mucorales
Fruit rot, various hosts	*Botrytis cinerea*	Deuteromycetes: Moniliales
Fruit tree red spider mite	*Panonychus ulmi*	Acari: Tetranychidae
Fruit tree red spider mite predator	*Amblyseius degenerans*	Mesostigmata: Phytoseiidae
Fruit tree red spider mite predator	*Typhlodromus pyri*	Mesostigmata: Phytoseiidae
Fruitworm, tomato	*Helicoverpa zea*	Lepidoptera: Noctuidae
Fuchsia	*Fuchsia hybrida*	Onagraceae
Fungal vector of rhizomania virus	*Polymyxa betae*	Chromista: Oomycetes: Plasmodiophorales
Fungus gnat predator	*Hypoaspis aculeifer*	Mesostigmata: Laelapidae
Fusarium foot and root rots, various hosts	*Fusarium culmorum*	Deuteromycetes: Moniliales
Fusarium wilt, various hosts	*Fusarium oxysporum*	Deuteromycetes: Moniliales
Gangrene, potatoes	*Phoma exigua*	Deuteromycetes: Sphaeropsidales
Garden chafer	*Phyllopertha horticola*	Coleoptera: Scarabaeidae
Garden wireworms	*Athous*	Coleoptera: Elateridae
Giant knotweed	*Reynoutria sachalinensis (= Polygonum sachalinense)*	Polygonaceae
Glasshouse potato aphid	*Aulacorthum solani*	Hemiptera: Aphididae
Glasshouse whitefly	*Trialeurodes vaporariorum*	Hemiptera: Aleyrodidae
Glasshouse whitefly parasitoid	*Encarsia formosa*	Hymenoptera: Aphelinidae
Gloeosporium rot, apples	*Gloeosporium*	Deuteromycetes: Sphaeriales
Gloeosporium rot, apples	*Glomerella cingulata*	Deuteromycetes: Sphaeriales
Glume blotch, wheat	*Leptosphaeria nodorum (= Septoria nodorum)*	Ascomycetes: Sphaeriales
Golden chalcid	*Aphytis melinus*	Hymenoptera: Aphelinidae
Gooseberry bryobia mite	*Bryobia ribis*	Acari: Tetranychidae
Goosegrass	*Eleusine indica*	Poaceae

English	Latin	Order and/or Family
Gorse	Ulex europaeus	Fabaceae
Grain aphid	Sitobium avenae	Hemiptera: Aphididae
Grain beetles	Cryptolestes	Coleoptera: Cucujidae
Grape berry moth	Eupoecilia ambiguella	Lepidoptera: Cochylidae
Grass and cereal flies	Opomyza	Diptera: Opomyzidae
Grass moth	Chrysoteuchia caliginosellus (= Crambus caliginosellus)	Lepidoptera: Pyralidae
Grass moth	Crambus caliginosellus	Lepidoptera: Pyralidae
Green leafhopper	Empoasca fabae	Hemiptera: Cicadellidae
Green leafhoppers	Nephotettix	Hemiptera: Cicadellidae
Green mould, citrus	Penicillium digitatum	Deuteromycetes: Moniliales
Green rice leafhopper	Nephotettix cincticeps	Hemiptera: Cicadellidae
Green rice leafhopper	Nephotettix impicticepts	Hemiptera: Cicadellidae
Gypsy moth	Lymantria dispar	Lepidoptera: Lymantriidae
Hairy nightshade	Solanum sarrachoides	Solanaceae
Halo blight, beans	Pseudomonas phaseolicola	Pseudomonadales: Pseudomonadaceae
Hay bacillus	Bacillus subtilis	Schizomycetes: Eubacteriales
Head smut, maize	Sphacelotheca reiliana	Basidiomycetes: Ustilaginales
Helmet scale	Saissetia coffeae	Hemiptera: Coccidae
Helminthosporium blight, rice	Helminthosporium oryzae	Deuteromycetes: Moniliales
Hemispherical scale	Saissetia coffeae	Hemiptera: Coccidae
Hemp sesbania	Sesbania exaltata	Fabaceae
Hickory shuckworm	Cydia caryana	Lepidoptera: Tortricidae
Honey fungus	Armillaria mellea	Basidiomycetes: Agaricales
Hornweed, common	Ceratophyllum demersum	Ceratophyllaceae
Housefly	Musca domestica	Diptera: Muscidae
House mosquito	Culex fatigans (= C. quinquefasciatus)	Diptera
Hyperparasite of fungi	Ampelomyces quisqualis	Deuteromycetes: Sphaeropsidales
Indian meal moth	Plodia interpunctella	Lepidoptera: Pyralidae
Ivy	Hedera helix	Araliaceae
Japanese beetle	Popillia japonicus	Coleoptera: Scarabaeidae
Japanese bullrush	Scirpus juncoides	Cyperaceae
Japanese knotweed	Reynoutria japonica (= Polygonum cuspidatum)	Polygonaceae
Jimson weed	Datura stramonium	Solanaceae
Johnson grass	Sorghum halepense	Poaceae
Joint vetches	Aeschynomene	Fabaceae
Kanzawa spider mite	Tetranychus kanzawai	Acari: Tetranychidae
Knapweed, diffuse	Centaurea diffusa	Asteraceae

English	Latin	Order and/or Family
Knapweed, spotted	*Centaurea maculosa*	Asteraceae
Knapweeds	*Centaurea*	Asteraceae
Knot grass	*Polygonum aviculare*	Polygonaceae
Knotweeds	*Polygonum*	Polygonaceae
Koa seed worm	*Crytophlebia illepida*	Lepidoptera: Tortricidae
Kyllinga, green	*Cyperus brevifolius*	Cyperaceae
Ladybird	*Harmonia axyridis*	Coleoptera: Coccinellidae
Ladybird	*Hippodamia convergens*	Coleoptera: Coccinellidae
Ladybug	*Harmonia axyridis*	Coleoptera: Coccinellidae
Ladybug	*Hippodamia convergens*	Coleoptera: Coccinellidae
Large white butterfly	*Pieris brassicae*	Lepidoptera: Pieridae
Late blight, potato, tomato	*Phytophthora infestans*	Chromista: Oomycetes: Peronosporales
Late-flowering cyperus	*Cyperus serotinus*	Cyperaceae
Leaf and glume spots, various hosts	*Septoria*	Sphaeropsidales
Leaf and pod spot, peas	*Ascochyta pinodes*	Deuteromycetes: Sphaeropsidales
Leaf and pod spot, peas	*Ascochyta pisi*	Deuteromycetes: Sphaeropsidales
Leaf blight, rice	*Xanthomonas oryzae*	Pseudomonadales: Pseudomonadaceae
Leaf blotch, barley and rye	*Rhynchosporium secalis*	Deuteromycetes: Moniliales
Leaf blotches etc., various hosts	*Marssonina*	Melanconiales
Leaf miners	*Agromyza*	Diptera: Agromyzidae
Leaf miners	*Liriomyza*	Diptera: Agromyzidae
Leaf miners	*Phytomyza*	Diptera: Agromyzidae
Leaf-mining moths	*Leucoptera*	Lepidoptera: Lyonetiidae
Leaf-mining moths	*Phyllonorycter*	Lepidoptera: Gracillariidae
Leaf mould, tomato	*Fulvia fulva*	Deuteromycetes: Hyphales
Leaf moulds	*Fulvia*	Deuteromycetes: Hyphales
Leaf roller	*Archips podanus*	Lepidoptera: Tortricidae
Leaf roller	*Pandemis heparana*	Lepidoptera: Tortricidae
Leaf roller	*Platynota stultana*	Lepidoptera: Tortricidae
Leaf scorch, apples	*Gymnosporangium*	Basidiomycetes: Uredinales
Leaf scorch, strawberry	*Diplocarpon earliana*	Ascomycetes: Helotiales
Leaf smuts, various hosts	*Urocystis*	Basidiomycetes: Ustilaginales
Leaf spot	*Alternaria alternata*	Deuteromycetes: Moniliales
Leaf spot	*Curvularia*	Moniliales; Dematiaceae
Leaf spot, beans	*Ascochyta fabae*	Deuteromycetes: Sphaeropsidales
Leaf spot, beet crops	*Cercospora beticola*	Deuteromycetes: Moniliales
Leaf spot, beet crops	*Ramularia beticola*	Deuteromycetes: Moniliales
Leaf spot, currants and gooseberry	*Pseudopeziza ribis*	Ascomycetes: Helotiales
Leaf spot, melon	*Corynespora melonis*	Deuteromycetes: Moniliales
Leaf spot, soybean	*Cercosporidium* (C. sojinum = Cercospora sojina)	Deuteromycetes: Moniliales

English	Latin	Order and/or Family
Leaf spot and black rot, apple	*Botryosphaeria obtusa* (= *Physalospora obtusa*)	Ascomycetes: Sphaeriales
Leaf spot and stem canker, sunflowers	*Diaporthe helianthi*	Ascomycetes: Sphaeriales
Leaf spot diseases, various hosts	*Mycosphaerella*	Ascomycetes: Sphaeriales
Leaf spots, grasses	*Rhynchosporium*	Hyphales
Leaf spots, various	*Alternaria*	Deuteromycetes: Moniliales
Leaf spots, various	*Cercospora*	Deuteromycetes: Moniliales
Leaf spots, various	*Ramularia*	Deuteromycetes: Moniliales
Leaf spots, various hosts	*Ascochyta*	Deuteromycetes: Sphaeropsidales
Leaf stripe, barley	*Pyrenophora graminea*	Ascomycetes: Sphaeriales
Leafhopper	*Empoasca decipiens and Hauptidia maraccana*	Hemiptera: Cicadellidae
Leafhopper egg parasitoid	*Anagrus atomus*	Hymenoptera: Mymaridae
Leafy spurge	*Euphorbia esula*	Euphorbiaceae
Leatherjackets	*Tipula*	Diptera: Tipulidae
Lemon-shaped cyst nematodes	*Heterodera*	Nematoda: Heteroderidae
Leopard moth	*Zeuzera pyrina*	Lepidoptera: Cossidae
Lepidopteran egg parasites	*Trichogramma*	Hymenoptera: Trichogrammatidae
Lepidopteran parasitic wasp	*Cotesia*	Hymenoptera: Aphidiidae
Lesser armyworm	*Spodoptera exigua*	Lepidoptera: Noctuidae
Lesser grain borer	*Rhyzopertha dominica*	Coleoptora: Bostrichidae
Lesser peach tree borer	*Synanthedon pictipes*	Lepidoptera: Aegeriidae
Light leaf spot, brassicas	*Pyrenopeziza brassicae*	Ascomycetes: Helotiales
Long-tailed mealybug	*Pseudococcus longispinus*	Hemiptera: Pseudococcidae
Loose silky-bent	*Apera spica-venti*	Poaceae
Loose smut, barley and wheat	*Ustilago nuda*	Basidiomycetes: Ustilaginales
Macadamia nut borer	*Cryptophlebia ombrodelta*	Lepidoptera: Tortricidae
Macrotermites	*Macrotermes, Microtermes and Odontotermes*	Uniramia: Macrotermitinae
Maggot, apple	*Rhagoletis pomonella*	Diptera: Tephritidae
Maggots	*Delia and Hylemya*	Diptera: Anthomyiidae
Maize stalk borer	*Diatraea saccharalis*	Lepidoptera: Crambidae
Maize stalk rot	*Gibberella fujikuroi*	Deuteromycetes: Hypocreales
Maize weevil	*Sitophilus zeamais*	Coleoptera: Curculionidae
Mallow	*Malva.*	Malvaceae
Mangold flea beetle	*Chaetocnema concinna*	Coleoptera: Chrysomelidae
Mangold fly	*Pegomya hyoscamni*	Diptera: Anthomyiidae
Mangrove	*Rhizophora mangle*	Rhizophoraceae
Mayweed, scentless	*Tripleurospermum maritimum* (= *Matricaria inodora*)	Asteraceae
Mayweeds (some)	*Chamomilla*	Asteraceae

English	Latin	Order and/or Family
Mayweeds (some)	*Matricaria*	Asteraceae
McDaniel's spider mite	*Tetranychus mcdanieli*	Acari: Tetranychidae
Meadow grass, annual	*Poa annua*	Poaceae
Meadow grass, rough	*Poa trivialis*	Poaceae
Meadow grasses	*Poa*	Poaceae
Mealworms	*Alphitobius*	Coleoptera: Tenebrionidae
Mealybug, citrus	*Planococcus citri*	Hemiptera: Pseudococcidae
Mealybug, long-tailed	*Pseudococcus longispinus*	Hemiptera: Pseudococcidae
Mealybug, obscure	*Pseudococcus affinis*	Hemiptera: Pseudococcidae
Mealybug parasite	*Leptomastix dactylopii*	Hymenoptera: Encrytidae
Mealybug predator	*Cryptolaemus montrouzieri*	Coleoptera: Coccinelidae
Mealybugs	*Pseudococcus*	Hemiptera: Pseudococcidae
Mediterranean black scale	*Saissetia oleae*	Hemiptera: Coccidae
Mediterranean fruit fly	*Ceratitis capitata*	Diptera: Tephritidae
Melanosis, citrus	*Diaporthe citri*	Ascomycetes: Sphaeriales
Melon and cotton aphid	*Aphis gossypii*	Hemiptera: Aphididae
Melon fly	*Dacus cucurbitae*	Diptera: Tephritidae
Mexican bean beetle	*Epilachna varivestis*	Coleoptera: Coccinellidae
Milkweed vine	*Morrenia odorata*	Asclepiadaceae
Millet, Texas	*Panicum texanum*	Poaceae
Minute black ladybird	*Stethorus punctum*	Coleoptera: Coccinellidae
Minute pirate bugs	*Orius*	Hemiptera: Anthocoridae
Mite	*Aculops*	Acari: Eriophyidae
Mite	*Calepitrimerus*	Acari: Eriophyidae
Mite	*Eriophyes*	Acari: Eriophyidae
Mite	*Phyllocoptes*	Acari: Eriophyidae
Mite	*Vasates*	Acari: Eriophyidae
Mite, avocado brown	*Oligonychus punicae*	Acari: Tetranychidae
Mite, carmine spider	*Tetranychus cinnabarinus*	Acari: Tetranychidae
Mite, McDaniel's spider	*Tetranychus mcdanieli*	Acari: Tetranychidae
Mite, persea	*Oligonychus perseae*	Acari: Tetranychidae
Mite predator	*Feltiella acarisuga*	Diptera: Cecidomyiidae
Mite predator	*Galendromus occidentalis*	Acarina: Phytoseiidae
Mite, six-spotted	*Eotetranychus sexmaculatus*	Acari: Tetranychidae
Mite, southern red	*Oligonychus illicis*	Acari: Tetranychidae
Mite, spruce spider	*Oligonychus ununguis*	Acari: Tetranychidae
Mite, storage	*Acarus siro*	Acarina: Acaridae
Mite, storage	*Lepidoglyphus (Glycyphagus) destructor*	Acarina: Acaridae
Mite, strawberry	*Tetranychus atlanticus*	Acari: Tetranychidae
Mite, two-spotted spider	*Tetranychus urticae*	Acari: Tetranychidae
Mite, willamette	*Eotetranychus willamettei*	Acari: Tetranychidae
Mites, spider	*Tetranychidae*	Acari
Mites, spider	*Tetranychus spp.*	Acari: Tetranychidae
Mock cypress	*Kochia scoparia*	Chenopodiaceae
Mole cricket	*Gryllotalpa gryllotalpa*	Orthoptera: Gryllotalpidae

English	Latin	Order and/or Family
Mole cricket	*Scapteriscus vicinus*	Orthoptera: Gryllotalpidae
Mole crickets	*Gryllotalpa*	Orthoptera: Gryllotalpidae
Monilinia leaf blight, apple	*Monilinia mali*	Deuteromycetes: Moniliales
Morning glory, ivyleaf	*Ipomoea hederacea*	Convolvulaceae
Morning glory, tall	*Ipomoea purpurea*	Convolvulaceae
Mosquitoes	*Aedes, Anopheles, Coquillettidea, Culex, Culiseta, Deinocerites, Eretmapodites, Haemagogus, Mansonia, Opifex, Orthopodomyia, Psorophora, Sabethes, Uranotaenia and Wyeomyia species*	Diptera: Culicideae
Moss	*Ceratodon purpureus*	Bryophyta
Mountain pine beetle	*Dendroctonus ponderosae*	Coleoptera: Scolytidae
Mugwort	*Artemisia vulgaris*	Asteraceae
Mushroom cecid	*Heteropeza pygmaea*	Diptera: Cecidomyiidae
Mushroom sciarid	*Lycoriella auripila*	Diptera: Sciaridae
N-fixing bacteria	*Nitrosomonas*	Bacteria
Navel orangeworm	*Amylois transitella*	Lepidoptera: Pyralidae
Neck rot, lettuce	*Sclerotinia minor*	Ascomycetes: Helotiales
Neck rot, onions	*Botrytis allii*	Deuteromycetes: Moniliales
Needle nematodes	*Longidorus*	Nematoda
Neem tree	*Azadirachta indica*	Meliaceae
Net blotch, barley	*Pyrenophora teres*	Ascomycetes: Sphaeriales
Nettle, common	*Urtica dioica*	Urticaceae
Nettle, small	*Urtica urens*	Urticaceae
Nettles	*Urtica*	Urticaceae
New Zealand grass grub	*Costelytra zealandrica*	Coleoptera: Scarabaeidae
Nightshade, black	*Solanum nigrum*	Solanaceae
Nightshade, hairy	*Solanum sarrachoides*	Solanaceae
Nipplewort	*Lapsana communis*	Asteraceae
Northern joint vetch	*Aeschynomene virginica*	Fabaceae
Northern leaf blight, maize	*Helminthosporium turcicum*	Deuteromycetes: Moniliales
Nutgrass	*Cyperus rotundus*	Cyperaceae
Nutsedges	*Cyperus*	Cyperaceae
Oat, sterile	*Avena sterilis*	Poaceae
Oats (wild and cultivated)	*Avena*	Poaceae
Old World bollworm	*Helicoverpa armigera*	Lepidoptera: Noctuidae
Olive fruit fly	*Bactrocera oleae (= Dacus oleae)*	Diptera: Tephritidae
Olive moth	*Prays oleae*	Lepidoptera: Yponomeutidae
Onion couch	*Arrhenatherum elatius*	Poaceae
Onion thrips	*Thrips tabaci*	Thysanoptera: Thripidae
Orangeworm, navel	*Amylois transitella*	Lepidoptera: Pyralidae
Oriental fruit moth	*Grapholitha molesta*	Lepidoptera: Tortricidae

English	Latin	Order and/or Family
Oriental fruit tree moth	*Carposina niponensis*	Lepidoptera: Tortricidae
Oriental red scale	*Aonidiella orientalis*	Hemiptera: Diaspididae
Oriental tobacco budworm	*Helicoverpa assulta*	Lepidoptera: Noctuidae
Pale persicaria	*Polygonum lapathifolium*	Polygonaceae
Palm scale	*Chrysomphalus dictyospermi*	Hemiptera: Diaspididae
Panic grasses	*Panicum spp.*	Poaceae
Parasitic yeasts	*Candida*	Ascomycetes: Endomycetales
Pea and bean weevils	*Sitona*	Coleoptera: Curculionidae
Pea aphid	*Acyrthosiphon pisum*	Hemiptera: Aphididae
Pea cyst nematode	*Heterodera goettingiana*	Nematoda: Heteroderidae
Pea moth	*Cydia nigicana*	Lepidoptera: Tortricidae
Peach leaf-curl	*Taphrina deformans*	Ascomycetes: Taphrinales
Peach-potato aphid	*Myzus persicae*	Hemiptera: Aphididae
Peach tree borer	*Anarsia lineatella*	Lepidoptera: Gelechiidae
Peach tree borer	*Sanninoidea exitiosa*	Lepidoptera: Aegeriidae
Pear leaf blister moth	*Leucoptera malifoliella*	Lepidoptera: Lyonetiidae
Pear psylla	*Cacopsylla pyri and Cacopsylla pyricola*	Hemiptera: Psyllidae
Pear rust	*Gymnosporangium fuscum*	Basidiomycetes: Uredinales
Pear rust mite	*Epitrimerus pyri*	Acari: Eriophyidae
Pearly green lacewing	*Chrysoperla carnea*	Neuroptera: Chrysopidae
Penicillium rots	*Penicillium*	Deuteromycetes: Moniliales
Persicaria	*Polygonum persicaria*	Polygonaceae
Petunia	*Petunia*	Solanaceae
Pharaoh's ant	*Monomorium pharaonis*	Hymenoptera: Formicidae
Pickerel weed	*Monochoria vaginalis*	Pontederiaceae
Pimpernel, false	*Lindernia procumbens*	Scrophulariaceae
Pine false webworms	*Acantholyda erythrocephala*	Hymenoptera: Pamphiliidae
Pine processionary caterpillar	*Thaumetopoea pityocampa*	Lepidoptera: Thaumetopoeidae
Pine sawfly	*Neodiprion lecontei*	Hymenoptera: Diprionidae
Pine wood nematode	*Bursaphelenchus xylophilus*	Nematoda: Aphelenchoididae
Pinewood nematode	*Bursaphelenchus xylophilus*	Nematoda: Aphelenchoididae
Pink bollworm	*Pectinophora gossypiella*	Lepidoptera: Gelechiidae
Pink citrus rust mite	*Aculops pelekassi*	Acari: Eriophyidae
Pitch canker disease	*Fusarium moniliforme var. subglutinans*	Deuteromycetes: Moniliales
Plantains	*Plantago*	Plantaginaceae
Planthopper, citrus	*Metcalfa pruinosa*	Hemiptera: Fulgoroidea
Planthoppers	*Nilaparvata*	Hemiptera: Delphacidae
Plum rust	*Tranzschelia pruni-spinosae*	Basidiomycetes: Uredinales
Pod rot, cocoa	*Monilia roreri*	Deuteromycetes: Moniliales

English	Latin	Order and/or Family
Pollen beetle	*Meligethes aeneus*	Coleoptera: Nitidulidae
Pollen beetles	*Meligethes*	Coleoptera: Nitidulidae
Pondweed, American	*Potamogeton distinctus*	Potamogetonaceae
Pondweeds	*Potamogeton*	Potamogetonaceae
Poplars	*Populus*	Salicaceae
Poppies	*Papaver*	Papaveraceae
Post-harvest rots	*Rhizopus*	Mucorales
Post-harvest rots, various hosts	*Sclerotium*	Deuteromycetes: Agonomycetales
Potato aphid	*Macrosiphum euphorbiae*	Hemiptera: Aphididae
Potato cyst nematodes	*Globodera*	Nematoda: Heteroderidae
Potato early blight	*Alternaria solani*	Deuteromycetes: Moniliales
Potato leafhopper	*Eupterycyba jucunda*	Hemiptera: Cicadellidae
Potato moth	*Phthorimaea operculella*	Lepidoptera: Gelechiidae
Powdery mildew, apple	*Podosphaera leucotricha*	Ascomycetes: Erysiphales
Powdery mildew, beet crops	*Erysiphe betae*	Ascomycetes: Erysiphales
Powdery mildew, cereals, grasses	*Blumeria graminis*	Ascomycetes: Erysiphales
Powdery mildew, cucurbits	*Erysiphe cichoracearum*	Ascomycetes: Erysiphales
Powdery mildew, cucurbits	*Sphaerotheca fuliginea*	Ascomycetes: Erysiphales
Powdery mildew, grapevines	*Uncinula necator*	Ascomycetes: Erysiphales
Powdery mildew, peppers	*Leveillula*	Ascomycetes: Erysiphales
Powdery mildew, rose	*Sphaerotheca pannosa*	Ascomycetes: Erysiphales
Powdery mildew, various hosts	*Erysiphe*	Ascomycetes: Erysiphales
Powdery mildew, various hosts	*Phyllactinia*	Ascomycetes: Erysiphales
Powdery mildew, various hosts	*Podosphaera*	Ascomycetes: Erysiphales
Powdery mildew, various hosts	*Sphaerotheca*	Ascomycetes: Erysiphales
Predatory mite	*Amblyseius barkeri*	Mesostigmata: Phytoseiidae
Predatory mite	*Amblyseius californicus*	Mesostigmata: Phytoseiidae
Predatory mite	*Amblyseius cucumeris*	Mesostigmata: Phytoseiidae
Predatory mite	*Typhlodromus occidentalis*	Mesostigmata: Phytoseiidae
Prickly pear cactus	*Opuntia lindheimerii*	Cactaceae
Psyllid, eucalyptus	*Ctenarytaina eucalypti*	Hemiptera: Psyllidae
Psyllids	*Psylla*	Hemiptera: Psyllidae
Puncture vine	*Tribulus terrestris*	Zygophyllaceae
Purslane	*Portulaca*	Portulacaceae
Purslane	*Portulaca oleracea*	Portulacaceae
Pygmy mangold beetle	*Atomaria linearis*	Coleoptera: Cryptophagidae
Pyrethrum daisy	*Tanacetum cinerariaefolium*	Asteraceae
Quackgrass	*Elytrigia repens (= Elymus repens)*	Poaceae
Ragweed, common	*Ambrosia artemisifolia*	Asteraceae
Ragwort, common or tansy	*Senecio jacobaea*	Asteraceae
Rape	*Brassica napus*	Brassicaceae
Ray blight, chrysanthemum	*Didymella ligulicola*	Ascomycetes: Sphaeriales
Red core, strawberry	*Phytophthora fragariae*	Phycomycetes: Peronosporales

English	Latin	Order and/or Family
Red crevice tea mite	*Brevipalpus phoenicis*	Acari: Tenuipalpidae
Red dead-nettle	*Lamium purpureum*	Labiateae
Redheaded cockchafer	*Adoryphorus couloni*	Coleoptera: Scarabaeidae
Coleoptera: Scarabaeida		
Redroot pigweed	*Amaranthus retroflexus*	Amaranthaceae
Red scale	*Aspidiotus nerii*	Hemiptera: Diaspididae
Red scale parasite	*Aphytis lignanensis*	Hymenoptera: Aphelinidae
Redshank	*Polygonum persicaria*	Polygonaceae
Red spider mites	*Panonychus*	Acari: Tetranychidae
Reniform nematode	*Rotylenchulus reniformis*	Nematoda
Rhododendron	*Rhododendron ponticum*	Ericaceae
Rice blast (imperfect stage)	*Pyricularia oryzae*	Deuteromycetes: Moniliales
Rice blast (perfect stage)	*Magnaporthe grisea*	Ascomycetes
Rice brown planthopper	*Nilaparvata lugens*	Hemiptera: Delphacidae
Rice leaf beetle	*Oulema oryzae*	Coleoptera: Chrysomelidae
Rice leaf roller	*Cnaphalocrocis medinalis*	Lepidoptera: Pyralidae
Rice leaf scald	*Monographella nivalis*	Ascomycetes: Sphaeriales
Rice sheath blight	*Pellicularia sasakii*	Basidiomycetes: Agaricales
Rice stalk borer	*Chilo suppressalis*	Lepidoptera: Pyralidae
Rice stem borer	*Chilo plejadellus*	Lepidoptera: Pyralidae
Rice stem borer	*Chilo suppressalis*	Lepidoptera: Pyralidae
Rice water weevil	*Lissorhoptrus oryzophilus*	Coleoptera: Curculionidae
Rice weevil	*Sitophilus oryzae*	Coleoptera: Curculionidae
Rice weevil	*Sitophilus zeamais*	Coleoptera: Curculionidae
Ring-spot, brassicas	*Mycosphaerella brassicicola*	Ascomycetes: Sphaeriales
Root and stem rots, various	*Phoma*	Deuteromycetes
Root flies	*Delia (= Hylemya)*	Diptera: Anthomyiidae
Root-knot nematodes	*Meloidogyne*	Nematoda
Root-lesion nematodes	*Pratylenchus*	Nematoda: Hoplolaimidae
Root rot, brassicas	*Phytophthora megasperma*	Chromista: Oomycetes: Peronosporales
Root rot, tomato	*Colletotrichum coccodes*	Ascomycetes: Melanconiales
Root rots, various	*Pythium*	Chromista: Oomycetes: Peronosporales
Rose aphid	*Macrosiphum rosae*	Hemiptera: Aphididae
Rose thrips	*Thrips fuscipennis*	Thysanoptera: Thripidae
Rot, various crops	*Phytophthora palmivora*	Chromista: Oomycetes: Peronosporales
Rots, damping off, etc. various hosts	*Pellicularia*	Basidiomycetes: Agaricales
Rots, ear blights and wilts, various hosts (Imperfect fungi with perfect stages in various genera)	*Fusarium*	Deuteromycetes: Moniliales
Rots, various hosts	*Sclerotium rolfsii*	Deuteromycetes: Agonomycetales
Rots of stems, storage organs, etc., various crops	*Sclerotinia sclerotiorum*	Ascomycetes: Helotiales

English	Latin	Order and/or Family
Rough cocklebur	*Xanthium strumarium*	Asteraceae
Roughseed bullrush	*Scirpus mucronatus*	Cyperaceae
Rubbery rot, potatoes	*Geotrichum candidum*	Deuteromycetes: Moniliales
Runch	*Raphanus raphanistrum*	Brassicaceae
Rush, flowering	*Butomus umbellatus*	Butomaceae
Rush skeleton weed	*Chondrilla juncea*	Asteraceae
Russian thistle	*Salsola kali*	Chenopodiaceae
Rust, beet crops	*Uromyces betae*	Basidiomycetes: Uredinales
Rust, roses	*Phragmidium mucronatum*	Basidiomycetes: Uredinales
Rust, soybean	*Phakopsora pachyrhizi*	Basidiomycetes: Uredinales
Rust, various hosts	*Gymnosporangium*	Basidiomycetes: Uredinales
Rust, various hosts	*Puccinia*	Basidiomycetes: Uredinales
Rust mite, apple	*Aculus schlechtendali*	Acari: Eriophyidae
Rust mite, pink citrus	*Aculops pelekassi*	Acari: Eriophyidae
Rust mites	*Aculus*	Acari: Eriophyidae
Rust mites	*Phyllocoptruta*	Acari: Eriophyidae
Rusts, various crops	*Uromyces*	Basidiomycetes: Uredinales
Ryegrass, annual	*Lolium rigidum*	Poaceae
Ryegrass, Italian	*Lolium multiflorum*	Poaceae
Ryegrass, perennial	*Lolium perenne*	Poaceae
Ryegrass, Wimmera	*Lolium rigidum*	Poaceae
Ryegrasses	*Lolium*	Poaceae
St John's Wort	*Hypericum perforatum*	Guttiferae
St John's Wort, pale	*Hypericum montanum*	Guttiferae
San José scale	*Quadraspidiotus perniciosus*	Hemiptera: Coccidae
Saramatta grass	*Ischaemum rugosum*	Poaceae
Sawflies	*Diprion*	Hymenoptera: Diprionidae
Scab, apples	*Venturia inaequalis*	Ascomycetes: Sphaeriales
Scab, cereals	*Gibberella* (= *Fusarium*)	Deuteromycetes: Hypocreales
Scab, cereals	*Gibberella zeae*	Deuteromycetes: Hypocreales
Scab, citrus	*Elsinoe fawcettii*	Ascomycetes: Myriangiales
Scab, pears	*Venturia pirina*	Ascomycetes: Sphaeriales
Scale, citrus black	*Parlatoria ziziphi*	Hemiptera: Diaspididae
Scale, cottony cushion	*Icerya purchasi*	Hemiptera: Margarodidae
Scale, euonymus	*Unaspis euonymi*	Hemiptera: Coccidae
Scale, siebold	*Lepidosaphes yangicola*	Hemiptera: Coccidae
Scale insect	*Didesmococcus brevipes*	Hemiptera: Coccidae
Scale insect parasite	*Metaphycus bartletti*	Hymenoptera: Encrytidae
Scale insect parasite	*Metaphycus helvolus*	Hymenoptera: Encrytidae
Scale insects	*Coccus*	Hemiptera: Coccidae
Scentless mayweed	*Matricaria perforata* (= *M. inodora*)	Asteraceae
Sciarid flies	*Bradysia*	Diptera: Sciaridae

English	Latin	Order and/or Family
Sciarid flies	*Lycoriella*	Diptera: Sciaridae
Sciarid flies	*Sciara*	Diptera: Sciaridae
Sciarid fly predator	*Hypoaspis miles*	Mesostigmata: Laelapidae
Sclerotinia rots, various hosts	*Sclerotinia*	Ascomycetes: Helotiales
Scotchbroom	*Sarothamnus scoparius*	Fabaceae
Scuttle flies	*Megaselia*	Diptera: Phoridae
Scuttle flies	*Phoridae*	Diptera
Sea club-rush	*Scirpus maritimus*	Cyperaceae
Sea-rush	*Juncus maritimus*	Juncaceae
Sedges	*Carex*	Cyperaceae
Seed-eating ants	*Monomorium*	Hymenoptera: Formicidae
Septoria leaf spot, wheat	*Mycosphaerella graminicola*	Ascomycetes: Sphaeriales
Sharp eyespot, cereals	*Ceratobasidium cereale*	Basidiomycetes: Tulasnellales
Shattercane	*Sorghum bicolor*	Poaceae
Shepherd's purse	*Capsella bursa-pastoris*	Brassicaceae
Shoot borer, Eastern pine	*Eucosma gloriola*	Lepidoptera: Tortricidae
Shothole, prunus	*Stigmina carpophila*	Hyphales
Siam weed	*Chromolaena odorata*	Asteraceae
Siam weed	*Eupatorium odoratum (= Chromolaena odorata)*	Asteraceae
Sickle pod	*Cassia obtusifolia*	Fabaceae
Sigatoka	*Mycosphaerella musicola*	Ascomycetes: Sphaeriales
Silver leaf	*Chondrostereum purpureum*	Agaricales: Agaricaceae
Silver scurf, potatoes	*Helminthosporium solani*	Deuteromycetes: Moniliales
Silvery moth	*Plusia*	Lepidoptera: Noctuidae
Six-spined ips	*Ips sexdentatus*	Coleoptera: Ipidae
Six-toothed spruce bark beetle	*Pityogenes chalcographus*	Coleoptera: Scolytidae
Skin spot, potatoes	*Polyscytalum pustulans*	Hyphales
Slugs and snails	*Gastropoda*	Mollusca
Small brown planthopper	*Laodelphax striatella*	Hemiptera: Delphacidae
Small white butterfly	*Pieris rapae*	Lepidoptera: Pieridae
Smaller European elm bark beetle	*Scolytus multistratus*	Coleoptera: Scolytidae
Smartweed	*Polygonum persicaria*	Polygonaceae
Smooth sowthistle	*Sonchus oleraceus*	Asteraceae
Smooth witchgrass	*Panicum dichotomiflorum*	Poaceae
Smut, various hosts	*Tilletia*	Basidiomycetes: Ustilaginales
Smut, various hosts	*Ustilago*	Ustilaginales
Snow mould, grasses, cereals	*Microdochium nivalis*	Deuteromycetes: Moniliales
Snow rot, cereals	*Typhula incarnata*	Basidiomycetes: Agaricales
Sooty blotch, apple, pear and citrus	*Gloeodes pomigena*	Deuteromycetes: Sphaeropsidales
Sooty mould	*Cladosporium*	Deuteromycetes: Moniliales
Sorghum grasses	*Sorghum*	Poaceae
South American leaf miner	*Liriomyza huidobrensis*	Diptera: Agromyzidae
Southern pine beetle	*Dendroctonus frontalis*	Coleoptera: Scolytidae
Southern root-knot nematode	*Meloidogyne incognita*	Nematoda

English	Latin	Order and/or Family
Soybean brown spot	*Septoria glycines*	Ascomycetes: Sphaeriales
Soybean cyst nematode	*Heterodera glycines*	Nematoda: Heteroderidae
Soybean looper	*Anticarsia gemmatalis*	Lepidoptera: Noctuidae
Speedwell, common field	*Veronica persica*	Scrophulariaceae
Speedwell, ivy-leaved	*Veronica hederifolia*	Scrophulariaceae
Speedwell, slender	*Veronica filiformis*	Scrophulariaceae
Speedwells	*Veronica*	Scrophulariaceae
Spider mite predator	*Amblyseius fallacis*	Mesostigmata: Phytoseiidae
Spider mite, Kanzawa	*Tetranychus kanzawai*	Acari: Tetranychidae
Spider mites	*Tetranychus*	Acari: Tetranychidae
Spike rush	*Eleocharis acicularis*	Cyperaceae
Spiny sida	*Sida spinosa*	Malvaceae
Spiral nematodes	*Helicotylenchus*	Nematoda: Tylenchidae
Spotted knapweed	*Centaurea maculosa*	Asteraceae
Spotted spurge	*Euphorbia maculata*	Euphorbiaceae
Sprangletop, bearded	*Leptochloa fascicularis* (= *Diplachne fascicularis*)	Poaceae
Sprangletop, red	*Leptochloa chinensis*	Poaceae
Sprangletop grasses	*Leptochloa*	Poaceae
Spruce beetle	*Dendroctonus rufipennis*	Coleoptera: Scolytidae
Spur blight, cane fruit	*Didymella applanata*	Ascomycetes: Sphaeriales
Spurge, leafy	*Euphorbia esula*	Euphorbiaceae
Stalk rots, various hosts	*Diplodia*	Deuteromycetes: Sphaeropsidales
Star grasses	*Cynodon*	Poaceae
Star-thistle, purple	*Centaurea calcitrapa*	Asteraceae
Star-thistle, yellow	*Centaurea solstitialis*	Asteraceae
Stem borers	*Chilo*	Lepidoptera: Pyralidae
Stem canker fungi, various hosts	*Diaporthe*	Ascomycetes: Sphaeriales
Stem nematode	*Ditylenchus dipsaci*	Nematoda: Tylenchidae
Sting nematode	*Belonolaimus longicausatus*	Nematoda
Stinking smut	*Tilletia caries*	Basidiomycetes: Ustilaginales
Storage fungi	*Aspergillus*	Deuteromycetes: Moniliales
Storage mite	*Acarus siro*	Acarina: Acaridae
Storage mite	*Lepidoglyphus* (*Glycyphagus*) *destructor*	Acarina: Acaridae
Strangler vine	*Morrenia odorata*	Asclepiadaceae
Stubby-root nematodes	*Trichodorus*	Nematoda
Sugar beet root maggot	*Tetanops myopaeformis*	Diptera: Otitidae
Sugar cane borer	*Diatreae saccharalis*	Lepidoptera: Pyralidae
Sugar cane rootstalk borer	*Diaprepes abbreviatus*	Coleoptera: Curculionidae
Sugarcane spittle bug	*Mahanarva postica*	Hemiptera: Cercopidae
Sumac	*Rhus aromatica*	Anacardiacea
Summer fruit tortrix moth	*Adoxophyes orana*	Lepidoptera: Tortricidae
Sunflower	*Helianthus annuus*	Asteraceae
Symphilids	*Symphyla*	Myriapoda

English	Latin	Order and/or Family
Tan spot, wheat	*Pyrenophora tritici-vulgaris*	Ascomycetes: Sphaeriales
Tarnished plant bug	*Lygus pabulinus*	Hemiptera: Miridae
Tarnished plant bug, Southern	*Lygus lineolaris*	Hemiptera: Miridae
Tarnished plant bug, Western	*Lygus herperus*	Hemiptera: Miridae
Tarsonemid mites	*Tarsonemus (= Phytonemus)*	Acari: Tarsonemidae
Tea leaf roller	*Caloptilia theivora*	Lepidoptera: Gracillariidae
Tea tortrix	*Homona magnanima*	Lepidoptera: Tortricidae
Tetranychid mites	*Eutetranychus*	Acari: Tetranychidae
Texas citrus mite	*Eutetranychus banksi*	Acari: Tetranychidae
Thistle, creeping	*Cirsium arvense*	Asteraceae
Thistles	*Carduus*	Asteraceae
Thorn apple	*Datura stramonium*	Solanaceae
Thrips	*Thrips*	Thysanoptera: Thripidae
Thrips	*Thrips palmi*	Thysanoptera: Thripidae
Thrips, avocado	*Scirtothrips persea*	Thysanoptera: Thripidae
Thrips, flower	*Frankliniella intonsa*	Thysanoptera: Thripidae
Thrips, legume flower	*Megalurothrips sjostedti*	Thysanoptera: Thripidae
Thrips, New Zealand flower	*Thrips obscuratus*	Thysanoptera: Thripidae
Thrips, palm	*Parthenothrips dracaenae*	Thysanoptera: Thripidae
Thrips, Western flower	*Frankliniella occidentalis*	Thysanoptera: Thripidae
Tobacco	*Nicotiana rustica*	Solanaceae
Tobacco aphid	*Myzus nicotianae*	Hemiptera: Aphididae
Tobacco budworm	*Heliothis virescens*	Lepidoptera: Noctuidae
Tobacco flea beetle	*Epitrix hirtipennis*	Coleoptera: Chrysomelidae
Tobacco whitefly	*Bemisia tabaci*	Hemiptera: Aleyrodidae
Tomato canker	*Clavibacter michiganensis*	Eubacteriales
Tomato fruitworm	*Helicoverpa zea*	Lepidoptera: Noctuidae
Tomato leaf miner	*Liriomyza bryoniae*	Diptera: Agromyzidae
Tomato looper	*Chrysodeixis chalcites*	Lepidoptera: Noctuidae
Tomato pinworm	*Keiferia lycopersicella*	Lepidoptera: Gelechiidae
Tomato powdery mildew	*Leveillula taurica*	Ascomycetes: Erysiphales
Tomato powdery mildew	*Oidiopsis taurica*	Deuteromycetes: Moniliales
Tortrix moths	*Tortrix*	Lepidoptera: Tortricidae
Tortrix moths and leaf rollers	*Adoxophyes*	Lepidoptera: Tortricidae
Tortrix moths and leaf rollers	*Homona*	Lepidoptera: Tortricidae
Tropical green rice leafhopper	*Nephotettix nigropictus*	Hemiptera: Cicadellidae
Tufted apple moth	*Platynota idaeusalis*	Lepidoptera: Tortricidae
Turnip gall weevil	*Ceutorhynchus pleurostigmata*	Coleoptera: Curculionidae
Turnip moth	*Agrotis segetum*	Lepidoptera: Noctuidae
Tussock moth, whitemarked	*Orgyia leucostigma*	Lepidoptera: Lymantriidae
Two-spotted spider mite	*Tetranychus urticae*	Acari: Tetranychidae
Two-spotted spider mite predator	*Phytoseiulus persimilis*	Mesostigmata: Phytoseiidae
Umbrella plant	*Cyperus difformis*	Cyperaceae
Valsa canker of apple	*Valsa ceratosperma*	Ascomycetes: Sphaeriales

English	Latin	Order and/or Family
Various rots	*Monilia*	Deuteromycetes: Moniliales
Velvet bean caterpillar	*Anticarsia gemmatalis*	Lepidoptera: Noctuidae
Velvetleaf	*Abutilon theophrasti*	Malvaceae
Verticillium wilt, various hosts	*Verticillium*	Deuteromycetes: Moniliales
Vine weevil	*Otiorhynchus sulcatus*	Coleoptera: Curculionidae
Wandering Jew	*Commelina*	Commelinaceae
Warehouse moth	*Ephestia elutella*	Lepidoptera: Pyralidae
Water duckweed	*Pistia stratiotes*	Araceae
Water hyacinth	*Eichhornia crassipes*	Pontederiaceae
Water primroses	*Jussiaea*	Onagraceae
Water purslane	*Ludwigia peploides*	Onagraceae
Welsh chafer	*Hoplia philanthus*	Coleoptera: Scarabaeidae
Western balsam bark beetle	*Dryocoetes confusus*	Coleoptera: Scolytidae
Western flower thrips	*Frankliniella occidentalis*	Thysanoptera: Thripidae
Wheat bulb fly	*Delia coarctata*	Diptera: Anthomyiidae
White-backed planthopper	*Sogatella furcifera*	Hemiptera: Delphacidae
White blister	*Albugo candida*	Chromista: Oomycetes: Peronosporales
Whitefly	*Bemisia*	Hemiptera: Aleyrodidae
Whitefly, glasshouse	*Trialeurodes vaporariorum*	Hemiptera: Aleyrodidae
Whitefly, silverleaf	*Bemisia argentifolii*	Hemiptera: Aleyrodidae
Whitefly, tobacco	*Bemisia tabaci*	Hemiptera: Aleyrodidae
White grubs	*Hoplochelis marginalis*	Coleoptera: Scarabaeidae
White leaf spot, oilseed rape	*Pseudocercosporella capsellae*	Deuteromycetes: Hyphales
White leaf spot, strawberry	*Mycosphaerella fragariae*	Ascomycetes: Sphaeriales
White mould, mushrooms	*Mycogone perniciosa*	Deuteromycetes: Moniliales
White muscardine	*Beauvaria bassiana*	Deuteromycetes; Moniliales
White mustard	*Sinapis alba*	Brassicaceae
White rot, onion	*Sclerotium cepivorum*	Deuteromycetes: Agonomycetales
White rot, timber	*Ganoderma*	Basidiomycetes: Agaricales
Whiteflies	*Bemisia*	Hemiptera: Aleyrodidae
Whitefly	*Aleurothrixus floccosus*	Hemiptera: Aleyrodidae
Whitefly, citrus	*Dialeurodes citri*	Hemiptera: Aleyrodidae
Whitefly parasite	*Eretmocerus sp. nr. californicus* (= *Eretmocerus eremicus*)	Hymenoptera: Aphelinidae
Whitefly predatory beetle	*Delphasus pusillus*	Coleoptera: Coccinellidae
Wild fire of tobacco and soybean	*Pseudomonas tabaci*	Pseudomonadales: Pseudomonadaceae
Wild oat	*Avena fatua*	Poaceae
Wild oat, winter	*Avena ludoviciana*	Poaceae
Wild pansies	*Viola*	Violaceae
Wild radish	*Raphanus raphanistrum*	Brassicaceae
Wilt	*Verticillium albo-atrum*	Deuteromycetes: Moniliales
Wimmera ryegrass	*Lolium rigidum*	Poaceae

English	Latin	Order and/or Family
Wireworms	*Agriotes*	Coleoptera: Elateridae
Witches' broom	*Crinipellis perniciosa*	Basidiomycetes: Agaricales
Woolly aphid	*Eriosoma lanigerum*	Hemiptera: Pemphigidae
Wormwood	*Artemesia vulgaris*	Asteraceae
Yellow birch	*Betula lutea*	Betulaceae
Yellow cereal fly	*Opomyza florum*	Diptera: Opomyzidae
Yellow fever mosquito	*Aedes aegypti*	Diptera: Culicidae
Yellow-headed spruce sawfly	*Pikonema alaskensis*	Hymenoptera: Tenthredinidae
Yellow nutsedge	*Cyperus esculentus*	Cyperaceae
Yellow rust, cereals	*Puccinia striiformis*	Basidiomycetes: Uredinales
Yellow underwing moth	*Noctua pronuba*	Lepidoptera: Noctuidae
Yew	*Taxus baccata*	Taxaceae

Directory of companies

Parts of names given in bold represent the short form of the company name which is used in the text of Main Entries.

Abbott Laboratories
See **Valent BioSciences** Inc.

Aberdeen Road Co.
PO Box 435
Emigsville
PA 17318
USA

AEF Global
855 rue Pepin
Sherbrooke
Quebec
J1L 2P8
CANADA
Tel: 1 819 348 9461
Fax: 1 819 348 9465
e-mail: aefglobal@qc.aira.com

Afrasa
See Industrias **Afrasa** sa

AgBio
9915 Raleigh Street
Westminster
CO 80030
USA
Tel: 1 303 469 9221
Fax: 1 303 469 9598

AgBioChem Inc.
3 Fleetwood Court
Orinda
CA 94563
USA
Tel: 1 510 254 0789
Fax: 1 925 254 0186

AGC MicroBio Ltd
See **Becker Underwood** Inc.

Agralan Ltd
The Old Brickyard
Ashton Keynes
Swindon
Wiltshire SN6 6QR
UK
Tel: 44 1285 860015
Fax: 44 1285 860056
e-mail: agralan@cybermail.uk.com
Home page: www.agralan.co.uk

AgraQuest Inc.
1105 Kennedy Place No. 4
Davis
CA 95616-1272
USA
Tel: 1 530 750 0150
Fax: 1 530 750 0153
e-mail: agraquest@agraquest.com

AgResearch
Level 1A
6 Viaduct Harbour Avenue
Maritime Square
Private Bag 92080
Auckland
NEW ZEALAND
Tel: 64 9 968 9100
Fax: 64 9 968 9101
e-mail: agresearch@agresearch.co.nz
Home page: www.agresearch.co.nz

AgrEvo
See **Bayer CropScience**

Agrichem B.V.
4900 AG Oosterhout
NETHERLANDS
Tel: 31 162 431931
Fax: 31 162 456797
e-mail: info@agrichem.nl
Home page: www.agrichem.net/english

Agricola el Sol
30 calle 11-42 zona 12 01012
Ciudad de Guatemala
GUATEMALA
Fax: 502 2 760496
e-mail: restrada@guate.net

Agricultural Research Initiatives Inc.
700 Research Center Blvd
Fayetteville
AR 72701
USA
e-mail: dkellyc@alltel.net

Agriculture Sciences Inc.
3601 Garden Brook
Dallas
TX 75234
USA
Tel: 1 972 243 8930
Fax: 1 972 406 1125

Agriculture Solutions
PO Box 141
Strong
ME 04983
USA
Tel: 1 207-684-3939
e-mail: info@agriculturesolutions.com

Agridyne Technologies Inc.
See **Certis** USA LLC

Agrifutur s.r.l.
via Countrywomen 8
I-25020 Alfianello
Brescia
ITALY
Tel: 39 30 99 34 776
Fax: 39 30 99 34 777
e-mail: rkm@agrifutur.it

Agri Life
1162 Mayuri Nagar
Miyapur
Hyderabad 500 049
INDIA
Tel: 91 40 5664 6278
Fax: 91 40 5532 6278
e-mail: dr_venkatesh@yahoo.com
Home page: www.webspawner.comusers/
drvenkatesh/index.html

Agrimm Biologicals Ltd
231 Fitzgerald Avenue
PO Box 13-245
Christchurch
NEW ZEALAND
Tel: 64 3 366 8671
Fax: 64 3 365 1859
e-mail: j.hunt@agrimm.co.nz

AgriSense-BCS Ltd
Treforest Industrial Estate
Pontypridd CF37 5SU
UK
Tel: 44 1443 841155
Fax: 44 1443 841152
e-mail: mail@agrisense.demon.co.uk
Home page: www.agrisense.co.uk
See also **Certis** and **Mitsui** Chemicals Inc.

AgriSystems International
125 West Seventh Street
Wind Gap
PA 18091
USA
Tel: 1 610 863 6700
Fax: 1 610 863 4622

AgriVir LLC
1901 L Street NW
Suite 250
Washington DC 20036
USA

Agrobiologicos of the Northwest
(**AGROBIONSA**)
River Mocorito No. 575
The West
Col. Guadalupe
80220 Culiacán
Sinaloa
MEXICO
Tel/Fax: 91 67 13 17 39
e-mail: agrobion@docs.ccs.net.mx

AGROBIONSA
See Agrobiologicos of the Northwest
(**AGROBIONSA**)

Agro-Kanesho Co. Ltd
Akasaka Shasta-East 7th Fl.
4-2-19 Akasaka
Minato-Ku
Tokyo 107-0052
JAPAN
Tel: 81 3 5570 4711
Fax: 81 3 5570 4708
Home page: www.agrokanesho.co.jp

Agro Logistics
PO Box 5799
Diamond Bar
CA 91765
USA
Tel: 1 909 594 7222
Fax: 1 909 594 7330
e-mail: info@agrologistic.com

Agropharm Ltd
Buckingham House
Church Road, Penn
High Wycombe
Bucks HP10 8LN
UK
Tel: 44 1494 816575
Fax: 44 1494 816578

Agros Holding S.A.
ul. Chałubińskiego 8
00-613 Warsaw
POLAND
Tel.: 48 22 830 95 66
Fax: 48 22 597 11 84
e-mail: agros@agros.com.pl

AGSCI
See **Agriculture Sciences** Inc.

Agtrol Chemical Products
7322 Southwest Freeway
Suite 1400
Houston
TX 77074
USA
Tel: 1 713 995 0111
Fax: 1 713 995 9505

AgVenture Inc.
PO Box 29
Kentland
IN 47951-0029
USA
Tel: 1 218 474 5557
Fax: 1 219 474 5533

A.H. **Hoffman** Inc.
Lancaster
NY 14086
USA

Aimco Pesticides Ltd
Akhand Jyoti Block 1 & 3
8th Road
Santacruz (East)
PO Box 6822
Mumbai 400 055
INDIA
Tel: 91 22 618 3042
Fax: 91 22 611 6736

Akzo Nobel B.V.
See **Nufarm** B.V.

American Biophysics Corp.
140 Frenchtown Road
North Kingstown
RI 02852
USA

American **Cyanamid** Co.
See **BASF** AG

American Insectaries Inc.
30805 Rodriguez Road
Escondido
CA 92026-5312
USA
Tel: 1 760 751 1436
Fax: 1 760 749 7061
e-mail: jdavis@mailhost2.csusm.edu

Amvac Chemical Corp.
4695 MacArthur Court
Suite 1250
Newport Beach
CA 92660
USA
Tel/Fax: 1 949 260 1212
e-mail: alfredop@amvac-chemical.com
Home page: www.amvac-chemical.com

Amycel Spawn Mate
260 Westgate Drive
Watsonville
CA 95076
USA
Tel: 1 831 763 5300
Fax: 1 831 763 0700
e-mail: PKooker@montmush.com
Home page: www.amycel.com

Anchor Bio-Technologies
Cape Town
Western Cape
SOUTH AFRICA
Tel: 27 21 534 1351
Fax: 27 21 534 4819
e-mail: Consultant@AnchorBioTechnologies

Andermatt Biocontrol AG
Stahlermatten 6
CH-6146 Grossdietwil
SWITZERLAND
Tel: 41 62 917 50 00
Fax: 41 62 917 50 01
e-mail: sales@biocontrol.ch
Home page: www.biocontrol.ch

Apicure Inc.
8 Meader Road
Greenwich
NY 12834
USA

Applied Bio-Nomics Ltd
11074 W Saanich Rd
Sidney
BC
V8L 5P5
CANADA
Tel: 1 604 656 2123
Fax: 1 604 656 3844

Applied BioPest
3310 Net Place
Oxnard
CA 93035
USA
Tel: 1 805 984 9224
FAX: 1 805 984 1517
e-mail: biopest@jetlink.net

Applied Chemicals (**Thailand**) Co. Ltd
1575/15 Phaholyothin Road
Samsennai Nai Phrayathai
Bangkok 10400
THAILAND
Tel: 66 2 279 2615
Fax: 66 2 278 1343

Aqua-10 Corp.
PO Box 818
Beaufort
NC 28516
USA

Aragonesas Agro S.A.
Po Recoletos 27
28004 Madrid
SPAIN
Tel: 34 91 5853800
Fax: 34 91 5852310

Arbico Inc.
See **Arizona Biological Control Inc.**

Aries Agro-Vet Industries Ltd
Aries House
24 Deonar
Govandi E
Mumbai
Maharashtra 400 043
INDIA
Tel: 91 22 556 4052
Fax: 91 22 556 4054

Arizona Biological Control Inc. (Arbico)
PO Box 4247 CRB
Tucson
AZ 85738-1247
USA
Tel: 1 520 825 9785
Fax: 1 520 825 2038
Home page: www.usit.net/BICONET

Arizona Cotton Research and Protection
Council
3721 East Weir Avenue
Phoenix
AZ 85040
USA

Armatron International Inc.
15 Highland Avenue
Malden
MA 02148-6603
USA
Tel: 1 781 321 2300
Fax: 1 781 324 8403

Arvesta Corp.
100 First Street
Suite 1700
San Francisco
CA 94105
USA

Arysta LifeScience Corporation
St Luke's Tower 38.39 Fl.
8-1 Akashicho
Chuo-ku
Tokyo 104-6591
JAPAN
Tel: 81 3 3547 4500
Fax: 81 3 3547 4699
Home page: www.arysta-ls.com
Agricultural products: www.arysta-ls.comengproducts/agri.html
See also **Nichimen** Corporation

Asahi Chemical Mnfg. Co. Ltd
500 Takayasu
Ikaruga-cho
Ikoma-gun
Nara Pref.
JAPAN
Tel: 81 7457 4 1131
Fax: 81 7457 4 1961

Atlantic and Pacific Research Inc.
PO Box 1336
Hendersonville
NC 28793
USA
Fax: 1 704 693 0071

Australian Technological Innovation Corporation Pty. Ltd
Innovation House
124 Gymnasium Drive
Macquarie University
Sydney
NSW 2109
AUSTRALIA
Tel: 61 2 9850 8216
Fax: 61 2 9884 7290

Auxien
See **Emerald** BioAgriculture Corp.
AVA Chemical Ventures LLC
80 Rochester Avenue
Suite 214
Portsmouth
NH 03801
USA

Aventis
See **Bayer CropScience**

Babolna Bioenvironmental Centre Ltd
1107 Budapest R.
Szallas U.6
HUNGARY

Bactec Corp.
2020 Holmes Road
Houston
TX 77045
USA
Tel: 1 713 797 0406
Fax: 1 713 795 4665

BASF AG
Crop Protection Division
Agricultural Center
Postfach 120
D-67114 Limburgerhof
GERMANY
Tel: 49 621 60 0
Fax: 49 621 60 27144
Home page: www.basf.com
Agricultural products: www.basf.de/
enprodukte/gesundheit/pflanzen

Bayer Corporation
Agriculture Div.
8400 Hawthorn Rd
PO Box 4913
Kansas City
MO 64120-0013
USA
Tel: 1 816 242 2000
Fax: 1 816 242 2738

Bayer CropScience
Business Group Crop Protection
Development/Regulatory Affairs
Agrochemical Center Monheim
D-51368 Leverkusen
GERMANY
Tel: 49 2173 38 3280
Fax: 49 2173 38 3564
Home page: www.bayer.com
Agricultural products: www.bayer.comen/tk/
cropscience.php

Bayer Garden
230 Cambridge Science Park
Milton Road
Cambridge CB4 0WB
UK
Tel: 44 845 345 4100
Home page: www.bayergarden.co.uk

BCP
See **Biological Crop Protection**

BCS
See **Biosystemes** France

Becker Microbial Products Inc.
9464 NW 11th Street
Plantation
FL 33322
USA
Tel: 1 954 474 7590
Fax: 1 954 474 2463

Becker Underwood Inc.
801 Dayton Avenue
Ames
IA 50010
USA
Tel: 1 515 232 5907
Fax: 1 515 232 5961
e-mail: info@beckerunderwood.com
Home page: www.beckerunderwood.com

Bedoukian Research Inc.
(Chemical Manufacturer)
21 Finance Drive, Danbury
CT 06810-4192
USA
Tel: 1 203 830 4000
Fax: 1 203 830 4010

Beneficial Insectary
14751 Oak Run Rd, Oak Run
CA 96069
USA
Tel: 1 916 472 3715
Fax: 1 916 472 3523
Home page: www.insectary.com

The **Beneficial Insect Company**
See **The Beneficial Insect Company**

BioAgri AB
PO Box 914
Dag Hammarskjolds 180
SE-751 09 Uppsala
SWEDEN
Tel: 46 18 67 49 00
Fax: 46 18 67 49 01

Bioagro
Ninos Heroes No. 105
Rio Bravo, Tamaulipas
MEXICO
Tel: 52 893 45556

Bio-Agronomics
PO Box 1013
Clovis
CA 93616
USA
Tel: 1 209 297 9288

Biobest
Ilse Velden 18
B-2260 Westerlo
BELGIUM
Tel: 32 14 231701
Fax: 32 14 231831
Home page: www.biobest.be

Bio-Care Technology Pty. Ltd
RMB 1084 Pacific Highway
Somersby
NSW 2250
AUSTRALIA
Tel: 61 2 4340 2246
Fax: 61 2 4340 2243
e-mail: gary@bio-care.com.au
Home page: www.bio-care.com.au

Bio Collect
5841 Crittenden Street
Oakland
CA 94601
USA
Tel: 1 510 436 8052
Fax: 1 510 532 0288

Biocontrol Ltd
PO Box 515
Warwick
Queensland 4370
AUSTRALIA
Tel: 61 76 61 4488
Fax: 61 76 61 7211

Biocontrol Ltd
400 East Evergreen Blvd
Suite 205
Vancouver
WA 98660
USA

Biocontrol Network
5116 Williamsburg Rd
Brentwood
TN 37027
USA
Tel: 1 615 370 4301
Fax: 1 615 370 0662
e-mail: ebugs@biconet.com

Biofac Inc.
PO Box 87, Mathis
TX 78368
USA
Tel: 1 512 547 3259
Fax: 1 512 547 9660

Bioganic Safety Brands Inc.
318 Seaboard Lane
Suite 202
Franklin
TN 37067
USA
Tel: 1 888 326 7233

Bio Huma Netics Inc.
201 South Roosevelt
Chandler
AZ 85226
USA
Tel: 1 480 961 1220
Fax: 1 480 961 3501
e-mail: info@biohumanetics.com
Home page: www.biohumanetics.com

Bio-Innovation AB
Bredholmen Box 56
SE 545 02 Algaras
SWEDEN
Tel: 46 506 42005
Fax: 46 506 42072

Bio Italia
See Intrachem **Bio Italia** S.p.A.

BioLogic
Springtown Rd
PO Box 177
Willow Hill
PA 17271
USA
Tel: 1 717 349 2789
Fax: 1 717 349 2922

Biological Control of Weeds Inc.
1418 Maple Drive
Bozeman
MT 59715
USA
Tel: 1 406 586 5111
Fax: 1 406 586 1679
e-mail: bugs@bio-control.com

Biological Control Products S.A. Ltd
PO Box 1561
Pinetown 3600
SOUTH AFRICA
e-mail: infor@biocontrol.co.za
Home page: www.biocontrol.co.za

Biological Crop Protection
Occupation Rd
Wye, Ashford
Kent TN25 5AH
UK
Tel: 44 1233 813240
Fax: 44 1233 813383
Home page: www.biological-crop-protection.
co.uk
See also **Certis** Europe B.V. and **Mitsui**
Chemicals Inc.

Biological Services
PO Box 501
Loxton
South Australia 5333
AUSTRALIA
Tel: 61 85 846 977
Fax: 61 85 845 057

Bioplanet Soc. Coop
Via Masiera 1a 1195
I-47020 Martorano di Cesena (FC)
ITALY
Tel: 39 0547 632212
Fax: 39 0547 632244
e-mail: info@bioplanet.it

Biopreparaty Ltd
Tylisovska 1
160 00 Prague 6
CZECH REPUBLIC
Tel: 420 2 311 42 98
Fax: 420 2 3332 12 17

Biopreventa
Blahoslavova 1
674 01 Trebic
CZECH REPUBLIC
Tel: 420 618 6774

Bioprodex Inc.
8520 Nw 2nd Pl.
Gainesville
FL 32607-1423
USA
Tel: 1 352 332 8198

Bio Protection Pty. Ltd
PO Box 35
Warwick
Queensland 4370
AUSTRALIA
Tel: 61 76 661 590
Fax: 61 76 661 639

BioSafer
99/220 Tessabansongkraoh Rd
Ladyao
Jatujak
Bangkok 10900
THAILAND
Tel: 662 9543120 6
Fax: 662 9543128 (& 5802178)

Biosensory Inc.
322 Main Street
Building 1
2nd Floor
Willimantic
CT 06226
USA

BioSmith Pest Management Service
385 West Shaw Avenue #121
Fresno
CA 93704
USA
Tel: 1 209 265 0266

Biostadt Agrisciences
New India Centre
17 Cooperage Road
Mumbai 400001
INDIA
Tel: 91 20 20676
Fax: 91 20 27858
e-mail: biostadt@bom5.vsnl.net.in

Biosys
See **Certis** USA LLC

Biosystemes France
Parc d'Activités des Bellevues
BP 227
95614 Cergy-Pontoise CEDEX
FRANCE
Tel: 33 34 48 99 26
Fax: 33 34 48 99 27

Biotactics Inc.
425 La Cadena Drive
West #12 Riverside
CA 92501
USA
Tel: 1 909 320 1366
Fax: 1 909 781 6572
e-mail: benemite@biohaven.com
Home page: www.biohaven.combus/biotactics

Biotech International
PO Box 1539
9109 Main Street
Needville
TX 77461
USA
Tel: 1 979 793 78 80
Fax: 1 979 793 49 11
e-mail: info@biotechintl.com
Home page: www.biotechintl.com

Biotepp Inc.
895 Chemin Benoit Mont
St-Hilaire
Quebec
J3G 4S6
CANADA

Biotepp Research
2229 Avenue C North
Saskatoon
SK
S7L 5Z2
CANADA
Tel: 1 306 242 0417
Fax: 1 306 242 0872

Biotop
Biocity, Turku
Tykistökatu 6B 3
20520
FINLAND
Tel: 358 2 241 0099
Fax: 358 2 241 0111

Biovéd Bt.
1223 Budapest
Közgazdász u. 10
HUNGARY
Tel: 36 20 951 8151

Biowise
Hoyle Depot
Graffham, Petworth
West Sussex GU28 0LR
UK
Tel/Fax: 44 1798 867574

BioWorks Inc.
345 Woodcliff Drive
Fairport
NY 14450
USA
Tel: 1 800 877 9443
Fax: 1 800 903 2377
Home page: www.bioworksinc.com

Boehringer Inc.
See **BASF** AG

Bog Madsen
Garta Odense
Torveporten 10
DK-2500
Valby
DENMARK
Tel: 45 63 17 24 53
Fax: 45 66 17 28 80

Bonide Products Inc.
2 Wurz Avenue
Yorkville
NY 13495
USA
Tel: 1 315 736 8231

Brandt Consolidated Inc.
2935 So. Koke Mill Road
Springfield
IL 62711
USA

Brinkman
See Royal **Brinkman** B.V.

Buckman Laboratories Inc.
1256 N McLean Blvd
Memphis
TN 38108
USA
Tel: 1 901 278 0330
Fax: 1 901 276 5970

Buena Biosytems
PO Box 4008
Ventura
CA 93007
USA
Tel: 1 805 525 2525
Fax: 1 805 525 6058
Home page: www.buenabiosystems.com

BugLogical Control Systems
PO Box 32046
Tucson
AZ 85751-2046
USA
Tel: 1 520 298 4400
Fax: 1 520 298 4400
e-mail: www.buglogical.com

Bugs for Bugs
28 Orton Street
Mundubbera 4626
AUSTRALIA
Tel: 61 71 654576
Fax: 61 71 654626

Burlington Bio-Medical & Scientific Corp.
222 Sherwood Avenue
Farmingdale
NY 11735
USA
Tel: 1 516 694 9000
Fax: 1 516 694 9177

Caffaro S.p.A.
See Industrie Chimiche **Caffaro** S.p.A.

Calliope S.A.
N.P.P.
Route d'Artix
BP 80
64150 Nogueres
FRANCE
Tel: 33 5 59 60 92 92
Fax: 33 5 59 60 92 99
e-mail: calliope-nogueres@calliope-sa.com
Home page: www.calliope-sa.comAccueil.html
See also **Arysta** LifeScience Corporation and
Nichimen Corporation

Caltec Agri Marketing Services
PO Box 576 155
Modesto
CA 95357
USA
Tel: 1 209 575 1295
Fax: 1 209 575 0366

Canadian Forest Service
Natural Resources Canada
580 Booth Street 8th Floor
Ottawa
Ontario
K1A 0E4
CANADA
Tel: 1 613 947 7341
Fax: 1 613 947 7397
Home page: www.nrcan.gc.ca/cfs-scf/national

Cardinal Laboratories Inc.
710 South Ayon Avenue
Azusa
CA 91702
USA

CCT Corp.
5115 Avenida Encinas
Suite A, Carlsbad
CA 92008
USA
Tel: 1 760 929 9228
Fax: 1 760 929 9522

CEPLAC/CEPEC
Centro de Pesquisas do Cacau
Rod. Ilheus - Itabuna km 22 CP 07
Itabuna 45600-000, Ba
BRAZIL
Tel/Fax: 55 73 214 3215

Cequisa
Muntaner 322 1°
08021 Barcelona
SPAIN
Tel: 34 93 200 0322
Fax: 34 93 200 5648

Cerexagri Inc.
630 Freedom Business Center
Suite 402, King of Prussia
PA 19406
USA
Tel: 1 610 491 2800
Fax: 1 610 491 2801
e-mail: contact@cerexagri.com
Home page: www.cerexagri.com

Certis Europe B.V.
Boulevard de la Woluwe 60
B-1200 Brussels
BELGIUM
Tel: 32 2 331 3894
Fax: 32 2 331 3860
e-mail: Waalkens@proagro.com
Home page: www.certiseurope.com

Certis USA LLC
9145 Guilford Road Suite 175
Columbia
MD 21046
USA
Tel/Fax: 1 301 483 3807
Home page: www.certisusa.comindex.html
See also **Mitsui** Chemicals Inc.

CFPI
28 Boulevard Camelinat
BP 75
92233 Gennevilliers
FRANCE
Tel: 33 1 40 85 5050
Fax: 33 1 47 92 2545

Champon Millennium Chemicals Inc.
570 Herndon Parkway
Suite 500, Herndon
VA 20170
USA
Tel: 1 703 766 3787
e-mail: champon@ix.netcom.com
Home page: www.champon.com

Chase Organics
River Dene Estate
Molesey Rd, Hersham
Surrey KT12 4RG
UK
Tel: 44 1932 253666
Fax: 44 1932 252707
Home page: www.organiccatalog.com

Cheminova Agro A/S
PO Box 9
7620 Lemvig
DENMARK
Tel: 45 96 90 96 90
Fax: 45 96 90 96 91
e-mail: info@cheminova.dk
Home page: www.cheminova.com
Agricultural products: www.cheminova.
comhtml/agro/html

ChemTica Internacional
Apdo. 159-2150
San José
COSTA RICA
Tel: 506 261 2424
Fax: 506 261 5397
e-mail: info@pheroshop.com
Home page: www.pheroshop.com

Chemvet Advanced Agrochemicals &
Veterinary Products Industrial Co.
PO Box 294
Industrial Estate – Sahab
Amman
JORDAN
Tel: 962 6 40226 40
Fax: 962 6 40223 40
e-mail: chemvet@go.com.jo

Ciba Bunting
See **Syngenta Bioline** Ltd

Ciba-Geigy
See **Syngenta**

Ciech SA Agro-Sulphur Group
PO Box 271
Powazkowska Street 46/50
00-950 Warsaw
POLAND
Tel: 48 22 639 1580
Fax: 48 22 639 1598

Cillus Technology Inc.
8801 Long Street
Lenexa
KS 66215
USA

Circle One Global Inc.
1 Industrial Park Drive
PO Box 408
Cuthbert
GA 39840-0408
USA
Tel: 1 **229 732 3101**
Fax: 1 **229 732 3191**
e-mail: info@circleoneglobal.com
Home page: www.circleoneglobal.com

CK **Witco** Corp./Uniroyal Chemical
Benson Road
Middlebury
CT 06749
USA
Tel: 1 203 573 2000
Fax: 1 203 573 3394

CNICM
Havana
CUBA

Coated Seed Ltd
See **Wrightson Seeds** Ltd

Consep Inc.
213 SW Columbia Street
Bend
OR 97703
USA
Tel: 1 503 388 3688
Fax: 1 503 388 3705

Countrymark Cooperative Inc.
PO Box 2500
Bloomington
IL 61702-2500
USA
Tel: 1 309 557 6399
Fax: 1 309 557 6860

Crompton Corporation
199 Benson Road
Middlebury
CT 06749
USA
Tel: 1 203 573 2000
Fax: 1 203 573 3394
Home page: www.cromptoncorp.com

Crop Genetics International
See **Certis** USA

Croplan Genetics
PO Box 64089
MS 690 St Paul
MN 55164-0089
USA
Tel: 1 612 451 5458

CTT Corp.
2776 Loker Avenue
West Carlsbad
CA 92008
USA
Tel: 1 619 929 9228

Cyanamid
See **BASF** AG

Cyclo International S. A. de C.V.
Calle Laurel
Rosarita #10
Baja California 22710
MEXICO
Tel: 52 66 123 209
Fax: 52 66 121 976

Dainippon Ink & Chemicals Inc.
7-20 Nihonbashi 3-chome
Chuo-ku
Tokyo 103
JAPAN
Tel: 81 3 3272 4511
Fax: 81 3 3281 8589

DCM
See De Ceuster Meststoffen NV (**DCM**)

De Ceuster Meststoffen NV (**DCM**)
Forstsesteenweg 30
B-2860 St-Katelijne-Waver
BELGIUM
Tel: 32 15 31 22 57
Fax: 32 15 31 36 15

Defenders Ltd
Occupation Road
Wye
Ashford
Kent TN25 5EN
UK
Tel: 44 1233 813121
Fax: 44 1233 813633
e-mail: help@defenders.co.uk
Home page: www.defenders.co.uk

Defensa S.A.
Industria de Defensivos Agricolas S
Rua Padra Chagas 79 - 7th Fl.
90570-080 Porto Alegre RS
BRAZIL
Tel: 55 51 346 2121
Fax: 55 51 346 1844

Delicia
See Frunol **Delicia** GmbH

Deltapine Seed
Box 157
Scott
MS 38732
USA
Tel: 1 601 742 4000
Fax: 1 601 742 4055

Denka International B.V.
PO Box 337
3770 A H Barneveld
NETHERLANDS
Tel: 31 342 455455
Fax: 31 342 490587
e-mail: moskal@euronet.nl
Home page: www.denka.nl

De Sangosse
Bonnel - BP 5
47480 Pont-du-Casse
FRANCE
Tel: 33 5 53 69 36 30
Fax: 33 5 53 66 30 65
Home page: www.desangosse.com

Diatect International
875 S. Industrial Parkway
Heber City
UT 84032
USA
Tel: 1 435 654 4370
Fax: 1 435 657 9794
e-mail: info@diatect.com
Home page: www.diatect.com

Dong Bang Agrochemical Corp
Dongbang Agro Bldg.
1055-2 Namhyeon-dong
Gwanak-gu 151-802
Seoul
SOUTH KOREA
Tel: 82 2 580 3600
Fax: 82 2 523 3505
e-mail: admin@dongbangagro.co.kr
Home page: www.dongbangagro.co.kr

Dow AgroScience
9330 Zionsville Rd
Indianapolis
IN 46268-1054
USA
Tel: 1 317 337 4974
Fax: 1 317 337 7344
e-mail: info@dowagro.com
Home page: www.dowagro.com

DowElanco
See **Dow AgroScience**

Dunhill Chemical Co.
3026 Muscatel Avenue
Rosmead
CA 91770
USA
Tel: 1 818 288 1271
Fax: 1 818 288 3930

Du Pont
See E. I. **du Pont** de Nemours

Earth BioScience Inc.
451 Orange Street
New Haven
CT 06511
USA
Tel: 1 203 562 7773
Fax: 1 203 562 2207
Home page: www.taensa.com

Ecogen Inc.
2005 Cabot Boulevard West
PO Box 3023
Langhorne
PA 19047-1810
USA
Tel: 1 215 757 1590
Fax: 1 215 757 2956
(Ecogen is no longer trading)

Ecogen Inc.
PO Box 4309
Jerusalem
ISRAEL
Tel: 972 2 733212
Fax: 972 2 733265

EcoScience Corp.
Head Office
10 Alvin Court
East Brunswick
NJ 08816
USA

EcoScience Corp.
Produce Systems Division
PO Box 3228
Orlando
FL 32802-3228
USA
Tel: 1 407 872 2224
Fax: 1 407 872 2261

Ecosense Labs (I) Pvt. Ltd
54 Yogendra Bhavan
J.B. Nagar
Andheri (E)
Mumbai 400 059
INDIA
Tel: 91 36 830 0967
Fax: 91 22 822 8016

EcoSMART Technologies Inc.
318 Seaboard Lane
Suite 202
Franklin
TN 37067
USA
Tel: 1 888 326 7233

Eco Soil Systems Inc.
10740 Thornmint Road
San Diego
CA 92127
USA

Ecostyle A/S
Rugarddsvej 877B
Hindervad
5471 Sonderso
DENMARK
Tel: 45 7022 7067
Fax: 45 7022 7068
e-mail: info@ecostyle.dk

Ecoval Inc.
293 Church Street
Oakville
Ontario
I6J 1N9
CANADA

EDEN Bioscience Corp.
See Plant Health Care Inc.

E.I.D. Parry (India) Ltd
Pesticides Division
Dare House
234 N.S.C. Bose Rd
Chennai 600 001
INDIA
Tel: 91 44 5340251
Fax: 91 44 5340858
e-mail: NeemAzal@murugappa.co.in
Home page: www.parrysaza.com

E. I. du Pont de Nemours
Du Pont Agricultural Products
Walker's Mill
Barley Mill Plaza
Wilmington
DE 19880
USA
Tel: 1 800 441 7515
Fax: 1 302 992 6470
Home page: www.dupont.comindex.html
Agricultural products: www.dupont.comag

Eikou Kasei Co. Ltd
Agrochemicals Division
Violet Akihabara Bldg.
18-1 Kanda Matsunaga-cho
Chiyoda-ku
Tokyo 101
JAPAN
Tel: 81 3 5256 3861/2
Fax: 81 3 5256 3864

Elf Atochem Agri S.A.
See Cerexagri Inc.

EMBRAPA
Park Biological Station - PqEB s/n°
Brasilia DF
CEP 70770-901
BRAZIL
Tel: 55 61 448 4433
Fax: 55 61 347 1041

Emerald BioAgriculture Corp.
Lansing
MI 48909
USA
Tel: 1 517 882 7370
Fax: 1 517 882 7560
e-mail: info@emeraldbio.com
Home page: www.emeraldbio.com

EM Industries Inc.
7 Skyline Drive
Hawthorne
NY
USA

Encore Technologies LLC
2950 Xenium Lane Suite 110
Minneapolis
MN 55447
USA
Tel: 1 763 577 5958
Fax: 1 763 577 1936
Home page: www.encoretechllc.com

English Woodlands Biocontrol
See **Biowise**

Entocare CV
Haagsteeg 4
PO Box 162
6700 AD Wageningen
NETHERLANDS
Tel: 31 317 411188
Fax: 31 317 413166
e-mail: info@entocare.nl

Environmental Biocontrol International
3521 Silverside Rd Suite 1-L
Wilmington
DE 19810
USA
Tel: 1 302 695 5781
Fax: 1 302 695 5763
e-mail: nobirds@flightcontrol.com

Eric Schweizer Samen AG
Postfach 150
3602 Thun
SWITZERLAND
Tel: 33 227 57 57
Fax: 33 227 57 58
Home page: www.schweizerseeds.ch

Fairfax Biological Laboratory Inc.
PO Box 300
Clinton Corners
NY 12514
USA
Tel: 1 914 266 3705
Fax: 1 914 266 4892

Fargro Ltd
Toddington Lane
Littlehampton
West Sussex BN17 7PP
UK
Tel: 44 1903 721591
Fax: 44 1903 730737
e-mail: promos-fargro@btinternet.com

Farmers Cooperative Co.
PO Box 208
109 Railroad Street
Bayard
IA 50029
USA
Tel: 1 712 651 2091

Fine Agrochemicals Ltd
3 The Bull Ring
Worcester WR2 5AA
UK
Tel: 44 1905 748444
Fax: 44 1905 748440
e-mail: enquire@fine-agrochemicals.com

F. Joh. **Kwizda** GmbH
Kwizda AGRO
Laaer Bundesstrasse
2100 Leobendorf
AUSTRIA
Tel: 43 2262 735 40 40
Fax: 43 2262 735 40 49
e-mail: lw.leobdf.@kwizda.co.at
Home page: www.kwizda.co.at
Agricultural products: www.kwizda-agro.at

Flora Nützlinge
Wulkower Weg (Gärtnerei)
15518 Hangelsberg
Germany
Tel: 49 33632 59363
Fax: 49 3363 59364
E-mail: info@floranuetzlinge.de
Home page: www.floranuetzlinge.de

Florida Silvics Inc.
D/B/A Tree Tech. Microinjection Systems
950 S.E. 215th Ave.
Morriston
FL 32668
USA

FMC Corp.
Agricultural Products Group
1735 Market Street
Philadelphia
PA 19103
USA
Tel: 1 215 299 6661
Fax: 1 215 299 6256

Fortune Biotech
6-6-125 Annam Gardens
Kavidiguda
Secunderabad 500 380
INDIA
Tel: 91 40 817978
Fax: 91 40 843945
e-mail: info@fortunebiotech.com
Home page: www.fortunebiotech.com

Forward International Ltd
PO Box 81-249
5/F No. 112 Tun Hua Road
Tiapei
TAIWAN
Tel: 886 2 2545 1592
Fax: 886 2 2718 2614 (& 2094)
e-mail: thailand@forwardinter.com
Home page: www.forwardinter.com

Frost Technology Corporation
Plant Health Technologies
PO Box 198
Lathrop
CA 95330
USA

Frunol **Delicia** GmbH
Duebener Str. 137
04509 Delitzsch
GERMANY
Tel: 49 23032 53600
Fax: 49 23032 52650

Fujisawa Pharmaceutical
Chemicals Group
3-4-6 Nihonbashi Honcho
Chuo-ku
Tokyo 103
JAPAN
Tel: 81 3 3279 0882
Fax: 81 3 3241 5805

FZB Biotechnik GmbH
Glienicker Weg 185
D-12489 Berlin
GERMANY
Tel: 49 30 670570
Fax: 49 30 67057233
Home page: www.fzb-biotechnik.de/

Garlic Research Labs.
624 Ruberta Avenue
Glendale
CA 91201-2335
USA
Tel: 1 818 247 9600

Gartnersenteret **Lier Fruktlager** Handel AS
Ringeriksveien 107
3400 Lier
NORWAY
Tel: 47 32 22 85 65
Fax: 47 32 22 85 60
Home page: www.lierfrukt.no

Gharda Chemicals Ltd
B-27-29
MIDC
Dombivli (E) 421 203
Dist. Thane
INDIA
Tel: 91 251 471215
Fax: 91 251 472777
e-mail: gclrnd@giasbm01.vsnl.net.in

Gilmore Marketing & Development Inc.
152 Collins Street
Memphis
TN 38112
USA
Tel: 1 901 323 5870
Fax: 1 901 454 0295
e-mail: development@gmdinc.com
Home page: www.gmdinc.com

Gist-Brocades B.V.
Wateringseweg 1
Postbus 1
2600 MA Delft
NETHERLANDS
Tel: 31 15 2799111
Fax: 31 15 2793200

GlycoGenesys Inc.
Park Square Building 31
St James Avenue 8th Floor
Boston
MA 02116
USA
Tel: 1 617 422 0674
Fax: 1 617 422 0675
Home page: www.glycogenesys.com

Goemar
La Madeleine
Avenue du Général Patton
BP 55
35413 Saint-Malo CEDEX
FRANCE
Tel: 33 2 99 21 53 70
Fax: 33 2 99 82 56 17
Home page: www.goemar.com

Goemar USA (Agrimar Corp.)
PO Box 1419
5634 Atlanta Highway – Suite 200
Flowery Branch
GA 30542
USA
Tel: 1 770 965 9063
Fax: 1 770 965 9766

Gowan Co.
PO Box 5569, Yuma
AZ 85366-5569
USA
Tel: 1 928 783 8844
Fax: 1 928 343 9255
Home page: www.gowanco.com

Grace
See **Certis** USA

Greenfire Inc.
347 Nord Avenue #1
Chico
CA 95926
USA
Tel: 1 916 895 8301
Fax: 1 916 895 8317

Green Spot Ltd
93 Priest Rd
Nottingham
UK
Tel: 44 603 942 8925
Fax: 44 603 942 8932
e-mail: Info@GreenMethods.com
Home page: www.GreenMethods.com

Greenville Farms
1689 N. 1200 E.
North Logan
UT 84341
USA

Griffin Corp.
PO Box 1847
Rocky Ford Road, Valdosta
GA 31603
USA
Tel: 1 912 242 8635
Fax: 1 912 244 5813
Home page: griffinllc.com

Growmark Inc.
PO Box 2500
Bloomington
IL 61702-2500
USA
Tel: 1 309 557 6399
Fax: 1 309 557 6860

Growth Products Ltd
PO Box 1252
White Plains
NY 10602
USA
Tel: 1 914 428 1316
Fax: 1 914 428 2780
Home page: www.growthproducts.com

Guardian Spray
900 Lancer Way
Lebec
CA 93242
USA

Gustafson Inc.
1400 Preston Road
Suite 400, Plano
TX 75093
USA
Tel: 1 972 985 8877
Fax: 1 972 985 1696

Harmony Farm Supply
1740 N Research Parkway
Sebastopol
CA 95472
USA
Tel: 1 707 823 9125
Fax: 1 707 823 1734
e-mail: info@harmonyfarm.com

Hasel Tarim Co.
Sirinyali Mh. Sinanoglu Cd.
Cag Koop. Siteleri A Blok No: 46/6
Antalya
TURKEY
Tel: 90 242 316 50 25
Fax: 90 242 316 50 27
e-mail: info@haseltarim.com
Home page: www.haseltarim.com

Hawkesbury Integrated Pest Management
Service
PO Box 436
Richmond
NSW 2753
AUSTRALIA
Tel: 61 45 701331
Fax: 61 45 701314

Heads Up Plant Protectants
PO Box 519
428 3rd Street
Kamsack, SK
SOA 1SO
CANADA

Helena Chemical Co.
Corporate Center
225 Schilling Blvd
Collierville
TN 38017
USA
Tel: 1 901 761 0050
Fax: 1 901 761 5754
e-mail:
ProprietaryProducts@helenachemical.com
Home page: www.helenachemical.com

Henry Doubleday Research Association
Sales Ltd
Ryton Organic Gardens
Ryton on Dunsmore
Coventry
Warwickshire CV8 3LG
UK
Tel: 44 24 7630 3517
Fax: 44 24 7663 9229
e-mail: enquiry@hdra.org.uk

Hercon Environmental Corp.
PO Box 467
Aberdeen Road
Emigsville
PA 17318-0467
USA
Tel: 1 717 764 1191
Fax: 1 717 767 1016
e-mail: pmaclean@herconenviron.com
Home page: www.herconenviron.com

Hindustan Insecticides Ltd
Scope Complex Core 6 2nd Floor
7 Lodi Road
New Delhi 1100 03
INDIA
Tel: 91 112 436 2165 4549
Fax: 91 112 436 2116
e-mail: hilhq@nde.vsnl.net.in
Home page: www.hil-india.com

Hodogaya Chemical Co. Ltd
66-2 Horikawa-cho
Saiwai-ku
Kawasaki
Kanagawa 210
JAPAN
Tel: 81 44 549 6600
Fax: 81 44 549 6630
e-mail: nouyaka.hcc@hodogaya.co.jp

Hoechst Schering **AgrEvo** GmbH
See **Bayer CropScience**

Hokko Chemical Industry Co. Ltd
Mitsui Building No. 2 4-20
Nihonbashi Hongoku-cho 4-Chome
Chuo-ku
Tokyo 103-8341
JAPAN
Tel: 81 3 3279 5361
Fax: 81 3 3279 5165
e-mail: fort@hokkochem.co.jp
Home page: www.hokkochem.co.jp

Holy Terra Products Inc.
115 Fox Street
Portland
ME 04101
USA

Hortichem Ltd
14 Edison Rd, Churchfields Industrial Estate
Salisbury
Wilts SP2 7NU
UK
Tel: 44 1722 320133
Fax: 44 1722 326799
e-mail: hortichem@hortichem.co.uk

Hot Pepper Wax Inc.
305 Third Street
Greenville
PA 16125
USA
Tel: 1 724 646 2300
Fax: 1 724 646 2302
e-mail: lindag@hotpepperwax.com

Hubei **Sanonda** Co. Ltd
1 East Beijing Road
Shashi
Hubei 434001
CHINA
Tel: 86 716 8316975
Fax: 86 716 8315265
e-mail: sanonda@public.jn.hb.cn

Hydro-Gardens Inc.
PO Box 25845
Colorado Springs
CO 80936
USA
Tel: 1 719 495 2266
Fax: 1 719 531 0506
e-mail: hgi@usa.net

ICIPE
See **The International Centre of Insect
Physiology and Ecology (ICIPE)**

Idemitsu Kosan Co. Ltd
3-1-1 Marunouchi, Chiyoda-ku
Tokyo 100
JAPAN
Tel: 81 3 3213 3115
Fax: 81 3 3213 9354
Home page: www.idemitsu.co.jpe/index.html

Igene Biotechnology Inc.
9110 Red Branch Road
Columbia
MD 21045-2024
USA
Tel: 1 410 997 2599
Fax: 1 410 730 0540
e-mail: igene@igene.com

Ihara Chemical Industry Co. Ltd
1-4-26 Ikenohata
Taito-ku
Tokyo 110
JAPAN
Tel: 81 3 3822 5233
Fax: 81 3 3828 9887
Home page: www.iharachem.co.jp
See also **Kumiai** Chemical Industry Co. Ltd

IIBC
See **International Institute of Biological
Control**

Industrias **Afrasa** sa
Ciudad de Sevilla 53
Polígono Industrial Fuente del Jarro
E-46988 Paterna
Valencia
SPAIN
Tel: 34 96 132 17 00
Fax: 34 96 132 17 16
e-mail: afrasa@afrasa.es
Home page: www.afrasa.es

Industrias Quimicas del Vallés SA
Av. Rafael de Casanova 81
08100 Mollet del Vallés
Barcelona
SPAIN
Tel: 34 93 579 66 77
Fax: 34 93 593 80 11
e-mail: iqv@iqv-valles.com
Home page: www.iqv-valles.com

Industrie Chimiche **Caffaro** S.p.A.
Via Friuli 55
20031 Cesano Maderno
Milan
ITALY
Tel: 39 362 51 4266
Fax: 39 362 51 4454
e-mail: caffaro.chem@caffaro.it
Home page: www.caffarochem.comcaffaro/
itaindex.html
See also **Isagro** S.p.A. **Sipcam** and **Sipcam-
Oxon Group**

Insect Biotechnology Inc.
PO Box 2311
Chapel Hill
NC 27515-2311
USA
Tel: 1 919 475 8821

Integrated BioControl Systems Inc.
PO Box 96
Aurora
IN 47001
USA
Tel: 1 812 537 6652
e-mail: goodbug@nematodary.com

Integrated Pest Management
See **Bugs for Bugs**

Integrated Weed Control
4027 Bridger Canyon Road
Bozeman
MT 59715-8433
USA
Tel: 1 888 319 1632
Fax: 1 406 587 1989
e-mail: iwc@integratedweedcontrol.com

International Centre of Insect Physiology and
Ecology
See **The International Centre of Insect
Physiology and Ecology (ICIPE)**

International Institute of Biological Control
Silwood Park
Ascot
Berks SL5 7PY
UK
Tel: 44 1344 872999
Fax: 44 1344 872901

International Specialty Products
II Fairview Plaza
5950 Fairview Rd Suite 612
Charlotte
NC 28210
USA
Tel: 1 704 554 9332
Fax: 1 704 554 0911
e-mail: info@ispcorp.com
Home page: www.ispcorp.comhome.html
Agricultural products: http://ecom.ispcorp.
comcgi-bin/ncommerce3/ExecMacro/
ISPChemicals/main.d2w/report

International Technology Services
PO Box 75
Lafayette
CO 80026
USA
Tel: 1 303 661 9546
Fax: 1 303 661 9543
e-mail: sales@greenhouseinfo.com

Intrachem Bio (International) S.A.
34 Quai de Cologny
Cologny
CH-1223 Geneva
SWITZERLAND
Tel: 41 22 736 78 87
Fax: 41 22 736 24 10
Home page: www.intrachembio.int.ch

Intrachem **Bio Italia** S.p.A.
Via XXV Aprile 44
Grassobbio (BG) 24050
ITALY
Tel: 39 35 335 313
Fax: 39 35 335 334
e-mail: info@intrachem.it
Home page: www.intrachem.it

IPM Laboratories Inc.
Main Street
Locke
NY 13092-0300
USA
Tel: 1 315 497 2063
Fax: 1 315 497 3129

IPM of Alaska
PO Box 875006
Rocco
Moschetti
Wasilla
AK 99687-5006
USA
Tel: 1 907 745 7233
e-mail: ipmofak@mtaonline.net
Home page: www.ipmofalaska.com

IPM Technologies Inc.
4134 North Vancouver Avenue
Suite 305
Portland
OR 97217
USA
Tel: 1 503 288 2493
Fax: 1 503 288 1887

IQV
See **Industrias Quimicas** del Vallés SA

Isagro S.p.A.
Centro Uffici San Siro
Fabricato D
ala 3 - Via Caldera 21
20153 Milan
ITALY
Tel: 39 02 40901 1
Fax: 39 02 40901 287
e-mail: Isagro@Isagro.it
Home page: www.isagro.com

ISCA Technologies Inc.
PO Box 5266
Riverside
CA 92517
USA
Tel: 1 951 686 5008
Fax: 1 815 346 1722
e-mail: info@iscatech.com
Home page: www.iscatech.com

Ishihara Sangyo Kaisha Ltd
Biosciences Business Headquarters
3-15 Edobori 2-chome
Nishi-ku
Osaka 550-0002
JAPAN
Tel: 81 6 6444 7154
Fax: 81 6 6444 7156
Home page: www.iskweb.co.jp

Japan Tobacco Inc.
Agribusiness Division
Akasaka 1-chome Center Bldg. 13th Floor
11-30
Akasaka 1-chome
Minato-ku
Tokyo 107-0052
JAPAN
Tel: 81 3 5571 6648
Fax: 81 3 5571 6715
Home page: www.jti.co.jpJTI_E/Welcome.html

J B Chemicals and Pharmaceuticals
Neelam Centre 4th Floor B-Wing
Hind Cycle Road
Worli
Mumbai
Maharashtra 400025
INDIA

J F McKenna Ltd
66 Cathedral Rd
Armagh BT60 4BF
UK
Tel: 44 28 3752 4800
Fax: 44 28 3752 2227
Home page: www.jfmckenna.com

J H Biotech
4951 Olivas Park Drive
Ventura
CA 93003
USA
Tel: 1 805 650 8933
Fax: 1 805 650 8942
e-mail: biotech@rain.org

J. J. **Mauget** Co.
2810 North Figueroa Street
Los Angeles
CA 90065
USA

Johnson
See S C **Johnson** & Son Inc.

Kaken Pharmaceutical Co. Ltd
2-28-8 Honkomagome
Bunkyo-ku
Tokyo 113-8650
JAPAN
Tel: 81 3 5977 5002
Fax: 81 3 5977 5131
e-mail: ag_am@kaken.co.jp
Home page: www.kaken.co.jp

Karapur Agro Pvt. Ltd
Vaibhav Apts. 1st Floor
Vidya Nagar Colony
Miramar
Panjim
Goa 403 001
INDIA
Tel: 91 832 224844
Home page: www.karapur.com

Kemira Fine Chemicals Oy
Porkkalankatu 3
PO Box 330
FIN-00101 Helsinki
FINLAND
Tel: 358 10 862 1552
Fax: 358 10 862 1624
e-mail: reijo.partanen@kemira.com
Home page: http://fine.kemira.com
Agricultural products: http://fine.kemira.
comshowproducts.cfm?division=4

KFZB Biotechnik GmbH
Glienicker Weg 185
D-12489 Berlin
GERMANY
Tel: 49 30 670570
Fax: 49 30 67057233

Koppert B.V.
Veilingweg 17
PO Box 155
2650 AD Berkel en Rodenrijs
NETHERLANDS
Tel: 31 10 5140444
Fax: 31 10 5115203
e-mail: info@koppert.nl
Home page: www.koppert.nl

Koppert Biological Systems
2856 South Main
Ann Arbor
MI 48103
USA
Tel: 1 313 998 589
Fax: 1 313 998 5557

Krishi Rasayan (Bihar)
FMC Fortuna
Block No. A11 4th Fl.
234/3A Acharya Jagadish Chandra Bose Rd
Calcutta 700 020
INDIA
Tel: 91 33 247 5719/37
Fax: 91 33 247 1436
e-mail: krishi.rasayan@gems.vsnl.net.in

Kubota Corp.
1-2-47 Shikitsuhigashi
Naniwa-ku
Osaka 556-91
JAPAN
Tel: 81 6 648 2111
Fax: 81 6 648 3826

Kuida Ag. Supply Co.
PO Box 2598
Salinas
CA 93902
USA
Tel: 1 408 758 9914; 1 209 674 3730; 1 310 378 5137
Fax: 1 408 758 9933

Kumiai Chemical Industry Co. Ltd
4-26 Ikenohata 1-chome
Taitoh-ku
Tokyo 110-8782
JAPAN
Tel: 81 3 3822 5036
Fax: 81 3 3822 6830
e-mail: soumu@kumiai-chem.co.jp
Home page: www.kumiai-chem.co.jp
See also **Ihara** Chemical Industry Co. Ltd

Kunafin
Rte 1 Box 39
Quemado
TX 78877
USA
Tel: 1 800 832 1113
Fax: 1 512 757 1468

Kureha Chemical Industry Co. Ltd
1-9-11 Nihonbashi Horidome-cho
Chuo-ku
Tokyo 103-8552
JAPAN
Tel: 81 3 3249 4632
Fax: 81 3 3249 4745
Home page: www.kureha.co.jp

Kwizda AGRO
See F. Joh. **Kwizda** GmbH

Ladda Co. Ltd
GPO Box 2562
99/220 Tessabarnsongkroah Road
Ladyao
Jatujak
Bangkok 10900
THAILAND
Tel: 66 2 954 3120 6
Fax: 66 2 580 2178 (& 954 3128)

Latham Seeds
131 180th Street
Alexander
IA 50420
USA
Tel: 1 515 692 3258
Fax: 1 515 692 3250

Lier Fruktlager
See Gartnersenteret **Lier Fruktlager** Handel
AS

Live Systems Technology SA
COLOMBIA
Home page: http://lstsa.comhtml/LST

Loveland Products Inc.
419 18th Street
Greeley
CO 80632-1286
USA
Tel: 1 970 356 4400
Home Page: www.uap.comuap

LT Biosyn Inc.
11324 17th Avenue Court NW
Gig Harbor
WA 98332
USA

3M Canada
(Pesticide Producer)
3M Center
St Paul
MN 55144
USA

M&R Durango Inc.
PO Box 886
Bayfield
CO 81122
USA
Tel: 1 303 259 3521
Fax: 1 303 259 3857
e-mail: mail@goodbug.com
Home page: www.goodbug.com

Makhteshim-Agan
See **Makhteshim** Chemical Works Ltd

Makhteshim Chemical Works Ltd
PO Box 60
84100 Beer-Sheva
ISRAEL
Tel: 972 7 6469837
Fax: 972 7 6469846
e-mail: main@main.co.il

Mann Lake Ltd
501 South 1st Street
Hackensack
MN 5645
USA

Mauget
See J. J. **Mauget** Co.

Mauri Foods
67 Epping Road
North Ryde
AUSTRALIA

McLaughlin **G**ormley **K**ing Co.
8810 Tenth Avenue North
Minneapolis
MN 55427
USA
Tel: 1 612 544 0341
Fax: 1 612 544 6437

Meiji Seika Kaisha Ltd
Agrochemical Department
2-4-16 Kyobashi 2-chome
Chuo-ku
Tokyo 104
JAPAN
Tel: 81 3 3272 6511
Fax: 81 3 3281 4058

Merck & Co. Inc.
See **Syngenta**

MGK
See **M**cLaughlin **G**ormley **K**ing Co.

MicroBio
See **Becker Underwood** Inc.

Micro Flo Company LLC
530 Oak Court
Memphis
TN 38117
USA
Tel: 1 901 432 5000
Fax: 1 901 432 5100
Home page: www.microflocompany.com
See also **BASF** AG

Microgen Bioproducts Ltd
1 Admiralty Way
Camberley
Surrey GU15 3DT
UK
Tel: 44 1276 600081
Fax: 44 1276 600151
Home page: www.microgenbioproducts.com

Minrav Infrastructures
The Biological Division
Jerusalem 91291
ISRAEL

Mitsubishi Chemical Corp.
Mitsubishi Building
5-2 Marunouchi 2-chome
Chiyoda-ku
Tokyo 100
JAPAN
Tel: 81 3 3283 6274
Fax: 81 3 3283 6287

Mitsui Chemicals Inc.
Agri Science Division
3-2-5 Kasumigaseki Bldg.
Chiyoda-ku
Tokyo 100-6070
JAPAN
Tel: 81 3 3592 4852
Fax: 81 3 3592 4282
e-mail: chiaki.sakurama@mitsui-chem.co.jp
Home page: www.mitsui.co.jp

Monsanto Co.
Crop Protection
800 N. Lindbergh Blvd
St Louis
MO 63167
USA
Tel: 1 314 694 1000
Fax: 1 314 694 7625
Home page: www.monsanto.com

Monterey Chemical Company
PO Box 5317
Fresno
CA 93755
USA
Tel: 1 209 499 2100
Fax: 1 209 499 1015
e-mail: info@montereychemical.com

Monterey Laboratories
Watsonville
CA 95076
USA

Morse Enterprises Ltd Inc.
151 South East 15 Road
Brickell East
Floor Ten
Miami
FL 33129
USA

Mstrs Technologies Inc.
2501 North Loop Drive
Suite 1614 A
Ames
IA 50010
USA
Tel: 1 515 296 6332
Fax: 1 515 296 9910

Myco-Forestis Corp.
801 Route 344
Casier postal 3158
L'Assomption
Quebec
J5W 4M9
CANADA

Mycogen Corp.
See **Dow AgroScience**

Mycogen Crop Protection
See **Dow AgroScience**

Mycologic Inc.
University of Victoria IDC
PO Box 3075 STN CSC
R-Hut McKenzie Avenue
Victoria, BC
CANADA
Tel: 1 250 721 6500
Fax: 1 250 721 6497
e-mail: idc@uvic.ca
Home page: http://web.uvic.ca/idc/mycologic/
myco

Mycontrol Ltd
Alon Hagalil MP
Nazereth Elit 17920
ISRAEL
Tel/Fax: 972 4 9861827

Mycotech Corp.
See **Emerald** BioAgriculture Corp.

Nagarjuna Agrichem Ltd
Auto Plaza First floor
Road No. 3
Banjara Hills
Hyderabad 500 034
INDIA
Tel: 91 40 318217
Fax: 91 40 319234
e-mail: cropprotection@nagarjunagroup.com

National Business Corp.
8 Lorong Bakar Batu
Singapore 1334
SINGAPORE
Tel: 65 748 8555
Fax: 65 747 0936

Natural Animal Health Products Inc.
PO Box 1177
St Augustine
FL 32085
USA
Fax: 1 904 824 5100
e-mail: sales@naturalanimal.com

Natural Industries Inc.
6223 Theall Road
Houston
TX 77066
USA

Natural Insect Control
RR#2
Stevensville
Ontario
L0S 1S0
CANADA
Tel: 1 905 382 2904
Fax: 1 905 382 4418
e-mail: nic@niagara.com
Home page: www.natural-insect-control.com

Natural Pest Control
8864 Little Creek Drive
Orangevale
CA 95662
USA
Tel: 1 916 726 0855
Fax: 1 916 726 0855
e-mail: natpestco@cwnet.com

Natural Plant Protection
See **NPP**

NatureMark Potatoes
300 East Mallard Drive
Suite 220
Boise
ID 83706
USA
Tel: 1 208 389 2236
Fax: 1 208 309 2280

Nature's Alternative Insectary Ltd
Box 19
Dawson Rd
Nanoose Bay
BC
V0R 2R0
CANADA
Tel/Fax: 1 604 468 7912

Nature's Control
PO Box 35, Medford
OR 97501
USA
Tel: 1 541 245 6033
Fax: 1 541 245 6944

Nematech
Plant Nematology Lab.
Centre for Plant Sciences LIBA
Irene Manton Building, University of Leeds
Leeds LS2 9JT
UK
e-mail: pbiu@leeds.ac.uk

Neudorff
See W. **Neudorff** GmbH KG

New BioProducts Inc.
4737 N.W. Elmwood Drive
Corvallis
OR 97330
USA
Tel: 1 503 752 2045

Nichimen Corporation
4-1-23 Shiba, Minato-ku
Tokyo 108
JAPAN
Tel: 81 3 5446 1111
Fax: 81 3 5446 1010
Home page: www.nichimen.co.jp

Nicobrand
See **The Nicobrand Company**

Nihon Nohyaku Co. Ltd
5th Floor Eitaro Building
2-5 Nihonbashi 1-chome, Chuo-ku
Tokyo 103-8236
JAPAN
Tel: 81 3 3274 3415
Fax: 81 3 3271 2443
e-mail: nnc-overseasdiv@nichino.co.jp
Home page: www.nichino.co.jp
Agricultural products: www.nichino.co.jpeng/
ag/index.html

NIJHOF BGB
Vogelzangsteeg 19
9479 TE Noordlaren
The Netherlands
Tel: 31 504 062 817
Fax: 31 504 062 819
E-mail: info@nijhofbgb.nl
Home page: www.nijhofbgb.nl

Nippon Kayaku Co. Ltd
Agrochemicals Division
11-2 Fujimi 1-chome
Chiyoda-ku
Tokyo
JAPAN
Tel: 81 3 3237 5219
Fax: 81 3 3237 5089
e-mail: nk-agro@magical3.egg.or.jp

Nippon Soda Co. Ltd
Agrochemicals Division
2-1 Ohtemachi 2-chome
Chiyoda-ku
Tokyo 100
JAPAN
Tel: 81 3 3245 6168
Fax: 81 3 3245 6287

Nissan Chemical Industries Ltd
Kowa-Hitotsubashi Building
7-1 3-chome Kanda-nishiki-cho
Chiyoda-ku
Tokyo 101
JAPAN
Tel: 81 3 3296 8151
Fax: 81 3 3296 8016
Home page: www.nissanchem.co.jp
Agricultural products: www.nissan-nouyaku.
net/english

Nod Apiary Products US Inc.
46763 US Interstate 81
Alexandria Bay
NY 13607
USA

NOR-AM
See **Bayer CropScience**

Northwest Agricultural Products
821 S. Chestnut Street
Pasco
WA 99301
USA

Novartis BCM
See **Syngenta Bioline** Ltd

Novartis Crop Protection AG
See **Syngenta**

Novozymes Biologicals Inc.
111 Kesler Mill Road
Salem
VA 24153
USA
Tel: 1 540 389 9361

NPO Vector
633159 Koltsovo
Novosibirsk Region
RUSSIA
Tel: 7 3832 64 01 40
Fax: 7 3832 32 88 31
e-mail: vector@vector.nsk.su

NPP
Route d'Artix
B.P. 80
64150 Nogueres
FRANCE
Tel: 33 559 60 92 92
Fax: 33 559 60 92 19

Nufarm Americas Inc.
1333 Burr Ridge Parkway Suite 125A
Burr Ridge
IL 60527
USA
Tel: 1 708 754 3330
Fax: 1 708 754 0314
Home page: www.nufarm.com
Agricultural products: www.nufarm.us.com

Nufarm B.V.
Welplaatweg 12
Rotterdam Botlek 3197
NETHERLANDS
Tel: 31 10 438 9545
Fax: 31 10 472 2826
See also **Nufarm** Ltd

Nufarm Ltd
103-105 Pipe Rd, Laverton North
Victoria 3026
AUSTRALIA
Tel: 61 39 282 1000
Fax: 61 39 282 1001
e-mail: corporate.information@au.nufarm.com
Home page: www.nufarm.com

Nu-Gro Professional and Consumer Group
2270 Speers Road, Oakville
Ontario
L6L 2X8
CANADA
Tel: 1 905 825 8418
Fax: 1 905 825 8368
e-mail: marwolff@nu-gro.ca

Nutrachem Inc.
123 Horn Road
Rayville
LA 71269
USA

Nutribiotic
133 Copeland Street
Suite C
Petaluma
CA 94952
USA

Oecos Ltd
130 High Street
Kimpton, Hitchin
Herts SG4 8QP
UK
Tel: 44 1438 832481
Fax: 44 1438 832157
e-mail: sales@oecos.co.uk

Olive Trees
134 Davis Street
Santa Paula
CA 93060
USA
Tel: 1 805 921 3900
e-mail: sales@olive-trees.net

OmniLytics
5450 W. Wiley Post Way
Salt Lake City
UT 84116
USA
Tel: 1 801 746 3600
Fax: 1 801 746 3461
Home page: www.omnilytics.com

Organica Corp.
705 General Washington Avenue Suite 500
Norristown
PA 19403
USA
Tel: 1 888 244 7336
Fax: 1 610 539 8383
e-mail: info@organica.net

Ozone Biotech
Plot No. 6
14/3 Mathura Road
Faridabad
Haryana - 121003
INDIA
Tel: 91 129 4047601/2/3
Fax: 91 129 4047604
e-mail: neem@ozonebiotech.com

Ozone Biotech
8132 Bent Tree Springs Drive
Plano
TX 75025
USA
Tel: 1 972 208 7768
e-mail: goyal_pawan@hotmail.com

Pacific Biocontrol Corp.
400 E. Evergreen Blvd #306
Vancouver
WA 98660
USA
Tel: 1 206 693 2866
Fax: 1 206 693 3088

Paushak Ltd
Alembic Road
Baroda 390 003
Gujarat
INDIA
Tel/Fax: 91 265 380 371

PBI/Gordon
1217 W. 12ᵗʰ Street
Kansas City
MO 64101
USA
Tel: 1 816 421 4070
Fax: 1 816 474 0462

PBT Inc.
HC 66, PO Box 74
Deming
NM 88030
USA

Perifleur Products Ltd
Hangleton Lane
Ferring
Worthing
West Sussex BN12 6PP
UK

PGS
See **Plant Genetic Systems**

PheroAsia
1845 Cathedral Glen
Escondido
CA 92029
USA
Tel: 1 760 855 1246
Fax: 1 760 758 9951
e-mail: dloring3@cox.net

Phero Tech Inc.
Gulf Road
Point Robert
WA 98281
USA

Philom Bios
318-111 Research Drive
Saskatoon
SK
S7N 3R2
CANADA
Tel: 1 306 668 8220
Fax: 1 306 975 1215

Plant Genetic Systems
Josef Plateaustraat 22
B-9000 Ghent
BELGIUM
Tel: 32 9 235 8411
Fax: 32 9 224 0694

Plant Health Care Inc
440 William Pitt Way
Pittsburgh
PA 15238
USA
Tel: 1 800 421 9051

Plant Health Technologies
PO Box 15057
Boise
ID 83715
USA
Tel: 1 208 345 1021
Fax: 1 208 345 1032

Plant Impact plc
2 Lockside Business Park
Lockside Road
Riversway
Preston
Lancashire PR2 2YS
UK
Tel: 44 1772 333 874
Fax: 44 1772 733 838
Home page: www.plantimpact.com

Plant Products Co. Ltd
314 Orenda Road
Brampton
ON
L6T 1G1
CANADA
Tel: 1 905 793 7000
Fax: 1 905 793 9632
e-mail: sales@plantprod.com

Plant Research International
PO Box 16
6700 AA Wageningen
NETHERLANDS
Tel: 31 0317 486001
Fax: 31 0317 418094
e-mail: info.plant@wur.nl
Home page: www.pri.wur.nl

Plato Industries Inc.
2020 Holmes Road
Houston
TX 77045
USA
Tel: 1 713 797 0406
Fax: 1 713 795 4665
e-mail: plato@nol.net

Point Enterprises
PO Box 48
12 rue des Marchandises
CH-1260 Nyon
SWITZERLAND
Tel: 41 22 362 5553
Fax: 41 22 362 5557

Pokon and Chrysal B.V.
Gooimeer 7
1411 DD Naarden
NETHERLANDS

Poulenger USA Inc.
3705 Century Blvd #3
Lakeland
FL 33811
USA
e-mail: info@poulengerusa.com
Home page: www.poulengerusa.com

PPRI Weed Pathology Unit
Stellenbosch
SOUTH AFRICA

Praxis
2723 116th Avenue
Allegan
MI 49010
USA
Tel/Fax: 1 616 673 2793
e-mail: praxis@datawise.net

Precision Herbs
9804 Township Road 89
Killbuck
OH 44637
USA
Tel: 1 330 276 6511
Fax: 1 330 276 0411
e-mail: Sales@PrecisionHerbs.com
Home page: www.precisionherbs.com

Premier Horticulture Inc.
127 South Fifth Street
Suite 300
Quakertown
PA 18951
USA

Prentiss Inc.
21 Vernon Street CB 2000
Floral Park
NY 11001
USA
Tel: 1 516 326 1919
Fax: 1 516 326 2312
e-mail: info@prentiss.com

Probelte S.A.
Ctra Madrid Km 384 - DP 4579
30100 Espinardo
Murcia
SPAIN
Tel: 34 968 307250
Fax: 34 968 305432
e-mail: andressanchez@probelte.es
Home page: www.probelte.es

Productos OSA
See **Reposo** SA/C

Proguard Inc.
6111 Lambie Road
Suisun City
CA 94585-9789
USA
Tel: 1 707 426 2500
Fax: 1 707 429 8960

Prophyta Biologischer Pflanzenschutz GmbH
Inselstrasse 12
D-23999 Malchow/Poel
GERMANY
Tel: 49 38425 230
Fax: 49 38425 2323

Pyrethrum Board of Kenya
PO Box 591
Nakuru
KENYA
Tel: 254 9037211 567
Fax: 254 903745 274

Rallis India Ltd
Agrochemical Res. Station
21/22 Peenya Industrial Area
PO Box 5813
Bangalore 560 058
Karnataka
INDIA
Tel: 91 80 839 4959
Fax: 91 80 839 4015

Reanal Fine Chemical Co. Rt.
H-1147 Budapest
Telepos u. 53
HUNGARY
Tel: 36 1 4677 500
Fax: 36 1 384 3102
e-mail: reanex@reanal.hu
Home page: www.reanal.hu

Remeslo ssro
Plant Protection Institute
Tylisovska 1/722
Prague 6 16000
CZECH REPUBLIC

Reposo SA/C
Avenue Santa Fe 1578 1er A
1640 Martinez
Pcia. de Buenos Aires
ARGENTINA
Tel: 54 11 733 3007
Fax: 54 11 798 6797

Reuter
See Theodore E **Reuter**

Rhône-Poulenc Secteur Agro
See **Bayer CropScience**

Riken Green Co. Ltd
See **Kumiai** Chemical Industry Co. Ltd and
Ihara Chemical Industry Co. Ltd

Rimi Chemicals Co. Ltd
23 Shenkar Street
Kiryat Arye
PO Box 4002
Petah Tikva 49130
ISRAEL
Tel: 972 3 9221551
Fax: 972 3 9221550
e-mail: mail@rimi.co.il

Rincon-Vitova Insectaries Inc.
PO Box 1555
Ventura
CA 93002
USA
Tel: 1 805 643 5407
Fax: 1 805 643 6267
e-mail: bugnet@west.net
Home page: www.rain.org/~sals/rincon.html

Rohm & Haas
See **Dow AgroSciences**

Rotam Group
7/F Cheung Tat Centre
18 Cheung Lee Street
Chai Wan
Hong Kong
CHINA
Tel: 852 2896 5608
Fax: 852 2558 6577
e-mail: rotam@hkstar.com

Royal **Brinkman** B.V.
PO Box 2
2690 AA 's-Gravenzande
NETHERLANDS
Tel: 31 174 411 333
Fax: 31 174 414 301

RPG Life Sciences Ltd
415-419 Himalaya House 4th Floor
79 Palton Road
Mumbai 400 001
INDIA
Tel: 91 22 2679778
Fax: 91 22 2679748

Russell Fine Chemicals Ltd
111 Garden Lane
Chester
Cheshire CH1 4EY
UK
Tel: 44 1244 371 821
Fax: 44 1244 372 048

Russell IPM Ltd
68 Third Avenue
Deeside Industrial Park
Deeside
Flintshire CH5 2LA
UK
Tel: 44 1244 281 333
Fax: 44 1244 281 878
e-mail: info@russellipm.com
Home page: www.russellipm.com

SafeScience
See **GlycoGenesys** Inc.

Samabiol
La Grande Marine
84800 L'Isle-sur-la-Sorgue
FRANCE
Tel: 33 4 90 21 44 44
Fax: 33 4 90 38 10 55
e-mail: samabiol@samabiol.com
Home page: http://perso.wanadoo.fr/samabiol

Sandoz Agro Ltd
See **Syngenta**

Sanex
See **Nu-Gro** Professional and Consumer
Group

Sankyo Co. Ltd
Agrochemicals Division
7-12 Ginza 2-chome
Chuo-ku
Tokyo 104
JAPAN
Tel: 81 3 3562 7524
Fax: 81 3 3562 7525

Sanonda
See Hubei **Sanonda** Co. Ltd

Sarritor
Montreal
CANADA
e-mail: info@sarritor.ca

Sautter & Stepper GmbH
Rosenstr. 19
D-72119 Ammerbuch 5
Altingen
GERMANY
Tel: 49 7032 75501
Fax: 49 7032 74199

Scarletts Plant Care
Nayland Road
West Bergholt
Colchester
Essex CO6 3DH
UK
Tel: 44 1206 242533
Fax: 44 1206 242530
Home page: www.scarletts.co.uk

Scent-Off
See **The Scent-Off Corp.**

Scentry Biologicals Inc.
610 Central Avenue
Billings
MT 59102
USA
Tel: 1 406 248 5856
Fax: 1 406 245 2790
e-mail: scentry@scentry.com
Home page: www.scentry.com

S C **Johnson** & Son Inc.
1525 Howe Street
Racine
WI 53403-2236
USA

Scotts
See **The Scotts Company**

SDS Biotech KK
5-6 Shiba 2-chome
Minato-ku
Tokyo 105-0014
JAPAN
Tel: 81 3 5427 2411
Fax: 81 3 5427 2430
See also **Showa Denko KK**

Searle (India)
See **RPG** Life Sciences Ltd

SEDQ
See **S**ociedad **E**spanola de **D**esarrolos
Quimicos S.A. (**SEDQ**)

Sesaco Corp.
700 West Center Street
Paris
TX 75461
USA

Sespe Creek Insectary
PO Box 176
Lindsay
CA 93247
USA
Tel: 1 559 562 6464
Fax: 1 559 562 5565

SGB Pty Ltd
10 Bradford Street
PO Box 387
Wodonga
Victoria 3689
AUSTRALIA
Tel: 61 2 6024 1438
Fax: 61 2 6024 3557

Shin-Etsu Chemical Co. Ltd
Fine Chemicals Dept.
2-6-1 Ohtemachi
Chiyoda-ku
Tokyo 100
JAPAN
Tel: 81 3 3246 502484
Fax: 81 3 3246 504877
Home page: www.shinetsu.co.jp
Agricultural products: www.shinetsu.co.jpe/
product/cellulose.shtml#cellulose

Showa Denko KK
1-13-9 Shiba Daimon
Minato-ku
Tokyo 105
JAPAN
Tel: 81 3 5470 3462
Fax: 81 3 3436 4668
Home page: www.sdk.co.jpindex_e.htm

SIAPA
See Industrie Chimiche **Caffaro** S.p.A., **Isagro**
S.p.A. **Sipcam** and **Sipcam-Oxon Group**

Siebens Hybrids Inc.
633 North College Avenue
Geneseo
IL 61254
USA
Tel: 1 309 944 5131
Fax: 1 309 944 6090

Sime Darby
See **ChemTica** Internacional

Sipcam
Via Sempione 195
20016 Pero
Milan
ITALY
Tel: 39 2 3537 8400
Fax: 39 2 3390 275
e-mail: sipcam@sipcam.it
Home page: www.sipcam.it

Sipcam-Oxon Group
Via Sempione 195
20016 Pero
Milan
ITALY
Tel: 39 2 3537 8400
Fax: 39 2 339 10876
e-mail: oxon@oxon.it
Home page: www.sipcam-oxon.com

Sociedad **E**spanola de **D**esarrolos **Q**uimicos
S.A. (**SEDQ**)
Avda. Diagonal 352 entlo
08013 Barcelona
SPAIN
Tel: 34 93 458 85 00
Fax: 34 93 458 40 07

Soil Technologies Corp.
2103 185th Street
Fairfield
IA 52556
USA
Tel: 1 641 472 3963
Fax: 1 641 472 6189

SOM Phytopharma (India) Ltd
154 A/5 S V Co-op Indl. Estate
IDA BOLLARAM
502 325 District: Medak
Andhra Pradesh
INDIA
Tel: 91 8458 79474

Source Technology Biologicals Inc.
3355 Hiawatha Ave.
Suite 222
Minneapolis
MN 55406
USA
Tel: 1 612 724 7102
Fax: 1 612 724 1642

Spectrum
A Division of United Industries Corp.
PO Box 142642
St Louis
MO 63114
USA

Spiess-Urania Chemicals GmbH
Heidenkampsweg 77
200 97 Hamburg
GERMANY
Tel: 49 40 23 65 20
Fax: 49 40 23 65 22 55
e-mail: mail@spiess-urania.com
Home page: www.spiess-urania.com

Stanes
See T **Stanes** & Co. Ltd

St Gabriel Laboratories
14044 Litchfield Drive
Orange
VA 22960
USA
Tel: 1 800 801 0061

Stine Microbial Products
6613 Haskins
Shawnee
KS 66216
USA
Tel/Fax: 1 913 268 7504

Stoller Enterprises Inc.
4001 W. Sam Houston Parkway
N. Suite 100
Houston
TX 77043
USA
Tel: 1 713 461 1493
Fax: 1 713 461 4467
Home page: www.stollerusa.com

Sumika Technoservice
See **Sumitomo** Chemical Company Ltd

Sumitomo Chemical Company Ltd
5-33 Kitahama 4-chome
Chuo-ku
Osaka 541
JAPAN
Tel: 81 6 220 3683
Fax: 81 6 220 3342

Sumitomo Chemical **Takeda** Agro Company
Ltd
2-13-10 Nihonbashi
Chuo-ku
Tokyo
JAPAN
See also **Sumitomo** Chemical Company Ltd

Sun Moon Chemical Co. Ltd
K.W.T.C.
PO Box 7
Seoul
SOUTH KOREA
Tel: 82 2 565 1653
Fax: 82 2 565 1654

Suterra
213 SW Columbia Street
Bend
OR 97702
USA
Tel: 1 541 388 3688
Fax: 1 541 388 3705

Svenska Predator AB
Box 14017
250 14 Helsingborg
SWEDEN
Tel: 46 42 20 11 30
Fax: 46 42 20 09 05

Sylvan Spawn Laboratory
West Hills Industrial Park
Kittanning
PA 16201
USA
Tel: 1 412 543 2242

Synexus
Avenue de Tervuren 270 272
1150 Brussels
BELGIUM
Tel: 32 2 776 4111
Fax: 32 2 776 4385

Syngenta
CH-4002
Basel
SWITZERLAND
Tel: 41 61 323 1111
Fax: 41 61 323 1212
Home page: www.syngenta.com

Syngenta Bioline Ltd
Telstar Nursery
Holland Road
Little Clacton
Clacton-on-Sea
Essex CO16 9QG
UK
Tel: 44 1255 863200
Fax: 44 1255 863206
e-mail: syngenta.bioline@syngenta.com
Home page: www.syngenta-bioline.co.uk

Syngenta Bioline North America
PO Box 2430
Oxnard
CA 93034
USA
Tel: 1 805 986 8265
Fax: 1 805 986 8267

Syngenta Crop Protection Inc.
PO Box 18300
Greensboro
NC 27419
USA

Taensa
See **Earth BioScience** Inc.

Tagros Chemicals India Ltd
Jhaver Centre
Rajah Annamalai Building
19 Marshalls Road
Egmore
Chennai 600008
INDIA
Tel: 91 44 8587880
Fax: 91 44 8587573
e-mail: tagros1@md3.vsnl.net.in

Takeda Chemical Industries Ltd
See Sumitomo Chemical **Takeda** Agro
Company Ltd

Tecomag Srl
via Quattro Passi 108
41043 Formigine (Modena)
ITALY
Tel: 39 59 57 37 45
Fax: 39 59 57 21 70
e-mail: inc@tecomag.com

Terminex
See **The Terminex International Company L.P.**

Terra Industries Inc.
PO Box 6000, Sioux City
IA 51102-6000
USA
Tel: 1 712 277 1340
Fax: 1 712 233 3648

TGT Inc.
See **BioWorks** Inc.

The **Beneficial Insect Company**
244 Forest Street, Fort Mill
SC 29715
USA
Tel: 1 803 547 2301

The International Centre of Insect Physiology
and Ecology (**ICIPE**)
PO Box 30772
00100 Nairobi
KENYA
Tel: 254 20 861680 4
Fax: 254 20 806330/860110
e-mail: dg@icipe.org

The **Nicobrand** Company
189 Castleroe Rd, Coleraine
Northern Ireland BT51 3RP
UK
Tel: 44 1265 868733
Fax: 44 1265 868735

Theodore E **Reuter**
1404 Litchfield Drive, Orange
VA 22960
USA

Thermo Trilogy
See **Certis** USA

The **Scent-Off** Corp.
4978 61 Avenue South
St Petersburg
FL 33715
USA

The **Scotts** Company
14111 Scottslawn Road
Marysville
OH 43041
USA
Tel: 1 937 644 0011
Fax: 1 937 644 7679
Home page: www.scotts.com

The **Scotts** Co. Ltd
Salisbury House
Weyside Park
Colleshall Lane
Godalming
Surrey GU7 1XE
UK
Tel: 44 870 5301010
Home page: lovethegarden.com

The **Terminex** International Company L.P.
860 Ridge Lake Blvd
Memphis
TN 38120
USA
Tel: 1 800 8376 4639
Home page: www.terminex.com

Tide International Co. Ltd
486-26 Jian Guo Bei Road
Hangzhou 310004
CHINA
Tel: 86 571 5181421
Fax: 86 571 5181422

Tifa (C.I.) Ltd
Tifa Square, Millington
NJ 07946
USA
Tel: 1 908 647 2517
Fax: 1 908 647 7338

Tomono Agrica Co. Ltd
2-12-25 Kasuga
Shizuoka City 420
JAPAN
Tel: 81 54 254 6261
Fax: 81 54 254 6263

Transagra International Inc.
155 N. Michigan Street
Suite 728
Chicago
IL 60601
USA

Trece Inc.
7569 Highway 28 West
Adair
OK 74330
USA
Tel: 1 918 785 3061
Fax: 1 918 785 3063
e-mail: custserv@trece.com
Home page: www.trece.com

Troy Biosciences Inc.
2620 N. 37th Drive
Phoenix
AZ 85009
USA
Tel: 1 602 233 9047
Fax: 1 602 272 4155
e-mail: info@troybiosciences.com
Home page: www.troybiosciences.com

T **Stanes** & Co. Ltd
8/23-24 Race Course Road
PO Box 3709
Coimbatore 641 018
INDIA
Tel: 91 422 211 514
Fax: 91 422 217 857

Turf Science Laboratories Inc.
2121 Hoover Avenue
National City
CA 91950
USA

Ube Industries Ltd
See **SDS Biotech KK** and **Showa Denko KK**

Uniroyal
See CK **Witco** Corp./Uniroyal Chemical

United Phosphorus Ltd
Readymoney Terrace
167 Dr Annie Besant Rd
Worli, Mumbai
Maharashtra 400 018
INDIA
Tel: 91 22 493 0681/49
Fax: 91 22 493 7331
e-mail: uniphos@bom3.vsnl.net.in

USDA Forest Service
180 Canfield Street
Morgantown
WV 26505
USA

Valent BioSciences Inc.
870 Technology Way
Libertyville
IL 60048
USA
Tel: 1 847 935 3570
Fax: 1 847 937 3679
e-mail: vbcwebmaster@valent.com
Home page: www.valentbiosciences.com
Agricultural products: www.valentbiosciences.
comagricultural_products/agricultural_
products_1.asp
See also **Sumitomo** Chemical Company Ltd

Vapco - Veterinary & Agricultural Products
Mfg. Co. Ltd
PO Box 17058
Amman 11195
JORDAN
Tel: 962 5 694991
Fax: 962 5 694998
e-mail: vapco@vapco.net
Home page: www.vapco.net

Verdera Oy
PO Box 1, FI-02201 Espoo
FINLAND
Tel: 358 10 217 3724
Fax: 358 10 217 3711
e-mail: marina.niemi@verdera.fi
Home page: www.verdera.fi

Vipesco – Vietnam Pesticide Co.
102 Nguyen Dinh Chieu Street
District 1, Ho Chi Minh City
VIETNAM
Tel: 84 8 8230751
Fax: 84 8 8230752
e-mail: vipesco@hcm.vnn.vn
Home page: www.vipesco.com

Vita (Europe) Ltd
21/23 Wote Street
Basingstoke
Hants RG21 7NE
UK

Vital Earth Resources
706 East Broadway Avenue
Gladewater
TX 75647
USA
Tel: 1 903 845 4397
Fax: 1 903 845 5144

Vitax Ltd
Owen St, Coalville
Leics LE6 2DE
UK
Tel: 44 1530 510060
Fax: 44 1530 510299

Westgro Sales Inc.
7430 Hopcott Road Delta
BC
V4G 1B6
CANADA
Tel: 1 604 940 0290
Fax: 1 800 663 2552
e-mail: westgro@westgro.com

W F **Young** Inc.
302 Benton Drive
East Longmeadow
MA 01028
USA

Wilbur-Ellis Co.
191 West Shaw Avenue, Suite 107
Fresno, CA 93704-2876
USA
Tel: 1 209 226 1934
Fax: 1 209 226 7630

Witco
See CK **Witco** Corp./Uniroyal Chemical

W. **Neudorff** GmbH KG
Abt. Nutzorgardsmen
An der Muhle 3, Postfach 1209
D-31857 Emmerthal
GERMANY
Tel: 49 5155 62460
Fax: 49 5155 62457
e-mail: w.neudorff@t-online.de

Woodstream Inc.
PO Box 327, 69 N. Locust Street
Lititz
PA 17543-0327
USA
Tel: 1 717 626 2125
Fax: 1 717 626 1912
e-mail: consumercare@woodstream.com
Home page: www.woodstream.com

Wrightson Seeds Ltd
Wrightson Research
Kimihia Research Centre
Tancreds Road, Lincoln
PO Box 939
Christchurch
NEW ZEALAND
Tel: 64 3 325 3158
Fax: 64 3 325 2417

Young
See W F **Young** Inc.

Zeneca Agrochemicals
See **Syngenta**

Abbreviations and codes

Throughout the Main Entries of *The Manual of Biocontrol Agents*, many standard abbreviations and codes have been used. For those unfamiliar with these abbreviations and codes, they are listed in the following pages under four headings:

1   Formulation codes

2   WHO (World Health Organization) and EPA (Environmental Protection Agency) toxicity classification

3   IOBC codes for beneficial species

4   General abbreviations

1 Formulation codes

The following standard codes are used. For further details, see *Catalogue of Pesticide Formulation Types and International Coding System*, Technical Monograph No. 2, 5th edn, March 2002, CropLife International, Brussels, Belgium.

Code	Term	Code	Term
AB	Grain bait	DT	Tablet for direct application
AE	Aerosol dispenser	EC	Emulsifiable concentrate
AI	Active ingredient	ED	Electrochargeable liquid
AL	Other liquids to be applied undiluted	EG	Emulsifiable granule
		EO	Emulsion, water in oil
AP	Other powder	EP	Emulsifiable powder
BB	Block bait	ES	Emulsion for seed treatment
BR	Briquette	EW	Emulsion, oil in water
CB	Bait concentrate	FD	Smoke tin
CF	Capsule Suspension for seed treatment	FG	Fine granule
		FK	Smoke candle
CG	Encapsulated granule	[FL]	Flowable; either SC or OF[†]
CL	Contact liquid or gel	FP	Smoke cartridge
CP	Contact powder	FR	Smoke rodlet
CS	Capsule suspension	FS	Flowable concentrate for seed treatment
DC	Dispersible concentrate		
[DL]	Driftless formulation[#]	FT	Smoke tablet
DP	Dustable powder	FU	Smoke generator
DS	Powder for dry seed treatment	FW	Smoke pellet

Code	Term	Code	Term
GA	Gas	SC	Suspension concentrate (= flowable concentrate)
GB	Granular bait		
GE	Gas generating product	SD	Suspension concentrate for direct application
GF	Gel for seed treatment		
GG	Macrogranule	SE	Suspo-emulsion
GL	Emulsifiable gel	SG	Water soluble granules
GP	Flo-dust	SL	Soluble concentrate
GR	Granule	SO	Spreading oil
GS	Grease	SP	Water soluble powder
GW	Water soluble gel	SS	Water soluble powder for seed treatment
HN	Hot fogging concentrate		
KK	Combi-pack solid/liquid	ST	Water soluble tablet
KL	Combi-pack liquid/liquid	SU	Ultra-low volume (ULV) suspension
KN	Cold fogging concentrate		
KP	Combi-pack solid/solid	TB	Tablet
LA	Lacquer	TC	Technical material
LS	Solution for seed treatment	TK	Technical concentrate
LV	Liquid vaporiser	(TP)	Tracking powder*
MC	Mosquito coil	UL	Ultra-low volume (ULV) liquid
ME	Micro-emulsion	VP	Vapour releasing product
MG	Microgranule	WG	Water dispersible granules
MV	Vaporising mat	WP	Wettable powder
OD	Oil dispersion	WS	Water dispersible powder for slurry treatment
OF	Oil miscible flowable concentrate (oil miscible suspension)		
		WT	Water dispersible tablet
OL	Oil miscible liquid	XX	Others
OP	Oil dispersible powder	ZC	A mixed formulation of CS and SC
PA	Paste		
PB	Plate bait	ZE	A mixed formulation of CS and SE
PC	Gel or paste concentrate		
PO	Pour-on	ZW	A mixed formulation of CS and EW
PR	Plant rodlet		
PS	Seed coated with a pesticide		
RB	Bait (ready for use)		
SA	Spot-on		
SB	Scrap bait		

Japan; not a CropLife International code.
† Not a CropLife International code.
* Discontinued term; see CP.

The following are additional formulation types:

Beads	A coacervate with a polymer to form a slow release complex
Coating agent	Solution, paste or powder used for treating plant roots or cut trees. Could be coded AL or AP
Controlled release strips	These are often coacervates with PVC formed into strips and work by vapour action. Similar to VP
Fumigant	Fumigant material, sometimes pelletised, for application to glasshouses etc. by fumigation application equipment
Hollow fibre	Slow release mechanism using hollow absorbent fibres
Impregnated strip	Insecticide impregnated in PVC and used as strips for vapour action (see Controlled release strips)
Membrane dispensers	Pheromone sex attractant device
Micro flake	Pheromone liquid absorbed onto solid substrate as small flakes
Pasteboard sticks	The a.i. is impregnated into an inert matrix, then formed into sticks for pot plant soil application
Pellets	Pelleted formulation for aquatic application
Plastic flakes	Insect pheromone flakes for aerial application, to disrupt mating
Plastic laminates	Same as plastic flakes, but used for insect monitoring on the ground
Plastic sachets	Insect pheromone in a plastic sachet, from which a.i. slowly evaporates
Polymer matrix	Similar to plastic laminates
PVC strips	Insect pheromone for slow release. Similar to Controlled release strips and Impregnated strips
Slow release capsules	Slow release containers special slow release CS formulation in bubble cap packs
Slow release containers	Special slow release CS formulation in bubble cap packs
Spirals	Vapour-releasing coils for insect semiochemicals
Tube	Mechanical device for dispensing insect semiochemicals

2 WHO and EPA toxicity classification

WHO toxicity classification

The World Health Organization classification for estimating the acute toxicity of pesticides.

| Description | Class | LD$_{50}$ for the rat (mg/kg b.w.) | | | |
| | | Oral | | Dermal | |
		Solids	Liquids	Solids	Liquids
Extremely hazardous	Ia	≤5	≤20	≤10	≤40
Highly hazardous	Ib	5–50	20–200	10–100	40–400
Moderately hazardous	II	50–500	200–2000	100–1000	400–4000
Slightly hazardous	III	≥501	≥2001	≥1001	≥4001
Product unlikely to present acute hazard in normal use	U	≥2000	≥3000	–	–
Not classified; believed obsolete	O				
Fumigants, not classified under WHO	FM				

EPA toxicity classification

| Class | Acute toxicity to rat | | | Eye effects | Skin effects |
	Oral LD$_{50}$ (mg/kg)	Dermal LD$_{50}$ (mg/kg)	Inhalation LC$_{50}$ (mg/l)		
I	≤50	≤200	≤0.2	Corrosive; corneal opacity not reversible within 7 days	Corrosive
II	50–500	200–2000	0.2–2.0	Corneal opacity reversible within 7 days; irritation persisting for 7 days	Severe irritation at 72 hours
III	500–5000	2000–20000	2.0–20	No corneal opacity; irritation reversible within 7 days	Moderate irritation at 72 hours
IV	≥5000	≥20000	≥20	No irritation	Mild or slight irritation at 72 hours

3 IOBC codes for beneficial species

The following scale is used to describe the effects of pesticides on beneficial species in a wide range of tests. See S A Hassan *et al.* 1994. *Entomophaga*, **39**(1), 107–19 for details. The figures relate to percentage mortality.

Laboratory studies:

1	harmless	<30%
2	slightly harmful	30–79%
3	moderately harmful	80–99%
4	harmful	>99%

Semi-field and field studies:

1	harmless	<25%
2	slightly harmful	25–50%
3	moderately harmful	51–75%
4	harmful	>75%

4 General abbreviations

The following abbreviations have been used, some being SI units.

a	acre
ACCase	acetyl CoA carboxylase
ACGIH	The American Conference of Governmental Industrial Hygienists, Inc.
ACS	American Chemical Society
ADI	acceptable daily intake
a.e.	acid equivalent (active ingredient expressed in terms of parent acid)
AFPP	Association Française de Protection des Plantes
AG	Aktiengesellschaft (Company)
AHAS	acetohydroxyacid synthase
a.i.	active ingredient
ALC_{50}	approximate concentration required to kill 50% of test organisms
ALS	acetolactate synthase
ANPP	Association Nationale pour la Protection des Plantes (now AFPP)
ANSI	American National Standards Institute
AOAC	Association of Official Analytical Chemists
AOAC Methods	Official Methods of Analysis of The Association of Official Analytical Chemists
ave.	average

BAN	British Approved Name (by British Pharmacopoeia Commission)
BBA	Biologische Bundesanstalt Abteilung
BCPC	British Crop Protection Council
BGA	Bundesgesundheitsamt (former Federal Health Office, Germany), now BgVV (Bundesinstitut für gesundheitlichen Verbraucherschutz und Veterinärmedizin – Federal Institute for Health Protection of Consumers and Veterinary Medicine)
BIOS	British Intelligence Objective Sub-Committee (former)
BS	British Standard
BSI	British Standards Institution
B.V.	Beperkt Vennootschap (Limited)
b.w.	body weight
c.	*circa* (about)
C.A.	*Chemical Abstracts*
calc.	calculated
CAS RN	Chemical Abstracts Services Registry Number
cbi	carotenoid biosynthesis inhibitor
cdi	cell division inhibitor
cf.	compare
cfu	colony-forming units
CHO	Chinese hamster ovary
9CI	9th collective index period (1972–1976) of *Chemical Abstracts*
CIPAC	Collaborative International Pesticides Analytical Council Limited
COLUMA	Comite de Lutte Contre les Mauvaises Herbes
concn.	concentration
cwt	hundredweight
d	day(s)
dat	days after treatment
Defra	Department for Environment, Food & Rural Affairs (England and Wales)
DT_{50}	time for 50% loss; half-life
EAC	ecologically acceptable concentration
EC_{50}	median effective concentration
EbC_{50}	median effective concentration (biomass, e.g. of algae)
ErC_{50}	median effective concentration (growth rate, e.g. of algae)
ECETOC	European Chemical Industry Ecotoxicology and Toxicology Centre
Edn	edition
ED_{50}	median effective dose
E-ISO	ISO name (English spelling)
EPA	Environmental Protection Agency (of USA)

EPPO	European and Mediterranean Plant Protection Organisation
ESA	Entomological Society of America
ESCORT	European Standard Characteristics of Beneficials Regulatory Testing
est.	estimated
EU	European Union
EUP	Experimental Use Permit (US EPA)
EWRC	European Weed Research Council (pre-1975)
EWRS	European Weed Research Society (since 1975)
FAO	Food and Agricultural Organisation (of the United Nations)
F-ISO	ISO name (French spelling)
g	gram (hence also ng, μg, mg, kg, etc.)
GABA	gamma-aminobutyric acid
gc	gas chromatography
gc-ms	combined gas chromatography-mass spectrometry
GCPF	Global Crop Protection Federation (formerly GIFAP)
GIFAP	Groupement International des Associations Nationales de Fabricants de Produits Agrochimiques (now known as GCPF)
glc	gas–liquid chromatography
GmbH	Gesellschaft mit beschränkter Haftung (limited liability company, Germany)
GUS	groundwater ubiquity score
GV	granulovirus (formerly granulosis virus)
h	hour(s)
ha	hectare(s) $(10^4 \, m^2)$
hl	hectolitre (100 l)
HMSO	Her Majesty's Stationery Office (UK) (now TSO – The Stationery Office)
IARC	International Agency for Research on Cancer
ibid.	in the journal last mentioned
IC_{50}	concentration that produces 50% inhibition
ICM	integrated crop management
idem	by the author(s) last mentioned
i.e.	that is
INRA	Institut National de la Recherche Agronomique
IOBC	International Organisation for Biological Control
i.p.	intraperitoneal
IPCS	International Programme on Chemical Safety
IPM	integrated pest management
i.r.	infrared
ISO	International Organisation for Standardisation
i.u.	international unit (measure of activity of micro-organisms)

IUPAC	International Union of Pure and Applied Chemistry
i.v.	intravenous
JECFA	Joint FAO/WHO Expert Committee on Food Additives
JMAF	Japanese Ministry for Agriculture, Forestry and Fisheries (*formerly* Japanese Ministry for Agriculture and Forestry)
JMPR	Joint meeting of the FAO Panel of Experts on Pesticide Residues and the Environment and the WHO Expert Group on Pesticide Residues
j.v.	joint venture
k	kilo, multiplier (1000) for SI units
kg	kilogram(s)
l	litre (hence also μl, ml, etc.)
lb	pound(s)
lb/a	pounds per acre
lc	liquid chromatography
LC_{50}	concentration required to kill 50% of test organisms
LD_{50}	dose required to kill 50% of test organisms
LOEC	lowest observed effect concentration
LR_{50}	the application rate causing 50% mortality
Ltd	Limited
m	metre (hence also nm, mm, etc.)
m	milli, multiplier (10^{-3}) for SI units
M	mega, multiplier (10^{6}) for SI units
MAFF	Ministry of Agriculture Fisheries and Food (UK, now Defra, Department for Environment, Food & Rural Affairs)
MATC	maximum acceptable toxicant concentration
mg	milligram(s), (0.001 g)
min	minute(s)
mm	millimetre(s), (0.001 m)
m/m	proportion by mass
MNPV	multiple-embedded nucleopolyhedrovirus (formerly multiple-embedded nuclear polyhedrosis virus)
mo	month(s)
ms	mass spectroscopy
n	nano, multiplier (10^{-9}) for SI units
ng	nanogram (10^{-9} g)
nm	nanometre (10^{-9} m)
nmr	nuclear magnetic resonance
NOAEL	no observed adverse effect level
NOEL	no observed effect level

nPa	nanopascal, (10^{-9} Pa)
NPV	nucleopolyhedrovirus (formerly nuclear polyhedrosis virus)
NRDC	National Research and Development Corporation (former, of UK)
N.V.	Naamloze Vennootschap (Limited)
OECD	Organization for Economic Co-operation and Development
o.m.	organic matter
oz	ounce(s)
p	pico, multiplier (10^{-12}) for SI units
Pa	pascal (SI unit for pressure)
pH	$-\log_{10}$ hydrogen ion concentration
PIB	polyhedral inclusion body
plc	Public Limited Company
post-em.	after emergence
ppb	parts per billion
ppi	pre-plant incorporated
ppm	parts per million
pre-em.	before emergence
q.v.	quod vide (which see)
ref.	reference
RfD	Reference Dose: an estimate of a daily oral exposure that is likely to be without an appreciable risk of deleterious effects during a lifetime.
r.h.	relative humidity
rplc	reverse phase liquid chromatography
s	second(s)
S.A.	Société Anonyme (Company)
SAR	structure–activity relationship(s)
SI	International System of Units
sp.	species (singular)
S.p.A.	Societé par Actions (Company)
spp.	species (plural)
sq. m.	square metre (m^2)
t	tonne, 1000 kg
tech.	technical grade
TLm	median tolerance limit
TSO	The Stationery Office (formerly HMSO – Her Majesty's Stationery Office)
UNEP	United Nations Environment Programme
USDA	United States Department of Agriculture
u.v.	ultraviolet
viz.	namely

v.p.	vapour pressure
w	week(s)
WHO	World Health Organization (of the United Nations)
WIPO	World Intellectual Property Organization
WSSA	Weed Science Society of America
wt	weight
y	year(s)
μ	micro, multiplier (10^{-6}) for SI units
>	greater than
≥	greater than or equal to
<	less than
≤	less than or equal to

Index 1

Chemical Abstracts Service (CAS) Registry Numbers

Indexes

CAS RN	Entry no.	CAS RN	Entry no.
[8000-34-8]	2:152; 2:217; 2:218	[28111-33-3]	2:221
[8000-48-4]	2:207	[29804-22-6]	4:387
[8000-78-0]	2:169; 2:178	[30820-22-5]	4:408
[8001-29-4]	2:169	[31654-77-0]	4:441
[8003-34-7]	2:213	[33189-72-9]	4:446; 4:448; 4:449
[8006-82-4]	2:159	[33446-70-7]	2:170
[8007-46-3]	2:206; 2:218; 2:233	[33956-49-9]	4:382
[8007-70-3]	2:207	[34010-21-4]	4:412; 4:414
[8030-53-3]	2:180	[35148-19-7]	4:391
[8047-13-0]	2:220	[35153-18-5]	4:449
[8051-02-3]	2:221	[35153-16-3]	4:416
[8071-03-2]	2:180	[35153-15-2]	4:444
[9012-76-4]	2:208	[35396-61-5]	2:209
[10052-59-2]	2:170; 2:238	[35597-43-4]	2:158
[11002-92-9]	2:160	[35628-05-8]	4:420
[11113-80-7]	2:211; 2:212	[35628-00-3]	4:420; 4:421
[11104-05-5]	4:408	[35857-62-6]	4:426
[11141-17-6]	2:156	[35900-26-6]	4:402
[12002-53-8]	4:452	[37248 -47-8]	2:236
[12673-85-7]	1:30; 1:31; 1:32; 1:33;	[38401-84-2]	4:397
	1:34; 1:35	[38421-90-8]	4:383
[12767-55-4]	2:160	[38363-29-0]	4:390
[13255-47-5]	2:170	[38665-10-0]	2:236
[13466-78-9]	4:404	[40642-40-8]	4:390
[14959-86-5]	4:388; 4:445	[42822-86-6]	2:194
[14979-39-6]	4:385	[42922-74-7]	2:231
[15662-33-6]	2:220	[49669-74-1]	2:153
[16725-53-4]	4:441; 4:445	[50767-79-8]	4:439
[16974-11-1]	4:392	[50933-33-0]	4:407
[18309-32--5]	4:451	[51596-11-3]	2:196
[19396-06-6]	2:211	[51596-10-2]	2:196
[19408-46-9]	2:189	[52207-99-5]	4:407
[20261-85-2]	2:209	[52304-36-6]	2:161
[20711-10-8]	4:446; 4:447; 4:448	[53042-79-8]	4:407
[22083-74-5]	2:200	[53120-27-7]	4:430; 4:431; 4:432
[22976-86-9]	2:212	[53120-26-6]	4:430; 4:432
[25402-06-6]	2:214	[53880-51-6]	4:395
[26532-95-6]	4:447	[53939-27-8]	4:413
[26532-22-9]	4:408	[53939-28-9]	4:409; 4:410; 4:411;
[26532-25-2]	4:408		4:412; 4:413
[26532-24-1]	4:408	[54364-62-4]	4:394
[26532-23-0]	4:408	[54364-63-5]	4:394
[27519-02-4]	4:426	[54815-06-4]	4:385
[28079-04-1]	4:389; 4:390	[54910-51-9]	4:387

CAS RN	Entry no.	CAS RN	Entry no.
[54910-52-0]	4:387	[77518-55-9]	4:384
[55512-33-9]	2:169	[78617-58-0]	4:418
[56196-53-3]	4:434	[83212-30-0]	4:436
[56219-04-6]	4:410; 4:411	[84776-33-0]	2:176
[56578-18-8]	4:383	[86252-74-6]	4:429
[57497-78-6]	2:197	[89997-47-7]	2:175
[58064-47-4]	2:231	[93091-95-3]	4:425
[58594-45-9]	4:411	[94513-55-0]	2:220
[58917-26-3]	4:423	[97424-51-6]	4:425
[59014-03-8]	4:385	[105700-87-6]	4:438
[60018-04-4]	4:399; 4:400	[106049-29-0]	4:411
[60478-96-8]	4:403; 4:404; 4:405	[108175-15-1]	2:231
[62532-53-0]	4:399; 4:404	[111192-53-1]	4:384
[62628-54-0]	1:30; 1:31; 1:32; 1:33; 1:34; 1:35	[112025-60-2]	2:192
		[122616-64-2]	4:407
[63408-44-6]	4:419	[129900-07-8]	2:170; 2:238
[64726-93-8]	4:384	[131929-60-7]	2:229
[64726-91-6]	4:384	[131929-63-0]	2:229
[65128-96-3]	4:443	[133494-64-1]	4:384
[65195-55-3]	2:150	[137512-74-4]	2:174
[65195-56-4]	2:150	[143447-72-7]	1:21
[65954-19-0]	4:450	[146659-78-1]	2:212
[66900-48-9]	4:436	[149745-51-7]	4:438
[67383-05-5]	1:30; 1:31; 1:32; 1:33; 1:34; 1:35	[152045-16-4]	4:402
		[152203-43-5]	4:402
[67527-71-3]	2:197	[152783-91-0]	4:402
[67983 -11-3]	2:197	[155569-91-8]	2:174
[68038 -70-0]	1:22	[162334-33-0]	4:437
[68038-71-1]	1:30; 1:31; 1:32; 1:33; 1:34; 1:35	[162428-75-3]	4:437
		[162428-76-4]	4:437
[68990-67-0]	2:222	[162490-88-2]	4:437
[69744-44-1]	4:397	[168316-95-8]	2:229
[69744-43-0]	4:397	[178152-25-5]	4:398
[71048-99-2]	2:158	[179607-18-2]	2:174
[71751-41-2]	2:150	[187166-40-1]	2:228
[71899-16-6]	4:422; 4:423	[187237-90-7]	2:221
[72194-83-3]	4:409	[187166-15-0]	2:228
[72269-48-8]	4:450	[308064-23-5]	2:170
[73360-07-3]	2:153	[404589-23-7]	2:222
[75202-10-7]	2:200		

Index 2

Approved names, common names, code numbers, tradenames, chemical names, Latin names and major classes

This index lists alphabetically all approved names, common names, Latin names, code numbers, tradenames and chemical names. In addition, there are entries that guide the reader to major classes [*Predators of aphids*; *Biological insecticides (bacterial)*]. These major classes list all entries within a broad classification, enabling the reader to locate all related products rapidly. They appear in bold italics. Some entries may appear twice, particularly those beginning with a number or other symbol; one takes account of the symbol and the other does not [e.g. (+)-grandisol appears at the beginning and also under grandisol].

Latin names appear in italics (e.g. *Bacillus thuringiensis*) and tradenames appear in single quotation marks (e.g. 'Tracer').

All are referred to by entry number.

Indexes

Name	Entry no.	Name	Entry no.
'Advanced 3 Hour Weedkiller'	2:203	'Allityn Insect Repellent'	2:178
'Advanced Moss Killer'	2:203	allo-ocimenol	2:191
'Affirm'	2:150; 2:174	4-allyl anisole	4:381
'Afla-guard'	1:11	4-allyl-2-methoxyphenol	2:152
AfMNPV	1:08	alpha-ionone	2:185
AfNPV	1:08	*Alternaria cassiae*	1:04
African weaver ant	3:340	*Alternaria cuscutacidae*	1:05
Agapeta zoegana	3:241	*Alternaria destruens*	1:06
Agapeta zoegana'	3:241	'AM 301'	1:105
Agasicles hygrophila	3:242	'Amazin 3% EC'	2:156
AgMNPV	1:09	amber disease	1:121
AgNPV	1:09	amber spurge flea beetle	3:269
'Agree'	1:33; 1:35	'AmbliPAK50000'	3:247
'Agrept'	2:230	'Amblyline cal'	3:246
'Agriblend'	2:170	'Amblyline crs-wp'	3:247
Agrilus hyperici	3:243	'Amblyline cu'	3:247
'Agrilus hyperici'	3:243	'Amblyline d'	3:248
'Agrimec'	2:150	'Amblyline flo'	3:247
'Agri-Mek'	2:150	'Amblyline m'	3:250
'Agrimycin 17'	2:230	*Amblyseius andersoni*	3:244
'AgriPhage'	1:149	'Amblyseius andersoni'	3:244
'Agrispon'	2:205	*Amblyseius barkeri*	3:245
'Agrobac'	1:33	'Amblyseius barkeri'	3:245; 3:247
Agrobacterium radiobacter isolate K84	1:03	'Amblyseius Breeding System'	3:247
		Amblyseius californicus	3:246
Agrobacterium radiobacter isolate K89	1:03	*Amblyseius chilensis*	3:246
		Amblyseius cucumeris	3:245; 3:247
Agrobacterium radiobacter isolate K1026	1:03	'Amblyseius cucumeris mite'	3:247
		'Amblyseius cucumeris'	3:247
'Agrocare'	1:24	*Amblyseius degenerans*	3:248
'Agro-Gibb'	2:179	'Amblyseius degenerans'	3:248
'Agromec'	2:150	*Amblyseius fallacis*	3:348; 3:249; 3:362
'Agroneem'	2:156		
'Agtrol 6-BA'	2:157	'Amblyseius fallacis'	3:249
AIA	2:184	*Amblyseius mackenziei*	3:245
aker-tuba	2:219	*Amblyseius montdorensis*	3:250
'Alfadex'	2:213	*Amblyseius mungeri*	3:246
alfalfa weevil parasite	3:332	*Amblyseius swirskii*	3:251
alfalfa weevil parasitoid	3:279	'Amblyseius System'	3:247
Algal-derived fungicide	2:190	'Amblyseius Thrips Predators'	3:247
'Align '	2:156	'Amblyseius-C'	3:247
alligatorweed flea beetle	3:242	'Amblyseius-Vermiculite-System'	3:247

Indexes

Indexes

Name	Entry no.	Name	Entry no.
'Bactospeine'	1:33	beetle repellent	4:424
'Bactucide'	1:33	'Beetleball Technical'	4:381
'Bag-A-Bug Japanese Beetle Trap'	2:152; 2:172; 2:204	'Belpromec'	2:150
		Beneficial bacterium	1:03
'Baits Motel Stay Awhile - Rest Forever'	1:37	*Beneficial fungus*	1:11
		beneficial nematodes	1:125; 1:126; 1:128; 1:129; 1:130; 1:67; 1:68; 1:129
'balEnce'	1:39		
banana corm weevil aggregation pheromone	4:437		
banana weevil aggregation pheromone	4:398	6-benzyladenine	2:157
		6-benzylaminopurine	2:157; 2:180
Bangasternus orientalis	3:278	'Berelex'	2:179
'Bangasternus orientalis'	3:278	bergaptene-free bergamot oils	2:168
'Banker-E-System'	3:266	bertha armyworm nucleopolyhedrovirus	1:74
BAP	2:157		
Barbasco	2:219	beta-ionone	2:185
BAS 114 UBF	2:198	'Betel'	1:40
BAS 280 I	4:447	'B-Green'	1:67
BAS 281 I	4:392	bialaphos	2:158
BAS 284 I	4:390	big-eyed bug	3:312
BAS 286 I	4:431	bilanafos	2:158
BAS 287 I	4:383	bilanafos-sodium	2:158
Bathyplectes curculionis	3:279	'BINAB T Pellets'	1:141
'Bathyplectes curculionis'	3:279	'BINAB T Vector'	1:141
'Baturad'	1:33	'BINAB T WP'	1:141
Bayeria capitigena	3:361	'BINAB TF WP'	1:141
BcS-3	2:160	bindweed gall mite	3:239
'Beanin'	2:157	BIO 1020	1:76
Beauveria bassiana isolate 447	1:37	'Bio AquaGuard'	1:104; 1:144
Beauveria bassiana isolate ATCC 74040	1:36	'Bio Bug Insect Repellent'	2:168
		'Bio Safe WG'	1:125
Beauveria bassiana isolate Bb 147	1:36	'BioAct WG'	1:89
Beauveria bassiana isolate GHA	1:36; 1:38	'Biobac'	1:23
Beauveria bassiana isolate HF23	1:39	'Biobit'	1:33
Beauveria bassiana isolate Stanes	1:36	'BioBlast Biological Termiticide'	1:76
Beauveria brongniartii isolate 95.041	1:40	'BioCane'	1:78
Beauveria brongniartii isolate Bb96	1:40	'Bio-Catch'	1:71
Beauveria brongniartii isolate IMBST 95.031	1:40	'Bio-Catch M'	1:76
		'Biochon'	1:46
Beauveria tenella sensu	1:40	'Biocot'	1:33
'Bedoukian Octenol Technical'	2:201	'Bio-cure F'	1:144
beet armyworm sex pheromone	4:441; 4:444	'Biocure'	1:44; 1:104

Name	Entry no.	Name	Entry no.
'Biofox C'	1:62	*Biological herbicide (fungus)*	1:01; 1:04; 1:05;
'Biofungicide Green-Releaf'	1:16		1:06; 1:45; 1:46;
'Bio-Fungus'	1:138		1:48; 1:49; 1:50;
'Bioganic Crawling Insect Killer'	2:216		1:51; 1:52; 1:53;
'Bioganic Dust'	2:152		1:58; 1:59; 1:60;
'Bioganic Flying Insect Killer'	2:152; 2:204		1:75; 1:97; 1:98;
'Bioganic Lawn and Garden Spray'	2:152		1:99; 1:109; 1:110; 1:111;
'Bioganic Weed and Grass Killer'	2:152		1:112; 1:113; 1:114; 1:115;
'BioGreen'	1:81		1:116; 1:117;
'BioGuard-V'	1:100		1:118; 1:119;
'Biok'	2:150		1:120; 1:122;
'BioKeeper'	1:61		1:145
Biological acaricide (fungus)	1:69	*Biological herbicide (virus)*	1:136
Biological bactericide (bacteriophage)	1:107; 1:149	*Biological insecticide (bacterium)*	1:18; 1:21; 1:27; 1:28; 1:29; 1:30;
Biological bactericide (bacterium)	1:61; 1:91; 1:92		1:31; 1:32; 1:33; 1:34; 1:35; 1:90;
Biological fungicide (bacterium)	1:16; 1:17; 1:19; 1:20; 1:22; 1:23;		1:121
	1:24; 1:25; 1:26; 1:101; 1:102;	*Biological insecticide (fungus)*	1:36; 1:37; 1:38; 1:39; 1:40; 1:41;
	1:103; 1:131; 1:132		1:70; 1:71; 1:76; 1:77; 1:78; 1:79;
Biological fungicide (fungus)	1:07; 1:43; 1:44; 1:47; 1:54; 1:55;		1:80; 1:81
	1:56; 1:62; 1:63; 1:64; 1:83; 1:94;	*Biological insecticide (microsporidium)*	1:86; 1:146
	1:96; 1:108; 1:134; 1:135;	*Biological insecticide/acaricide (fungus)*	1:88
	1:137; 1:138; 1:139; 1:140;	*Biological nematicide (bacterium)*	1:15; 1:93
	1:141; 1:142; 1:143; 1:144	*Biological nematicide (fungus)*	1:10; 1:84; 1:89
Biological fungicide/bactericide (bacterium)	1:26; 1:104	*Biological plant growth promoter (bacterium)*	1:13
Biological fungicide/nematicide (bacterium)	1:42	*Biological plant growth regulator*	1:14
Biological fungicide/plant growth regulator (bacterium)	1:82	'Biolug'	1:95
		'BioLure Japanese Beetle Trap'	4:384
Biological herbicide (bacterium)	1:148	'BioMal'	1:53
		'Biomite'	2:172
		'Biomonas'	1:104
		'BioNeem'	2:156
		'Bio-neem'	2:156
		'BioNem'	1:15
		'Bio-Path'	1:76
		'Bio-Power'	1:36

Indexes

Name	Entry no.	Name	Entry no.
cypermethrin	4:426	'Delphastus-A'	3:297
Cyrtopeltis tenuis	3:338	*Delphastus catalinae*	3:296
'Cytex'	2:170; 2:238	'Delphastus catalinae'	3:296
'Cytogro Hormone Biostimulant'	2:170	*Delphastus pusillus*	3:297
'Cytokin Bioregulator Concentrate'	2:170	'Delphastus pusillus'	3:297
		'Delphastus pusillus predator beetle'	3:297
cytokinin	2:170; 2:179		
'Cytoplex HMS'	2:170; 2:179	'Delphastus-System'	3:297
'Cyzer'	2:170	'Delsure (c)'	3:296
'Dacdigline si'	3:295; 3:300	'Delsure (p)'	3:297
'Dacline s'	3:295	'Delvolan'	2:199
'Dacnusa'	3:295	'Denim'	2:174
Dacnusa sibirica	3:295; 3:300	'Deny'	1:42
'Dacnusa sibirica'	3:295	*Deraeocoris brevis*	3:298
'Dacnusa sibirica – Schlupfwespen'	3:295	'Deraeocoris brevis adults'	3:298
		'Derisom'	2:187
'Dacnusa/Diglyphus (225:25)'	3:295	derris	2:219
'Dacnusa-Mix-System'	3:295; 3:300	derris root	2:219
'Dacnusa-System'	3:295	'Detur'	2:186
'Dacsure (si)'	3:295; 3:300	'DeVine'	1:99
Dacus cucurbitae pheromone	4:435	'Devour WG'	1:129
'Dagger'	1:104	diamond-back moth parasitoid	3:334
'Dantotsupadanvalida'	2:236	diamondback moth sex pheromone	4:412
DAZA	2:171		
'DAZA'	2:171	*Diaporthe parasitica*	1:55
'Dazitol'	2:163	'Diatect II'	2:213
DE-105	2:229	'Diatect III'	2:213
DE-175	2:228	'Diatect V'	2:213
(*E*)-dec-5-enol	4:383; 4:417	'Dickmaulrüsslernematoden'	1:68
(*E*)-dec-5-enyl acetate	4:383; 4:417	'Dicyphus h'	3:299
(*R,Z*)-5-(1-decenyl)dihydro-2(3*H*)-furanone	4:384	*Dicyphus hesperus*	3:299
		'Dicyphus hesperus'	3:299
decollate snail	3:357	*Dicyphus nocivus*	3:338
'Decollate Snails'	3:357	'Dicyphus System'	3:299
'Decoy EPSM'	4:392	9,21-didehydroryanodine	2:220
'DeCoy PBW Band'	4:407	'Diegall'	1:03
'Degenerans L'	3:248	'Digline i'	3:300
'Degenerans-System'	3:248	'DigliPAK 100'	3:300
'Delegate WG'	2:228	*Diglyphus isaea*	3:295; 3:300
'Delfin'	1:33	'Diglyphus isaea'	3:300
'Delivery'	1:33	'Diglyphus/Dacnusa'	3:295; 3:300
'Delphaline p'	3:297	'Diglyphus-System'	3:300

Indexes

Name	Entry no.	Name	Entry no.
extract of Mexican tea	2:175	'Foliar Triggrr MFG'	2:170
F-7744	1:36	'Foliar Triggrr'	2:170
'Falgro'	2:179	'Foli-Zyme GA'	2:170; 2:179
fall armyworm sex pheromone	4:445	'Foray'	1:33
farnesol	4:401	'Forgibbs'	2:179
fatty acids	2:176; 2:203	formic acid	2:177
'Felsure (a)'	3:307	'Formic Acid Gel'	2:177
Feltiella acarisuga	3:307	'For-Mite'	2:177
'Feltiella acarisuga'	3:307	'Fortune Aza'	2:156
'Feltiella LV-System'	3:307	'Fortune Biotech'	2:156
'Feltiella-System'	3:307	'Forwarbit'	1:33
'Feltiline a'	3:307	*Franklinothrips orizabensis*	3:308
fenbutatin oxide	2:209	Franklinothrips orizabensis'	3:308
fenitrothion	2:188; 2:189; 2:236	frontalin	4:403; 4:404; 4:405
fenobucarb	2:189	'Frostban A'	1:104; 1:106
fenobucarb	2:209	'Frostban D'	1:104; 1:106
'Ferag OR'	4:434	fruit tree predatory mite	3:374
'Ferag RF'	4:402	'FruitGuard-V'	1:100
ferimzone	2:236	'FruitPlus'	1:56
'Fermone Naturalis L-225'	1:36	'Frustrate PBW'	4:407
ferrolure+	4:402	'Frustrate TPW'	4:450
'Ferrolure+'	4:402	**Fungal-derived fungicide**	2:237
ferrugineol	4:402	**Fungal-derived herbicide**	2:151
ferruginone	4:402	**Fungal-derived plant growth regulator**	2:179; 2:180
'Fersex CCh'	4:388		
'Fersex CF'	4:389	'Fungastop'	2:167
'Fersex ChS'	4:409; 4:411	fungus fly predator	3:319
'Fersex PO'	4:443	fungus gnat predator	3:318; 3:319
'Fersex SN'	4:414	'Fusaclean G'	1:62
'Fersex TP'	4:418	'Fusaclean L'	1:62
'Fersex ZP'	4:429; 4:430	*Fusarium oxysporum* isolate Fo 47	1:62
filth fly parasitoid	3:336	GA₃	2:179
'FitoPAK 2000'	3:348	GA4	2:157
'Flight Control'	2:154	GA7	2:157
'Florbac'	1:35	GABA	2:181
flour moth sex pheromone	4:440	*Galendromus annectens*	3:309
flower bug	3:262	'Galendromus annectens'	3:309
'Flowtron Octenol Mosquito Attractant'	2:201	*Galendromus helveolus*	3:310
		'Galendromus helveolus'	3:310
'Flybait'	4:426	*Galendromus occidentalis*	3:311
'Foil'	1:33	'Galendromus occidentalis'	3:311

Name	Entry no.	Name	Entry no.
Galendromus pyri	3:374	'Goldben'	4:426
gall midge	3:268	golden chalcid	3:272; 3:273
Gallobelicus crassicornis	3:338	golden eye	3:288
'Galltrol-A'	1:03	'Golden Malrin Fly Bait'	4:426
'Garden Dust'	2:213; 2:219	'Golden Pest Spray Oil'	2:207
'Garlic Barrier'	2:178	'Goldengro TM R'	2:170
'Garlic Grow'	2:178	*Goniozus legneri*	3:314
garlic oil	2:169	'Goniozus legneri'	3:314
garlic oil extract	2:178	gorse seed weevil	3:275
'GB 34 Concentrate Biological Fungicide'	1:19	gossyplure	4:407
		grain itch mite	3:351
'Gbm Rope'	4:392; 4:449	grandisol	4:408
GC-91	1:35	(+)-grandisol	4:408
'GemStar'	1:66	grandlure	4:408
Geocoris punctipes	3:312	grandlure I	4:408
'Geocoris punctipes'	3:312	grandlure II	4:408
Geolaelaps miles	3:319	grandlure III	4:408
'Geolaelaps sp. (= Hypoaspis sp.) – Fungus Gnat Destroyer'	3:319	grandlure IV	4:408
		'Grandlure'	4:408
geraniol	2:152; 2:172; 2:185; 2:204	'Grantico'	2:203
		'Granupom'	1:57
geranyl alcohol	2:172	grape leafhopper parasitoid	3:254
German cockroach pheromone	4:406	*Grapholitha molesta* sex pheromone	4:390
'GF-120'	2:229		
'GIB'	2:179	'Grasshopper Control Semaspore Bait'	1:86
gibberellic acid	2:170; 2:179		
gibberellin A₃	2:179	'Green Ban'	2:168
gibberellin A₄	2:180	'Green Guard'	1:77
gibberellin A₄ plus gibberellin A₇	2:180	green lacewing	3:289
gibberellin A₇	2:180	'Green Lacewing'	3:288
'Gibbex'	2:179	green muscardine fungus	1:76; 1:77; 1:79; 1:80
'GibGro'	2:179		
'Gibrel'	2:179	'Green Muscle'	1:77
'Gilmectin'	2:150	'Green Sol 48'	2:170; 2:179
glasshouse thrips parasitoid	3:365	'Green Sol 70'	2:170
glasshouse whitefly parasitoid	3:302	greenbug parasitoid	3:324
Gliocladium catenulatum	1:47; 1:63	greenhouse thrips parasitoid	3:365
Gliocladium virens	1:64; 1:143	'Grocel'	2:179
'Glissade'	1:83	'Grubsure LT'	1:128
Glyptapanteles liparidis	3:313	guaniol	2:172
'Glyptapanteles liparidis'	3:313	'Guardian Spray'	2:178
'Gnatrol'	1:31	'Guardian'	1:125

Indexes

Indexes

Name	Entry no.	Name	Entry no.
'Lysiphlebus testaceipes'	3:324	McNPV	1:74
'M/C'	1:29	'Mealikil'	1:71
Macleaya extract	2:192	mealybug destroyer	3:288
Macrocentrus ancylivorus	3:325	'Mealybug Destroyer'	3:288
'Macrocentrus ancylivorus'	3:325	'Mealybug Parasite'	3:320
'Macroline c'	3:326	mealybug predator	3:288
Macrolophus caliginosus	3:326	'MEC Eastern Pine Shoot Borer'	4:391
'Macrolophus-N-System'	3:326	'MEC.CM'	4:382
'Macrolophus-System'	3:326	'MEC.LRX'	4:447
'Macropolis'	3:347	'MEC.OFM'	4:390
'Madex'	1:57	Mediterranean fruit fly attractant	4:452
'Magnet Oli'	4:386	'MegaGro L'	2:170
'Magnet'	1:126	'MeliPAK5000'	3:273
maize oil	2:169	'MeloConTM WG'	1:89
malathion	4:408	'Melocont-Pilzgerste'	1:40
'Mallet WP'	1:52	melon fly attractant	4:435
Mamestra brassicae MNPV	1:73	melon fly pheromone	4:435
Mamestra brassicae NPV	1:73	'MepPlus'	1:14
Mamestra brassicae nucleopolyhedrovirus	1:73	'Mesa'	2:196
		Mesoseiulus longipes	3:326
Mamestra configurata MNPV	1:74	'Mesoseiulus longipes'	3:326
Mamestra configurata NPV	1:74	'Messenger'	2:183
Mamestra configurata nucleopolyhedrovirus	1:74	metalaxyl	1:23
		Metaphycus bartletti	3:328
'Mamestrin'	1:73	'Metaphycus bartletti'	3:328
maneb	2:230	*Metaphycus flavus*	3:329
maple lactone	2:193	'Metaphycus flavus'	3:329
Maravalia cryptostegiae	1:75	*Metaphycus helvolus*	3:330
'Margosom'	2:156	'Metaphycus helvolus wasps'	3:330
marmalade hoverfly	3:303	'Metaphycus helvolus'	3:330
'Matran'	2:152	*Metarhizium anisopliae* isolate ESF1	1:76
'Matsuguard'	2:196	*Metarhizium anisopliae* isolate F52	1:76
'Mattch'	1:29	*Metarhizium anisopliae* isolate ICIPE 30	1:79
'Maxon II'	2:179		
'Maxon'	2:170	*Metarhizium anisopliae* isolate ICIPE 69	1:80
MBI 600	1:23		
MbMNPV	1:73	*Metarhizium anisopliae* var. acridium isolate FI-985	1:77
MbNPV	1:73		
'M-C KM'	1:35	*Metarhizium anisopliae* var. acridium isolate IMI 330189	1:77
MCH	4:424		
'MCH'	4:424	*Metarhizium anisopliae* var. anisopliae	1:77
McMNPV	1:74		

Indexes

Indexes

Name	Entry no.	Name	Entry no.
'Natur'l Oil'	2:227	'NeemPlus Liquid'	2:156
Natural acaricide	2:177	nekoe	2:219
Natural plant growth regulator, fungicide and plant metabolic primer	2:181	'NemaChek'	1:89
		'Nema-Q'	2:222
'Naturalis'	1:36	'Nemaslug'	1:95
'Naturalis-L'	1:36	'Nemastop'	2:205
'Naturalis-O'	1:36	'Nemastroy'	2:163
'Naturalis-T'	1:36	'Nemasys F'	1:126
'Naturalyte'	2:229	'Nemasys G'	1:68
'Nature's Glory Weed and Grass Killer Ready to Use'	2:151	'Nemasys H'	1:68
		'Nemasys L'	1:128
'Nature's Glory Weed and Grass Killer'	2:151	'Nemasys M'	1:126
		'Nemasys'	1:126
'Naturell WK Herbicide'	2:176	'Nematac C'	1:125
'Naturell WK Mosskiller'	2:176	'Nematac S'	1:130
'Naturell'	2:176	'Nematrol'	2:225
'Natur-Gro R-50'	2:220	'Nemitol'	2:207
'Natur-Gro Triple Plus'	2:220	'NemoPAK H'	1:67
navel orangeworm parasitoid	3:314; 3:346	'NemoPAK S'	1:126
navel orangeworm sex pheromone	4:417	*Neoaplectana bibionis*	1:126
		Neoaplectana carpocapsae	1:125
'N-Cap'	1:30	*Neoaplectana feltiae*	1:126
n-decanol	4:449	*Neoaplectana leucaniae*	1:126
n-dodecanol	4:382; 4:449	'Neochek-S'	1:85
Neem	2:156	*Neodiprion lecontei* MNPV	1:85
'Neem Cake'	2:156	*Neodiprion lecontei* NPV	1:85
neem oil	2:156	*Neodiprion lecontei* nucleopolyhedrovirus	1:85
'Neem Oil 70%'	2:156		
'Neem Suraksha'	2:156	*Neodiprion sertifer* MNPV	1:85
'Neem Wave'	2:156	*Neodiprion sertifer* NPV	1:85
'Neem'	2:156	*Neodiprion sertifer* nucleopolyhedrovirus	1:85
'Neemactin'	2:156		
'NeemAzad'	2:156	*Neodryinus typhlocybae*	3:337
'Neemazad'	2:156	Neodryinus typhlocybae'	3:337
'Neemazal T/S 1.2% EC'	2:156	'Neo-Fat'	2:176
'Neememulsion'	2:156	*Neoseiulus barkeri*	3:245
'Neemgard'	2:156	*Neoseiulus californicus*	3:246
'Neemitaf'	2:156	'Neoseiulus californicus'	3:246
'Neemitox'	2:156	*Neoseiulus cucumeris*	3:247
'Neemix'	2:156	'Neoseiulus cucumeris'	3:247
'Neemolin'	2:156	*Neoseiulus fallacis*	3:249
		'Neoseiulus fallacis'	3:249

Name	Entry no.	Name	Entry no.
nerolidol	4:401	*n*-tetradecanol	4:382
Nesidiocoris tenuis	3:338	'Nusyn-Noxfish'	2:219
'Ness-A'	1:123	'NutGuard-V'	1:100
'Ness-E'	1:123	*Oberea erythrocephala*	3:339
'Neu 1128'	2:176	'Oberea erythrocephala'	3:339
'New BIO 1020'	1:76	(*Z*)-octadec-9-enal	4:409; 4:447
'Niblecidine'	2:156	(*Z*)-octadec-13-enal	4:409; 4:411; 4:447
nicotine	2:200		
'Nicotine 40% Shreds'	2:200	(*E,Z*)-2,13-octadecadienal	4:428
nicotine sulfate	2:200	(*E,Z*)-octadeca-2,13-dienyl acetate	4:429; 4:430
'Nimbecidine'	2:156		
'Nitrozyme'	2:170	'(*E,Z*)-2,13-Octadecadienyl Acetate'	4:429
'N-Large'	2:179		
NIMNPV	1:85	(*E,Z*)-octadeca-3,13-dienyl acetate	4:429; 4:430; 4:432
NINPV	1:85		
'No-Fid'	2:200	'(*E,Z*)-Octadeca-3,13-dienyl acetate'	4:430
'Nogall'	1:03		
'NOLO Bait'	1:86	(*Z,Z*)-octadeca-3,13-dienyl acetate	4:431; 4:432
'Nolo Bait'	1:86		
'No-Mate BAW'	4:441; 4:444	(*Z*)-9-octadecenoic acid	2:176
'NoMate CM'	4:382	oct-1-en-3-ol	2:201
'No-Mate DBM'	4:409; 4:412; 4:414	1-octen-3-ol	2:201; 4:433
		octenol	2:201
'No-Mate FAW'	4:388; 4:441; 4:445	*Oecophylla longinoda*	3:340
		'OFF! Citronella Candle'	2:168
'Nomate LRX'	4:449	oil of anise	2:207
'NoMate OLR'	4:446; 4:449	oil of mustard	2:207
'Nomate PBW'	4:407	oils and stearates	2:168
'No-Mate PTB'	4:432	OK-8905	2:176
'NoMate TABM'	4:449	'Oleate'	2:176
'NoMate TPW'	4:450	oleic acid	2:176
(2*S*,7*S*)-nonadiyl dibutyrate	4:427	oleoresin	2:163
nonanoic acid	2:203	olive fly sex pheromone	4:386
nonoic acid	2:203	'Olive Moth Pheromone'	4:443
'Norbac 84C'	1:03	olive moth sex pheromone	4:443
Nosema locustae	1:86	'Ontrol'	2:224
'Novagib'	2:179; 2:180	'OPM Disrupt'	4:389
'Novodor'	1:30	OpMNPV	1:87
'Novosol FC'	1:33	OpNPV	1:87
'Noxfish'	2:219	'OpNPV'	1:87
NsMNPV	1:85	'Organic Triggrr'	2:170
NsNPV	1:85	'Organigard'	1:38

Name	Entry no.	Name	Entry no.
Parasitoid of navel orangeworm	3:314; 3:346	'Pentalitomastix plethorica'	3:346
Parasitoid of oriental fruit moth	3:325	peppermint	2:168
Parasitoid of plant bugs	3:258	'Periplan'	3:276
Parasitoid of planthoppers	3:337	'Perlan'	2:157; 2:180
Parasitoid of psyllids	3:350	'PERMATROL'	2:186
Parasitoid of red scale	3:270	permethrin	4:382; 4:407
Parasitoid of scale insects	3:272; 3:328;	persea mite predator	3:309
	3:329; 3:330	'Pesticidal Spray Oil'	2:207
Parasitoid of thrips	3:365	PFR 97	1:88
Parasitoid of weevils	3:279; 3:332	'PFR 97'	1:88
Parasitoid of whiteflies	3:282; 3:301;	'PGR Plus'	2:170
	3:302; 3:304;	'PGR-IV'	2:179
	3:305; 3:306	'Phagus'	1:107
Parasitoid of woolly aphids	3:252; 3:264	Phasmarhabditis hermaphrodita	1:95
'Para-strip'	3:302	'Phasmarhabditis-System'	1:95
Pasteuria penetrans	1:93	2-phenylethyl propionate	2:152; 2:172;
'Pasteuria Wettable Powder'	1:93		2:204
Patasson gonipteri	3:259	3-phenyl-2-propenal	2:165
'Paturyl'	2:157	'PHEROBANK sweet potato	4:396
'Paushamycin'	2:230	weevil pheromone'	
'Pavois'	1:57	'PHEROBANK tomato moth	4:442
'PB Rope'	4:407	pheromone'	
'PB-Nox'	2:219	'Pheromone trap for Brinjal	4:415
'PBW Rope-L'	4:407	shoot and fruit borer'	
PBW sex pheromone	4:407	'Pheropax'	4:421
'Pea Moth Monitoring System'	4:395	Phlebia gigantea	1:96
pea moth pheromone	4:395	Phlebiopsis gigantea	1:94; 1:96
'Pea Moth Pheromone'	4:395	Phloeospora mimosae-pigrae	1:122
peach fruit moth sex	4:419	Phomopsis amaranthicola	1:97
pheromone		phosphinothricylalanyl-alanine	2:158
'Peach Leafminer Pheromone'	4:425	Phragmidium violaceum	1:98
peach miner sex pheromone	4:425	phthalide	2:188; 2:189;
'Peach tree borer sex	4:383		2:236
pheromone		'Phyta-Guard EC'	2:217
pearly green lacewing	3:288	'Phytoline p'	3:348
Pectobacterium carotovorum	1:61	'Phytomycin'	2:202
Pectone'	4:407	Phytophagous beetle	3:241; 3:242;
Pediobius foveolatus	3:345		3:243; 3:269;
'Pediobius foveolatus mummies'	3:345		3:286; 3:287;
pelargonic acid	2:203		3:322; 3:333;
Penicillium anisopliae	1:76; 1:79; 1:80		3:339; 3:360
Peniophora gigantea	1:94; 1:96	Phytophagous caterpillar	3:281; 3:317;
Pentalitomastix plethorica	3:346		3:321; 3:376

Indexes

Name	Entry no.	Name	Entry no.
Phytophagous midge	3:361	*Plant-derived fungicide*	2:169; 2:176;
Phytophagous mite	3:239		2:186; 2:192;
Phytophagous seedhead fly	3:377		2:198; 2:207;
Phytophagous weevil	3:274; 3:275;		2:217; 2:218;
	3:278; 3:352;		2:222; 2:233
	3:373	*Plant-derived fungicide and*	2:167; 2:169;
Phytophthora palmivora	1:99	*bactericide*	2:218
Phytoseiulus chanti	3:347	*Plant-derived fungicide and*	2:186; 2:223;
Phytoseiulus longipes	3:327	*insecticide*	2:233
Phytoseiulus macropilis	3:347	*Plant-derived fungicide, insect*	2:165; 2:167
Phytoseiulus persimilis	3:348; 3:249;	*attractant and animal repellent*	
	3:362	*Plant-derived fungicide, insecticide*	2:233
'Phytoseiulus persimilis'	3:348	*and nematicide*	
Phytoseiulus riegeli	3:348	*Plant-derived herbicide and plant*	2:203
Phytoseiulus speyeri	3:347	*growth regulator*	
Phytoseiulus tardi	3:348	*Plant-derived herbicide, fungicide*	2:176
'Phytoseiulus-System'	3:348	*and insecticide*	
'Phytosure (p)'	3:348	*Plant-derived insect and small*	2:163; 2:165
'Phytosure (pt)'	3:348	*mammal repellent*	
pimaricin	2:199	*Plant-derived insect attractant*	2:193; 2:201;
α-pinene	4:404; 4:405;		2:204
	4:421; 4:423	*Plant-derived insect repellent*	2:161; 2:168;
pin-2-en-2-one	4:451		2:178; 2:194
pine processionary moth sex	4:418	*Plant-derived insecticide*	2:152; 2:156;
pheromone			2:162; 2:166;
pink bollworm sex pheromone	4:407		2:171; 2:172;
pink lady beetle	3:290		2:175; 2:176;
pink spotted lady beetle	3:290		2:185; 2:186;
piperonyl butoxide	2:213; 2:219		2:187; 2:191;
pirate bug	3:262		2:200; 2:204;
'Pityolure'	4:418		2:206; 2:207;
'PL Plus'	1:89		2:216; 2:219;
plant essential oil	2:227; 2:233;		2:221; 2:226;
	2:167; 2:169;		2:227; 2:233
	2:206; 2:207;	*Plant-derived insecticide and*	2:187; 2:213;
	2:217; 2:218;	*acaricide*	2:214; 2:215;
	2:223		2:219; 2:232
plant extract 620	2:205	*Plant-derived insecticide and*	2:204
'Plant Extract 620'	2:205	*attractant*	
Plant-derived acaricide	2:234	*Plant-derived insecticide and*	2:223; 2:225
Plant-derived animal repellent	2:159; 2:163;	*fungicide*	
	2:165; 2:207	*Plant-derived insecticide and*	2:231
Plant-derived bird repellent	2:154	*nematicide*	
		Plant-derived insecticide, fungicide	2:207
		and animal repellent	

Name	Entry no.	Name	Entry no.
Plant-derived nematicide	2:173; 2:233	*Predator of aphids*	3:240; 3:288; 3:289; 3:303; 3:315
Plant-derived plant growth regulator	2:157; 2:170; 2:184; 2:238		
Plant-derived plant growth regulator and nematicide	2:170; 2:173; 2:184; 2:205	*Predator of aphids and whiteflies*	3:298
planthopper parasitoid	3:337	predator of armoured scales	3:285
'Plantomycin'	2:230	*Predator of bark beetles*	3:353
'PlantShield HC'	1:137	predator of black scale	3:285
Plodia interpunctella granulosis virus	1:100	*Predator of caterpillars*	3:349
		Predator of euonymus scale	3:285; 3:294
Plodia interpunctella GV	1:100	*Predator of insects*	3:290; 3:312; 3:316; 3:351; 3:378
plum fruit moth sex pheromone	4:389		
Plutella xylostella sex pheromone	4:412	*Predator of mealybugs*	3:293
p-menthane-3,8-diol	2:194	*Predator of mites*	3:244; 3:246; 3:249; 3:358; 3:374; 3:375
'Podibug'	3:349		
'Podiline m'	3:349		
Podisus maculiventris	3:349	predator of oleander scale	3:285
'Podisus maculiventris'	3:349	predator of olive scale	3:285
'Point Acigib'	2:179	predator of oriental scale	3:285
'Poison-Free Ant Killer'	2:207	*Predator of persea mite*	3:309
'Pol-Gibrescol'	2:179	*Predator of psyllids and aphids*	3:262
'Polybelin'	2:211	predator of red scale	3:285
'Polycore SKL'	4:386	*Predator of scale insects*	3:285; 3:354; 3:355; 3:356
poly-D-glucosamine	2:208		
'Polygandron'	1:117	*Predator of sciarid flies*	3:318; 3:319
'Polygen'	1:09	*Predator of slugs and snails*	3:357
poly-N-acetyl-D-glucosamine	2:210	*Predator of spider mites*	3:307; 3:310; 3:311; 3:347; 3:348; 3:362
polynactins	2:209		
'Polyoxin AL'	2:211	*Predator of thrips*	3:245; 3:247; 3:248; 3:250; 3:308; 3:341; 3:342; 3:343; 3:344
polyoxin B	2:211		
polyoxin D	2:212		
'Polyoxin Z'	2:212		
polyoxins	2:211; 2:212	*Predator of two-spotted spider mite*	3:327
polyoxorim	2:212		
'Polyversum'	1:117	predator of white louse scale	3:285
porphyrin-derivatives	2:205	*Predator of whiteflies*	3:296; 3:297; 3:299; 3:359
potassium salts of plant oils.	2:226		
PP761	4:407	*Predator of whiteflies and spider mites*	3:326
praying mantis	3:363		
'Praying Mantis'	3:363	*Predator of whiteflies and thrips*	3:338
Prays oleae sex pheromone	4:443	*Predator of whitefly*	3:251

Indexes

Name	Entry no.	Name	Entry no.
Predatory ant	3:340	*Pseudomonas fluorescens*	1:104; 1:106; 1:144
Predatory beetle	3:277; 3:283		
Predatory insect	3:363	*Pseudomonas gladioli* pv. *gladioli*	1:105
predatory ladybird	3:315	*Pseudomonas syringae*	1:104
'Predatory Minute Pirate Bug'	3:342	*Pseudomonas syringae* isolate ESC-10	1:106
predatory mirid	3:299		
Predatory mite	3:284	*Pseudomonas syringae* isolate ESC-11	1:106
predatory stink bug	3:349		
'PreFeRal'	1:88	*Pseudomonas syringae* pv. *tomato* bacteriophage	1:149
'Premium Pyganic 175'	2:213		
'Prenfish'	2:219	*Pseudomonas tolassii* bacteriophage	1:107
'Prentox Pyrethrum Extract'	2:213		
'Pré-Phytoseiulus-System'	3:348	*Pseudozyma flocculosa* isolate PF-A22 UL	1:108
'Prestop'	1:47		
prickly pear moth	3:281	*Psyllaephagus pilosus*	3:350
'Primastop'	1:47	'Psyllaephagus pilosus'	3:350
'Priority'	1:88	*Puccinia abrupta* var. *partheniicola*	1:109
'ProAct'	2:182	*Puccinia canaliculata*	1:110
'Proclaim'	2:174	*Puccinia carduorum*	1:111
'Procone'	2:180	*Puccinia chondrillina*	1:112
'Proecol'	1:36	*Puccinia evadens*	1:113
profenfos	4:408	*Puccinia jaceae* var. *solstitialis* isolate FDWSRU 84-71	1:114
'ProGibb'	2:179		
'Promalin'	2:157; 2:180	*Puccinia jaceae* var. *solstitialis* isolate TR 84-96	1:114
'Promax'	2:233		
'Pro-mix with Biofungicide'	1:25	*Puccinia myrsiphylli*	1:115
'Promot'	1:138	*Puccinia thlaspeos*	1:116
'Proneem'	2:156	*Puccinia violacea*	1:98
3-propylphenol	4:433	'Puffer CM'	4:382
'ProShear'	2:157	'Puffer NOW'	4:417
'Protus WG'	1:134	puncture vine seed weevil	3:333
'Proud 3'	2:233	purple scale parasitoid	3:271
'ProVide'	2:180	'Purple Scale Predator'	3:355
Pseudomonas aureofaciens isolate Tx-1	1:101	'Pycon'	2:213
		Pyemotes tritici	3:351
Pseudomonas campestris	1:148	'Pyemotes tritici'	3:351
Pseudomonas cepacia	1:102	'Pyganic Crop Protection EC 1.4'	2:213
Pseudomonas cepacia Wisconsin isolate	1:42	'Pyganic Crop Protection EC 5.0'	2:213
		'Pyratyp'	3:366
Pseudomonas cerasi	1:106	'Pyrellin'	2:219
Pseudomonas chlororaphis isolate 63-28	1:103	pyrethrates	2:215
		pyrèthres	2:213
		pyrethrin I	2:214

Indexes

Indexes

Name	Entry no.
'Streptrol'	2:230
striped rice stem borer sex pheromone	4:411
'Strong'	2:179
'Stumpout'	1:58
'Subtilex'	1:25
'Success'	2:229
sucrose octanoate esters	2:231
sugar octanoate esters	2:231
'Sukasiba-con'	4:430
sulcatol	4:423
sulfate de nicotine	2:200
sulfur	2:213; 2:219
'Super Lagniappe'	2:170; 2:179
'Superneem 4.5-B'	2:156
'Suppress'	2:170
'Surefire Japanese Beetle Trap'	2:152; 2:172; 2:204
'Sweet potato weevil sex pheromone	4:396
'Sweet Potato Whitefly Predator'	3:302
'Swirskii-System'	3:251
'Swirskiline as'	3:251
Synanthedon pictipes sex pheromone	4:432
synergol	2:232
Syngrapha falcifera nucleopolyhedrovirus	1:133
'Synpren Fish'	2:219
'System 3'	1:23
'T. lineatum Lure'	4:423
'T-22 Planter Box'	1:137
'T-22 PlantShield'	1:137
TAB-1	3:302
'TAE-022'	1:22
'Taegro'	1:22
'Taenure Granular Bioinsecticide'	1:76
'Taerain'	1:77
Talaromyces flavus isolate V117b	1:134
Talaromyces flavus isolate Y-9401	1:135
tansy ragwort root beetle	3:322
tea tortrix sex pheromone	4:446

Name	Entry no.
'Teknar'	1:31
tennectin	2:199
Tenodera aridifolia sinensis	3:363
terramitsin	2:202
'Terranal'	1:22
tert-butyl-2-methyl-4(or 5)-chlorocyclohexanecarboxylate	4:452
(9*Z*,12*E*)-tetradecadien-l-ol	4:440
(9*Z*,12*E*)-tetradecadien-l-ol acetate	4:440
(3*E*,8*Z*)-tetradecadien-1-yl acetate	4:442
(*E*,*Z*)-tetradeca-4,10-dienyl acetate	4:389; 4:416; 4:419; 4:438
(3*E*,8*Z*,11*Z*)-tetradecatrien-1-yl acetate	4:442
(*Z*,*E*)-tetradeca-9,11-dienyl acetate	4:439
(*Z*,*E*)-tetradeca-9,12-dienyl acetate	4:414; 4:441; 4:444
tetradecan-1-ol	4:382; 4:447
(*Z*)-tetradec-7-enal	4:443
(*Z*)-tetradec-9-enal	4:413
(*Z*)-tetradec-9-enol	4:414; 4:441; 4:444
(*E*)-tetradec-11-enol	4:449
(*Z*)-tetradec-9-enyl acetate	4:388; 4:389; 4:392; 4:419; 4:441; 4:444; 4:445; 4:449
tetradec-11-enyl acetate	4:382; 4:389; 4:392; 4:416; 4:419; 4:425; 4:438; 4:441; 4:446
(*Z*)-tetradec-11-enyl acetate	4:382; 4:447; 4:448; 4:449
(*E*)-tetradec-11-enyl acetate	4:448; 4:449
(*Z*)-9-tetradecenyl acetate	4:382; 4:447
(*Z*)-11-tetradecenyl acetate	4:382
tetranactin	2:209
'Tetrastichus parasitoids'	3:364
Tetrastichus species	3:364
TF-138	2:197

Indexes

Indexes